Detection, Estimation, and Modulation Theory

Detection, Estimation, and Modulation Theory

PART I Detection, Estimation, and Linear Modulation Theory

HARRY L. VAN TREES

Massachusetts Institute of Technology

John Wiley and Sons, Inc.   *New York · London · Sydney*

Library of Congress Catalog Card Number: 67–23331

GB 471 89955X

Printed in the United States of America

To Diane

and Stephen, Mark, Kathleen, Patricia,
Eileen, Harry, and Julia

Preface

The area of detection and estimation theory that we shall study in this book represents a combination of the classical techniques of statistical inference and the random process characterization of communication, radar, sonar, and other modern data processing systems. The two major areas of statistical inference are decision theory and estimation theory. In the first case we observe an output that has a random character and decide which of two possible causes produced it. This type of problem was studied in the middle of the eighteenth century by Thomas Bayes [1]. In the estimation theory case the output is related to the value of some parameter of interest, and we try to estimate the value of this parameter. Work in this area was published by Legendre [2] and Gauss [3] in the early nineteenth century. Significant contributions to the classical theory that we use as background were developed by Fisher [4] and Neyman and Pearson [5] more than 30 years ago. In 1941 and 1942 Kolmogoroff [6] and Wiener [7] applied statistical techniques to the solution of the optimum linear filtering problem. Since that time the application of statistical techniques to the synthesis and analysis of all types of systems has grown rapidly. The application of these techniques and the resulting implications are the subject of this book.

This book and the subsequent volume, Detection, Estimation, and Modulation Theory, Part II, are based on notes prepared for a course entitled "Detection, Estimation, and Modulation Theory," which is taught as a second-level graduate course at M.I.T. My original interest in the material grew out of my research activities in the area of analog modulation theory. A preliminary version of the material that deals with modulation theory was used as a text for a summer course presented at M.I.T. in 1964. It turned out that our viewpoint on modulation theory could best be understood by an audience with a clear understanding of modern detection and estimation theory. At that time there was no suitable text available to cover the material of interest and emphasize the points that I felt were important, so I started writing notes. It was clear that in order to present the material to graduate students in a reasonable amount of time it would be necessary to develop a unified presentation of the three topics: detection, estimation, and modulation theory, and exploit the fundamental ideas that connected them. As the development proceeded, it grew in size until the material that was originally intended to be background for modulation theory occupies the entire contents of this book. The original material on modulation theory starts at the beginning of the second book. Collectively, the two books provide a unified coverage of the three topics and their application to many important physical problems.

For the last three years I have presented successively revised versions of the material in my course. The audience consists typically of 40 to 50 students who have completed a graduate course in random processes which covered most of the material in Davenport and Root [8]. In general, they have a good understanding of random process theory and a fair amount of practice with the routine manipulation required to solve problems. In addition, many of them are interested in doing research in this general area or closely related areas. This interest provides a great deal of motivation which I exploit by requiring them to develop many of the important ideas as problems. It is for this audience that the book is primarily intended. The appendix contains a detailed outline of the course.

On the other hand, many practicing engineers deal with systems that have been or should have been designed and analyzed with the techniques developed in this book. I have attempted to make the book useful to them. An earlier version was used successfully as a text for an in-plant course for graduate engineers.

From the standpoint of specific background little advanced material is required. A knowledge of elementary probability theory and second moment characterization of random processes is assumed. Some familiarity with matrix theory and linear algebra is helpful but certainly not necessary. The level of mathematical rigor is low, although in most sections the results could be rigorously proved by simply being more careful in our derivations. We have adopted this approach in order not to obscure the important ideas with a lot of detail and to make the material readable for the kind of engineering audience that will find it useful. Fortunately, in almost all cases we can verify that our answers are intuitively logical. It is worthwhile to observe that this ability to check our answers intuitively would be necessary even if our derivations were rigorous, because our ultimate objective is to obtain an answer that corresponds to some physical system of interest. It is easy to find physical problems in which a plausible mathematical model and correct mathematics lead to an unrealistic answer for the original problem.

We have several idiosyncrasies that it might be appropriate to mention. In general, we look at a problem in a fair amount of detail. Many times we look at the same problem in several different ways in order to gain a better understanding of the meaning of the result. Teaching students a number of ways of doing things helps them to be more flexible in their approach to new problems. A second feature is the necessity for the reader to solve problems to understand the material fully. Throughout the course and the book we emphasize the development of an ability to work problems. At the end of each chapter are problems that range from routine manipulations to significant extensions of the material in the text. In many cases they are equivalent to journal articles currently being published. Only by working a fair number of them is it possible to appreciate the significance and generality of the results. Solutions for an individual problem will be supplied on request, and a book containing solutions to about one third of the problems is available to faculty members teaching the course. We are continually generating new problems in conjunction with the course and will send them to anyone who is using the book as a course text. A third issue is the abundance of block diagrams, outlines, and pictures. The diagrams are included because most engineers (including myself) are more at home with these items than with the corresponding equations.

One problem always encountered is the amount of notation needed to cover the large range of subjects. We have tried to choose the notation in a logical manner and to make it mnemonic. All the notation is summarized in the glossary at the end of the book. We have tried to make our list of references as complete as possible and to acknowledge any ideas due to other people.

A number of people have contributed in many ways and it is a pleasure to acknowledge them. Professors W. B. Davenport and W. M. Siebert have provided continual encouragement and technical comments on the various chapters. Professors Estil Hoversten and Donald Snyder of the M.I.T. faculty and Lewis Collins, Arthur Baggeroer, and Michael Austin, three of my doctoral students, have carefully read and criticized the various chapters. Their suggestions have improved the manuscript appreciably. In addition, Baggeroer and Collins contributed a number of the problems in the various chapters and Baggeroer did the programming necessary for many of the graphical results. Lt. David Wright read and criticized Chapter 2. L. A. Frasco and H. D. Goldfein, two of my teaching assistants, worked all of the problems in the book. Dr. Howard Yudkin of Lincoln Laboratory read the entire manuscript and offered a number of important criticisms. In addition, various graduate students taking the course have made suggestions which have been incorporated. Most of the final draft was typed by Miss Aina Sils. Her patience with the innumerable changes is sincerely appreciated. Several other secretaries, including Mrs. Jarmila Hrbek, Mrs. Joan Bauer, and Miss Camille Tortorici, typed sections of the various drafts.

As pointed out earlier, the books are an outgrowth of my research interests. This research is a continuing effort, and I shall be glad to send our current work to people working in this area on a regular reciprocal basis. My early work in modulation theory was supported by Lincoln Laboratory as a summer employee and consultant in groups directed by Dr. Herbert Sherman and Dr. Barney Reiffen. My research at M.I.T. was partly supported by the Joint Services and the National Aeronautics and Space Administration under the auspices of the Research Laboratory of Electronics. This support is gratefully acknowledged.

Harry L. Van Trees

Cambridge, Massachusetts
October, 1967.

REFERENCES

[1] Thomas Bayes, "An Essay Towards Solving a Problem in the Doctrine of Chances," *Phil. Trans*, **53**, 370–418 (1764).
[2] A. M. Legendre, *Nouvelles Méthodes pour La Détermination ces Orbites des Comètes*, Paris, 1806.
[3] K. F. Gauss, *Theory of Motion of the Heavenly Bodies Moving About the Sun in Conic Sections*, reprinted by Dover, New York, 1963.
[4] R. A. Fisher, "Theory of Statistical Estimation," *Proc. Cambridge Philos. Soc.*, **22**, 700 (1925).
[5] J. Neyman and E. S. Pearson, "On the Problem of the Most Efficient Tests of Statistical Hypotheses," *Phil. Trans. Roy. Soc. London*, **A 231**, 289, (1933).
[6] A. Kolmogoroff, "Interpolation and Extrapolation von Stationären Zufälligen Folgen," *Bull. Acad. Sci. USSR, Ser. Math.* **5**, 1941.
[7] N. Wiener, *Extrapolation, Interpolation, and Smoothing of Stationary Time Series*, Tech. Press of M.I.T. and Wiley, New York, 1949 (originally published as a classified report in 1942).
[8] W. B. Davenport and W. L. Root, *Random Signals and Noise*, McGraw-Hill, New York, 1958.

Contents

1

Introduction

In these two books, we shall study three areas of statistical theory, which we have labeled detection theory, estimation theory, and modulation theory. The goal is to develop these theories in a common mathematical framework and to demonstrate how they can be used to solve a wealth of practical problems in many diverse physical situations.

In this chapter we present three outlines of the material. The first is a topical outline in which we develop a qualitative understanding of the three areas by examining some typical problems of interest. The second is a logical outline in which we explore the various methods of attacking the problems. The third is a chronological outline in which we explain the structure of the books.

1.1 TOPICAL OUTLINE

An easy way to explain what is meant by detection theory is to examine several physical situations that lead to detection theory problems.

A simple digital communication system is shown in Fig. 1.1. The source puts out a binary digit every T seconds. Our object is to transmit this sequence of digits to some other location. The channel available for transmitting the sequence depends on the particular situation. Typically, it could be a telephone line, a radio link, or an acoustical channel. For

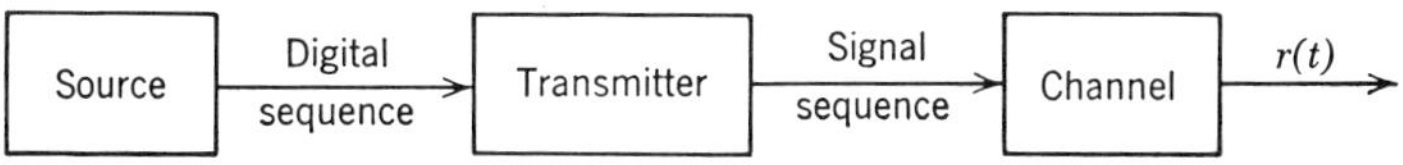

Fig. 1.1 Digital communication system.

purposes of illustration, we shall consider a radio link. In order to transmit the information, we must put it into a form suitable for propagating over the channel. A straightforward method would be to build a device that generates a sine wave,

$$s_1(t) = \sin \omega_1 t, \tag{1}$$

for T seconds if the source generated a "one" in the preceding interval, and a sine wave of a different frequency,

$$s_0(t) = \sin \omega_0 t, \tag{2}$$

for T seconds if the source generated a "zero" in the preceding interval. The frequencies are chosen so that the signals $s_0(t)$ and $s_1(t)$ will propagate over the particular radio link of concern. The output of the device is fed into an antenna and transmitted over the channel. Typical source and transmitted signal sequences are shown in Fig. 1.2. In the simplest kind of channel the signal sequence arrives at the receiving antenna attenuated but essentially undistorted. To process the received signal we pass it through the antenna and some stages of rf-amplification, in the course of which a thermal noise $n(t)$ is added to the message sequence. Thus in any T-second interval we have available a waveform $r(t)$ in which

$$r(t) = s_1(t) + n(t), \qquad 0 \le t \le T, \tag{3}$$

if $s_1(t)$ was transmitted, and

$$r(t) = s_0(t) + n(t), \qquad 0 \le t \le T, \tag{4}$$

if $s_0(t)$ was transmitted. We are now faced with the problem of deciding which of the two possible signals was transmitted. We label the device that does this a decision device. It is simply a processor that observes $r(t)$ and guesses whether $s_1(t)$ or $s_0(t)$ was sent according to some set of rules. This is equivalent to guessing what the source output was in the preceding interval. We refer to designing and evaluating the processor as a detection

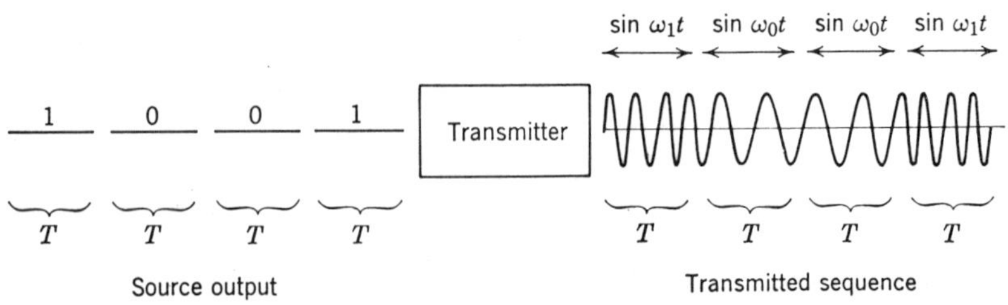

Fig. 1.2 Typical sequences.

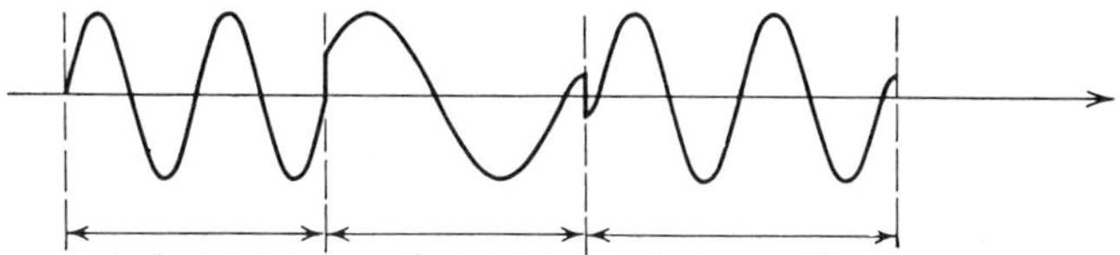

Fig. 1.3 Sequence with phase shifts.

theory problem. In this particular case the only possible source of error in making a decision is the additive noise. If it were not present, the input would be completely known and we could make decisions without errors. We denote this type of problem as the *known signal in noise problem.* It corresponds to the lowest level (i.e., simplest) of the detection problems of interest.

An example of the next level of detection problem is shown in Fig. 1.3. The oscillators used to generate $s_1(t)$ and $s_0(t)$ in the preceding example have a phase drift. Therefore in a particular T-second interval the received signal corresponding to a "one" is

$$r(t) = \sin(\omega_1 t + \theta_1) + n(t), \qquad 0 \le t \le T, \tag{5}$$

and the received signal corresponding to a "zero" is

$$r(t) = \sin(\omega_0 t + \theta_0) + n(t), \qquad 0 \le t \le T, \tag{6}$$

where θ_0 and θ_1 are unknown constant phase angles. Thus even in the absence of noise the input waveform is not completely known. In a practical system the receiver may include auxiliary equipment to measure the oscillator phase. If the phase varies slowly enough, we shall see that essentially perfect measurement is possible. If this is true, the problem is the same as above. However, if the measurement is not perfect, we must incorporate the signal uncertainty in our model.

A corresponding problem arises in the radar and sonar areas. A conventional radar transmits a pulse at some frequency ω_c with a rectangular envelope:

$$s_t(t) = \sin \omega_c t, \qquad 0 \le t \le T. \tag{7}$$

If a target is present, the pulse is reflected. Even the simplest target will introduce an attenuation and phase shift in the transmitted signal. Thus the signal available for processing in the interval of interest is

$$\begin{aligned} r(t) &= V_r \sin[\omega_c(t-\tau) + \theta_r] + n(t), && \tau \le t \le \tau + T, \\ &= n(t), && 0 \le t < \tau, \tau + T < t < \infty, \end{aligned} \tag{8}$$

if a target is present and

$$r(t) = n(t), \qquad 0 \le t < \infty, \tag{9}$$

if a target is absent. We see that in the absence of noise the signal still contains three unknown quantities: V_r, the amplitude, θ_r, the phase, and τ, the round-trip travel time to the target.

These two examples represent the second level of detection problems. We classify them as *signal with unknown parameters in noise problems.*

Detection problems of a third level appear in several areas. In a passive sonar detection system the receiver listens for noise generated by enemy vessels. The engines, propellers, and other elements in the vessel generate acoustical signals that travel through the ocean to the hydrophones in the detection system. This composite signal can best be characterized as a sample function from a random process. In addition, the hydrophone generates self-noise and picks up sea noise. Thus a suitable model for the detection problem might be

$$r(t) = s_{\Omega}(t) + n(t) \tag{10}$$

if the target is present and

$$r(t) = n(t) \tag{11}$$

if it is not. In the absence of noise the signal is a sample function from a random process (indicated by the subscript Ω).

In the communications field a large number of systems employ channels in which randomness is inherent. Typical systems are tropospheric scatter links, orbiting dipole links, and chaff systems. A common technique is to transmit one of two signals separated in frequency. (We denote these frequencies as ω_1 and ω_0.) The resulting received signal is

$$r(t) = s_{\Omega_1}(t) + n(t) \tag{12}$$

if $s_1(t)$ was transmitted and

$$r(t) = s_{\Omega_0}(t) + n(t) \tag{13}$$

if $s_0(t)$ was transmitted. Here $s_{\Omega_1}(t)$ is a sample function from a random process centered at ω_1, and $s_{\Omega_0}(t)$ is a sample function from a random process centered at ω_0. These examples are characterized by the lack of any deterministic signal component. Any decision procedure that we design will have to be based on the difference in the statistical properties of the two random processes from which $s_{\Omega_0}(t)$ and $s_{\Omega_1}(t)$ are obtained. This is the third level of detection problem and is referred to as a *random signal in noise problem.*

In our examination of representative examples we have seen that detection theory problems are characterized by the fact that we must decide which of several alternatives is true. There were only two alternatives in the examples cited; therefore we refer to them as binary detection problems. Later we will encounter problems in which there are M alternatives available (the M-ary detection problem). Our hierarchy of detection problems is presented graphically in Fig. 1.4.

There is a parallel set of problems in the estimation theory area. A simple example is given in Fig. 1.5, in which the source puts out an analog message $a(t)$ (Fig. 1.5*a*). To transmit the message we first sample it every T seconds. Then, every T seconds we transmit a signal that contains

	Detection theory
Level 1. Known signals in noise	1. Synchronous digital communication 2. Pattern recognition problems
Level 2. Signals with unknown parameters in noise	1. Conventional pulsed radar or sonar, target detection 2. Target classification (orientation of target unknown) 3. Digital communication systems without phase reference 4. Digital communication over slowly-fading channels
Level 3. Random signals in noise	1. Digital communication over scatter link, orbiting dipole channel, or chaff link 2. Passive sonar 3. Seismic detection system 4. Radio astronomy (detection of noise sources)

Fig. 1.4 Detection theory hierarchy.

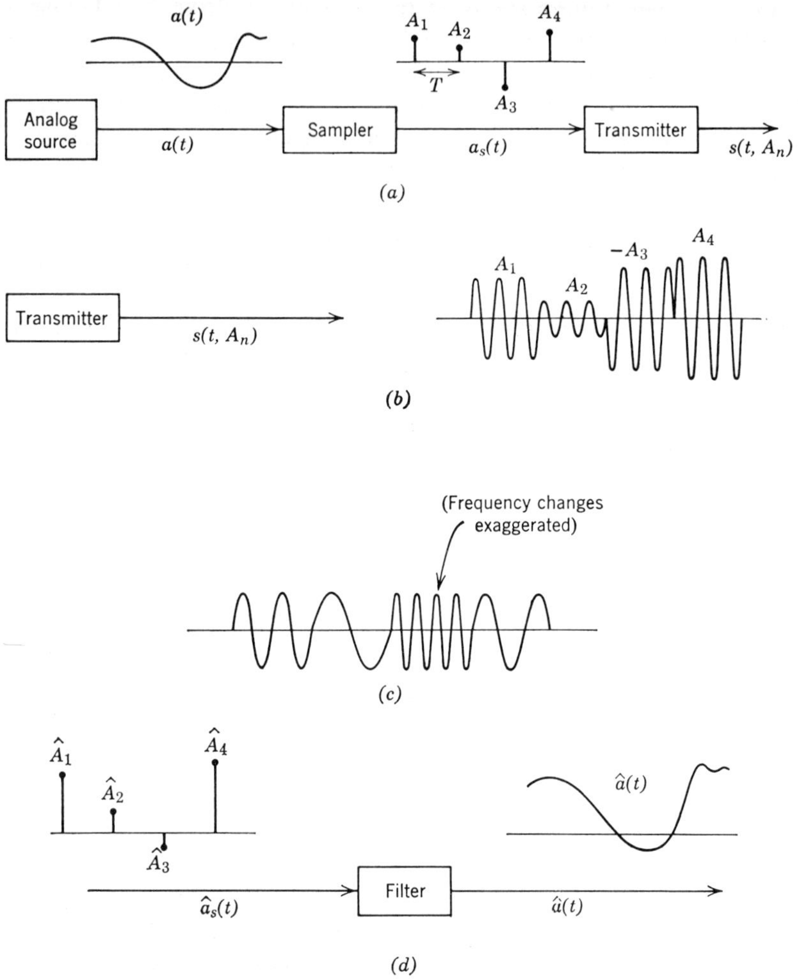

Fig. 1.5 **(*a*) Sampling an analog source; (*b*) pulse-amplitude modulation; (*c*) pulse-frequency modulation; (*d*) waveform reconstruction.**

a parameter which is uniquely related to the last sample value. In Fig. 1.5*b* the signal is a sinusoid whose amplitude depends on the last sample. Thus, if the sample at time nT is A_n, the signal in the interval $[nT, (n+1)T]$ is

$$s(t, A_n) = A_n \sin \omega_c t, \qquad nT \le t \le (n+1)T. \tag{14}$$

A system of this type is called a pulse amplitude modulation (PAM) system. In Fig. 1.5*c* the signal is a sinusoid whose frequency in the interval differs from the reference frequency ω_c by an amount proportional to the preceding sample value,

$$s(t, A_n) = \sin(\omega_c t + A_n t), \qquad nT \le t \le (n+1)T. \tag{15}$$

A system of this type is called a pulse frequency modulation (PFM) system. Once again there is additive noise. The received waveform, given that A_n was the sample value, is

$$r(t) = s(t, A_n) + n(t), \qquad nT \le t \le (n+1)T. \tag{16}$$

During each interval the receiver tries to estimate A_n. We denote these estimates as $\hat{A}_n$. Over a period of time we obtain a sequence of estimates, as shown in Fig. 1.5*d*, which is passed into a device whose output is an estimate of the original message $a(t)$. If $a(t)$ is a band-limited signal, the device is just an ideal low-pass filter. For other cases it is more involved.

If, however, the parameters in this example were known and the noise were absent, the received signal would be completely known. We refer to problems in this category as *known signal in noise problems.* If we assume that the mapping from A_n to $s(t, A_n)$ in the transmitter has an inverse, we see that if the noise were not present we could determine A_n unambiguously. (Clearly, if we were allowed to design the transmitter, we should always choose a mapping with an inverse.) The *known signal in noise problem* is the first level of the estimation problem hierarchy.

Returning to the area of radar, we consider a somewhat different problem. We assume that we know a target is present but do not know its range or velocity. Then the received signal is

$$\begin{aligned} r(t) &= V_r \sin[(\omega_c + \omega_d)(t - \tau) + \theta_r] + n(t), && \tau \le t \le \tau + T, \\ &= n(t), && 0 \le t < \tau, \tau + T < t < \infty, \end{aligned} \tag{17}$$

where ω_d denotes a Doppler shift caused by the target's motion. We want to estimate τ and ω_d. Now, even if the noise were absent and τ and ω_d were known, the signal would still contain the unknown parameters V_r and θ_r. This is a typical second-level estimation problem. As in detection theory, we refer to problems in this category as *signal with unknown parameters in noise problems.*

At the third level the signal component is a random process whose statistical characteristics contain parameters we want to estimate. The received signal is of the form

$$r(t) = s_\Omega(t, A) + n(t), \tag{18}$$

where $s_\Omega(t, A)$ is a sample function from a random process. In a simple case it might be a stationary process with the narrow-band spectrum shown in Fig. 1.6. The shape of the spectrum is known but the center frequency

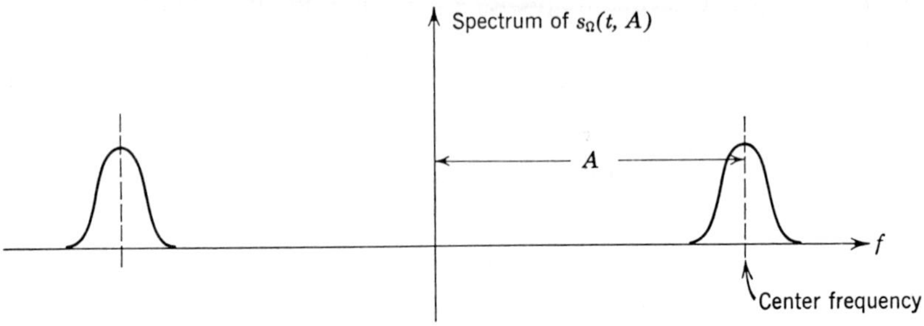

Fig. 1.6 Spectrum of random signal.

	Estimation Theory
Level 1. Known signals in noise	1. PAM, PFM, and PPM communication systems with phase synchronization 2. Inaccuracies in inertial systems (e.g., drift angle measurement)
Level 2. Signals with unknown parameters in noise	1. Range, velocity, or angle measurement in radar/sonar problems 2. Discrete time, continuous amplitude communication system (with unknown amplitude or phase in channel)
Level 3. Random signals in noise	1. Power spectrum parameter estimation 2. Range or Doppler spread target parameters in radar/sonar problem 3. Velocity measurement in radio astronomy 4. Target parameter estimation: passive sonar 5. Ground mapping radars

Fig. 1.7 Estimation theory hierarchy.

is not. The receiver must observe $r(t)$ and, using the statistical properties of $s_{\Omega}(t, A)$ and $n(t)$, estimate the value of A. This particular example could arise in either radio astronomy or passive sonar. The general class of problem in which the signal containing the parameters is a sample function from a random process is referred to as the *random signal in noise problem.* The hierarchy of estimation theory problems is shown in Fig. 1.7.

We note that there appears to be considerable parallelism in the detection and estimation theory problems. We shall frequently exploit these parallels to reduce the work, but there is a basic difference that should be emphasized. In binary detection the receiver is either "right" or "wrong." In the estimation of a continuous parameter the receiver will seldom be exactly right, but it can try to be close most of the time. This difference will be reflected in the manner in which we judge system performance.

The third area of interest is frequently referred to as modulation theory. We shall see shortly that this term is too narrow for the actual problems. Once again a simple example is useful. In Fig. 1.8 we show an analog message source whose output might typically be music or speech. To convey the message over the channel, we transform it by using a modulation scheme to get it into a form suitable for propagation. The transmitted signal is a continuous waveform that depends on $a(t)$ in some deterministic manner. In Fig. 1.8 it is an amplitude modulated waveform:

$$s[t, a(t)] = [1 + ma(t)] \sin(\omega_c t). \tag{19}$$

(This is conventional double-sideband AM with modulation index m.) In Fig. 1.8*c* the transmitted signal is a frequency modulated (FM) waveform:

$$s[t, a(t)] = \sin\left[\omega_c t + \int_{-\infty}^{t} a(u)\, du\right]. \tag{20}$$

When noise is added the received signal is

$$r(t) = s[t, a(t)] + n(t). \tag{21}$$

Now the receiver must observe $r(t)$ and put out a continuous estimate of the message $a(t)$, as shown in Fig. 1.8. This particular example is a first-level modulation problem, for if $n(t)$ were absent and $a(t)$ were known the received signal would be completely known. Once again we describe it as a *known signal in noise problem.*

Another type of physical situation in which we want to estimate a continuous function is shown in Fig. 1.9. The channel is a time-invariant linear system whose impulse response $h(\tau)$ is unknown. To estimate the impulse response we transmit a known signal $x(t)$. The received signal is

$$r(t) = \int_0^{\infty} h(\tau)\, x(t - \tau)\, d\tau + n(t). \tag{22}$$

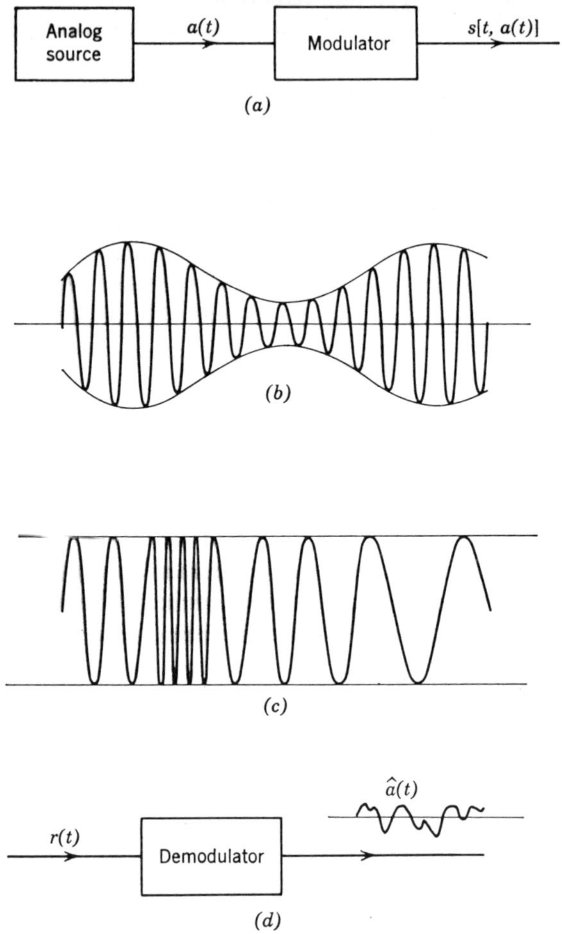

Fig. 1.8 A modulation theory example: (*a*) analog transmission system; (*b*) amplitude modulated signal; (*c*) frequency modulated signal; (*d*) demodulator.

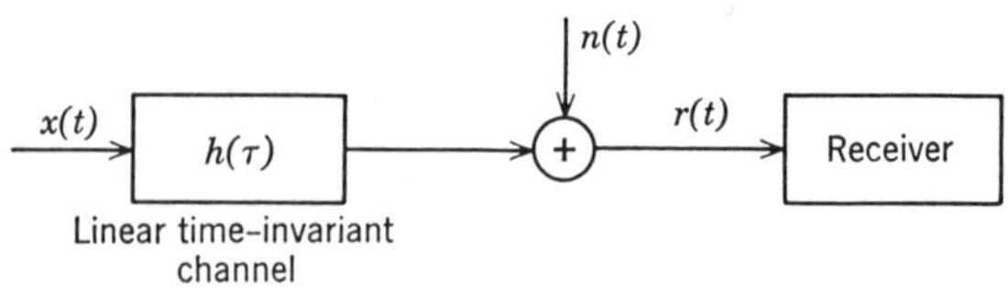

Fig. 1.9 Channel measurement.

The receiver observes $r(t)$ and tries to estimate $h(\tau)$. This particular example could best be described as a continuous estimation problem. Many other problems of interest in which we shall try to estimate a continuous waveform will be encountered. For convenience, we shall use the term *modulation theory* for this category, even though the term continuous waveform estimation might be more descriptive.

The other levels of the modulation theory problem follow by direct analogy. In the amplitude modulation system shown in Fig. 1.8*b* the receiver frequently does not know the phase angle of the carrier. In this case a suitable model is

$$r(t) = (1 + ma(t)) \sin (\omega_c t + \theta) + n(t), \tag{23}$$

	Modulation Theory (Continuous waveform estimation)
1. Known signals in noise	1. Conventional communication systems such as AM (DSB-AM, SSB), FM, and PM with phase synchronization 2. Optimum filter theory 3. Optimum feedback systems 4. Channel measurement 5. Orbital estimation for satellites 6. Signal estimation in seismic and sonar classification systems 7. Synchronization in digital systems
2. Signals with unknown parameters in noise	1. Conventional communication systems without phase synchronization 2. Estimation of channel characteristics when phase of input signal is unknown
3. Random signals in noise	1. Analog communication over randomly varying channels 2. Estimation of statistics of time-varying processes 3. Estimation of plant characteristics

Fig. 1.10 Modulation theory hierarchy.

where θ is an unknown parameter. This is an example of a *signal with unknown parameter problem* in the modulation theory area.

A simple example of a third-level problem (*random signal in noise*) is one in which we transmit a frequency-modulated signal over a radio link whose gain and phase characteristics are time-varying. We shall find that if we transmit the signal in (20) over this channel the received waveform will be

$$r(t) = V(t) \sin \left[\omega_c t + \int_{-\infty}^{t} a(u)\, du + \theta(t)\right] + n(t), \tag{24}$$

where $V(t)$ and $\theta(t)$ are sample functions from random processes. Thus, even if $a(u)$ were known and the noise $n(t)$ were absent, the received signal would still be a random process. An over-all outline of the problems of interest to us appears in Fig. 1.10. Additional examples included in the table to indicate the breadth of the problems that fit into the outline are discussed in more detail in the text.

Now that we have outlined the areas of interest it is appropriate to determine how to go about solving them.

1.2 POSSIBLE APPROACHES

From the examples we have discussed it is obvious that an inherent feature of all the problems is randomness of source, channel, or noise (often all three). Thus our approach must be statistical in nature. Even assuming that we are using a statistical model, there are many different ways to approach the problem. We can divide the possible approaches into two categories, which we denote as "structured" and "nonstructured." Some simple examples will illustrate what we mean by a structured approach.

Example 1. The input to a linear time-invariant system is $r(t)$:

$$\begin{aligned} r(t) &= s(t) + w(t) \qquad 0 \le t \le T, \\ &= 0, \qquad \text{elsewhere.} \end{aligned} \tag{25}$$

The impulse response of the system is $h(\tau)$. The signal $s(t)$ is a known function with energy E_s,

$$E_s = \int_0^T s^2(t)\, dt, \tag{26}$$

and $w(t)$ is a sample function from a zero-mean random process with a covariance function:

$$K_w(t, u) = \frac{N_0}{2}\, \delta(t-u). \tag{27}$$

We are concerned with the output of the system at time T. The output due to the signal is a deterministic quantity:

$$s_o(T) = \int_0^T h(\tau)\, s(T - \tau)\, d\tau. \tag{28}$$

The output due to the noise is a random variable:

$$n_o(T) = \int_0^T h(\tau)\, n(T - \tau)\, d\tau. \tag{29}$$

We can define the output signal-to-noise ratio at time T as

$$\frac{S}{N} \triangleq \frac{s_o^2(T)}{E[n_o^2(T)]}, \tag{30}$$

where $E(\cdot)$ denotes expectation.

Substituting (28) and (29) into (30), we obtain

$$\frac{S}{N} = \frac{\left[\int_0^T h(\tau)\, s(T - \tau)\, d\tau\right]^2}{E\left[\iint_0^T h(\tau)\, h(u)\, n(T - \tau)\, n(T - u)\, d\tau\, du\right]}. \tag{31}$$

By bringing the expectation inside the integral, using (27), and performing the integration with respect to u, we have

$$\frac{S}{N} = \frac{\left[\int_0^T h(\tau)\, s(T - \tau)\, d\tau\right]^2}{N_0/2 \int_0^T h^2(\tau)\, d\tau}. \tag{32}$$

The problem of interest is to choose $h(\tau)$ to maximize the signal-to-noise ratio. The solution follows easily, but it is not important for our present discussion. (See Problem 3.3.1.)

This example illustrates the three essential features of the structured approach to a statistical optimization problem:

Structure. The processor was required to be a linear time-invariant filter. We wanted to choose the best system in this class. Systems that were not in this class (e.g., nonlinear or time-varying) were not allowed.

Criterion. In this case we wanted to maximize a quantity that we called the signal-to-noise ratio.

Information. To write the expression for S/N we had to know the signal shape and the covariance function of the noise process.

If we knew more about the process (e.g., its first-order probability density), we could not use it, and if we knew less, we could not solve the problem. Clearly, if we changed the criterion, the information required might be different. For example, to maximize x

$$x = \frac{s_o^4(T)}{E[n_o^4(T)]}, \tag{33}$$

the covariance function of the noise process would not be adequate. Alternatively, if we changed the structure, the information required might change. Thus the three ideas of structure, criterion, and information are closely related. It is important to emphasize that the structured approach does not imply a linear system, as illustrated by Example 2.

Example 2. The input to the nonlinear no-memory device shown in Fig. 1.11 is $r(t)$, where

$$r(t) = s(t) + n(t), \qquad -\infty < t < \infty. \tag{34}$$

At any time t, $s(t)$ is the value of a random variable s with known probability density $p_s(S)$. Similarly, $n(t)$ is the value of a statistically independent random variable n with known density $p_n(N)$. The output of the device is $y(t)$, where

$$y(t) = a_0 + a_1[r(t)] + a_2[r(t)]^2 \tag{35}$$

is a quadratic no-memory function of $r(t)$. [The adjective no-memory emphasizes that the value of $y(t_0)$ depends *only* on $r(t_0)$.] We want to choose the coefficients a_0, a_1, and a_2 so that $y(t)$ is the minimum mean-square error estimate of $s(t)$. The mean-square error is

$$\begin{aligned} \xi(t) &\triangleq E\{[y(t) - s(t)^2]\} \\ &= E(\{a_0 + a_1[r(t)] + a_2[r^2(t)] - s(t)\}^2) \end{aligned} \tag{36}$$

and a_0, a_1, and a_2 are chosen to minimize $\xi(t)$. The solution to this particular problem is given in Chapter 3.

The technique for solving structured problems is conceptually straightforward. We allow the structure to vary within the allowed class and choose the particular system that maximizes (or minimizes) the criterion of interest.

An obvious advantage to the structured approach is that it usually requires only a partial characterization of the processes. This is important because, in practice, we must measure or calculate the process properties needed.

An obvious disadvantage is that it is often impossible to tell if the structure chosen is correct. In Example 1 a simple nonlinear system might

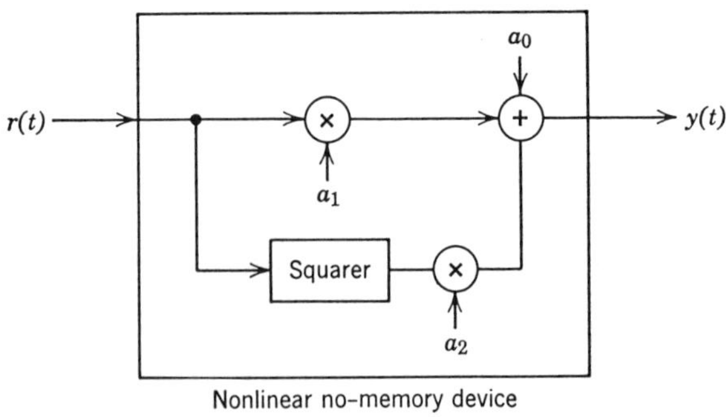

Fig. 1.11 A structured nonlinear device.

be far superior to the best linear system. Similarly, in Example 2 some other nonlinear system might be far superior to the quadratic system. Once a class of structure is chosen we are committed. A number of trivial examples demonstrate the effect of choosing the wrong structure. We shall encounter an important practical example when we study frequency modulation in Chapter II-2.

At first glance it appears that one way to get around the problem of choosing the proper strucutre is to let the structure be an arbitrary nonlinear time-varying system. In other words, the class of structure is chosen to be so large that every possible system will be included in it. The difficulty is that there is no convenient tool, such as the convolution integral, to express the output of a nonlinear system in terms of its input. This means that there is no convenient way to investigate all possible systems by using a structured approach.

The alternative to the structured approach is a nonstructured approach. Here we refuse to make any a priori guesses about what structure the processor should have. We establish a criterion, solve the problem, and implement whatever processing procedure is indicated.

A simple example of the nonstructured approach can be obtained by modifying Example 2. Instead of assigning characteristics to the device, we denote the estimate by $y(t)$. Letting

$$\xi(t) \triangleq E\{[y(t) - s(t)]^2\}, \tag{37}$$

we solve for the $y(t)$ that is obtained from $r(t)$ in *any* manner to minimize ξ. The obvious advantage is that if we can solve the problem we know that our answer, is *with respect to the chosen criterion*, the best processor of all possible processors. The obvious disadvantage is that we must completely characterize all the signals, channels, and noises that enter into the problem. Fortunately, it turns out that there are a large number of problems of practical importance in which this complete characterization is possible. Throughout both books we shall emphasize the nonstructured approach.

Our discussion up to this point has developed the topical and logical basis of these books. We now discuss the actual organization.

1.3 ORGANIZATION

The material covered in this book and Volume II can be divided into five parts. The first can be labeled *Background* and consists of Chapters 2 and 3. In Chapter 2 we develop in detail a topic that we call Classical Detection and Estimation Theory. Here we deal with problems in which the observations are sets of random variables instead of random waveforms. The theory needed to solve problems of this type has been studied by statisticians for many years. We therefore use the adjective classical to describe it. The purpose of the chapter is twofold: first, to derive all the basic statistical results we need in the remainder of the chapters; second, to provide a general background in detection and estimation theory that can be extended into various areas that we do not discuss in detail. To accomplish the second purpose we keep the discussion as general as possible. We consider in detail the binary and M-ary hypothesis testing problem, the problem of estimating random and nonrandom variables, and the composite hypothesis testing problem. Two more specialized topics, the general Gaussian problem and performance bounds on binary tests, are developed as background for specific problems we shall encounter later.

The next step is to bridge the gap between the classical case and the waveform problems discussed in Section 1.1. Chapter 3 develops the necessary techniques. The key to the transition is a suitable method for characterizing random processes. When the observation interval is finite, the most useful characterization is by a series expansion of the random process which is a generalization of the conventional Fourier series. When the observation interval is infinite, a transform characterization, which is a generalization of the usual Fourier transform, is needed. In the process of developing these characterizations, we encounter integral equations and we digress briefly to develop methods of solution. Just as in Chapter 2, our discussion is general and provides background for other areas of application.

With these two chapters in the first part as background, we are prepared to work our way through the hierarchy of problems outlined in Figs. 1.4, 1.7, and 1.10. The second part of the book (Chapter 4) can be labeled *Elementary Detection and Estimation Theory*. Here we develop the first two levels described in Section 1.1. (This material corresponds to the upper two levels in Figs. 1.4 and 1.7.) We begin by looking at the simple binary digital communication system described in Fig. 1.1 and then proceed to more complicated problems in the communications, radar, and sonar area involving M-ary communication, random phase channels, random amplitude and phase channels, and colored noise interference. By exploiting the parallel nature of the estimation problem, results are obtained easily for the estimation problem outlined in Fig. 1.5 and other more complex systems. The extension of the results to include the multiple channel (e.g., frequency diversity systems or arrays) and multiple parameter (e.g., range and Doppler) problems completes our discussion. The results in this chapter are fundamental to the understanding of modern communication and radar/sonar systems.

The third part, which can be labeled *Modulation Theory or Continuous Estimation Theory*, consists of Chapters 5 and 6 and Chapter 2 of Volume II. In Chapter 5 we formulate a quantitative model for the first two levels of the continuous waveform estimation problem and derive a set of integral equations whose solution is the optimum estimate of the message. We also derive equations that give bounds on the performance of the estimators. In order to study solution techniques, we divide the estimation problem into two categories, linear and nonlinear.

In Chapter 6 we study linear estimation problems in detail. In the first section of the chapter we discuss the relationships between various criteria, process characteristics, and the structure of the processor. In the next section we discuss the special case in which the processes are stationary and the infinite past is available. This case, the Wiener problem, leads to straightforward solution techniques. The original work of Wiener is extended to obtain some important closed-form error expressions. In the next section we discuss the case in which the processes can be characterized by using state-variable techniques. This case, the Kalman-Bucy problem, enables us to deal with nonstationary, finite-interval problems and adds considerable insight to the results of the preceding section.

The material in Chapters 1 through 6 has two characteristics:

1. In almost all cases we can obtain *explicit*, *exact* solutions to the problems that we formulate.

2. Most of the topics discussed are of such fundamental interest that everyone concerned with the statistical design of communication, radar, or sonar systems should be familiar with them.

As soon as we try to solve the nonlinear estimation problem, we see a sharp departure. To obtain useful results we must resort to approximate solution techniques. To decide what approximations are valid, however, we must consider specific nonlinear modulation systems. Thus the precise quantitative results are only applicable to the specific system. In view of this departure, we pause briefly in our logical development and summarize our results in Chapter 7.

After a brief introduction we return to the nonlinear modulation problem in Chapter 2 of Volume II and consider angle modulation systems in great detail. After an approximation to the optimum processor is developed, its performance and possible design modification are analyzed both theoretically and experimentally. More advanced techniques from Markov process theory and information theory are used to obtain significant results.

In the fourth part we revisit the problems of detection, estimation, and modulation theory at the third level of the hierarchy described in Section 1.1. Looking at the bottom boxes in Figs. 1.4, 1.7, and 1.10, we see that this is the *Random Signals in Noise* problem. Chapter II-3 studies it in detail. We find that the linear processors developed in Chapter I-6 play a fundamental role in the random signal problem. This result, coupled with the corresponding result in Chapter II-2, emphasizes the fundamental importance of the results in Chapter I-6. They also illustrate the inherent unity of the various problems. Specific topics such as power-spectrum parameter estimation and analog transmission over time-varying channels are also developed.

The fifth part is labeled *Applications* and includes Chapters II-4 and II-5. Throughout the two books we emphasize applications of the theory to models of practical problems. In most of them the relation of the actual physical situation can be explained in a page or two. The fifth part deals with physical situations in which developing the model from the physical situation is a central issue. Chapter II-4 studies the radar/sonar problem in depth. It builds up a set of target and channel models, starting with slowly fluctuating point targets and culminating in deep targets that fluctuate at arbitrary rates. This set of models enables us to study the signal design problem for radar and sonar, the resolution problem in mapping radars, the effect of reverberation on sonar-system performance, estimation of parameters of spread targets, communication over spread channels, and other important problems.

In Chapter II-5 we study various multidimensional problems such as multiplex communication systems and multivariable processing problems encountered in continuous receiving apertures and optical systems. The primary emphasis in the chapter is on optimum array processing in sonar (or seismic) systems. Both active and passive sonar systems are discussed; specific processor configurations are developed and their performance is analyzed.

Finally, in Chapter II-6 we summarize some of the more important results, mention some related topics that have been omitted, and suggest areas of future research.

2

Classical Detection and Estimation Theory

2.1 INTRODUCTION

In this chapter we develop in detail the basic ideas of classical detection and estimation theory. The first step is to define the various terms.

The basic components of a simple decision-theory problem are shown in Fig. 2.1. The first is a *source* that generates an output. In the simplest case this output is one of two choices. We refer to them as hypotheses and label them H_0 and H_1 in the two-choice case. More generally, the output might be one of M hypotheses, which we label $H_0, H_1, \ldots, H_{M-1}$. Some typical source mechanisms are the following:

1. A digital communication system transmits information by sending ones and zeros. When "one" is sent, we call it H_1, and when "zero" is sent, we call it H_0.
2. In a radar system we look at a particular range and azimuth and try

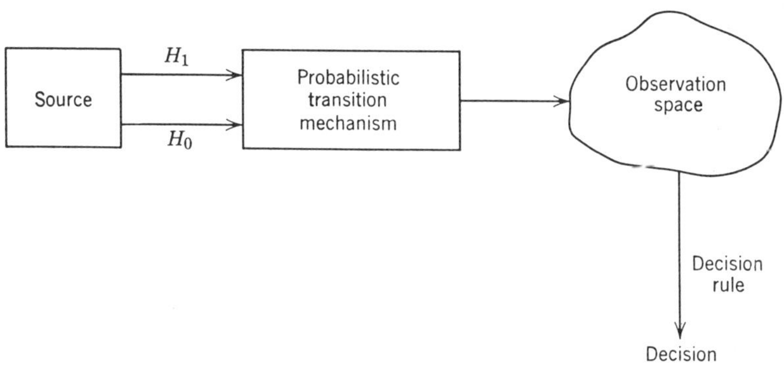

Fig. 2.1 Components of a decision theory problem.

to decide whether a target is present; H_1 corresponds to the presence of a target and H_0 corresponds to no target.

3. In a medical diagnosis problem we examine an electrocardiogram. Here H_1 could correspond to the patient having had a heart attack and H_0 to the absence of one.

4. In a speaker classification problem we know the speaker is German, British, or American and either male or female. There are six possible hypotheses.

In the cases of interest to us we do not know which hypothesis is true.

The second component of the problem is a *probabilistic transition mechanism*; the third is an *observation space*. The transition mechanism

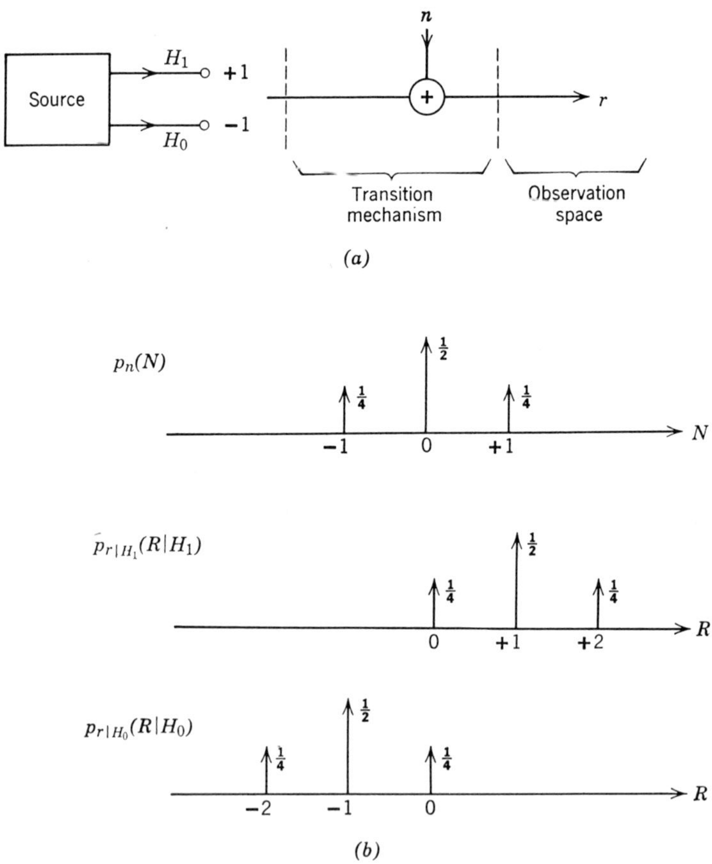

Fig. 2.2 A simple decision problem: (*a*) model; (*b*) probability densities.

can be viewed as a device that knows which hypothesis is true. Based on this knowledge, it generates a point in the observation space according to some probability law.

A simple example to illustrate these ideas is given in Fig. 2.2. When H_1 is true, the source generates $+1$. When H_0 is true, the source generates -1. An independent discrete random variable n whose probability density is shown in Fig. 2.2*b* is added to the source output. The sum of the source output and n is the observed variable r.

Under the two hypotheses, we have

$$\begin{aligned} H_1{:}r &= 1 + n, \\ H_0{:}r &= -1 + n. \end{aligned} \tag{1}$$

The probability densities of r on the two hypotheses are shown in Fig. 2.2*b*. The observation space is one-dimensional, for any output can be plotted on a line.

A related example is shown in Fig. 2.3*a* in which the source generates two numbers in sequence. A random variable n_1 is added to the first number and an independent random variable n_2 is added to the second. Thus

$$\begin{aligned} H_1{:}r_1 &= 1 + n_1 \\ r_2 &= 1 + n_2, \\ H_0{:}r_1 &= -1 + n_1 \\ r_2 &= -1 + n_2. \end{aligned} \tag{2}$$

The joint probability density of r_1 and r_2 when H_1 is true is shown in Fig. 2.3*b*. The observation space is two-dimensional and any observation can be represented as a point in a plane.

In this chapter we confine our discussion to problems in which the observation space is finite-dimensional. In other words, the observations consist of a set of N numbers and can be represented as a point in an N-dimensional space. This is the class of problem that statisticians have treated for many years. For this reason we refer to it as the *classical* decision problem.

The fourth component of the detection problem is a *decision* rule. After observing the outcome in the observation space we shall guess which hypothesis was true, and to accomplish this we develop a decision rule that assigns each point to one of the hypotheses. Suitable choices for decision rules will depend on several factors which we discuss in detail later. Our study will demonstrate how these four components fit together to form the total decision (or hypothesis-testing) problem.

The classical estimation problem is closely related to the detection problem. We describe it in detail later.

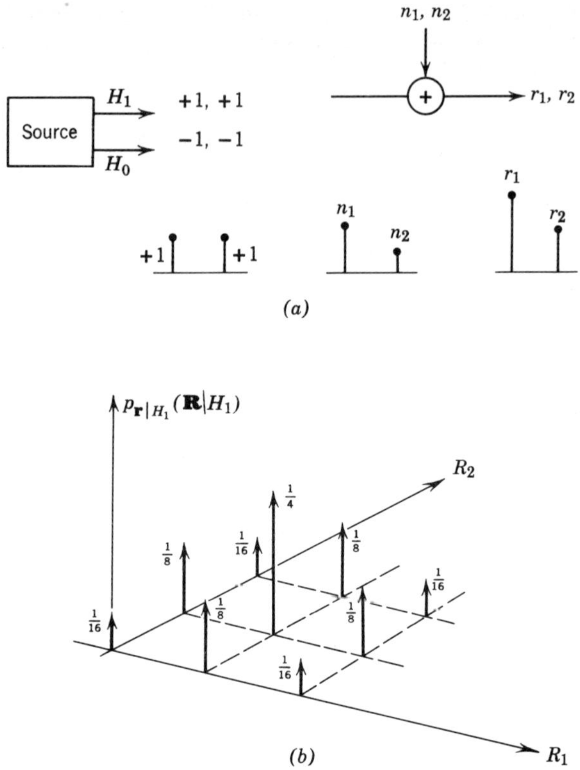

Fig. 2.3 A two-dimensional problem: (a) model; (b) probability density.

Organization. This chapter is organized in the following manner. In Section 2.2 we study the binary hypothesis testing problem. Then in Section 2.3 we extend the results to the case of M hypotheses. In Section 2.4 classical estimation theory is developed.

The problems that we encounter in Sections 2.2 and 2.3 are characterized by the property that each source output corresponds to a different hypothesis. In Section 2.5 we shall examine the composite hypothesis testing problem. Here a number of source outputs are lumped together to form a single hypothesis.

All of the developments through Section 2.5 deal with arbitrary probability transition mechanisms. In Section 2.6 we consider in detail a special class of problems that will be useful in the sequel. We refer to it as the general Gaussian class.

In many cases of practical importance we can develop the "optimum" decision rule according to certain criteria but cannot evaluate how well the test will work. In Section 2.7 we develop bounds and approximate expressions for the performance that will be necessary for some of the later chapters.

Finally, in Section 2.8 we summarize our results and indicate some of the topics that we have omitted.

2.2 SIMPLE BINARY HYPOTHESIS TESTS

As a starting point we consider the decision problem in which each of two source outputs corresponds to a hypothesis. Each hypothesis maps into a point in the observation space. We assume that the observation space corresponds to a set of N observations: $r_1, r_2, r_3, \ldots, r_N$. Thus each set can be thought of as a point in an N-dimensional space and can be denoted by a vector $\mathbf{r}$:

$$\mathbf{r} \triangleq \begin{bmatrix} r_1 \\ r_2 \\ \vdots \\ r_N \end{bmatrix} \tag{3}$$

The probabilistic transition mechanism generates points in accord with the two known conditional probability densities $p_{\mathbf{r}|H_1}(\mathbf{R}|H_1)$ and $p_{\mathbf{r}|H_0}(\mathbf{R}|H_0)$. The object is to use this information to develop a suitable decision rule. To do this we must look at various criteria for making decisions.

2.2.1 Decision Criteria

In the binary hypothesis problem we know that either H_0 or H_1 is true. We shall confine our discussion to decision rules that are required to make a choice. (An alternative procedure would be to allow decision rules with three outputs (a) H_0 true, (b) H_1 true, (c) don't know.) Thus each time the experiment is conducted one of four things can happen:

1. H_0 true; choose H_0.
2. H_0 true; choose H_1.
3. H_1 true; choose H_1.
4. H_1 true; choose H_0.

The first and third alternatives correspond to correct choices. The second and fourth alternatives correspond to errors. The purpose of a decision criterion is to attach some relative importance to the four possible courses of action. It might be expected that the method of processing the received data (**r**) would depend on the decision criterion we select. In this section we show that for the two criteria of most interest, the Bayes and the Neyman–Pearson, the operations on **r** are identical.

Bayes Criterion. A Bayes test is based on two assumptions. The first is that the source outputs are governed by probability assignments, which are denoted by P_1 and P_0, respectively, and called the a priori probabilities. These probabilities represent the observer's information about the source before the experiment is conducted. The second assumption is that a cost is assigned to each possible course of action. We denote the cost for the four courses of action as C_{00}, C_{10}, C_{11}, C_{01}, respectively. The first subscript indicates the hypothesis chosen and the second, the hypothesis that was true. Each time the experiment is conducted a certain cost will be incurred. We should like to design our decision rule so that *on the average* the cost will be as small as possible. To do this we first write an expression for the expected value of the cost. We see that there are two probabilities that we must average over; the a priori probability and the probability that a particular course of action will be taken. Denoting the expected value of the cost as the risk $\mathcal{R}$, we have:

$$\begin{aligned}\mathcal{R} = {} & C_{00}P_0 \Pr(\text{say } H_0|H_0 \text{ is true}) \\ & + C_{10}P_0 \Pr(\text{say } H_1|H_0 \text{ is true}) \\ & + C_{11}P_1 \Pr(\text{say } H_1|H_1 \text{ is true}) \\ & + C_{01}P_1 \Pr(\text{say } H_0|H_1 \text{ is true}).\end{aligned} \tag{4}$$

Because we have assumed that the decision rule must say either H_1 or H_0, we can view it as a rule for dividing the total observation space Z into two parts, Z_0 and Z_1, as shown in Fig. 2.4. Whenever an observation falls in Z_0 we say H_0, and whenever an observation falls in Z_1 we say H_1.

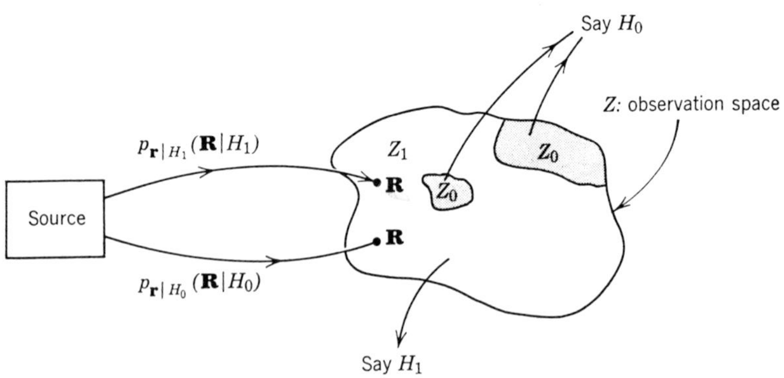

Fig. 2.4 Decision regions.

We can now write the expression for the risk in terms of the transition probabilities and the decision regions:

$$\begin{aligned}\mathcal{R} = {} & C_{00}P_0 \int_{Z_0} p_{\mathbf{r}|H_0}(\mathbf{R}|H_0)\, d\mathbf{R} \\ & + C_{10}P_0 \int_{Z_1} p_{\mathbf{r}|H_0}(\mathbf{R}|H_0)\, d\mathbf{R} \\ & + C_{11}P_1 \int_{Z_1} p_{\mathbf{r}|H_1}(\mathbf{R}|H_1)\, d\mathbf{R} \\ & + C_{01}P_1 \int_{Z_0} p_{\mathbf{r}|H_1}(\mathbf{R}|H_1)\, d\mathbf{R}. \end{aligned} \tag{5}$$

For an N-dimensional observation space the integrals in (5) are N-fold integrals.

We shall assume throughout our work that the cost of a wrong decision is higher than the cost of a correct decision. In other words,

$$\begin{aligned} C_{10} &> C_{00}, \\ C_{01} &> C_{11}. \end{aligned} \tag{6}$$

Now, to find the Bayes test we must choose the decision regions Z_0 and Z_1 in such a manner that the risk will be minimized. Because we require that a decision be made, this means that we must assign each point $\mathbf{R}$ in the observation space Z to Z_0 or Z_1.

Thus

$$Z = Z_0 + Z_1 \triangleq Z_0 \cup Z_1. \tag{7}$$

Rewriting (5), we have

$$\begin{aligned}\mathcal{R} = {} & P_0C_{00} \int_{Z_0} p_{\mathbf{r}|H_0}(\mathbf{R}|H_0)\, d\mathbf{R} + P_0C_{10} \int_{Z-Z_0} p_{\mathbf{r}|H_0}(\mathbf{R}|H_0)\, d\mathbf{R} \\ & + P_1C_{01} \int_{Z_0} p_{\mathbf{r}|H_1}(\mathbf{R}|H_1)\, d\mathbf{R} + P_1C_{11} \int_{Z-Z_0} p_{\mathbf{r}|H_1}(\mathbf{R}|H_1)\, d\mathbf{R}. \end{aligned} \tag{8}$$

Observing that

$$\int_Z p_{\mathbf{r}|H_0}(\mathbf{R}|H_0)\, d\mathbf{R} = \int_Z p_{\mathbf{r}|H_1}(\mathbf{R}|H_1)\, d\mathbf{R} = 1, \tag{9}$$

(8) reduces to

$$\begin{aligned}\mathcal{R} = {} & P_0C_{10} + P_1C_{11} \\ & + \int_{Z_0} \{[P_1(C_{01} - C_{11})p_{\mathbf{r}|H_1}(\mathbf{R}|H_1)] \\ & - [P_0(C_{10} - C_{00})p_{\mathbf{r}|H_0}(\mathbf{R}|H_0)]\}\, d\mathbf{R}. \end{aligned} \tag{10}$$

The first two terms represent the fixed cost. The integral represents the cost controlled by those points $\mathbf{R}$ that we assign to Z_0. The assumption in (6) implies that the two terms inside the brackets are positive. Therefore all values of $\mathbf{R}$ where the second term is larger than the first should be included in Z_0 because they contribute a negative amount to the integral. Similarly, all values of $\mathbf{R}$ where the first term is larger than the second should be excluded from Z_0 (assigned to Z_1) because they would contribute a positive amount to the integral. Values of $\mathbf{R}$ where the two terms are equal have no effect on the cost and may be assigned arbitrarily. We shall assume that these points are assigned to H_1 and ignore them in our subsequent discussion. Thus the decision regions are defined by the statement: If

$$P_1(C_{01} - C_{11})p_{\mathbf{r}|H_1}(\mathbf{R}|H_1) \geq P_0(C_{10} - C_{00})p_{\mathbf{r}|H_0}(\mathbf{R}|H_0), \tag{11}$$

assign $\mathbf{R}$ to Z_1 and consequently say that H_1 is true. Otherwise assign $\mathbf{R}$ to Z_0 and say H_0 is true.

Alternately, we may write

$$\frac{p_{\mathbf{r}|H_1}(\mathbf{R}|H_1)}{p_{\mathbf{r}|H_0}(\mathbf{R}|H_0)} \underset{H_0}{\overset{H_1}{\gtrless}} \frac{P_0(C_{10} - C_{00})}{P_1(C_{01} - C_{11})}. \tag{12}$$

The quantity on the left is called the *likelihood ratio* and denoted by $\Lambda(\mathbf{R})$

$$\boxed{\Lambda(\mathbf{R}) \triangleq \frac{p_{\mathbf{r}|H_1}(\mathbf{R}|H_1)}{p_{\mathbf{r}|H_0}(\mathbf{R}|H_0)}.} \tag{13}$$

Because it is the ratio of two functions of a random variable, it is a random variable. We see that regardless of the dimensionality of $\mathbf{R}$, $\Lambda(\mathbf{R})$ is a one-dimensional variable.

The quantity on the right of (12) is the threshold of the test and is denoted by η:

$$\eta \triangleq \frac{P_0(C_{10} - C_{00})}{P_1(C_{01} - C_{11})}. \tag{14}$$

Thus Bayes criterion leads us to a *likelihood ratio test* (LRT)

$$\Lambda(\mathbf{R}) \underset{H_0}{\overset{H_1}{\gtrless}} \eta. \tag{15}$$

We see that all the data processing is involved in computing $\Lambda(\mathbf{R})$ and is not affected by a priori probabilities or cost assignments. This invariance of the data processing is of considerable practical importance. Frequently the costs and a priori probabilities are merely educated guesses. The result in (15) enables us to build the entire processor and leave η as a variable threshold to accommodate changes in our estimates of a priori probabilities and costs.

Because the natural logarithm is a monotonic function, and both sides of (15) are positive, an equivalent test is

$$\ln \Lambda(\mathbf{R}) \underset{H_0}{\overset{H_1}{\gtrless}} \ln \eta. \tag{16}$$

Two forms of a processor to implement a likelihood ratio test are shown in Fig. 2.5.

Before proceeding to other criteria, we consider three simple examples.

Example 1. We assume that under H_1 the source output is a constant voltage m. Under H_0 the source output is zero. Before observation the voltage is corrupted by an additive noise. We sample the output waveform each second and obtain N samples. Each noise sample is a zero-mean Gaussian random variable n with variance σ^2. The noise samples at various instants are independent random variables and are independent of the source output. Looking at Fig. 2.6, we see that the observations under the two hypotheses are

$$\begin{aligned} H_1: r_i &= m + n_i \qquad i = 1, 2, \ldots, N, \\ H_0: r_i &= \phantom{m +{}} n_i \qquad i = 1, 2, \ldots, N, \end{aligned} \tag{17}$$

and

$$p_{n_i}(X) = \frac{1}{\sqrt{2\pi}\,\sigma} \exp\left(-\frac{X^2}{2\sigma^2}\right), \tag{18}$$

because the noise samples are Gaussian.

The probability density of r_i under each hypothesis follows easily:

$$p_{r_i|H_1}(R_i|H_1) = p_{n_i}(R_i - m) = \frac{1}{\sqrt{2\pi}\,\sigma} \exp\left(-\frac{(R_i - m)^2}{2\sigma^2}\right) \tag{19}$$

and

$$p_{r_i|H_0}(R_i|H_0) = p_{n_i}(R_i) = \frac{1}{\sqrt{2\pi}\,\sigma} \exp\left(-\frac{R_i^2}{2\sigma^2}\right). \tag{20}$$

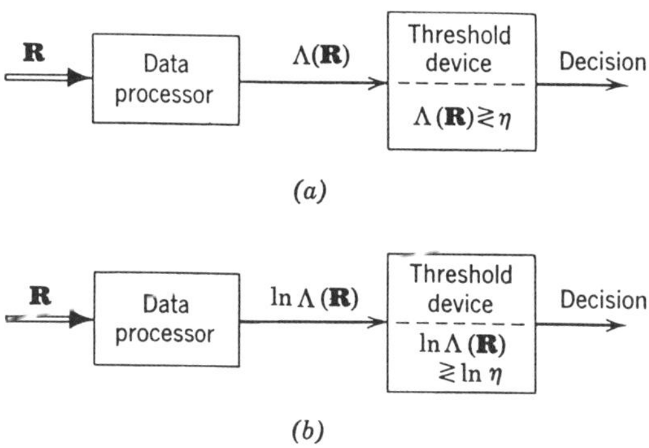

Fig. 2.5 Likelihood ratio processors.

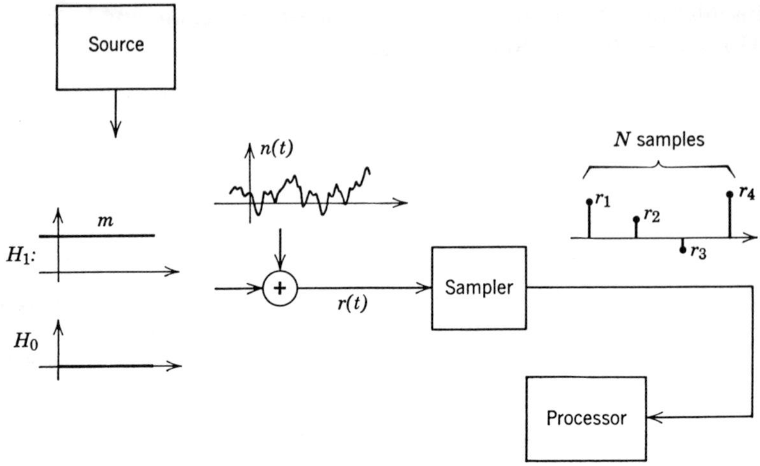

Fig. 2.6 Model for Example 1.

Because the n_i are statistically independent, the joint probability density of the r_i (or, equivalently, of the vector **r**) is simply the product of the individual probability densities. Thus

$$p_{\mathbf{r}|H_1}(\mathbf{R}|H_1) = \prod_{i=1}^{N} \frac{1}{\sqrt{2\pi}\,\sigma} \exp\left(-\frac{(R_i - m)^2}{2\sigma^2}\right), \tag{21}$$

and

$$p_{\mathbf{r}|H_0}(\mathbf{R}|H_0) = \prod_{i=1}^{N} \frac{1}{\sqrt{2\pi}\,\sigma} \exp\left(-\frac{R_i^2}{2\sigma^2}\right). \tag{22}$$

Substituting into (13), we have

$$\Lambda(\mathbf{R}) = \frac{\displaystyle\prod_{i=1}^{N} \frac{1}{\sqrt{2\pi}\,\sigma} \exp\left(-\frac{(R_i - m)^2}{2\sigma^2}\right)}{\displaystyle\prod_{i=1}^{N} \frac{1}{\sqrt{2\pi}\,\sigma} \exp\left(-\frac{R_i^2}{2\sigma^2}\right)}. \tag{23}$$

After canceling common terms and taking the logarithm, we have

$$\ln \Lambda(\mathbf{R}) = \frac{m}{\sigma^2} \sum_{i=1}^{N} R_i - \frac{Nm^2}{2\sigma^2}. \tag{24}$$

Thus the likelihood ratio test is

$$\frac{m}{\sigma^2} \sum_{i=1}^{N} R_i - \frac{Nm^2}{2\sigma^2} \underset{H_0}{\overset{H_1}{\gtrless}} \ln \eta \tag{25}$$

or, equivalently,

$$\sum_{i=1}^{N} R_i \underset{H_0}{\overset{H_1}{\gtrless}} \frac{\sigma^2}{m} \ln \eta + \frac{Nm}{2} \triangleq \gamma. \tag{26}$$

We see that the processor simply *adds* the observations and compares them with a threshold.

In this example the only way the data appear in the likelihood ratio test is in a sum. This is an example of a *sufficient statistic*, which we denote by $l(\mathbf{R})$ (or simply l when the argument is obvious). It is just a function of the received data which has the property that $\Lambda(\mathbf{R})$ can be written as a function of l. In other words, when making a decision, knowing the value of the sufficient statistic is just as good as knowing $\mathbf{R}$. In Example 1, l is a linear function of the R_i. A case in which this is not true is illustrated in Example 2.

Example 2. Several different physical situations lead to the mathematical model of interest in this example. The observations consist of a set of N values: $r_1, r_2, r_3, \ldots, r_N$. Under both hypotheses, the r_i are independent, identically distributed, zero-mean Gaussian random variables. Under H_1 each r_i has a variance σ_1^2. Under H_0 each r_i has a variance σ_0^2. Because the variables are independent, the joint density is simply the product of the individual densities. Therefore

$$p_{\mathbf{r}|H_1}(\mathbf{R}|H_1) = \prod_{i=1}^{N} \frac{1}{\sqrt{2\pi}\,\sigma_1} \exp\left(-\frac{R_i^2}{2\sigma_1^2}\right) \tag{27}$$

and

$$p_{\mathbf{r}|H_0}(\mathbf{R}|H_0) = \prod_{i=1}^{N} \frac{1}{\sqrt{2\pi}\,\sigma_0} \exp\left(-\frac{R_i^2}{2\sigma_0^2}\right). \tag{28}$$

Substituting (27) and (28) into (13) and taking the logarithm, we have

$$\frac{1}{2}\left(\frac{1}{\sigma_0^2} - \frac{1}{\sigma_1^2}\right) \sum_{i=1}^{N} R_i^2 + N \ln \frac{\sigma_0}{\sigma_1} \underset{H_0}{\overset{H_1}{\gtrless}} \ln \eta. \tag{29}$$

In this case the sufficient statistic is the sum of the squares of the observations

$$l(\mathbf{R}) = \sum_{i=1}^{N} R_i^2, \tag{30}$$

and an equivalent test for $\sigma_1^2 > \sigma_0^2$ is

$$l(\mathbf{R}) \underset{H_0}{\overset{H_1}{\gtrless}} \frac{2\sigma_0^2\sigma_1^2}{\sigma_1^2 - \sigma_0^2}\left(\ln \eta - N \ln \frac{\sigma_0}{\sigma_1}\right) \triangleq \gamma. \tag{31}$$

For $\sigma_1^2 < \sigma_0^2$ the inequality is reversed because we are multiplying by a negative number:

$$l(\mathbf{R}) \underset{H_1}{\overset{H_0}{\gtrless}} \frac{2\sigma_0^2\sigma_1^2}{\sigma_0^2 - \sigma_1^2}\left(N \ln \frac{\sigma_0}{\sigma_1} - \ln \eta\right) \triangleq \gamma'; \qquad (\sigma_1^2 < \sigma_0^2). \tag{32}$$

These two examples have emphasized Gaussian variables. In the next example we consider a different type of distribution.

Example 3. The Poisson distribution of events is encountered frequently as a model of shot noise and other diverse phenomena (e.g., [1] or [2]). Each time the experiment is conducted a certain number of events occur. Our observation is just this number which ranges from 0 to ∞ and obeys a Poisson distribution on both hypotheses; that is,

$$\Pr(n \text{ events}) = \frac{(m_i)^n}{n!} e^{-m_i}, \qquad n = 0, 1, 2 \ldots, i = 0, 1, \tag{33}$$

where m_i is the parameter that specifies the average number of events:

$$E(n) = m_i. \tag{34}$$

It is this parameter m_i that is different in the two hypotheses. Rewriting (33) to emphasize this point, we have for the two Poisson distributions

$$H_1: \Pr(n \text{ events}) = \frac{m_1^n}{n!} e^{-m_1}, \qquad n = 0, 1, 2, \ldots, \tag{35}$$

$$H_0: \Pr(n \text{ events}) = \frac{m_0^n}{n!} e^{-m_0}, \qquad n = 0, 1, 2, \ldots. \tag{36}$$

Then the likelihood ratio test is

$$\Lambda(n) = \left(\frac{m_1}{m_0}\right)^n \exp[-(m_1 - m_0)] \underset{H_0}{\overset{H_1}{\gtrless}} \eta \tag{37}$$

or, equivalently,

$$n \underset{H_0}{\overset{H_1}{\gtrless}} \frac{\ln \eta + m_1 - m_0}{\ln m_1 - \ln m_0}, \qquad \text{if } m_1 > m_0,$$

$$n \underset{H_1}{\overset{H_0}{\gtrless}} \frac{\ln \eta + m_1 - m_0}{\ln m_1 - \ln m_0}, \qquad \text{if } m_0 > m_1. \tag{38}$$

This example illustrates how the likelihood ratio test which we originally wrote in terms of probability densities can be simply adapted to accommodate observations that are discrete random variables. We now return to our general discussion of Bayes tests.

There are several special kinds of Bayes test which are frequently used and which should be mentioned explicitly.

If we assume that C_{00} and C_{11} are zero and $C_{01} = C_{10} = 1$, the expression for the risk in (8) reduces to

$$\mathcal{R} = P_0 \int_{Z_1} p_{\mathbf{r}|H_0}(\mathbf{R}|H_0)\, d\mathbf{R} + P_1 \int_{Z_0} p_{\mathbf{r}|H_1}(\mathbf{R}|H_1)\, d\mathbf{R}. \tag{39}$$

We see that (39) is just the total probability of making an error. Therefore for this cost assignment the Bayes test is minimizing the total probability of error. The test is

$$\ln \Lambda(\mathbf{R}) \underset{H_0}{\overset{H_1}{\gtrless}} \ln \frac{P_0}{P_1} = \ln P_0 - \ln(1 - P_0). \tag{40}$$

When the two hypotheses are equally likely, the threshold is zero. This assumption is normally true in digital communication systems. These processors are commonly referred to as minimum probability of error receivers.

A second special case of interest arises when the a priori probabilities are unknown. To investigate this case we look at (8) again. We observe that once the decision regions Z_0 and Z_1 are chosen, the values of the ntegrals are determined. We denote these values in the following manner:

$$P_F = \int_{Z_1} p_{\mathbf{r}|H_0}(\mathbf{R}|H_0)\,d\mathbf{R},$$

$$P_D = \int_{Z_1} p_{\mathbf{r}|H_1}(\mathbf{R}|H_1)\,d\mathbf{R}, \tag{41}$$

$$P_M = \int_{Z_0} p_{\mathbf{r}|H_1}(\mathbf{R}|H_1)\,d\mathbf{R} = 1 - P_D.$$

We see that these quantities are *conditional probabilities*. The subscripts are mnemonic and chosen from the radar problem in which hypothesis H_1 corresponds to the presence of a target and hypothesis H_0 corresponds to its absence. P_F is the probability of a *false alarm* (i.e., we say the target is present when it is not); P_D is the probability of *detection* (i.e., we say the target is present when it is); P_M is the probability of a *miss* (we say the target is absent when it is present). Although we are interested in a much larger class of problems than this notation implies, we shall use it for convenience.

For any choice of decision regions the risk expression in (8) can be written in the notation of (41):

$$\mathcal{R} = P_0C_{10} + P_1C_{11} + P_1(C_{01} - C_{11})P_M - P_0(C_{10} - C_{00})(1 - P_F). \tag{42}$$

Because

$$P_0 = 1 - P_1, \tag{43}$$

(42) becomes

$$\mathcal{R}(P_1) = C_{00}(1 - P_F) + C_{10}P_F + P_1[(C_{11} - C_{00}) + (C_{01} - C_{11})P_M - (C_{10} - C_{00})P_F]. \tag{44}$$

Now, if all the costs and a priori probabilities are known, we can find a Bayes test. In Fig. 2.7*a* we plot the Bayes risk, $\mathcal{R}_B(P_1)$, as a function of P_1. Observe that as P_1 changes the decision regions for the Bayes test change and therefore P_F and P_M change.

Now consider the situation in which a certain P_1 (say $P_1 = P_1^*$) is *assumed* and the corresponding Bayes test designed. We now fix the threshold and assume that P_1 is allowed to change. We denote the risk for this fixed threshold test as $\mathcal{R}_F(P_1^*, P_1)$. Because the threshold is fixed, P_F and P_M are fixed, and (44) is just a straight line. Because it is a Bayes test for $P_1 = P_1^*$, it touches the $\mathcal{R}_B(P_1)$ curve at that point. Looking at (14), we see that the threshold changes continuously with P_1. Therefore, whenever $P_1 \neq P_1^*$, the threshold in the Bayes test will be different. Because the Bayes test minimizes the risk,

$$\mathcal{R}_F(P_1^*, P_1) \geq \mathcal{R}_B(P_1). \tag{45}$$

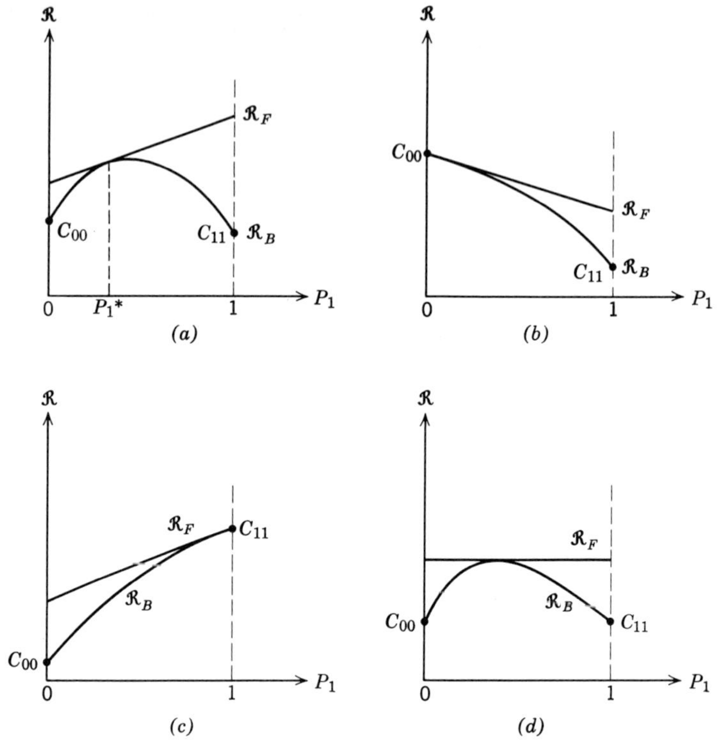

Fig. 2.7 Risk curves: (*a*) fixed risk versus typical Bayes risk; (*b*) maximum value of $\mathcal{R}_1$ at $P_1 = 0$.

If Λ is a continuous random variable with a probability distribution function that is strictly monotonic, then changing η always changes the risk. $\mathcal{R}_B(P_1)$ is strictly concave downward and the inequality in (45) is strict. This case, which is one of particular interest to us, is illustrated in Fig. 2.7*a*. We see that $\mathcal{R}_F(P_1^*, P_1)$ is tangent to $\mathcal{R}_B(P_1)$ at $P_1 = P_1^*$. These curves demonstrate the effect of incorrect knowledge of the a priori probabilities.

An interesting problem is encountered if we assume that the a priori probabilities are chosen to make our performance as bad as possible. In other words, P_1 is chosen to maximize our risk $\mathcal{R}_F(P_1^*, P_1)$. Three possible examples are given in Figs. 2.7*b*, *c*, and *d*. In Fig. 2.7*b* the maximum of $\mathcal{R}_B(P_1)$ occurs at $P_1 = 0$. To minimize the maximum risk we use a Bayes test designed assuming $P_1 = 0$. In Fig. 2.7*c* the maximum of $\mathcal{R}_B(P_1)$ occurs at $P_1 = 1$. To minimize the maximum risk we use a Bayes test designed assuming $P_1 = 1$. In Fig. 2.7*d* the maximum occurs inside the interval

[0, 1], and we choose $\mathcal{R}_F$ to be the horizontal line. This implies that the coefficient of P_1 in (44) must be zero:

$$(C_{11} - C_{00}) + (C_{01} - C_{11})P_M - (C_{10} - C_{00})P_F = 0. \tag{46}$$

A Bayes test designed to minimize the maximum possible risk is called a *minimax test.* Equation 46 is referred to as the minimax equation and is useful whenever the maximum of $\mathcal{R}_B(P_1)$ is interior to the interval.

A special cost assignment that is frequently logical is

$$C_{00} = C_{11} = 0 \tag{47}$$

(This guarantees the maximum is interior.)

Denoting,

$$\begin{aligned} C_{01} &= C_M, \\ C_{10} &= C_F. \end{aligned} \tag{48}$$

the risk is,

$$\begin{aligned} \mathcal{R}_F &= C_F P_F + P_1(C_M P_M - C_F P_F) \\ &= P_0 C_F P_F + P_1 C_M P_M \end{aligned} \tag{49}$$

and the minimax equation is

$$C_M P_M = C_F P_F. \tag{50}$$

Before continuing our discussion of likelihood ratio tests we shall discuss a second criterion and prove that it also leads to a likelihood ratio test.

Neyman–Pearson Tests. In many physical situations it is difficult to assign realistic costs or a priori probabilities. A simple procedure to bypass this difficulty is to work with the *conditional probabilities* P_F and P_D. In general, we should like to make P_F as small as possible and P_D as large as possible. For most problems of practical importance these are conflicting objectives. An obvious criterion is to constrain one of the probabilities and maximize (or minimize) the other. A specific statement of this criterion is the following:

Neyman–Pearson Criterion. Constrain $P_F = \alpha' \le \alpha$ and design a test to maximize P_D (or minimize P_M) under this constraint.

The solution is obtained easily by using Lagrange multipliers. We construct the function F,

$$F = P_M + \lambda [P_F - \alpha'], \tag{51}$$

or

$$F = \int_{Z_0} p_{\mathbf{r}|H_1}(\mathbf{R}|H_1)\, d\mathbf{R} + \lambda \left[\int_{Z_1} p_{\mathbf{r}|H_0}(\mathbf{R}|H_0)\, d\mathbf{R} - \alpha'\right], \tag{52}$$

Clearly, if $P_F = \alpha'$, then minimizing F minimizes P_M.

or

$$F = \lambda(1 - \alpha') + \int_{Z_0} [p_{\mathbf{r}|H_1}(\mathbf{R}|H_1) - \lambda p_{\mathbf{r}|H_0}(\mathbf{R}|H_0)]\, d\mathbf{R}. \tag{53}$$

Now observe that for any positive value of λ an LRT will minimize F. (A negative value of λ gives an LRT with the inequalities reversed.)

This follows directly, because to minimize F we assign a point $\mathbf{R}$ to Z_0 only when the term in the bracket is negative. This is equivalent to the test

$$\frac{p_{\mathbf{r}|H_1}(\mathbf{R}|H_1)}{p_{\mathbf{r}|H_0}(\mathbf{R}|H_0)} < \lambda, \qquad \text{assign point to } Z_0 \text{ or say } H_0. \tag{54}$$

The quantity on the left is just the likelihood ratio. Thus F is minimized by the likelihood ratio test

$$\Lambda(\mathbf{R}) \underset{H_0}{\overset{H_1}{\gtrless}} \lambda. \tag{55}$$

To satisfy the constraint we choose λ so that $P_F = \alpha'$. If we denote the density of Λ when H_0 is true as $p_{\Lambda|H_0}(\Lambda|H_0)$, then we require

$$P_F = \int_{\lambda}^{\infty} p_{\Lambda|H_0}(\Lambda|H_0)\, d\Lambda = \alpha'. \tag{56}$$

Solving (56) for λ gives the threshold. The value of λ given by (56) will be non-negative because $p_{\Lambda|H_0}(\Lambda|H_0)$ is zero for negative values of λ. Observe that decreasing λ is equivalent to increasing Z_1, the region where we say H_1. Thus P_D increases as λ decreases. Therefore we decrease λ until we obtain the largest possible $\alpha' \leq \alpha$. In most cases of interest to us P_F is a continuous function of λ and we have $P_F = \alpha$. We shall assume this continuity in all subsequent discussions. Under this assumption the Neyman–Pearson criterion leads to a likelihood ratio test. On p. 41 we shall see the effect of the continuity assumption not being valid.

Summary. In this section we have developed two ideas of fundamental importance in hypothesis testing. The first result is the demonstration that for a Bayes or a Neyman–Pearson criterion the optimum test consists of processing the observation $\mathbf{R}$ to find the likelihood ratio $\Lambda(\mathbf{R})$ and then comparing $\Lambda(\mathbf{R})$ to a threshold in order to make a decision. Thus, regardless of the dimensionality of the observation space, the decision space is one-dimensional.

The second idea is that of a sufficient statistic $l(\mathbf{R})$. The idea of a sufficient statistic originated when we constructed the likelihood ratio and saw that it depended explicitly only on $l(\mathbf{R})$. If we actually construct $\Lambda(\mathbf{R})$ and then recognize $l(\mathbf{R})$, the notion of a sufficient statistic is perhaps of secondary value. A more important case is when we can recognize $l(\mathbf{R})$ directly. An easy way to do this is to examine the geometric interpretation of a sufficient statistic. We considered the observations $r_1, r_2, \ldots, r_N$ as a point $\mathbf{r}$ in an N-dimensional space, and one way to describe this point is to use these coordinates. When we choose a sufficient statistic, we are simply describing the point in a coordinate system that is more useful for the decision problem. We denote the first coordinate in this system by l, the sufficient statistic, and the remaining $N - 1$ coordinates which will not affect our decision by the $(N - 1)$-dimensional vector $\mathbf{y}$. Thus

$$\Lambda(\mathbf{R}) = \Lambda(L, \mathbf{Y}) = \frac{p_{l,\mathbf{y}|H_1}(L, \mathbf{Y}|H_1)}{p_{l,\mathbf{y}|H_0}(L, \mathbf{Y}|H_0)}. \tag{57}$$

Now the expression on the right can be written as

$$\Lambda(L, \mathbf{Y}) = \frac{p_{l|H_1}(L|H_1)p_{\mathbf{y}|l,H_1}(\mathbf{Y}|L, H_1)}{p_{l|H_0}(L|H_0)p_{\mathbf{y}|l,H_0}(\mathbf{Y}|L, H_0)}. \tag{58}$$

If l is a sufficient statistic, then $\Lambda(\mathbf{R})$ must reduce to $\Lambda(L)$. This implies that the second terms in the numerator and denominator must be equal. In other words,

$$p_{\mathbf{y}|l,H_0}(\mathbf{Y}|L, H_0) = p_{\mathbf{y}|l,H_1}(\mathbf{Y}|L, H_1) \tag{59}$$

because the density of $\mathbf{y}$ cannot depend on which hypothesis is true. We see that choosing a sufficient statistic simply amounts to picking a coordinate system in which one coordinate contains all the information necessary to making a decision. The other coordinates contain no information and can be disregarded for the purpose of making a decision.

In Example 1 the new coordinate system could be obtained by a simple rotation. For example, when $N = 2$,

$$\begin{aligned} L &= \frac{1}{\sqrt{2}}(R_1 + R_2), \\ Y &= \frac{1}{\sqrt{2}}(R_1 - R_2). \end{aligned} \tag{60}$$

In Example 2 the new coordinate system corresponded to changing to polar coordinates. For $N = 2$

$$\begin{aligned} L &= R_1^2 + R_2^2, \\ Y &= \tan^{-1}\frac{R_2}{R_1}. \end{aligned} \tag{61}$$

Notice that the vector $\mathbf{y}$ can be chosen in order to make the demonstration of the condition in (59) as simple as possible. The only requirement is that the pair $(l, \mathbf{y})$ must describe any point in the observation space. We should also observe that the condition

$$p_{\mathbf{y}|H_1}(\mathbf{Y}|H_1) = p_{\mathbf{y}|H_0}(\mathbf{Y}|H_0) \tag{62}$$

does *not* imply (59) unless l and $\mathbf{y}$ are independent under H_1 and H_0. Frequently we will choose $\mathbf{y}$ to obtain this independence and then use (62) to verify that l is a sufficient statistic.

2.2.2 Performance: Receiver Operating Characteristic

To complete our discussion of the simple binary problem we must evaluate the performance of the likelihood ratio test. For a Neyman–Pearson test the values of P_F and P_D completely specify the test performance. Looking at (42) we see that the Bayes risk $\mathcal{R}_B$ follows easily if P_F and P_D are known. Thus we can concentrate our efforts on calculating P_F and P_D.

We begin by considering Example 1 in Section 2.2.1.

Example 1. From (25) we see that an equivalent test is

$$l = \frac{1}{\sqrt{N}\,\sigma} \sum_{i=1}^{N} R_i \underset{H_0}{\overset{H_1}{\gtrless}} \frac{\sigma}{\sqrt{N}\,m} \ln \eta + \frac{\sqrt{N}\,m}{2\sigma}. \tag{63}$$

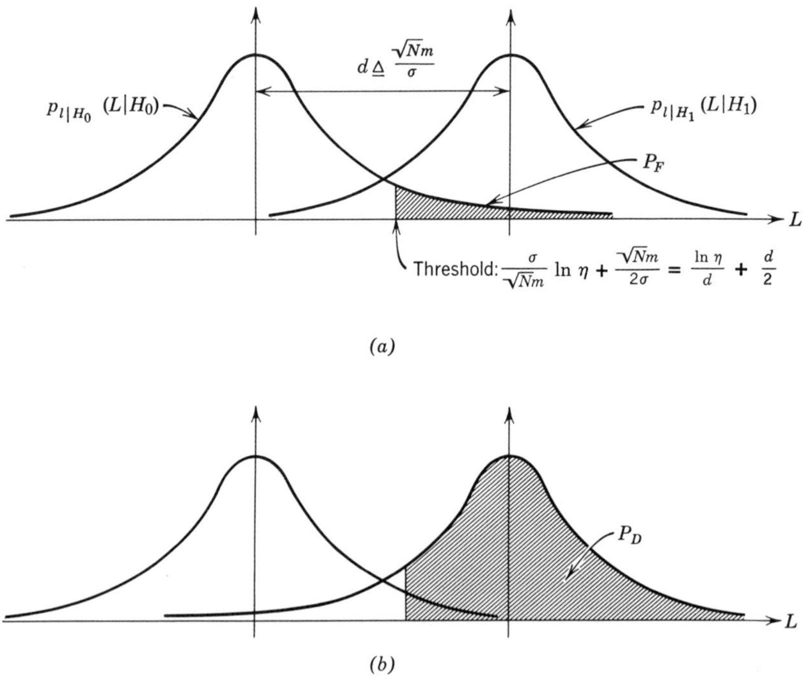

Fig. 2.8 Error probabilities: (*a*) P_F calculation; (*b*) P_D calculation.

We have multiplied (25) by $\sigma/\sqrt{N}\,m$ to normalize the next calculation. Under H_0, l is obtained by adding N independent zero-mean Gaussian variables with variance σ^2 and then dividing by $\sqrt{N}\,\sigma$. Therefore l is $N(0, 1)$.

Under H_1, l is $N(\sqrt{N}\,m/\sigma, 1)$. The probability densities on the two hypotheses are sketched in Fig. 2.8*a*. The threshold is also shown. Now, P_F is simply the integral of $p_{l|H_0}(L|H_0)$ to the right of the threshold.

Thus

$$P_F = \int_{(\ln \eta)/d + d/2}^{\infty} \frac{1}{\sqrt{2\pi}} \exp\left(-\frac{x^2}{2}\right) dx, \tag{64}$$

where $d \triangleq \sqrt{N}\,m/\sigma$ is the distance between the means of the two densities. The integral in (64) is tabulated in many references (e.g., [3] or [4]).

We generally denote

$$\text{erf}_* (X) \triangleq \int_{-\infty}^{X} \frac{1}{\sqrt{2\pi}} \exp\left(-\frac{x^2}{2}\right) dx, \tag{65}$$

where erf_* is an abbreviation for the error function† and

$$\text{erfc}_* (X) \triangleq \int_{X}^{\infty} \frac{1}{\sqrt{2\pi}} \exp\left(-\frac{x^2}{2}\right) dx \tag{66}$$

is its complement. In this notation

$$P_F = \text{erfc}_* \left(\frac{\ln \eta}{d} + \frac{d}{2}\right). \tag{67}$$

Similarly, P_D is the integral of $p_{l|H_1}(L|H_1)$ to the right of the threshold, as shown in Fig. 2.8*b*:

$$\begin{aligned} P_D &= \int_{(\ln \eta)/d + d/2}^{\infty} \frac{1}{\sqrt{2\pi}} \exp\left[-\frac{(x - d)^2}{2}\right] dx \\ &= \int_{(\ln \eta)/d - d/2}^{\infty} \frac{1}{\sqrt{2\pi}} \exp\left(-\frac{y^2}{2}\right) dy \triangleq \text{erfc}_* \left(\frac{\ln \eta}{d} - \frac{d}{2}\right). \end{aligned} \tag{68}$$

In Fig. 2.9*a* we have plotted P_D versus P_F for various values of d with η as the varying parameter. For $\eta = 0$, $\ln \eta = -\infty$, and the processor always guesses H_1. Thus $P_F = 1$ and $P_D = 1$. As η increases, P_F and P_D decrease. When $\eta = \infty$, the processor always guesses H_0 and $P_F = P_D = 0$.

As we would expect from Fig. 2.8, the performance increases monotonically with d. In Fig. 2.9*b* we have replotted the results to give P_D versus d with P_F as a parameter on the curves. For a particular d we can obtain any point on the curve by choosing η appropriately $(0 \le \eta \le \infty)$.

The result in Fig. 2.9*a* is referred to as the receiver operating characteristic (ROC). It completely describes the performance of the test as a function of the parameter of interest.

A special case that will be important when we look at communication systems is the case in which we want to minimize the total probability of error

$$\Pr(\epsilon) \triangleq P_0 P_F + P_1 P_M. \tag{69a}$$

† The function that is usually tabulated is $\text{erf}(X) = \sqrt{2/\pi} \int_0^X \exp(-y^2)\, dy$, which is related to (65) in an obvious way.

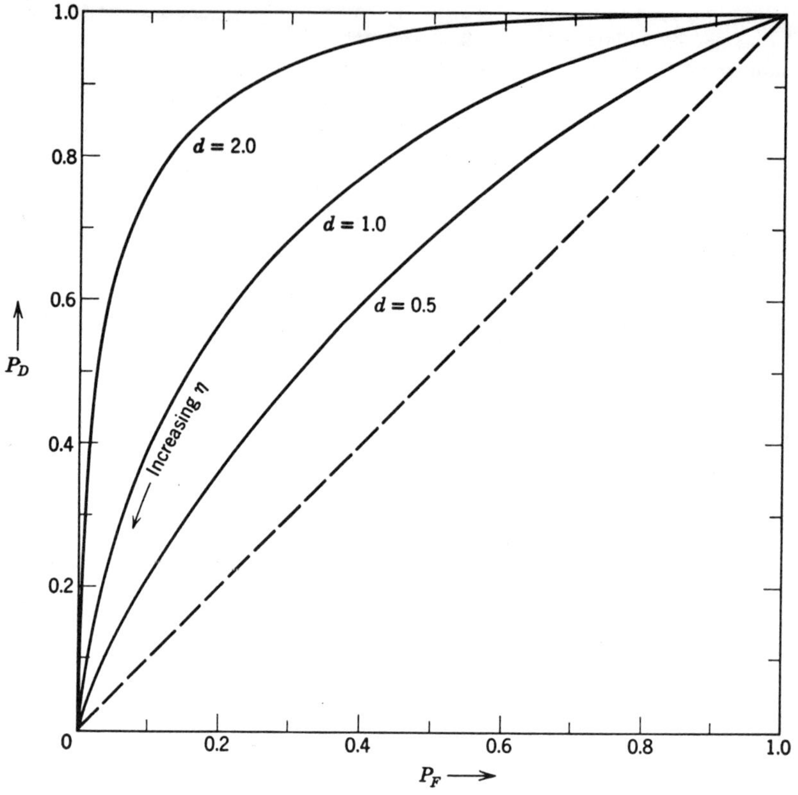

(a)

Fig. 2.9 (a) Receiver operating characteristic: Gaussian variables with unequal means.

The threshold for this criterion was given in (40). For the special case in which $P_0 = P_1$ the threshold η equals one and

$$\Pr(\epsilon) = \tfrac{1}{2}(P_F + P_M). \tag{69b}$$

Using (67) and (68) in (69), we have

$$\Pr(\epsilon) = \int_{+d/2}^{\infty} \frac{1}{\sqrt{2\pi}} \exp\left(-\frac{x^2}{2}\right) dx = \operatorname{erfc}_*\left(+\frac{d}{2}\right). \tag{70}$$

It is obvious from (70) that we could also obtain the Pr (ϵ) from the ROC. However, if this is the only threshold setting of interest, it is generally easier to calculate the Pr (ϵ) directly.

Before calculating the performance of the other two examples, it is worthwhile to point out two simple bounds on $\operatorname{erfc}_*(X)$. They will enable us to discuss its approximate behavior analytically. For $X > 0$

$$\frac{1}{\sqrt{2\pi}\,X}\left(1 - \frac{1}{X^2}\right)\exp\left(-\frac{X^2}{2}\right) < \operatorname{erfc}_*(X) < \frac{1}{\sqrt{2\pi}\,X}\exp\left(-\frac{X^2}{2}\right). \quad (71)$$

This can be derived by integrating by parts. (See Problem 2.2.15 or Feller [30].) A second bound is

$$\operatorname{erfc}_*(X) < \tfrac{1}{2}\exp\left(-\frac{X^2}{2}\right), \qquad x > 0, \quad (72)$$

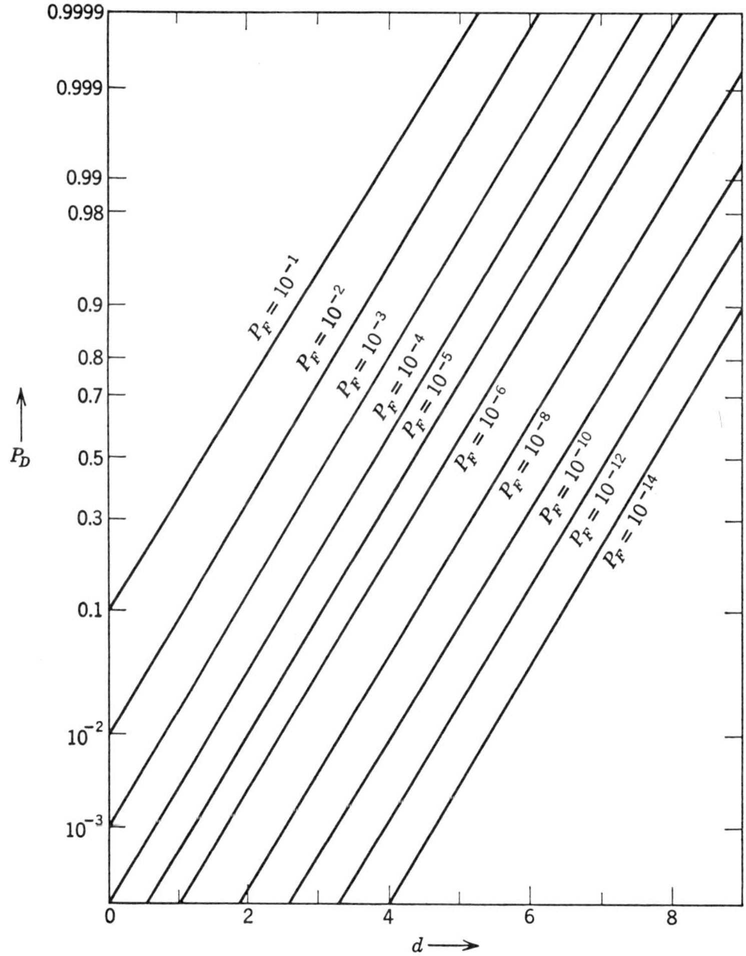

(b)

Fig. 2.9 **(b) detection probability versus *d*.**

which can also be derived easily (see Problem 2.2.16). The four curves are plotted in Fig. 2.10. We note that $\text{erfc}_*(X)$ decreases exponentially.

The receiver operating characteristics for the other two examples are also of interest.

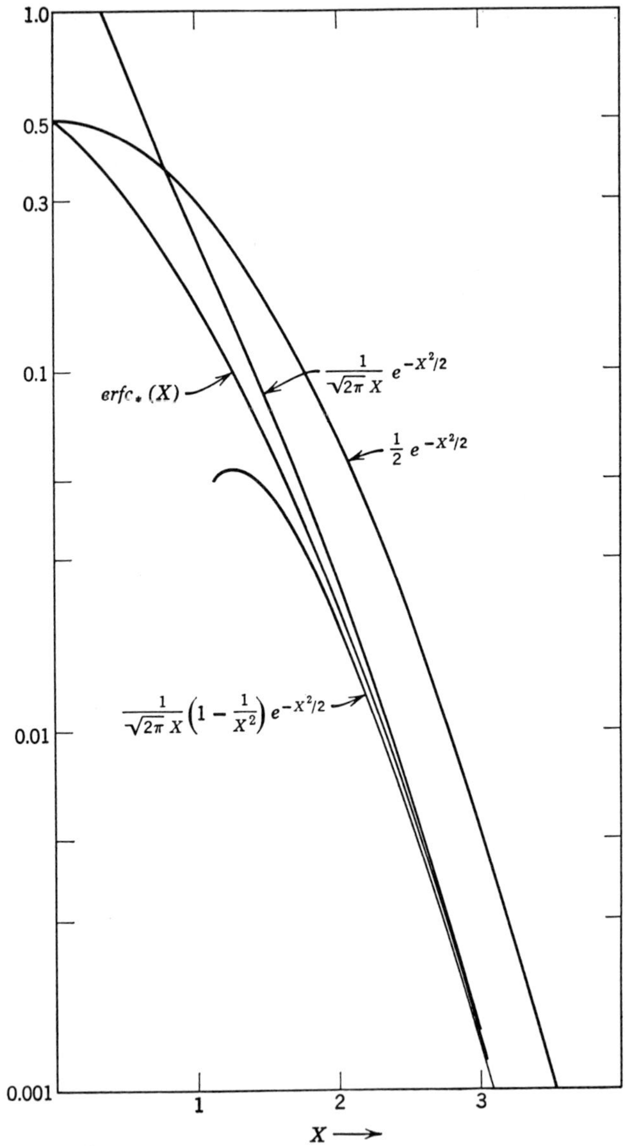

Fig. 2.10 Plot of $\text{erfc}_*(X)$ and related functions.

Example 2. In this case the test is

$$l(\mathbf{R}) = \sum_{i=1}^{N} R_i^2 \underset{H_0}{\overset{H_1}{\gtrless}} \frac{2\sigma_0^2\sigma_1^2}{\sigma_1^2 - \sigma_0^2}\left(\ln \eta - N \ln \frac{\sigma_0}{\sigma_1}\right) = \gamma, \qquad (\sigma_1 > \sigma_0). \tag{73}$$

The performance calculation for arbitrary N is somewhat tedious, so we defer it until Section 2.6. A particularly simple case appearing frequently in practice is $N = 2$. Under H_0 the r_i are independent zero-mean Gaussian variables with variances equal to σ_0^2:

$$P_F = \Pr(l \geq \gamma | H_0) = \Pr(r_1^2 + r_2^2 \geq \gamma | H_0). \tag{74}$$

To evaluate the expression on the right, we change to polar coordinates:

$$\begin{aligned} r_1 &= z \cos\theta, \qquad & z &= \sqrt{r_1^2 + r_2^2} \\ r_2 &= z \sin\theta, \qquad & \theta &= \tan^{-1}\frac{r_2}{r_1} \end{aligned} \tag{75}$$

Then

$$\Pr(z^2 \geq \gamma | H_0) = \int_0^{2\pi} d\theta \int_{\sqrt{\gamma}}^{\infty} Z \frac{1}{2\pi\sigma_0^2} \exp\left(-\frac{Z^2}{2\sigma_0^2}\right) dZ. \tag{76}$$

Integrating with respect to θ, we have

$$P_F = \int_{\sqrt{\gamma}}^{\infty} Z \frac{1}{\sigma_0^2} \exp\left(-\frac{Z^2}{2\sigma_0^2}\right) dZ. \tag{77}$$

We observe that l, the sufficient statistic, equals z^2. Changing variables, we have

$$P_F = \int_{\gamma}^{\infty} \frac{1}{2\sigma_0^2} \exp\left(-\frac{L}{2\sigma_0^2}\right) dL = \exp\left(-\frac{\gamma}{2\sigma_0^2}\right). \tag{78}$$

(Note that the probability density of the sufficient statistic is exponential.)

Similarly,

$$P_D = \exp\left(-\frac{\gamma}{2\sigma_1^2}\right). \tag{79}$$

To construct the ROC we can combine (78) and (79) to eliminate the threshold γ. This gives

$$P_D = (P_F)^{\sigma_0^2/\sigma_1^2}. \tag{80}$$

In terms of logarithms

$$\ln P_D = \frac{\sigma_0^2}{\sigma_1^2} \ln P_F. \tag{81}$$

As expected, the performance improves monotonically as the ratio σ_1^2/σ_0^2 increases. We shall study this case and its generalizations in more detail in Section 2.6.

The two Poisson distributions are thc third example.

Example 3. From (38), the likelihood ratio test is

$$n \underset{H_0}{\overset{H_1}{\gtrless}} \frac{\ln \eta + m_1 - m_0}{\ln m_1 - \ln m_0} = \gamma, \qquad (m_1 > m_0). \tag{82}$$

Because n takes on only integer values, it is more convenient to rewrite (82) as

$$n \underset{H_0}{\overset{H_1}{\gtrless}} \gamma_I, \qquad \gamma_I = 0, 1, 2, \ldots, \tag{83}$$

where γ_I takes on only integer values. Using (35),

$$P_D = 1 - e^{-m_1} \sum_{n=0}^{\gamma_I - 1} \frac{(m_1)^n}{n!}, \qquad \gamma_I = 0, 1, 2, \ldots, \tag{84}$$

and from (36)

$$P_F = 1 - e^{-m_0} \sum_{n=0}^{\gamma_I - 1} \frac{(m_0)^n}{n!}, \qquad \gamma_I = 0, 1, 2, \ldots. \tag{85}$$

The resulting ROC is plotted in Fig. 2.11*a* for some representative values of m_0 and m_1.

We see that it consists of a series of points and that P_F goes from 1 to $1 - e^{-m_0}$ when the threshold is changed from 0 to 1. Now suppose we wanted P_F to have an intermediate value, say $1 - \frac{1}{2}e^{-m_0}$. To achieve this performance we proceed in the following manner. Denoting the LRT with $\gamma_I = 0$ as LRT No. 0 and the LRT with $\gamma_I = 1$ as LRT No. 1, we have the following table:

LRT	γ_I	P_F	P_D
0	0	1	1
1	1	$1 - e^{-m_0}$	$1 - e^{-m_1}$

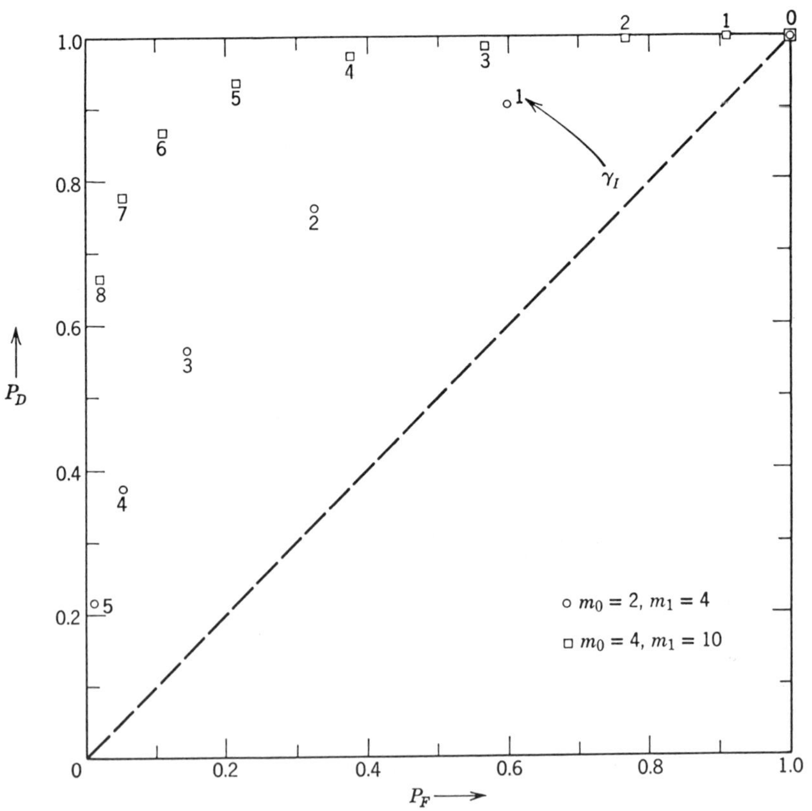

Fig. 2.11 (*a*) Receiver operating characteristic, Poisson problem.

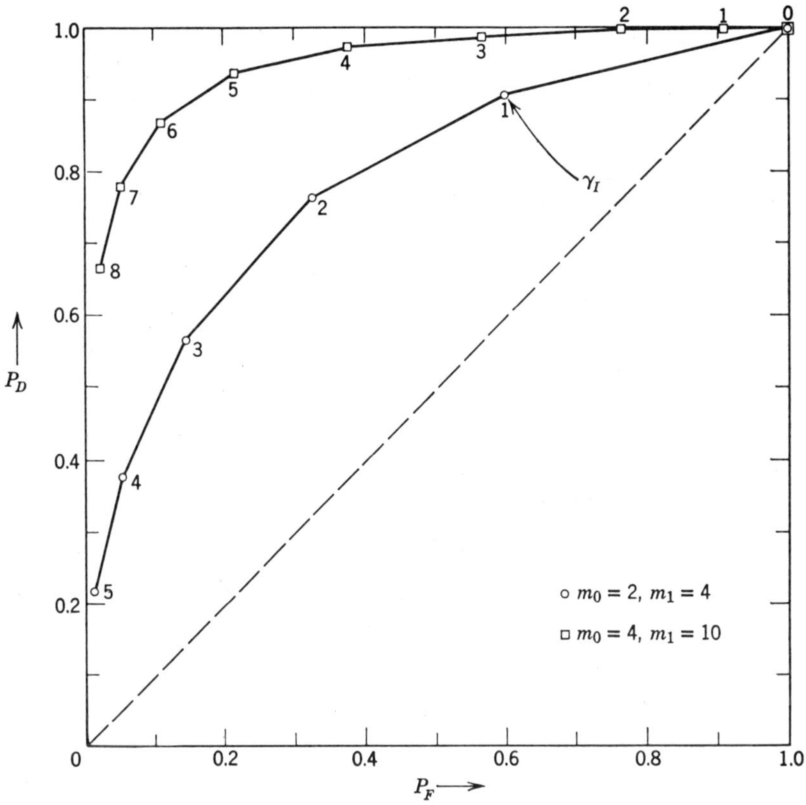

Fig. 2.11 (*b*) Receiver operating characteristic with randomized decision rule.

To get the desired value of P_F we use LRT No. 0 with probability $\frac{1}{2}$ and LRT No. 1 with probability $\frac{1}{2}$. The test is

$$\begin{aligned} \text{If } n = 0, \quad & \text{say } H_1 \text{ with probability } \tfrac{1}{2}, \\ & \text{say } H_0 \text{ with probability } \tfrac{1}{2}, \\ n \geq 1 \quad & \text{say } H_1. \end{aligned}$$

This procedure, in which we mix two likelihood ratio tests in some probabilistic manner, is called a *randomized decision rule*. The resulting P_D is simply a weighted combination of detection probabilities for the two tests.

$$P_D = 0.5(1) + 0.5(1 - e^{-m_1}) = (1 - 0.5\,e^{-m_1}). \tag{86}$$

We see that the ROC for randomized tests consists of straight lines which connect the points in Fig. 2.11*a*, as shown in Fig. 2.11*b*. The reason that we encounter a randomized test is that the observed random variables are discrete. Therefore $\Lambda(\mathbf{R})$ is a discrete random variable and, using an ordinary likelihood ratio test, only certain values of P_F are possible.

Looking at the expression for P_F in (56) and denoting the threshold by η, we have

$$P_F(\eta) = \int_{\eta}^{\infty} p_{\Lambda|H_0}(X|H_0)\,dX. \tag{87}$$

If $P_F(\eta)$ is a continuous function of η, we can achieve a desired value from 0 to 1 by a suitable choice of η and a randomized test will never be needed. This is the only case of interest to us in the sequel (see Prob. 2.2.12).

With these examples as a background, we now derive a few general properties of receiver operating characteristics. We confine our discussion to continuous likelihood ratio tests.

Two properties of *all* ROC's follow immediately from this example.

Property 1. All continuous likelihood ratio tests have ROC's that are concave downward. If they were not, a randomized test would be better. This would contradict our proof that a LRT is optimum (see Prob. 2.2.12).

Property 2. All continuous likelihood ratio tests have ROC's that are above the $P_D = P_F$ line. This is just a special case of Property 1 because the points ($P_F = 0, P_D = 0$) and ($P_F - 1, P_D = 1$) are contained on all ROC's.

Property 3. The slope of a curve in a ROC at a particular point is equal to the value of the threshold η required to achieve the P_D and P_F of that point.

Proof.

$$P_D = \int_{\eta}^{\infty} p_{\Lambda|H_1}(\Lambda|H_1)\,d\Lambda,$$

$$P_F = \int_{\eta}^{\infty} p_{\Lambda|H_0}(\Lambda|H_0)\,d\Lambda. \tag{88}$$

Differentiating both expressions with respect to η and writing the results as a quotient, we have

$$\frac{dP_D/d\eta}{dP_F/d\eta} = \frac{-p_{\Lambda|H_1}(\eta|H_1)}{-p_{\Lambda|H_0}(\eta|H_0)} = \frac{dP_D}{dP_F}. \tag{89}$$

We now show that

$$\frac{p_{\Lambda|H_1}(\eta|H_1)}{p_{\Lambda|H_0}(\eta|H_0)} = \eta. \tag{90}$$

Let

$$\Omega(\eta) \triangleq \{\mathbf{R}|\Lambda(\mathbf{R}) \geq \eta\} = \left[\mathbf{R}\left|\frac{p_{\mathbf{r}|H_1}(\mathbf{R}|H_1)}{p_{\mathbf{r}|H_0}(\mathbf{R}|H_0)} \geq \eta\right.\right]. \tag{91}$$

Then

$$P_D(\eta) \triangleq \Pr\{\Lambda(\mathbf{R}) \geq \eta|H_1\} = \int_{\Omega(\eta)} p_{\mathbf{r}|H_1}(\mathbf{R}|H_1)\,d\mathbf{R}$$

$$= \int_{\Omega(\eta)} \Lambda(\mathbf{R})p_{\mathbf{r}|H_0}(\mathbf{R}|H_0)\,d\mathbf{R}, \tag{92}$$

where the last equality follows from the definition of the likelihood ratio. Using the definition of $\Omega(\eta)$, we can rewrite the last integral

$$P_D(\eta) = \int_{\Omega(\eta)} \Lambda(\mathbf{R}) p_{\mathbf{r}|H_0}(\mathbf{R}|H_0)\, d\mathbf{R} = \int_\eta^\infty X p_{\Lambda|H_0}(X|H_0)\, dX. \quad (93)$$

Differentiating (93) with respect to η, we obtain

$$\frac{dP_D(\eta)}{d\eta} = -\eta p_{\Lambda|H_0}(\eta|H_0). \quad (94)$$

Equating the expression for $dP_D(n)/d\eta$ in the numerator of (89) to the right side of (94) gives the desired result.

We see that this result is consistent with Example 1. In Fig. 2.9*a*, the curves for nonzero d have zero slope at $P_F = P_D = 1$ $(\eta = 0)$ and infinite slope at $P_F = P_D = 0$ $(\eta = \infty)$.

Property 4. Whenever the maximum value of the Bayes risk is interior to the interval (0, 1) on the P_1 axis, the minimax operating point is the intersection of the line

$$(C_{11} - C_{00}) + (C_{01} - C_{11})(1 - P_D) - (C_{10} - C_{00})P_F = 0 \quad (95)$$

and the appropriate curve of the ROC (see 46). In Fig. 2.12 we show the special case defined by (50),

$$C_F P_F = C_M P_M = C_M(1 - P_D), \quad (96)$$

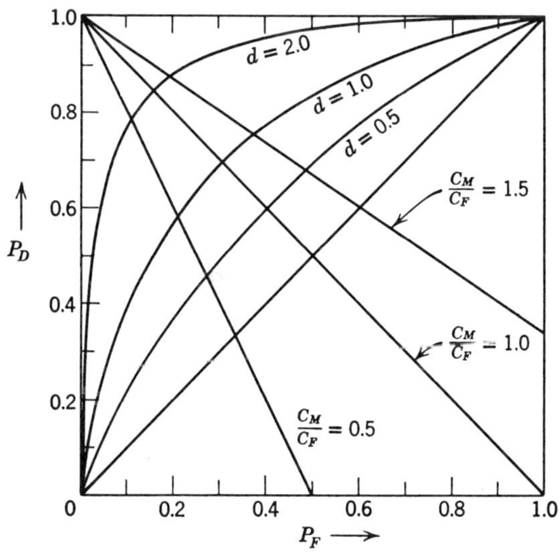

Fig. 2.12 Determination of minimax operating point.

superimposed on the ROC of Example 1. We see that it starts at the point $P_F = 0$, $P_D = 1$, and intersects the $P_F = 1$ line at

$$P_F = 1 - \frac{C_F}{C_M}. \tag{97}$$

This completes our discussion of the binary hypothesis testing problem. Several key ideas should be re-emphasized:

1. Using either a Bayes criterion or a Neyman–Pearson criterion, we find that the optimum test is a likelihood ratio test. Thus, regardless of the dimensionality of the observation space, the test consists of comparing a scalar variable $\Lambda(\mathbf{R})$ with a threshold. (We assume $P_F(\eta)$ is continuous.)
2. In many cases construction of the LRT can be simplified if we can identify a sufficient statistic. Geometrically, this statistic is just that coordinate in a suitable coordinate system which describes the observation space that contains *all* the information necessary to make a decision.
3. A complete description of the LRT performance was obtained by plotting the conditional probabilities P_D and P_F as the threshold η was varied. The resulting ROC could be used to calculate the Bayes risk for any set of costs. In many cases only one value of the threshold is of interest and a complete ROC is not necessary.

A number of interesting binary tests are developed in the problems.

2.3 M HYPOTHESES

The next case of interest is one in which we must choose one of M hypotheses. In the simple binary hypothesis test there were two source outputs, each of which corresponded to a single hypothesis. In the simple M-ary test there are M source outputs, each of which corresponds to one of M hypotheses. As before, we assume that we are forced to make a decision. Thus there are M^2 alternatives that may occur each time the experiment is conducted. The Bayes criterion assigns a cost to each of these alternatives, assumes a set of a priori probabilities $P_0, P_1, \ldots, P_{M-1}$, and minimizes the risk. The generalization of the Neyman–Pearson criterion to M hypotheses is also possible. Because it is not widely used in practice, we shall discuss only the Bayes criterion in the text.

Bayes Criterion. To find a Bayes test we denote the cost of each course of action as C_{ij}. The first subscript signifies that the ith hypothesis is chosen. The second subscript signifies that the jth hypothesis is true. We denote the region of the observation space in which we choose H_i as Z_i

and the a priori probabilities are P_i. The model is shown in Fig. 2.13. The expression for the risk is

$$\mathcal{R} = \sum_{i=0}^{M-1} \sum_{j=0}^{M-1} P_j C_{ij} \int_{Z_i} p_{\mathbf{r}|H_j}(\mathbf{R}|H_j)\, d\mathbf{R}. \tag{98}$$

To find the optimum Bayes test we simply vary the Z_i to minimize $\mathcal{R}$. This is a straightforward extension of the technique used in the binary case. For simplicity of notation, we shall only consider the case in which $M = 3$ in the text.

Noting that $Z_0 = Z - Z_1 - Z_2$, because the regions are disjoint, we obtain

$$\begin{aligned}
\mathcal{R} = {} & P_0 C_{00} \int_{Z-Z_1-Z_2} p_{\mathbf{r}|H_0}(\mathbf{R}|H_0)\, d\mathbf{R} + P_0 C_{10} \int_{Z_1} p_{\mathbf{r}|H_0}(\mathbf{R}|H_0)\, d\mathbf{R} \\
& + P_0 C_{20} \int_{Z_2} p_{\mathbf{r}|H_0}(\mathbf{R}|H_0)\, d\mathbf{R} + P_1 C_{11} \int_{Z-Z_0-Z_2} p_{\mathbf{r}|H_1}(\mathbf{R}|H_1)\, d\mathbf{R} \\
& + P_1 C_{01} \int_{Z_0} p_{\mathbf{r}|H_1}(\mathbf{R}|H_1)\, d\mathbf{R} + P_1 C_{21} \int_{Z_2} p_{\mathbf{r}|H_1}(\mathbf{R}|H_1)\, d\mathbf{R} \\
& + P_2 C_{22} \int_{Z-Z_0-Z_1} p_{\mathbf{r}|H_2}(\mathbf{R}|H_2)\, d\mathbf{R} + P_2 C_{02} \int_{Z_0} p_{\mathbf{r}|H_2}(\mathbf{R}|H_2)\, d\mathbf{R} \\
& + P_2 C_{12} \int_{Z_1} p_{\mathbf{r}|H_2}(\mathbf{R}|H_2)\, d\mathbf{R}.
\end{aligned} \tag{99}$$

This reduces to

$$\begin{aligned}
\mathcal{R} = {} & P_0 C_{00} + P_1 C_{11} + P_2 C_{22} \\
& + \int_{Z_0} [P_2(C_{02} - C_{22}) p_{\mathbf{r}|H_2}(\mathbf{R}|H_2) + P_1(C_{01} - C_{11}) p_{\mathbf{r}|H_1}(\mathbf{R}|H_1)]\, d\mathbf{R} \\
& + \int_{Z_1} [P_0(C_{10} - C_{00}) p_{\mathbf{r}|H_0}(\mathbf{R}|H_0) + P_2(C_{12} - C_{22}) p_{\mathbf{r}|H_2}(\mathbf{R}|H_2)]\, d\mathbf{R} \\
& + \int_{Z_2} [P_0(C_{20} - C_{00}) p_{\mathbf{r}|H_0}(\mathbf{R}|H_0) + P_1(C_{21} - C_{11}) p_{\mathbf{r}|H_1}(\mathbf{R}|H_1)]\, d\mathbf{R}.
\end{aligned} \tag{100}$$

As before, the first three terms represent the fixed cost and the integrals represent the variable cost that depends on our choice of Z_0, Z_1, and Z_2. Clearly, we assign each $\mathbf{R}$ to the region in which the value of the integrand is the smallest. Labeling these integrands $I_0(\mathbf{R})$, $I_1(\mathbf{R})$, and $I_2(\mathbf{R})$, we have the following rule:

$$\begin{aligned}
&\text{if } I_0(\mathbf{R}) < I_1(\mathbf{R}) \text{ and } I_2(\mathbf{R}), \text{ choose } H_0, \\
&\text{if } I_1(\mathbf{R}) < I_0(\mathbf{R}) \text{ and } I_2(\mathbf{R}), \text{ choose } H_1, \\
&\text{if } I_2(\mathbf{R}) < I_0(\mathbf{R}) \text{ and } I_1(\mathbf{R}), \text{ choose } H_2.
\end{aligned} \tag{101}$$

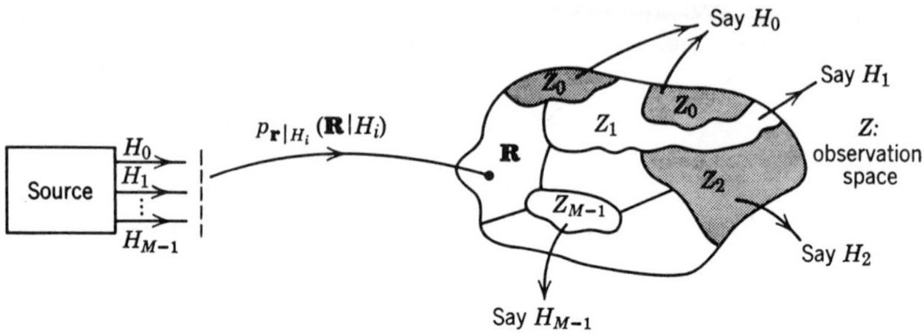

Fig. 2.13 ***M* hypothesis problem.**

We can write these terms in terms of likelihood ratios by defining

$$\Lambda_1(\mathbf{R}) \triangleq \frac{p_{\mathbf{r}|H_1}(\mathbf{R}|H_1)}{p_{\mathbf{r}|H_0}(\mathbf{R}|H_0)},$$
$$\Lambda_2(\mathbf{R}) \triangleq \frac{p_{\mathbf{r}|H_2}(\mathbf{R}|H_2)}{p_{\mathbf{r}|H_0}(\mathbf{R}|H_0)}. \tag{102}$$

Using (102) in (100) and (101), we have

$$P_1(C_{01} - C_{11})\,\Lambda_1(\mathbf{R}) \underset{H_0 \text{ or } H_2}{\overset{H_1 \text{ or } H_2}{\gtrless}} P_0(C_{10} - C_{00}) + P_2(C_{12} - C_{02})\,\Lambda_2(\mathbf{R}) \tag{103}$$

$$P_2(C_{02} - C_{22})\,\Lambda_2(\mathbf{R}) \underset{H_0 \text{ or } H_1}{\overset{H_2 \text{ or } H_1}{\gtrless}} P_0(C_{20} - C_{00}) + P_1(C_{21} - C_{01})\,\Lambda_1(\mathbf{R}), \tag{104}$$

$$P_2(C_{12} - C_{22})\,\Lambda_2(\mathbf{R}) \underset{H_1 \text{ or } H_0}{\overset{H_2 \text{ or } H_0}{\gtrless}} P_0(C_{20} - C_{10}) + P_1(C_{21} - C_{11})\,\Lambda_1(\mathbf{R}). \tag{105}$$

We see that the decision rules correspond to three lines in the Λ_1, Λ_2 plane. It is easy to verify that these lines intersect at a common point and therefore uniquely define three decision regions, as shown in Fig. 2.14. The decision space is two-dimensional for the three-hypothesis problem. It is easy to verify that M hypotheses *always* lead to a decision space which has, at most, $(M - 1)$ dimensions.

Several special cases will be useful in our later work. The first is defined by the assumptions

$$\begin{aligned} C_{00} = C_{11} &= C_{22} = 0, \\ C_{ij} &= 1, \qquad i \neq j. \end{aligned} \tag{106}$$

These equations indicate that any error is of equal importance. Looking at (98), we see that this corresponds to minimizing the total probability of error.

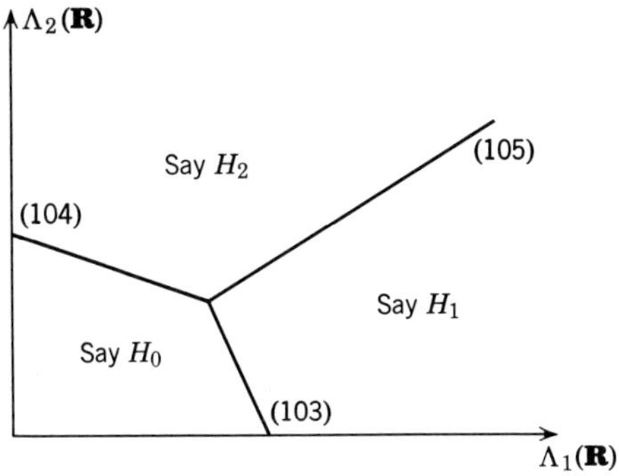

Fig. 2.14 Decision space.

Substituting into (103)–(105), we have

$$
\begin{aligned}
P_1\Lambda_1(\mathbf{R}) &\underset{H_0 \text{ or } H_2}{\overset{H_1 \text{ or } H_2}{\gtrless}} P_0,\\
P_2\Lambda_2(\mathbf{R}) &\underset{H_0 \text{ or } H_1}{\overset{H_2 \text{ or } H_1}{\gtrless}} P_0,\\
P_2\Lambda_2(\mathbf{R}) &\underset{H_1 \text{ or } H_0}{\overset{H_2 \text{ or } H_0}{\gtrless}} P_1\Lambda_1(\mathbf{R}).
\end{aligned}
\tag{107}
$$

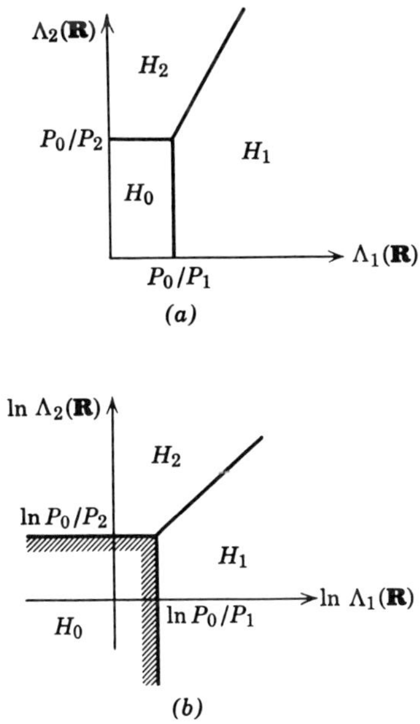

Fig. 2.15 Decision spaces.

The decision regions in the (Λ_1, Λ_2) plane are shown in Fig. 2.15*a*. In this particular case, the transition to the $(\ln \Lambda_1, \ln \Lambda_2)$ plane is straightforward (Fig. 2.15*b*). The equations are

$$\begin{aligned}
\ln \Lambda_1(\mathbf{R}) &\underset{H_0 \text{ or } H_2}{\overset{H_1 \text{ or } H_2}{\gtrless}} \ln \frac{P_0}{P_1}, \\
\ln \Lambda_2(\mathbf{R}) &\underset{H_0 \text{ or } H_1}{\overset{H_1 \text{ or } H_2}{\gtrless}} \ln \frac{P_0}{P_2}, \\
\ln \Lambda_2(\mathbf{R}) &\underset{H_0 \text{ or } H_1}{\overset{H_0 \text{ or } H_2}{\gtrless}} \ln \Lambda_1(\mathbf{R}) + \ln \frac{P_1}{P_2}.
\end{aligned} \tag{108}$$

The expressions in (107) and (108) are adequate, but they obscure an important interpretation of the processor. The desired interpretation is obtained by a little manipulation.

Substituting (102) into (103–105) and multiplying both sides by $p_{\mathbf{r}|H_0}(\mathbf{R}|H_0)$, we have

$$\begin{aligned}
P_1 p_{\mathbf{r}|H_1}(\mathbf{R}|H_1) &\underset{H_0 \text{ or } H_2}{\overset{H_1 \text{ or } H_2}{\gtrless}} P_0 p_{\mathbf{r}|H_0}(\mathbf{R}|H_0), \\
P_2 p_{\mathbf{r}|H_2}(\mathbf{R}|H_2) &\underset{H_0 \text{ or } H_1}{\overset{H_2 \text{ or } H_1}{\gtrless}} P_0 p_{\mathbf{r}|H_0}(\mathbf{R}|H_0), \\
P_2 p_{\mathbf{r}|H_2}(\mathbf{R}|H_2) &\underset{H_1 \text{ or } H_0}{\overset{H_2 \text{ or } H_0}{\gtrless}} P_1 p_{\mathbf{r}|H_1}(\mathbf{R}|H_1).
\end{aligned} \tag{109}$$

Looking at (109), we see that an equivalent test is to compute the a posteriori probabilities $\Pr[H_0|\mathbf{R}]$, $\Pr[H_1|\mathbf{R}]$, and $\Pr[H_2|\mathbf{R}]$ and choose the largest. (Simply divide both sides of each equation by $p_{\mathbf{r}}(\mathbf{R})$ and examine the resulting test.) For this reason the processor for the minimum probability of error criterion is frequently referred to as a *maximum a posteriori probability computer*. The generalization to M hypotheses is straightforward.

The next two topics deal with degenerate tests. Both results will be useful in later applications. A case of interest is a degenerate one in which we combine H_1 and H_2. Then

$$C_{12} = C_{21} = 0, \tag{110}$$

and, for simplicity, we can let

$$C_{01} = C_{10} = C_{20} = C_{02} \tag{111}$$

and

$$C_{00} = C_{11} = C_{22} = 0. \tag{112}$$

Then (103) and (104) both reduce to

$$P_1 \Lambda_1(\mathbf{R}) + P_2 \Lambda_2(\mathbf{R}) \underset{H_0}{\overset{H_1 \text{ or } H_2}{\gtrless}} P_0 \tag{113}$$

and (105) becomes an identity.

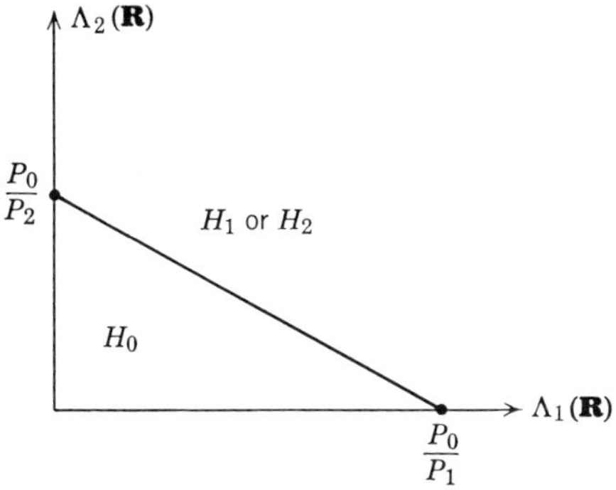

Fig. 2.16 Decision spaces.

The decision regions are shown in Fig. 2.16. Because we have eliminated all of the cost effect of a decision between H_1 and H_2, we have reduced it to a binary problem.

We next consider the dummy hypothesis technique. A simple example illustrates the idea. The actual problem has two hypotheses, H_1 and H_2, but occasionally we can simplify the calculations by introducing a dummy hypothesis H_0 which occurs with zero probability. We let

$$P_0 = 0, \qquad P_1 + P_2 = 1,$$

and (114)

$$C_{12} = C_{02}, C_{21} = C_{01}.$$

Substituting these values into (103–105), we find that (103) and (104) imply that we always choose H_1 or H_2 and the test reduces to

$$P_2(C_{12} - C_{22})\, \Lambda_2(\mathbf{R}) \underset{H_1}{\overset{H_2}{\gtrless}} P_1(C_{21} - C_{11})\, \Lambda_1(\mathbf{R}). \tag{115}$$

Looking at (12) and recalling the definition of $\Lambda_1(\mathbf{R})$ and $\Lambda_2(\mathbf{R})$, we see that this result is exactly what we would expect. [Just divide both sides of (12) by $p_{\mathbf{r}|H_0}(\mathbf{R}|H_0)$.] On the surface this technique seems absurd, but it will turn out to be useful when the ratio

$$\frac{p_{\mathbf{r}|H_2}(\mathbf{R}|H_2)}{p_{\mathbf{r}|H_1}(\mathbf{R}|H_1)}$$

is difficult to work with and the ratios $\Lambda_1(\mathbf{R})$ and $\Lambda_2(\mathbf{R})$ can be made simple by a proper choice of $p_{\mathbf{r}|H_0}(\mathbf{R}|H_0)$.

In this section we have developed the basic results needed for the M-hypothesis problem. We have not considered any specific examples because the details involved in constructing the likelihood ratios are the same as those in the binary case. Typical examples are given in the problems. Several important points should be emphasized.

1. The minimum dimension of the decision space is no more than $M - 1$. The boundaries of the decision regions are hyperplanes in the $(\Lambda_1, \dots, \Lambda_{M-1})$ plane.
2. The optimum test is straightforward to find. We shall find however, when we consider specific examples that the error probabilities are frequently difficult to compute.
3. A particular test of importance is the minimum total probability of error test. Here we compute the a posteriori probability of each hypothesis $\Pr(H_i|\mathbf{R})$ and choose the largest.

These points will be appreciated more fully as we proceed through various applications.

These two sections complete our discussion of simple hypothesis tests. A case of importance that we have not yet discussed is the one in which several source outputs are combined to give a single hypothesis. To study this detection problem, we shall need some ideas from estimation theory. Therefore we defer the composite hypothesis testing problem until Section 2.5 and study the estimation problem next.

2.4 ESTIMATION THEORY

In the last two sections we have considered a problem in which one of several hypotheses occurred. As the result of a particular hypothesis, a vector random variable $\mathbf{r}$ was observed. Based on our observation, we shall try to choose the true hypothesis.

In this section we discuss the problem of *parameter estimation*. Before formulating the general problem, let us consider a simple example.

Example 1. We want to measure a voltage a at a single time instant. From physical considerations, we know that the voltage is between $-V$ and $+V$ volts. The measurement is corrupted by noise which may be modeled as an independent additive zero-mean Gaussian random variable n. The observed variable is r. Thus

$$r = a + n. \tag{116}$$

The probability density governing the observation process is $p_{r|a}(R|A)$. In this case

$$p_{r|a}(R|A) = p_n(R - A) = \frac{1}{\sqrt{2\pi}\,\sigma_n} \exp\left(-\frac{(R - A)^2}{2\sigma_n^2}\right). \tag{117}$$

The problem is to observe r and estimate a.

This example illustrates the basic features of the estimation problem.

A model of the general estimation problem is shown in Fig. 2.17. The model has the following four components:

Parameter Space. The output of the source is a parameter (or variable). We view this output as a point in a parameter space. For the single-parameter case, which we shall study first, this will correspond to segments of the line $-\infty < A < \infty$. In the example considered above the segment is $(-V, V)$.

Probabilistic Mapping from Parameter Space to Observation Space. This is the probability law that governs the effect of a on the observation.

Observation Space. In the classical problem this is a finite-dimensional space. We denote a point in it by the vector $\mathbf{R}$.

Estimation Rule. After observing $\mathbf{R}$, we shall want to estimate the value of a. We denote this estimate as $\hat{a}(\mathbf{R})$. This mapping of the observation space into an estimate is called the estimation rule. The purpose of this section is to investigate various estimation rules and their implementations.

The second and third components are familiar from the detection problem. The new features are the parameter space and the estimation rule. When we try to describe the parameter space, we find that two cases arise. In the first, the parameter is a random variable whose behavior is governed by a probability density. In the second, the parameter is an unknown quantity but not a random variable. These two cases are analogous to the

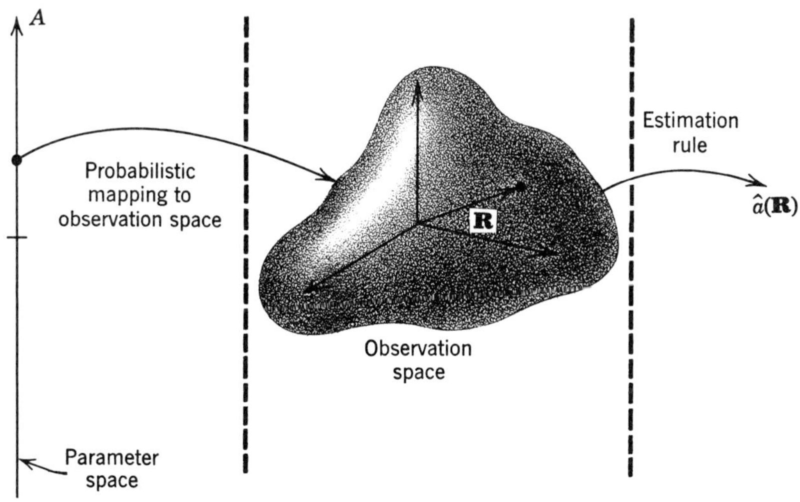

Fig. 2.17 Estimation model.

source models we encountered in the hypothesis-testing problem. To correspond with each of these models of the parameter space, we shall develop suitable estimation rules. We start with the random parameter case.

2.4.1 Random Parameters: Bayes Estimation

In the Bayes detection problem we saw that the two quantities we had to specify were the set of costs C_{ij} and the a priori probabilities P_i. The cost matrix assigned a cost to each possible course of action. Because there were M hypotheses and M possible decisions, there were M^2 costs. In the estimation problem a and $\hat{a}(\mathbf{R})$ are continuous variables. Thus we must assign a cost to all pairs $[a, \hat{a}(\mathbf{R})]$ over the range of interest. This is a function of two variables which we denote as $C(a, \hat{a})$. In many cases of interest it is realistic to assume that the cost depends only on the error of the estimate. We define this error as

$$a_\epsilon(\mathbf{R}) \triangleq \hat{a}(\mathbf{R}) - a. \tag{118}$$

The cost function $C(a_\epsilon)$ is a function of a single variable. Some typical cost functions are shown in Fig. 2.18. In Fig. 2.18*a* the cost function is simply the square of the error:

$$C(a_\epsilon) = a_\epsilon^2. \tag{119}$$

This cost is commonly referred to as the squared error cost function. We see that it accentuates the effects of large errors. In Fig. 2.18*b* the cost function is the absolute value of the error:

$$C(a_\epsilon) = |a_\epsilon|. \tag{120}$$

In Fig. 2.18*c* we assign zero cost to all errors less than $\pm\Delta/2$. In other words, an error less than $\Delta/2$ in magnitude is as good as no error. If $a_\epsilon > \Delta/2$, we assign a uniform value:

$$\begin{aligned} C(a_\epsilon) &= 0, \qquad |a_\epsilon| \le \frac{\Delta}{2}, \\ &= 1, \qquad |a_\epsilon| > \frac{\Delta}{2}. \end{aligned} \tag{121}$$

In a given problem we choose a cost function to accomplish two objectives. First, we should like the cost function to measure user satisfaction adequately. Frequently it is difficult to assign an analytic measure to what basically may be a subjective quality.

Our goal is to find an estimate that minimizes the expected value of the cost. Thus our second objective in choosing a cost function is to assign one that results in a tractable problem. In practice, cost functions are usually some compromise between these two objectives. Fortunately, in many

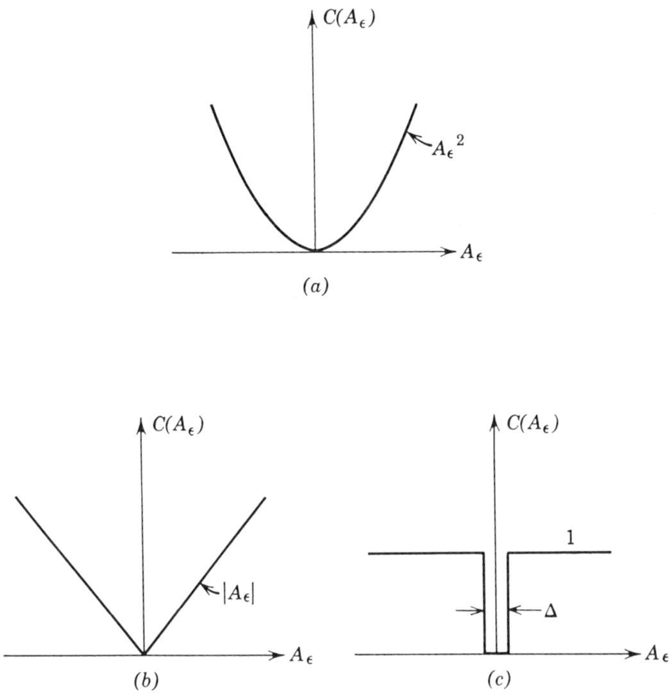

Fig. 2.18 Typical cost functions: (*a*) mean-square error; (*b*) absolute error; (*c*) uniform cost function.

problems of interest the same estimate will be optimum for a large class of cost functions.

Corresponding to the a priori probabilities in the detection problem, we have an a priori probability density $p_a(A)$ in the random parameter estimation problem. In all of our discussions we assume that $p_a(A)$ is known. If $p_a(A)$ is not known, a procedure analogous to the minimax test may be used.

Once we have specified the cost function and the a priori probability, we may write an expression for the risk:

$$\mathcal{R} \triangleq E\{C[a, \hat{a}(\mathbf{R})]\} = \int_{-\infty}^{\infty} dA \int_{-\infty}^{\infty} C[A - \hat{a}(\mathbf{R})] p_{a,\mathbf{r}}(A, \mathbf{R})\, d\mathbf{R}. \quad (122)$$

The expectation is over the random variable a and the observed variables $\mathbf{r}$. For costs that are functions of one variable only (122) becomes

$$\mathcal{R} = \int_{-\infty}^{\infty} dA \int_{-\infty}^{\infty} C[A - \hat{a}(\mathbf{R})] p_{a,\mathbf{r}}(A, \mathbf{R})\, d\mathbf{R}. \quad (123)$$

The Bayes estimate is the estimate that minimizes the risk. It is straightforward to find the Bayes estimates for the cost functions in Fig. 2.18. For the cost function in Fig. 2.18*a*, the risk corresponds to mean-square error. We denote the risk for the mean-square error criterion as $\mathcal{R}_{\text{ms}}$. Substituting (122) into (123), we have

$$\mathcal{R}_{\text{ms}} = \int_{-\infty}^{\infty} dA \int_{-\infty}^{\infty} d\mathbf{R}[A - \hat{a}(\mathbf{R})]^2 p_{a,\mathbf{r}}(A, \mathbf{R}). \tag{124}$$

The joint density can be rewritten as

$$p_{a,\mathbf{r}}(A, \mathbf{R}) = p_{\mathbf{r}}(\mathbf{R}) p_{a|\mathbf{r}}(A|\mathbf{R}). \tag{125}$$

Using (125) in (124), we have

$$\mathcal{R}_{\text{ms}} = \int_{-\infty}^{\infty} d\mathbf{R}\, p_{\mathbf{r}}(\mathbf{R}) \int_{-\infty}^{\infty} dA[A - \hat{a}(\mathbf{R})]^2 p_{a|\mathbf{r}}(A|\mathbf{R}). \tag{126}$$

Now the inner integral and $p_{\mathbf{r}}(\mathbf{R})$ are non-negative. Therefore we can minimize $\mathcal{R}_{\text{ms}}$ by minimizing the inner integral. We denote this estimate $\hat{a}_{\text{ms}}(\mathbf{R})$. To find it we differentiate the inner integral with respect to $\hat{a}(\mathbf{R})$ and set the result equal to zero:

$$\begin{aligned} \frac{d}{d\hat{a}} \int_{-\infty}^{\infty} dA[A - \hat{a}(\mathbf{R})]^2 p_{a|\mathbf{r}}(A|\mathbf{R}) \\ = -2 \int_{-\infty}^{\infty} A p_{a|\mathbf{r}}(A|\mathbf{R})\, dA + 2\hat{a}(\mathbf{R}) \int_{-\infty}^{\infty} p_{a|\mathbf{r}}(A|\mathbf{R})\, dA. \end{aligned} \tag{127}$$

Setting the result equal to zero and observing that the second integral equals 1, we have

$$\hat{a}_{\text{ms}}(\mathbf{R}) = \int_{-\infty}^{\infty} dA\, A p_{a|\mathbf{r}}(A|\mathbf{R}). \tag{128}$$

This is a unique minimum, for the second derivative equals two. The term on the right side of (128) is familiar as the mean of the a posteriori density (or the conditional mean).

Looking at (126), we see that if $\hat{a}(\mathbf{R})$ is the conditional mean the inner integral is just the a posteriori variance (or the conditional variance). Therefore the minimum value of $\mathcal{R}_{\text{ms}}$ is just the average of the conditional variance over all observations $\mathbf{R}$.

To find the Bayes estimate for the absolute value criterion in Fig. 2.18*b* we write

$$\mathcal{R}_{\text{abs}} = \int_{-\infty}^{\infty} d\mathbf{R}\, p_{\mathbf{r}}(\mathbf{R}) \int_{-\infty}^{\infty} dA[|A - \hat{a}(\mathbf{R})|] p_{a|\mathbf{r}}(A|\mathbf{R}). \tag{129}$$

To minimize the inner integral we write

$$I(\mathbf{R}) = \int_{-\infty}^{\hat{a}(\mathbf{R})} dA[\hat{a}(\mathbf{R}) - A]\, p_{a|\mathbf{r}}(A|\mathbf{R}) + \int_{\hat{a}(\mathbf{R})}^{\infty} dA[A - \hat{a}(\mathbf{R})]\, p_{a|\mathbf{r}}(A|\mathbf{R}). \tag{130}$$

Differentiating with respect to $\hat{a}(\mathbf{R})$ and setting the result equal to zero, we have

$$\int_{-\infty}^{\hat{a}_{abs}(\mathbf{R})} dA\, p_{a|\mathbf{r}}(A|\mathbf{R}) = \int_{\hat{a}_{abs}(\mathbf{R})}^{\infty} dA\, p_{a|\mathbf{r}}(A|\mathbf{R}). \tag{131}$$

This is just the definition of the median of the a posteriori density.

The third criterion is the uniform cost function in Fig. 2.18*c*. The risk expression follows easily:

$$\mathcal{R}_{unf} = \int_{-\infty}^{\infty} d\mathbf{R}\, p_{\mathbf{r}}(\mathbf{R})\left[1 - \int_{\hat{a}_{unf}(\mathbf{R})-\Delta/2}^{\hat{a}_{unf}(\mathbf{R})+\Delta/2} p_{a|\mathbf{r}}(A|\mathbf{R})\, dA\right]. \tag{132}$$

To minimize this equation we maximize the inner integral. Of particular interest to us is the case in which Δ is an arbitrarily small but nonzero number. A typical a posteriori density is shown in Fig. 2.19. We see that for small Δ the best choice for $\hat{a}(\mathbf{R})$ is the value of A at which the a posteriori density has its maximum. We denote the estimate for this special case as $\hat{a}_{map}(\mathbf{R})$, the *maximum a posteriori* estimate. In the sequel we use $\hat{a}_{map}(\mathbf{R})$ without further reference to the uniform cost function.

To find $\hat{a}_{map}$ we must have the location of the maximum of $p_{a|\mathbf{r}}(A|\mathbf{R})$. Because the logarithm is a monotone function, we can find the location of the maximum of $\ln p_{a|\mathbf{r}}(A|\mathbf{R})$ equally well. As we saw in the detection problem, this is frequently more convenient.

If the maximum is interior to the allowable range of A and $\ln p_{a|\mathbf{r}}(A|\mathbf{R})$ has a continuous first derivative then a necessary, but not sufficient, condition for a maximum can be obtained by differentiating $\ln p_{a|\mathbf{r}}(A|\mathbf{R})$ with respect to A and setting the result equal to zero:

$$\left.\frac{\partial \ln p_{a|\mathbf{r}}(A|\mathbf{R})}{\partial A}\right|_{A=\hat{a}(\mathbf{R})} = 0. \tag{133}$$

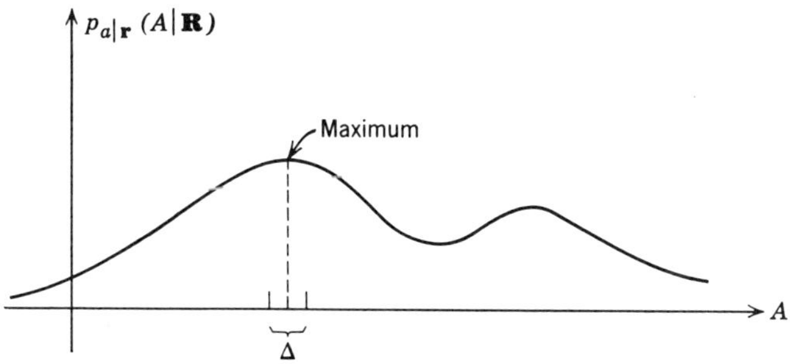

Fig. 2.19 An a posteriori density.

We refer to (133) as the MAP equation. In each case we must check to see if the solution is the absolute maximum.

We may rewrite the expression for $p_{a|\mathbf{r}}(A|\mathbf{R})$ to separate the role of the observed vector $\mathbf{R}$ and the a priori knowledge:

$$p_{a|\mathbf{r}}(A|\mathbf{R}) = \frac{p_{\mathbf{r}|a}(\mathbf{R}|A)p_a(A)}{p_{\mathbf{r}}(\mathbf{R})}. \tag{134}$$

Taking logarithms,

$$\ln p_{a|\mathbf{r}}(A|\mathbf{R}) = \ln p_{\mathbf{r}|a}(\mathbf{R}|A) + \ln p_a(A) - \ln p_{\mathbf{r}}(\mathbf{R}). \tag{135}$$

For MAP estimation we are interested only in finding the value of A where the left-hand side is maximum. Because the last term on the right-hand side is not a function of A, we can consider just the function

$$l(A) \triangleq \ln p_{\mathbf{r}|a}(\mathbf{R}|A) + \ln p_a(A). \tag{136}$$

The first term gives the probabilistic dependence of $\mathbf{R}$ on A and the second describes a priori knowledge.

The MAP equation can be written as

$$\left.\frac{\partial l(A)}{\partial A}\right|_{A=\hat{a}(\mathbf{R})} = \left.\frac{\partial \ln p_{\mathbf{r}|a}(\mathbf{R}|A)}{\partial A}\right|_{A=\hat{a}(\mathbf{R})} + \left.\frac{\partial \ln p_a(A)}{\partial A}\right|_{A=\hat{a}(\mathbf{R})} = 0. \tag{137}$$

Our discussion in the remainder of the book emphasizes minimum mean-square error and maximum a posteriori estimates.

To study the implications of these two estimation procedures we consider several examples.

Example 2. Let

$$r_i = a + n_i, \qquad i = 1, 2, \ldots, N. \tag{138}$$

We assume that a is Gaussian, $N(0, \sigma_a)$, and that the n_i are each independent Gaussian variables $N(0, \sigma_n)$. Then

$$\begin{aligned} p_{\mathbf{r}|a}(\mathbf{R}|A) &= \prod_{i=1}^{N} \frac{1}{\sqrt{2\pi}\,\sigma_n} \exp\left(-\frac{(R_i - A)^2}{2\sigma_n^2}\right), \\ p_a(A) &= \frac{1}{\sqrt{2\pi}\,\sigma_a} \exp\left(-\frac{A^2}{2\sigma_a^2}\right). \end{aligned} \tag{139}$$

To find $\hat{a}_{ms}(\mathbf{R})$ we need to know $p_{a|\mathbf{r}}(A|\mathbf{R})$. One approach is to find $p_{\mathbf{r}}(\mathbf{R})$ and substitute it into (134), but this procedure is algebraically tedious. It is easier to observe that $p_{a|\mathbf{r}}(A|\mathbf{R})$ is a probability density with respect to a for any $\mathbf{R}$. Thus $p_{\mathbf{r}}(\mathbf{R})$ just contributes to the constant needed to make

$$\int_{-\infty}^{\infty} p_{a|\mathbf{r}}(A|\mathbf{R})\, dA = 1. \tag{140}$$

(In other words, $p_{\mathbf{r}}(\mathbf{R})$ is simply a normalization constant.) Thus

$$p_{a|\mathbf{r}}(A|\mathbf{R}) = \left[\frac{\left(\prod_{i=1}^{N} \frac{1}{\sqrt{2\pi}\,\sigma_n}\right) \frac{1}{\sqrt{2\pi}\,\sigma_a}}{p_{\mathbf{r}}(\mathbf{R})}\right] \exp\left\{-\frac{1}{2}\left[\frac{\sum_{i=1}^{N}(R_i - A)^2}{\sigma_n^2} + \frac{A^2}{\sigma_a^2}\right]\right\}. \tag{141}$$

Rearranging the exponent, completing the square, and absorbing terms depending only on R_i^2 into the constant, we have

$$p_{a|\mathbf{r}}(A|\mathbf{R}) = k(\mathbf{R}) \exp\left\{-\frac{1}{2\sigma_p^2}\left[A - \frac{\sigma_a^2}{\sigma_a^2 + \sigma_n^2/N}\left(\frac{1}{N}\sum_{i=1}^{N} R_i\right)\right]^2\right\}, \tag{142}$$

where

$$\sigma_p^2 \triangleq \left(\frac{1}{\sigma_a^2} + \frac{N}{\sigma_n^2}\right)^{-1} = \frac{\sigma_a^2\sigma_n^2}{N\sigma_a^2 + \sigma_n^2} \tag{143}$$

is the a posteriori variance.

We see that $p_{a|\mathbf{r}}(A|\mathbf{R})$ is just a Gaussian density. The estimate $\hat{a}_{ms}(\mathbf{R})$ is just the conditional mean

$$\hat{a}_{ms}(\mathbf{R}) = \frac{\sigma_a^2}{\sigma_a^2 + \sigma_n^2/N}\left(\frac{1}{N}\sum_{i=1}^{N} R_i\right). \tag{144}$$

Because the a posteriori variance is not a function of $\mathbf{R}$, the mean-square risk equals the a posteriori variance (see (126)).

Two observations are useful:

1. The R_i enter into the a posteriori density only through their sum. Thus

$$l(\mathbf{R}) = \sum_{i=1}^{N} R_i \tag{145}$$

is a *sufficient statistic*. This idea of a sufficient statistic is identical to that in the detection problem.

2. The estimation rule uses the information available in an intuitively logical manner. If $\sigma_a^2 \ll \sigma_n^2/N$, the a priori knowledge is much better than the observed data and the estimate is very close to the a priori mean. (In this case, the a priori mean is zero.) On the other hand, if $\sigma_a^2 \gg \sigma_n^2/N$, the a priori knowledge is of little value and the estimate uses primarily the received data. In the limit $\hat{a}_{ms}$ is just the arithmetic average of the R_i.

$$\lim_{\frac{\sigma_n^2}{N\sigma_a^2} \to 0} \hat{a}_{ms}(\mathbf{R}) = \frac{1}{N}\sum_{i=1}^{N} R_i. \tag{146}$$

The MAP estimate for this case follows easily. Looking at (142), we see that because the density is Gaussian the maximum value of $p_{a|\mathbf{r}}(A|\mathbf{R})$ occurs at the conditional mean. Thus

$$\hat{a}_{map}(\mathbf{R}) = \hat{a}_{ms}(\mathbf{R}). \tag{147}$$

Because the conditional median of a Gaussian density occurs at the conditional mean, we also have

$$\hat{a}_{abs}(\mathbf{R}) = \hat{a}_{ms}(\mathbf{R}). \tag{148}$$

Thus we see that for this particular example all three cost functions in Fig. 2.18 lead to the same estimate. This invariance to the choice of a cost function is obviously a useful feature because of the subjective judgments that are frequently involved in choosing $C(a_\epsilon)$. Some conditions under which this invariance holds are developed in the next two properties.†

† These properties are due to Sherman [20]. Our derivation is similar to that given by Viterbi [36].

Property 1. We assume that the cost function $C(a_\epsilon)$ is a symmetric, convex-upward function and that the a posteriori density $p_{a|\mathbf{r}}(A|\mathbf{R})$ is symmetric about its conditional mean; that is,

$$C(a_\epsilon) = C(-a_\epsilon) \qquad \text{(symmetry)}, \quad (149)$$

$$C(bx_1 + (1 - b)x_2) \le bC(x_1) + (1 - b)\, C(x_2) \qquad \text{(convexity)} \quad (150)$$

for any b inside the range (0, 1) and for all x_1 and x_2. Equation 150 simply says that all chords lie above or on the cost function.

This condition is shown in Fig. 2.20*a*. If the inequality is strict whenever $x_1 \neq x_2$, we say the cost function is strictly convex (upward). Defining

$$z \triangleq a - \hat{a}_{\text{ms}} = a - E[a|\mathbf{R}] \qquad (151)$$

the symmetry of the a posteriori density implies

$$p_{z|\mathbf{r}}(Z|\mathbf{R}) = p_{z|\mathbf{r}}(-Z|\mathbf{R}). \qquad (152)$$

The estimate $\hat{a}$ that minimizes any cost function in this class is identical to $\hat{a}_{\text{ms}}$ (which is the conditional mean).

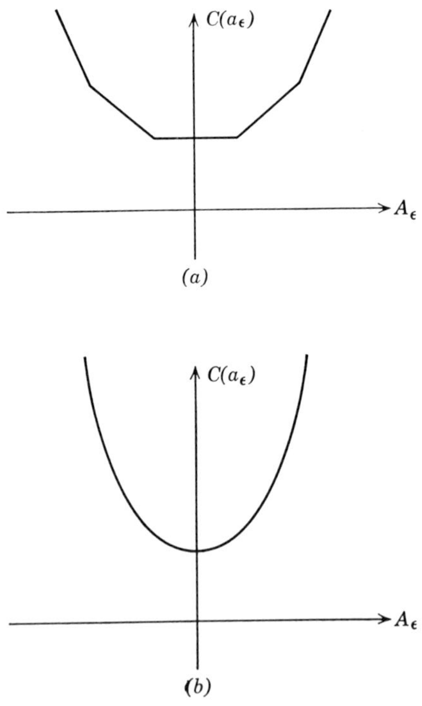

Fig. 2.20 Symmetric convex cost functions: (*a*) convex; (*b*) strictly convex.

Proof. As before we can minimize the conditional risk [see (126)]. Define

$$\mathcal{R}_B(\hat{a}|\mathbf{R}) \triangleq E_a[C(\hat{a} - a)|\mathbf{R}] = E_a[C(a - \hat{a})|\mathbf{R}], \tag{153}$$

where the second equality follows from (149). We now write four equivalent expressions for $\mathcal{R}_B(\hat{a}|\mathbf{R})$:

$$\mathcal{R}_B(\hat{a}|\mathbf{R}) = \int_{-\infty}^{\infty} C(\hat{a} - \hat{a}_{\text{ms}} - Z)p_{z|\mathbf{r}}(Z|\mathbf{R})\,dZ \tag{154}$$

[Use (151) in (153)]

$$= \int_{-\infty}^{\infty} C(\hat{a} - \hat{a}_{\text{ms}} + Z)p_{z|\mathbf{r}}(Z|\mathbf{R})\,dZ \tag{155}$$

[(152) implies this equality]

$$= \int_{-\infty}^{\infty} C(\hat{a}_{\text{ms}} - \hat{a} - Z)p_{z|\mathbf{r}}(Z|\mathbf{R})\,dZ \tag{156}$$

[(149) implies this equality]

$$= \int_{-\infty}^{\infty} C(\hat{a}_{\text{ms}} - \hat{a} + Z)p_{z|\mathbf{r}}(Z|\mathbf{R})\,dZ \tag{157}$$

[(152) implies this equality].

We now use the convexity condition (150) with the terms in (155) and (157):

$$\begin{aligned}\mathcal{R}_B(\hat{a}|\mathbf{R}) &= \tfrac{1}{2}E(\{C[Z + (\hat{a}_{\text{ms}} - \hat{a})] + C[Z - (\hat{a}_{\text{ms}} - \hat{a})]\}|\mathbf{R})\\ &\geq E\{C[\tfrac{1}{2}(Z + (\hat{a}_{\text{ms}} - \hat{a})) + \tfrac{1}{2}(Z - (\hat{a}_{\text{ms}} - \hat{a}))]|\mathbf{R}\}\\ &= E[C(Z)|\mathbf{R}].\end{aligned} \tag{158}$$

Equality will be achieved in (158) if $\hat{a}_{\text{ms}} = \hat{a}$. This completes the proof. If $C(a_\epsilon)$ is strictly convex, we will have the additional result that the minimizing estimate $\hat{a}$ is unique and equals $\hat{a}_{\text{ms}}$.

To include cost functions like the uniform cost functions which are not convex we need a second property.

Property 2. We assume that the cost function is a symmetric, nondecreasing function and that the a posteriori density $p_{a|\mathbf{r}}(A|\mathbf{R})$ is a symmetric (about the conditional mean), unimodal function that satisfies the condition

$$\lim_{x \to \infty} C(x)p_{a|\mathbf{r}}(x|\mathbf{R}) = 0.$$

The estimate $\hat{a}$ that minimizes any cost function in this class is identical to $\hat{a}_{\text{ms}}$. The proof of this property is similar to the above proof [36].

The significance of these two properties should not be underemphasized. Throughout the book we consider only minimum mean-square and maximum a posteriori probability estimators. Properties 1 and 2 ensure that whenever the a posteriori densities satisfy the assumptions given above the estimates that we obtain will be optimum for a large class of cost functions. Clearly, if the a posteriori density is Gaussian, it will satisfy the above assumptions.

We now consider two examples of a different type.

Example 3. The variable a appears in the signal in a nonlinear manner. We denote this dependence by $s(A)$. Each observation r_i consists of $s(A)$ plus a Gaussian random variable n_i, $N(0, \sigma_n)$. The n_i are statistically independent of each other and a. Thus

$$r_i = s(A) + n_i. \tag{159}$$

Therefore

$$p_{a|\mathbf{r}}(A|\mathbf{R}) = k(\mathbf{R}) \exp\left(-\frac{1}{2}\left\{\frac{\sum_{i=1}^{N}[R_i - s(A)]^2}{\sigma_n^2} + \frac{A^2}{\sigma_a^2}\right\}\right). \tag{160}$$

This expression cannot be further simplified without specifying $s(A)$ explicitly.

The MAP equation is obtained by substituting (160) into (137)

$$\hat{a}_{\text{map}}(\mathbf{R}) = \frac{\sigma_a^2}{\sigma_n^2}\sum_{i=1}^{N}[R_i - s(A)]\left.\frac{\partial s(A)}{\partial A}\right|_{A=\hat{a}_{\text{map}}(\mathbf{R})}. \tag{161}$$

To solve this explicitly we must specify $s(A)$. We shall find that an analytic solution is generally not possible when $s(A)$ is a nonlinear function of A.

Another type of problem that frequently arises is the estimation of a parameter in a probability density.

Example 4. The number of events in an experiment obey a Poisson law with mean value a. Thus

$$\Pr(n \text{ events} \mid a = A) = \frac{A^n}{n!}\exp(-A), \qquad n = 0, 1, \ldots. \tag{162}$$

We want to observe the number of events and estimate the parameter a of the Poisson law. We shall assume that a is a random variable with an exponential density

$$p_a(A) = \begin{cases} \lambda \exp(-\lambda A), & A > 0, \\ 0, & \text{elsewhere.} \end{cases} \tag{163}$$

The a posteriori density of a is

$$p_{a|n}(A|N) = \frac{\Pr(n = N \mid a = A)p_a(A)}{\Pr(n = N)}. \tag{164}$$

Substituting (162) and (163) into (164), we have

$$p_{a|n}(A|N) = k(N)[A^N \exp(-A(1 + \lambda))], \qquad A \geq 0, \tag{165}$$

where

$$k(N) = \frac{(1 + \lambda)^{N+1}}{N!} \tag{166}$$

in order for the density to integrate to 1. (As already pointed out, the constant is unimportant for MAP estimation but is needed if we find the MS estimate by integrating over the conditional density.)

The mean-square estimate is the conditional mean:

$$\hat{a}_{ms}(N) = \frac{(1+\lambda)^{N+1}}{N!}\int_0^\infty A^{N+1}\exp[-A(1+\lambda)]\,dA$$

$$= \frac{(1+\lambda)^{N+1}}{(1+\lambda)^{N+2}}(N+1) = \left(\frac{1}{\lambda+1}\right)(N+1). \tag{167}$$

To find $\hat{a}_{map}$ we take the logarithm of (165)

$$\ln p_{a|n}(A|N) = N\ln A - A(1+\lambda) + \ln k(N). \tag{168}$$

By differentiating with respect to A, setting the result equal to zero, and solving, we obtain

$$\hat{a}_{map}(N) = \frac{N}{1+\lambda}. \tag{169}$$

Observe that $\hat{a}_{map}$ is not equal to $\hat{a}_{ms}$.

Other examples are developed in the problems. The principal results of this section are the following:

1. The minimum mean-square error estimate (MMSE) is always the mean of the a posteriori density (the conditional mean).
2. The maximum a posteriori estimate (MAP) is the value of A at which the a posteriori density has its maximum.
3. For a large class of cost functions the optimum estimate is the conditional mean whenever the a posteriori density is a unimodal function which is symmetric about the conditional mean.

These results are the basis of most of our estimation work. As we study more complicated problems, the only difficulty we shall encounter is the actual evaluation of the conditional mean or maximum. In many cases of interest the MAP and MMSE estimates will turn out to be equal.

We now turn to the second class of estimation problems described in the introduction.

2.4.2 Real (Nonrandom) Parameter Estimation†

In many cases it is unrealistic to treat the unknown parameter as a random variable. The problem formulation on pp. 52–53 is still appropriate. Now, however, the parameter is assumed to be nonrandom, and we want to design an estimation procedure that is good in some sense.

† The beginnings of classical estimation theory can be attributed to Fisher [5, 6, 7, 8]. Many discussions of the basic ideas are now available (e.g., Cramér [9]), Wilks [10], or Kendall and Stuart [11]).

A logical first approach is to try to modify the Bayes procedure in the last section to eliminate the average over $p_a(A)$. As an example, consider a mean-square error criterion,

$$\mathcal{R}(A) \triangleq \int_{-\infty}^{\infty} [\hat{a}(\mathbf{R}) - A]^2 \, p_{\mathbf{r}|a}(\mathbf{R}|A) \, d\mathbf{R}, \tag{170}$$

where the expectation is only over $\mathbf{R}$, for it is the only random variable in the model. Minimizing $\mathcal{R}(A)$, we obtain

$$\hat{a}_{ms}(\mathbf{R}) = A. \tag{171}$$

The answer is correct, but not of any value, for A is the unknown quantity that we are trying to find. Thus we see that this direct approach is not fruitful. A more useful method in the nonrandom parameter case is to examine other possible measures of quality of estimation procedures and then to see whether we can find estimates that are good in terms of these measures.

The first measure of quality to be considered is the expectation of the estimate

$$E[\hat{a}(\mathbf{R})] \triangleq \int_{-\infty}^{+\infty} \hat{a}(\mathbf{R}) \, p_{\mathbf{r}|a}(\mathbf{R}|A) \, d\mathbf{R}. \tag{172}$$

The possible values of the expectation can be grouped into three classes

1. If $E[\hat{a}(\mathbf{R})] = A$, for all values of A, we say that the estimate is *unbiased.* This statement means that the average value of the estimates equals the quantity we are trying to estimate.
2. If $E[\hat{a}(\mathbf{R})] = A + B$, where B is not a function of A, we say that the estimate has a *known bias.* We can always obtain an unbiased estimate by subtracting B from $\hat{a}(\mathbf{R})$.
3. If $E[\hat{a}(\mathbf{R})] = A + B(A)$, we say that the estimate has an *unknown bias.* Because the bias depends on the unknown parameter, we cannot simply subtract it out.

Clearly, even an unbiased estimate may give a bad result on a particular trial. A simple example is shown in Fig. 2.21. The probability density of the estimate is centered around A, but the variance of this density is large enough that big errors are probable.

A second measure of quality is the variance of estimation error:

$$\text{Var}\,[\hat{a}(\mathbf{R}) - A] = E\{[\hat{a}(\mathbf{R}) - A]^2\} - B^2(A). \tag{173}$$

This provides a measure of the spread of the error. In general, we shall try to find unbiased estimates with small variances. There is no straightforward minimization procedure that will lead us to the minimum variance unbiased estimate. Therefore we are forced to try an estimation procedure to see how well it works.

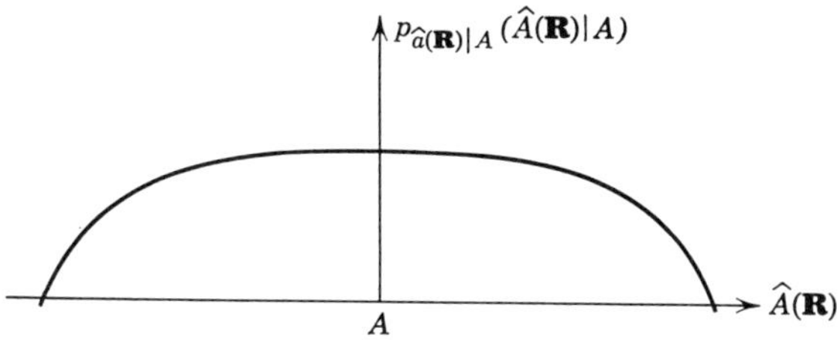

Fig. 2.21 Probability density for an estimate.

Maximum Likelihood Estimation. There are several ways to motivate the estimation procedure that we shall use. Consider the simple estimation problem outlined in Example 1. Recall that

$$r = A + n, \tag{174}$$

$$p_{r|a}(R|A) = (\sqrt{2\pi}\,\sigma_n)^{-1} \exp\left[-\tfrac{1}{2}(R - A)^2\right]. \tag{175}$$

We choose as our estimate the value of A that most likely caused a given value of R to occur. In this simple additive case we see that this is the same as choosing the most probable value of the noise ($N = 0$) and subtracting it from R. We denote the value obtained by using this procedure as a maximum likelihood estimate.

$$\hat{a}_{ml}(R) = R. \tag{176}$$

In the general case we denote the function $p_{\mathbf{r}|a}(\mathbf{R}|A)$, viewed as a function of A, as the *likelihood function*. Frequently we work with the logarithm, $\ln p_{\mathbf{r}|a}(\mathbf{R}|A)$, and denote it as the *log likelihood function*. The maximum likelihood estimate $\hat{a}_{ml}(\mathbf{R})$ is that value of A at which the likelihood function is a maximum. If the maximum is interior to the range of A, and $\ln p_{\mathbf{r}|a}(\mathbf{R}|A)$ has a continuous first derivative, then a necessary condition on $\hat{a}_{ml}(\mathbf{R})$ is obtained by differentiating $\ln p_{\mathbf{r}|a}(\mathbf{R}|A)$ with respect to A and setting the result equal to zero:

$$\left.\frac{\partial \ln p_{\mathbf{r}|a}(\mathbf{R}|A)}{\partial A}\right|_{A=\hat{a}_{ml}(\mathbf{R})} = 0. \tag{177}$$

This equation is called the *likelihood equation*. Comparing (137) and (177), we see that the ML estimate corresponds mathematically to the limiting case of a MAP estimate in which the a priori knowledge approaches zero.

In order to see how effective the ML procedure is we can compute the bias and the variance. Frequently this is difficult to do. Rather than approach the problem directly, we shall first derive a lower bound on the variance on *any* unbiased estimate. Then we shall see how the variance of $\hat{a}_{ml}(\mathbf{R})$ compares with this lower bound.

Cramér-Rao Inequality: Nonrandom Parameters. We now want to consider the variance of *any* estimate $\hat{a}(\mathbf{R})$ of the real variable A. We shall prove the following statement.

Theorem. (a) If $\hat{a}(\mathbf{R})$ is *any* unbiased estimate of A, then

$$\operatorname{Var}[\hat{a}(\mathbf{R}) - A] \geq \left(E\left\{\left[\frac{\partial \ln p_{\mathbf{r}|a}(\mathbf{R}|A)}{\partial A}\right]^2\right\}\right)^{-1} \tag{178}$$

or, equivalently,

(b)

$$\operatorname{Var}[\hat{a}(\mathbf{R}) - A] \geq \left\{-E\left[\frac{\partial^2 \ln p_{\mathbf{r}|a}(\mathbf{R}|A)}{\partial A^2}\right]\right\}^{-1}, \tag{179}$$

where the following conditions are assumed to be satisfied:

(c)

$$\frac{\partial p_{\mathbf{r}|a}(\mathbf{R}|A)}{\partial A} \quad \text{and} \quad \frac{\partial^2 p_{\mathbf{r}|a}(\mathbf{R}|A)}{\partial A^2}$$

exist and are absolutely integrable.

The inequalities were first stated by Fisher [6] and proved by Dugué [31]. They were also derived by Cramér [9] and Rao [12] and are usually referred to as the Cramér-Rao bound. Any estimate that satisfies the bound with an equality is called an *efficient* estimate.

The proof is a simple application of the Schwarz inequality. Because $\hat{a}(\mathbf{R})$ is unbiased,

$$E[\hat{a}(\mathbf{R}) - A] \triangleq \int_{-\infty}^{\infty} p_{\mathbf{r}|a}(\mathbf{R}|A)[\hat{a}(\mathbf{R}) - A]\, d\mathbf{R} = 0. \tag{180}$$

Differentiating both sides with respect to A, we have

$$\frac{d}{dA}\int_{-\infty}^{\infty} p_{\mathbf{r}|a}(\mathbf{R}|A)[\hat{a}(\mathbf{R}) - A]\, d\mathbf{R} = \int_{-\infty}^{\infty} \frac{\partial}{\partial A}\{p_{\mathbf{r}|a}(\mathbf{R}|A)[\hat{a}(\mathbf{R}) - A]\}\, d\mathbf{R} = 0, \tag{181}$$

where condition (c) allows us to bring the differentiation inside the integral. Then

$$-\int_{-\infty}^{\infty} p_{\mathbf{r}|a}(\mathbf{R}|A)\, d\mathbf{R} + \int_{-\infty}^{\infty} \frac{\partial p_{\mathbf{r}|a}(\mathbf{R}|A)}{\partial A}[\hat{a}(\mathbf{R}) - A]\, d\mathbf{R} = 0. \tag{182}$$

The first integral is just $+1$. Now observe that

$$\frac{\partial p_{\mathbf{r}|a}(\mathbf{R}|A)}{\partial A} = \frac{\partial \ln p_{\mathbf{r}|a}(\mathbf{R}|A)}{\partial A}\, p_{\mathbf{r}|a}(\mathbf{R}|A). \tag{183}$$

Substituting (183) into (182), we have

$$\int_{-\infty}^{\infty} \frac{\partial \ln p_{\mathbf{r}|a}(\mathbf{R}|A)}{\partial A} p_{\mathbf{r}|a}(\mathbf{R}|A)[\hat{a}(\mathbf{R}) - A]\, d\mathbf{R} = 1. \tag{184}$$

Rewriting, we have

$$\int_{-\infty}^{\infty} \left[\frac{\partial \ln p_{\mathbf{r}|a}(\mathbf{R}|A)}{\partial A} \sqrt{p_{\mathbf{r}|a}(\mathbf{R}|A)}\right] \left[\sqrt{p_{\mathbf{r}|a}(\mathbf{R}|A)}\, [\hat{a}(\mathbf{R}) - A]\right] d\mathbf{R} = 1, \tag{185}$$

and, using the Schwarz inequality, we have

$$\left\{\int_{-\infty}^{\infty} \left[\frac{\partial \ln p_{\mathbf{r}|a}(\mathbf{R}|A)}{\partial A}\right]^2 p_{\mathbf{r}|a}(\mathbf{R}|A)\, d\mathbf{R}\right\} \times \left\{\int_{-\infty}^{\infty} [\hat{a}(\mathbf{R}) - A]^2\, p_{\mathbf{r}|a}(\mathbf{R}|A)\, d\mathbf{R}\right\} \geq 1, \tag{186}$$

where we recall from the derivation of the Schwarz inequality that equality holds if and only if

$$\frac{\partial \ln p_{\mathbf{r}|a}(\mathbf{R}|A)}{\partial A} = [\hat{a}(\mathbf{R}) - A]\, k(A), \tag{187}$$

for all $\mathbf{R}$ and A. We see that the two terms of the left side of (186) are the expectations in statement (*a*) of (178). Thus,

$$E\{[\hat{a}(\mathbf{R}) - A]^2\} \geq \left\{E\left[\frac{\partial \ln p_{\mathbf{r}|a}(\mathbf{R}|A)}{\partial A}\right]^2\right\}^{-1}. \tag{188}$$

To prove statement (b) we observe

$$\int_{-\infty}^{\infty} p_{\mathbf{r}|a}(\mathbf{R}|A)\, d\mathbf{R} = 1. \tag{189}$$

Differentiating with respect to A, we have

$$\int_{-\infty}^{\infty} \frac{\partial p_{\mathbf{r}|a}(\mathbf{R}|A)}{\partial A}\, d\mathbf{R} = \int_{-\infty}^{\infty} \frac{\partial \ln p_{\mathbf{r}|a}(\mathbf{R}|A)}{\partial A} p_{\mathbf{r}|a}(\mathbf{R}|A)\, d\mathbf{R} = 0. \tag{190}$$

Differentiating again with respect to A and applying (183), we obtain

$$\int_{-\infty}^{\infty} \frac{\partial^2 \ln p_{\mathbf{r}|a}(\mathbf{R}|A)}{\partial A^2} p_{\mathbf{r}|a}(\mathbf{R}|A)\, d\mathbf{R} + \int_{-\infty}^{\infty} \left(\frac{\partial \ln p_{\mathbf{r}|a}(\mathbf{R}|A)}{\partial A}\right)^2 p_{\mathbf{r}|a}(\mathbf{R}|A)\, d\mathbf{R} = 0 \tag{191}$$

or

$$E\left[\frac{\partial^2 \ln p_{\mathbf{r}|a}(\mathbf{R}|A)}{\partial A^2}\right] = -E\left[\frac{\partial \ln p_{\mathbf{r}|a}(\mathbf{R}|A)}{\partial A}\right]^2, \tag{192}$$

which together with (188) gives condition (b).

Several important observations should be made about this result.

1. It shows that any unbiased estimate must have a variance greater than a certain number.

2. If (187) is satisfied, the estimate $\hat{a}_{ml}(\mathbf{R})$ will satisfy the bound with an equality. We show this by combining (187) and (177). The left equality is the maximum likelihood equation. The right equality is (187):

$$0 = \frac{\partial \ln p_{\mathbf{r}|a}(\mathbf{R}|A)}{\partial A}\bigg|_{A=\hat{a}_{ml}(\mathbf{R})} = (\hat{a}(\mathbf{R}) - A)\, k(A)\bigg|_{A=\hat{a}_{ml}(\mathbf{R})}. \tag{193}$$

In order for the right-hand side to equal zero either

$$\hat{a}(\mathbf{R}) = \hat{a}_{ml}(\mathbf{R}) \tag{194}$$

or

$$k(\hat{a}_{ml}) = 0. \tag{195}$$

Because we want a solution that depends on the data, we eliminate (195) and require (194) to hold.

Thus, *if* an efficient estimate exists, it is $\hat{a}_{ml}(\mathbf{R})$ and can be obtained as a unique solution to the likelihood equation.

3. If an efficient estimate *does not* exist [i.e., $\partial \ln p_{\mathbf{r}|a}(\mathbf{R}|A)/\partial A$ cannot be put into the form of (187)], we do not know how good $\hat{a}_{ml}(\mathbf{R})$ is. Further, we do not know how close the variance of any estimate will approach the bound.

4. In order to use the bound, we must verify that the estimate of concern is unbiased. Similar bounds can be derived simply for biased estimates (Problem 2.4.17).

We can illustrate the application of ML estimation and the Cramér–Rao inequality by considering Examples 2, 3, and 4. The observation model is identical. We now assume, however, that the parameters to be estimated are nonrandom variables.

Example 2. From (138) we have

$$r_i = A + n_i, \qquad i = 1, 2, \ldots, N. \tag{196}$$

Taking the logarithm of (139) and differentiating, we have

$$\frac{\partial \ln p_{\mathbf{r}|a}(\mathbf{R}|A)}{\partial A} = \frac{N}{\sigma_n^2}\left(\frac{1}{N}\sum_{i=1}^{N} R_i - A\right). \tag{197}$$

Thus

$$\hat{a}_{ml}(\mathbf{R}) = \frac{1}{N}\sum_{i=1}^{N} R_i. \tag{198}$$

To find the bias we take the expectation of both sides,

$$E[\hat{a}_{ml}(\mathbf{R})] = \frac{1}{N}\sum_{i=1}^{N} E(R_i) = \frac{1}{N}\sum_{i=1}^{N} A = A, \tag{199}$$

so that $\hat{a}_{ml}(\mathbf{R})$ is unbiased.

Because the expression in (197) has the form required by (187), we know that $\hat{a}_{ml}(\mathbf{R})$ is an efficient estimate. To evaluate the variance we differentiate (197):

$$\frac{\partial^2 \ln p_{\mathbf{r}|a}(\mathbf{R}|A)}{\partial A^2} = -\frac{N}{\sigma_n^2}. \tag{200}$$

Using (179) and the efficiency result, we have

$$\text{Var}\,[\hat{a}_{ml}(\mathbf{R}) - A] = \frac{\sigma_n^2}{N}. \tag{201}$$

Skipping Example 3 for the moment, we go to Example 4.

Example 4. Differentiating the logarithm of (162), we have

$$\frac{\partial \ln \Pr\,(n = N|A)}{\partial A} = \frac{\partial}{\partial A}(N \ln A - A - \ln N!)$$

$$= \frac{N}{A} - 1 = \frac{1}{A}(N - A). \tag{202}$$

The ML estimate is

$$\hat{a}_{ml}(N) = N. \tag{203}$$

It is clearly unbiased and efficient. To obtain the variance we differentiate (202):

$$\frac{\partial^2 \ln \Pr\,(n = N|A)}{\partial A^2} = -\frac{N}{A^2}. \tag{204}$$

Thus

$$\text{Var}\,[\hat{a}_{ml}(N) - A] = \frac{A^2}{E(N)} = \frac{A^2}{A} = A. \tag{205}$$

In both Examples 2 and 4 we see that the ML estimates could have been obtained from the MAP estimates [let $\sigma_a \to \infty$ in (144) and recall that $\hat{a}_{ms}(\mathbf{R}) = \hat{a}_{map}(\mathbf{R})$ and let $\lambda \to 0$ in (169)].

We now return to Example 3.

Example 3. From the first term in the exponent in (160), we have

$$\frac{\partial \ln p_{\mathbf{r}|a}(\mathbf{R}|A)}{\partial A} = \frac{1}{\sigma_n^2}\sum_{i=1}^{N}[R_i - s(A)]\frac{\partial s(A)}{\partial A}. \tag{206}$$

In general, the right-hand side cannot be written in the form required by (187), and therefore an unbiased efficient estimate does not exist.

The likelihood equation is

$$\left[\frac{\partial s(A)}{\partial A}\frac{1}{\sigma_n^2}\right]\left[\frac{1}{N}\sum_{i=1}^{N} R_i - s(A)\right]\Bigg|_{A = \hat{a}_{ml}(\mathbf{R})} = 0. \tag{207}$$

If the range of $s(A)$ includes $(1/N)\sum_{i=1}^{N} R_i$, a solution exists:

$$s[\hat{a}_{ml}(\mathbf{R})] = \frac{1}{N}\sum_{i=1}^{N} R_i. \tag{208}$$

If (208) can be satisfied, then

$$\hat{a}_{ml}(\mathbf{R}) = s^{-1}\left(\frac{1}{N}\sum_{i=1}^{N} R_i\right). \tag{209}$$

[Observe that (209) tacitly assumes that $s^{-1}(\cdot)$ exists. If it does not, then even in the absence of noise we shall be unable to determine A unambiguously. If we were designing a system, we would always choose an $s(\cdot)$ that allows us to find A unambiguously in the absence of noise.] If the range of $s(a)$ does not include $(1/N)\sum_{i=1}^{N} R_i$, the maximum is at an end point of the range.

We see that the maximum likelihood estimate commutes over nonlinear operations. (This is *not* true for MS or MAP estimation.) If it is unbiased, we evaluate the bound on the variance by differentiating (206):

$$\frac{\partial^2 \ln p_{\mathbf{r}|a}(\mathbf{R}|A)}{\partial A^2} = \frac{1}{\sigma_n^2}\sum_{i=1}^{N}[R_i - s(A)]\frac{\partial^2 s(A)}{\partial A^2} - \frac{N}{\sigma_n^2}\left[\frac{\partial s(A)}{\partial A}\right]^2. \tag{210}$$

Observing that

$$E[r_i - s(A)] = E(n_i) = 0, \tag{211}$$

we obtain the following bound for any unbiased estimate,

$$\text{Var}\,[\hat{a}(\mathbf{R}) - A] \geq \frac{\sigma_n^2}{N[\partial s(A)/\partial A]^2}. \tag{212}$$

We see that the bound is exactly the same as that in Example 2 except for a factor $[\partial s(A)/\partial A]^2$. The intuitive reason for this factor and also some feeling for the conditions under which the bound will be useful may be obtained by inspecting the typical function shown in Fig. 2.22. Define

$$Y = s(A). \tag{213}$$

Then

$$r_i = Y + n_i. \tag{214}$$

The variance in estimating Y is just σ_n^2/N. However, if y_ϵ, the error in estimating Y, is small enough so that the slope is constant, then

$$A_\epsilon \simeq \frac{Y_\epsilon}{\left.\dfrac{\partial s(A)}{\partial A}\right|_{A=\hat{a}(\mathbf{R})}} \tag{215}$$

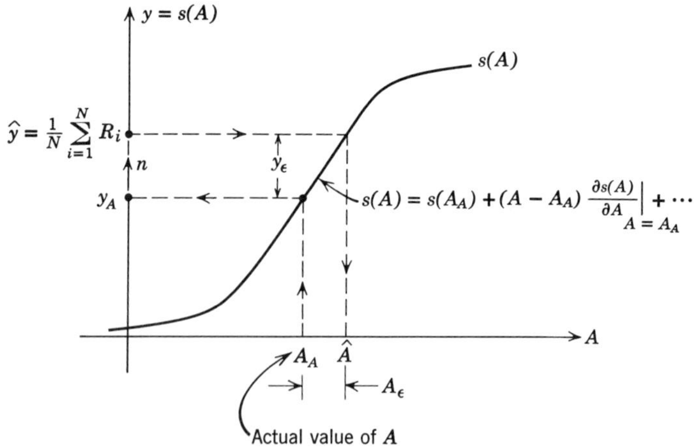

Fig. 2.22 Behavior of error variance in the presence of small errors.

and

$$\text{Var}(a_\epsilon) \cong \frac{\text{Var}(y_\epsilon)}{[\partial s(A)/\partial A]^2} = \frac{\sigma_n^2}{N[\partial s(A)/\partial A]^2}. \tag{216}$$

We observe that if y_ϵ is large there will no longer be a simple linear relation between y_ϵ and a_ϵ. This tells us when we can expect the Cramér-Rao bound to give an accurate answer in the case in which the parameter enters the problem in a nonlinear manner. Specifically, whenever the estimation error is small, relative to $A\ \partial^2 s(A)/\partial A^2$, we should expect the actual variance to be close to the variance bound given by the Cramér-Rao inequality.

The properties of the ML estimate which are valid when the error is small are generally referred to as asymptotic. One procedure for developing them formally is to study the behavior of the estimate as the number of independent observations N approaches infinity. Under reasonably general conditions the following may be proved (e.g., Cramér [9], pp. 500–504).

1. The solution of the likelihood equation (177) converges in probability to the correct value of A as $N \to \infty$. *Any* estimate with this property is called consistent. Thus the ML estimate is consistent.
2. The ML estimate is asymptotically efficient; that is,

$$\lim_{N\to\infty} \frac{\text{Var}\,[\hat{a}_{ml}(\mathbf{R}) - A]}{\left(-E\left[\dfrac{\partial^2 \ln p_{\mathbf{r}|a}(\mathbf{R}|A)}{\partial A^2}\right]\right)^{-1}} = 1.$$

3. The ML estimate is asymptotically Gaussian, $N(A, \sigma_{a_\epsilon})$.

These properties all deal with the behavior of ML estimates for large N. They provide some motivation for using the ML estimate even when an efficient estimate does not exist.

At this point a logical question is: "Do better estimation procedures than the maximum likelihood procedure exist?" Certainly if an efficient estimate does not exist, there may be unbiased estimates with lower variances. The difficulty is that there is no general rule for finding them. In a particular situation we can try to improve on the ML estimate. In almost all cases, however, the resulting estimation rule is more complex, and therefore we emphasize the maximum likelihood technique in all of our work with real variables.

A second logical question is: "Do better lower bounds than the Cramér–Rao inequality exist?" One straightforward but computationally tedious procedure is the Bhattacharyya bound. The Cramér–Rao bound uses $\partial^2 p_{\mathbf{r}|a}(\mathbf{R}|A)/\partial A^2$. Whenever an efficient estimate does not exist, a larger bound which involves the higher partial derivatives can be obtained. Simple derivations are given in [13] and [14] and in Problems 2.4.23–24. For the cases of interest to us the computation is too involved to make the bound of much practical value. A second bound is the Barankin bound (e.g. [15]). Its two major advantages are that it does not require the probability density to be differentiable and it gives the greatest lower bound. Its disadvantages are that it requires a maximization over a function to obtain the bound and the procedure for finding this maximum is usually not straightforward. Some simple examples are given in the problems (2.4.18–19). In most of our discussions, we emphasize the Cramér–Rao bound.

We now digress briefly to develop a similar bound on the mean-square error when the parameter is random.

Lower Bound on the Minimum Mean-Square Error in Estimating a Random Parameter. In this section we prove the following theorem.

Theorem. Let a be a random variable and $\mathbf{r}$, the observation vector. The mean-square error of any estimate $\hat{a}(\mathbf{R})$ satisfies the inequality

$$\begin{aligned} E\{[\hat{a}(\mathbf{R}) - a]^2\} &\geq \left(E\left\{\left[\frac{\partial \ln p_{\mathbf{r},a}(\mathbf{R}, A)}{\partial A}\right]^2\right\}\right)^{-1} \\ &= \left\{-E\left[\frac{\partial^2 \ln p_{\mathbf{r},a}(\mathbf{R}, A)}{\partial A^2}\right]\right\}^{-1}. \end{aligned} \tag{217}$$

Observe that the probability density is a joint density and that the expectation is over both a and $\mathbf{r}$. The following conditions are assumed to exist:

1. $\dfrac{\partial p_{\mathbf{r},a}(\mathbf{R}, A)}{\partial A}$ is absolutely integrable with respect to $\mathbf{R}$ and A.

2. $\dfrac{\partial^2 p_{\mathbf{r},a}(\mathbf{R}, A)}{\partial A^2}$ is absolutely integrable with respect to $\mathbf{R}$ and A.

3. The conditional expectation of the error, given A, is

$$B(A) = \int_{-\infty}^{\infty} [\hat{a}(\mathbf{R}) - A]\, p_{\mathbf{r}|a}(\mathbf{R}|A)\, d\mathbf{R}. \tag{218}$$

We assume that

$$\lim_{A \to \infty} B(A)\, p_a(A) = 0, \tag{219}$$

$$\lim_{A \to -\infty} B(A)\, p_a(A) = 0. \tag{220}$$

The proof is a simple modification of the one on p. 66. Multiply both sides of (218) by $p_a(A)$ and then differentiate with respect to A:

$$\begin{aligned} \frac{d}{dA}[p_a(A)\, B(A)] = &-\int_{-\infty}^{\infty} p_{\mathbf{r},a}(\mathbf{R}, A)\, d\mathbf{R} \\ &+ \int_{-\infty}^{\infty} \frac{\partial p_{\mathbf{r},a}(\mathbf{R}, A)}{\partial A} [\hat{a}(\mathbf{R}) - A]\, d\mathbf{R}. \end{aligned} \tag{221}$$

Now integrate with respect to A:

$$p_a(A)\,B(A)\Big|_{-\infty}^{+\infty} = -1 + \int_{-\infty}^{\infty}\int_{-\infty}^{\infty} \frac{\partial p_{\mathbf{r},a}(\mathbf{R}, A)}{\partial A}\,[\hat{a}(\mathbf{R}) - A]\,dA\,d\mathbf{R}. \quad (222)$$

The assumption in Condition 3 makes the left-hand side zero. The remaining steps are identical. The result is

$$E\{[\hat{a}(\mathbf{R}) - a]^2\} \geq \left\{E\left[\left(\frac{\partial \ln p_{\mathbf{r},a}(\mathbf{R}, A)}{\partial A}\right)^2\right]\right\}^{-1} \quad (223)$$

or, equivalently,

$$E\{[\hat{a}(\mathbf{R}) - a]^2\} \geq \left\{-E\left[\frac{\partial^2 \ln p_{\mathbf{r}|a}(\mathbf{R}|A)}{\partial A^2}\right] - E\left[\frac{\partial^2 \ln p_a(A)}{\partial A^2}\right]\right\}^{-1} \quad (224)$$

with equality if and only if

$$\frac{\partial \ln p_{\mathbf{r},a}(\mathbf{R}, A)}{\partial A} = k[\hat{a}(\mathbf{R}) - A], \quad (225)$$

for all $\mathbf{R}$ and all A. (In the nonrandom variable case we used the Schwarz inequality on an integral over $\mathbf{R}$ so that the constant $k(A)$ could be a function of A. Now the integration is over both $\mathbf{R}$ and A so that k cannot be a function of A.) Differentiating again gives an equivalent condition

$$\frac{\partial^2 \ln p_{\mathbf{r},a}(\mathbf{R}, A)}{\partial A^2} = -k. \quad (226)$$

Observe that (226) may be written in terms of the a posteriori density,

$$\frac{\partial^2 \ln p_{a|\mathbf{r}}(A|\mathbf{R})}{\partial A^2} = -k. \quad (227)$$

Integrating (227) twice and putting the result in the exponent, we have

$$p_{a|\mathbf{r}}(A|\mathbf{R}) = \exp\left(-kA^2 + C_1 A + C_2\right) \quad (228)$$

for all $\mathbf{R}$ and A; but (228) is simply a statement that the a posteriori probability density of a must be Gaussian for all $\mathbf{R}$ in order for an efficient estimate to exist. (Note that C_1 and C_2 are functions of $\mathbf{R}$).

Arguing as in (193)–(195), we see that if (226) is satisfied the MAP estimate will be efficient. Because the minimum MSE estimate cannot have a larger error, this tells us that $\hat{a}_{\text{ms}}(\mathbf{R}) = \hat{a}_{\text{map}}(\mathbf{R})$ whenever an efficient estimate exists. As a matter of technique, when an efficient estimate does exist, it is usually computationally easier to solve the MAP equation than it is to find the conditional mean. When an efficient estimate does not exist, we do not know how closely the mean-square error, using either $\hat{a}_{\text{ms}}(\mathbf{R})$ or $\hat{a}_{\text{map}}(\mathbf{R})$, approaches the lower bound. Asymptotic results similar to those for real variables may be derived.

2.4.3 Multiple Parameter Estimation

In many problems of interest we shall want to estimate more than one parameter. A familiar example is the radar problem in which we shall estimate the range and velocity of a target. Most of the ideas and techniques can be extended to this case in a straightforward manner. The model is shown in Fig. 2.23. If there are K parameters, $a_1, a_2, \ldots, a_K$, we describe them by a parameter vector $\mathbf{a}$ in a K-dimensional space. The other elements of the model are the same as before. We shall consider both the case in which $\mathbf{a}$ is a random parameter vector and that in which $\mathbf{a}$ is a real (or nonrandom) parameter vector. Three issues are of interest. In each the result is the vector analog to a result in the scalar case.

1. Estimation procedures.
2. Measures of error.
3. Bounds on performance.

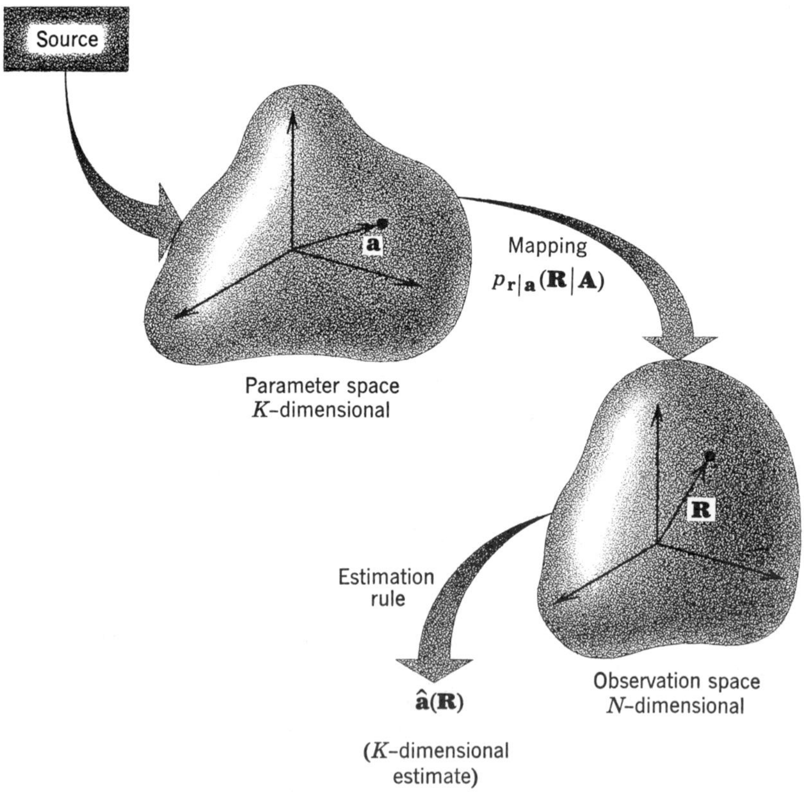

Fig. 2.23 Multiple parameter estimation model.

Estimation Procedure. For random variables we could consider the general case of Bayes estimation in which we minimize the risk for some arbitrary scalar cost function $C(\mathbf{a}, \hat{\mathbf{a}})$, but for our purposes it is adequate to consider only cost functions that depend on the error. We define the error vector as

$$\mathbf{a}_\epsilon(\mathbf{R}) = \begin{bmatrix} \hat{a}_1(\mathbf{R}) - a_1 \\ \hat{a}_2(\mathbf{R}) - a_2 \\ \vdots \\ \hat{a}_K(\mathbf{R}) - a_K \end{bmatrix} = \hat{\mathbf{a}}(\mathbf{R}) - \mathbf{a}. \tag{229}$$

For a mean-square error criterion, the cost function is simply

$$C(\mathbf{a}_\epsilon(\mathbf{R})) \triangleq \sum_{i=1}^{K} a_{\epsilon_i}{}^2(\mathbf{R}) = \mathbf{a}_\epsilon{}^T(\mathbf{R})\, \mathbf{a}_\epsilon(\mathbf{R}). \tag{230}$$

This is just the sum of the squares of the errors. The risk is

$$\mathcal{R}_{\text{ms}} = \iint_{-\infty}^{\infty} C(\mathbf{a}_\epsilon(\mathbf{R}))\, p_{\mathbf{r},\mathbf{a}}(\mathbf{R}, \mathbf{A})\, d\mathbf{R}\, d\mathbf{A} \tag{231}$$

or

$$\mathcal{R}_{\text{ms}} = \int_{-\infty}^{\infty} p_{\mathbf{r}}(\mathbf{R})\, d\mathbf{R} \int_{-\infty}^{\infty} \left[\sum_{i=1}^{K} (\hat{a}_i(\mathbf{R}) - A_i)^2\right] p_{\mathbf{a}|\mathbf{r}}(\mathbf{A}|\mathbf{R})\, d\mathbf{A}. \tag{232}$$

As before, we can minimize the inner integral for each $\mathbf{R}$. Because the terms in the sum are positive, we minimize them separately. This gives

$$\hat{a}_{\text{ms}_i}(\mathbf{R}) = \int_{-\infty}^{\infty} A_i p_{\mathbf{a}|\mathbf{r}}(\mathbf{A}|\mathbf{R})\, d\mathbf{A} \tag{233}$$

or

$$\hat{\mathbf{a}}_{\text{ms}}(\mathbf{R}) = \int_{-\infty}^{\infty} \mathbf{A} p_{\mathbf{a}|\mathbf{r}}(\mathbf{A}|\mathbf{R})\, d\mathbf{A}. \tag{234}$$

It is easy to show that mean-square estimation commutes over *linear transformations*. Thus, if

$$\mathbf{b} = \mathbf{D}\mathbf{a}, \tag{235}$$

where $\mathbf{D}$ is a $L \times K$ matrix, and we want to minimize

$$E[\mathbf{b}_\epsilon{}^T(\mathbf{R})\, \mathbf{b}_\epsilon(\mathbf{R})] = E\left[\sum_{i=1}^{L} b_{\epsilon_i}{}^2(\mathbf{R}),\right] \tag{236}$$

the result will be,

$$\hat{\mathbf{b}}_{\text{ms}}(\mathbf{R}) = \mathbf{D}\hat{\mathbf{a}}_{\text{ms}}(\mathbf{R}) \tag{237}$$

[see Problem 2.4.20 for the proof of (237)].

For MAP estimation we must find the value of $\mathbf{A}$ that maximizes $p_{\mathbf{a}|\mathbf{r}}(\mathbf{A}|\mathbf{R})$. If the maximum is interior and $\partial \ln p_{\mathbf{a}|\mathbf{r}}(\mathbf{A}|\mathbf{R})/\partial A_i$ exists at the maximum then a necessary condition is obtained from the MAP equations. By analogy with (137) we take the logarithm of $p_{\mathbf{a}|\mathbf{r}}(\mathbf{A}|\mathbf{R})$, differentiate with respect to each parameter A_i, $i = 1, 2, \ldots, K$, and set the result equal to zero. This gives a set of K simultaneous equations:

$$\left.\frac{\partial \ln p_{\mathbf{a}|\mathbf{r}}(\mathbf{A}|\mathbf{R})}{\partial A_i}\right|_{\mathbf{A}=\hat{\mathbf{a}}_{\text{map}}(\mathbf{R})} = 0, \qquad i = 1, 2, \ldots, K. \tag{238}$$

We can write (238) in a more compact manner by defining a partial derivative matrix operator

$$\nabla_{\mathbf{A}} \triangleq \begin{bmatrix} \dfrac{\partial}{\partial A_1} \\ \dfrac{\partial}{\partial A_2} \\ \vdots \\ \dfrac{\partial}{\partial A_K} \end{bmatrix}. \tag{239}$$

This operator can be applied only to $1 \times m$ matrices; for example,

$$\nabla_{\mathbf{A}}\, \mathbf{G} = \begin{bmatrix} \dfrac{\partial G_1}{\partial A_1} & \dfrac{\partial G_2}{\partial A_1} & \cdots & \dfrac{\partial G_m}{\partial A_1} \\ \vdots & & & \\ \dfrac{\partial G_1}{\partial A_K} & & & \dfrac{\partial G_m}{\partial A_K} \end{bmatrix}. \tag{240}$$

Several useful properties of $\nabla_{\mathbf{A}}$ are developed in Problems 2.4.27–28. In our case (238) becomes a single vector equation,

$$\nabla_{\mathbf{A}}[\ln p_{\mathbf{a}|\mathbf{r}}(\mathbf{A}|\mathbf{R})]|_{\mathbf{A}=\hat{\mathbf{a}}_{\text{map}}(\mathbf{R})} = \mathbf{0}. \tag{241}$$

Similarly, for ML estimates we must find the value of $\mathbf{A}$ that maximizes $p_{\mathbf{r}|\mathbf{a}}(\mathbf{R}|\mathbf{A})$. If the maximum is interior and $\partial \ln p_{\mathbf{r}|\mathbf{a}}(\mathbf{R}|\mathbf{A})|\partial A_i$ exists at the maximum then a necessary condition is obtained from the likelihood equations:

$$\nabla_{\mathbf{A}}[\ln p_{\mathbf{r}|\mathbf{a}}(\mathbf{R}|\mathbf{A})]|_{\mathbf{A}=\hat{\mathbf{a}}_{\text{ml}}(\mathbf{R})} = \mathbf{0}. \tag{242}$$

In both cases we must verify that we have the absolute maximum.

Measures of Error. For nonrandom variables the first measure of interest is the bias. Now the bias is a vector,

$$\mathbf{B}(\mathbf{A}) \triangleq E[\mathbf{a}_\epsilon(\mathbf{R})] = E[\hat{\mathbf{a}}(\mathbf{R})] - \mathbf{A}. \tag{243}$$

If each component of the bias vector is zero for every $\mathbf{A}$, we say that the estimate is unbiased.

In the single parameter case a rough measure of the spread of the error was given by the variance of the estimate. In the special case in which $a_\epsilon(\mathbf{R})$ was Gaussian this provided a complete description:

$$p_{a_\epsilon}(A_\epsilon) = \frac{1}{\sqrt{2\pi}\,\sigma_{a_\epsilon}} \exp\left(-\frac{A_\epsilon^2}{2\sigma_{a_\epsilon}^2}\right). \tag{244}$$

For a vector variable the quantity analogous to the variance is the covariance matrix

$$E[(\mathbf{a}_\epsilon - \bar{\mathbf{a}}_\epsilon)(\mathbf{a}_\epsilon^T - \bar{\mathbf{a}}_\epsilon^T)] \triangleq \mathbf{\Lambda}_\epsilon, \tag{245}$$

where

$$\bar{\mathbf{a}}_\epsilon \triangleq E(\mathbf{a}_\epsilon) = \mathbf{B}(\mathbf{A}). \tag{246}$$

The best way to determine how the covariance matrix provides a measure of spread is to consider the special case in which the a_{ϵ_i} are jointly Gaussian. For algebraic simplicity we let $E(\mathbf{a}_\epsilon) = \mathbf{0}$. The joint probability density for a set of K jointly Gaussian variables is

$$p_{\mathbf{a}_\epsilon}(\mathbf{A}_\epsilon) = (|2\pi|^{K/2}|\mathbf{\Lambda}_\epsilon|^{1/2})^{-1} \exp\left(-\tfrac{1}{2}\mathbf{A}_\epsilon^T\mathbf{\Lambda}_\epsilon^{-1}\mathbf{A}_\epsilon\right) \tag{247}$$

(e.g., p. 151 in Davenport and Root [1]).

The probability density for $K = 2$ is shown in Fig. 2.24*a*. In Figs. 2.24*b*,*c* we have shown the equal-probability contours of two typical densities.From (247) we observe that the equal-height contours are defined by the relation,

$$\mathbf{A}_\epsilon^T\mathbf{\Lambda}_\epsilon^{-1}\mathbf{A}_\epsilon = C^2, \tag{248}$$

which is the equation for an ellipse when $K = 2$. The ellipses move out monotonically with increasing C. They also have the interesting property that the probability of being inside the ellipse is only a function of C^2.

Property. For $K = 2$, the probability that the error vector lies inside an ellipse whose equation is

$$\mathbf{A}_\epsilon^T\mathbf{\Lambda}_\epsilon^{-1}\mathbf{A}_\epsilon = C^2, \tag{249}$$

is

$$P = 1 - \exp\left(-\frac{C^2}{2}\right). \tag{250}$$

Proof. The area inside the ellipse defined by (249) is

$$\mathcal{A} = |\mathbf{\Lambda}_\epsilon|^{1/2}\pi C^2. \tag{251}$$

The differential area between ellipses corresponding to C and $C + dC$ respectively is

$$d\mathcal{A} = |\mathbf{\Lambda}_\epsilon|^{1/2}2\pi C\,dC. \tag{252}$$

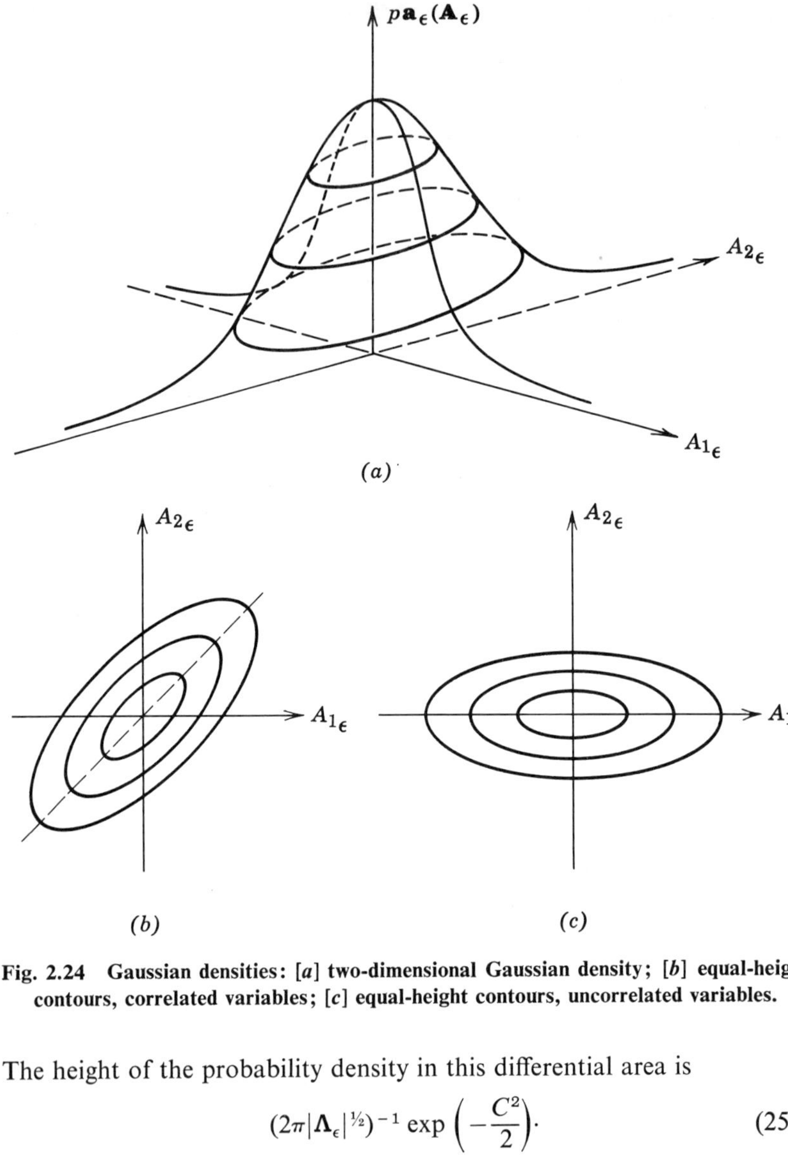

Fig. 2.24 Gaussian densities: [*a*] two-dimensional Gaussian density; [*b*] equal-height contours, correlated variables; [*c*] equal-height contours, uncorrelated variables.

The height of the probability density in this differential area is

$$(2\pi|\mathbf{\Lambda}_\epsilon|^{1/2})^{-1} \exp\left(-\frac{C^2}{2}\right). \tag{253}$$

We can compute the probability of a point lying outside the ellipse by multiplying (252) by (253) and integrating from C to ∞.

$$1 - P = \int_C^\infty X \exp\left(-\frac{X^2}{2}\right) dX = \exp\left(-\frac{C^2}{2}\right), \tag{254}$$

which is the desired result.

For this reason the ellipses described by (248) are referred to as *concentration ellipses* because they provide a measure of the concentration of the density.

A similar result holds for arbitrary K. Now, (248) describes an *ellipsoid*. Here the differential volume† in K-dimensional space is

$$dv = |\mathbf{\Lambda}_\epsilon|^{1/2} \frac{\pi^{K/2}}{\Gamma(K/2 + 1)} KC^{K-1}\, dC. \tag{255}$$

The value of the probability density on the ellipsoid is

$$[(2\pi)^{K/2}|\mathbf{\Lambda}_\epsilon|^{1/2}]^{-1} \exp\left(-\frac{C^2}{2}\right). \tag{256}$$

Therefore

$$1 - P = \frac{K}{(2)^{K/2}\Gamma(K/2 + 1)} \int_C^\infty X^{K-1} e^{-X^2/2}\, dX, \tag{257}$$

which is the desired result. We refer to these ellipsoids as *concentration ellipsoids*.

When the probability density of the error is *not* Gaussian, the concentration ellipsoid no longer specifies a unique probability. This is directly analogous to the one-dimensional case in which the variance of a non-Gaussian zero-mean random variable does not determine the probability density. We can still interpret the concentration ellipsoid as a rough measure of the spread of the errors. When the concentration ellipsoids of a given density lie wholly outside the concentration ellipsoids of a second density, we say that the second density is more concentrated than the first. With this motivation, we derive some properties and bounds pertaining to concentration ellipsoids.

Bounds on Estimation Errors: Nonrandom Variables. In this section we derive two bounds. The first relates to the variance of an individual error; the second relates to the concentration ellipsoid.

Property 1. Consider *any* unbiased estimate of A_i. Then

$$\sigma_{\epsilon_i}{}^2 \triangleq \text{Var}\,[\hat{a}_i(\mathbf{R}) - A_i] \geq J^{ii}, \tag{258}$$

where J^{ii} is the iith element in the $K \times K$ square matrix $\mathbf{J}^{-1}$. The elements in $\mathbf{J}$ are

$$\begin{aligned} J_{ij} &\triangleq E\left[\frac{\partial \ln p_{\mathbf{r}|\mathbf{a}}(\mathbf{R}|\mathbf{A})}{\partial A_i} \cdot \frac{\partial \ln p_{\mathbf{r}|\mathbf{a}}(\mathbf{R}|\mathbf{A})}{\partial A_j}\right] \\ &= -E\left[\frac{\partial^2 \ln p_{\mathbf{r}|\mathbf{a}}(\mathbf{R}|\mathbf{A})}{\partial A_i\, \partial A_j}\right] \end{aligned} \tag{259}$$

† e.g., Cramér [9], p. 120, or Sommerfeld [32].

or

$$\mathbf{J} \triangleq E(\{\nabla_{\mathbf{A}}[\ln p_{\mathbf{r}|\mathbf{a}}(\mathbf{R}|\mathbf{A})]\}\{\nabla_{\mathbf{A}}[\ln p_{\mathbf{r}|\mathbf{a}}(\mathbf{R}|\mathbf{A})]\}^T) \\ = -E[\nabla_{\mathbf{A}}(\{\nabla_{\mathbf{A}}[\ln p_{\mathbf{r}|\mathbf{a}}(\mathbf{R}|\mathbf{A})]\}^T)]. \tag{260}$$

The $\mathbf{J}$ matrix is commonly called *Fisher's information matrix*. The equality in (258) holds if and only if

$$\hat{a}_i(\mathbf{R}) - A_i = \sum_{j=1}^{K} k_{ij}(\mathbf{A}) \frac{\partial \ln p_{\mathbf{r}|\mathbf{a}}(\mathbf{R}|\mathbf{A})}{\partial A_j} \tag{261}$$

for all values of A_i and $\mathbf{R}$.

In other words, the estimation error can be expressed as the weighted sum of the partial derivatives of $\ln p_{\mathbf{r}|\mathbf{a}}(\mathbf{R}|\mathbf{A})$ with respect to the various parameters.

Proof. Because $\hat{a}_i(\mathbf{R})$ is unbiased,

$$\int_{-\infty}^{\infty} [\hat{a}_i(\mathbf{R}) - A_i]\, p_{\mathbf{r}|\mathbf{a}}(\mathbf{R}|\mathbf{A})\, d\mathbf{R} = 0 \tag{262}$$

or

$$\int_{-\infty}^{\infty} \hat{a}_i(\mathbf{R})\, p_{\mathbf{r}|\mathbf{a}}(\mathbf{R}|\mathbf{A})\, d\mathbf{R} = A_i. \tag{263}$$

Differentiating both sides with respect to A_j, we have

$$\int_{-\infty}^{\infty} \hat{a}_i(\mathbf{R}) \frac{\partial p_{\mathbf{r}|\mathbf{a}}(\mathbf{R}|\mathbf{A})}{\partial A_j}\, d\mathbf{R} \\ = \int_{-\infty}^{\infty} \hat{a}_i(\mathbf{R}) \frac{\partial \ln p_{\mathbf{r}|\mathbf{a}}(\mathbf{R}|\mathbf{A})}{\partial A_j}\, p_{\mathbf{r}|\mathbf{a}}(\mathbf{R}|\mathbf{A})\, d\mathbf{R} = \delta_{ij}. \tag{264}$$

We shall prove the result for $i = 1$. We define a $K + 1$ vector

$$\mathbf{x} = \begin{bmatrix} \hat{a}_1(\mathbf{R}) - A_1 \\ \dfrac{\partial \ln p_{\mathbf{r}|\mathbf{a}}(\mathbf{R}|\mathbf{A})}{\partial A_1} \\ \vdots \\ \dfrac{\partial \ln p_{\mathbf{r}|\mathbf{a}}(\mathbf{R}|\mathbf{A})}{\partial A_K} \end{bmatrix}. \tag{265}$$

The covariance matrix is

$$E[\mathbf{x}\mathbf{x}^T] = \begin{bmatrix} \sigma_{\epsilon_1}{}^2 & 1 & 0 & 0 & 0 \\ 1 & J_{11} & J_{12} & \cdots & J_{1K} \\ 0 & \vdots & & \ddots & \vdots \\ 0 & J_{K1} & & & J_{KK} \end{bmatrix}. \tag{266}$$

[The ones and zeroes in the matrix follow from (264).] Because it is a covariance matrix, it is nonnegative definite, which implies that the determinant of the entire matrix is greater than or equal to zero. (This condition is only necessary, not sufficient, for the matrix to be nonnegative definite.)

Evaluating the determinant using a cofactor expansion, we have

$$\sigma_{\epsilon_1}{}^2|\mathbf{J}| - \text{cofactor } J_{11} \geq 0. \tag{267}$$

If we assume that $\mathbf{J}$ is nonsingular, then

$$\sigma_{\epsilon_1}{}^2 \geq \frac{\text{cofactor } J_{11}}{|\mathbf{J}|} = J^{11}, \tag{268}$$

which is the desired result. The modifications for the case when $\mathbf{J}$ is singular follow easily for any specific problem.

In order for the determinant to equal zero, the term $\hat{A}_1(\mathbf{R}) - A_1$ must be expressible as a linear combination of the other terms. This is the condition described by (261). The second line of (259) follows from the first line in a manner exactly analogous to the proof in (189)–(192). The proof for $i \neq 1$ is an obvious modification.

Property 2. Consider *any* unbiased estimate of $\mathbf{A}$. The concentration ellipse

$$\mathbf{A}_\epsilon{}^T\mathbf{\Lambda}_\epsilon{}^{-1}\mathbf{A}_\epsilon = C^2 \tag{269}$$

lies either outside or on the bound ellipse defined by

$$\mathbf{A}_\epsilon{}^T\mathbf{J}\mathbf{A}_\epsilon = C^2. \tag{270}$$

Proof. We shall go through the details for $K = 2$. By analogy with the preceding proof, we construct the covariance matrix of the vector.

$$\mathbf{x} = \begin{bmatrix} \hat{a}_1(\mathbf{R}) - A_1 \\ \hat{a}_2(\mathbf{R}) - A_2 \\ \dfrac{\partial \ln p_{\mathbf{r}|\mathbf{a}}(\mathbf{R}|\mathbf{A})}{\partial A_1} \\ \dfrac{\partial \ln p_{\mathbf{r}|\mathbf{a}}(\mathbf{R}|\mathbf{A})}{\partial A_2} \end{bmatrix}. \tag{271}$$

Then

$$E[\mathbf{x}\mathbf{x}^T] = \left[\begin{array}{cc|cc} \sigma_{1_\epsilon}{}^2 & \rho\sigma_{1_\epsilon}\sigma_{2_\epsilon} & 1 & 0 \\ \rho\sigma_{1_\epsilon}\sigma_{2_\epsilon} & \sigma_{2_\epsilon}{}^2 & 0 & 1 \\ \hline 1 & 0 & J_{11} & J_{12} \\ 0 & 1 & J_{21} & J_{22} \end{array}\right] = \left[\begin{array}{c|c} \mathbf{\Lambda}_\epsilon & \mathbf{I} \\ \hline \mathbf{I} & \mathbf{J} \end{array}\right]. \tag{272}$$

The second equality defines a partition of the 4×4 matrix into four 2×2 matrices. Because it is a covariance matrix, it is nonnegative definite. Using a formula for the determinant of a partitioned matrix,† we have

$$|\mathbf{\Lambda}_\epsilon \mathbf{J} - \mathbf{I}| \geq 0 \tag{273}$$

or, assuming that $\mathbf{\Lambda}_\epsilon$ is nonsingular and applying the product rule for determinants,

$$|\mathbf{\Lambda}_\epsilon|\ |\mathbf{J} - \mathbf{\Lambda}_\epsilon^{-1}| \geq 0. \tag{274}$$

This implies

$$|\mathbf{J} - \mathbf{\Lambda}_\epsilon^{-1}| \geq 0. \tag{275}$$

Now consider the two ellipses. The intercept on the A_{ϵ_1} axis is

$$A_{1_\epsilon}^{\ 2}\Big|_{A_{2\epsilon}=0} = C^2 \frac{|\mathbf{\Lambda}_\epsilon|}{\sigma_2^{\ 2}} \tag{276}$$

for the actual concentration ellipse and

$$A_{1_\epsilon}^{\ 2}\Big|_{A_{2\epsilon}=0} = C^2 \frac{1}{J_{11}} \tag{277}$$

for the bound ellipse.

We want to show that the actual intercept is greater than or equal to the bound intercept. This requires

$$J_{11}|\mathbf{\Lambda}_\epsilon| \geq \sigma_2^{\ 2}. \tag{278}$$

This inequality follows because the determinant of the 3×3 matrix in the upper left corner of (272) is greater than or equal to zero. (Otherwise the entire matrix is not nonnegative definite, e.g. [16] or [18].) Similarly, the actual intercept on the A_{2_ϵ} axis is greater than or equal to the bound intercept. Therefore the actual ellipse is either always outside (or on) the bound ellipse *or* the two ellipses intersect.

If they intersect, we see from (269) and (270) that there must be a solution, $\mathbf{A}_\epsilon$, to the equation

$$\mathbf{A}_\epsilon^T \mathbf{\Lambda}_\epsilon^{-1} \mathbf{A}_\epsilon = \mathbf{A}_\epsilon^T \mathbf{J} \mathbf{A}_\epsilon \tag{279}$$

or

$$\mathbf{A}_\epsilon^T [\mathbf{J} - \mathbf{\Lambda}_\epsilon^{-1}] \mathbf{A}_\epsilon \triangleq \mathbf{A}_\epsilon^T \mathbf{D} \mathbf{A}_\epsilon = 0. \tag{280}$$

In scalar notation

$$A_{1_\epsilon}^{\ 2} D_{11} + 2A_{1_\epsilon} A_{2_\epsilon} D_{12} + A_{2_\epsilon}^{\ 2} D_{22} = 0 \tag{281}$$

or, equivalently,

$$\left(\frac{A_{1_\epsilon}}{A_{2_\epsilon}}\right)^2 D_{11} + 2\left(\frac{A_{1_\epsilon}}{A_{2_\epsilon}}\right) D_{12} + D_{22} = 0. \tag{282}$$

† Bellman [16], p. 83.

Solving for $A_{1_\epsilon}/A_{2_\epsilon}$, we would obtain real roots only if the discriminant were greater than or equal to zero. This requires

$$|\mathbf{J} - \mathbf{\Lambda}_\epsilon^{-1}| \leq 0. \tag{283}$$

The inequality is a contradiction of (275). One possibility is $|\mathbf{J} - \mathbf{\Lambda}_\epsilon^{-1}| = 0$, but this is true only when the ellipses coincide. In this case all the estimates are efficient.

For arbitrary K we can show that $\mathbf{J} - \mathbf{\Lambda}_\epsilon^{-1}$ is nonnegative definite. The implications with respect to the concentration ellipsoids are the same as for $K = 2$.

Frequently we want to estimate functions of the K basic parameters rather than the parameters themselves. We denote the desired estimates as

$$\begin{aligned} d_1 &= g_{d_1}(\mathbf{A}), \\ d_2 &= g_{d_2}(\mathbf{A}), \\ &\vdots \\ d_M &= g_{d_M}(\mathbf{A}). \end{aligned} \tag{284}$$

or

$$\mathbf{d} = \mathbf{g_d}(\mathbf{A})$$

The number of estimates M is not related to K in general. The functions may be nonlinear. The estimation error is

$$\hat{d}_i - g_i(\mathbf{A}) \triangleq d_{\epsilon_i}. \tag{285}$$

If we assume that the estimates are unbiased and denote the error covariance matrix as $\mathbf{\Lambda}_\epsilon$, then by using methods identical to those above we can prove the following properties.

Property 3. The matrix

$$\mathbf{\Lambda}_\epsilon - \{\nabla_{\mathbf{A}}[\mathbf{g_d}^T(\mathbf{A})]\}^T \mathbf{J}^{-1}\{\nabla_{\mathbf{A}}[\mathbf{g_d}^T(\mathbf{A})]\} \tag{286}$$

is nonnegative definite.

This implies the following property (just multiply the second matrix out and recall that all diagonal elements of nonnegative definite matrix are nonnegative):

Property 4.

$$\text{Var}\,(d_{\epsilon_l}) \geq \sum_i^K \sum_j^K \frac{\partial g_{d_l}(\mathbf{A})}{\partial A_i} J^{ij} \frac{\partial g_{d_l}(\mathbf{A})}{\partial A_j}. \tag{287}$$

For the special case in which the desired functions are linear, the result in (287) can be written in a simpler form.

Property 5. Assume that

$$\mathbf{g_d}(\mathbf{A}) \triangleq \mathbf{G_d}\mathbf{A}, \tag{288}$$

where $\mathbf{G_d}$ is an $M \times K$ matrix. If the estimates are unbiased, then

$$\mathbf{\Lambda}_\epsilon - \mathbf{G_d}\mathbf{J}^{-1}\mathbf{G_d}^T$$

is nonnegative definite.

Property 6. Efficiency commutes with linear transformations but does not commute with nonlinear transformations. In other words, if $\hat{\mathbf{A}}$ is efficient, then $\mathbf{g_d}(\mathbf{A})$ will be efficient if and only if $\mathbf{g_d}(\mathbf{A})$ is a linear transformation.

Bounds on Estimation Errors: Random Parameters. Just as in the single parameter case, the bound for random parameters is derived by a straightforward modification of the derivation for nonrandom parameters. The information matrix now consists of two parts:

$$\mathbf{J}_T \triangleq \mathbf{J}_D + \mathbf{J}_P. \tag{289}$$

The matrix $\mathbf{J}_D$ is the information matrix defined in (260) and represents information obtained from the *data*. The matrix $\mathbf{J}_P$ represents the a priori information. The elements are

$$\begin{aligned} J_{P_{ij}} &\triangleq E\left[\frac{\partial \ln p_{\mathbf{a}}(\mathbf{A})}{\partial A_i}\frac{\partial \ln p_{\mathbf{a}}(\mathbf{A})}{\partial A_j}\right] \\ &= -E\left[\frac{\partial^2 \ln p_{\mathbf{a}}(\mathbf{A})}{\partial A_i\, \partial A_j}\right]. \end{aligned} \tag{290}$$

The *correlation matrix* of the errors is

$$\mathbf{R}_\epsilon \triangleq E(\mathbf{a}_\epsilon \mathbf{a}_\epsilon^T). \tag{291}$$

The diagonal elements represent the mean-square errors and the off-diagonal elements are the cross correlations. Three properties follow easily:

Property No. 1.

$$E[a_{\epsilon_i}^2] \geq J_T^{ii}. \tag{292}$$

In other words, the diagonal elements in the inverse of the total information matrix are lower bounds on the corresponding mean-square errors.

Property No. 2. The matrix

$$\mathbf{J}_T - \mathbf{R}_\epsilon^{-1}$$

is nonnegative definite. This has the same physical interpretation as in the nonrandom parameter problem.

Property No. 3. If $\mathbf{J}_T = \mathbf{R}_\epsilon^{-1}$, all of the estimates are efficient. A necessary and sufficient condition for this to be true is that $p_{\mathbf{a}|\mathbf{r}}(\mathbf{A}|\mathbf{R})$ be Gaussian for all $\mathbf{R}$. This will be true iff $\mathbf{J}$ is constant. [Modify (261), (228)].

A special case of interest occurs when the a priori density is a Kth-order Gaussian density. Then

$$\mathbf{J}_P = \mathbf{\Lambda_a}^{-1}, \tag{293}$$

where $\mathbf{\Lambda_a}$ is the covariance matrix of the random parameters.

An even simpler case arises when the variables are independent Gaussian variables. Then

$$J_{P_{ij}} = \frac{1}{\sigma_{a_i}^{\ 2}}\,\delta_{ij}, \tag{294}$$

Under these conditions only the diagonal terms of $\mathbf{J}_T$ are affected by the a priori information.

Results similar to Properties 3 to 6 for nonrandom parameters can be derived for the random parameter case.

2.4.4 Summary of Estimation Theory

In this section we developed the estimation theory results that we shall need for the problems of interest. We began our discussion with Bayes estimation of random parameters. The basic quantities needed in the model were the a priori density $p_a(A)$, the probabilistic mapping to the observation space $p_{\mathbf{r}|a}(\mathbf{R}|A)$, and a cost function $C(A_\epsilon)$. These quantities enabled us to find the risk. The estimate which minimized the risk was called a Bayes estimate and the resulting risk, the Bayes risk. Two types of Bayes estimate, the MMSE estimate (which was the mean of the a posteriori density) and the MAP estimate (the mode of the a posteriori density), were emphasized. In Properties 1 and 2 (pp. 60–61) we saw that the conditional mean was the Bayes estimate for a large class of cost functions when certain conditions on the cost function and a posteriori density were satisfied.

Turning to nonrandom parameter estimation, we introduced the idea of bias and variance as two separate error measures. The Cramér-Rao inequality provided a bound on the variance of any unbiased estimate. Whenever an efficient estimate existed, the maximum likelihood estimation procedure gave this estimate. This property of the ML estimate, coupled with its asymptotic properties, is the basis for our emphasis on ML estimates.

The extension to multiple parameter estimation involved no new concepts. Most of the properties were just multidimensional extensions of the corresponding scalar result.

It is important to emphasize the close relationship between detection and estimation theory. Both theories are based on a likelihood function or likelihood ratio, which, in turn, is derived from the probabilistic transition mechanism. As we proceed to more difficult problems, we shall find that a large part of the work is the manipulation of this transition mechanism. In many cases the mechanism will not depend on whether the problem is one of detection or estimation. Thus the difficult part of the problem will be applicable to either problem. This close relationship will become even more obvious as we proceed. We now return to the detection theory problem and consider a more general model.

2.5 COMPOSITE HYPOTHESES

In Sections 2.2 and 2.3 we confined our discussion to the decision problem in which the hypotheses were simple. We now extend our discussion to the case in which the hypotheses are composite. The term composite is most easily explained by a simple example.

Example 1. Under hypothesis 0 the observed variable r is Gaussian with zero mean and variance σ^2. Under hypothesis 1 the observed variable r is Gaussian with mean m and variance σ^2. The value of m can be anywhere in the interval $[M_0, M_1]$. Thus

$$\begin{aligned} H_0: p_{r|H_0}(R|H_0) &= \frac{1}{\sqrt{2\pi}\,\sigma} \exp\left(-\frac{R^2}{2\sigma^2}\right), \\ H_1: p_{r|H_1}(R|H_1) &= \frac{1}{\sqrt{2\pi}\,\sigma} \exp\left(-\frac{(R-M)^2}{2\sigma^2}\right), \qquad M_0 \le M \le M_1. \end{aligned} \tag{295}$$

We refer to H_1 as a composite hypothesis because the parameter value M, which characterizes the hypothesis, ranges over a set of values. A model of this decision problem is shown in Fig. 2.25*a*. The output of the source is a parameter value M, which we view as a point in a parameter space χ. We then define the hypotheses as subspaces of χ. In this case H_0 corresponds to the point $M = 0$ and H_1 corresponds to the interval $[M_0, M_1.]$ We assume that the probability density governing the mapping from the parameter space to the observation space, $p_{r|m}(R|M)$, is known for all values of M in χ.

The final component is a decision rule that divides the observation space into two parts which correspond to the two possible decisions. It is important to observe that we are interested *solely* in making a decision and that the actual value of M is not of interest to us. For this reason the parameter M is frequently referred to as an "unwanted" parameter.

The extension of these ideas to the general composite hypothesis-testing problem is straightforward. The model is shown in Fig. 2.25*b*. The output of the source is a set of parameters. We view it as a point in a parameter space χ and denote it by the vector $\boldsymbol{\theta}$. The hypotheses are subspaces of χ. (In Fig. 2.25*b* we have indicated nonoverlapping spaces for convenience.) The probability density governing the mapping from the parameter space to the observation space is denoted by $p_{\mathbf{r}|\boldsymbol{\theta}}(\mathbf{R}|\boldsymbol{\theta})$ and is assumed to be known for all values of $\boldsymbol{\theta}$ in χ. Once again, the final component is a decision rule.

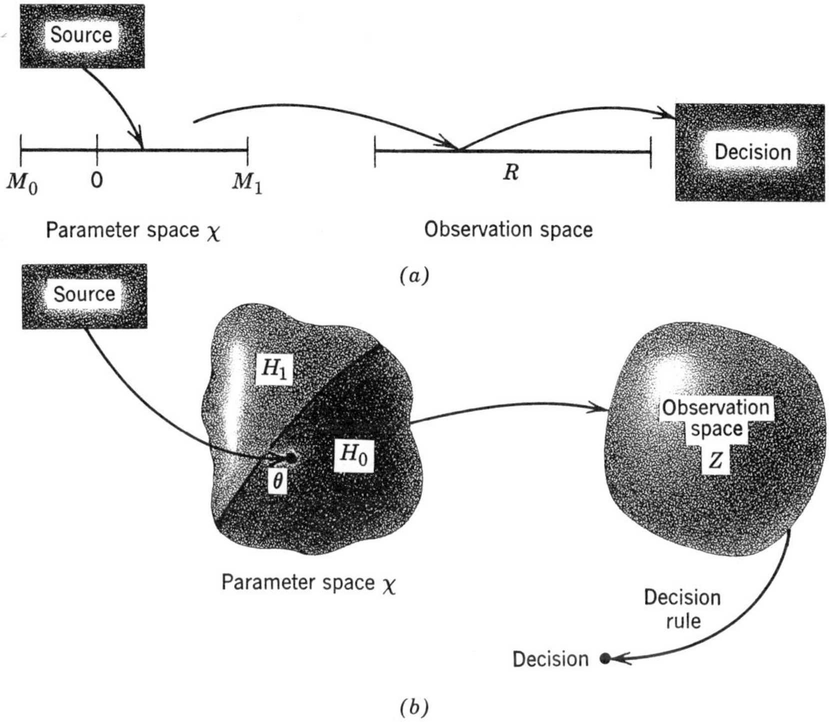

Fig. 2.25 ***a.*** **Composite hypothesis testing problem for single-parameter example.** ***b.*** **Composite hypothesis testing problem.**

To complete the formulation, we must characterize the parameter $\boldsymbol{\theta}$. Just as in the parameter estimation case the parameter $\boldsymbol{\theta}$ may be a nonrandom or random variable. If $\boldsymbol{\theta}$ is a random variable with a known probability density, the procedure is straightforward. Denoting the probability density of $\boldsymbol{\theta}$ on the two hypotheses as $p_{\boldsymbol{\theta}|H_0}(\boldsymbol{\theta}|H_0)$ and $p_{\boldsymbol{\theta}|H_1}(\boldsymbol{\theta}|H_1)$, the likelihood ratio is

$$\Lambda(\mathbf{R}) \triangleq \frac{p_{\mathbf{r}|H_1}(\mathbf{R}|H_1)}{p_{\mathbf{r}|H_0}(\mathbf{R}|H_0)} = \frac{\int_{\chi} p_{\mathbf{r}|\boldsymbol{\theta}}(\mathbf{R}|\boldsymbol{\theta}) p_{\boldsymbol{\theta}|H_1}(\boldsymbol{\theta}|H_1)\, d\boldsymbol{\theta}}{\int_{\chi} p_{\mathbf{r}|\boldsymbol{\theta}}(\mathbf{R}|\boldsymbol{\theta}) p_{\boldsymbol{\theta}|H_0}(\boldsymbol{\theta}|H_0)\, d\boldsymbol{\theta}}. \tag{296}$$

The reason for this simplicity is that the known probability density on $\boldsymbol{\theta}$ enables us to reduce the problem to a simple hypothesis-testing problem by integrating over $\boldsymbol{\theta}$. We can illustrate this procedure for the model in Example 1.

Example 1 (*continued.*) We assume that the probability density governing m on H_1 is

$$p_{m|H_1}(M|H_1) = \frac{1}{\sqrt{2\pi}\,\sigma_m} \exp\left(-\frac{M^2}{2\sigma_m^2}\right), \qquad -\infty < M < \infty, \tag{297}$$

Then (296) becomes

$$\Lambda(R) = \frac{\displaystyle\int_{-\infty}^{\infty} \frac{1}{\sqrt{2\pi}\,\sigma} \exp\left(-\frac{(R-M)^2}{2\sigma^2}\right) \cdot \frac{1}{\sqrt{2\pi}\,\sigma_m} \exp\left(-\frac{M^2}{2\sigma_m^2}\right) dM}{\displaystyle\frac{1}{\sqrt{2\pi}\,\sigma} \exp\left(-\frac{R^2}{2\sigma^2}\right)} \underset{H_0}{\overset{H_1}{\gtrless}} \eta. \tag{298}$$

Integrating and taking the logarithm of both sides, we obtain

$$R^2 \underset{H_0}{\overset{H_1}{\gtrless}} \frac{2\sigma^2(\sigma^2 + \sigma_m^2)}{\sigma_m^2} \left[\ln \eta + \frac{1}{2} \ln\left(1 + \frac{\sigma_m^2}{\sigma^2}\right)\right]. \tag{299}$$

This result is equivalent to Example 2 on p. 29 because the density used in (297) makes the two problems identical.

As we expected, the test uses only the magnitude of R because the mean m has a symmetric probability density.

For the general case given in (296) the actual calculation may be more involved, but the desired procedure is well defined.

When $\boldsymbol{\theta}$ is a random variable with an unknown density, the best test procedure is not clearly specified. One possible approach is a minimax test over the unknown density. An alternate approach is to try several densities based on any partial knowledge of $\boldsymbol{\theta}$ that is available. In many cases the test structure will be insensitive to the detailed behavior of the probability density.

The second case of interest is the case in which $\boldsymbol{\theta}$ is a nonrandom variable. Here, just as in the problem of estimating nonrandom variables, we shall try a procedure and investigate the results. A first observation is that, because $\boldsymbol{\theta}$ has no probability density over which to average, a Bayes test is not meaningful. Thus we can devote our time to Neyman-Pearson tests.

We begin our discussion by examining what we call a *perfect measurement* bound on the test performance. We illustrate this idea for the problem in Example 1.

Example 2. In this case $\theta = M$.
From (295)

$$H_1: p_{r|m}(R|M) = \frac{1}{\sqrt{2\pi}\,\sigma} \exp\left(-\frac{(R-M)^2}{2\sigma^2}\right), \qquad (M_0 \le M \le M_1),$$

and

$$H_0: p_{r|m}(R|M) = \frac{1}{\sqrt{2\pi}\,\sigma} \exp\left(-\frac{R^2}{2\sigma^2}\right). \tag{300}$$

where M is an unknown nonrandom parameter.

It is clear that whatever test we design can never be better than a hypothetical test in which the receiver first measures M perfectly (or, alternately, it is told M) and then designs the optimum likelihood ratio test. Thus we can bound the ROC of any test by the ROC of this fictitious perfect measurement test. For this example we could use the ROC's in Fig. 2.9*a* by letting $d^2 = M^2/\sigma^2$. Because we are interested in the behavior versus M, the format in Fig. 2.9*b* is more useful. This is shown in Fig. 2.26. Such a curve is called a *power function*. It is simply a plot of P_D for all values of M (more generally $\boldsymbol{\theta}$) for various values of P_F. Because $H_0 = H_1$ for $M = 0$, $P_D = P_F$. The curves in Fig. 2.26 represent a bound on how well any test could do. We now want to see how close the actual test performance comes to this bound.

The best performance we could achieve would be obtained if an actual test's curves equaled the bound for all $M \in \chi$. We call such tests *uniformly most powerful* (UMP). In other words, for a given P_F a UMP test has a P_D greater than or equal to any other test for *all* $M \in \chi$. The conditions for a UMP test to exist can be seen in Fig. 2.27.

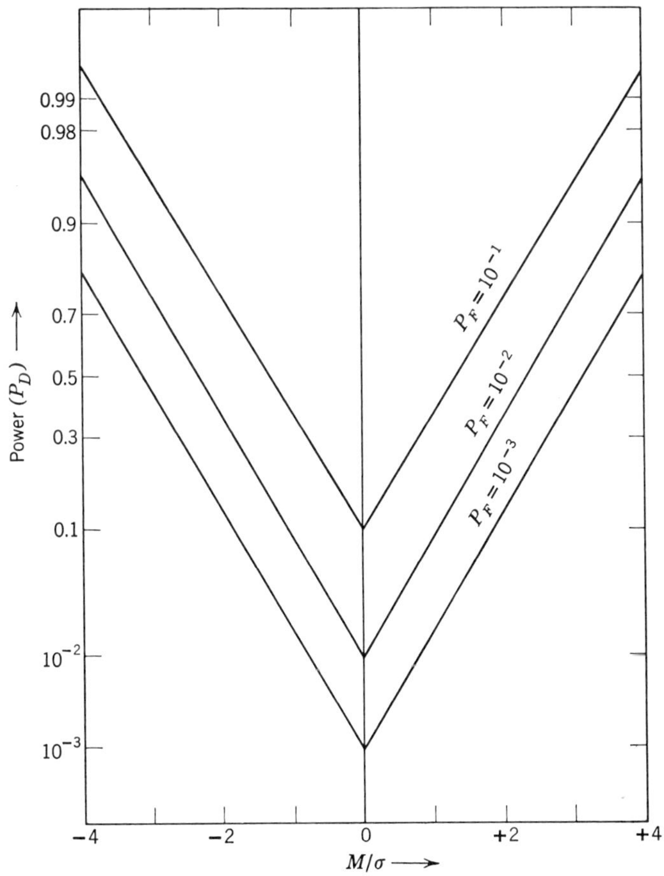

Fig. 2.26 Power function for perfect measurement test.

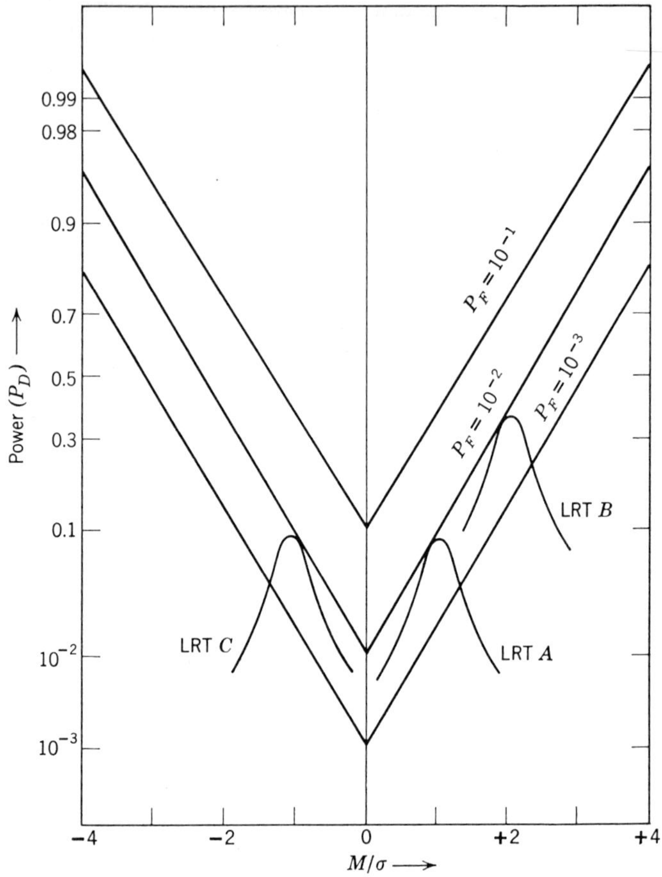

Fig. 2.27 Power functions for various likelihood ratio tests.

We first construct the perfect measurement bound. We next consider other possible tests and their performances. Test A is an ordinary likelihood ratio test designed under the assumption that $M = 1$. The first observation is that the power of this test equals the bound at $M = 1$, which follows from the manner in which we constructed the bound. For other values of M the power of test A may or may not equal the bound. Similarly, test B is a likelihood ratio test designed under the assumption that $M = 2$, and test C is a likelihood ratio test designed under the assumption that $M = -1$. In each case their power equals the bound at their design points. (The power functions in Fig. 2.27 are drawn to emphasize this and are not quantitatively correct away from the design point. The quantitatively correct curves are shown in Fig. 2.29.) They may also equal the bound at other points. The conditions for a UMP test are now obvious. We must be able to design a complete likelihood ratio test (including the threshold) for every $M \in \chi$ without knowing M.

The analogous result for the general case follows easily.

It is clear that in general the bound can be reached for any particular $\boldsymbol{\theta}$ simply by designing an ordinary LRT for that particular $\boldsymbol{\theta}$. Now a UMP test must be as good as any other test for every $\boldsymbol{\theta}$. This gives us a necessary and sufficient condition for its existence.

Property. A UMP test exists if and only if the likelihood ratio test for every $\boldsymbol{\theta} \in \chi$ can be completely defined (including threshold) without knowledge of $\boldsymbol{\theta}$.

The "if" part of the property is obvious. The "only if" follows directly from our discussion in the preceding paragraph. If there exists some $\boldsymbol{\theta} \in \chi$ for which we cannot find the LRT without knowing $\boldsymbol{\theta}$, we should have to use some other test, because we do not know $\boldsymbol{\theta}$. This test will necessarily be inferior for that particular $\boldsymbol{\theta}$ to a LRT test designed for that particular θ and therefore is not *uniformly* most powerful.

Returning to our example and using the results in Fig. 2.8, we know that the likelihood ratio test is

$$R \underset{H_0}{\overset{H_1}{\gtrless}} \gamma^+, \tag{301}$$

and

$$P_F = \int_{\gamma^+}^{\infty} \frac{1}{\sqrt{2\pi}\,\sigma} \exp\left(-\frac{R^2}{2\sigma^2}\right) dR, \qquad \text{if } M > 0. \tag{302}$$

(The superscript + emphasizes the test assumes $M > 0$. The value of γ^+ may be negative.) This is shown in Fig. 2.28*a*.

Similarly, for the case in which $M < 0$ the likelihood ratio test is

$$R \underset{H_1}{\overset{H_0}{\gtrless}} \gamma^-, \tag{303}$$

where

$$P_F = \int_{-\infty}^{\gamma^-} \frac{1}{\sqrt{2\pi}\,\sigma} \exp\left(-\frac{R^2}{2\sigma^2}\right) dR, \qquad M < 0. \tag{304}$$

This is shown in Fig. 2.28*b*. We see that the threshold is just the negative of the threshold for $M > 0$. This reversal is done to get the largest portion of $p_{r|H_1}(R|H_1)$ inside the H_1 region (and therefore maximize P_D).

Thus, with respect to Example 1, we draw the following conclusions:

1. If M can take on *only* nonnegative values (i.e., $M_0 \geq 0$), a UMP test exists [use (301)].
2. If M can take on *only* nonpositive values (i.e., $M_1 \leq 0$), a UMP test exists [use (303)].
3. If M can take on both negative and positive values (i.e., $M_0 < 0$ and $M_1 > 0$), then a UMP test does not exist. In Fig. 2.29 we show the power function for a likelihood ratio test designed under the assumption that M was positive. For negative values of M, P_D is less than P_F because the threshold is on the wrong side.

Whenever a UMP test exists, we use it, and the test works as well as if we knew $\boldsymbol{\theta}$. A more difficult problem is presented when a UMP test does

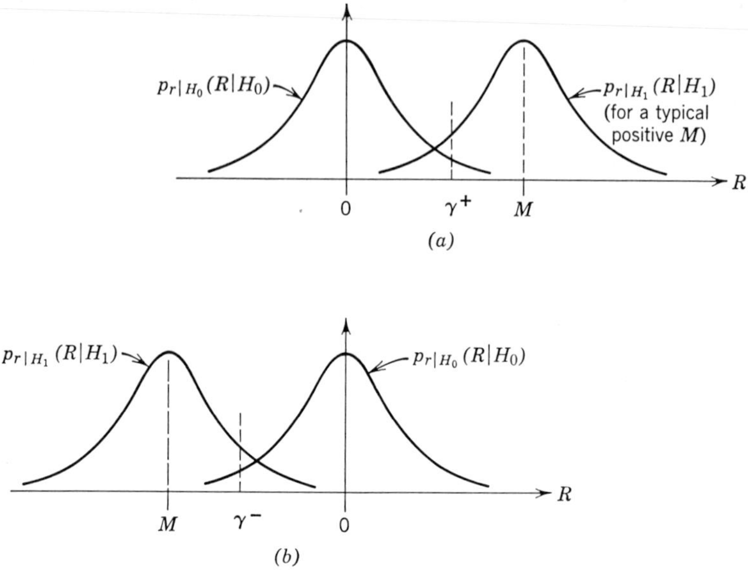

Fig. 2.28 Effect of sign of M: [a] threshold for positive M; [b] threshold for negative M.

not exist. The next step is to discuss other possible tests for the cases in which a UMP test does not exist. We confine our discussion to one possible test procedure. Others are contained in various statistics texts (e.g., Lehmann [17]) but seem to be less appropriate for the physical problems of interest in the sequel.

The perfect measurement bound suggests that a logical procedure is to estimate $\boldsymbol{\theta}$ assuming H_1 is true, then estimate $\boldsymbol{\theta}$ assuming H_0 is true, and use these estimates in a likelihood ratio test as if they were correct. If the maximum likelihood estimates discussed on p. 65 are used, the result is called a *generalized likelihood ratio test*. Specifically,

$$\Lambda_g(\mathbf{R}) = \frac{\max\limits_{\boldsymbol{\theta}_1} p_{\mathbf{r}|\boldsymbol{\theta}_1}(\mathbf{R}|\boldsymbol{\theta}_1)}{\max\limits_{\boldsymbol{\theta}_0} p_{\mathbf{r}|\boldsymbol{\theta}_0}(\mathbf{R}|\boldsymbol{\theta}_0)} \underset{H_0}{\overset{H_1}{\gtrless}} \gamma, \tag{305}$$

where $\boldsymbol{\theta}_1$ ranges over all $\boldsymbol{\theta}$ in H_1 and $\boldsymbol{\theta}_0$ ranges over all $\boldsymbol{\theta}$ in H_0. In other words, we make a ML estimate of $\boldsymbol{\theta}_1$, assuming that H_1 is true. We then evaluate $p_{\mathbf{r}|\boldsymbol{\theta}_1}(\mathbf{R}|\boldsymbol{\theta}_1)$ for $\boldsymbol{\theta}_1 = \hat{\boldsymbol{\theta}}_1$ and use this value in the numerator. A similar procedure gives the denominator.

A simple example of a generalized LRT is obtained by using a slightly modified version of Example 1.

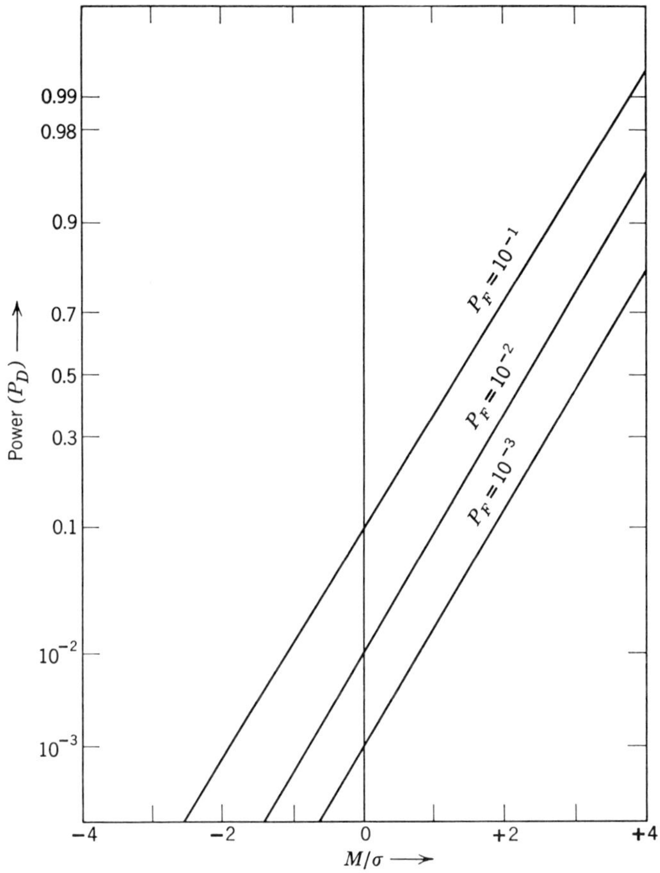

Fig. 2.29 Performance of LRT assuming positive *M*.

Example 2. The basic probabilities are the same as in Example 1. Once again, $\boldsymbol{\theta} = M$. Instead of one, we have N independent observations, which we denote by the vector $\mathbf{R}$. The probability densities are,

$$p_{\mathbf{r}|m,H_1}(\mathbf{R}|M,H_1) = \prod_{i=1}^{N} \frac{1}{\sqrt{2\pi}\,\sigma} \exp\left(-\frac{(R_i - M)^2}{2\sigma^2}\right),$$
$$p_{\mathbf{r}|m,H_0}(R|M,H_0) = \prod_{i=1}^{N} \frac{1}{\sqrt{2\pi}\,\sigma} \exp\left(-\frac{R_i^2}{2\sigma^2}\right). \tag{306}$$

In this example H_1 is a composite hypothesis and H_0, a simple hypothesis. From (198)

$$\hat{M}_1 = \frac{1}{N} \sum_{i=1}^{N} R_i. \tag{307}$$

Then

$$\Lambda_g(\mathbf{R}) = \frac{\prod_{i=1}^{N} \frac{1}{\sqrt{2\pi}\,\sigma} \exp\left\{-\frac{[R_i - (1/N)\sum_{j=1}^{N} R_j]^2}{2\sigma^2}\right\}}{\prod_{i=1}^{N} \frac{1}{\sqrt{2\pi}\,\sigma} \exp\left(-R_i^2/2\sigma^2\right)} \underset{H_0}{\overset{H_1}{\gtrless}} \gamma. \tag{308}$$

Canceling common terms and taking the logarithm, we have

$$\ln \Lambda_g(\mathbf{R}) = \frac{1}{2\sigma^2 N}\left(\sum_{i=1}^{N} R_i\right)^2 \underset{H_0}{\overset{H_1}{\gtrless}} \ln \gamma. \tag{309}$$

The left side of (309) is always greater than or equal to zero. Thus, γ can always be chosen greater than or equal to one. Therefore, an equivalent test is

$$\left(\frac{1}{N^{1/2}} \sum_{i=1}^{N} R_i\right)^2 \underset{H_0}{\overset{H_1}{\gtrless}} \gamma_1^2 \tag{310}$$

where $\gamma_1 \geq 0$. Equivalently,

$$|z| \triangleq \left|\frac{1}{N^{1/2}} \sum_{i=1}^{N} R_i\right| \underset{H_0}{\overset{H_1}{\gtrless}} \gamma_1. \tag{311}$$

The power function of this test follows easily. The variable z has a variance equal

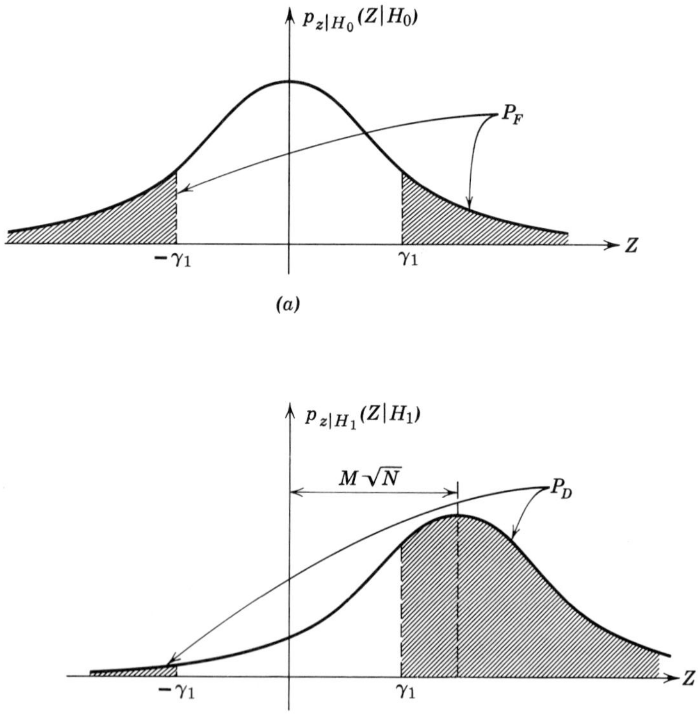

Fig. 2.30 Errors in generalized likelihood ratio test: [a] P_F calculation; [b] P_D calculation.

to σ^2. On H_0 its mean is zero and on H_1 its mean is $M\sqrt{N}$. The densities are sketched in Fig. 2.30.

$$P_F = \int_{-\infty}^{-\gamma_1} \frac{1}{\sqrt{2\pi}\,\sigma} \exp\left(-\frac{Z^2}{2\sigma^2}\right) dZ + \int_{\gamma_1}^{\infty} \frac{1}{\sqrt{2\pi}\,\sigma} \exp\left(-\frac{Z^2}{2\sigma^2}\right) dZ$$

$$= 2\,\text{erfc}_*\left(\frac{\gamma_1}{\sigma}\right) \tag{312}$$

and

$$P_D(M) = \int_{-\infty}^{-\gamma_1} \frac{1}{\sqrt{2\pi}\,\sigma} \exp\left[-\frac{(Z - M\sqrt{N})^2}{2\sigma^2}\right] dZ$$

$$+ \int_{\gamma_1}^{\infty} \frac{1}{\sqrt{2\pi}\,\sigma} \exp\left[-\frac{(Z - M\sqrt{N})^2}{2\sigma^2}\right] dZ$$

$$= \text{erfc}_*\left[\frac{\gamma_1 + M\sqrt{N}}{\sigma}\right] + \text{erfc}_*\left[\frac{\gamma_1 - M\sqrt{N}}{\sigma}\right]. \tag{313}$$

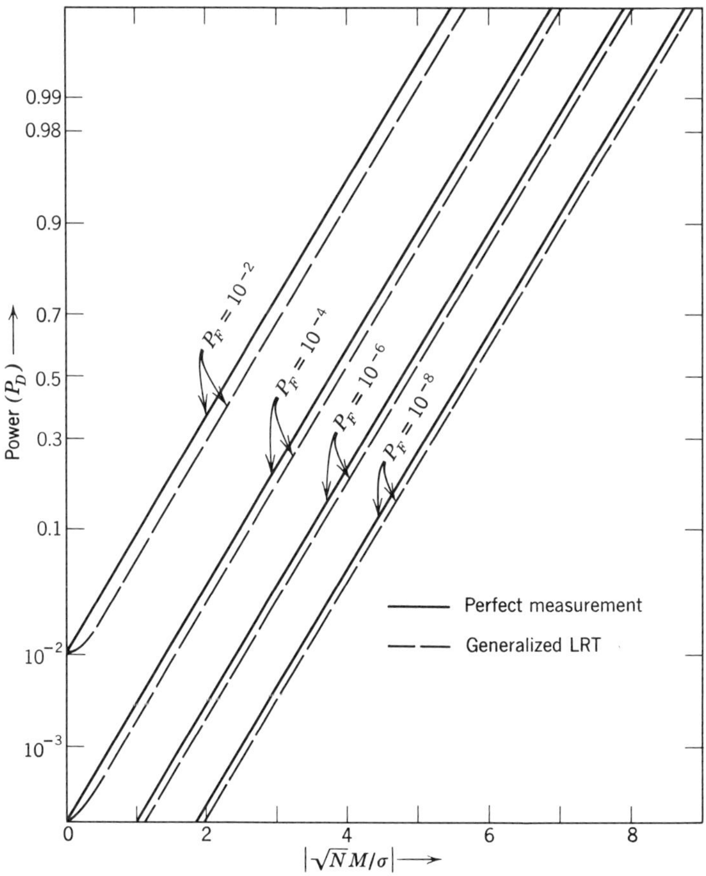

Fig. 2.31 Power function: generalized likelihood ratio tests.

The resulting power function is plotted in Fig. 2.31. The perfect measurement bound is shown for comparison purposes. As we would expect from our discussion of ML estimates, the difference approaches zero as $\sqrt{N}\, M/\sigma \to \infty$.

Just as there are cases in which the ML estimates give poor results, there are others in which the generalized likelihood ratio test may give bad results. In these cases we must look for other test procedures. Fortunately, in most of the physical problems of interest to us either a UMP test will exist or a generalized likelihood ratio test will give satisfactory results.

2.6 THE GENERAL GAUSSIAN PROBLEM

All of our discussion up to this point has dealt with arbitrary probability densities. In the binary detection case $p_{\mathbf{r}|H_1}(\mathbf{R}|H_1)$ and $p_{\mathbf{r}|H_0}(\mathbf{R}|H_0)$ were not constrained to have any particular form. Similarly, in the estimation problem $p_{\mathbf{r}|\mathbf{a}}(\mathbf{R}|\mathbf{A})$ was not constrained. In the classical case, constraints are not particularly necessary. When we begin our discussion of the waveform problem, we shall find that most of our discussions concentrate on problems in which the conditional density of $\mathbf{r}$ is Gaussian. We discuss this class of problem in detail in this section. The material in this section and the problems associated with it lay the groundwork for many of the results in the sequel. We begin by defining a Gaussian random vector and the general Gaussian problem.

Definition. A set of random variables $r_1, r_2, \ldots, r_N$ are defined as jointly Gaussian if all their linear combinations are Gaussian random variables.

Definition. A vector $\mathbf{r}$ is a Gaussian random vector when its components $r_1, r_2, \ldots, r_N$ are jointly Gaussian random variables.

In other words, if

$$z = \sum_{i=1}^{N} g_i r_i \triangleq \mathbf{G}^T \mathbf{r} \tag{314}$$

is a Gaussian random variable for all finite $\mathbf{G}^T$, then $\mathbf{r}$ is a Gaussian vector.

If we define

$$E(\mathbf{r}) = \mathbf{m} \tag{315}$$

and

$$\text{Cov}\,(\mathbf{r}) = E[(\mathbf{r} - \mathbf{m})(\mathbf{r}^T - \mathbf{m}^T)] \triangleq \mathbf{\Lambda}, \tag{316}$$

then (314) implies that the characteristic function of $\mathbf{r}$ is

$$M_{\mathbf{r}}(j\mathbf{v}) \triangleq E[e^{j\mathbf{v}^T\mathbf{r}}] = \exp\,(+j\mathbf{v}^T\mathbf{m} - \tfrac{1}{2}\mathbf{v}^T\mathbf{\Lambda}\mathbf{v}) \tag{317}$$

and assuming $\boldsymbol{\Lambda}$ is nonsingular the probability density of $\mathbf{r}$ is

$$p_{\mathbf{r}}(\mathbf{R}) = [(2\pi)^{N/2}|\boldsymbol{\Lambda}|^{1/2}]^{-1} \exp\left[-\tfrac{1}{2}(\mathbf{R}^T - \mathbf{m}^T)\boldsymbol{\Lambda}^{-1}(\mathbf{R} - \mathbf{m})\right]. \tag{318}$$

The proof is straightforward (e.g., Problem 2.6.20).

Definition. A hypothesis testing problem is called a general Gaussian problem if $p_{\mathbf{r}|H_i}(\mathbf{R}|H_i)$ is a Gaussian density on all hypotheses. An estimation problem is called a general Gaussian problem if $p_{\mathbf{r}|\mathbf{a}}(\mathbf{R}|\mathbf{A})$ has a Gaussian density for all $\mathbf{A}$.

We discuss the binary hypothesis testing version of the general Gaussian problem in detail in the text. The M-hypothesis and the estimation problems are developed in the problems. The basic model for the binary detection problem is straightforward. We assume that the observation space is N-dimensional. Points in the space are denoted by the N-dimensional vector (or column matrix) $\mathbf{r}$:

$$\mathbf{r} = \begin{bmatrix} r_1 \\ r_2 \\ \vdots \\ r_N \end{bmatrix}. \tag{319}$$

Under the first hypothesis H_1 we assume that $\mathbf{r}$ is a Gaussian random vector, which is completely specified by its mean vector and covariance matrix. We denote these quantities as

$$E[\mathbf{r}|H_1] = \begin{bmatrix} E(r_1|H_1) \\ E(r_2|H_1) \\ \vdots \\ E(r_N|H_1) \end{bmatrix} \triangleq \begin{bmatrix} m_{11} \\ m_{12} \\ \vdots \\ m_{1N} \end{bmatrix} \triangleq \mathbf{m}_1. \tag{320}$$

The covariance matrix is

$$\mathbf{K}_1 \triangleq E[(\mathbf{r} - \mathbf{m}_1)(\mathbf{r}^T - \mathbf{m}_1{}^T)|H_1]$$

$$= \begin{bmatrix} {}_1K_{11} & {}_1K_{12} & {}_1K_{13} & \cdots & {}_1K_{1N} \\ {}_1K_{21} & {}_1K_{22} & \ddots & & \\ \vdots & \vdots & & & \vdots \\ {}_1K_{N1} & & & & {}_1K_{NN} \end{bmatrix}. \tag{321}$$

We define the inverse of $\mathbf{K}_1$ as $\mathbf{Q}_1$

$$\mathbf{Q}_1 \triangleq \mathbf{K}_1^{-1} \tag{322}$$

$$\mathbf{Q}_1\mathbf{K}_1 = \mathbf{K}_1\mathbf{Q}_1 = \mathbf{I}, \tag{323}$$

where $\mathbf{I}$ is the identity matrix (ones on the diagonal and zeroes elsewhere). Using (320), (321), (322), and (318), we may write the probability density of $\mathbf{r}$ on H_1,

$$p_{\mathbf{r}|H_1}(\mathbf{R}|H_1) = [(2\pi)^{N/2}|\mathbf{K}_1|^{1/2}]^{-1} \exp[-\tfrac{1}{2}(\mathbf{R}^T - \mathbf{m}_1^T)\mathbf{Q}_1(\mathbf{R} - \mathbf{m}_1)]. \quad (324)$$

Going through a similar set of definitions for H_0, we obtain the probability density

$$p_{\mathbf{r}|H_0}(\mathbf{R}|H_0) = [(2\pi)^{N/2}|\mathbf{K}_0|^{1/2}]^{-1} \exp[-\tfrac{1}{2}(\mathbf{R}^T - \mathbf{m}_0^T)\mathbf{Q}_0(\mathbf{R} - \mathbf{m}_0)]. \quad (325)$$

Using the definition in (13), the likelihood ratio test follows easily:

$$\Lambda(\mathbf{R}) \triangleq \frac{p_{\mathbf{r}|H_1}(\mathbf{R}|H_1)}{p_{\mathbf{r}|H_0}(\mathbf{R}|H_0)} = \frac{|\mathbf{K}_0|^{1/2} \exp[-\frac{1}{2}(\mathbf{R}^T - \mathbf{m}_1^T)\mathbf{Q}_1(\mathbf{R} - \mathbf{m}_1)]}{|\mathbf{K}_1|^{1/2} \exp[-\frac{1}{2}(\mathbf{R}^T - \mathbf{m}_0^T)\mathbf{Q}_0(\mathbf{R} - \mathbf{m}_0)]} \mathop{\gtrless}_{H_0}^{H_1} \eta. \quad (326)$$

Taking logarithms, we obtain

$$\boxed{\begin{aligned}&\tfrac{1}{2}(\mathbf{R}^T - \mathbf{m}_0^T)\,\mathbf{Q}_0(\mathbf{R} - \mathbf{m}_0) - \tfrac{1}{2}(\mathbf{R}^T - \mathbf{m}_1^T)\,\mathbf{Q}_1(\mathbf{R} - \mathbf{m}_1)\\ &\qquad \mathop{\gtrless}_{H_0}^{H_1} \ln\eta + \tfrac{1}{2}\ln|\mathbf{K}_1| - \tfrac{1}{2}\ln|\mathbf{K}_0| \triangleq \gamma^*.\end{aligned}} \quad (327)$$

We see that the test consists of finding the difference between two *quadratic forms*. The result in (327) is basic to many of our later discussions. For this reason we treat various cases of the general Gaussian problem in some detail. We begin with the simplest.

2.6.1 Equal Covariance Matrices

The first special case of interest is the one in which the covariance matrices on the two hypotheses are equal,

$$\mathbf{K}_1 = \mathbf{K}_0 \triangleq \mathbf{K}, \quad (328)$$

but the means are different.

Denote the inverse as $\mathbf{Q}$:

$$\mathbf{Q} = \mathbf{K}^{-1}. \quad (329)$$

Substituting into (327), multiplying the matrices, canceling common terms, and using the symmetry of $\mathbf{Q}$, we have

$$(\mathbf{m}_1^T - \mathbf{m}_0^T)\mathbf{Q}\mathbf{R} \mathop{\gtrless}_{H_0}^{H_1} \ln\eta + \tfrac{1}{2}(\mathbf{m}_1^T\mathbf{Q}\mathbf{m}_1 - \mathbf{m}_0^T\mathbf{Q}\mathbf{m}_0) \triangleq \gamma_*'. \quad (330)$$

We can simplify this expression by defining a vector corresponding to the difference in the mean value vectors on the two hypotheses:

$$\Delta\mathbf{m} \triangleq \mathbf{m}_1 - \mathbf{m}_0. \quad (331)$$

Then (327) becomes

$$\boxed{l(\mathbf{R}) \triangleq \boldsymbol{\Delta}\mathbf{m}^T\mathbf{Q}\mathbf{R} \underset{H_0}{\overset{H_1}{\gtrless}} \gamma'_*} \tag{332}$$

or, equivalently,

$$l(\mathbf{R}) \triangleq \mathbf{R}^T\mathbf{Q}\, \boldsymbol{\Delta}\mathbf{m} \underset{H_0}{\overset{H_1}{\gtrless}} \gamma'_*. \tag{333}$$

The quantity on the left is a *scalar* Gaussian random variable, for it was obtained by a linear transformation of jointly Gaussian random variables. Therefore, as we discussed in Example 1 on pp. 36–38, we can completely characterize the performance of the test by the quantity d^2. In that example, we defined d as the distance between the means on the two hypothesis when the variance was normalized to equal one. An identical definition is,

$$d^2 \triangleq \frac{[E(l|H_1) - E(l|H_0)]^2}{\text{Var}\,(l|H_0)}. \tag{334}$$

Substituting (320) into the definition of l, we have

$$E(l|H_1) = \boldsymbol{\Delta}\mathbf{m}^T\mathbf{Q}\mathbf{m}_1 \tag{335}$$

and

$$E(l|H_0) = \boldsymbol{\Delta}\mathbf{m}^T\mathbf{Q}\mathbf{m}_0. \tag{336}$$

Using (332), (333), and (336) we have

$$\text{Var}\,[l|H_0] = E\{[\boldsymbol{\Delta}\mathbf{m}^T\mathbf{Q}(\mathbf{R} - \mathbf{m}_0)][(\mathbf{R}^T - \mathbf{m}_0{}^T)\mathbf{Q}\,\boldsymbol{\Delta}\mathbf{m}]\}. \tag{337}$$

Using (321) to evaluate the expectation and then (323), we have

$$\text{Var}\,[l|H_0] = \boldsymbol{\Delta}\mathbf{m}^T\mathbf{Q}\,\boldsymbol{\Delta}\mathbf{m}. \tag{338}$$

Substituting (335), (336), and (338) into (334), we obtain

$$\boxed{d^2 = \boldsymbol{\Delta}\mathbf{m}^T\mathbf{Q}\,\boldsymbol{\Delta}\mathbf{m}}. \tag{339}$$

Thus the performance for the equal covariance Gaussian case is completely determined by the quadratic form in (339). We now interpret it for some cases of interest.

Case 1. Independent Components with Equal Variance. Each r_i has the same variance σ^2 and is statistically independent. Thus

$$\mathbf{K} = \sigma^2\mathbf{I} \tag{340}$$

and

$$\mathbf{Q} = \frac{1}{\sigma^2}\mathbf{I}. \tag{341}$$

Substituting (341) into (339), we obtain

$$d^2 = \mathbf{\Delta m}^T \frac{1}{\sigma^2} \mathbf{I}\, \mathbf{\Delta m} = \frac{1}{\sigma^2} \mathbf{\Delta m}^T \mathbf{\Delta m} = \frac{1}{\sigma^2} |\mathbf{\Delta m}|^2 \tag{342}$$

or

$$\boxed{d = \frac{|\mathbf{\Delta m}|}{\sigma}}. \tag{343}$$

We see that d corresponds to the *distance* between the two mean-value vectors divided by the standard deviation of R_i. This is shown in Fig. 2.32. In (332) we see that

$$l = \frac{1}{\sigma^2} \mathbf{\Delta m}^T \mathbf{R}. \tag{344}$$

Thus the sufficient statistic is just the dot (or scalar) product of the observed vector $\mathbf{R}$ and the mean difference vector $\mathbf{\Delta m}$.

Case 2. Independent Components with Unequal Variances. Here the r_i are statistically independent but have unequal variances. Thus

$$\mathbf{K} = \begin{bmatrix} {\sigma_1}^2 & & & 0 \\ & {\sigma_2}^2 & & \\ & & \ddots & \\ 0 & & & {\sigma_N}^2 \end{bmatrix} \tag{345}$$

and

$$\mathbf{Q} = \begin{bmatrix} \dfrac{1}{{\sigma_1}^2} & & & 0 \\ & \dfrac{1}{{\sigma_2}^2} & & \\ & & \ddots & \\ 0 & & & \dfrac{1}{{\sigma_N}^2} \end{bmatrix}. \tag{346}$$

Substituting into (339) and performing the multiplication, we have

$$d^2 = \sum_{i=1}^{N} \frac{(\Delta m_i)^2}{{\sigma_i}^2}. \tag{347}$$

Now the various difference components contribute to d^2 with weighting that is inversely proportional to the variance along that coordinate. We can also interpret the result as distance in a new coordinate system.

Let

$$\Delta\mathbf{m}' = \begin{bmatrix} \dfrac{1}{\sigma_1} m_1 \\ \dfrac{1}{\sigma_2} m_2 \\ \vdots \\ \dfrac{1}{\sigma_N} m_N \end{bmatrix} \tag{348}$$

and

$$R_i' = \frac{1}{\sigma_i} R_i. \tag{349}$$

This transformation changes the scale on each axis so that the variances are all equal to one. We see that d corresponds exactly to the difference vector in this "scaled" coordinate system.

The sufficient statistic is

$$l(\mathbf{R}) = \sum_{i=1}^{N} \frac{\Delta m_i \cdot R_i}{\sigma_i^2}. \tag{350}$$

In the scaled coordinate system it is the dot product of the two vectors

$$l(\mathbf{R}') = \Delta\mathbf{m}'^T\mathbf{R}'. \tag{351}$$

Case 3. This is the general case. A satisfactory answer for l and d is already available in (332) and (339):

$$l(\mathbf{R}) = \Delta\mathbf{m}^T\mathbf{Q}\mathbf{R} \tag{352}$$

and

$$d^2 = \Delta\mathbf{m}^T\mathbf{Q}\,\Delta\mathbf{m}. \tag{353}$$

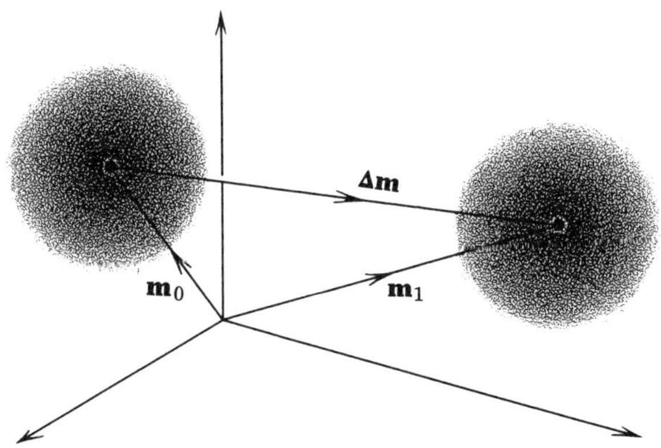

Fig. 2.32 Mean-value vectors.

Valuable insight into the important features of the problem can be gained by looking at it in a different manner.

The key to the simplicity in Cases 1 and 2 is the diagonal covariance matrix. This suggests that we try to represent $\mathbf{R}$ in a new coordinate system in which the components are statistically independent random variables. In Fig. 2.33*a* we show the observation in the original coordinate system. In Fig. 2.33*b* we show a new set of coordinate axes, which we denote by the orthogonal unit vectors $\boldsymbol{\phi}_1, \boldsymbol{\phi}_2, \ldots, \boldsymbol{\phi}_N$:

$$\boldsymbol{\phi}_i^T \boldsymbol{\phi}_j = \delta_{ij}. \tag{354}$$

We denote the observation in the new coordinate system by $\mathbf{r}'$. We want to choose the orientation of the new system so that the components r_i' and r_j' are uncorrelated (and therefore statistically independent, for they are Gaussian) for all $i \neq j$. In other words,

$$E[(r_i' - m_i')(r_j' - m_j')] = \lambda_i \delta_{ij}, \tag{355}$$

where

$$m_i' \triangleq E(r_i') \tag{356}$$

and

$$\text{Var}\,[r_i'] \triangleq \lambda_i. \tag{357}$$

Now the components of $\mathbf{r}'$ can be expressed simply in terms of the dot product of the original vector $\mathbf{r}$ and the various unit vectors

$$r_i' = \mathbf{r}^T \boldsymbol{\phi}_i = \boldsymbol{\phi}_i^T \mathbf{r}. \tag{358}$$

Using (358) in (355), we obtain

$$E[\boldsymbol{\phi}_i^T(\mathbf{r} - \mathbf{m})(\mathbf{r}^T - \mathbf{m}^T)\boldsymbol{\phi}_j] = \lambda_i \delta_{ij}. \tag{359}$$

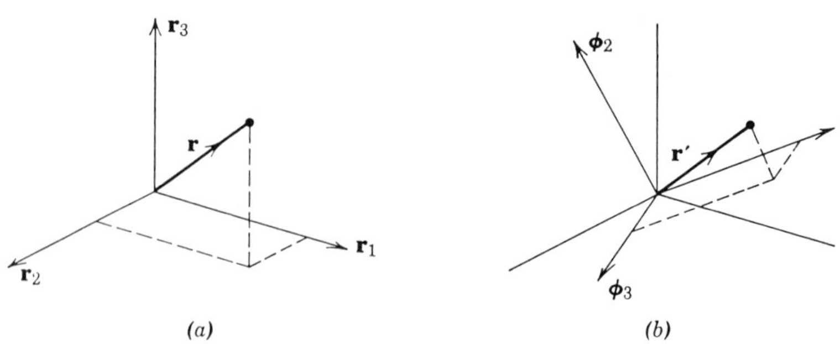

Fig. 2.33 Coordinate systems: [*a*] original coordinate system; [*b*] new coordinate system.

The expectation of the random part is just **K** [see (321)]. Therefore (359) becomes

$$\lambda_i \delta_{ij} = \boldsymbol{\phi}_i^T \mathbf{K} \boldsymbol{\phi}_j. \tag{360}$$

This will be satisfied if and only if

$$\lambda_j \boldsymbol{\phi}_j = \mathbf{K} \boldsymbol{\phi}_j \qquad \text{for } j = 1, 2, \ldots, N. \tag{361}$$

To check the "if" part of this result, substitute (361) into (360):

$$\lambda_i \delta_{ij} = \boldsymbol{\phi}_i^T \lambda_j \boldsymbol{\phi}_j = \lambda_j \delta_{ij}, \tag{362}$$

where the right equality follows from (354). The "only if" part follows using a simple proof by contradiction. Now (361) can be written with the j subscript suppressed:

$$\boxed{\lambda \boldsymbol{\phi} = \mathbf{K} \boldsymbol{\phi}.} \tag{363}$$

We see that the question of finding the proper coordinate system reduces to the question of whether we can find N solutions to (363) that satisfy (354).

It is instructive to write (363) out in detail. Each $\boldsymbol{\phi}$ is a vector with N components:

$$\boldsymbol{\phi} = \begin{bmatrix} \phi_1 \\ \phi_2 \\ \phi_3 \\ \vdots \\ \phi_N \end{bmatrix}. \tag{364}$$

Substituting (364) into (363), we have

$$\begin{aligned} K_{11}\phi_1 + K_{12}\phi_2 + \cdots + K_{1N}\phi_N &= \lambda\phi_1 \\ K_{21}\phi_1 + K_{22}\phi_2 + \cdots + K_{2N}\phi_N &= \lambda\phi_2 \\ \vdots \qquad\qquad & \\ K_{N1}\phi_1 + K_{N2}\phi_2 + \cdots + K_{NN}\phi_N &= \lambda\phi_N \end{aligned} \tag{365}$$

We see that (365) corresponds to a set of N homogeneous simultaneous equations. A nontrivial solution will exist if and only if the determinant of the coefficient matrix is zero. In other words, if and only if

$$|\mathbf{K} - \lambda\mathbf{I}| = \begin{vmatrix} K_{11} - \lambda & K_{12} & K_{13} & \cdots & \\ K_{21} & K_{22} - \lambda & K_{23} & & \\ K_{31} & K_{32} & \ddots & & \\ \vdots & \vdots & & & K_{NN} - \lambda \end{vmatrix} = 0. \tag{366}$$

We see that this is an Nth-order polynomial in λ. The N roots, denoted by $\lambda_1, \lambda_2, \ldots, \lambda_N$, are called the *eigenvalues* of the covariance matrix **K**. It can be shown that the following properties are true (e.g., [16] or [18]):

1. Because **K** is symmetric, the eigenvalues are real.
2. Because **K** is a covariance matrix, the eigenvalues are nonnegative. (Otherwise we would have random variables with negative variances.)

For each λ_i we can find a solution $\boldsymbol{\phi}_i$ to (363). Because there is an arbitrary constant associated with each solution to (363), we may choose the $\boldsymbol{\phi}_i$ to have unit length

$$\boldsymbol{\phi}_i^T \boldsymbol{\phi}_i = 1. \tag{367}$$

These solutions are called the normalized *eigenvectors* of **K**. Two other properties may also be shown for symmetric matrices.

3. If the roots λ_i are distinct, the corresponding eigenvectors are orthogonal.
4. If a particular root λ_j is of multiplicity M, the M associated eigenvectors are linearly independent. They can be chosen to be orthonormal.

We have now described a coordinate system in which the observations are statistically independent. The mean difference vector can be expressed as

$$\begin{aligned} \Delta m_1' &= \boldsymbol{\phi}_1^T \, \boldsymbol{\Delta}\mathbf{m} \\ \Delta m_2' &= \boldsymbol{\phi}_2^T \, \boldsymbol{\Delta}\mathbf{m} \\ &\vdots \\ \Delta m_N' &= \boldsymbol{\phi}_N^T \, \boldsymbol{\Delta}\mathbf{m} \end{aligned} \tag{368}$$

or in vector notation

$$\boldsymbol{\Delta}\mathbf{m}' = \begin{bmatrix} \boldsymbol{\phi}_1^T \\ \hdashline \boldsymbol{\phi}_2^T \\ \hdashline \vdots \\ \hdashline \boldsymbol{\phi}_N^T \end{bmatrix} \boldsymbol{\Delta}\mathbf{m} \triangleq \mathbf{W}\,\boldsymbol{\Delta}\mathbf{m}. \tag{369}$$

The resulting sufficient statistic in the new coordinate system is

$$l(\mathbf{R}') = \sum_{i=1}^{N} \frac{\Delta m_i' \cdot R_i'}{\lambda_i} \tag{370}$$

and d^2 is

$$\boxed{d^2 = \sum_{i=1}^{N} \frac{(\Delta m_i')^2}{\lambda_i}.} \tag{371}$$

The derivation leading to (371) has been somewhat involved, but the result is of fundamental importance, for it demonstrates that there always exists a coordinate system in which the random variables are uncorrelated and that the new system is related to the old system by a linear transformation. To illustrate the technique we consider a simple example.

Example. For simplicity we let $N = 2$ and $\mathbf{m}_0 = \mathbf{0}$. Let

$$\mathbf{K} = \begin{bmatrix} 1 & \rho \\ \rho & 1 \end{bmatrix} \tag{372}$$

and

$$\mathbf{m}_1 = \begin{bmatrix} m_{11} \\ m_{12} \end{bmatrix}. \tag{373}$$

To find the eigenvalues we solve

$$\begin{vmatrix} 1 - \lambda & \rho \\ \rho & 1 - \lambda \end{vmatrix} = 0 \tag{374}$$

or

$$(1 - \lambda^2) - \rho^2 = 0. \tag{375}$$

Solving,

$$\begin{aligned} \lambda_1 &= 1 + \rho, \\ \lambda_2 &= 1 - \rho. \end{aligned} \tag{376}$$

To find $\boldsymbol{\phi}_1$ we substitute λ_1 into (365),

$$\begin{bmatrix} 1 & \rho \\ \rho & 1 \end{bmatrix} \begin{bmatrix} \phi_{11} \\ \phi_{12} \end{bmatrix} = \begin{bmatrix} (1 + \rho)\phi_{11} \\ (1 + \rho)\phi_{12} \end{bmatrix} \tag{377}$$

Solving, we obtain

$$\phi_{11} = \phi_{12}. \tag{378}$$

Normalizing gives

$$\boldsymbol{\phi}_1 = \begin{bmatrix} +\frac{1}{\sqrt{2}} \\ \\ +\frac{1}{\sqrt{2}} \end{bmatrix}. \tag{379}$$

Similarly,

$$\boldsymbol{\phi}_2 = \begin{bmatrix} +\frac{1}{\sqrt{2}} \\ \\ -\frac{1}{\sqrt{2}} \end{bmatrix}. \tag{380}$$

The old and new axes are shown in Fig. 2.34. The transformation is

$$\mathbf{W} = \left[\begin{array}{c:c} +\frac{1}{\sqrt{2}} & +\frac{1}{\sqrt{2}} \\ \hdashline +\frac{1}{\sqrt{2}} & -\frac{1}{\sqrt{2}} \end{array}\right] \tag{381}$$

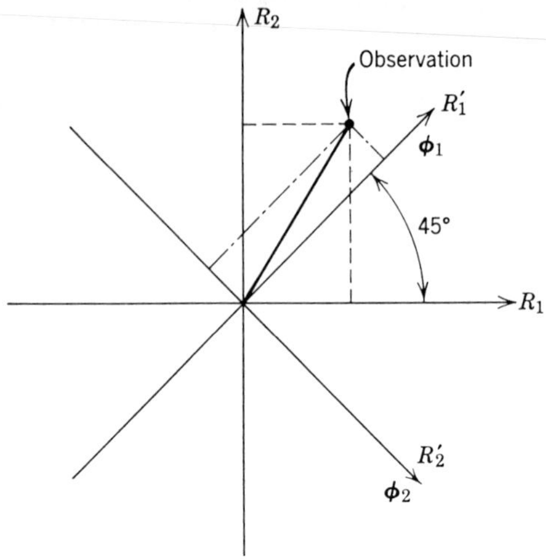

Fig. 2.34 Rotation of axes.

$$\begin{aligned} R_1' &= \frac{R_1 + R_2}{\sqrt{2}}, \\ R_2' &= \frac{R_1 - R_2}{\sqrt{2}}, \\ m_{11}' &= \frac{m_{11} + m_{12}}{\sqrt{2}}, \\ m_{12}' &= \frac{m_{11} - m_{12}}{\sqrt{2}}. \end{aligned} \tag{382}$$

The sufficient statistic is obtained by using (382) in (370),

$$l(\mathbf{R}') = \frac{1}{1+\rho}\frac{(R_1 + R_2)(m_{11} + m_{12})}{2} + \frac{1}{1-\rho}\frac{(R_1 - R_2)(m_{11} - m_{12})}{2} \tag{383}$$

and d^2 is

$$d^2 = \frac{(m_{11} + m_{12})^2}{2(1+\rho)} + \frac{(m_{11} - m_{12})^2}{2(1-\rho)} = \frac{(m_{11}')^2}{(1+\rho)} + \frac{(m_{12}')^2}{(1-\rho)}. \tag{384}$$

To illustrate a typical application in which the transformation is important we consider a simple optimization problem. The length of the mean vector is constrained,

$$|\mathbf{m}_1|^2 = 1. \tag{385}$$

We want to choose m_{11} and m_{12} to maximize d^2. Because our transformation is a rotation, it preserves lengths

$$|\mathbf{m}_1'|^2 = 1. \tag{386}$$

Looking at (384), we obtain the solution by inspection:

If $\rho > 0$, choose $m_{11}' = 0$ and $m_{12}' = 1$.

If $\rho < 0$, choose $m_{11}' = 1$ and $m_{12}' = 0$.

If $\rho = 0$, all vectors satisfying (385) give the same d^2.

We see that this corresponds to choosing the mean-value vector to be equal to the eigenvector with the smallest eigenvalue. This result can be easily extended to N dimensions.

The result in this example is characteristic of a wide class of optimization problems in which the solution corresponds to an eigenvector (or the waveform analog to it).

In this section, we have demonstrated that when the covariance matrices on the two hypotheses are equal the sufficient statistic $l(\mathbf{R})$ is a Gaussian random variable obtained by a linear transformation of $\mathbf{R}$. The performance for any threshold setting is determined by using the value of d^2 given by (339) on the ROC in Fig. 2.9. Because the performance improves monotonically with increasing d^2, we can use any freedom in the parameters to maximize d^2 without considering the ROC explicitly.

2.6.2 Equal Mean Vectors

In the second special case of interest the mean-value vectors on the two hypotheses are equal. In other words,

$$\mathbf{m}_1 = \mathbf{m}_0 \triangleq \mathbf{m}. \tag{387}$$

Substituting (387) into (327), we have

$$\tfrac{1}{2}(\mathbf{R}^T - \mathbf{m}^T)(\mathbf{Q}_0 - \mathbf{Q}_1)(\mathbf{R} - \mathbf{m}) \underset{H_0}{\overset{H_1}{\gtrless}} \ln \eta + \tfrac{1}{2} \ln \frac{|\mathbf{K}_1|}{|\mathbf{K}_0|} = \gamma^*. \tag{388}$$

Because the mean-value vectors contain no information that will tell us which hypothesis is true, the likelihood test subtracts them from the received vector. Therefore, without loss of generality, we may assume that $\mathbf{m} = \mathbf{0}$.

We denote the difference of the inverse matrices as $\Delta\mathbf{Q}$:

$$\Delta\mathbf{Q} \triangleq \mathbf{Q}_0 - \mathbf{Q}_1. \tag{389}$$

The likelihood ratio test may be written as

$$\boxed{l(\mathbf{R}) \triangleq \mathbf{R}^T\, \Delta\mathbf{Q}\mathbf{R} \underset{H_0}{\overset{H_1}{\gtrless}} 2\gamma^* \triangleq \gamma'.} \tag{390}$$

Note that $l(\mathbf{R})$ is the dot product of two Gaussian vectors, $\mathbf{R}^T$ and $\Delta\mathbf{Q}\mathbf{R}$. Thus, $l(\mathbf{R})$ is not a Gaussian random variable.

We now consider the behavior of this test for some interesting special cases.

Case 1. Diagonal Covariance Matrix on H_0: Equal Variances. Here the R_i on H_0 are statistically independent variables with equal variances:

$$\mathbf{K}_0 = \sigma_n^2\mathbf{I}. \tag{391}$$

We shall see later that (391) corresponds to the physical situation in which there is "noise" only on H_0. The following notation is convenient:

$$r_i = n_i, \qquad H_0. \tag{392}$$

On H_1 the r_i contain the same variable as on H_0, plus additional signal components that may be correlated:

$$\begin{aligned} r_i &= s_i + n_i, \qquad H_1, \\ \mathbf{K}_1 &= \mathbf{K}_s + \sigma_n^2\mathbf{I}, \end{aligned} \tag{393}$$

where the matrix $\mathbf{K}_s$ represents the covariance matrix of the signal components. Then

$$\mathbf{Q}_0 = \frac{1}{\sigma_n^2}\mathbf{I} \tag{394}$$

and

$$\mathbf{Q}_1 = \frac{1}{\sigma_n^2}\left(\mathbf{I} + \frac{1}{\sigma_n^2}\mathbf{K}_s\right)^{-1}. \tag{395}$$

It is convenient to write (395) as

$$\mathbf{Q}_1 = \frac{1}{\sigma_n^2}[\mathbf{I} - \mathbf{H}], \tag{396}$$

which implies

$$\mathbf{H} = (\sigma_n^2\mathbf{I} + \mathbf{K}_s)^{-1}\mathbf{K}_s = \mathbf{K}_s(\sigma_n^2\mathbf{I} + \mathbf{K}_s)^{-1} = \mathbf{Q}_0 - \mathbf{Q}_1 = \Delta\mathbf{Q}. \tag{397}$$

The $\mathbf{H}$ matrix has an important interpretation which we shall develop later. We take the first expression in (397) as its definition. Substituting (397) into (389) and the result into (390), we have

$$\boxed{l(\mathbf{R}) = \frac{1}{\sigma_n^2}\mathbf{R}^T\mathbf{H}\mathbf{R} \underset{H_0}{\overset{H_1}{\gtrless}} \gamma'.} \tag{398}$$

Several subcases are important.

Case 1A. Uncorrelated, Identically Distributed Signal Components. In this case the signal components s_i are independent variables with identical variances:

$$\mathbf{K}_1 = \sigma_s^2\mathbf{I}. \tag{399}$$

Then

$$\mathbf{H} = (\sigma_n^2\mathbf{I} + \sigma_s^2\mathbf{I})^{-1}\sigma_s^2\mathbf{I}, \tag{400}$$

or

$$\mathbf{H} = \frac{\sigma_s^2}{\sigma_n^2 + \sigma_s^2}\mathbf{I} \tag{401}$$

and

$$l(\mathbf{R}) = \frac{1}{\sigma_n^2}\frac{\sigma_s^2}{\sigma_n^2+\sigma_s^2}\mathbf{R}^T\mathbf{R} = \frac{1}{\sigma_n^2}\frac{\sigma_s^2}{\sigma_n^2+\sigma_s^2}\sum_{i=1}^{N} R_i^2. \tag{402}$$

The constant can be incorporated in the threshold to give

$$l(\mathbf{R}) \triangleq \sum_{i=1}^{N} R_i^2 \underset{H_0}{\overset{H_1}{\gtrless}} \gamma''. \tag{403}$$

We now calculate the performance of the test. On both hypotheses $l(\mathbf{R})$ is the sum of the squares of N Gaussian variables. The difference in the hypotheses is in the variance of the Gaussian variables. For simplicity, we shall assume that N is an even integer.

To find $p_{l|H_0}(L|H_0)$ we observe that the characteristic function of each R_i^2 is

$$\begin{aligned} M_{R_i^2|H_0}(jv) &\triangleq \int_{-\infty}^{\infty} e^{jvR_i^2}\frac{1}{\sqrt{2\pi}\,\sigma_n}e^{-R_i^2/2\sigma_n^2}\,dR_i \\ &= (1-2jv\sigma_n^2)^{-1/2}. \end{aligned} \tag{404}$$

Because of the independence of the variables, $M_{l|H_0}(jv)$ can be written as a product. Therefore

$$M_{l|H_0}(jv) = (1-2jv\sigma_n^2)^{-N/2}. \tag{405}$$

Taking the inverse transform, we obtain $p_{l|H_0}(L|H_0)$:

$$\begin{aligned} p_{l|H_0}(L|H_0) &= \frac{L^{N/2-1}e^{-L/2\sigma_n^2}}{2^{N/2}\sigma_n^N\Gamma\left(\frac{N}{2}\right)}, &\quad L \ge 0, \\ &= 0, &\quad L < 0, \end{aligned} \tag{406}$$

which is familiar as the χ^2 (chi-square) density function with N degrees of freedom. It is tabulated in several references (e.g., [19] or [3]). For $N = 2$ it is easy to check that it is the simple exponential on p. 41. Similarly,

$$\begin{aligned} p_{l|H_1}(L|H_1) &= \frac{L^{N/2-1}e^{-L/2\sigma_1^2}}{2^{N/2}\sigma_1^N\Gamma\left(\frac{N}{2}\right)}, &\quad L \ge 0, \\ &= 0, &\quad L < 0, \end{aligned} \tag{407}$$

where $\sigma_1^2 \triangleq \sigma_s^2 + \sigma_n^2$.

The expressions for P_D and P_F are,

$$P_D = \int_{\gamma''}^{\infty} [2^{N/2}\sigma_1^N\Gamma(N/2)]^{-1}L^{N/2-1}e^{-L/2\sigma_1^2}\,dL \tag{408}$$

and

$$P_F = \int_{\gamma''}^{\infty} [2^{N/2}\sigma_n^N\Gamma(N/2)]^{-1}L^{N/2-1}e^{-L/2\sigma_n^2}\,dL. \tag{409}$$

Construction of the ROC requires an evaluation of the two integrals. We see that for $N = 2$ we have the same problem as Example 2 on p. 41 and (408) and (409) reduce to

$$P_D = \exp\left(-\frac{\gamma''}{2\sigma_1^2}\right),$$
$$P_F = \exp\left(-\frac{\gamma''}{2\sigma_n^2}\right), \tag{410}$$

and

$$P_F = (P_D)^{(1+\sigma_s^2/\sigma_n^2)}. \tag{411}$$

For the general case there are several methods of proceeding. First, let $M = N/2 - 1$ and $\gamma''' = \gamma''/2\sigma_n^2$. Then write

$$P_F = 1 - \int_0^{\gamma'''} \frac{x^M}{M!} e^{-x}\, dx. \tag{412}$$

The integral, called the incomplete Gamma function, has been tabulated by Pearson [21]:

$$I_\Gamma(u, M) \triangleq \int_0^{u\sqrt{M+1}} \frac{x^M}{M!} e^{-x}\, dx, \tag{413}$$

and

$$P_F = 1 - I_\Gamma\left(\frac{\gamma'''}{\sqrt{M+1}}, M\right). \tag{414}$$

These tables are most useful for $P_F \geq 10^{-6}$ and $M \leq 50$.

In a second approach we integrate by parts M times. The result is

$$P_F = \exp(-\gamma''') \sum_{k=0}^{M} \frac{(\gamma''')^k}{k!}. \tag{415}$$

For *small* P_F, γ''' is large and we can approximate the series by the last few terms,

$$P_F = \frac{(\gamma''')^M e^{-\gamma'''}}{M!}\left[1 + \frac{M}{\gamma'''} + \frac{M(M-1)}{(\gamma''')^2} + \cdots\right]. \tag{416}$$

Furthermore, we can approximate the bracket as $(1 - M/\gamma''')^{-1}$. This gives

$$P_F \cong \frac{(\gamma''')^M e^{-\gamma'''}}{M!(1 - M/\gamma''')}. \tag{417}$$

A similar expression for P_D follows in which γ''' is replaced by $\gamma^{iv} \triangleq \gamma''/2\sigma_1^2$. The approximate expression in (417) is useful for manual calculation. In actual practice, we use (415) and calculate the ROC numerically. In Fig. 2.35*a* we have plotted the receiver operating characteristic for some representative values of N and σ_s^2/σ_n^2.

Two particularly interesting curves are those for $N = 8$, $\sigma_s^2/\sigma_n^2 = 1$ and $N = 2$, $\sigma_s^2/\sigma_n^2 = 4$. In both cases the product $N\sigma_s^2/\sigma_n^2 = 8$. We see that when the desired P_F is greater than 0.3, P_D is higher if the available "signal strength" is divided into more components. This suggests that for each P_F

and product $N\sigma_s^2/\sigma_n^2$ there should be an optimum N. In Chapter 4 we shall see that this problem corresponds to optimum diversity in communication systems and the optimum energy per pulse in radar. In Figs. 2.35*b* and *c* we have sketched P_M as a function of N for $P_F = 10^{-2}$ and 10^{-4}, respectively, and various $N\sigma_s^2/\sigma_n^2$ products. We discuss the physical implications of these results in Chapter 4.

Case 1B. Independent Signal Components: Unequal Variances. In this case the signal components s_i are independent variables with variances $\sigma_{s_i}^2$:

$$\mathbf{K}_1 = \begin{bmatrix} \sigma_{s_1}^2 & & & 0 \\ & \sigma_{s_2}^2 & & \\ 0 & & \ddots & \\ & & & \sigma_{s_N}^2 \end{bmatrix}. \tag{418}$$

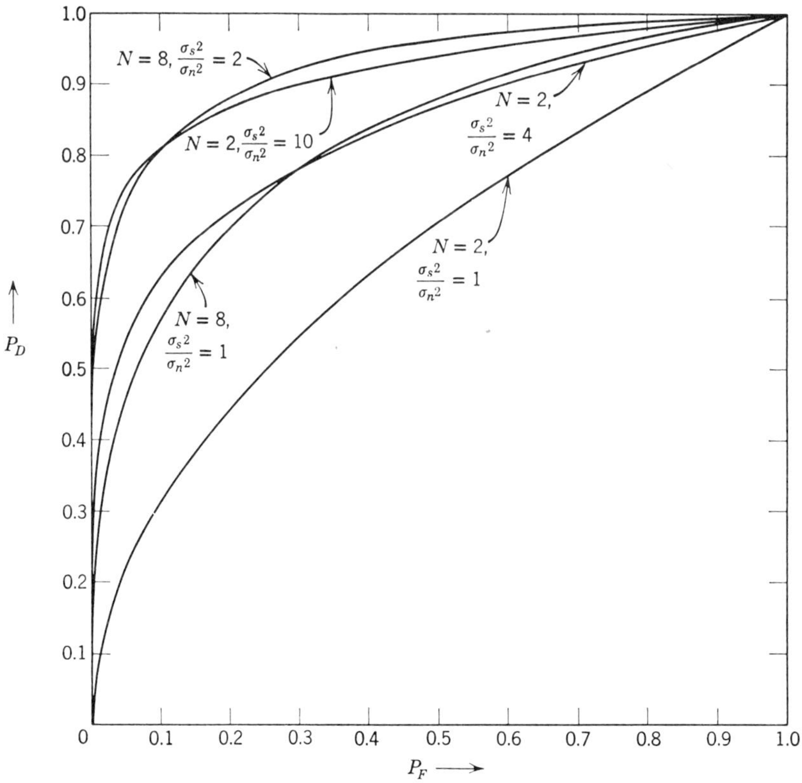

Fig. 2.35 ***a.*** **Receiver operating characteristic: Gaussian variables with identical means and unequal variances on the two hypotheses.**

Then

$$\mathbf{H} = \begin{bmatrix} \dfrac{\sigma_{s_1}{}^2}{\sigma_n{}^2 + \sigma_{s_1}{}^2} & & & 0 \\ & \dfrac{\sigma_{s_2}{}^2}{\sigma_n{}^2 + \sigma_{s_2}{}^2} & & \\ & & \ddots & \\ 0 & & & \dfrac{\sigma_{s_N}{}^2}{\sigma_n{}^2 + \sigma_{s_N}{}^2} \end{bmatrix} \tag{419}$$

and

$$l(\mathbf{R}) = \frac{1}{\sigma_n{}^2} \sum_{i=1}^{N} \frac{\sigma_{s_i}{}^2}{\sigma_n{}^2 + \sigma_{s_i}{}^2} R_i{}^2 \underset{H_0}{\overset{H_1}{\gtrless}} \gamma'. \tag{420}$$

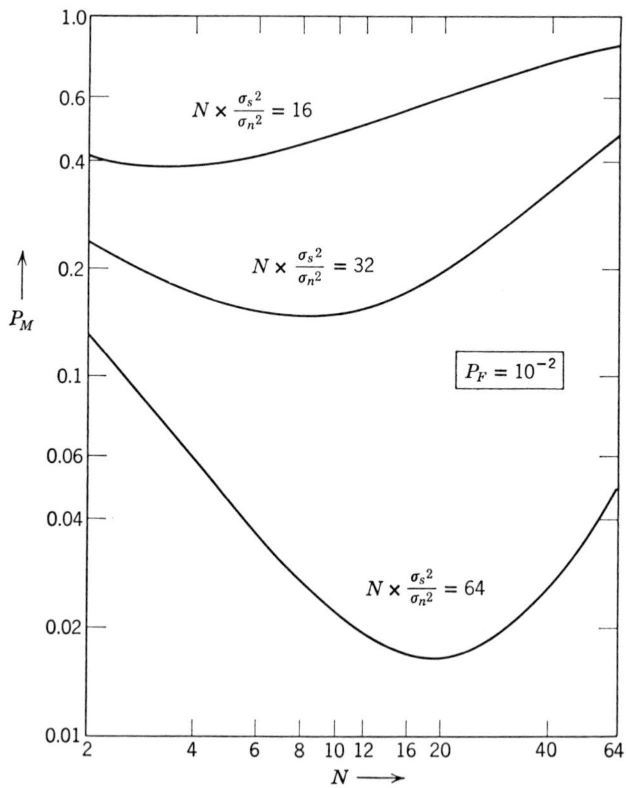

Fig. 2.35 ***b.*** **P_M as a function of N [$P_F = 10^{-2}$].**

The characteristic function of $l(\mathbf{R})$ follows easily, but the calculation of P_F and P_D is difficult. In Section 2.7 we derive approximations to the performance that lead to simpler expressions.

Case 1C. Arbitrary Signal Components. This is, of course, the general case. We revisit it merely to point out that it can always be reduced to Case 1B by an orthogonal transformation (see discussion on pp. 102–106).

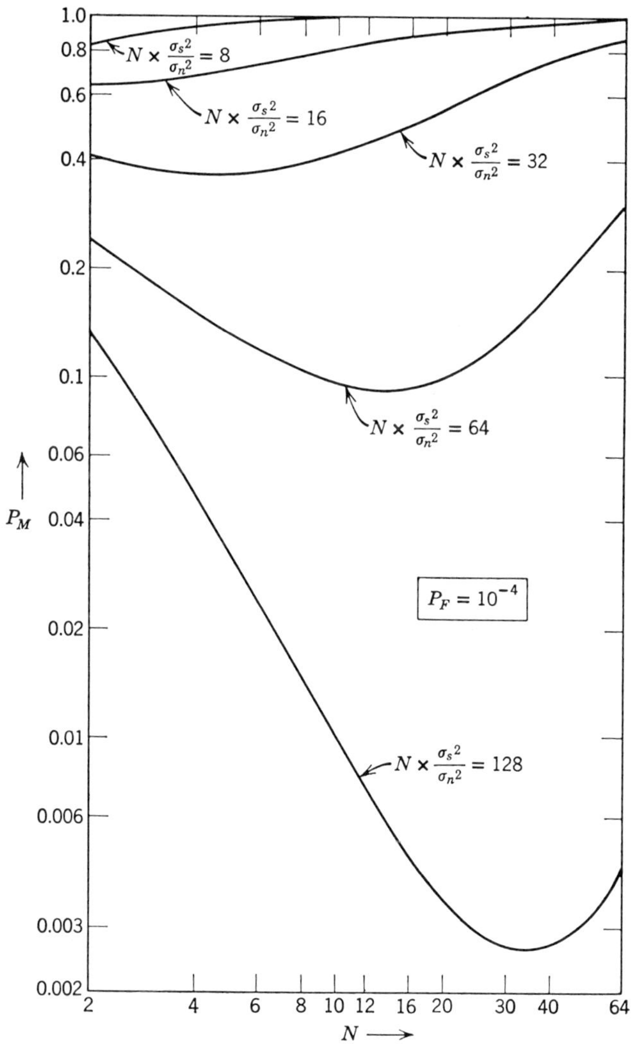

Fig. 2.35 ***c.*** **P_M as a function of N [$P_F = 10^{-4}$].**

Case 2. Symmetric Hypotheses, Uncorrelated Noise. Case 1 was unsymmetric because of the noise-only hypothesis. Here we have the following hypotheses:

$$\begin{aligned} H_1: r_i &= s_i + n_i \qquad & i &= 1, \ldots, N \\ & \quad\;\; n_i & i &= N+1, \ldots, 2N, \\ H_0: r_i &= \quad\;\; n_i & i &= 1, \ldots, N \\ & s_i + n_i & i &= N+1, \ldots, 2N, \end{aligned} \tag{421}$$

where the n_i are independent variables with variance σ_n^2 and the s_i have a covariance matrix $\mathbf{K}_s$. Then

$$\mathbf{K}_1 = \left[\begin{array}{c|c} \sigma_n^2\mathbf{I} + \mathbf{K}_s & \mathbf{0} \\ \hline \mathbf{0} & \sigma_n^2\mathbf{I} \end{array}\right] \tag{422}$$

and

$$\mathbf{K}_0 = \left[\begin{array}{c|c} \sigma_n^2\mathbf{I} & \mathbf{0} \\ \hline \mathbf{0} & \sigma_n^2\mathbf{I} + \mathbf{K}_s \end{array}\right], \tag{423}$$

where we have partitioned the $2N \times 2N$ matrices into $N \times N$ submatrices. Then

$$\Delta\mathbf{Q} = \left[\begin{array}{c|c} \dfrac{1}{\sigma_n^2}\mathbf{I} & \mathbf{0} \\ \hline \mathbf{0} & (\sigma_n^2\mathbf{I} + \mathbf{K}_s)^{-1} \end{array}\right] - \left[\begin{array}{c|c} (\sigma_n^2\mathbf{I} + \mathbf{K}_s)^{-1} & \mathbf{0} \\ \hline \mathbf{0} & \dfrac{1}{\sigma_n^2}\mathbf{I} \end{array}\right]. \tag{424}$$

Using (397), we have

$$\Delta\mathbf{Q} = \frac{1}{\sigma_n^2}\left[\begin{array}{c|c} \mathbf{H} & \mathbf{0} \\ \hline \mathbf{0} & -\mathbf{H} \end{array}\right], \tag{425}$$

where, as previously defined in (397), $\mathbf{H}$ is

$$\mathbf{H} \triangleq (\sigma_n^2\mathbf{I} + \mathbf{K}_s)^{-1}\mathbf{K}_s. \tag{426}$$

If we partition $\mathbf{R}$ into two $N \times 1$ matrices,

$$\mathbf{R} = \left[\begin{array}{c} \mathbf{R}_1 \\ \hline \mathbf{R}_2 \end{array}\right], \tag{427}$$

then

$$l(\mathbf{R}) = \frac{1}{\sigma_n^2}\,\mathbf{R}_1^T\mathbf{H}\mathbf{R}_1 - \mathbf{R}_2^T\mathbf{H}\mathbf{R}_2 \underset{H_0}{\overset{H_1}{\gtrless}} \gamma'. \tag{428}$$

The special cases analogous to 1A and 1B follow easily.

Case 2A. Uncorrelated, Identically Distributed Signal Components. Let

$$\mathbf{K}_s = \sigma_s^2 \mathbf{I}; \tag{429}$$

then

$$l(\mathbf{R}) = \sum_{i=1}^{N} R_i^2 - \sum_{i=N+1}^{2N} R_i^2 \underset{H_0}{\overset{H_1}{\gtrless}} \gamma^v. \tag{430}$$

If the hypotheses are equally likely and the criterion is minimum $\Pr(\epsilon)$, the threshold η in the LRT is unity (see 69). From (388) and (390) we see that this will result in $\gamma^v = 0$. This case occurs frequently and leads to a simple error calculation. The test then becomes

$$l_1(\mathbf{R}) \triangleq \sum_{i=1}^{N} R_i^2 \underset{H_0}{\overset{H_1}{\gtrless}} \sum_{i=N+1}^{2N} R_i^2 \triangleq l_0(\mathbf{R}). \tag{431}$$

The probability of error given that H_1 is true is the probability that $l_0(\mathbf{R})$ is greater than $l_1(\mathbf{R})$. Because the test is symmetric with respect to the two hypotheses,

$$\Pr(\epsilon) = \tfrac{1}{2}\Pr(\epsilon|H_1) + \tfrac{1}{2}\Pr(\epsilon|H_0) = \Pr(\epsilon|H_1). \tag{432a}$$

Thus

$$\Pr(\epsilon) = \int_0^\infty dL_1 p_{l_1|H_1}(L_1|H_1) \int_{L_1}^\infty p_{l_0|H_1}(L_0|H_1)\, dL_0. \tag{432b}$$

Substituting (406) and (407) in (432*b*), recalling that N is even, and evaluating the inner integral, we have

$$\Pr(\epsilon) = \int_0^\infty \frac{1}{2^{N/2}\sigma_1^N \Gamma(N/2)} L_1^{N/2-1} e^{-L_1/2\sigma_1^2} \times \left[e^{-L_1/2\sigma_n^2} \sum_{k=0}^{N/2-1} \frac{(L_1/2\sigma_n^2)^k}{k!}\right] dL_1. \tag{432c}$$

Defining

$$\alpha = \frac{\sigma_n^2}{\sigma_1^2 + \sigma_n^2}, \tag{433}$$

and integrating, (432*c*) reduces to

$$\Pr(\epsilon) = \alpha^{N/2} \sum_{j=0}^{N/2-1} \binom{\frac{N}{2} + j - 1}{j} (1-\alpha)^j. \tag{434}$$

This result is due to Pierce [22]. It is a closed-form expression but it is tedious to use. We delay plotting (434) until Section 2.7, in which we derive an approximate expression for comparison.

Case 2B. Uncorrelated Signal Components: Unequal Variances. Now,

$$\mathbf{K}_s = \begin{bmatrix} \sigma_{s_1}^2 & & & 0 \\ & \sigma_{s_2}^2 & & \\ & & \ddots & \\ 0 & & & \sigma_{s_N}^2 \end{bmatrix}. \tag{435}$$

It follows easily that

$$l(\mathbf{R}) = \frac{1}{\sigma_n^2}\left[\sum_{i=1}^{N} \frac{\sigma_{s_i}^2}{\sigma_n^2 + \sigma_{s_i}^2} R_i^2 - \sum_{i=N+1}^{2N} \frac{\sigma_{s_{i-N}}^2}{\sigma_n^2 + \sigma_{s_{i-N}}^2} R_i^2\right] \underset{H_0}{\overset{H_1}{\gtrless}} \gamma'. \tag{436}$$

As in Case 1B, the performance is difficult to evaluate. The approximations developed in Section 2.7 are also useful for this case.

2.6.3 Summary

We have discussed in detail the general Gaussian problem and have found that the sufficient statistic was the difference between two quadratic forms:

$$l(\mathbf{R}) = \tfrac{1}{2}(\mathbf{R}^T - \mathbf{m}_0^T)\mathbf{Q}_0(\mathbf{R} - \mathbf{m}_0) - \tfrac{1}{2}(\mathbf{R}^T - \mathbf{m}_1^T)\mathbf{Q}_1(\mathbf{R} - \mathbf{m}_1). \tag{437}$$

A particularly simple special case was the one in which the covariance matrices on the two hypotheses were equal. Then

$$l(\mathbf{R}) = \tfrac{1}{2}\, \Delta\mathbf{m}^T\mathbf{Q}\mathbf{R}, \tag{438}$$

and the performance was completely characterized by the quantity d^2:

$$d^2 = \Delta\mathbf{m}^T\mathbf{Q}\,\Delta\mathbf{m}. \tag{439}$$

When the covariance matrices are unequal, the implementation of the likelihood ratio test is still straightforward but the performance calculations are difficult (remember that d^2 is no longer applicable because $l(\mathbf{R})$ is not Gaussian). In the simplest case of diagonal covariance matrices with equal elements, exact error expressions were developed. In the general case, exact expressions are possible but are too unwieldy to be useful. This inability to obtain tractable performance expressions is the motivation for discussion of performance bounds and approximations in the next section.

Before leaving the general Gaussian problem, we should point out that similar results can be obtained for the M-hypothesis case and for the estimation problem. Some of these results are developed in the problems.

2.7 PERFORMANCE BOUNDS AND APPROXIMATIONS

Up to this point we have dealt primarily with problems in which we could derive the structure of the optimum receiver and obtain relatively simple expressions for the receiver operating characteristic or the error probability.

In many cases of interest the optimum test can be derived but an exact performance calculation is impossible. For these cases we must resort to bounds on the error probabilities or approximate expressions for these probabilities. In this section we derive some simple bounds and approximations which are useful in many problems of practical importance. The basic results, due to Chernoff [28], were extended initially by Shannon [23]. They have been further extended by Fano [24], Shannon, Gallager, and Berlekamp [25], and Gallager [26] and applied to a problem of interest to us by Jacobs [27]. Our approach is based on the last two references. Because the latter part of the development is heuristic in nature, the interested reader should consult the references given for more careful derivations. From the standpoint of use in later sections, we shall not use the results until Chapter II-3 (the results are also needed for some of the problems in Chapter 4).

The problem of interest is the general binary hypothesis test outlined in Section 2.2. From our results in that section we know that it will reduce to a likelihood ratio test. We begin our discussion at this point.

The likelihood ratio test is

$$l(\mathbf{R}) \triangleq \ln \Lambda(\mathbf{R}) = \ln \left[\frac{p_{\mathbf{r}|H_1}(\mathbf{R}|H_1)}{p_{\mathbf{r}|H_0}(\mathbf{R}|H_0)}\right] \underset{H_0}{\overset{H_1}{\gtrless}} \gamma. \tag{440}$$

The variable $l(\mathbf{R})$ is a random variable whose probability density depends on which hypothesis is true. In Fig. 2.36 we show a typical $p_{l|H_1}(L|H_1)$ and $p_{l|H_0}(L|H_0)$.

If the two densities are known, then P_F and P_D are given by

$$P_D = \int_\gamma^\infty p_{l|H_1}(L|H_1)\, dL, \tag{441}$$

$$P_F = \int_\gamma^\infty p_{l|H_0}(L|H_0)\, dL. \tag{442}$$

The difficulty is that it is often hard to find $p_{l|H_i}(L|H_i)$, and even if it can be found it is cumbersome. Typical of this complexity is Case 1A

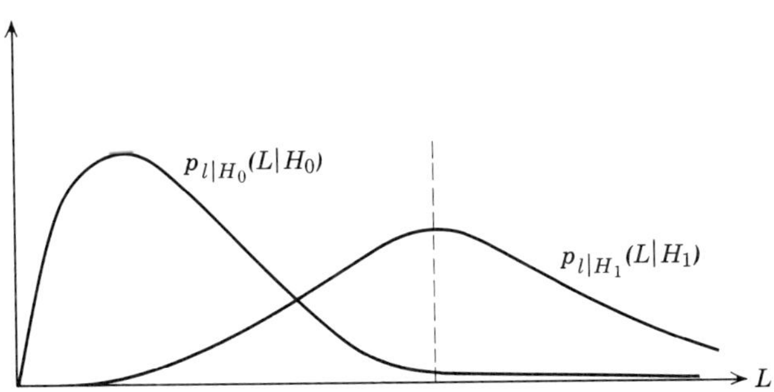

Fig. 2.36 Typical densities.

on p. 108, in which there are N Gaussian variables with equal variances making up the signal. To analyze a given system, the errors may be evaluated numerically. On the other hand, if we set out to synthesize a system, it is inefficient (if not impossible) to try successive systems and evaluate each numerically. Therefore we should like to find some simpler approximate expressions for the error probabilities.

In this section we derive some simple expressions that we shall use in the sequel. We first focus our attention on cases in which $l(\mathbf{R})$ is a sum of independent random variables. This suggests that its characteristic function may be useful, for it will be the product of the individual characteristic functions of the R_i. Similarly, the moment-generating function will be the product of individual moment-generating functions. Therefore an approximate expression based on one of these functions should be relatively easy to evaluate. The first part of our discussion develops bounds on the error probabilities in terms of the moment-generating function of $l(\mathbf{R})$.

In the second part we consider the case in which $l(\mathbf{R})$ is the sum of a *large* number of independent random variables. By the use of the central limit theorem we improve on the results obtained in the first part of the discussion.

We begin by deriving a simple upper bound on P_F in terms of the moment-generating function. The moment-generating function of $l(\mathbf{R})$ on hypothesis H_0 is

$$\phi_{l|H_0}(s) \triangleq E(e^{sl}|H_0) = \int_{-\infty}^{\infty} e^{sL} p_{l|H_0}(L|H_0)\, dL, \tag{443}$$

where s is a *real* variable. (The range of s corresponds to those values for which the integral exists.) We shall see shortly that it is more useful to write

$$\phi_{l|H_0}(s) \triangleq \exp\,[\mu(s)], \tag{444}$$

so that

$$\mu(s) = \ln \int_{-\infty}^{\infty} e^{sL} p_{l|H_0}(L|H_0)\, dL. \tag{445}$$

We may also express $\mu(s)$ in terms of $p_{\mathbf{r}|H_1}(\mathbf{R}|H_1)$ and $p_{\mathbf{r}|H_0}(\mathbf{R}|H_0)$. Because l is just a function of $\mathbf{r}$, we can write (443) as

$$\phi_{l|H_0}(s) = \int_{-\infty}^{\infty} e^{sl(\mathbf{R})} p_{\mathbf{r}|H_0}(\mathbf{R}|H_0)\, d\mathbf{R}. \tag{446}$$

Then

$$\mu(s) = \ln \int_{-\infty}^{\infty} e^{sl(\mathbf{R})} p_{\mathbf{r}|H_0}(\mathbf{R}|H_0)\, d\mathbf{R}. \tag{447}$$

Using (440),

$$\mu(s) = \ln \int_{-\infty}^{\infty} \left(\frac{p_{\mathbf{r}|H_1}(\mathbf{R}|H_1)}{p_{\mathbf{r}|H_0}(\mathbf{R}|H_0)}\right)^s p_{\mathbf{r}|H_0}(\mathbf{R}|H_0)\, d\mathbf{R}, \tag{448}$$

or

$$\boxed{\mu(s) = \ln \int_{-\infty}^{\infty} [p_{\mathbf{r}|H_1}(\mathbf{R}|H_1)]^s [p_{\mathbf{r}|H_0}(\mathbf{R}|H_0)]^{1-s}\, d\mathbf{R}.} \tag{449}$$

The function $\mu(s)$ plays a central role in the succeeding discussion. It is now convenient to rewrite the error expressions in terms of a new random variable whose mean is in the vicinity of the threshold. The reason for this step is that we shall use the central limit theorem in the second part of our derivation. It is most effective near the mean of the random variable of interest. Consider the simple probability density shown in Fig. 2.37*a*. To get the new family of densities shown in Figs. 2.37*b*–*d* we multiply $p_x(X)$ by e^{sX} for various values of s (and normalize to obtain a unit area). We see that for $s > 0$ the mean is shifted to the right. For the moment we leave s as a parameter. We see that increasing s "tilts" the density more.

Denoting this new variable as x_s, we have

$$p_{x_s}(X) \triangleq \frac{e^{sX} p_{l|H_0}(X|H_0)}{\int_{-\infty}^{\infty} e^{sL} p_{l|H_0}(L|H_0)\, dL} = \frac{e^{sX} p_{l|H_0}(X|H_0)}{e^{\mu(s)}}. \tag{450}$$

Observe that we define x_s in terms of its density function, for that is what we are interested in. Equation 450 is a general definition. For the density shown in Fig. 2.37, the limits would be $(-A, A)$.

We now find the mean and variance of x_s:

$$E(x_s) = \int_{-\infty}^{\infty} X p_{x_s}(X)\, dX = \frac{\int_{-\infty}^{\infty} X e^{sX} p_{l|H_0}(X|H_0)\, dX}{\int_{-\infty}^{\infty} e^{sL} p_{l|H_0}(L|H_0)\, dL}. \tag{451}$$

Comparing (451) and (445), we see that

$$E(x_s) = \frac{d\mu(s)}{ds} \triangleq \dot{\mu}(s). \tag{452}$$

Similarly, we find

$$\operatorname{Var}(x_s) = \ddot{\mu}(s). \tag{453}$$

[Observe that (453) implies that $\mu(s)$ is convex.]

We now rewrite P_F in terms of this tilted variable x_s:

$$P_F = \int_\gamma^\infty p_{l|H_0}(L|H_0)\,dL = \int_\gamma^\infty e^{\mu(s)-sX} p_{x_s}(X)\,dX$$
$$= e^{\mu(s)} \int_\gamma^\infty e^{-sX} p_{x_s}(X)\,dX. \tag{454}$$

We can now find a simple upper bound on P_F. For values of $s \geq 0$,

$$e^{-sX} \leq e^{-s\gamma}, \qquad \text{for } X \geq \gamma. \tag{455}$$

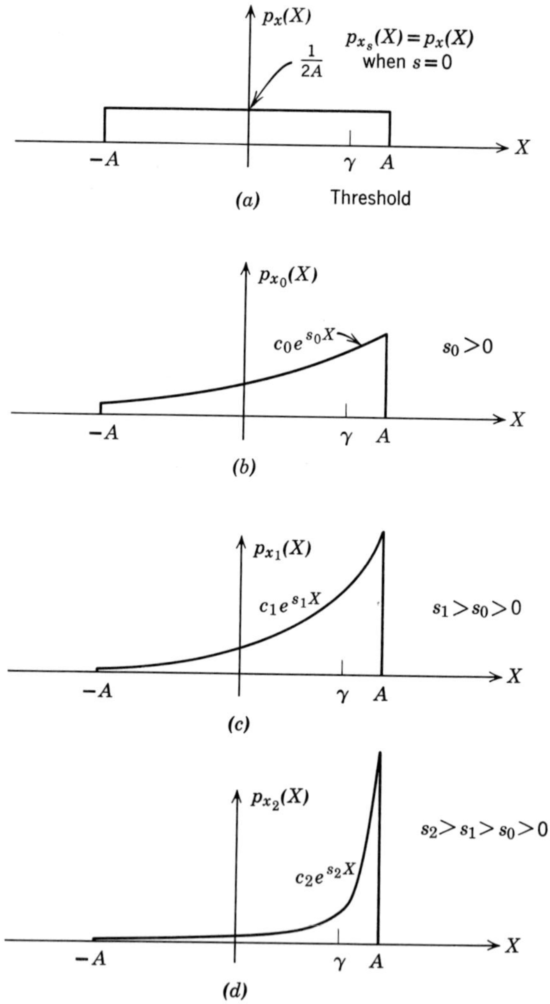

Fig. 2.37 Tilted probability densities.

Thus

$$P_F \le e^{\mu(s) - s\gamma} \int_{\gamma}^{\infty} p_{x_s}(X)\, dX, \qquad s \ge 0. \tag{456}$$

Clearly the integral is less than one. Thus

$$P_F \le e^{\mu(s) - s\gamma}, \qquad s \ge 0. \tag{457}$$

To get the best bound we minimize the right-hand side of (457) with respect to s. Differentiating the exponent and setting the result equal to zero, we obtain

$$\dot{\mu}(s) = \gamma. \tag{458}$$

Because $\ddot{\mu}(s)$ is nonnegative, a solution will exist if

$$\dot{\mu}(0) \le \gamma \le \dot{\mu}(\infty). \tag{459}$$

Because

$$\dot{\mu}(0) = E(l \mid H_0), \tag{460}$$

the left inequality implies that the threshold must be to the right of the mean of l on H_0. Assuming that (459) is valid, we have the desired result:

$$P_F \le \exp\,[\mu(s) - s\dot{\mu}(s)], \qquad s \ge 0, \tag{461}$$

where s satisfies (458). (We have assumed $\mu(s)$ exists for the desired s.)

Equation 461 is commonly referred to as the Chernoff bound [28]. Observe that s is chosen so that the mean of the tilted variable x_s is at the threshold.

The next step is find a bound on P_M, the probability of a miss:

$$P_M = \int_{-\infty}^{\gamma} p_{l \mid H_1}(X \mid H_1)\, dX, \tag{462}$$

which we want to express in terms of the tilted variable x_s.

Using an argument identical to that in (88) through (94), we see that

$$p_{l \mid H_1}(X \mid H_1) = e^{X} p_{l \mid H_0}(X \mid H_0). \tag{463}$$

Substituting (463) into the right side of (450), we have

$$p_{l \mid H_1}(X \mid H_1) = e^{\mu(s) + (1-s)X} p_{x_s}(X). \tag{464}$$

Substituting into (462),

$$P_M = e^{\mu(s)} \int_{-\infty}^{\gamma} e^{(1-s)X} p_{x_s}(X)\, dX. \tag{465}$$

For $s \le 1$

$$e^{(1-s)X} \le e^{(1-s)\gamma}, \qquad \text{for } X \le \gamma. \tag{466}$$

Thus

$$\begin{aligned} P_M &\le e^{\mu(s) + (1-s)\gamma} \int_{-\infty}^{\gamma} p_{x_s}(X)\, dX \\ &\le e^{\mu(s) + (1-s)\gamma}, \qquad s \le 1. \end{aligned} \tag{467}$$

Once again the bound is minimized for

$$\gamma = \dot{\mu}(s) \tag{468}$$

if a solution exists for $s \leq 1$. Observing that

$$\dot{\mu}(1) = E(l|H_1), \tag{469}$$

we see that this requires the threshold to be to the left of the mean of l on H_1.

Combining (461) and (467), we have

$$\boxed{\begin{aligned} P_F &\leq \exp\,[\mu(s) - s\dot{\mu}(s)] \\ & \qquad\qquad\qquad\qquad 0 \leq s \leq 1 \\ P_M &\leq \exp\,[\mu(s) + (1 - s)\dot{\mu}(s)] \end{aligned}} \tag{470}$$

and

$$\gamma = \dot{\mu}(s)$$

is the threshold that lies *between* the means of l on the two hypotheses. Confining s to [0, 1] is not too restrictive because if the threshold is not between the means the error probability will be large on one hypothesis (greater than one half if the median coincides with the mean). If we are modeling some physical system this would usually correspond to unacceptable performance and necessitate a system change.

As pointed out in [25], the exponents have a simple graphical interpretation. A typical $\mu(s)$ is shown in Fig. 2.38. We draw a tangent at the point at which $\dot{\mu}(s) = \gamma$. This tangent intersects vertical lines at $s = 0$ and $s = 1$. The value of the intercept at $s = 0$ is the exponent in the P_F bound. The value of the intercept at $s = 1$ is the exponent in the P_M bound.

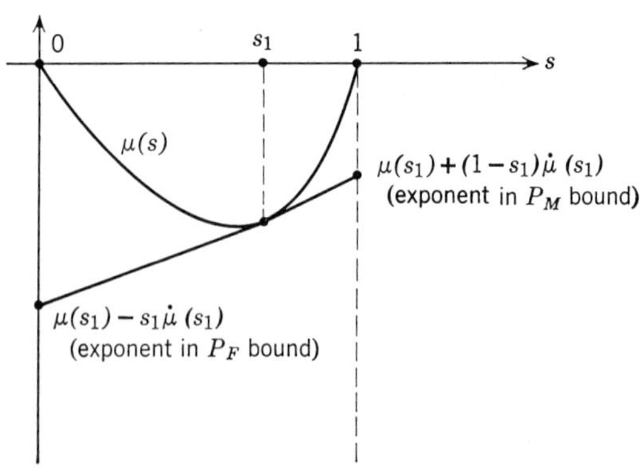

Fig. 2.38 Exponents in bounds.

For the special case in which the hypotheses are equally likely and the error costs are equal we know that $\gamma = 0$. Therefore to minimize the bound we choose that value of s where $\dot{\mu}(s) = 0$.

The probability of error $\Pr(\epsilon)$ is

$$\Pr(\epsilon) = \tfrac{1}{2}P_F + \tfrac{1}{2}P_M. \tag{471}$$

Substituting (456) and (467) into (471) and denoting the value s for which $\dot{\mu}(s) = 0$ as s_m, we have

$$\Pr(\epsilon) \leq \tfrac{1}{2}e^{\mu(s_m)} \int_0^{\infty} p_{x_s}(X)\, dX + \tfrac{1}{2}e^{\mu(s_m)} \int_{-\infty}^{0} p_{x_s}(X)\, dX, \tag{472}$$

or

$$\boxed{\Pr(\epsilon) \leq \tfrac{1}{2}e^{\mu(s_m)}.} \tag{473}$$

Up to this point we have considered arbitrary binary hypothesis tests. The bounds in (470) and (473) are always valid if $\mu(s)$ exists. In many cases of interest $l(\mathbf{R})$ consists of a sum of a large number of independent random variables, and we can obtain a simple approximate expression for P_F and P_M that provides a much closer estimate of their actual value than the above bounds. The exponent in this expression is the same, but the multiplicative factor will often be appreciably smaller than unity.

We start the derivation with the expression for P_F given in (454). Motivated by our result in the bound derivation (458), we choose s so that

$$\dot{\mu}(s) = \gamma.$$

Then (454) becomes

$$P_F = e^{\mu(s)} \int_{\dot{\mu}(s)}^{\infty} e^{-sX} p_{x_s}(X)\, dX. \tag{474}$$

This can be written as

$$P_F = e^{\mu(s) - s\dot{\mu}(s)} \int_{\dot{\mu}(s)}^{\infty} e^{+s[\dot{\mu}(s) - X]} p_{x_s}(X)\, dX. \tag{475}$$

The term outside is just the bound in (461). We now use a central limit theorem argument to evaluate the integral. First define a standardized variable:

$$y \triangleq \frac{x_s - E(x_s)}{(\mathrm{Var}\,[x_s])^{1/2}} = \frac{x_s - \dot{\mu}(s)}{\sqrt{\ddot{\mu}(s)}}. \tag{476}$$

Substituting (476) into (475), we have

$$P_F = e^{\mu(s) - s\dot{\mu}(s)} \int_0^{\infty} e^{-s\sqrt{\ddot{\mu}(s)}\, Y} p_y(Y)\, dY. \tag{477}$$

In many cases the probability density governing $\mathbf{r}$ is such that y approaches a Gaussian random variable as N (the number of components of $\mathbf{r}$) approaches infinity.† A simple case in which this is true is the case in which the r_i are independent, identically distributed random variables with finite means and variances. In such cases, y approaches a zero-mean Gaussian random variable with unit variance and the integral in (477) can be evaluated by substituting the limiting density.

$$\int_0^\infty e^{-s\sqrt{\ddot{\mu}(s)}\,Y} \frac{1}{\sqrt{2\pi}} e^{-(Y^2/2)}\, dY = e^{s^2\ddot{\mu}(s)/2} \operatorname{erfc}_*\left(s\sqrt{\ddot{\mu}(s)}\right). \tag{478}$$

Then

$$P_F \simeq \left\{\exp\left[\mu(s) - s\dot{\mu}(s) + \frac{s^2}{2}\ddot{\mu}(s)\right]\right\} \operatorname{erfc}_*\left[s\sqrt{\ddot{\mu}(s)}\right]. \tag{479}$$

The approximation arises because y is only approximately Gaussian for finite N. For values of $s\sqrt{\ddot{\mu}(s)} > 3$ we can approximate $\operatorname{erfc}_*(\cdot)$ by the upper bound in (71). Using this approximation,

$$\boxed{P_F \simeq \frac{1}{\sqrt{2\pi s^2\ddot{\mu}(s)}} \exp\left[\mu(s) - s\dot{\mu}(s)\right], \qquad s \geq 0.} \tag{480}$$

It is easy to verify that the approximate expression in (480) can also be obtained by letting

$$p_y(Y) \simeq p_y(0) \simeq \frac{1}{\sqrt{2\pi}}. \tag{481}$$

Looking at Fig. 2.39, we see that this is valid when the exponential function decreases to a small value while $Y \ll 1$.

In exactly the same manner we obtain

$$P_M \simeq \left\{\exp\left[\mu(s) + (1-s)\,\dot{\mu}(s) + \frac{(s-1)^2}{2}\ddot{\mu}(s)\right]\right\} \operatorname{erfc}_*\left[(1-s)\sqrt{\ddot{\mu}(s)}\right]. \tag{482}$$

For $(1-s)\sqrt{\ddot{\mu}(s)} > 3$, this reduces to

$$\boxed{P_M \simeq \frac{1}{\sqrt{2\pi(1-s)^2\ddot{\mu}(s)}} \exp\left[\mu(s) + (1-s)\dot{\mu}(s)\right], \qquad s \leq 1.} \tag{483}$$

Observe that the exponents in (480) and (483) are identical to those obtained by using the Chernoff bound. The central limit theorem argument has provided a multiplicative factor that will be significant in many of the applications of interest to us.

† An excellent discussion is contained in Feller [33], pp. 517–520.

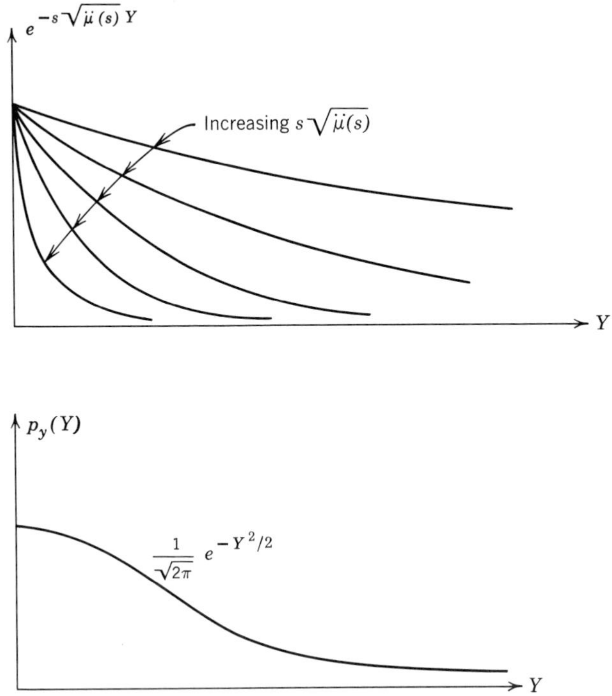

Fig. 2.39 Behavior of functions.

For the case in which $\Pr(\epsilon)$ is the criterion and the hypotheses are equally likely we have

$$
\begin{aligned}
\Pr(\epsilon) &= \tfrac{1}{2}P_F + \tfrac{1}{2}P_M \\
&= \tfrac{1}{2}\exp\left[\mu(s_m) + \frac{s_m^{\ 2}}{2}\,\ddot{\mu}(s_m)\right] \operatorname{erfc}_*\left[s_m\sqrt{\ddot{\mu}(s_m)}\right] \\
&\quad + \tfrac{1}{2}\exp\left[\mu(s_m) + \frac{(1-s_m)^2}{2}\,\ddot{\mu}(s_m)\right] \operatorname{erfc}_*\left[(1-s_m)\sqrt{\ddot{\mu}(s_m)}\right],
\end{aligned}
\tag{484}
$$

where s_m is defined in the statement preceding (472) [i.e., $\dot{\mu}(s_m) = 0 = \gamma$]. When both $s_m\sqrt{\ddot{\mu}(s_m)} > 3$ and $(1 - s_m)\sqrt{\ddot{\mu}(s_m)} > 3$, this reduces to

$$
\boxed{\Pr(\epsilon) \simeq \frac{1}{[2(2\pi\ddot{\mu}(s_m))^{1/2}s_m(1-s_m)]}\exp\mu(s_m).}
\tag{485}
$$

We now consider several examples to illustrate the application of these ideas. The first is one in which the exact performance is known. We go through the bounds and approximations to illustrate the manipulations involved.

Example 1. In this example we consider the simple Gaussian problem first introduced on p. 27:

$$p_{\mathbf{r}|H_1}(\mathbf{R}|H_1) = \prod_{i=1}^{N} \frac{1}{\sqrt{2\pi}\,\sigma} \exp\left[-\frac{(R_i - m)^2}{2\sigma^2}\right] \tag{486}$$

and

$$p_{\mathbf{r}|H_0}(\mathbf{R}|H_0) = \prod_{i=1}^{N} \frac{1}{\sqrt{2\pi}\,\sigma} \exp\left(-\frac{R_i^2}{2\sigma^2}\right). \tag{487}$$

Then, using (449)

$$\mu(s) = \ln \int_{-\infty}^{\infty} \cdots \int_{-\infty}^{\infty} \prod_{i=1}^{N} \frac{1}{\sqrt{2\pi}\,\sigma} \exp\left[-\frac{(R_i - m)^2 s + R_i^2(1 - s)}{2\sigma^2}\right] dR_i. \tag{488a}$$

Because all the integrals are identical,

$$\mu(s) = N \ln \int_{-\infty}^{\infty} \frac{1}{\sqrt{2\pi}\,\sigma} \exp\left[-\frac{(R - m)^2 s + R^2(1 - s)}{2\sigma^2}\right] dR. \tag{488b}$$

Integrating we have

$$\mu(s) = Ns(s - 1)\frac{m^2}{2\sigma^2} \triangleq \frac{s(s - 1)d^2}{2}, \tag{489}$$

where d^2 was defined in the statement after (64). The curve is shown in Fig. 2.40:

$$\dot{\mu}(s) = \frac{(2s - 1)d^2}{2}. \tag{490}$$

Using the bounds in (470), we have

$$\left.\begin{aligned} P_F &\leq \exp\left(\frac{-s^2 d^2}{2}\right) \\ P_M &\leq \exp\left[-\frac{(1 - s)^2 d^2}{2}\right] \end{aligned}\right\} \quad 0 \leq s \leq 1. \tag{491}$$

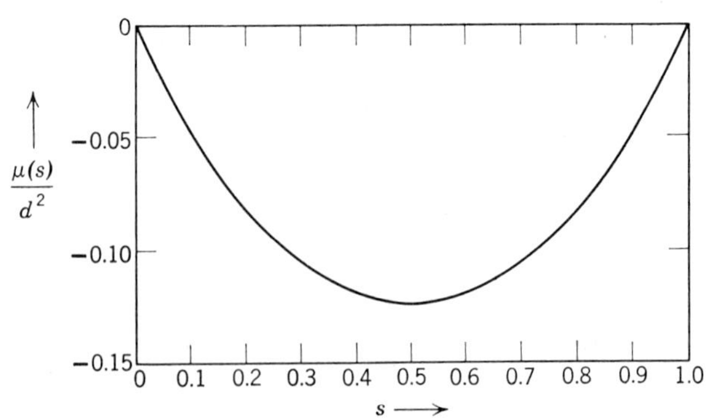

Fig. 2.40 $\mu(s)$ for Gaussian variables with unequal means.

Because $l(\mathbf{R})$ is the sum of Gaussian random variables, the expressions in (479) and (482) are exact. Evaluating $\ddot{\mu}(s)$, we obtain

$$\ddot{\mu}(s) = d^2. \tag{492}$$

Substituting into (479) and (482), we have

$$P_F = \text{erfc}_* \, [s\sqrt{\ddot{\mu}(s)}] = \text{erfc}_* \, (sd) \tag{493}$$

and

$$P_M = \text{erfc}_* \, [(1 - s)\sqrt{\ddot{\mu}(s)}] = \text{erfc}_* \, [(1 - s)d]. \tag{494}$$

These expressions are identical to (64) and (68) (let $s = (\ln \eta)/d^2 + \frac{1}{2}$).

An even simpler case is one in which the total probability of error is the criterion. Then we choose an s_m such as $\dot{\mu}(s_m) = 0$. From Fig. 2.40, we see that $s_m = \frac{1}{2}$. Using (484) and (485) we have

$$\Pr(\epsilon) = \text{erfc}_* \left(\frac{d}{2}\right) \simeq \left(\frac{2}{\pi d^2}\right)^{1/2} \exp\left(-\frac{d^2}{8}\right), \tag{495}$$

where the approximation is very good for $d > 6$.

This example is a special case of the binary symmetric hypothesis problem in which $\mu(s)$ is symmetric about $s = \frac{1}{2}$. When this is true *and* the criterion is minimum $\Pr(\epsilon)$, then $\mu(\frac{1}{2})$ is the important quantity.

$$\mu(\tfrac{1}{2}) = \ln \int_{-\infty}^{\infty} [p_{\mathbf{r}|H_1}(\mathbf{R}|H_1)]^{1/2}[p_{\mathbf{r}|H_0}(\mathbf{R}|H_0)]^{1/2} \, d\mathbf{R}. \tag{496}$$

The negative of this quantity is frequently referred to as the Bhattacharyya distance (e.g., [29]). It is important to note that it is the significant quantity only when $s_m = \frac{1}{2}$.

In our next example we look at a more interesting case.

Example 2. This example is Case 1A of the general Gaussian problem described on p. 108:

$$\begin{aligned} p_{\mathbf{r}|H_1}(\mathbf{R}|H_1) &= \prod_{i=1}^{N} \frac{1}{\sqrt{2\pi}\,\sigma_1} \exp\left(-\frac{R_i^2}{2\sigma_1^2}\right), \\ p_{\mathbf{r}|H_0}(\mathbf{R}|H_0) &= \prod_{i=1}^{N} \frac{1}{\sqrt{2\pi}\,\sigma_0} \exp\left(-\frac{R_i^2}{2\sigma_0^2}\right). \end{aligned} \tag{497}$$

Substituting (497) into (499) gives,

$$\mu(s) = N \ln \int_{-\infty}^{\infty} \frac{1}{(\sqrt{2\pi}\,\sigma_1{}^s\sigma_0^{1-s})} \exp\left[-\frac{sR^2}{2\sigma_1^2} - \frac{(1-s)R^2}{2\sigma_0^2}\right] dR \tag{498}$$

or

$$\mu(s) = \frac{N}{2} \ln \left[\frac{(\sigma_0^2)^s(\sigma_1^2)^{1-s}}{s\sigma_0^2 + (1 - s)\sigma_1^2}\right]. \tag{499}$$

A case that will be of interest in the sequel is

$$\begin{aligned} \sigma_1^2 &= \sigma_n^2 + \sigma_s^2, \\ \sigma_0^2 &= \sigma_n^2. \end{aligned} \tag{500}$$

Substituting (500) into (499) gives

$$\frac{\mu(s)}{N/2} = \left\{(1 - s)\ln\left(1 + \frac{\sigma_s^2}{\sigma_n^2}\right) - \ln\left[1 + (1 - s)\frac{\sigma_s^2}{\sigma_n^2}\right]\right\}. \tag{501}$$

This function is shown in Fig. 2.41.

$$\dot{\mu}(s) = \frac{N}{2}\left[-\ln\left(1 + \frac{\sigma_s^2}{\sigma_n^2}\right) + \frac{\sigma_s^2/\sigma_n^2}{1 + (1 - s)\sigma_s^2/\sigma_n^2}\right] \tag{502}$$

and

$$\ddot{\mu}(s) = \frac{N}{2}\left[\frac{\sigma_s^2/\sigma_n^2}{1 + (1 - s)(\sigma_s^2/\sigma_n^2)}\right]^2. \tag{503}$$

By substituting (501), (502), and (503) into (479) and (482) we can plot an approximate receiver operating characteristic. This can be compared with the exact ROC in Fig. 2.35*a* to estimate the accuracy of the approximation. In Fig. 2.42 we show the comparison for $N = 4$ and 8, and $\sigma_s^2/\sigma_n^2 = 1$. The lines connect the equal threshold points. We see that the approximation is good. For larger N the exact and approximate ROC are identical for all practical purposes.

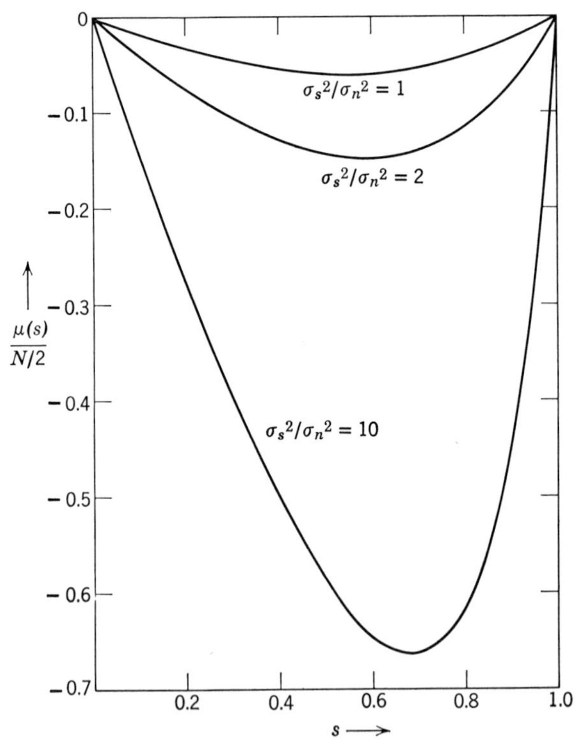

Fig. 2.41 $\mu(s)$ **for Gaussian variables with unequal variances.**

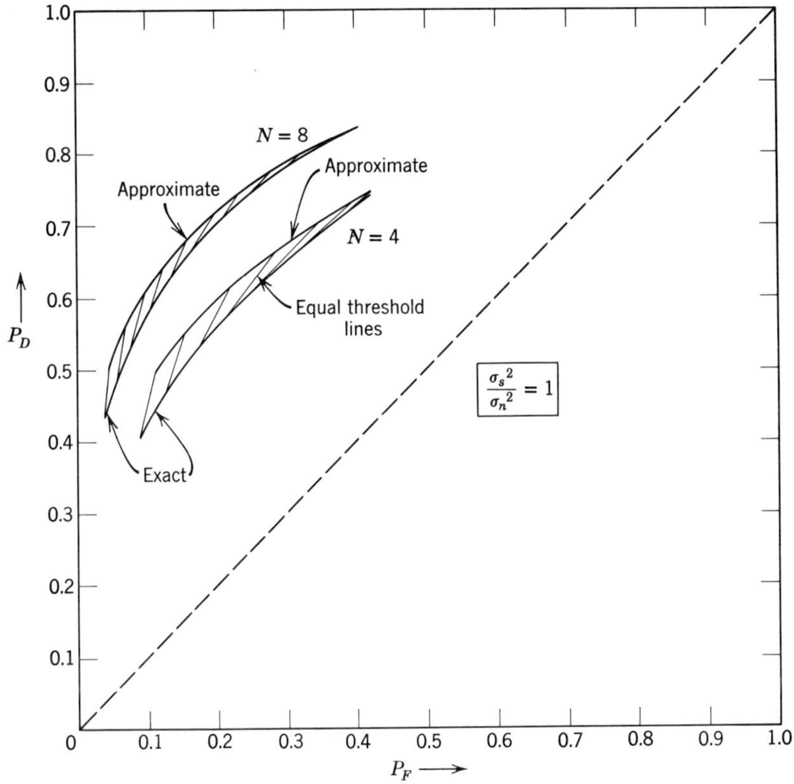

Fig. 2.42 Approximate receiver operating characteristics.

Example 3. In this example we consider first the simplified version of the symmetric hypothesis situation described in Case 2A (p. 115) in which $N = 2$.

$$p_{\mathbf{r}|H_1}(\mathbf{R}|H_1) = \frac{1}{(2\pi)^2\sigma_1^2\sigma_0^2} \exp\left(-\frac{R_1^2 + R_2^2}{2\sigma_1^2} - \frac{R_3^2 + R_4^2}{2\sigma_0^2}\right) \tag{504}$$

and

$$p_{\mathbf{r}|H_0}(\mathbf{R}|H_0) = \frac{1}{(2\pi)^2\sigma_1^2\sigma_0^2} \exp\left(-\frac{R_1^2 + R_2^2}{2\sigma_0^2} - \frac{R_3^2 + R_4^2}{2\sigma_1^2}\right), \tag{505}$$

where

$$\begin{aligned} \sigma_1^2 &= \sigma_s^2 + \sigma_n^2 \\ \sigma_0^2 &= \sigma_n^2. \end{aligned} \tag{506}$$

Then

$$\begin{aligned} \mu(s) = {} & s \ln \sigma_n^2 + (1 - s) \ln (\sigma_n^2 + \sigma_s^2) - \ln (\sigma_n^2 + \sigma_s^2 s) \\ & + (1 - s) \ln \sigma_n^2 + s \ln (\sigma_n^2 + \sigma_s^2) - \ln [\sigma_n^2 + \sigma_s^2(1 - s)] \\ = {} & \ln\left(1 + \frac{\sigma_s^2}{\sigma_n^2}\right) - \ln\left[\left(1 + \frac{s\sigma_s^2}{\sigma_n^2}\right)\left(1 + \frac{(1 - s)\sigma_s^2}{\sigma_n^2}\right)\right]. \end{aligned} \tag{507}$$

The function $\mu(s)$ is plotted in Fig. 2.43*a*. The minimum is at $s = \frac{1}{2}$. This is the point of interest at which minimum Pr (ϵ) is the criterion.

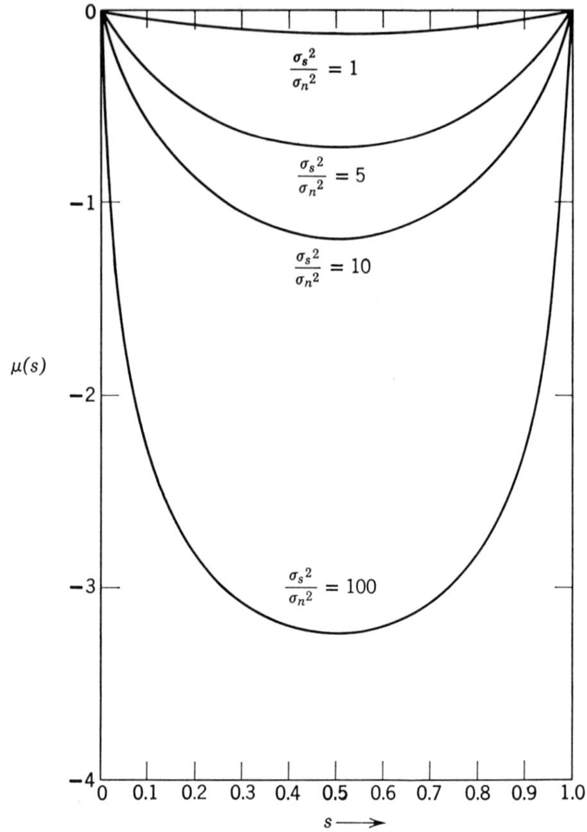

Fig. 2.43*a* $\mu(s)$ for the binary symmetric hypothesis problem.

Thus from (473), a *bound* on the error is,

$$\Pr(\epsilon) \leq \frac{1}{2}\,\frac{(1 + \sigma_s^2/\sigma_n^2)}{(1 + \sigma_s^2/2\sigma_n^2)^2}. \tag{508}$$

The bound in (508) is plotted in Fig. 2.43*b*.

Example 3A. An interesting extension of Example 3 is the problem in which

$$\mathbf{K}_s = \begin{bmatrix} \sigma_1^2 & & & & & & & \\ & \sigma_1^2 & & & & 0 & & \\ & & \sigma_2^2 & & & & & \\ & & & \sigma_2^2 & & & & \\ & & & & \sigma_3^2 & & & \\ & & & & & \ddots & & \\ & 0 & & & & & \sigma_{N/2}^2 & \\ & & & & & & & \sigma_{N/2}^2 \end{bmatrix}. \tag{509}$$

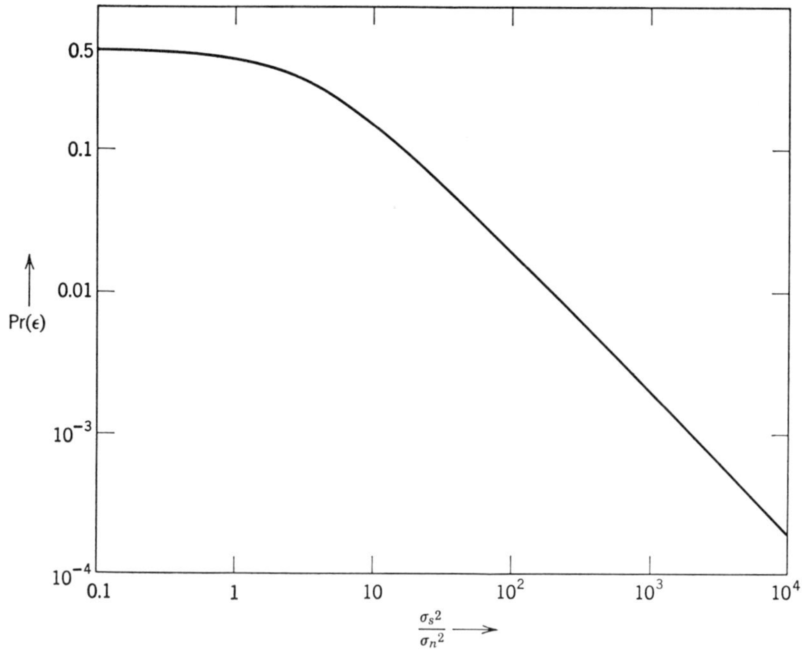

Fig. 2.43*b* Bound on the probability of error (Pr(ε)).

The r_i's are independent variables and their variances are pairwise equal. This is a special version of Case 2B on p. 115. We shall find later that it corresponds to a physical problem of appreciable interest.

Because of the independence, $\mu(s)$ is just the sum of the $\mu(s)$ for each pair, but each pair corresponds to the problem in Example 3. Therefore

$$\mu(s) = \sum_{i=1}^{N/2} \ln\left(1 + \frac{\sigma_{s_i}{}^2}{\sigma_n{}^2}\right) - \sum_{i=1}^{N/2} \ln\left\{\left(1 + s\frac{\sigma_{s_i}{}^2}{\sigma_n{}^2}\right)\left(1 + (1-s)\frac{\sigma_{s_i}{}^2}{\sigma_n{}^2}\right)\right\}. \tag{510}$$

Then

$$\dot{\mu}(s) = -\sum_{i=1}^{N/2}\left[\frac{\sigma_{s_i}{}^2}{\sigma_n{}^2 + s\sigma_{s_i}{}^2} - \frac{\sigma_{s_i}{}^2}{\sigma_n{}^2 + (1-s)\sigma_{s_i}{}^2}\right] \tag{511}$$

and

$$\ddot{\mu}(s) = \sum_{i=1}^{N/2}\left\{\frac{\sigma_{s_i}{}^4}{(\sigma_n{}^2 + s\sigma_{s_i}{}^2)^2} + \frac{\sigma_{s_i}{}^4}{[\sigma_n{}^2 + (1-s)\sigma_{s_i}{}^2]^2}\right\}. \tag{512}$$

For a minimum probability of error criterion it is obvious from (511) that $s_m = \frac{1}{2}$. Using (485), we have

$$\Pr(\epsilon) \simeq \left[\pi \sum_{i=1}^{N/2} \frac{\sigma_{s_i}{}^4}{(\sigma_n{}^2 + \frac{1}{2}\sigma_{s_i}{}^2)^2}\right]^{-1/2} \exp\left[\sum_{i=1}^{N/2} \ln\left(1 + \frac{\sigma_{s_i}{}^2}{\sigma_n{}^2}\right) - 2\sum_{i=1}^{N/2} \ln\left(1 + \frac{\sigma_{s_i}{}^2}{2\sigma_n{}^2}\right)\right] \tag{513}$$

or

$$\Pr(\epsilon) \simeq \left[\pi \sum_{i=1}^{N/2} \frac{\sigma_{s_i}{}^4}{(\sigma_n{}^2 + \frac{1}{2}\sigma_{s_i}{}^2)^2}\right]^{-1/2} \prod_{i=1}^{N/2} \frac{\left(1 + \frac{\sigma_{s_i}{}^2}{\sigma_n{}^2}\right)}{\left(1 + \frac{\sigma_{s_i}{}^2}{2\sigma_n{}^2}\right)^2}. \tag{514}$$

For the special case in which the variances are equal

$$\sigma_{s_1}^2 = \sigma_s^2 \tag{515}$$

and (514) reduces to

$$\Pr(\epsilon) \simeq \sqrt{\frac{2}{\pi N}} \frac{(1 + \sigma_s^2/\sigma_n^2)^{N/2}}{(\sigma_s^2/\sigma_n^2)(1 + \sigma_s^2/2\sigma_n^2)^{N-1}}. \tag{516}$$

Alternately, we can use the approximation given by (484). For this case it reduces to

$$\Pr(\epsilon) \simeq \left[\frac{1 + \sigma_s^2/\sigma_n^2}{(1 + \sigma_s^2/2\sigma_n^2)^2}\right]^{N/2} \exp\left[\frac{N}{8}\left(\frac{\sigma_s^2/\sigma_n^2}{1 + \sigma_s^2/2\sigma_n^2}\right)^2\right] \operatorname{erfc}_*\left[\left(\frac{N}{4}\right)^{1/2}\left(\frac{\sigma_s^2/\sigma_n^2}{1 + \sigma_s^2/2\sigma_n^2}\right)\right]. \tag{517}$$

In Fig. 2.44 we have plotted the approximate $\Pr(\epsilon)$ using (517) and exact $\Pr(\epsilon)$ which was given by (434). We see that the approximation is excellent.

The principal results of this section were the bounds on P_F and P_M given in (470) and (473) and the approximate error expressions given in (479), (480), (482), (483), (484), and (485). These expressions will enable us to find performance results for a number of cases of physical interest.

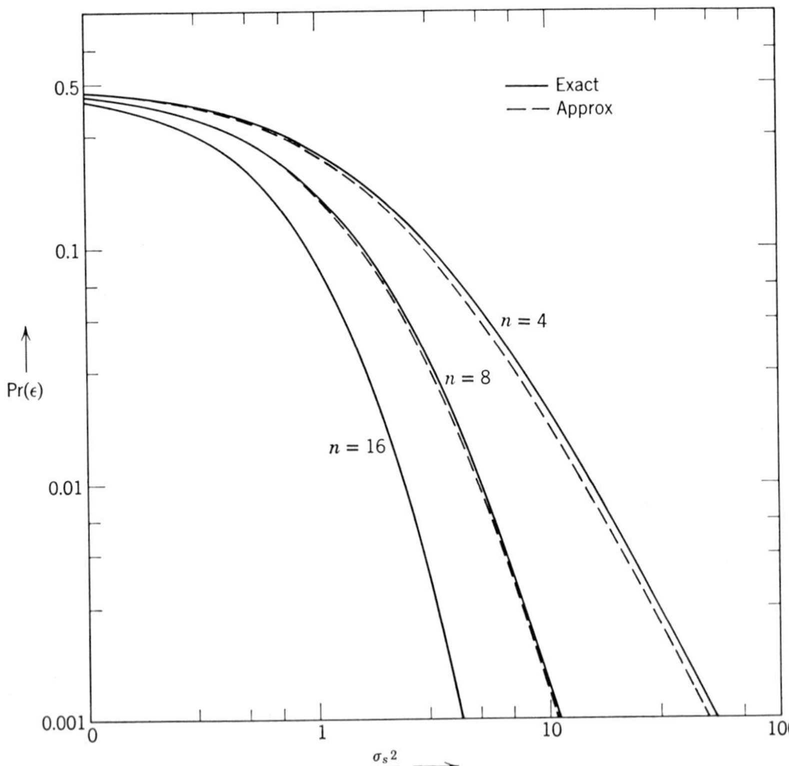

Fig. 2.44 Exact and approximate error expressions for the binary symmetric hypothesis case.

Results for some other cases are given in Yudkin [34] and Goblick [35] and the problems. In Chapter II-3 we shall study the detection of Gaussian signals in Gaussian noise. Suitable extensions of the above bounds and approximations will be used to evaluate the performance of the optimum processors.

2.8 SUMMARY

In this chapter we have derived the essential detection and estimation theory results that provide the basis for our work in the remainder of the book.

We began our discussion by considering the simple binary hypothesis testing problem. Using either a Bayes or a Neyman-Pearson criterion, we were led to a likelihood ratio test, whose performance was described by a receiver operating characteristic. Similarly, the M-hypothesis problem led to the construction of a set of likelihood ratios. This criterion-invariant reduction of the observation to a single number in the binary case or to $M - 1$ numbers in the M hypothesis case is the key to our ability to solve the detection problem when the observation is a waveform.

The development of the necessary estimation theory results followed a parallel path. Here, the fundamental quantity was a likelihood function. As we pointed out in Section 2.4, its construction is closely related to the construction of the likelihood ratio, a similarity that will enable us to solve many parallel problems by inspection. The composite hypothesis testing problem showed further how the two problems were related.

Our discussion through Section 2.5 was deliberately kept at a general level and for that reason forms a broad background of results applicable to many areas in addition to those emphasized in the remainder of the book. In Section 2.6 we directed our attention to the general Gaussian problem, a restriction that enabled us to obtain more specific results than were available in the general case. The waveform analog to this general Gaussian problem plays the central role in most of the succeeding work.

The results in the general Gaussian problem illustrated that although we can always find the optimum processor the exact performance may be difficult to calculate. This difficulty motivated our discussion of error bounds and approximations in Section 2.7. These approximations will lead us to useful results in several problem areas of practical importance.

2.9 PROBLEMS

The problems are divided into sections corresponding to the major sections in the chapter. For example, section P2.2 pertains to text material in Section 2.2. In sections in which it is appropriate the problems are divided into topical groups.

As pointed out in the Preface, solutions to individual problems are available on request.

P2.2 Binary Hypothesis Tests

Simple Binary Tests

Problem 2.2.1. Consider the following binary hypothesis testing problem:

$$H_1: r = s + n,$$
$$H_0: r = n,$$

where s and n are independent random variables.

$$p_s(S) = \begin{cases} ae^{-aS} & S \geq 0, \\ 0 & S < 0, \end{cases}$$

$$p_n(N) = \begin{cases} be^{-bN} & N \geq 0, \\ 0 & N < 0. \end{cases}$$

1. Prove that the likelihood ratio test reduces to

$$R \underset{H_0}{\overset{H_1}{\gtrless}} \gamma.$$

2. Find γ for the optimum Bayes test as a function of the costs and a priori probabilities.

3. Now assume that we need a Neyman-Pearson test. Find γ as a function of P_F, where

$$P_F \triangleq \Pr(\text{say } H_1 | H_0 \text{ is true}).$$

Problem 2.2.2. The two hypotheses are

$$H_1: p_r(R) = \frac{1}{2} \exp(-|R|)$$

$$H_0: p_r(R) = \frac{1}{\sqrt{2\pi}} \exp\left(-\frac{1}{2} R^2\right)$$

1. Find the likelihood ratio $\Lambda(R)$.
2. The test is

$$\Lambda(R) \underset{H_0}{\overset{H_1}{\gtrless}} \eta.$$

Compute the decision regions for various values of η.

Problem 2.2.3. The random variable x is $N(0, \sigma)$. It is passed through one of two nonlinear transformations.

$$H_1: y = x^2,$$
$$H_0: y = x^3.$$

Find the LRT.

Problem 2.2.4. The random variable x is $N(m, \sigma)$. It is passed through one of two nonlinear transformations.

$$H_1: y = e^x,$$
$$H_0: y = x^2.$$

Find the LRT.

Problem 2.2.5. Consider the following hypothesis-testing problem. There are K independent observations.

$$H_1: r_i \text{ is Gaussian, } N(0, \sigma_1), \qquad i = 1, 2, \ldots, K,$$
$$H_0: r_i \text{ is Gaussian, } N(0, \sigma_0), \qquad i = 1, 2, \ldots, K,$$

where $\sigma_0 < \sigma_1$.

1. Compute the likelihood ratio.
2. Assume that the threshold is η:

$$\Lambda(\mathbf{R}) \underset{H_0}{\overset{H_1}{\gtrless}} \eta.$$

Show that a sufficient statistic is $l(\mathbf{R}) = \sum_{i=1}^{K} R_i^2$. Compute the threshold γ for the test

$$l(\mathbf{R}) \underset{H_0}{\overset{H_1}{\gtrless}} \gamma$$

in terms of η, σ_0, σ_1.

3. Define

$$P_F = \Pr\,(\text{choose } H_1 |\, H_0 \text{ is true}),$$
$$P_M = \Pr\,(\text{choose } H_0 |\, H_1 \text{ is true}).$$

Find an expression for P_F and P_M.

4. Plot the ROC for $K = 1$, $\sigma_1^2 = 2$, $\sigma_0^2 = 1$.
5. What is the threshold for the minimax criterion when $C_M = C_F$ and $C_{00} = C_{11} = 0$?

Problem 2.2.6. The observation r is defined in the following manner:

$$r = bm_1 + n : H_1,$$
$$r = n \qquad\quad : H_0,$$

where b and n are independent zero-mean Gaussian variables with variances σ_b^2 and σ_n^2, respectively

1. Find the LRT and draw a block diagram of the optimum processor.
2. Draw the ROC.
3. Assume that the two hypotheses are equally likely. Use the criterion of minimum probability of error. What is the $\Pr(\epsilon)$?

Problem 2.2.7. One of two possible sources supplies the inputs to the simple communication channel as shown in the figure.

Both sources put out either 1 or 0. The numbers on the line are the channel transition probabilities; that is,

$$\Pr(a \text{ out} \mid 1 \text{ in}) = 0.7.$$

The source characteristics are

Source 1: $\Pr(1) = 0.5$, $\Pr(0) = 0.5$;
Source 2: $\Pr(1) = 0.6$, $\Pr(0) = 0.4$.

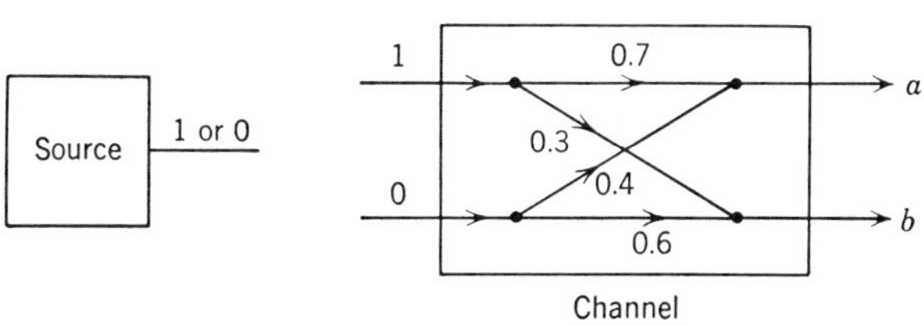

To put the problem in familiar notation, define

(a) false alarm—say source 2 when source 1 is present;
(b) detection—say source 2 when source 2 is present.

1. Compute the ROC of a test that maximizes P_D subject to the constraint that $P_F = \alpha$.
2. Describe the test procedure in detail for $\alpha = 0.25$.

Problem 2.2.8. The probability densities on the two hypotheses are

$$p_{x|H_i}(X|H_i) = \frac{1}{\pi[1 + (X - a_i)^2]} \qquad -\infty < X < \infty : H_i, \qquad i = 0, 1.$$

where $a_0 = 0$ and $a_1 = 1$.

1. Find the LRT.
2. Plot the ROC.

Problem 2.2.9. Consider a simple coin tossing problem:

$$\begin{aligned} &H_1: \text{ heads are up}, && \Pr[H_1] \triangleq P_1, \\ &H_0: \text{ tails are up}, && \Pr[H_0] \triangleq P_0 < P_1. \end{aligned}$$

N independent tosses of the coin are made. Show that the number of observed heads, N_H, is a sufficient statistic for making a decision between the two hypotheses.

Problem 2.2.10. A sample function of a simple Poisson counting process $N(t)$ is observed over the interval T:

$$\begin{aligned} &\text{hypothesis } H_1: && \text{the mean rate is } k_1 : \Pr(H_1) = \tfrac{1}{2}, \\ &\text{hypothesis } H_0: && \text{the mean rate is } k_0 : \Pr(H_0) = \tfrac{1}{2}. \end{aligned}$$

1. Prove that the number of events in the interval T is a "sufficient statistic" to choose hypothesis H_0 or H_1.
2. Assuming equal costs for the possible errors, derive the appropriate likelihood ratio test and the threshold.
3. Find an expression for the probability of error.

Problem 2.2.11. Let

$$y = \sum_{i=0}^{n} x_i,$$

where the x_i are statistically independent random variables with a Gaussian density $N(0, \sigma)$. The number of variables in the sum is a random variable with a Poisson distribution:

$$\Pr(n = k) = \frac{\lambda^k}{k!} e^{-\lambda}, \qquad k = 0, 1, \ldots.$$

We want to decide between the two hypotheses,

$$\begin{aligned} H_1 &: n \le 1, \\ H_0 &: n > 1. \end{aligned}$$

Write an expression for the LRT.

Problem 2.2.12. Randomized tests. Our basic model of the decision problem in the text (p. 24) did not permit randomized decision rules. We can incorporate them by assuming that at each point $\mathbf{R}$ in Z we say H_1 with probability $\phi(\mathbf{R})$ and say H_0 with probability $1 - \phi(\mathbf{R})$. The model in the text is equivalent to setting $\phi(\mathbf{R}) = 1$ for all $\mathbf{R}$ in Z_1 and $\phi(\mathbf{R}) = 0$ for all $\mathbf{R}$ in Z_0.

1. We consider the Bayes criterion first. Write the risk for the above decision model.
2. Prove that a LRT minimizes the risk and a randomized test is *never* necessary.

3. Prove that the risk is constant over the interior of any straight-line segment on an ROC. Because straight-line segments are generated by randomized tests, this is an alternate proof of the result in Part 2.

4. Consider the Neyman-Pearson criterion. Prove that the optimum test always consists of either

(i) an ordinary LRT with $P_F = \alpha$ or

(ii) a probabilistic mixture of *two* ordinary likelihood ratio tests constructed as follows: Test 1: $\Lambda(\mathbf{R}) \overset{H1}{\geq} \eta$ gives $P_F = \alpha^+$. Test 2: $\Lambda(\mathbf{R}) \overset{H1}{>} \eta$ gives $P_F = \alpha^-$, where $[\alpha^-, \alpha^+]$ is the smallest interval containing α. $\phi(\mathbf{R})$ is 0 or 1 except for those $\mathbf{R}$ where $\phi(\mathbf{R}) = \eta$. (Find $\phi(\mathbf{R})$ for this set.)

Mathematical Properties

Problem 2.2.13. The random variable $\Lambda(\mathbf{R})$ is defined by (13) and has a different probability density on H_1 and H_0. Prove the following:

1. $E(\Lambda^n|H_1) = E(\Lambda^{n+1}|H_0)$,
2. $E(\Lambda|H_0) = 1$,
3. $E(\Lambda|H_1) - E(\Lambda|H_0) = \text{Var}(\Lambda|H_0)$.

Problem 2.2.14. Consider the random variable Λ. In (94) we showed that

$$p_{\Lambda|H_1}(X|H_1) = X p_{\Lambda|H_0}(X|H_0).$$

1. Verify this relation by direct calculation of $p_{\Lambda|H_1}(\cdot)$ and $p_{\Lambda|H_0}(\cdot)$ for the densities in Example 1 [p. 27, (19) and (20)].

2. On page 37 we saw that the performance of the test in Example 1 was completely characterized by d^2. Show that

$$d^2 = \ln [1 + \text{Var}(\Lambda|H_0)].$$

Problem 2.2.15. The function $\text{erfc}_*(X)$ is defined in (66):

1. Integrate by parts to establish the bound

$$\frac{1}{\sqrt{2\pi}\, X}\left(1 - \frac{1}{X^2}\right) \exp\left(-\frac{X^2}{2}\right) < \text{erfc}_*(X) < \frac{1}{\sqrt{2\pi}\, X} \exp\left(-\frac{X^2}{2}\right), \qquad X > 0.$$

2. Generalize part 1 to obtain the asymptotic series

$$\text{erfc}_*(X) = \frac{1}{\sqrt{2\pi}\, X} e^{-X^2/2}\left[1 + \sum_{m=1}^{n} (-1)^m \frac{1\cdot 3\cdots(2m-1)}{X^{2m}} + R_n\right].$$

The remainder is less than the magnitude of the $n + 1$ term and is the same sign. *Hint.* Show that the remainder is

$$R_n = \left[(-1)^n \frac{1\cdot 3\cdots(2n-1)}{X^{2n}}\right]\theta,$$

where

$$\theta = \int_0^\infty e^{-t}\left(1 + \frac{2t}{X}\right)^{-n-\frac{1}{2}} dt \leq 1.$$

3. Assume that $X = 3$. Calculate a simple bound on the *percentage* error when $\text{erfc}_*(3)$ is approximated by the first n terms in the asymptotic series. Evaluate this percentage error for $n = 2, 3, 4$ and compare the results. Repeat for $X = 5$.

Problem 2.2.16.

1. Prove

$$\operatorname{erfc}_*(X) < \frac{1}{2}\exp\left(-\frac{X^2}{2}\right), \qquad X > 0.$$

Hint. Show

$$[\operatorname{erfc}_*(X)]^2 = \Pr(x \geq X, y \geq X) < \Pr(x^2 + y^2 \geq 2X^2),$$

where x and y are independent zero-mean Gaussian variables with unit variance.

2. For what values of X is this bound better than (71)?

Higher Dimensional Decision Regions

A simple binary test can always be reduced to a one-dimensional decision region. In many cases the results are easier to interpret in two or three dimensions. Some typical examples are illustrated in this section.

Problem 2.2.17.

$$H_1\!:p_{x_1,x_2|H_1}(X_1, X_2|H_1) = \frac{1}{4\pi\sigma_1\sigma_0}\left[\exp\left(-\frac{X_1^2}{2\sigma_1^2} - \frac{X_2^2}{2\sigma_0^2}\right) + \exp\left(-\frac{X_1^2}{2\sigma_0^2} - \frac{X_2^2}{2\sigma_1^2}\right)\right],$$

$$-\infty < X_1, X_2 < \infty.$$

$$H_0\!:p_{x_1,x_2|H_0}(X_1, X_2|H_0) = \frac{1}{2\pi\sigma_0^2}\exp\left(-\frac{X_1^2}{2\sigma_0^2} - \frac{X_2^2}{2\sigma_0^2}\right), \qquad -\infty < X_1, X_2 < \infty.$$

1. Find the LRT.
2. Write an exact expression for P_D and P_F. Upper and lower bound P_D and P_F by modifying the region of integration in the exact expression.

Problem 2.2.18. The joint probability density of the random variables x_i ($i = 1, 2, \ldots, M$) on H_1 and H_0 is

$$p_{\mathbf{x}|H_1}(\mathbf{X}|H_1) = \sum_{k=1}^{M} p_k \frac{1}{(2\pi\sigma^2)^{M/2}}\exp\left[-\frac{(X_k - m)^2}{2\sigma^2}\right]\prod_{i\neq k}^{M}\exp\left(-\frac{X_i^2}{2\sigma^2}\right),$$

where

$$\sum_{k=1}^{M} p_k = 1,$$

$$p_{\mathbf{x}|H_0}(\mathbf{X}|H_0) = \prod_{i=1}^{M}\frac{1}{\sqrt{2\pi}\,\sigma}\exp\left(-\frac{X_i^2}{2\sigma^2}\right) \qquad -\infty < X_i < \infty.$$

1. Find the LRT.
2. Draw the decision regions for various values of η in the X_1, X_2-plane for the special case in which $M = 2$ and $p_1 = p_2 = \frac{1}{2}$.
3. Find an upper and lower bound to P_F and P_D by modifying the regions of integration.

Problem 2.2.19. The probability density of r_i on the two hypotheses is

$$p_{r_i|H_k}(R_i|H_k) = \frac{1}{\sqrt{2\pi}\,\sigma_k}\exp\left[-\frac{(R_i - m_k)^2}{2\sigma_k^2}\right], \qquad \begin{array}{l} i = 1, 2, \ldots, N, \\ k = 0, 1. \end{array}$$

The observations are independent.

1. Find the LRT. Express the test in terms of the following quantities:

$$l_\alpha = \sum_{i=1}^{N} R_i,$$

$$l_\beta = \sum_{i=1}^{N} R_i^2.$$

2. Draw the decision regions in the l_α,l_β-plane for the case in which

$$2m_0 = m_1 > 0,$$
$$2\sigma_1 = \sigma_0.$$

Problem 2.2.20 **(*continuation*).**

1. Consider the special case

$$m_0 = 0,$$
$$\sigma_0 = \sigma_1.$$

Draw the decision regions and compute the ROC.

2. Consider the special case

$$m_0 = m_1 = 0,$$
$$\sigma_1^2 = \sigma_s^2 + \sigma_n^2,$$
$$\sigma_0 = \sigma_n.$$

Draw the decision regions.

Problem 2.2.21. A shell is fired at one of two targets: under H_1 the point of aim has coordinates x_1, y_1, z_1; under H_0 it has coordinates x_0, y_0, z_0. The distance of the actual landing point from the point of aim is a zero-mean Gaussian variable, $N(0, \sigma)$, in each coordinate. The variables are independent. We wish to observe the point of impact and guess which hypothesis is true.

1. Formulate this as a hypothesis-testing problem and compute the likelihood ratio. What is the simplest sufficient statistic? Is the ROC in Fig. 2.9*a* applicable? If so, what value of d^2 do we use?

2. Now include the effect of time. Under H_k the desired explosion time is t_k $(k = 1, 2)$. The distribution of the actual explosion time is

$$p_{\tau|H_k}(\tau) = \frac{1}{\sqrt{2\pi}\,\sigma_t} \exp\left(-\frac{(\tau - t_k)^2}{2\sigma_t^2}\right), \qquad -\infty < \tau < \infty, \quad k = 1, 2.$$

Find the LRT and compute the ROC.

P2.3 M-Hypothesis Tests

Problem 2.3.1.

1. Verify that the M-hypothesis Bayes test always leads to a decision space whose dimension is less than or equal to $M - 1$.

2. Assume that the coordinates of the decision space are

$$\Lambda_k(\mathbf{R}) \triangleq \frac{p_{\mathbf{r}|H_k}(\mathbf{R}|H_k)}{p_{\mathbf{r}|H_0}(\mathbf{R}|H_0)}, \qquad k = 1, 2, \ldots, M - 1.$$

Verify that the decision boundaries are hyperplanes.

Problem 2.3.2. The formulation of the M-hypothesis problem in the text leads to an efficient decision space but loses some of the symmetry.

1. Starting with (98) prove that an equivalent form of the Bayes test is the following:

Compute

$$\beta_i \triangleq \sum_{j=0}^{M-1} C_{ij} \Pr(H_j|\mathbf{R}), \qquad i = 0, 1, \ldots, M - 1,$$

and choose the *smallest*.

2. Consider the special cost assignment

$$\begin{aligned} C_{ii} &= 0, \qquad i = 0, 1, 2, \ldots, M-1, \\ C_{ij} &= C, \qquad i \neq j,\ i, j = 0, 1, 2, \ldots, M-1. \end{aligned}$$

Show that an equivalent test is the following:
Compute

$$\Pr(H_i|\mathbf{R}), \qquad i = 0, 1, 2, \ldots, M-1,$$

and choose the *largest*.

Problem 2.3.3. The observed random variable is Gaussian on each of five hypotheses.

$$p_{r|H_k}(R|H_k) = \frac{1}{\sqrt{2\pi}\,\sigma} \exp\left(-\frac{(R-m_k)^2}{2\sigma^2}\right), \qquad \begin{aligned} -\infty < R < \infty, \\ k = 1, 2, \ldots, 5, \end{aligned}$$

where

$$\begin{aligned} m_1 &= -2m, \\ m_2 &= -m, \\ m_3 &= 0, \\ m_4 &= m, \\ m_5 &= 2m. \end{aligned}$$

The hypotheses are equally likely and the criterion is minimum $\Pr(\epsilon)$.

1. Draw the decision regions on the R-axis.
2. Compute the error probability.

Problem 2.3.4. The observed random variable r has a Gaussian density on the three hypotheses,

$$p_{r|H_k}(R|H_k) = \frac{1}{\sqrt{2\pi}\,\sigma_k} \exp\left[-\frac{(R-m_k)^2}{2\sigma_k^2}\right], \qquad \begin{aligned} -\infty < R < \infty \\ k = 1, 2, 3, \end{aligned}$$

where the parameter values on the three hypotheses are,

$$\begin{aligned} H_1: m_1 &= 0, \quad \sigma_1 = \sigma_\alpha, \\ H_2: m_2 &= m, \quad \sigma_2 = \sigma_\alpha, \qquad (m > 0), \\ H_3: m_3 &= 0, \quad \sigma_3 = \sigma_\beta, \qquad (\sigma_\beta > \sigma_\alpha). \end{aligned}$$

The three hypotheses are equally likely and the criterion is minimum $\Pr(\epsilon)$.

1. Find the optimum Bayes test.
2. Draw the decision regions on the R-axis for the special case,

$$\begin{aligned} \sigma_\beta^2 &= 2\sigma_\alpha^2, \\ \sigma_\alpha &= m. \end{aligned}$$

3. Compute the $\Pr(\epsilon)$ for this special case.

Problem 2.3.5. The probability density of $\mathbf{r}$ on the three hypotheses is

$$p_{r_1, r_2|H_k}(R_1, R_2|H_k) = (2\pi\sigma_{1k}\sigma_{2k})^{-1} \exp\left[-\frac{1}{2}\left(\frac{R_1^2}{\sigma_{1k}^2} + \frac{R_2^2}{\sigma_{2k}^2}\right)\right], \qquad \begin{aligned} -\infty < R_1, R_2 < \infty, \\ k = 1, 2, 3, \end{aligned}$$

where

$$\begin{aligned} \sigma_{11}^2 &= \sigma_{21}^2 = \sigma_n^2, \\ \sigma_{12}^2 &= \sigma_s^2 + \sigma_n^2, \qquad \sigma_{22}^2 = \sigma_n^2, \\ \sigma_{13}^2 &= \sigma_n^2, \qquad \sigma_{23}^2 = \sigma_s^2 + \sigma_n^2. \end{aligned}$$

The cost matrix is

$$\begin{bmatrix} 0 & 1 & 1 \\ 1 & 0 & \alpha \\ 1 & \alpha & 0 \end{bmatrix},$$

where $0 \leq \alpha < 1$ and $\Pr(H_2) = \Pr(H_3) \triangleq p$. Define $l_1 = R_1^2$ and $l_2 = R_2^2$.

1. Find the optimum test and indicate the decision regions in the l_1, l_2-plane.
2. Write an expression for the error probabilities. (Do not evaluate the integrals.)
3. Verify that for $\alpha = 0$ this problem reduces to 2.2.17.

Problem 2.3.6. On H_k the observation is a value of a Poisson random variable

$$\Pr(r = n) = \frac{k_m^{\,n}}{n!} e^{-k_m}, \qquad m = 1, 2, \ldots, M,$$

where $k_m = mk$. The hypotheses are equally likely and the criterion is minimum $\Pr(\epsilon)$.

1. Find the optimum test.
2. Find a simple expression for the boundaries of the decision regions and indicate how you would compute the $\Pr(\epsilon)$.

Problem 2.3.7. Assume that the received vector on each of the three hypotheses is

$$\begin{aligned} H_0 &: \mathbf{r} = \mathbf{m}_0 + \mathbf{n}, \\ H_1 &: \mathbf{r} = \mathbf{m}_1 + \mathbf{n}, \\ H_2 &: \mathbf{r} = \mathbf{m}_2 + \mathbf{n}, \end{aligned}$$

where

$$\mathbf{r} \triangleq \begin{bmatrix} r_1 \\ r_2 \\ r_3 \end{bmatrix}, \qquad \mathbf{m}_i \triangleq \begin{bmatrix} m_{i1} \\ m_{i2} \\ m_{i3} \end{bmatrix}, \qquad \mathbf{n} \triangleq \begin{bmatrix} n_1 \\ n_2 \\ n_3 \end{bmatrix}.$$

The $\mathbf{m}_i$ are known vectors, and the components of $\mathbf{n}$ are statistically independent, zero-mean Gaussian random variables with variance σ^2.

1. Using the results in the text, express the Bayes test in terms of two sufficient statistics.

$$l_1 = \sum_{i=1}^{3} c_i r_i,$$

$$l_2 = \sum_{i=1}^{3} d_i r_i.$$

Find explicit expressions for c_i and d_i. Is the solution unique?

2. Sketch the decision regions in the l_1, l_2-plane for the particular cost assignment,

$$\begin{aligned} C_{00} &= C_{11} = C_{22} = 0, \\ C_{12} &= C_{21} = C_{01} = C_{10} = \tfrac{1}{2}C_{02} = \tfrac{1}{2}C_{20} > 0. \end{aligned}$$

P2.4 Estimation

BAYES ESTIMATION

Problem 2.4.1. Let

$$r = ab + n,$$

where a, b, and n are independent zero-mean Gaussian variables with variances σ_a^2, σ_b^2, and σ_n^2.

1. What is $\hat{a}_{\text{map}}$?
2. Is this equivalent to simultaneously finding $\hat{a}_{\text{map}}, \hat{b}_{\text{map}}$?

3. Now consider the case in which

$$r = a + \sum_{i=1}^{k} b_i + n,$$

where the b_i are independent zero-mean Gaussian variables with variances $\sigma_{b_i}{}^2$.

(a) What is $\hat{a}_{\text{map}}$?
(b) Is this equivalent to simultaneously finding $\hat{a}_{\text{map}}$, $\hat{b}_{i,\text{map}}$?
(c) Explain intuitively why the answers to part 2 and part 3b are different.

Problem 2.4.2. The observed random variable is x. We want to estimate the parameter λ. The probability density of x as a function of λ is,

$$\begin{aligned} p_{x|\lambda}(X|\lambda) &= \lambda e^{-\lambda X}, \qquad & X \geq 0, \lambda > 0, \\ &= 0, & X < 0. \end{aligned}$$

The a priori density of λ depends on two parameters: n_*, l_*.

$$p_{\lambda|n_*,l_*}(\lambda|n_*, l_*) \triangleq \begin{cases} \dfrac{l_*^{n_*}}{\Gamma(n_*)} e^{-\lambda l_*} \lambda^{n_* - 1}, & \lambda \geq 0, \\ 0, & \lambda < 0. \end{cases}$$

1. Find $E(\lambda)$ and Var (λ) *before* any observations are made.
2. Assume that one observation is made. Find $p_{\lambda|x}(\lambda|X)$. What interesting property does this density possess? Find $\hat{\lambda}_{\text{ms}}$ and $E[(\hat{\lambda}_{\text{ms}} - \lambda)^2]$.
3. Now assume that n independent observations are made. Denote these n observations by the vector $\mathbf{x}$. Verify that

$$p_{\lambda|\mathbf{x}}(\lambda|\mathbf{X}) \triangleq \begin{cases} \dfrac{(l')^{n'}}{\Gamma(n')} e^{-\lambda l'} \lambda^{n' - 1}, & \lambda \geq 0, \\ 0, & \lambda < 0, \end{cases}$$

where

$$\begin{aligned} l' &= l + l_*, \\ n' &= n + n_*, \end{aligned}$$

and

$$l = \sum_{i=1}^{n} X_i.$$

Find $\hat{\lambda}_{\text{ms}}$ and $E[(\hat{\lambda}_{\text{ms}} - \lambda)^2]$.
4. Does $\hat{\lambda}_{\text{map}} = \hat{\lambda}_{\text{ms}}$?

Comment. Reproducing Densities. The reason that the preceding problem was simple was that the a priori and a posteriori densities had the same functional form. (Only the parameters changed.) In general,

$$p_{a|r}(A|R) = \frac{p_{r|a}(R|A)p_a(A)}{p_r(R)},$$

and we say that $p_a(A)$ is a *reproducing density* or a *conjugate prior* density [with respect to the transition density $p_{r|a}(R|A)$] if the a posteriori density is of the same form as $p_a(A)$. Because the choice of the a priori density is frequently somewhat arbitrary, it is convenient to choose a reproducing density in many cases. The next two problems illustrate other reproducing densities of interest.

Problem 2.4.3. Let

$$r = a + n,$$

where n is $N(0, \sigma_n)$. Then

$$p_{r|a}(R|A) = \frac{1}{\sqrt{2\pi}\,\sigma_n} \exp\left[-\frac{(R - A)^2}{2\sigma_n^2}\right].$$

1. Verify that a conjugate priori density for a is $N\left(m_0, \frac{\sigma_n}{k_0}\right)$ by showing that

$$p_{a|r}(A|R) = N(m_1, \sigma_1),$$

where

$$m_1 = \frac{m_0 k_0^2 + R}{(1 + k_0^2)}$$

and

$$\sigma_1^2 = \frac{\sigma_n^2}{1 + k_0^2}.$$

2. Extend this result to N independent observations by verifying that

$$p_{a|\mathbf{r}}(A|\mathbf{R}) = N(m_N, \sigma_N),$$

where

$$m_N = \frac{m_0 k_0^2 + Nl}{N + k_0^2},$$

$$\sigma_N^2 = \frac{\sigma_n^2}{N + k_0^2},$$

and

$$l \triangleq \frac{1}{N} \sum_{i=1}^{N} R_i.$$

Observe that the a priori parameter k_0 can be interpreted as an equivalent number of observations (fractional observations are allowed).

Problem 2.4.4. Consider the observation process

$$p_{r|a}(R|A) = \frac{A^{1/2}}{(2\pi)^{1/2}} \exp\left[-\frac{A}{2}(R - m)^2\right],$$

where m is known and A is the parameter of interest (it is the reciprocal of the standard deviation). We assume that N independent observations are available.

1. Verify that

$$p_a(A|k_1, k_2) = c(A^{k_1/2 - 1}) \exp\left(-\tfrac{1}{2}Ak_1k_2\right), \qquad A \geq 0, \quad k_1, k_2 > 0,$$

(c is a normalizing factor) is a conjugate prior density by showing that

$$p_{a|\mathbf{r}}(A|\mathbf{R}) = p_a(A|k_1', k_2'),$$

where

$$k_2' = \frac{1}{k_1'}(k_1k_2 + Nw),$$

$$k_1' = k_1 + N,$$

$$w = \frac{1}{N} \sum_{i=1}^{N} (R_i - m)^2.$$

Note that k_1, k_2 are simply the parameters in the a priori density which are chosen based on our a priori knowledge.

2. Find $\hat{a}_{ms}$.

Problem 2.4.5. We make K observations: $R_1, \ldots, R_K$, where

$$r_i = a + n_i.$$

The random variable a has a Gaussian density $N(0, \sigma_a)$. The n_i are independent Gaussian variables $N(0, \sigma_n)$.

1. Find the MMSE estimate $\hat{a}_{\text{ms}}$.
2. Find the MAP estimate $\hat{a}_{\text{map}}$.
3. Compute the mean-square error.
4. Consider an alternate procedure using the same r_i.

(a) Estimate a after each observation using a MMSE criterion.

This gives a sequence of estimates $\hat{a}_1(R_1)$, $\hat{a}_2(R_1, R_2) \ldots \hat{a}_j(R_1, \ldots R_j) \ldots \hat{a}_K(R_1, \ldots, R_K)$. Denote the corresponding variances as $\sigma_1^2, \sigma_2^2, \ldots, \sigma_K^2$.

(b) Express $\hat{a}_j$ as a function of $\hat{a}_{j-1}$, σ_{j-1}^2, and R_j.

(c) Show that

$$\frac{1}{\sigma_j^2} = \frac{1}{\sigma_a^2} + \frac{j}{\sigma_n^2}.$$

Problem 2.4.6. [36]. In this problem we outline the proof of Property 2 on p. 61. The assumptions are the following:

(a) The cost function is a symmetric, nondecreasing function. Thus

$$C(X) = C(-X)$$

$$C(X_1) \geq C(X_2) \quad \text{for} \quad X_1 \geq X_2 \geq 0, \tag{P.1}$$

which implies

$$\frac{dC(X)}{dX} \geq 0 \quad \text{for} \quad X \geq 0. \tag{P.2}$$

(b) The a posteriori probability density is symmetric about its conditional mean and is nonincreasing.

(c)

$$\lim_{X \to \infty} C(X) p_{x|\mathbf{r}}(X|\mathbf{R}) = 0. \tag{P.3}$$

We use the same notation as in Property 1 on p. 61. Verify the following steps:

1. The conditional risk using the estimate $\hat{a}$ is

$$\mathcal{R}(\hat{a}|\mathbf{R}) = \int_{-\infty}^{\infty} C(Z) p_{z|\mathbf{r}}(Z + \hat{a} - \hat{a}_{\text{ms}}|\mathbf{R})\, dZ. \tag{P.4}$$

2. The difference in conditional risks is

$$\Delta\mathcal{R} = \mathcal{R}(\hat{a}|\mathbf{R}) - \mathcal{R}(\hat{a}_{\text{ms}}|\mathbf{R}) = \int_0^{\infty} C(Z)[p_{z|\mathbf{r}}(Z + \hat{a} - \hat{a}_{\text{ms}}|\mathbf{R}) p_{z|\mathbf{r}}(Z - \hat{a} + \hat{a}_{\text{ms}}|\mathbf{R}) - 2p_{z|\mathbf{r}}(Z|\mathbf{R})]\, dZ. \tag{P.5}$$

3. For $\hat{a} > \hat{a}_{\text{ms}}$ the integral of the terms in the bracket with respect to Z from 0 to Z_0 is

$$\int_0^{\hat{a} - \hat{a}_{ms}} [p_{z|\mathbf{r}}(Z_0 + Y|\mathbf{R}) - p_{z|\mathbf{r}}(Z_0 - Y|\mathbf{R})]\, dY \triangleq g(Z_0) \tag{P.6}$$

4. Integrate (P.5) by parts to obtain

$$\Delta\mathcal{R} = C(Z)g(Z)\Big|_0^{\infty} - \int_0^{\infty} \frac{dC(Z)}{dZ} g(Z)\, dZ, \qquad \hat{a} > \hat{a}_{\text{ms}}. \tag{P.7}$$

5. Show that the assumptions imply that the first term is zero and the second term is nonnegative.
6. Repeat Steps 3 to 5 with appropriate modifications for $\hat{a} < \hat{a}_{\text{ms}}$.

7. Observe that these steps prove that $\hat{a}_{ms}$ minimizes the Bayes risk under the above assumptions. Under what conditions will the Bayes estimate be unique?

Nonrandom Parameter Estimation

Problem 2.4.7. We make n statistically independent observations: $r_1, r_2, \ldots, r_n$, with mean m and variance σ^2. Define the sample variance as

$$V = \frac{1}{n}\sum_{j=1}^{n}\left(R_j - \sum_{i=1}^{n}\frac{R_i}{n}\right)^2.$$

Is it an unbiased estimator of the actual variance?

Problem 2.4.8. We want to estimate a in a binomial distribution by using n observations.

$$\Pr(r \text{ events}|a) = \binom{n}{r}a^r(1-a)^{n-r}, \qquad r = 0, 1, 2, \ldots, n.$$

1. Find the ML estimate of a and compute its variance.
2. Is it efficient?

Problem 2.4.9.

1. Does an efficient estimate of the standard deviation σ of a zero-mean Gaussian density exist?
2. Does an efficient estimate of the variance σ^2 of a zero-mean Gaussian density exist?

Problem 2.4.10 (continuation). The results of Problem 2.4.9 suggest the general question. Consider the problem of estimating some function of the parameter A, say, $f_1(A)$. The observed quantity is R and $p_{r|a}(R|A)$ is known. Assume that A is a nonrandom variable.

1. What are the conditions for an efficient estimate $\hat{f_1}(A)$ to exist?
2. What is the lower bound on the variance of the error of any unbiased estimate of $f_1(A)$?
3. Assume that an efficient estimate of $f_1(A)$ exists. When can an efficient estimate of some other function $f_2(A)$ exist?

Problem 2.4.11. The probability density of r, given A_1 and A_2 is:

$$p_{r|a_1,a_2}(R|A_1, A_2) = (2\pi A_2)^{-1/2}\exp\left[-\frac{(R-A_1)^2}{2A_2}\right];$$

that is, A_1 is the mean and A_2 is the variance.

1. Find the joint ML estimates of A_1 and A_2 by using n independent observations.
2. Are they biased?
3. Are they coupled?
4. Find the error covariance matrix.

Problem 2.4.12. We want to transmit two parameters, A_1 and A_2. In a simple attempt to achieve a secure communication system we construct two signals to be transmitted over separate channels.

$$s_1 = x_{11}A_1 + x_{12}A_2,$$
$$s_2 = x_{21}A_1 + x_{22}A_2,$$

where x_{ij}, $i, j = 1, 2$, are known. The received variables are

$$r_1 = s_1 + n_1,$$
$$r_2 = s_2 + n_2.$$

The additive noises are independent, identically distributed, zero-mean Gaussian random variables, $N(0, \sigma_n)$. The parameters A_1 and A_2 are nonrandom.

1. Are the ML estimates $\hat{a}_1$ and $\hat{a}_2$ unbiased?
2. Compute the variance of the ML estimates $\hat{a}_1$ and $\hat{a}_2$.
3. Are the ML estimates efficient? In other words, do they satisfy the Cramér-Rao bound with equality?

Problem 2.4.13. Let

$$y = \sum_{i=1}^{N} x_i,$$

where the x_i are independent, zero-mean Gaussian random variables with variance σ_x^2. We observe y. In parts 1 through 4 treat N as a continuous variable.

1. Find the maximum likelihood estimate of N.
2. Is $\hat{n}_{ml}$ unbiased?
3. What is the variance of $\hat{n}_{ml}$?
4. Is $\hat{n}_{ml}$ efficient?
5. Discuss qualitatively how you would modify part 1 to take into account that N is discrete.

Problem 2.4.14. We observe a value of the discrete random variable x.

$$\Pr(x = i|A) = \frac{A^i}{i!} e^{-A}, \qquad i = 0, 1, 2, \ldots,$$

where A is nonrandom.

1. What is the lower bound on the variance of any unbiased estimate, $\hat{a}(x)$?
2. Assuming n independent observations, find an $\hat{a}(\mathbf{x})$ that is efficient.

Problem 2.4.15. Consider the Cauchy distribution

$$p_{x|a}(X|A) = \{\pi[1 + (X - A)^2]\}^{-1}.$$

Assume that we make n independent observations in order to estimate A.

1. Use the Cramér-Rao inequality to show that the variance of any unbiased estimate of A has a variance greater than $2/n$.
2. Is the sample mean a consistent estimate?
3. We can show that the sample median is asymptotically normal, $N(A, \pi/\sqrt{4n})$. (See pp. 367–369 of Cramér [9].) What is the asymptotic efficiency of the sample median as an estimator?

Problem 2.4.16. Assume that

$$p_{r_1,r_2|\rho}(R_1, R_2|\rho) = \frac{1}{2\pi(1 - \rho^2)^{1/2}} \exp\left\{-\frac{(R_1^2 - 2\rho R_1 R_2 + R_2^2)}{2(1 - \rho^2)}\right\}.$$

We want to estimate the correlation coefficient ρ by using n independent observations of (R_1, R_2).

1. Find the equation for the ML estimate $\hat{\rho}$.
2. Find a lower bound on the variance of any unbiased estimate of ρ.

Mathematical Properties

Problem 2.4.17. Consider the biased estimate $\hat{a}(\mathbf{R})$ of the *nonrandom* parameter A.

$$E(\hat{a}(\mathbf{R})) = A + B(A).$$

Show that

$$E[(\hat{a}(\mathbf{R}) - A)^2] \geq \frac{(1 + dB(A)/dA)^2}{E\left\{\left[\frac{\partial \ln p_{\mathbf{r}|a}(\mathbf{R}|A)}{\partial A}\right]^2\right\}}.$$

This is the Cramér-Rao inequality for biased estimates. Note that it is a bound on the mean-square error.

Problem 2.4.18. Let $p_{\mathbf{r}|a}(\mathbf{R}|A)$ be the probability density of $\mathbf{r}$, given A. Let h be an arbitrary random variable that is independent of $\mathbf{r}$ defined so that $A + h$ ranges over all possible values of A. Assume that $p_{h_1}(H)$ and $p_{h_2}(H)$ are *two* arbitrary probability densities for h. Assuming that $\hat{a}(\mathbf{R})$ is unbiased, we have

$$\int [\hat{a}(\mathbf{R}) - (A + H)] p_{\mathbf{r}|a}(\mathbf{R}|A + H)\, d\mathbf{R} = 0.$$

Multiplying by $p_{h_i}(H)$ and integrating over H, we have

$$\int dH\, p_{h_i}(H) \int [\hat{a}(\mathbf{R}) - (A + H)] p_{\mathbf{r}|a}(\mathbf{R}|A + H)\, d\mathbf{R} = 0.$$

1. Show that

$$\text{Var}\,[\hat{a}(R) - A] \geq \frac{[E_1(h) - E_2(h)]^2}{\displaystyle\int \left(\frac{p_{\mathbf{r}|a}(\mathbf{R}|A + H)\left\{\int [p_{h_1}(H) - p_{h_2}(H)]\, dH\right\}^2}{p_{\mathbf{r}|a}(\mathbf{R}|A)}\right) d\mathbf{R}}$$

for any $p_{h_1}(H)$ and $p_{h_2}(H)$. Observe that because this is true for all $p_{h_1}(H)$ and $p_{h_2}(H)$, we may write

$$\text{Var}\,[\hat{a}(R) - A] \geq \sup_{p_{h_1}, p_{h_2}} \text{(right-hand side of above equation)}.$$

Comment. Observe that this bound does not require any regularity conditions. Barankin [15] has shown that this is the greatest lower bound.

Problem 2.4.19 (***continuation***). We now derive two special cases.

1. First, let $p_{h_2}(H) = \delta(H)$. What is the resulting bound?
2. Second, let $p_{h_1}(H) = \delta(H - H_0)$, where $H_0 \neq 0$. Show that

$$\text{Var}\,[\hat{a}(\mathbf{R}) - A] \geq \left(\inf_{H_0} \left\{\frac{1}{H_0^2}\left[\int \frac{p^2_{\mathbf{r}|a}(\mathbf{R}|A + H_0)}{p_{\mathbf{r}|a}(\mathbf{R}|A)}\, d\mathbf{R} - 1\right]\right\}\right)^{-1}.$$

The infimum being over all $H_0 \neq 0$ such that $p_{\mathbf{r}|a}(\mathbf{R}|A) = 0$ implies

$$p_{\mathbf{r}|a}(\mathbf{R}|A + H_0) = 0.$$

3. Show that the bound given in part 2 is always as good as the Cramér-Rao inequality when the latter applies.

Problem 2.4.20. Let

$$\mathbf{a} = \mathbf{L}\mathbf{b},$$

where $\mathbf{L}$ is a nonsingular matrix and $\mathbf{a}$ and $\mathbf{b}$ are vector random variables. Prove that

$$\hat{\mathbf{a}}_{\text{map}} = \mathbf{L}\hat{\mathbf{b}}_{\text{map}} \quad \text{and} \quad \hat{\mathbf{a}}_{ms} = \mathbf{L}\hat{\mathbf{b}}_{ms}.$$

Problem 2.4.21. An alternate way to derive the Cramér-Rao inequality is developed in this problem. First, construct the vector $\mathbf{z}$.

$$\mathbf{z} \triangleq \begin{bmatrix} \hat{a}(\mathbf{R}) - A \\ \hdashline \dfrac{\partial \ln p_{\mathbf{r}|a}(\mathbf{R}|A)}{\partial A} \end{bmatrix}.$$

1. Verify that for unbiased estimates $E(\mathbf{z}) = \mathbf{0}$.

2. Assuming that $E(\mathbf{z}) = \mathbf{0}$, the covariance matrix is

$$\mathbf{\Lambda_z} = E(\mathbf{z}\mathbf{z}^T).$$

Using the fact that $\mathbf{\Lambda_z}$ is nonnegative definite, derive the Cramér-Rao inequality. If the equality holds, what does this imply about $|\mathbf{\Lambda_z}|$?

Problem 2.4.22. Repeat Problem 2.4.21 for the case in which a is a random variable. Define

$$\mathbf{z} = \begin{bmatrix} \hat{a}(\mathbf{R}) - a \\ \hdashline \dfrac{\partial \ln p_{\mathbf{r},a}(\mathbf{R}, A)}{\partial A} \end{bmatrix}$$

and proceed as before.

Problem 2.4.23. Bhattacharyya Bound. Whenever an efficient estimate does not exist, we can improve on the Cramér-Rao inequality. In this problem we develop a conceptually simple but algebraically tedious bound for unbiased estimates of nonrandom variables.

1. Define an $(N + 1)$-dimensional vector,

$$\mathbf{z} \triangleq \begin{bmatrix} \hat{a}(\mathbf{R}) - A \\ \hdashline \dfrac{1}{p_{\mathbf{r}|a}(\mathbf{R}|A)} \dfrac{\partial p_{\mathbf{r}|a}(\mathbf{R}|A)}{\partial A} \\ \hdashline \dfrac{1}{p_{\mathbf{r}|a}(\mathbf{R}|A)} \dfrac{\partial^2 p_{\mathbf{r}|a}(\mathbf{R}|A)}{\partial A^2} \\ \hdashline \vdots \\ \hdashline \dfrac{1}{p_{\mathbf{r}|a}(\mathbf{R}|A)} \dfrac{\partial^N p_{\mathbf{r}|a}(\mathbf{R}|A)}{\partial A^N} \end{bmatrix}.$$

Verify that

$$\mathbf{\Lambda_z} \triangleq E(\mathbf{z}\mathbf{z}^T) = \left[\begin{array}{c|c|c} \sigma_\epsilon^2 & 1 & \mathbf{0} \\ \hline 1 & \multicolumn{2}{c}{} \\ \cline{1-1} & \multicolumn{2}{c}{\tilde{\mathbf{J}}} \\ \mathbf{0} & \multicolumn{2}{c}{} \end{array}\right]$$

What are the elements in $\tilde{\mathbf{J}}$? Is $\mathbf{\Lambda_z}$ nonnegative definite? Assume that $\tilde{\mathbf{J}}$ is positive definite. When is $\mathbf{\Lambda_z}$ *not* positive definite?

2. Verify that the results in part 1 imply

$$\sigma_\epsilon^2 \geq \tilde{J}^{11}.$$

This is the Bhattacharyya bound. Under what conditions does the equality hold?

3. Verify that for $N = 1$ the Bhattacharyya bound reduces to Cramér-Rao inequality.

4. Does the Bhattacharyya bound always improve as N increases?

Comment. In part 2 the condition for equality is

$$\hat{a}(\mathbf{R}) - A = \sum_{i=1}^{N} c_i(A) \frac{1}{p_{\mathbf{r}|a}(\mathbf{R}|A)} \frac{\partial^i p_{\mathbf{r}|a}(\mathbf{R}|A)}{\partial A^i}.$$

This condition could be termed Nth-order efficiency but does not seem to occur in many problems of interest.

5. Frequently it is easier to work with

$$\frac{\partial^i \ln p_{\mathbf{r}|a}(\mathbf{R}|A)}{\partial A^i}.$$

Rewrite the elements $\tilde{J}_{ij}$ in terms of expectations of combinations of these quantities for $N = 1$ and 2.

***Problem 2.4.24* (*continuation*).** Let $N = 2$ in the preceding problem.

1. Verify that

$$\sigma_\epsilon^2 \geq \frac{1}{\tilde{J}_{11}} + \frac{\tilde{J}_{12}^2}{\tilde{J}_{11}(\tilde{J}_{11}\tilde{J}_{22} - \tilde{J}_{12}^2)}.$$

The second term represents the improvement in the bound.

2. Consider the case in which $\mathbf{r}$ consists of M independent observations with identical densities and finite conditional means and variances. Denote the elements of $\tilde{\mathbf{J}}$ due to M observations as $\tilde{J}_{ij}(M)$. Show that $\tilde{J}_{11}(M) = M\tilde{J}_{11}(1)$. Derive similar relations for $\tilde{J}_{12}(M)$ and $\tilde{J}_{22}(M)$. Show that

$$\sigma_\epsilon^2 \geq \frac{1}{M\tilde{J}_{11}(1)} + \frac{\tilde{J}_{12}^2(1)}{2M^2\tilde{J}_{11}^4(1)} + o\left(\frac{1}{M^2}\right).$$

Problem 2.4.25. [11] Generalize the result in Problem 2.4.23 to the case in which we are estimating a function of A, say $f(A)$. Assume that the estimate is unbiased. Define

$$\mathbf{z} = \begin{bmatrix} \hat{a}(\mathbf{R}) - f(A) \\ \hline k_1 \dfrac{1}{p_{\mathbf{r}|a}(\mathbf{R}|A)} \dfrac{\partial p_{\mathbf{r}|a}(\mathbf{R}|A)}{\partial A} \\ \hline k_2 \dfrac{1}{p_{\mathbf{r}|a}(\mathbf{R}|A)} \dfrac{\partial^2 p_{\mathbf{r}|a}(\mathbf{R}|A)}{\partial A^2} \\ \hline \vdots \\ \hline k_N \dfrac{1}{p_{\mathbf{r}|a}(\mathbf{R}|A)} \dfrac{\partial^N p_{\mathbf{r}|a}(\mathbf{R}|A)}{\partial A^N} \end{bmatrix}.$$

Let

$$y = [\hat{a}(\mathbf{R}) - f(A)] - \sum_{i=1}^{N} k_i \frac{1}{p_{\mathbf{r}|a}(\mathbf{R}|A)} \cdot \frac{\partial^i p_{\mathbf{r}|a}(\mathbf{R}|A)}{\partial A^i}.$$

1. Find an expression for $\xi_y = E[y^2]$. Minimize ξ_y by choosing the k_i appropriately.
2. Using these values of k_i, find a bound on $\text{Var}[\hat{a}(\mathbf{R}) - f(A)]$.
3. Verify that the result in Problem 2.4.23 is obtained by letting $f(A) = A$ in (2).

Problem 2.4.26.

1. Generalize the result in Problem 2.4.23 to establish a bound on the mean-square error in estimating a random variable.
2. Verify that the matrix of concern is

$$\mathbf{\Lambda_z} = \left[\begin{array}{c|c|c} E(a_\epsilon^2) & 1 & \mathbf{0} \\ \hline 1 & \multicolumn{2}{c}{\multirow{2}{*}{$\tilde{\mathbf{J}}_T$}} \\ \cline{1-1} \mathbf{0} & & \end{array}\right].$$

What are the elements in $\tilde{\mathbf{J}}_T$?

3. Find $\mathbf{\Lambda_z}$ for the special case in which a is $N(0, \sigma_a)$.

MULTIPLE PARAMETERS

Problem 2.4.27. In (239) we defined the partial derivative matrix $\nabla_{\mathbf{x}}$.

$$\nabla_{\mathbf{x}} \triangleq \begin{bmatrix} \dfrac{\partial}{\partial x_1} \\ \dfrac{\partial}{\partial x_2} \\ \vdots \\ \dfrac{\partial}{\partial x_n} \end{bmatrix}.$$

Verify the following properties.

1. The matrix $\mathbf{A}$ is $n \times 1$ and the matrix $\mathbf{B}$ is $n \times 1$. Show that

$$\nabla_{\mathbf{x}}(\mathbf{A}^T\mathbf{B}) = (\nabla_{\mathbf{x}}\mathbf{A}^T)\mathbf{B} + (\nabla_{\mathbf{x}}\mathbf{B}^T)\mathbf{A}.$$

2. If the $n \times 1$ matrix $\mathbf{B}$ is not a function of $\mathbf{x}$, show that

$$\nabla_{\mathbf{x}}(\mathbf{B}^T\mathbf{x}) = \mathbf{B}.$$

3. Let $\mathbf{C}$ be an $n \times m$ constant matrix,

$$\nabla_{\mathbf{x}}(\mathbf{x}^T\mathbf{C}) = \mathbf{C}.$$

4. $\nabla_{\mathbf{x}}(\mathbf{x}^T) = \mathbf{I}$.

Problem 2.4.28. A problem that occurs frequently is the differentiation of a quadratic form.

$$Q = \mathbf{A}^T(\mathbf{x})\,\mathbf{\Lambda A}(\mathbf{x}),$$

where $\mathbf{A}(\mathbf{x})$ is a $m \times 1$ matrix whose elements are a function of $\mathbf{x}$ and $\mathbf{\Lambda}$ is a symmetric nonnegative definite $m \times m$ matrix. Recall that this implies that we can write

$$\mathbf{\Lambda} = \mathbf{\Lambda}^{1/2}\mathbf{\Lambda}^{1/2}.$$

1. Prove

$$\nabla_{\mathbf{x}} Q = 2(\nabla_{\mathbf{x}}\mathbf{A}^T(\mathbf{x}))\,\mathbf{\Lambda A}(\mathbf{x})$$

2. For the special case

$$\mathbf{A}(\mathbf{x}) = \mathbf{Bx},$$

prove

$$\nabla_{\mathbf{x}} Q = 2\mathbf{B}^T\mathbf{\Lambda B x}.$$

3. For the special case

$$Q = \mathbf{x}^T\mathbf{\Lambda x},$$

prove

$$\nabla_{\mathbf{x}} Q = 2\mathbf{\Lambda x}.$$

Problem 2.4.29. Go through the details of the proof on p. 83 for arbitrary K.

Problem 2.4.30. As discussed in (284), we frequently estimate,

$$\mathbf{d} \triangleq \mathbf{g_d}(\mathbf{A}).$$

Assume the estimates are unbiased. Derive (286).

Problem 2.4.31. The cost function is a scalar-valued function of the vector $\mathbf{a}_\epsilon$, $C(\mathbf{a}_\epsilon)$. Assume that it is symmetric and convex,

1. $C(\mathbf{a}_\epsilon) = C(-\mathbf{a}_\epsilon)$,
2. $C(b\mathbf{x}_1 + (1 - b)\mathbf{x}_2) \leq bC(\mathbf{x}_1) + (1 - b)\, C(\mathbf{x}_2), \qquad 0 \leq b \leq 1.$

Assume that the a posteriori density is symmetric about its conditional mean. Prove that the conditional mean of **a** minimizes the Bayes risk.

Problem 2.4.32. Assume that we want to estimate K nonrandom parameters $A_1, A_2, \ldots,$ A_K, denoted by **A**. The probability density $p_{\mathbf{r}|\mathbf{a}}(\mathbf{R}|\mathbf{A})$ is known. Consider the biased estimates $\hat{\mathbf{a}}(\mathbf{R})$ in which

$$\mathrm{B}(a_i) \triangleq \int [\hat{a}_i(\mathbf{R}) - A_i] p_{\mathbf{r}|\mathbf{a}}(\mathbf{R}|\mathbf{A})\, d\mathbf{R}.$$

1. Derive a bound on the mean-square error in estimating A_i.
2. The error correlation matrix is

$$\mathbf{R}_\epsilon \triangleq E[(\hat{\mathbf{a}}(\mathbf{R}) - \mathbf{A})(\hat{\mathbf{a}}^T(\mathbf{R}) - \mathbf{A}^T)]$$

Find a matrix $\mathbf{J}_B$ such that, $\mathbf{J}_B - \mathbf{R}_\epsilon^{-1}$ is nonnegative definite.

MISCELLANEOUS

Problem 2.4.33. Another method of estimating nonrandom parameters is called the method of moments (Pearson [37]). If there are k parameters to estimate, the first k sample moments are equated to the actual moments (which are functions of the parameters of interest). Solving these k equations gives the desired estimates. To illustrate this procedure consider the following example. Let

$$\begin{aligned} p_{x|\lambda}(X|\lambda) &= \frac{1}{\Gamma(\lambda)} X^{\lambda-1} e^{-X}, \qquad & X \geq 0, \\ &= 0, & X < 0, \end{aligned}$$

where λ is a positive parameter. We have n independent observations of x.

1. Find a lower bound on the variance of any unbiased estimate.
2. Denote the method of moments estimate as $\hat{\lambda}_{mm}$. Show

$$\hat{\lambda}_{mm} = \frac{1}{n} \sum_{i=1}^{n} X_i,$$

and compute $E(\hat{\lambda}_{mm})$ and Var $(\hat{\lambda}_{mm})$.

Comment. In [9] the efficiency of $\hat{\lambda}_{mm}$ is computed. It is less than 1 and tends to zero as $n \to \infty$.

Problem 2.4.34. Assume that we have n independent observations from a Gaussian density $N(m, \sigma)$. Verify that the method of moments estimates of m and σ are identical to the maximum-likelihood estimates.

P2.5 Composite Hypotheses

Problem 2.5.1. Consider the following composite hypothesis testing problem,

$$H_0 : p_r(R) = \frac{1}{\sqrt{2\pi}\, \sigma_0} \exp\left(-\frac{R^2}{2\sigma_0^2}\right),$$

where σ_0 is known,

$$H_1 : p_r(R) = \frac{1}{\sqrt{2\pi}\, \sigma_1} \exp\left(-\frac{R^2}{2\sigma_1^2}\right),$$

where $\sigma_1 > \sigma_0$. Assume that we require $P_F = 10^{-2}$.

1. Construct an upper bound on the power function by assuming a perfect measurement scheme coupled with a likelihood ratio test.
2. Does a uniformly most powerful test exist?
3. If the answer to part 2 is negative, construct the power function of a generalized likelihood ratio test.

Problem 2.5.2. Consider the following composite hypothesis testing problem. *Two* statistically independent observations are received. Denote the observations as R_1 and R_2. Their probability densities on the two hypotheses are

$$H_0: p_{r_i}(R_i) = \frac{1}{\sqrt{2\pi}\,\sigma_0} \exp\left(-\frac{R_i^2}{2\sigma_0^2}\right), \qquad i = 1, 2,$$

where σ_0 is known,

$$H_1: p_{r_i}(R_i) = \frac{1}{\sqrt{2\pi}\,\sigma_1} \exp\left(-\frac{R_i^2}{2\sigma_1^2}\right), \qquad i = 1, 2,$$

where $\sigma_1 > \sigma_0$. Assume that we require a $P_F = \alpha$.

1. Construct an upper bound on the power function by assuming a perfect measurement scheme coupled with a likelihood ratio test.
2. Does a uniformly most powerful test exist?
3. If the answer to part 2 is negative, construct the power function of a generalized likelihood ratio test.

Problem 2.5.3. The observation consists of a set of values of the random variables, $r_1, r_2, \ldots, r_M$.

$$\begin{aligned} r_i &= s_i + n_i, \qquad & i &= 1, 2, \ldots, M, \qquad & &H_1, \\ r_i &= n_i, \qquad & i &= 1, 2, \ldots, M, \qquad & &H_0. \end{aligned}$$

The s_i and n_i are independent, identically distributed random variables with densities $N(0, \sigma_s)$ and $N(0, \sigma_n)$, respectively, where σ_n is known and σ_s is unknown.

1. Does a UMP test exist?
2. If the answer to part 1 is negative, find a generalized LRT.

Problem 2.5.4. The observation consists of a set of values of the random variables $r_1, r_2, \ldots, r_M$, which we denote by the vector $\mathbf{r}$. Under H_0 the r_i are statistically independent, with densities

$$p_{r_i}(R_i) = \frac{1}{\sqrt{2\pi\lambda_i^0}} \exp\left(-\frac{R_i^2}{2\lambda_i^0}\right)$$

in which the λ_i^0 are known. Under H_1 the r_i are statistically independent, with densities

$$p_{r_i}(R_i) = \frac{1}{\sqrt{2\pi\lambda_i^1}} \exp\left(-\frac{R_i^2}{2\lambda_i^1}\right)$$

in which $\lambda_i^1 > \lambda_i^0$ for all i. Repeat Problem 2.5.3.

Problem 2.5.5. Consider the following hypothesis testing problem. *Two* statistically independent observations are received. Denote the observations R_1 and R_2. The probability densities on the two hypotheses are

$$H_0: p_{r_i}(R_i) = \frac{1}{\sqrt{2\pi}\,\sigma} \exp\left(-\frac{R_i^2}{2\sigma^2}\right), \qquad i = 1, 2,$$

$$H_1: p_{r_i}(R_i) = \frac{1}{\sqrt{2\pi}\,\sigma} \exp\left[-\frac{(R_i - m)^2}{2\sigma^2}\right] \qquad i = 1, 2,$$

where m can be *any* nonzero number. Assume that we require $P_F = \alpha$.

1. Construct an upper bound on the power function by assuming a perfect measurement scheme coupled with a likelihood ratio test.
2. Does a uniformly most powerful test exist?
3. If the answer to part 2 is negative, construct the power function of a generalized likelihood ratio test.

Problem 2.5.6. Consider the following hypothesis-testing problem.

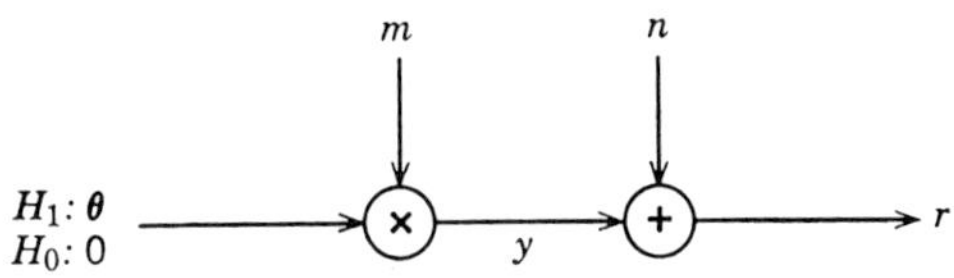

Under H_1 a nonrandom variable θ $(-\infty < \theta < \infty)$ is transmitted. It is multiplied by the random variable m. A noise n is added to the result to give r. Under H_0 nothing is transmitted, and the output is just n. Thus

$$H_1: r = m\theta + n,$$
$$H_0: r = n.$$

The random variables m and n are independent.

$$p_n(N) = \frac{1}{\sqrt{2\pi}\,\sigma_n} \exp\left(-\frac{N^2}{2\sigma_n^2}\right),$$

$$p_m(M) = \tfrac{1}{2}\,\delta(M - 1) + \tfrac{1}{2}\,\delta(M + 1).$$

1. Does a uniformly most powerful test exist? If it does, describe the test and give an expression for its power function? If it does not, indicate why.
2. Do one of the following:
   (a) If a UMP test exists for this example, derive a necessary and sufficient condition on $p_m(M)$ for a UMP test to exist. (The rest of the model is unchanged.)
   (b) If a UMP test does not exist, derive a generalized likelihood ratio test and an expression for its power function.

Problem 2.5.7 (CFAR receivers.) We have N independent observations of the variable x. The probability density on H_k is

$$p_{x_i|H_k}(X|H_k) = \frac{1}{\sqrt{2\pi}\,\sigma} \exp\left\{\frac{-(X_i - m_k)^2}{2\sigma^2}\right\} \qquad -\infty < X_i < \infty, \qquad \begin{aligned} i &= 1, 2, \dots N, \\ H_k: k &= 0,1, \\ m_0 &= 0. \end{aligned}$$

The variance σ^2 is *unknown*. Define

$$l_1 = \sum_{i=1}^{N} x_i$$

$$l_2 = \sum_{i=1}^{N} x_i^2$$

(a) Consider the test

$$l_1^2 \underset{H_0}{\overset{H_1}{\gtrless}} \alpha l_2$$

Verify that the P_F of this test does *not* depend on σ^2. (*Hint.* Use formula in Problem 2.4.6.)

(b) Find α as a function of P_F.
(c) Is this a UMP test?
(d) Consider the particular case in which $N = 2$ and $m_1 = m$. Find P_D as a function of P_F and m/σ. Compare your result with Figure 2.9*b* and see how much the lack of knowledge about the variance σ^2 has decreased the system performance.

Comment. Receivers of this type are called CFAR (constant false alarm rate) receivers in the radar/sonar literature.

Problem 2.5.8 (continuation). An alternate approach to the preceding problem would be a generalized LRT.

1. Find the generalized LRT and write an expression for its performance for the case in which $N = 2$ and $m_1 = m$.
2. How would you decide which test to use?

Problem 2.5.9. Under H_0, x is a Poisson variable with a known intensity k_0.

$$\Pr(x = n) = \frac{k_0^n}{n!} e^{-k_0}, \qquad n = 0, 1, 2, \ldots.$$

Under H_1, x is a Poisson variable with an unknown intensity k_1, where $k_1 > k_0$.

1. Does a UMP test exist?
2. If a UMP test does not exist, assume that M independent observations of x are available and construct a generalized LRT.

Problem 2.5.10. How are the results to Problem 2.5.2 changed if we know that $\sigma_0 < \sigma_c$ and $\sigma_1 > \sigma_c$ where σ_c is known. Neither σ_0 or σ_1, however, is known. If a UMP test does not exist, what test procedure (other than a generalized LRT) would be logical?

P2.6 General Gaussian Problem

Detection

Problem 2.6.1. The M-hypothesis, general Gaussian problem is

$$p_{\mathbf{r}|H_i}(\mathbf{R}|H_i) = [(2\pi)^{N/2}|\mathbf{K}_i|^{1/2}]^{-1} \exp[-\tfrac{1}{2}(\mathbf{R}^T - \mathbf{m}_i^T)\,\mathbf{Q}_i(\mathbf{R} - \mathbf{m}_i)], \qquad i = 1, 2, \ldots, M.$$

1. Use the results of Problem 2.3.2 to find the Bayes test for this problem.
2. For the particular case in which the cost of a correct decision is zero and the cost of any wrong decision is equal show that the test reduces to the following:

Compute

$$l_i(\mathbf{R}) = \ln P_i - \tfrac{1}{2}\ln|\mathbf{K}_i| - \tfrac{1}{2}(\mathbf{R}^T - \mathbf{m}_i^T)\,\mathbf{Q}_i(\mathbf{R} - \mathbf{m}_i)$$

and choose the largest.

Problem 2.6.2 (continuation). Consider the special case in which all $\mathbf{K}_i = \sigma_n^2\mathbf{I}$ and the hypotheses are equally likely. Use the costs in Part 2 of Problem 2.6.1.

1. What determines the dimension of the decision space? Draw some typical decision spaces to illustrate the various alternatives.
2. Interpret the processor as a minimum-distance decision rule.

Problem 2.6.3. Consider the special case in which $\mathbf{m}_i = 0$, $i = 1, 2, \ldots, M$, and the hypotheses are equally likely. Use the costs in Part 2 of Problem 2.6.1.

1. Show that the test reduces to the following:

Compute

$$l_i(\mathbf{R}) = \mathbf{R}^T\mathbf{Q}_i\mathbf{R} + \ln|\mathbf{K}_i|$$

and choose the *smallest*.

2. Write an expression for the Pr (ϵ) in terms of $p_{l|H_i}(\mathbf{L}|H_k)$, where

$$l \triangleq \begin{bmatrix} l_1 \\ l_2 \\ \vdots \\ l_M \end{bmatrix}.$$

Problem 2.6.4. Let

$$q_{\mathbf{B}} \triangleq \mathbf{x}^T\mathbf{B}\mathbf{x},$$

where $\mathbf{x}$ is a Gaussian vector $N(\mathbf{0}, \mathbf{I})$ and $\mathbf{B}$ is a symmetric matrix.

1. Verify that the characteristic function of $q_{\mathbf{B}}$ is

$$M_{q_{\mathbf{B}}}(jv) \triangleq E(e^{jvq_{\mathbf{B}}}) = \prod_{i=1}^{N} (1 - 2jv\lambda_{\mathbf{B}i})^{-1/2},$$

where $\lambda_{\mathbf{B}i}$ are the eigenvalues of $\mathbf{B}$.

2. What is $p_{q\mathbf{B}}(Q)$ when the eigenvalues are equal?

3. What is the form of $p_{q\mathbf{B}}(Q)$ when N is even and the eigenvalues are pair-wise equal but otherwise distinct; that is,

$$\lambda_{2i-1} = \lambda_{2i}, \qquad i = 1, 2, \ldots, \frac{N}{2},$$
$$\lambda_{2i} \neq \lambda_{2j}, \qquad \text{all } i \neq j.$$

Problem 2.6.5.

1. Modify the result of the preceding problem to include the case in which $\mathbf{x}$ is a Gaussian vector $N(\mathbf{0}, \mathbf{\Lambda_x})$, where $\mathbf{\Lambda_x}$ is positive definite.

2. What is $M_{q_{\mathbf{\Lambda_x}^{-1}}}(jv)$? Does the result have any interesting features?

Problem 2.6.6. Consider the M-ary hypothesis-testing problem. *Each* observation is a three-dimensional vector.

$$\begin{aligned} H_0&: \mathbf{r} = \mathbf{m}_0 + \mathbf{n}, \\ H_1&: \mathbf{r} = \mathbf{m}_1 + \mathbf{n}, \\ H_2&: \mathbf{r} = \mathbf{m}_2 + \mathbf{n}, \\ H_3&: \mathbf{r} = \mathbf{m}_3 + \mathbf{n}, \\ \mathbf{m}_1 &= +A, 0, B, \\ \mathbf{m}_2 &= 0, +A, B, \\ \mathbf{m}_3 &= -A, 0, B, \\ \mathbf{m}_4 &= 0, -A, B. \end{aligned}$$

The components of the noise vector are independent, identically distributed Gaussian variables, $N(0, \sigma)$. We have K independent observations. Assume a minimum Pr(ϵ) criterion and equally-likely hypotheses. Sketch the decision region and compute the Pr(ϵ).

Problem 2.6.7. Consider the following detection problem. Under either hypothesis the observation is a *two*-dimensional vector $\mathbf{r}$.

Under H_1:

$$\mathbf{r} \triangleq \begin{bmatrix} r_1 \\ r_2 \end{bmatrix} = \begin{bmatrix} x_1 \\ x_2 \end{bmatrix} + \begin{bmatrix} n_1 \\ n_2 \end{bmatrix} = \mathbf{x} + \mathbf{n}.$$

Under H_0:

$$\mathbf{r} \triangleq \begin{bmatrix} r_1 \\ r_2 \end{bmatrix} = \begin{bmatrix} y_1 \\ y_2 \end{bmatrix} + \begin{bmatrix} n_1 \\ n_2 \end{bmatrix} = \mathbf{y} + \mathbf{n}.$$

The signal vectors $\mathbf{x}$ and $\mathbf{y}$ are known. The length of the signal vector is constrained to equal $\sqrt{E}$ under both hypotheses; that is,

$$x_1^2 + x_2^2 = E,$$
$$y_1^2 + y_2^2 = E.$$

The noises are *correlated* Gaussian variables.

$$p_{n_1 n_2}(N_1, N_2) = \frac{1}{2\pi\sigma^2(1-\rho^2)^{1/2}} \exp\left(-\frac{N_1^2 - 2\rho N_1 N_2 + N_2^2}{2\sigma^2(1-\rho^2)}\right).$$

1. Find a sufficient statistic for a likelihood ratio test. Call this statistic $l(\mathbf{R})$. We have already shown that the quantity

$$d^2 = \frac{[E(l|H_1) - E(l|H_0)]^2}{\operatorname{Var}(l|H_0)}$$

characterizes the performance of the test in a monotone fashion.

2. Choose $\mathbf{x}$ and $\mathbf{y}$ to maximize d^2. Does the answer depend on ρ?
3. Call the d^2 obtained by using the best $\mathbf{x}$ and $\mathbf{y}$, d_0^2. Calculate d_0^2 for $\rho = -1, 0$, and draw a rough sketch of d_0^2 as ρ varies from -1 through 0 to 1.
4. Explain why the performance curve in part 3 is intuitively correct.

ESTIMATION

Problem 2.6.8. The observation is an N-dimensional vector

$$\mathbf{r} = \mathbf{a} + \mathbf{n},$$

where $\mathbf{a}$ is $N(\mathbf{0}, \mathbf{K_a})$, $\mathbf{n}$ is $N(\mathbf{0}, \mathbf{K_n})$, and $\mathbf{a}$ and $\mathbf{n}$ are statistically independent.

1. Find $\hat{\mathbf{a}}_{\text{map}}$. *Hint.* Use the properties of $\nabla_{\mathbf{a}}$ developed in Problems 2.4.27 and 2.4.28.
2. Verify that $\hat{\mathbf{a}}_{\text{map}}$ is efficient.
3. Compute the error correlation matrix

$$\mathbf{\Lambda}_\epsilon \triangleq E[(\hat{\mathbf{a}}_{\text{ms}} - \mathbf{a})(\hat{\mathbf{a}}_{\text{ms}} - \mathbf{a})^T].$$

Comment. Frequently this type of observation vector is obtained by sampling a random process $r(t)$ as shown below,

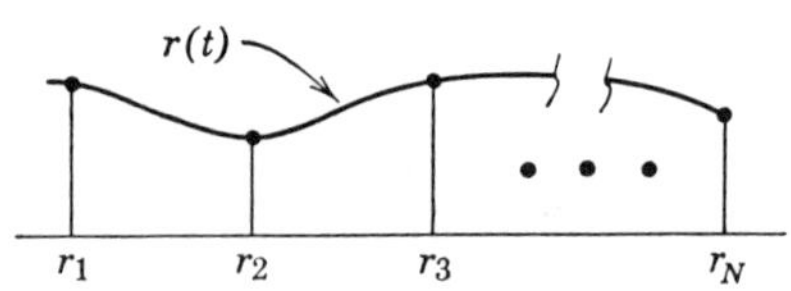

We denote the N samples by the vector $\mathbf{r}$. Using $\mathbf{r}$, we estimate the samples of $a(t)$ which are denoted by a_i. An error of interest is the sum of the squares of errors in estimating the a_i.

$$a_{\epsilon_i} = \hat{a}_i - a,$$

then

$$\xi_I \triangleq E\left[\sum_{i=1}^{N} (\hat{a}_i - a)^2\right] = E\left(\sum_{i=1}^{N} a_{\epsilon_i}^2\right) = E(\mathbf{a}_\epsilon^T \mathbf{a}_\epsilon) = \operatorname{Tr}(\mathbf{\Lambda}_\epsilon).$$

Problem 2.6.9 (continuation). Consider the special case

$$\mathbf{K}_n = \sigma_n^2 \mathbf{I}.$$

1. Verify that

$$\hat{\mathbf{a}}_{ms} = (\sigma_n^2\mathbf{I} + \mathbf{K}_a)^{-1}\mathbf{K}_a\mathbf{R}.$$

2. Now recall the detection problem described in Case 1 on p. 107. Verify that

$$l(\mathbf{R}) = \frac{1}{\sigma_n^2}\mathbf{R}^T\hat{\mathbf{a}}_{ms}.$$

Draw a block diagram of the processor. Observe that this is identical to the "unequal mean-equal covariance" case, except the mean $\mathbf{m}$ has been replaced by the mean-square estimate of the mean, $\hat{\mathbf{a}}_{ms}$.

3. What is the mean-square estimation error ξ_I?

Problem 2.6.10. Consider an alternate approach to Problem 2.6.8.

$$\mathbf{r} = \mathbf{a} + \mathbf{n},$$

where $\mathbf{a}$ is $N(\mathbf{0}, \mathbf{K_a})$ and $\mathbf{n}$ is $N(\mathbf{0}, \sigma_n^2\mathbf{I})$. Pass $\mathbf{r}$ through the matrix operation $\mathbf{W}$, which is defined in (369). The eigenvectors are those of $\mathbf{K}_a$.

$$\mathbf{r}' \triangleq \mathbf{W}\mathbf{r} = \mathbf{x} + \mathbf{n}'$$

1. Verify that $\mathbf{W}\mathbf{W}^T = \mathbf{I}$.
2. What are the statistics of $\mathbf{x}$ and $\mathbf{n}'$?
3. Find $\hat{\mathbf{x}}$. Verify that

$$\hat{x}_i = \frac{\lambda_i}{\lambda_i + \sigma_n^2} R_i',$$

where λ_i are the eigenvalues of $\mathbf{K_a}$.

4. Express $\hat{\mathbf{a}}$ in terms of a linear transformation of $\hat{\mathbf{x}}$. Draw a block diagram of the over-all estimator.
5. Prove

$$\xi_I \triangleq E[\mathbf{a}_\epsilon^T\mathbf{a}_\epsilon] = \sigma_n^2 \sum_{i=1}^{N} \frac{\lambda_i}{\lambda_i + \sigma_n^2}.$$

Problem 2.6.11 (Nonlinear Estimation). In the general Gaussian nonlinear estimation problem

$$\mathbf{r} = \mathbf{s}(\mathbf{A}) + \mathbf{n},$$

where $\mathbf{s}(\mathbf{A})$ is a nonlinear function of $\mathbf{A}$. The noise $\mathbf{n}$ is Gaussian $N(\mathbf{0}, \mathbf{K_n})$ and independent of $\mathbf{a}$.

1. Verify that

$$p_{\mathbf{r}|\mathbf{s}(\mathbf{A})}(\mathbf{R}|\mathbf{s}(\mathbf{A})) = [(2\pi)^{N/2}|\mathbf{K_n}|^{1/2}]^{-1} \exp[-\tfrac{1}{2}(\mathbf{R}^T - \mathbf{s}^T(\mathbf{A}))\mathbf{Q_n}(\mathbf{R} - \mathbf{s}(\mathbf{A}))].$$

2. Assume that $\mathbf{a}$ is a Gaussian vector $N(\mathbf{0}, \mathbf{K_a})$. Find an expression for $\ln p_{\mathbf{r},\mathbf{a}}(\mathbf{R}, \mathbf{A})$.
3. Using the properties of the derivative matrix $\nabla_{\mathbf{a}}$ derived in Problems 2.4.27 and 2.4.28, find the MAP equation.

Problem 2.6.12 (Optimum Discrete Linear Filter). Assume that we have a sequence of scalar observations $r_1, r_2, r_3, \ldots, r_K$, where $r_i = a_i + n_i$ and

$$E(a_i) = E(n_i) = 0,$$

$$E(\mathbf{r}\mathbf{r}^T) = \mathbf{\Lambda_r}, \qquad (N \times N),$$
$$E(\mathbf{r}a_i) = \mathbf{\Lambda}_{\mathbf{r}a_i}, \qquad (N \times 1).$$

We want to estimate a_K by using a realizable discrete linear filter. Thus

$$\hat{a}_K = \sum_{i=1}^{K} h_i R_i = \mathbf{h}^T\mathbf{R}.$$

Define the mean-square point estimation error as

$$\xi_P \triangleq E\{[\hat{a}_K(\mathbf{R}) - a_K]^2\}.$$

1. Use $\nabla_{\mathbf{h}}$ to find the discrete linear filter that minimizes ξ_P.
2. Find ξ_P for the optimum filter.
3. Consider the special case in which **a** and **n** are statistically independent. Find **h** and ξ_P.
4. How is $\hat{a}_K(\mathbf{R})$ for part 3 related to $\hat{\mathbf{a}}_{\text{map}}$ in Problem 2.6.8.

Note. No assumption about Gaussianness has been used.

Sequential Estimation

Problem 2.6.13. Frequently the observations are obtained in a time-sequence, $r_1, r_2, r_3, \ldots, r_N$. We want to estimate the k-dimensional parameter **a** in a sequential manner.

The ith observation is

$$r_i = \mathbf{C}\mathbf{a} + w_i, \qquad i = 1, 2, \ldots, N,$$

where **C** is a known $1 \times k$ matrix. The noises w_i are independent, identically distributed Gaussian variables $N(0, \sigma_n)$. The a priori knowledge is that **a** is Gaussian, $N(\mathbf{m}_0, \mathbf{\Lambda}_{\mathbf{a}})$.

1. Find $p_{\mathbf{a}|r_1}(\mathbf{A}|R_1)$.
2. Find the minimum mean-square estimate $\hat{\mathbf{a}}_1$ and the error correlation matrix $\mathbf{\Lambda}_{\epsilon_1}$. Put your answer in the form

$$p_{\mathbf{a}|r_1}(\mathbf{A}|R_1) = c \exp\left[-\tfrac{1}{2}(\mathbf{A} - \hat{\mathbf{a}}_1)^T\mathbf{\Lambda}_{\epsilon_1}^{-1}(\mathbf{A} - \hat{\mathbf{a}}_1)\right],$$

where

$$\mathbf{\Lambda}_{\epsilon_1}^{-1} = \mathbf{\Lambda}_{\mathbf{a}}^{-1} + \mathbf{C}^T\sigma_n^{-2}\mathbf{C}$$

and

$$\hat{\mathbf{a}}_1 = \mathbf{m}_0 + \frac{1}{\sigma_n^2}\mathbf{\Lambda}_{\epsilon_1}\mathbf{C}^T(R_1 - \mathbf{C}\mathbf{m}_0).$$

3. Draw a block diagram of the optimum processor.
4. Now proceed to the second observation R_2. What is the a priori density for this observation? Write the equations for $p_{\mathbf{a}|r_1,r_2}(\mathbf{A}|r_1, r_2)$, $\mathbf{\Lambda}_{\epsilon_2}^{-1}$, and $\hat{\mathbf{a}}_2$ in the same format as above.
5. Draw a block diagram of the sequential estimator and indicate exactly what must be stored at the end of each estimate.

Problem 2.6.14. Problem 2.6.13 can be generalized by allowing each observation to be an m-dimensional vector. The ith observation is

$$\mathbf{r}_i = \mathbf{C}\mathbf{a} + \mathbf{w}_i,$$

where **C** is a known $m \times k$ matrix. The noise vectors $\mathbf{w}_i$ are independent, identically distributed Gaussian vectors, $N(\mathbf{0}, \mathbf{\Lambda}_{\mathbf{w}})$, where $\mathbf{\Lambda}_{\mathbf{w}}$ is positive-definite.

Repeat Problem 2.6.13 for this model. Verify that

$$\hat{\mathbf{a}}_i = \hat{\mathbf{a}}_{i-1} + \mathbf{\Lambda}_{\epsilon_i}\mathbf{C}^T\mathbf{\Lambda}_{\mathbf{w}}^{-1}(\mathbf{R}_i - \mathbf{C}\hat{\mathbf{a}}_{i-1})$$

and

$$\mathbf{\Lambda}_{\epsilon_i}^{-1} = \mathbf{\Lambda}_{\epsilon_{i-1}}^{-1} + \mathbf{C}^T\mathbf{\Lambda}_{\mathbf{w}}^{-1}\mathbf{C}.$$

Draw a block diagram of the optimum processor.

Problem 2.6.15. **Discrete Kalman Filter.** Now consider the case in which the parameter $\mathbf{a}$ changes according to the equation

$$\mathbf{a}_{k+1} = \mathbf{\Phi a}_k + \mathbf{\Gamma u}_k, \qquad k = 1, 2, 3, \ldots,$$

where $\mathbf{a}_1$ is $N(\mathbf{m}_0, \mathbf{P}_0)$, $\mathbf{\Phi}$ is an $n \times n$ matrix (known), $\mathbf{\Gamma}$ is an $n \times p$ matrix (known), $\mathbf{u}_k$ is $N(\mathbf{0}, \mathbf{Q})$, and $\mathbf{u}_k$ is independent of $\mathbf{u}_j$ for $j \neq k$. The observation process is

$$\mathbf{r}_k = \mathbf{Ca}_k + \mathbf{w}_k, \qquad k = 1, 2, 3, \ldots,$$

where $\mathbf{C}$ is an $m \times n$ matrix, $\mathbf{w}_k$ is $N(\mathbf{0}, \mathbf{\Lambda_w})$ and the $\mathbf{w}_k$ are independent of each other and $\mathbf{u}_j$.

PART I. We first estimate $\mathbf{a}_1$, using a mean-square error criterion.

1. Write $p_{\mathbf{a}_1|\mathbf{r}_1}(\mathbf{A}_1|\mathbf{R}_1)$.
2. Use the $\nabla_{\mathbf{a}_1}$ operator to obtain $\hat{\mathbf{a}}_1$.
3. Verify that $\hat{\mathbf{a}}_1$ is efficient.
4. Use $\nabla_{\mathbf{a}_1}\{[\nabla_{\mathbf{a}_1}(\ln p_{\mathbf{a}_1|\mathbf{r}_1}(\mathbf{A}_1|\mathbf{R}_1))]^T\}$ to find the error covariance matrix $\mathbf{P}_1$,

where

$$\mathbf{P}_i \triangleq E[(\hat{\mathbf{a}}_i - \mathbf{a}_i)(\hat{\mathbf{a}}_i - \mathbf{a}_i)^T], \qquad i = 1, 2, \ldots.$$

Check.

$$\hat{\mathbf{a}}_1 = \mathbf{m}_0 + \mathbf{P}_1\mathbf{C}^T\mathbf{\Lambda_w}^{-1}[\mathbf{R} - \mathbf{Cm}_0]$$

and

$$\mathbf{P}_1^{-1} = \mathbf{P}_0^{-1} + \mathbf{C}^T\mathbf{\Lambda_w}^{-1}\mathbf{C}.$$

PART II. Now we estimate $\mathbf{a}_2$.

1. Verify that

$$p_{\mathbf{a}_2|\mathbf{r}_1,\mathbf{r}_2}(\mathbf{A}_2|\mathbf{R}_1, \mathbf{R}_2) = \frac{p_{\mathbf{r}_2|\mathbf{a}_2}(\mathbf{R}_2|\mathbf{A}_2)\, p_{\mathbf{a}_2|\mathbf{r}_1}(\mathbf{A}_2|\mathbf{R}_1)}{p_{\mathbf{r}_2|\mathbf{r}_1}(\mathbf{R}_2|\mathbf{R}_1)}.$$

2. Verify that $p_{\mathbf{a}_2|\mathbf{r}_1}(\mathbf{A}_2|\mathbf{R}_1)$ is $N(\mathbf{\Phi}\hat{\mathbf{a}}_1, \mathbf{M}_2)$, where

$$\mathbf{M}_2 \triangleq \mathbf{\Phi P}_1\mathbf{\Phi}^T + \mathbf{\Gamma Q \Gamma}^T.$$

3. Find $\hat{\mathbf{a}}_2$ and $\mathbf{P}_2$.

Check.

$$\hat{\mathbf{a}}_2 = \mathbf{\Phi}\hat{\mathbf{a}}_1 + \mathbf{P}_2\mathbf{C}^T\mathbf{\Lambda_w}^{-1}(\mathbf{R}_2 - \mathbf{C\Phi}\hat{\mathbf{a}}_1),$$
$$\mathbf{P}_2^{-1} = \mathbf{M}_2^{-1} + \mathbf{C}^T\mathbf{\Lambda_w}^{-1}\mathbf{C}.$$

4. Write

$$\mathbf{P}_2 = \mathbf{M}_2 - \mathbf{B}$$

and verify that $\mathbf{B}$ must equal

$$\mathbf{B} = \mathbf{M}_2\mathbf{C}^T(\mathbf{CM}_2\mathbf{C}^T + \mathbf{\Lambda_w})^{-1}\mathbf{CM}_2.$$

5. Verify that the answer to part 3 can be written as

$$\hat{\mathbf{a}}_2 = \mathbf{\Phi}\hat{\mathbf{a}}_1 + \mathbf{M}_2\mathbf{C}^T(\mathbf{CM}_2\mathbf{C}^T + \mathbf{\Lambda_w})^{-1}(\mathbf{R}_2 - \mathbf{C\Phi}\hat{\mathbf{a}}_1).$$

Compare the two forms with respect to ease of computation. What is the dimension of the matrix to be inverted?

PART III

1. Extend the results of Parts I and II to find an expression for $\hat{\mathbf{a}}_k$ and $\mathbf{P}_k$ in terms of $\hat{\mathbf{a}}_{k-1}$ and $\mathbf{M}_k$. The resulting equations are called the Kalman filter equations for discrete systems [38].
2. Draw a block diagram of the optimum processor.

PART IV. Verify that the Kalman filter reduces to the result in Problem 2.6.13 when $\mathbf{\Phi} = \mathbf{I}$ and $\mathbf{Q} = \mathbf{0}$.

Special Applications

A large number of problems in the areas of pattern recognition, learning systems, and system equalization are mathematically equivalent to the general Gaussian problem. We consider three simple problems (due to M. E. Austin) in this section. Other examples more complex in detail but not in concept are contained in the various references.

Problem 2.6.16. Pattern Recognition. A pattern recognition system is to be implemented for the classification of noisy samples taken from a set of M patterns. Each pattern may be represented by a set of parameters in which the mth pattern is characterized by the vector $\mathbf{s}_m$. In general, the $\mathbf{s}_m$ vectors are unknown. The samples to be classified are of the form

$$\mathbf{x} = \mathbf{s}_m + \mathbf{n},$$

where the $\mathbf{s}_m$ are assumed to be independent Gaussian random variables with mean $\bar{\mathbf{s}}_m$ and covariance $\mathbf{\Lambda}_m$ and $\mathbf{n}$ is assumed to be zero-mean Gaussian with covariance $\mathbf{\Lambda}_\mathbf{n}$ independent from sample to sample, and independent of $\mathbf{s}_m$.

1. In order to classify the patterns the recognition systems needs to know the pattern characteristics. We provide it with a "learning" sample:

$$\mathbf{x}_m = \mathbf{s}_m + \mathbf{n},$$

where the system knows that the mth pattern is present.

Show that if J learning samples, $\mathbf{x}_m^{(1)}, \mathbf{x}_m^{(2)}, \ldots, \mathbf{x}_m^{(J)}$, of the form $\mathbf{x}_m^{(j)} = \mathbf{s}_m + \mathbf{n}^{(j)}$ are available for each $m = 1, \ldots, M$, the pattern recognition system need store only the quantities

$$\mathbf{l}_m = \frac{1}{J}\sum_{j=1}^{J} \mathbf{x}_m{}^{(j)}$$

for use in classifying additional noisy samples; that is, show that the $\mathbf{l}_m$, $m = 1, \ldots M$' form a set of sufficient statistics extracted from the MJ learning samples.

2. What is the MAP estimate of $\mathbf{s}_m$? What is the covariance of this estimate as a function of J, the number of learning samples?

3. For the special case of two patterns ($M = 2$) characterized by unknown scalars s_1 and s_2, which have a priori densities $N(\bar{s}_1, \sigma)$ and $N(\bar{s}_2, \sigma)$, respectively, find the optimum decision rule for equiprobable patterns and observe that this approaches the decision rule of the "known patterns" classifier asymptotically with increasing number of learning samples J.

Problem 2.6.17. Intersymbol Interference. Data samples are to be transmitted over a known dispersive channel with an impulse response $h(t)$ in the presence of white Gaussian noise. The received waveform

$$r(t) = \sum_{k=-K}^{K} \xi_k h(t - kT) + n(t)$$

may be passed through a filter matched to the channel impulse response to give a set of numbers

$$a_j = \int r(t)\, h(t - jT)\, dt$$

for $j = 0, \pm 1, \pm 2, \ldots, \pm K$, which forms a set of sufficient statistics in the MAP estimation of the ξ_k. (This is proved in Chapter 4.) We denote the sampled channel autocorrelation function as

$$b_j = \int h(t)\, h(t - jT)\, dt$$

and the noise at the matched filter output as

$$n_j = \int n(t)\, h(t - jT)\, dt.$$

The problem then reduces to an estimation of the ξ_k, given a set of relations

$$a_j = \sum_{k=-K}^{K} \xi_k b_{j-k} + n_j \qquad \text{for } j, k = 0, \pm 1, \pm 2, \dots \pm K.$$

Using obvious notation, we may write these equations as

$$\mathbf{a} = \mathbf{B}\boldsymbol{\xi} + \mathbf{n}.$$

1. Show that if $n(t)$ has double-sided spectral height $\frac{1}{2}N_0$, that the noise vector $\mathbf{n}$ has a covariance matrix $\mathbf{\Lambda_n} = \frac{1}{2}N_0\mathbf{B}$.
2. If the ξ_k are zero-mean Gaussian random variables with covariance matrix $\mathbf{\Lambda}_{\boldsymbol{\xi}}$ show that the MAP estimate of $\boldsymbol{\xi}$ is of the form $\boldsymbol{\xi} = \mathbf{Ga}$ and therefore that $\hat{\boldsymbol{\xi}}_0 = \mathbf{g}^T\mathbf{a}$. Find $\mathbf{g}$ and note that the estimate of ξ_0 can be obtained by passing the sufficient statistics into a tapped delay line with tap gains equal to the elements of $\mathbf{g}$. This cascading of a matched filter followed by a sampler and a transversal filter is a well-known equalization method employed to reduce intersymbol interference in digital communication via dispersive media.

Problem 2.6.18. Determine the MAP estimate of ξ_0 in Problem 2.6.17; assuming further that the ξ_k are independent and that the ξ_k are known (say through a "teacher" or infallible estimation process) for $k < 0$. Show then that the weighting of the sufficient statistics is of the form

$$\hat{\xi}_0 = \sum_{j>0} g_j a_j - \sum_{j<0} f_j \xi_j$$

and find g_j and f_j. This receiver may be interpreted as passing the sampled matched-filter output through a transversal filter with tap gains g_j and subtracting the output from a second transversal filter whose input is the sequence of ξ_k which estimates have been made. Of course, in implementation such a receiver would be self-taught by using its earlier estimates as correct in the above estimation equation.

Problem No. 2.6.19. Let

$$z = \mathbf{G}^T\mathbf{r}$$

and assume that z is $N(m_z, \sigma_z)$ for all finite $\mathbf{G}$.

1. What is $M_z(jv)$? Express your result in terms of m_z and σ_z.
2. Rewrite the result in (1) in terms of $\mathbf{G}$, $\mathbf{m}$, and $\mathbf{\Lambda_r}$ [see (316)–(317) for definitions].
3. Observe that

$$M_z(ju) \triangleq E[e^{juz}] = E[e^{ju\mathbf{G}^T\mathbf{r}}]$$

and

$$M_{\mathbf{r}}(j\mathbf{v}) \triangleq E[e^{j\mathbf{v}^T\mathbf{r}}]$$

and therefore

$$M_z(ju) = M_{\mathbf{r}}(j\mathbf{v}) \quad \text{if} \quad \mathbf{G}u = \mathbf{v}.$$

Use these observations to verify (317).

Problem No. 2.6.20 (continuation).

(a) Assume that the $\mathbf{\Lambda_r}$ defined in (316) is positive definite. Verify that the expression for $p_{\mathbf{r}}(\mathbf{R})$ in (318) is correct. [*Hint.* Use the diagonalizing transformation $\mathbf{W}$ defined in (368).]

(b) How must (318) be modified if $\mathbf{\Lambda_r}$ is singular? What does this singularity imply about the components of $\mathbf{r}$?

P2.7 Performance Bounds and Approximations

Problem 2.7.1. Consider the binary test with N independent observations, r_i, where

$$p_{r_i|H_k} = N(m_k, \sigma_k), \qquad k = 0, 1, \quad i = 1, 2, \dots, N.$$

Find $\mu(s)$.

Problem 2.7.2 (continuation). Consider the special case of Problem 2.7.1 in which

$$m_0 = 0,$$
$$\sigma_0^2 = \sigma_n^2,$$

and

$$\sigma_1^2 = \sigma_s^2 + \sigma_n^2.$$

1. Find $\mu(s)$, $\dot{\mu}(s)$, and $\ddot{\mu}(s)$.
2. Assuming equally likely hypotheses, find an upper bound on the minimum $\Pr(\epsilon)$.
3. With the assumption in part 2, find an approximate expression for the $\Pr(\epsilon)$ that is valid for large N.

Problem 2.7.3. A special case of the binary Gaussian problem with $\mathbf{N}$ observations is

$$p_{\mathbf{r}|H_k}(\mathbf{R}|H_k) = \frac{1}{(2\pi)^{N/2}|\mathbf{K}_k|^{1/2}} \exp\left(-\frac{\mathbf{R}^T\mathbf{K}_k^{-1}\mathbf{R}}{2}\right), \qquad k = 0, 1.$$

1. Find $\mu(s)$.
2. Express it in terms of the eigenvalues of the appropriate matrices.

Problem 2.7.4 (continuation). Consider the special case in which

$$\mathbf{K}_0 = \sigma_n^2\mathbf{I}$$

and

$$\mathbf{K}_1 = \mathbf{K}_s + \mathbf{K}_0.$$

Find $\mu(s)$, $\dot{\mu}(s)$, $\ddot{\mu}(s)$.

Problem 2.7.5 (alternate continuation of 2.7.3). Consider the special case in which $\mathbf{K}_1$ and $\mathbf{K}_0$ are partitioned into the 4 $N \times N$ matrices given by (422) and (423).

1. Find $\mu(s)$.
2. Assume that the hypotheses are equally likely and that the criterion is minimum $\Pr(\epsilon)$. Find a bound on the $\Pr(\epsilon)$.
3. Find an approximate expression for the $\Pr(\epsilon)$.

Problem 2.7.6. The general binary Gaussian problem for N observations is

$$p_{\mathbf{r}|H_k}(\mathbf{R}|H_k) = \frac{1}{(2\pi)^{N/2}|\mathbf{K}_k|^{1/2}} \exp\left[-\frac{(\mathbf{R}^T - \mathbf{m}_k^T)\mathbf{K}_k^{-1}(\mathbf{R} - \mathbf{m}_k)}{2}\right], \qquad k = 0, 1.$$

Find $\mu(s)$.

Problem 2.7.7. Consider Example 3A on p. 130. A bound on the $\Pr(\epsilon)$ is

$$\Pr(\epsilon) \leq \tfrac{1}{2}\left[\frac{(1 + \sigma_s^2/\sigma_n^2)}{(1 + \sigma_s^2/2\sigma_n^2)^2}\right]^{N/2}$$

1. Constrain $N\sigma_s^2/\sigma_n^2 = x$. Find the value of N that minimizes the bound.
2. Evaluate the approximate expression in (516) for this value of N.

Problem 2.7.8. We derived the Chernoff bound in (461) by using tilted densities. This approach prepared us for the central limit theorem argument in the second part of our discussion. If we are interested only in (461), a much simpler derivation is possible.

1. Consider a function of the random variable x which we denote as $f(x)$. Assume

$$f(x) \geq 0, \qquad \text{all } x,$$

$$f(x) \geq f(X_0) > 0, \qquad \text{all } x \geq X_0.$$

Prove

$$\Pr[x \geq X_0] \leq \frac{E[f(x)]}{f(X_0)}.$$

2. Now let

$$f(x) = e^{sx}, \qquad s \geq 0,$$

and

$$X_0 = \gamma.$$

Use the result in (1) to derive (457). What restrictions on γ are needed to obtain (461)?

Problem 2.7.9. The reason for using tilted densities and Chernoff bounds is that a straightforward application of the central limit theorem gives misleading results when the region of interest is on the tail of the density. A trivial example taken from [4-18] illustrates this point.

Consider a set of statistically independent random variables x_i which assumes values 0 and 1 with equal probability. We are interested in the probability

$$\Pr\left[y_N = \frac{1}{N}\sum_{i=1}^{N} x_i \geq 1\right] \triangleq \Pr[A_N].$$

(a) Define a standardized variable

$$z \triangleq \frac{y_N - \bar{y}_N}{\sigma_{y_N}}.$$

Use a central limit theorem argument to estimate $\Pr[A_N]$. Denote this estimate as $\hat{\Pr}[A_N]$.

(b) Calculate $\Pr[A_N]$ exactly.

(c) Verify that the fractional error is,

$$\frac{\hat{\Pr}[A_N]}{\Pr[A_N]} \propto e^{0.19N}$$

Observe that the fractional error grows exponentially with N.

(d) Estimate $\Pr[A_N]$ using the Chernoff bound of Problem 2.7.8. Denote this estimate as $\Pr_c[A_N]$. Compute $\dfrac{\Pr_c[A_N]}{\Pr[A_N]}$.

REFERENCES

[1] W. B. Davenport and W. L. Root, *Random Signals and Noise*, McGraw-Hill, New York, 1958.

[2] A. T. Bharucha-Reid, *Elements of the Theory of Markov Processes and their Applications*, McGraw-Hill, New York, 1960.

[3] M. Abramovitz and I. A. Stegun (ed.), *Handbook of Mathematical Functions*, National Bureau of Standards, U.S. Government Printing Office, Washington, D.C., June 1964.

[4] J. A. Greenwood and H. O. Hartley, *Guides to Tables in Mathematical Statistics*, Princeton University Press, Princeton, New Jersey, 1962.

[5] R. A. Fisher, "Theory of Statistical Estimation," *Proc. Cambridge Phil. Soc.*, **22**, 700 (1925).

[6] R. A. Fisher, "On the Mathematical Foundations of Theoretical Statistics," *Phil. Trans. Roy. Soc., London*, **222**, 309 (1922).

[7] R. A. Fisher, "Two New Properties of Mathematical Likelihood," *Proc. Roy. Soc., London*, **144**, 285 (1934).

[8] R. A. Fisher, "The Logic of Inductive Inference," *J. Roy. Statist. Soc.*, **98**, 39 (1935).

[9] H. Cramér, *Mathematical Methods of Statistics*, Princeton University Press, Princeton, New Jersey, 1946.

[10] S. S. Wilks, *Mathematical Statistics*, Wiley, New York, 1962.

[11] M. G. Kendall and A. Stuart, *The Advanced Theory of Statistics, Vol. 2, Inference and Relationship*, Hafner, New York, 1961.

[12] C. R. Rao, "Information and Accuracy Attainable in the Estimation of Statistical Parameters," *Bull. Calcutta Math. Soc.*, **37**, 81–91 (1945).

[13] A. Bhattacharyya, "On Some Analogues of the Amount of Information and their Use in Statistical Estimation," *Sankhya*, **8**, 1, 201, 315 (1946, 1947, 1948).

[14] H. L. Van Trees, "A Generalized Bhattacharyya Bound," Internal Memo., Detection & Estimation Theory Group, MIT, 1966.

[15] E. W. Barankin, "Locally Best Unbiased Estimates," *Ann. Math. Stat.*, **20**, 477 (1949).

[16] R. Bellman, *Introduction to Matrix Analysis*, McGraw-Hill, New York, 1960.

[17] E. L. Lehmann, *Testing Statistical Hypotheses*, Wiley, New York, 1959.

[18] F. B. Hildebrand, *Methods of Applied Mathematics*, Prentice-Hall, Englewood Cliffs, New Jersey, 1952.

[19] R. Fisher and F. Yates, *Statistical Tables for Biological, Agricultural, and Medical Research*, Oliver & Boyd, Edinburgh, 1953.

[20] S. Sherman, "Non-mean-square Error Criteria," *IRE Trans. Information Theory*, **IT-4**, No. 3, 125–126 (1958).

[21] K. Pearson, *Tables of the Incomplete Γ-Function*, Cambridge University Press, Cambridge, 1934.

[22] J. N. Pierce, "Theoretical Diversity Improvement in Frequency-Shift Keying," *Proc. IRE*, **46**, 903–910 (May 1958).

[23] C. E. Shannon, "Seminar Notes for Seminar in Information Theory," MIT, 1956 (unpublished).

[24] R. M. Fano, *Transmission of Information*, MIT Press, Cambridge, Massachusetts, and Wiley, New York, 1961.

[25] C. E. Shannon, R. G. Gallager, and E. R. Berlekamp, "Lower Bounds to Error Probability for Coding on Discrete Memoryless Channels: I," *Information and Control*, Vol. **10**, No. 1, 65–103 (1967).

[26] R. G. Gallager, "Lower Bounds on the Tails of Probability Distributions," MIT, RLE, QPR 77, 277–291 (April 1965).

[27] I. M. Jacobs, "Probability-of-Error Bounds for Binary Transmission on the Slow Fading Rician Channel," *IEEE Trans. Information Theory*, Vol. **IT-12**, No. 4, Oct, 1966.

[28] H. Chernoff, "A Measure of Asymptotic Efficiency for Tests of a Hypothesi based on the Sum of Observations," *Annals Math. Stat.*, **23**, 493–507 (1962).

[29] A. Bhattacharyya, "On a Measure of Divergence Between Two Statistical Populations defined by their Probability Distributions," *Bull. Calcutta Math. Soc.*, **35**, No. 3, 99–110 (1943).

[30] W. Feller, *An Introduction to Probability Theory and Its Applications*, Vol. I, Wiley, New York, 1950, 1957.

[31] D. Dugué, "Application des Proprietes de la Limite au Sens du Calcul des Probabilities a L'etude des Diverses Questions D'estimation," *Ecol. Poly.*, **3**, No. 4, 305–372 (1937).

[32] A. Sommerfeld, *An Introduction to the Geometry of N Dimensions*, Dutton, New York, 1929.

[33] W. Feller, *An Introduction to Probability Theory and Its Applications*, Vol. II, Wiley, New York, 1966.

[34] H. L. Yudkin, "An Error Bound for Gaussian Signals in Gaussian Noise," MIT, RLE, QPR 73, April 15, 1964.

[35] T. J. Goblick, "Study of An Orbiting Dipole Belt Communication System," Lincoln Laboratory, Technical Report 369, December 22, 1964.

[36] A. J. Viterbi, *Principles of Coherent Communication*, McGraw-Hill, New York, 1966.

[37] K. Pearson, "On the Systematic Fitting of Curves to Observations and Measurements," *Biometrika*, **1**, 265 (1902).

[38] R. E. Kalman, "A New Approach to Linear Filtering and Prediction Problems," *J. Basic Eng.* (ASME Trans.) **82D**, 35–45 (1960).

3

Representations of Random Processes

3.1 INTRODUCTION

In this chapter we discuss briefly some of the methods of characterizing random processes that we need for the remainder of the book. The essential idea that we want to emphasize is straightforward. There are many alternate ways of characterizing waveforms and random processes, but the best depends heavily on the problem that we are trying to solve. An intelligent characterization frequently makes the problem solution almost trivial.

Several methods of characterizing signals come immediately to mind. The first is a time-domain characterization. A typical signal made up of pulses of various heights is shown in Fig. 3.1. A time-domain characterization describes the signal shape clearly.

Is it a good representation? To answer this question we must specify what we are going to do with the signal. In Fig. 3.2 we illustrate two possible cases. In the first we pass the signal through a limiter and want to

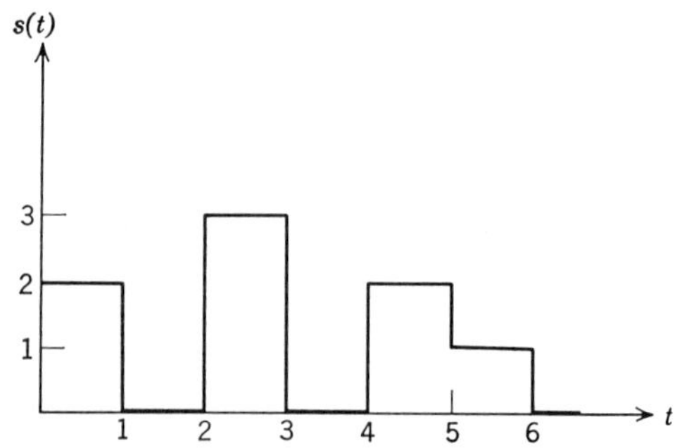

Fig. 3.1 A typical signal.

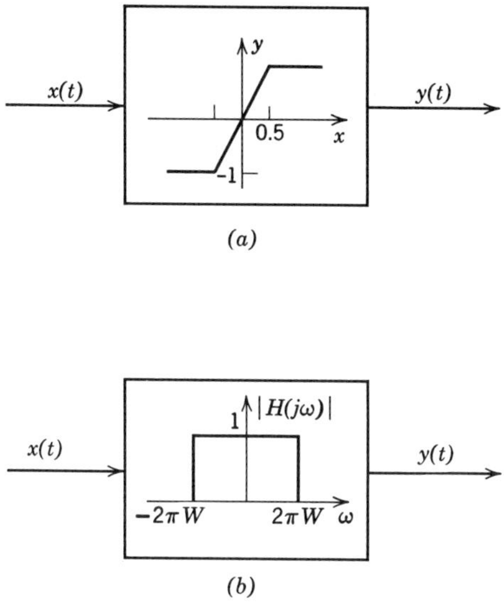

Fig. 3.2 Operations on signals

calculate the output. The time domain characterization enables us to find the output by inspection. In the second we pass the signal through an ideal low-pass filter and want to calculate the energy in the output. In this case a time-domain approach is difficult. If, however, we take the Fourier transform of $s(t)$,

$$S(j\omega) = \int_{-\infty}^{\infty} s(t)e^{-j\omega t}\,dt, \tag{1}$$

the resulting problem is straightforward. The energy in $y(t)$ is E_y, where

$$E_y = 2\int_0^{2\pi W} |S(j\omega)|^2 \frac{d\omega}{2\pi}. \tag{2}$$

Thus, as we well know, both the time-domain and frequency-domain descriptions play an important role in system analysis. The point of the example is that the most efficient characterization depends on the problem of interest.

To motivate another method of characterization consider the simple communication systems shown in Fig. 3.3. When hypothesis 1 is true, the deterministic signal $s_1(t)$ is transmitted. When hypothesis 0 is true, the signal $s_2(t)$ is transmitted. The particular transmitted waveforms are different in systems A, B, and C. The noise in each idealized system is constructed by multiplying the two deterministic waveforms by independent, zero-mean, Gaussian random variables and adding the resulting waveforms. The noise waveform will have a different shape in each system.

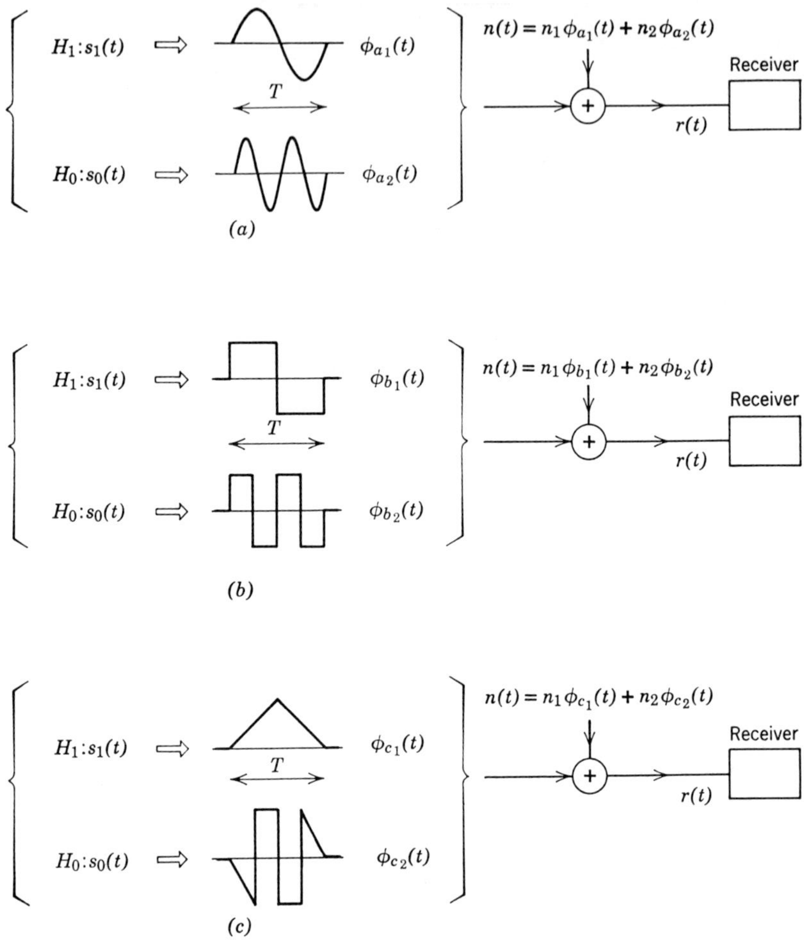

Fig. 3.3 Three hypothetical communication systems.

The receiver wants to decide which hypothesis is true. We see that the transmitted signal and additive noise are appreciably different waveforms in systems *A*, *B*, and *C*. In all cases, however, they can be written as

$$\begin{aligned} s_1(t) &= s_1\phi_1(t), & 0 \le t \le T,\\ s_2(t) &= s_2\phi_2(t), & 0 \le t \le T,\\ n(t) &= n_1\phi_1(t) + n_2\phi_2(t), & 0 \le t \le T, \end{aligned} \tag{3}$$

where the functions $\phi_1(t)$ and $\phi_2(t)$ are *orthonormal*; that is,

$$\int_0^T \phi_i(t)\,\phi_j(t)\,dt = \delta_{ij}, \qquad i, j = 1, 2. \tag{4}$$

The functions $\phi_1(t)$ and $\phi_2(t)$ are different in the three systems. It is clear that because

$$r(t) = (s_1 + n_1)\,\phi_1(t) + n_2\,\phi_2(t), \qquad 0 \le t \le T : H_1,$$
$$r(t) = n_1\,\phi_1(t) + (s_2 + n_2)\,\phi_2(t), \qquad 0 \le t \le T : H_0, \tag{5}$$

we must base our decision on the observed value of the coefficients of the two functions. Thus the test can be viewed as

$$\mathbf{r} \triangleq \begin{bmatrix} r_1 \\ r_2 \end{bmatrix} = \begin{bmatrix} s_1 \\ 0 \end{bmatrix} + \begin{bmatrix} n_1 \\ n_2 \end{bmatrix}, \qquad : H_1,$$
$$\mathbf{r} \triangleq \begin{bmatrix} r_1 \\ r_2 \end{bmatrix} = \begin{bmatrix} 0 \\ s_2 \end{bmatrix} + \begin{bmatrix} n_1 \\ n_2 \end{bmatrix}, \qquad : H_0. \tag{6}$$

This, however, is just a problem in classical detection that we encountered in Chapter 2.

The important observation is that any pair of orthonormal functions $\phi_1(t)$ and $\phi_2(t)$ will give the same detection performance. Therefore either a time-domain or frequency-domain characterization will tend to obscure the significant features of this particular problem. We refer to this third method of characterization as an *orthogonal series* representation.

We develop this method of characterizing both deterministic signals and random processes in this chapter. In the next section we discuss deterministic signals.

3.2 DETERMINISTIC FUNCTIONS: ORTHOGONAL REPRESENTATIONS

Consider the function $x(t)$ which is defined over the interval $[0, T]$ as shown in Fig. 3.4. We assume that the energy in the function has some finite value E_x.

$$E_x = \int_0^T x^2(t)\, dt < \infty. \tag{7}$$

Now the sketch implies one way of specifying $x(t)$. For every t we know the value of the function $x(t)$. Alternately, we may wish to specify $x(t)$ by a countable set of numbers.

The simple example in the last section suggests writing

$$x(t) = \sum_{i=1}^{\infty} x_i\, \phi_i(t), \tag{8†}$$

† Throughout most of our discussion we are concerned with expanding real waveforms using real orthonormal functions and real coefficients. The modifications to include complex orthonormal functions and coefficients are straightforward.

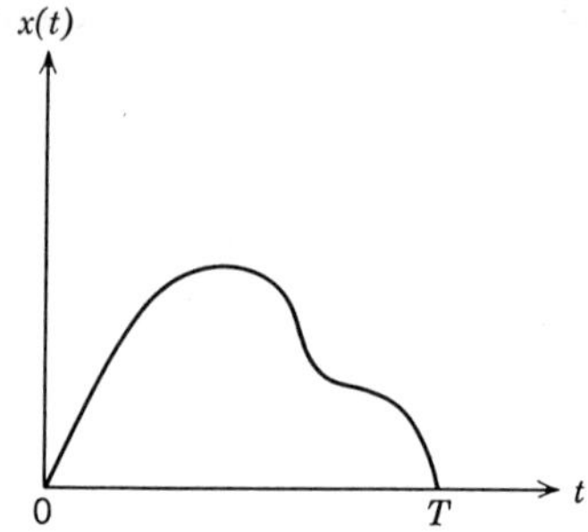

Fig. 3.4 A time-limited function.

where the $\phi_i(t)$ are some set of orthonormal functions. For example, we could choose a set of sines and cosines

$$\begin{aligned} \phi_1(t) &= \left(\frac{1}{T}\right)^{1/2}, \\ \phi_2(t) &= \left(\frac{2}{T}\right)^{1/2} \cos\left(\frac{2\pi}{T} t\right), \\ \phi_3(t) &= \left(\frac{2}{T}\right)^{1/2} \sin\left(\frac{2\pi}{T} t\right), \\ &\vdots \\ \phi_{2n}(t) &= \left(\frac{2}{T}\right)^{1/2} \cos\left(\frac{2\pi}{T} nt\right). \end{aligned} \qquad 0 \le t \le T. \tag{9}$$

Several mathematical and practical questions come to mind. The mathematical questions are the following:

1. Because it is only practical to use a finite number (N) of coefficients, how should we choose the coefficients to minimize the mean-square approximation (or representation) error?
2. As N increases, we would like the mean-square approximation error to go to zero. When does this happen?

The practical question is this:

If we receive $x(t)$ as a voltage waveform, how can we generate the coefficients experimentally?

First we consider the mathematical questions. The representation error is

$$e_N(t) = x(t) - \sum_{i=1}^{N} x_i\, \phi_i(t), \tag{10}$$

when we use N terms. The energy in the error is

$$E_e(N) \triangleq \int_0^T e_N^2(t)\, dt = \int_0^T \left[x(t) - \sum_{i=1}^{N} x_i\, \phi_i(t)\right]^2 dt. \tag{11}$$

We want to minimize this energy for any N by choosing the x_i appropriately. By differentiating with respect to some particular x_j, setting the result equal to zero, and solving, we obtain

$$x_j = \int_0^T x(t)\, \phi_j(t)\, dt. \tag{12}$$

Because the second derivative is a positive constant, the x_j given by (12) provides an absolute minimum. The choice of coefficient does not change as N is increased because of the orthonormality of the functions.

Finally, we look at the energy in the representation error as $N \to \infty$.

$$\begin{aligned} E_e(N) &\triangleq \int_0^T e_N^2(t)\, dt = \int_0^T \left[x(t) - \sum_{i=1}^N x_i\, \phi_i(t)\right]^2 dt \\ &= E_x - 2\sum_{i=1}^N \int_0^T x(t) x_i\, \phi_i(t)\, dt + \int_0^T \sum_{i=1}^N \sum_{j=1}^N x_i x_j\, \phi_i(t)\, \phi_j(t)\, dt \\ &= E_x - \sum_{i=1}^N x_i^2. \end{aligned} \tag{13}$$

Because the x_i^2 are nonnegative, the error is a monotone-decreasing function of N.

If,

$$\lim_{N\to\infty} E_e(N) = 0 \tag{14}$$

for all $x(t)$ with finite energy, we say that the $\phi_i(t)$, $i = 1, \ldots,$ are a *complete orthonormal* (CON) *set* over the interval $[0, T]$ for the class of functions with finite energy. The importance of completeness is clear. If we are willing to use more coefficients, the representation error decreases. In general, we want to be able to decrease the energy in the error to any desired value by letting N become large enough.

We observe that for CON sets

$$E_x = \sum_{i=1}^{\infty} x_i^2. \tag{15}$$

Equation 15 is just Parseval's theorem. We also observe that x_i^2 represents the energy in a particular component of the signal.

Two possible ways of generating the coefficients are shown in Fig. 3.5. In the first system we multiply $x(t)$ by $\phi_i(t)$ and integrate over $[0, T]$. This is referred to as a correlation operation. In the second we pass $x(t)$ into a set of linear filters with impulse responses $h_i(\tau) = \phi_i(T - \tau)$ and observe the outputs at time T. We see that the sampled output of the ith filter is

$$\int_0^T x(\tau)\, h_i(T - \tau)\, d\tau.$$

For the particular impulse response used this is x_i,

$$x_i = \int_0^T x(\tau)\, \phi_i(\tau)\, d\tau, \qquad i = 1, 2, \ldots, N. \tag{16}$$

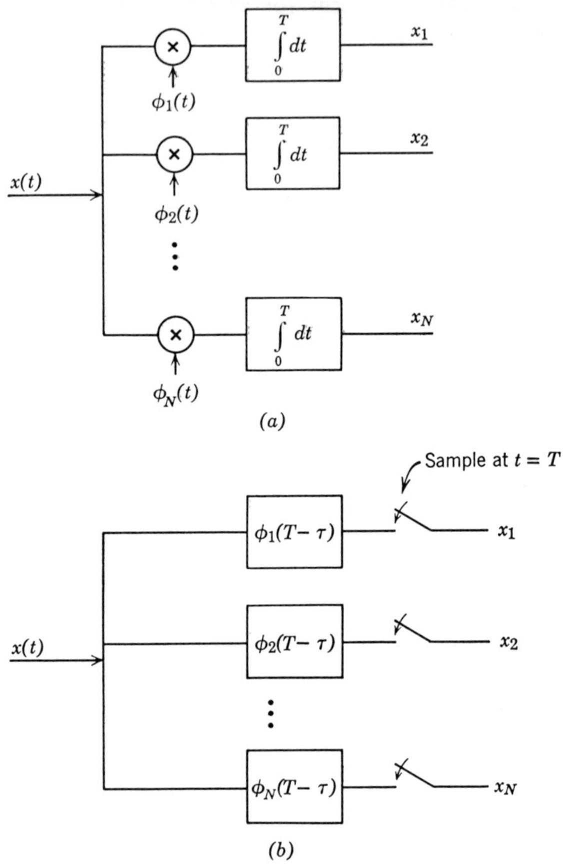

Fig. 3.5 Generation of expansion coefficients: (*a*) correlation operation; (*b*) filter operation.

In Chapter 2 we saw that it was convenient to consider N observations as a point in an N-dimensional space. We shall find that it is equally useful to think of the N coefficients as defining a point in a space. For arbitrary signals we may need an infinite dimensional space. Thus any finite energy signal can be represented as a vector. In Fig. 3.6 we show two signals—$s_1(t)$ and $s_2(t)$:

$$s_1(t) = \sum_{i=1}^{3} s_{1i}\, \phi_i(t),$$
$$s_2(t) = \sum_{i=1}^{3} s_{2i}\, \phi_i(t). \tag{17a}$$

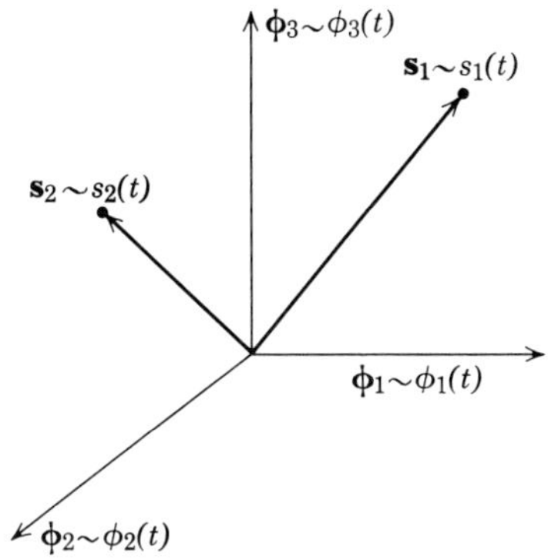

Fig. 3.6 Representation of a signal as a vector.

The corresponding signal vectors are

$$\mathbf{s}_1 \triangleq \begin{bmatrix} s_{11} \\ s_{12} \\ s_{13} \end{bmatrix} = \sum_{i=1}^{3} s_{1i}\boldsymbol{\phi}_i,$$

$$\mathbf{s}_2 \triangleq \begin{bmatrix} s_{21} \\ s_{22} \\ s_{23} \end{bmatrix} = \sum_{i=1}^{3} s_{2i}\boldsymbol{\phi}_i. \tag{17b}$$

Several observations follow immediately:

1. The length of the signal vector squared equals the energy in the signal.

$$\begin{aligned} |\mathbf{s}_1|^2 &= E_1, \\ |\mathbf{s}_2|^2 &= E_2. \end{aligned} \tag{18}$$

2. The correlation coefficient between two signals is defined as

$$\rho_{12} \triangleq \frac{\int_0^T s_1(t)\, s_2(t)\, dt}{\sqrt{E_1 E_2}}. \tag{19}$$

Substituting (17*a*) into (19), we have

$$\rho_{12} = \frac{\int_0^T \left[\sum_{i=1}^{3} s_{1i}\, \phi_i(t)\right]\left[\sum_{j=1}^{3} s_{2j}\, \phi_j(t)\right] dt}{\sqrt{E_1 E_2}}. \tag{20}$$

Using the orthonormality of the coordinate functions the integral reduces to

$$\rho_{12} = \frac{\sum_{i=1}^{3} s_{1i}s_{2i}}{\sqrt{E_1 E_2}}. \tag{21}$$

The numerator is just the dot product of $\mathbf{s}_1$ and $\mathbf{s}_2$. Using (18) in the denominator, we obtain,

$$\rho_{12} = \frac{\mathbf{s}_1 \cdot \mathbf{s}_2}{|\mathbf{s}_1|\,|\mathbf{s}_2|}. \tag{22}$$

The obvious advantage of the vector space interpretation is that it enables us to use familiar geometric ideas in dealing with waveforms.

We now extend these ideas to random waveforms.

3.3 RANDOM PROCESS CHARACTERIZATION

We begin our discussion in this section by reviewing briefly how random processes are conventionally defined and characterized.

3.3.1 Random Processes: Conventional Characterizations

The basic idea of a random process is familiar. Each time we conduct an experiment the outcome is a function over an interval of time instead of just a single number. Our mathematical model is illustrated in Fig. 3.7. Each point in the sample space Ω maps into a time function. We could write the function that came from ω_i as $x(t, \omega_i)$ to emphasize its origin, but it is easier to denote it simply as $x(t)$. The collection of waveforms generated from the points in Ω are referred to as an *ensemble*. If we look down the ensemble at any one time, say t_1, we will have a random variable $x_{t_1} \triangleq x(t_1, \omega)$. Similarly, at other times t_i we have random variables x_{t_i}.

Clearly, we could characterize any particular random variable x_{t_i} by its probability density. A more difficult question is how to characterize the entire process. There is an obvious property that this characterization should have. If we consider a set of times $t_1, t_2, \ldots, t_n$ in the interval in which the process is defined, there are n random variables $x_{t_1}, x_{t_2}, x_{t_3}, \ldots, x_{t_n}$. Any *complete* characterization should be able to specify the joint density $p_{x_{t_1}x_{t_2}\cdots x_{t_n}}(X_1, X_2, \ldots, X_n)$. Furthermore, it should be able to specify this density for any set of n times in the interval (for any finite n).

Unfortunately, it is not obvious that a characterization of this kind will be adequate to answer all questions of interest about a random process. Even if it does turn out to be adequate, there is a practical difficulty in actually specifying these densities for an arbitrary random process.

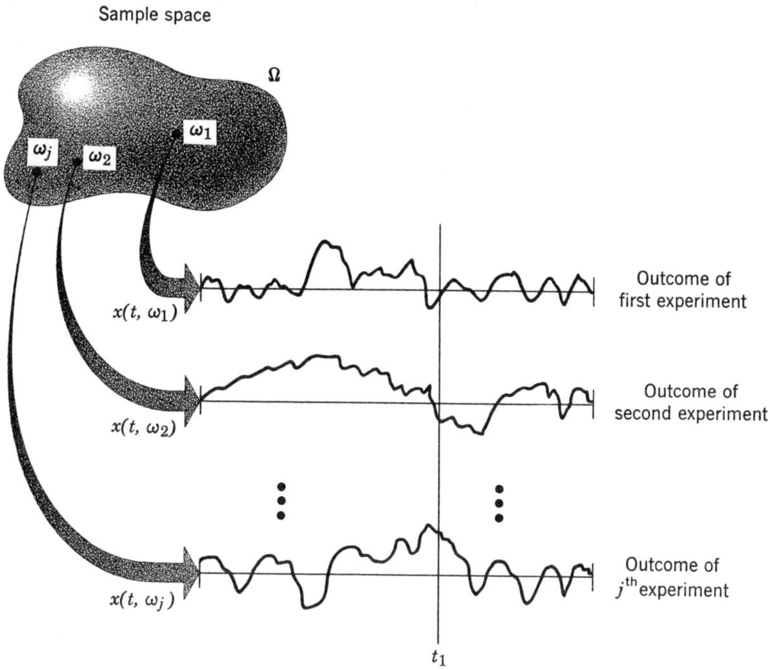

Fig. 3.7 An ensemble of sample functions.

There are two common ways of handling this difficulty in specifying the nth-order density.

Structured Processes. We consider *only* those processes in which any nth-order density has a certain structure that can be produced by using some low-order density and a known algorithm.

Example. Consider the probability density at the ordered set of times

$$t_1 < t_2 < t_3 < \cdots < t_{n-1} < t_n.$$

If

$$p_{x_{t_n}|x_{t_{n-1}}\cdots x_{t_1}}(X_{t_n}|X_{t_{n-1}}\cdots X_{t_1}) = p_{x_{t_n}|x_{t_{n-1}}}(X_{t_n}|X_{t_{n-1}}), \tag{23}$$

the process is called a Markov process. Here knowledge of the second-order density enables us to construct the nth order density (e.g., [2, p. 44] or Problems 3.3.9 and 3.3.10). Other structured processes will appear naturally as our discussion proceeds.

Partial Characterization. We now consider operations on the random process that can be studied without actually completely characterizing the process. For these operations we need only a partial characterization. A large number of partial characterizations are possible. Two of the most widely used are the following:

1. Single-time characterizations.
2. Second-moment characterizations.

In a single-time characterization we specify only $p_{x_t}(X)$, the first-order probability density at time t. In general, it will be a function of time. A simple example illustrates the usefulness of this characterization.

Example. Let

$$r(t) = x(t) + n(t). \tag{24}$$

Assume that x_t and n_t are statistically independent and $p_{x_t}(X)$ and $p_{n_t}(N)$ are known. We operate on $r(t)$ with a no-memory nonlinear device to obtain a minimum mean square error estimate of $x(t)$ which we denote by $\hat{x}(t)$.

From Chapter 2, $\hat{x}(t)$ is just the conditional mean. Because we are constrained to a no-memory operation, we can use only $r(t)$. Then

$$\hat{x}(t) = \int_{-\infty}^{\infty} X_t p_{x_t|r_t}(X_t|R_t)\, dX_t \triangleq f(R_t). \tag{25}$$

If x_t is Gaussian, $N(0, \sigma_x)$, and n_t is Gaussian, $N(0, \sigma_n)$, it is a simple exercise (cf. Problem 3.3.2) to show that

$$f(R_t) = \frac{\sigma_x^2}{\sigma_x^2 + \sigma_n^2} R_t, \tag{26}$$

so that the no-memory device happens to be linear. Observe that because we allowed only a no-memory device a complete characterization of the process was not necessary.

In a second-moment characterization we specify only the first and second moments of the process. We define the mean-value function of the process as

$$m_x(t) \triangleq E(x_t) = \int_{-\infty}^{\infty} X_t p_{x_t}(X_t)\, dX_t. \tag{27}$$

In general, this is a function of time. The correlation function is defined as

$$R_x(t, u) \triangleq E(x_t x_u) = \iint_{-\infty}^{\infty} X_t X_u p_{x_t x_u}(X_t, X_u)\, dX_t\, dX_u. \tag{28}$$

The covariance function is defined as

$$\begin{aligned} K_x(t, u) &\triangleq E\{[x_t - m_x(t)][x_u - m_x(u)]\} \\ &= R_x(t, u) - m_x(t)\, m_x(u). \end{aligned} \tag{29}$$

This partial characterization is well suited to *linear* operations on random processes. This type of application is familiar (e.g., [1], pp. 171–185).

The covariance function has several properties of interest to us. Looking at the definition in (29), we see that it is symmetric:

$$K_x(t, u) = K_x(u, t). \tag{30}$$

If we multiply a sample function $x(t)$ by some deterministic square-integrable function $f(t)$ and integrate over the interval $[0, T]$, we obtain a random variable:

$$x_f \triangleq \int_0^T x(t)\, f(t)\, dt. \tag{31}$$

The mean of this random variable is

$$E(x_f) = \bar{x}_f \triangleq E \int_0^T x(t)\, f(t)\, dt = \int_0^T m_x(t)\, f(t)\, dt, \tag{32}$$

and the variance is

$$\begin{aligned} \text{Var}\,(x_f) &\triangleq E[(x_f - \bar{x}_f)^2] \\ &= E\left\{\int_0^T [x(t) - m_x(t)]\, f(t)\, dt \int_0^T [x(u) - m_x(u)]\, f(u)\, du\right\}. \end{aligned} \tag{33}$$

Bringing the expectation inside the integral, we have

$$\text{Var}\,(x_f) = \iint_0^T f(t)\, K_x(t, u)\, f(u)\, dt\, du. \tag{34}$$

The variance must be greater than or equal to zero. Thus, we have shown that

$$\boxed{\iint_0^T f(t)\, K_x(t, u)\, f(u)\, dt\, du \geq 0} \tag{35}$$

for any $f(t)$ with finite energy. We call this property nonnegative definiteness. If the inequality is strict for every $f(t)$ with nonzero finite energy, we say that $K_x(t, u)$ is positive definite. We shall need the two properties in (30) and (35) in the next section.

If the process is defined over an infinite interval and the covariance function depends only on $|t - u|$ and not t or u individually, we say that the process is *covariance-stationary* and write

$$K_x(t, u) = K_x(t - u) = K_x(\tau). \tag{36†}$$

Similarly, if the correlation function depends only on $|t - u|$, we say that the process is *correlation-stationary* and write

$$R_x(t, u) = R_x(t - u) = R_x(\tau). \tag{37}$$

† It is important to observe that although $K_x(t, u)$ is a function of two variables and $K_x(\tau)$ of only one variable, we use the same notation for both. This economizes on symbols and should cause no confusion.

For stationary processes, a characterization using the power density spectrum $S_x(\omega)$ is equivalent to the correlation function characterization

$$S_x(\omega) \triangleq \int_{-\infty}^{\infty} R_x(\tau) e^{-j\omega\tau}\, d\tau$$

and

$$R_x(\tau) = \int_{-\infty}^{\infty} S_x(\omega) e^{j\omega\tau}\, \frac{d\omega}{2\pi}. \tag{38}$$

As already pointed out, these partial characterizations are useful only when the operations performed on the random process are constrained to have a certain form. A much more useful representation for the problems of interest to us is a characterization in terms of an orthogonal series expansion. In the next section we use a series expansion to develop a second-moment characterization. In the succeeding section we extend it to provide a complete characterization for a particular process of interest. It is worthwhile to observe that we have yet to commit ourselves in regard to a complete characterization of a random process.

3.3.2 Series Representation of Sample Functions of Random Processes

In Section 3.2 we saw how we could represent a deterministic waveform with finite energy in terms of a series expansion. We now want to extend these ideas to include sample functions of a random process. We start off by choosing an arbitrary complete orthonormal set: $\phi_1(t), \phi_2(t), \ldots$. For the moment we shall not specify the exact form of the $\phi_i(t)$. To expand $x(t)$ we write

$$x(t) = \lim_{N\to\infty} \sum_{i=1}^{N} x_i \phi_i(t), \qquad 0 \le t \le T, \tag{39}$$

where

$$x_i \triangleq \int_0^T x(t)\, \phi_i(t)\, dt. \tag{40}$$

We have not yet specified the type of convergence required of the sum on the right-hand side. Various types of convergence for sequences of random variables are discussed in the prerequisite references [1, p. 63] or [29].

An ordinary limit is not useful because this would require establishing conditions on the process to guarantee that *every* sample function could be represented in this manner.

A more practical type of convergence is mean-square convergence:

$$x(t) = \underset{N\to\infty}{\text{l.i.m.}} \sum_{i=1}^{N} x_i\, \phi_i(t), \qquad 0 \le t \le T. \tag{41}$$

The notation "l.i.m." denotes limit in the mean (e.g., [1, p. 63]) which is defined as,

$$\lim_{N\to\infty} E\left[\left(x_t - \sum_{i=1}^{N} x_i\,\phi_i(t)\right)^2\right] = 0, \qquad 0 \le t \le T. \tag{42}$$

For the moment we assume that we can find conditions on the process to guarantee the convergence indicated in (42).

Before doing so we discuss an appropriate choice for the orthonormal set. In our discussions of classical detection theory our observation space was finite dimensional and usually came with a built-in coordinate system. In Section 2.6 we found that problems were frequently easier to solve if we used a new coordinate system in which the random variables were uncorrelated (if they happened to be Gaussian variables, they were also statistically independent). In dealing with continuous waveforms we have the advantage that there is no specified coordinate system, and therefore we can choose one to suit our purposes. From our previous results a logical choice is a set of $\phi_i(t)$ that leads to *uncorrelated* coefficients.

If

$$E(x_i) \triangleq m_i, \tag{43}$$

we would like

$$E[(x_i - m_i)(x_j - m_j)] = \lambda_i \delta_{ij}. \tag{44}$$

For simplicity we assume that $m_i = 0$ for all i. Several observations are worthwhile:

1. The value x_i^2 has a simple physical interpretation. It corresponds to the *energy* along the coordinate function $\phi_i(t)$ in a particular sample function.
2. Similarly, $E(x_i^2) = \lambda_i$ corresponds to the *expected* value of the energy along $\phi_i(t)$, assuming that $m_i = 0$. Clearly, $\lambda_i \ge 0$ for all i.
3. If $K_x(t, u)$ is positive definite, every λ_i is greater than zero. This follows directly from (35). A little later it will be easy to show that if $K_x(t, u)$ is not positive definite, at least one λ_i must equal zero.

We now want to determine what the requirement in (44) implies about the complete orthogonal set. Substituting (40) into (44) and bringing the expectation inside the integral, we obtain

$$\begin{aligned}\lambda_i\delta_{ij} = E(x_i x_j) &= E\left[\int_0^T x(t)\,\phi_i(t)\,dt \int_0^T x(u)\,\phi_j(u)\,du\right]\\ &= \int_0^T \phi_i(t)\,dt \int_0^T K_x(t, u)\,\phi_j(u)\,du, \qquad \text{for all } i \text{ and } j.\end{aligned} \tag{45}$$

In order that (45) may hold for all choices of i and a particular j, it is necessary and sufficient that the inner integral equal $\lambda_j\, \phi_j(t)$:

$$\boxed{\lambda_j\, \phi_j(t) = \int_0^T K_x(t, u)\, \phi_j(u)\, du, \qquad 0 \le t \le T.} \tag{46}$$

The functions $\phi_i(t)$ are called eigenfunctions and the numbers λ_i are called eigenvalues.

Therefore we want to demonstrate that for some useful class of random processes there exist solutions to (46) with the desired properties. The form of (46) is reminiscent of the equation that specified the eigenvectors and eigenvalues in Section 2.6 (2–363),

$$\lambda \boldsymbol{\phi} = \mathbf{K}_x \boldsymbol{\phi}, \tag{47}$$

where $\mathbf{K}_x$ was a symmetric, nonnegative definite matrix. This was a set of N simultaneous homogeneous linear equations where N was the dimensionality of the observation space. Using results from linear equation theory, we saw that there were N real, nonnegative values of λ for which (47) had a nontrivial solution. Now the coordinate space is infinite and we have a homogeneous linear integral equation to solve.

The function $K_x(t, u)$ is called the kernel of the integral equation, and because it is a covariance function it is symmetric and nonnegative definite. We restrict our attention to processes with a finite mean-square value $[E(x^2(t)) < \infty]$. Their covariance functions satisfy the restriction

$$\iint_0^T K_x^2(t, u)\, dt\, du \le \int_0^T E[x^2(t)]\, dt < \infty, \tag{48}$$

where T is a finite number.

The restrictions in the last paragraph enable us to employ standard results from linear integral equation theory† (e.g., Courant and Hilbert [3], Chapter 3; Riesz and Nagy [4]; Lovitt [5]; or Tricomi [6]).

Properties of Integral Equations

1. There exist *at least one* square-integrable function $\phi(t)$ and real number $\lambda \neq 0$ that satisfy (46).

It is clear that there may not be more than one solution. For example,

$$K_x(t, u) = \sigma_f^2 f(t) f(u), \qquad 0 \le t, u \le T. \tag{49}$$

has only one nonzero eigenvalue and one normalized eigenfunction.

2. By looking at (46) we see that if $\phi_j(t)$ is a solution then $c\phi_j(t)$ is also a solution. Therefore we can always normalize the eigenfunctions.

† Here we follow Davenport and Root [1], p. 373.

3. If $\phi_1(t)$ and $\phi_2(t)$ are eigenfunctions associated with the same eigenvalue λ, then $c_1\phi_1(t) + c_2\phi_2(t)$ is also an eigenfunction associated with λ.
4. The eigenfunctions corresponding to different eigenvalues are orthogonal.
5. There is at most a countably infinite set of eigenvalues and all are bounded.
6. For any particular λ there is at most a *finite* number of linearly independent eigenfunctions. [Observe that we mean algebraic linear independence; $f(t)$ is linearly independent of the set $\phi_i(t)$, $i = 1, 2, \ldots, K$, if it cannot be written as a weighted sum of the $\phi_i(t)$.] These can always be orthonormalized (e.g., by the Gram-Schmidt procedure; see Problem 4.2.7 in Chapter 4).
7. Because $K_x(t, u)$ is nonnegative definite, the kernel $K_x(t, u)$ can be expanded in the series

$$K_x(t, u) = \sum_{i=1}^{\infty} \lambda_i \, \phi_i(t) \, \phi_i(u), \qquad 0 \le t, u \le T, \tag{50}$$

where the convergence is uniform for $0 \le t, u \le T$. (This is called Mercer's theorem.)
8. If $K_x(t, u)$ is positive definite, the eigenfunctions form a complete orthonormal set. From our results in Section 3.2 this implies that we can expand any deterministic function with finite energy in terms of the eigenfunctions.
9. If $K_x(t, u)$ is not positive definite, the eigenfunctions cannot form a complete orthonormal set. [This follows directly from (35) and (40).] Frequently, we augment the eigenfunctions with enough additional orthogonal functions to obtain a complete set. We occasionally refer to these additional functions as eigenfunctions with zero eigenvalues.
10. The sum of the eigenvalues is the expected value of the energy of the process in the interval $(0, T)$, that is,

$$E\left[\int_0^T x^2(t)\, dt\right] = \int_0^T K_x(t, t)\, dt = \sum_{i=1}^{\infty} \lambda_i.$$

(Recall that $x(t)$ is assumed to be zero-mean.)

These properties guarantee that we can find a set of $\phi_i(t)$ that leads to uncorrelated coefficients. It remains to verify the assumption that we made in (42). We denote the expected value of the error if $x(t)$ is approximated by the first N terms as $\xi_N(t)$:

$$\xi_N(t) \triangleq E\left[\left(x(t) - \sum_{i=1}^{N} x_i \, \phi_i(t)\right)^2\right]. \tag{51}$$

Evaluating the expectation, we have

$$\xi_N(t) = K_x(t, t) - 2E\left[x(t) \sum_{i=1}^{N} x_i \,\phi_i(t)\right] + E\left[\sum_{i=1}^{N} \sum_{j=1}^{N} x_i x_j \,\phi_i(t)\,\phi_j(t)\right], \tag{52}$$

$$\xi_N(t) = K_x(t, t) - 2E\left[x(t) \sum_{i=1}^{N} \left(\int_0^T x(u)\,\phi_i(u)\,du\right) \phi_i(t)\right] + \sum_{i=1}^{N} \lambda_i \,\phi_i(t)\,\phi_i(u), \tag{53}$$

$$\xi_N(t) = K_x(t, t) - 2\sum_{i=1}^{N} \left(\int_0^T K_x(t, u)\,\phi_i(u)\,du\right) \phi_i(t) + \sum_{i=1}^{N} \lambda_i \,\phi_i(t)\,\phi_i(u), \tag{54}$$

$$\xi_N(t) = K_x(t, t) - \sum_{i=1}^{N} \lambda_i \,\phi_i(t)\,\phi_i(t). \tag{55}$$

Property 7 guarantees that the sum will converge uniformly to $K_x(t, t)$ as $N \to \infty$. Therefore

$$\lim_{N \to \infty} \xi_N(t) = 0, \qquad 0 \le t \le T, \tag{56}$$

which is the desired result. (Observe that the convergence in Property 7 implies that for any $\epsilon > 0$ there exists an N_1 *independent of* t such that $\xi_N(t) < \epsilon$ for all $N > N_1$).

The series expansion we have developed in this section is generally referred to as the Karhunen-Loève expansion. (Karhunen [25], Loève [26], p. 478, and [30].) It provides a second-moment characterization in terms of uncorrelated random variables. This property, by itself, is not too important. In the next section we shall find that for a particular process of interest, the Gaussian random process, the coefficients in the expansion are statistically independent Gaussian random variables. It is in this case that the expansion finds its most important application.

3.3.3 Gaussian Processes

We now return to the question of a suitable complete characterization of a random process. We shall confine our attention to Gaussian random processes. To find a suitable definition let us recall how we defined jointly Gaussian random variables in Section 2.6. We said that the random variables $x_1, x_2, \ldots, x_N$ were jointly Gaussian if

$$y = \sum_{i=1}^{N} g_i x_i \tag{57}$$

was a Gaussian random variable for any set of g_i. In 2.6 N was finite and we required the g_i to be finite. If N is countably infinite, we require the g_i to be such that $E[y^2] < \infty$. In the random process, instead of a linear transformation on a set of random variables, we are interested in a linear functional of a random function. This suggests the following definition:

Definition. Let $x(t)$ be a random process defined over some interval $[T_\alpha, T_\beta]$ with a mean-value $m_x(t)$ and covariance function $K_x(t, u)$. If every linear functional of $x(t)$ is a Gaussian random variable, then $x(t)$ is a Gaussian random process. In other words, if

$$y = \int_{T_\alpha}^{T_\beta} g(u)\, x(u)\, du, \tag{58}$$

and $g(u)$ is any function such that $E[y^2] < \infty$. Then, in order for $x(u)$ to be a Gaussian random process, y must be a Gaussian random variable for every $g(u)$ in the above class.

Several properties follow immediately from this definition.

Property 1. The output of a linear system is a particular linear functional of interest. We denote the impulse response as $h(t, u)$, the output at time t due to a unit impulse input at time u. If the input is $x(t)$ which is a sample function from a Gaussian random process, the output $y(t)$ is also.

Proof:

$$y(t) = \int_{T_\alpha}^{T_\beta} h(t, u)\, x(u)\, du, \qquad T_\gamma \le t \le T_\Delta. \tag{59}$$

The interval $[T_\gamma, T_\Delta]$ is simply the range over which $y(t)$ is defined. We assume that $h(t, u)$ is such that $E[y^2(t)] < \infty$ for all t in $[T_\gamma, T_\Delta]$. From the definition it is clear that y_t is a Gaussian random variable. To show that $y(t)$ is a Gaussian random process we must show that any linear functional of it is a Gaussian random variable. Thus,

$$z \triangleq \int_{T_\gamma}^{T_\Delta} g_y(t)\, y(t)\, dt, \tag{60}$$

or

$$z = \int_{T_\gamma}^{T_\Delta} g_y(t)\, dt \int_{T_\alpha}^{T_\beta} h(t, u)\, x(u)\, du, \tag{61}$$

must be Gaussian for every $g_y(t)$ [such that $E[z^2] < \infty$]. Integrating with respect to t and defining the result as

$$g(u) \triangleq \int_{T_\gamma}^{T_\Delta} g_y(t)\, h(t, u)\, dt, \tag{62}$$

we have

$$z = \int_{T_\alpha}^{T_\beta} g(u)\, x(u)\, du, \tag{63}$$

which is Gaussian by definition.

Thus we have shown that if the input to a linear system is a Gaussian random process the output is a Gaussian random process.

Property 2. If

$$y_1 = \int_{T_\alpha}^{T_\beta} g_1(u)\, x(u)\, du \tag{64}$$

and

$$y_2 = \int_{T_\alpha}^{T_\beta} g_2(u)\, x(u)\, du, \tag{65}$$

where $x(u)$ is a Gaussian random process, then y_1 and y_2 are jointly Gaussian. (The proof is obvious in light of (57).)

Property 3. If

$$x_i = \int_{T_\alpha}^{T_\beta} \phi_i(u)\, x(u)\, du \tag{66}$$

and

$$x_j = \int_{T_\alpha}^{T_\beta} \phi_j(u)\, x(u)\, du, \tag{67}$$

where $\phi_i(u)$ and $\phi_j(u)$ are orthonormalized eigenfunctions of (46) [now the interval of interest is (T_α, T_β) instead of $(0, T)$] then x_i and x_j are statistically independent Gaussian random variables ($i \neq j$). Thus,

$$p_{x_i}(X_i) = \frac{1}{\sqrt{2\pi\lambda_i}} \exp\left[-\frac{(X_i - m_i)^2}{2\lambda_i}\right], \tag{68}$$

where

$$m_i \triangleq \int_{T_\alpha}^{T_\beta} m_x(t)\, \phi_i(t)\, dt. \tag{69}$$

This property follows from Property 2 and (45).

Property 4. For any set of times $t_1, t_2, t_3, \ldots, t_n$ in the interval $[T_\alpha, T_\beta]$ the random variables $x_{t_1}, x_{t_2}, \ldots, x_{t_n}$ are jointly Gaussian random variables.

Proof: If we denote the set by the vector $\mathbf{x}_t$,

$$\mathbf{x}_t \triangleq \begin{bmatrix} x_{t_1} \\ x_{t_2} \\ \vdots \\ x_{t_n} \end{bmatrix}, \tag{70}$$

whose mean is $\mathbf{m_x}$,

$$\mathbf{m_x} \triangleq E\begin{bmatrix} x_{t_1} \\ x_{t_2} \\ \vdots \\ x_{t_n} \end{bmatrix} = \begin{bmatrix} m_x(t_1) \\ m_x(t_2) \\ \vdots \\ m_x(t_n) \end{bmatrix}, \tag{71}$$

then the joint probability density is

$$p_{\mathbf{x}_t}(\mathbf{X}) = [(2\pi)^{n/2}|\mathbf{\Lambda}_{\mathbf{x}}|^{1/2}]^{-1} \exp\left[-\tfrac{1}{2}(\mathbf{X} - \mathbf{m}_{\mathbf{x}})^T\mathbf{\Lambda}_{\mathbf{x}}^{-1}(\mathbf{X} - \mathbf{m}_{\mathbf{x}})\right] \quad (72)$$

and the joint characteristic function is

$$M_{\mathbf{x}_t}(j\mathbf{v}) = \exp\left(j\mathbf{v}^T\mathbf{m}_{\mathbf{x}} - \tfrac{1}{2}\mathbf{v}^T\mathbf{\Lambda}_{\mathbf{x}}\mathbf{v}\right), \quad (73)$$

where $\mathbf{\Lambda}_{\mathbf{x}}$ is the covariance matrix of the random variables $x_{t_1}, x_{t_2}, \ldots, x_{t_n}$. (We assume $\mathbf{\Lambda}_{\mathbf{x}}$ is nonsingular.) The ij element is

$$\mathbf{\Lambda}_{\mathbf{x},ij} = E[(x_{t_i} - m_x(t_i))(x_{t_j} - m_x(t_j))]. \quad (74)$$

This property follows by using the function

$$g(u) = \sum_{i=1}^{n} g_i\, \delta(u - t_i) \quad (75)$$

in (58) and the result in (57). Thus we see that our definition has the desirable property suggested in Section 3.3.1, for it uniquely specifies the joint density at any set of times. Frequently Property 4 is used as the basic definition. The disadvantage of this approach is that it is more difficult to prove that our definition and Properties 1–3 follow from (72) than vice-versa.

The Gaussian process we have defined has two main virtues:

1. The physical mechanisms that produce many processes are such that a Gaussian model is appropriate.
2. The Gaussian process has many properties that make analytic results feasible.

Discussions of physical mechanisms that lead logically to Gaussian processes are available in [7] and [8]. Other properties of the Gaussian process which are not necessary for our main discussion, are developed in the problems (cf. Problems 3.3.12–3.3.18).

We shall encounter multiple processes that are jointly Gaussian. The definition is a straightforward extension of the preceding one.

Definition. Let $x_1(t), x_2(t), \ldots, x_N(t)$ be a set of random processes defined over the intervals $(T_{\alpha_1}, T_{\beta_1}), (T_{\alpha_2}, T_{\beta_2}), \ldots, (T_{\alpha_N}, T_{\beta_N})$, respectively. If every sum of arbitrary functionals of $x_i(t)$, $i = 1, \ldots, N$, is a Gaussian random variable, then the processes $x_1(t), x_2(t), \ldots, x_N(t)$ are defined to be jointly Gaussian random processes. In other words,

$$y = \sum_{i=1}^{N} \int_{T_{\alpha_i}}^{T_{\beta_i}} g_i(u)\, x_i(u)\, du$$

must be Gaussian for every set of $g_i(u)$ such that $E[y^2] < \infty$.

Other properties of jointly Gaussian processes are discussed in the problems.

Property 3 is the reason for our emphasis on the Karhunen-Loève expansion. It enables us to characterize a Gaussian process in terms of an at most countably infinite set of statistically independent Gaussian random variables. The significance of this will perhaps be best appreciated when we see how easy it makes our ensuing work. Observe that if we had chosen to emphasize Markov processes the orthogonal expansion method of characterization would not have been particularly useful. In Section 6.3 we discuss characterizations that emphasize the Markovian structure.

The Karhunen-Loève expansion is useful in two ways:

1. Many of our theoretical derivations use it as a tool. In the majority of these cases the eigenfunctions and eigenvalues do not appear in the final result. The integral equation that specifies them (46) need never be solved.
2. In other cases the result requires an explicit solution for one or more eigenfunctions and eigenvalues. Here we must be able to solve the equation exactly or find good approximate solutions.

In the next section we consider some useful situations in which solutions can be obtained.

3.4 HOMOGENEOUS INTEGRAL EQUATIONS AND EIGENFUNCTIONS

In this section we shall study in some detail the behavior of the solutions to (46). In addition to the obvious benefit of being able to solve for an eigenfunction when it is necessary, the discussion serves several other purposes:

1. By looking at several typical cases and finding the eigenvalues and eigenfunctions the idea of a coordinate expansion becomes somewhat easier to visualize.
2. In many cases we shall have to make approximations to get to the final result. We need to develop some feeling for what can be neglected and what is important.
3. We want to relate the behavior of the eigenvalues and eigenfunctions to more familiar ideas such as the power density spectrum.

In Section 3.4.1 we illustrate a technique that is useful whenever the random process is stationary and has a rational power density spectrum. In Section 3.4.2 we consider bandlimited stationary processes, and in Section 3.4.3 we look at an important nonstationary process. Next in Section 3.4.4 we introduce the idea of a "white" process. In Section 3.4.5, we derive the optimum linear filter for estimating a message corrupted by noise. Finally, in Section 3.4.6, we examine the asymptotic behavior of the eigenfunctions and eigenvalues for large time intervals.

3.4.1 Rational Spectra

The first set of random processes of interest are stationary and have spectra that can be written as a ratio of two polynomials in ω^2.

$$S_x(\omega) = \frac{N(\omega^2)}{D(\omega^2)}, \tag{76}$$

where $N(\omega^2)$ is a polynomial of order q in ω^2 and $D(\omega^2)$ is a polynomial of order p in ω^2. Because we assume that $x(t)$ has a finite mean-square value, $q < p$. We refer to these spectra as rational. There is a routine but tedious method of solution. The basic idea is straightforward. We convert the integral equation to a differential equation whose solution can be easily found. Then we substitute the solution back into the integral equation to satisfy the boundary conditions. We first demonstrate the technique by considering a simple example and then return to the general case and formalize the solution procedure. (Detailed discussions of similar problems are contained in Slepian [9], Youla [10], Davenport and Root [1], Laning and Battin [11], Darlington [12], Helstrom [13, 14], or Zadeh and Ragazzini [22].)

Example. Let

$$S_x(\omega) = \frac{2\alpha P}{\omega^2 + \alpha^2}, \qquad -\infty < \omega < \infty, \tag{77}$$

or

$$R_x(\tau) = P \exp(-\alpha|\tau|), \qquad -\infty < \tau < \infty. \tag{78}$$

The mean-square value of $x(t)$ is P. The integral equation of interest is

$$\int_{-T}^{T} P \exp(-\alpha|t-u|)\phi(u)\,du = \lambda\,\phi(t), \qquad -T \le t \le T. \tag{79}$$

(The algebra becomes less tedious with a symmetric interval.)

As indicated above, we solve the integral equation by finding the corresponding differential equation, solving it, and substituting it back into the integral equation. First, we rewrite (79) to eliminate the magnitude sign.

$$\lambda\,\phi(t) = \int_{-T}^{t} P \exp[-\alpha(t-u)]\,\phi(u)\,du + \int_{t}^{T} P \exp[-\alpha(u-t)]\,\phi(u)\,du. \tag{80}$$

Differentiating once, we have

$$\lambda\,\dot{\phi}(t) = -P\alpha e^{-\alpha t} \int_{-T}^{t} e^{+\alpha u}\,\phi(u)\,du + P\alpha e^{+\alpha t} \int_{t}^{T} e^{-\alpha u}\,\phi(u)\,du. \tag{81}$$

Differentiating a second time gives

$$\lambda \ddot{\phi}(t) = P\alpha^2 \int_{-T}^{T} e^{-\alpha|t-u|} \phi(u)\, du - 2P\alpha\, \phi(t); \tag{82}$$

but the first term on the right-hand side is just $\alpha^2 \lambda\, \phi(t)$. Therefore

$$\lambda \ddot{\phi}(t) = \alpha^2 \lambda\, \phi(t) - 2P\alpha\, \phi(t) \tag{83}$$

or, for $\lambda \neq 0$,

$$\ddot{\phi}(t) = \frac{\alpha^2(\lambda - 2P/\alpha)}{\lambda} \phi(t). \tag{84}$$

The solution to (83) has four possible forms corresponding to

$$\begin{aligned}
&\text{(i)} && \lambda = 0; \\
&\text{(ii)} && 0 < \lambda < \frac{2P}{\alpha}; \\
&\text{(iii)} && \lambda = \frac{2P}{\alpha}; \\
&\text{(iv)} && \lambda > \frac{2P}{\alpha}.
\end{aligned} \tag{85}$$

We can show that the integral equation cannot be satisfied for (i), (iii), and (iv). (Cf. Problem 3.4.1.)

For (ii) we may write

$$b^2 = \frac{-\alpha^2(\lambda - 2P/\alpha)}{\lambda}, \qquad 0 < b^2 < \infty. \tag{86}$$

Then

$$\phi(t) = c_1 e^{jbt} + c_2 e^{-jbt}. \tag{87}$$

Substituting (87) into (80) and performing the integration, we obtain

$$0 = e^{-\alpha t}\left[\frac{c_1 e^{-(\alpha + jb)T}}{\alpha + jb} + \frac{c_2 e^{-(\alpha - jb)T}}{\alpha - jb}\right] - e^{+\alpha t}\left[\frac{c_1 e^{-(\alpha - jb)T}}{-\alpha + jb} + \frac{c_2 e^{-(\alpha + jb)T}}{-\alpha - jb}\right]. \tag{88}$$

We can easily verify that if $c_1 \neq \pm c_2$, (88) cannot be satisfied for all time. For $c_1 = -c_2$ we require that $\tan bT = -b/\alpha$. For $c_1 = c_2$ we require $\tan bT = \alpha/b$. Combining these two equations, we have

$$\left(\tan bT + \frac{b}{\alpha}\right)\left(\tan bT - \frac{\alpha}{b}\right) = 0. \tag{89}$$

The values of b that satisfy (89) can be determined graphically as shown in Fig. 3.8. The upper set of intersections correspond to the second term in (89) and the lower set to the first term. The corresponding eigenvalues are

$$\lambda_i = \frac{2P\alpha}{\alpha^2 + b_i^2}, \qquad i = 1, 2, \ldots. \tag{90}$$

Observe that we have ordered the solutions to (89), $b_1 < b_2 < b_3 < \cdots$. From (90) we see that this orders the eigenvalues $\lambda_1 > \lambda_2 > \lambda_3 \cdots$. The odd-numbered solutions correspond to $c_1 = c_2$ and therefore

$$\phi_i(t) = \frac{1}{T^{1/2}\left(1 + \dfrac{\sin 2b_i T}{2b_i T}\right)^{1/2}} \cos b_i t, \qquad -T \leq t \leq T \qquad (i \text{ odd}). \tag{91}$$

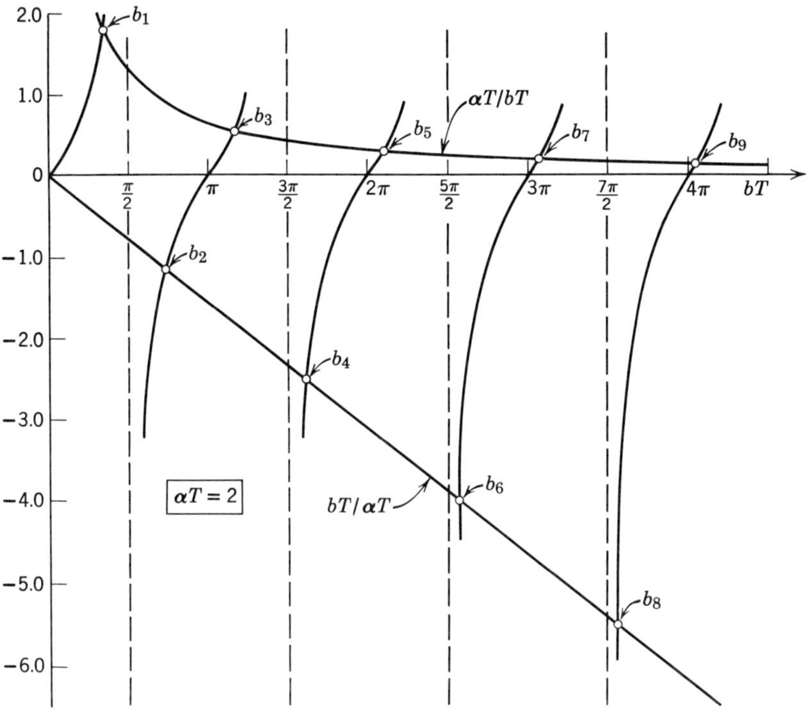

Fig. 3.8 Graphical solution of transcendental equation.

The even-numbered solutions correspond to $c_1 = -c_2$ and therefore

$$\phi_i(t) = \frac{1}{T^{1/2}\left(1 - \dfrac{\sin 2b_iT}{2b_iT}\right)^{1/2}} \sin b_i t, \qquad -T \le t \le T \qquad (i \text{ even}). \tag{92}$$

We see that the eigenfunctions are cosines and sines whose frequencies are *not* harmonically related.

Several interesting observations may be made with respect to this example:

1. The eigenvalue corresponding to a particular eigenfunction is equal to the height of the power density spectrum at that frequency.
2. As T increases, b_n decreases monotonically and therefore λ_n increases monotonically.
3. As bT increases, the upper intersections occur at approximately $(i - 1)\,\pi/2$ [i odd] and the lower intersections occur at approximately

$(i - 1)\,\pi/2$ [i even]. From (91) and (92) we see that the higher index eigenfunctions are approximately a set of periodic sines and cosines.

$$\phi_i(t) \cong \begin{cases} \dfrac{1}{T^{1/2}\left(1 + \dfrac{\sin 2b_iT}{2b_iT}\right)^{1/2}} \cos\left[\dfrac{(i-1)\pi}{2T}\,t\right], & -T \le t \le T \quad (i \text{ odd}). \\[2ex] \dfrac{1}{T^{1/2}\left(1 - \dfrac{\sin 2b_iT}{2b_iT}\right)^{1/2}} \sin\left[\dfrac{(i-1)\pi}{2T}\,t\right], & -T \le t \le T \quad (i \text{ even}). \end{cases}$$

This behavior is referred to as the asymptotic behavior.

The first observation is not true in general. In a later section (p. 204) we shall show that the λ_n are always monotonically increasing functions of T. We shall also show that the asymptotic behavior seen in this example is typical of stationary processes.

Our discussion up to this point has dealt with a particular spectrum. We now return to the general case.

It is easy to generalize the technique to arbitrary rational spectra. First we write $S_x(\omega)$ as a ratio of two polynomials,

$$S_x(\omega) = \frac{N(\omega^2)}{D(\omega^2)}. \tag{93}$$

Looking at (83), we see that the differential equation does *not* depend explicitly on T. This independence is true whenever the spectrum has the form in (93). Therefore we would obtain the same differential equation if we started with the integral equation

$$\lambda\,\phi(t) = \int_{-\infty}^{\infty} K_x(t-u)\,\phi(u)\,du, \qquad -\infty < t < \infty. \tag{94}$$

By use of Fourier transforms a formal solution to this equation follows immediately:

$$\lambda\Phi(j\omega) = S_x(\omega)\,\Phi(j\omega) = \frac{N(\omega^2)}{D(\omega^2)}\,\Phi(j\omega) \tag{95}$$

or

$$0 = [\lambda D(\omega^2) - N(\omega^2)]\,\Phi(j\omega). \tag{96}$$

There are $2p$ homogeneous solutions to the differential equation corresponding to (96) for every value of λ (corresponding to the roots of the polynomial in the bracket). We denote them as $\phi_{h_i}(t, \lambda)$, $i = 1, \ldots, 2p$. To find the solution to (46) we substitute

$$\phi(t) = \sum_{i=1}^{2p} a_i\phi_{h_i}(t, \lambda) \tag{97}$$

into the integral equation and solve for those values of λ and a_i that lead to a solution. There are no conceptual difficulties, but the procedure is tedious.†

One particular family of spectra serves as a useful model for many physical processes and also leads to tractable solutions to the integral equation for the problem under discussion. This is the family described by the equation

$$S_x(\omega:n) = \left(\frac{2nP}{\alpha}\right)\frac{\sin(\pi/2n)}{1 + (\omega/\alpha)^{2n}}. \tag{98}$$

It is referred to as the Butterworth family and is shown in Figure 3.9. When $n = 1$, we have the simple one-pole spectrum. As n increases, the attenuation versus frequency for $\omega > \alpha$ increases more rapidly. In the

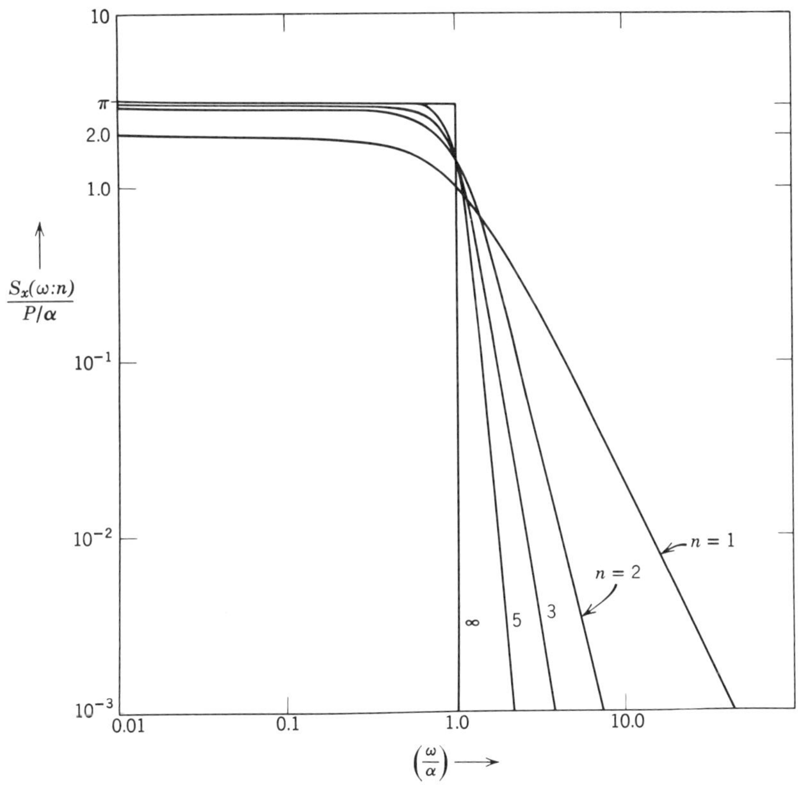

Fig. 3.9 Butterworth spectra.

† In Appendix I of Part II we shall develop a technique due to A. Baggeroer [32, 33] that is more efficient. At the present point in our discussion we lack the necessary background for the development.

limit, as $n \to \infty$, we have an ideal bandlimited spectrum. In the next section we discuss the eigenfunctions and eigenvalues for the bandlimited spectrum.

3.4.2 Bandlimited Spectra

When the spectrum is not rational, the differential equation corresponding to the integral equation will usually have time-varying coefficients. Fortunately, in many cases of interest the resulting differential equation is some canonical type whose solutions have been tabulated. An example in this category is the bandlimited spectrum shown in Fig. 3.10. In this case

$$S_x(\omega) = \begin{cases} \dfrac{\pi P}{\alpha}, & |\omega| \leq \alpha, \\ 0, & |\omega| > \alpha, \end{cases} \tag{99}$$

or, in cycles per second,

$$S_x(\omega) = \begin{cases} \dfrac{P}{2W}, & |f| \leq W, \\ 0, & |f| > W, \end{cases} \tag{100}$$

where

$$2\pi W = \alpha. \tag{101}$$

The corresponding covariance function is

$$K_x(t, u) = P \frac{\sin \alpha(t-u)}{\alpha(t-u)}. \tag{102}$$

The integral equation of interest becomes

$$\lambda\, \phi(t) = \int_{-T/2}^{+T/2} P \frac{\sin \alpha(t-u)}{\alpha(t-u)}\, \phi(u)\, du. \tag{103}$$

[This is just (46) with the interval shifted to simplify notation.]

Once again the procedure is to find a related differential equation and to examine its solution. We are, however, more interested in the results

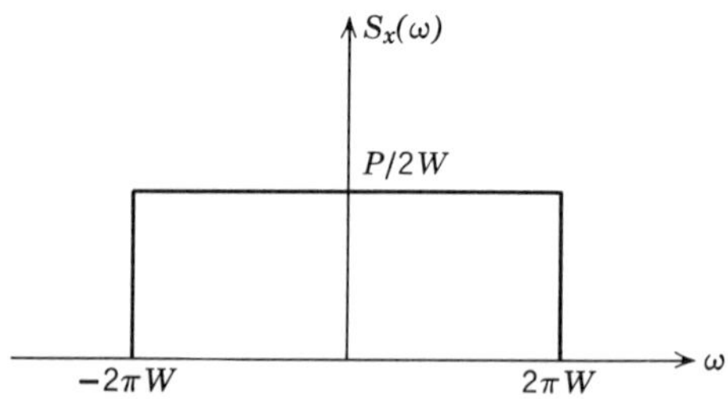

Fig. 3.10 Bandlimited spectrum.

than the detailed techniques; therefore we merely state them ([9], [15], [16], [17], and [18] are useful for further study).

The related differential equation over a normalized interval is

$$(1 - t^2)\ddot{f}(t) - 2t\dot{f}(t) + (\mu - c^2t^2) f(t) = 0, \qquad -1 < t < 1, \tag{104}$$

where

$$c = \frac{\alpha T}{2} = \pi WT \tag{105}$$

and μ is the eigenvalue. This equation has continuous solutions for certain values of $\mu(c)$. These solutions are called *angular prolate spheroidal functions* and are denoted by $S_{0n}(c, t)$, $n = 0, 1, 2, \ldots$. A plot of typical $S_{0n}(c, t)$ is contained in [15], [16], and [17]. These functions also satisfy the integral equation

$$2[R_{0n}^{(1)}(c, 1)]^2\, S_{0n}(c, t) = \int_{-1}^{+1} \frac{\sin c(t - u)}{c(t - u)}\, S_{0n}(c, u)\, du, \qquad -1 \le t \le 1, \tag{106}$$

or changing variables

$$PT\left[R_{0n}^{(1)}\left(\frac{\alpha T}{2}, 1\right)\right]^2 S_{0n}\left(\frac{\alpha T}{2}, \frac{2t}{T}\right) = \int_{-T/2}^{T/2} \frac{P \sin \alpha(t - u)}{\alpha(t - u)}\, S_{0n}\left(\frac{\alpha T}{2}, \frac{2u}{T}\right) du, \qquad -\frac{T}{2} \le t \le \frac{T}{2}, \tag{107}$$

where $R_{0n}^{(1)}(\alpha T/2, 1)$ is a *radial prolate spheroidal function*. Thus the eigenvalues are

$$\lambda_n = PT\left[R_{0n}^{(1)}\left(\frac{\alpha T}{2}, 1\right)\right]^2, \qquad n = 0, 1, 2, \ldots.$$

These functions are tabulated in several references (e.g. [18] or [19]).

The first several eigenvalues for various values of WT are shown in Figs. 3.11 and 3.12. We observe a very interesting phenomenon. For values of $n > (2WT + 1)$ the values of λ_n rapidly approach zero. We can check the total energy in the remaining eigenvalues, for

$$\sum_{i=0}^{\infty} \lambda_i = \int_{-T/2}^{+T/2} K_x(t, t)\, dt = PT. \tag{108}$$

In Fig. 3.11, $2WT = 2.55$ and the first four eigenvalues sum to $(2.54/2.55)$ PT. In Fig. 3.12, $2WT = 5.10$ and the first six eigenvalues sum to $(5.09/5.10)PT$. This behavior is discussed in detail in [17]. Our example suggests that the following statement is plausible. When a bandlimited process $[-W, W$ cps$]$ is observed over a T-second interval, there are only $(2TW + 1)$ significant eigenvalues. This result will be important to us in later chapters (specifically Chapters II-2 and II-3) when we obtain approximate solutions

$2WT = 2.55$
$\lambda_0 = 0.996 \frac{P}{2W}$
$\lambda_1 = 0.912 \frac{P}{2W}$
$\lambda_2 = 0.519 \frac{P}{2W}$
$\lambda_3 = 0.110 \frac{P}{2W}$
$\lambda_4 = 0.009 \frac{P}{2W}$
$\lambda_5 = 0.0004 \frac{P}{2W}$

Fig. 3.11 Eigenvalues for a bandlimited spectrum ($2WT = 2.55$).

by neglecting the higher eigenfunctions. More precise statements about the behavior are contained in [15], [16], and [17].

3.4.3 Nonstationary Processes

The process of interest is the simple Wiener process. It was developed as a model for Brownian motion and is discussed in detail in [20] and [21]. A typical sample function is shown in Fig. 3.13.

$2WT = 5.10$
$\lambda_0 = 1.000 \frac{P}{2W}$
$\lambda_1 = 0.999 \frac{P}{2W}$
$\lambda_2 = 0.997 \frac{P}{2W}$
$\lambda_3 = 0.961 \frac{P}{2W}$
$\lambda_4 = 0.748 \frac{P}{2W}$
$\lambda_5 = 0.321 \frac{P}{2W}$
$\lambda_6 = 0.061 \frac{P}{2W}$
$\lambda_7 = 0.006 \frac{P}{2W}$
$\lambda_8 = 0.0004 \frac{P}{2W}$

Fig. 3.12 Eigenvalues of a bandlimited spectrum ($2WT = 5.10$).

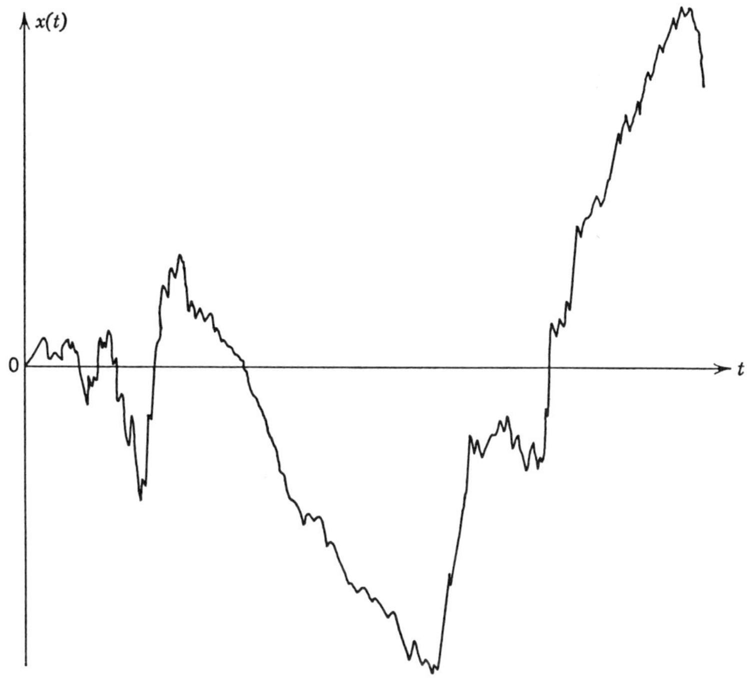

Fig. 3.13 Sample function of a Wiener process.

This process is defined for $t \geq 0$ and is characterized by the following properties:

$$x(0) = 0,$$

$$E[x(t)] = 0, \tag{109}$$

$$E[x^2(t)] = \sigma^2 t, \tag{110}$$

$$p_{x_t}(X_t) = \frac{1}{\sqrt{2\pi\sigma^2 t}} \exp\left(-\frac{X_t^{\,2}}{2\sigma^2 t}\right). \tag{111}$$

The increment variables are independent; that is, if $t_3 > t_2 > t_1$, then $(x_{t_3} - x_{t_2})$ and $(x_{t_2} - x_{t_1})$ are statistically independent. In the next example we solve (46) for the Wiener process.

Example. Wiener Process. Using the properties of the Wiener process, we can show that

$$K_x(t, u) = \sigma^2 \min(u,t) = \begin{cases} \sigma^2 u, & u \leq t, \\ \sigma^2 t & t \leq u, \end{cases} \tag{112}$$

In this case (46) becomes

$$\lambda\, \phi(t) = \int_0^T K_x(t, u)\, \phi(u)\, du, \qquad 0 \leq t \leq T. \tag{113}$$

Substituting (112) into (113)

$$\lambda\, \phi(t) = \sigma^2 \int_0^t u\, \phi(u)\, du + \sigma^2 t \int_t^T \phi(u)\, du. \tag{114}$$

Proceeding as in Section 3.4.1, we differentiate (114) and obtain,

$$\lambda\, \dot{\phi}(t) = \sigma^2 \int_t^T \phi(u)\, du. \tag{115}$$

Differentiating again, we obtain

$$\lambda \ddot{\phi}(t) = -\sigma^2\, \phi(t), \tag{116}$$

or, for $\lambda \neq 0$,

$$\ddot{\phi}(t) + \frac{\sigma^2}{\lambda}\, \phi(t) = 0. \tag{117}$$

There are three possible ranges for λ:

$$\begin{array}{lc} \text{(i)} & \lambda < 0, \\ \text{(ii)} & \lambda = 0, \\ \text{(iii)} & \lambda > 0. \end{array} \tag{118}$$

We can easily verify (cf. Problem 3.4.3) that (i) and (ii) do not provide solutions that will satisfy the integral equation. For $\lambda > 0$ we proceed exactly as in the preceding section and find

$$\lambda_n = \frac{\sigma^2 T^2}{(n - \frac{1}{2})^2 \pi^2}, \qquad n = 1, 2, \cdots \tag{119}$$

and

$$\phi_n(t) = \left(\frac{2}{T}\right)^{1/2} \sin\left[\left(n - \frac{1}{2}\right)\frac{\pi}{T} t\right] \qquad 0 \leq t \leq T. \tag{120}$$

Once again the eigenfunctions are sinusoids.

The Wiener process is important for several reasons.

1. A large class of processes can be transformed into the Wiener process
2. A large class of processes can be generated by passing a Wiener process through a linear or nonlinear system. (We discuss this in detail later.)

3.4.4 White Noise Processes

Another interesting process can be derived from the Wiener process. Using (41) and (119) we can expand $x(t)$ in a series.

$$x(t) = \underset{K \to \infty}{\text{l.i.m.}} \sum_{n=1}^{K} x_n \left(\frac{2}{T}\right)^{1/2} \sin\left[\left(n - \frac{1}{2}\right)\frac{\pi}{T} t\right], \tag{121}$$

where the mean-square value of the coefficient is given by (118):

$$E[x_n^2] = \frac{\sigma^2 T^2}{(n - \frac{1}{2})^2 \pi^2}. \tag{122}$$

We denote the K-term approximation as $x_K(t)$.

Now let us determine what happens when we differentiate $x_K(t)$:

$$\dot{x}_K(t) = \sum_{n=1}^{K} x_n\left(n - \frac{1}{2}\right)\frac{\pi}{T}\left\{\left(\frac{2}{T}\right)^{1/2} \cos\left[\left(n - \frac{1}{2}\right)\frac{\pi}{T}t\right]\right\}. \tag{123}$$

We see that the time function inside the braces is still normalized. Thus we may write

$$\dot{x}_K(t) = \sum_{n=1}^{K} w_n\left(\frac{2}{T}\right)^{1/2} \cos\left(n - \frac{1}{2}\right)\frac{\pi}{T}t, \tag{124}$$

where

$$E(w_n{}^2) = \sigma^2.$$

We observe that we have generated a process in which every eigenvalue is *equal.* Clearly, if we let $K \to \infty$, the series would not converge. If it did, it would correspond to a process with infinite energy over $[0, T]$.

We can *formally* obtain the covariance function of this resulting process by differentiating $K_x(t, u)$:

$$K_{\dot{x}}(t, u) = \frac{\partial^2}{\partial t\, \partial u} K_x(t, u) = \frac{\partial^2}{\partial t\, \partial u}\left[\sigma^2 \min(t, u)\right]$$

$$= \sigma^2\delta(t - u), \qquad 0 \le t, u \le T. \tag{125}$$

We see that the covariance function is an impulse. Still proceeding formally, we can look at the solution to the integral equation (46) for an impulse covariance function:

$$\lambda\, \phi(t) = \sigma^2 \int_0^T \delta(t - u)\, \phi(u)\, du, \qquad 0 < t < T. \tag{126}$$

The equation is satisfied for any $\phi(t)$ with $\lambda = \sigma^2$. Thus *any* set of orthonormal functions is suitable for decomposing this process. The reason for the nonuniqueness is that the impulse kernel is not square-integrable. The properties stated on pp. 180–181 assumed square-integrability.

We shall find the resulting process to be a useful artifice for many models. We summarize its properties in the following definitions.

Definition. A *Gaussian* white noise process is a Gaussian process whose covariance function is $\sigma^2\, \delta(t - u)$. It may be decomposed over the interval $[0, T]$ by using *any* set of orthonormal functions $\phi_i(t)$. The coefficients along *each* coordinate function are statistically independent Gaussian variables with equal variance σ^2.

Some related notions follow easily.

Property. We can write, formally

$$\sigma^2\,\delta(t-u) = \sum_{i=1}^{\infty} \sigma^2\,\phi_i(t)\,\phi_i(u), \qquad 0 \le t, u \le T. \tag{127}$$

or, equivalently,

$$\delta(t-u) = \sum_{i=1}^{\infty} \phi_i(t)\,\phi_i(u), \qquad 0 \le t, u \le T. \tag{128}$$

Property. If the coefficients are uncorrelated, with equal variances, but *not* Gaussian, the process is referred to as a white process.

Property. If the process is defined over the infinite interval, its spectrum is

$$S_x(\omega) = \sigma^2; \tag{129}$$

that is, it is constant over all frequencies. The value of each eigenvalue corresponds to the spectral height σ^2.

The utility of a white noise process is parallel to that of an impulse input in the analysis of linear systems. Just as we can observe an impulse only after it has been through a system with some finite bandwidth, we can observe white noise only after it has passed through a similar system. Therefore, as long as the bandwidth of the noise is appreciably larger than that of the system, it can be considered as having an infinite bandwidth.

To illustrate a typical application of eigenfunction expansions we consider a simple problem.

3.4.5 The Optimum Linear Filter

In this section we consider the problem of trying to estimate a message in the presence of interfering noise. Our treatment at this point is reasonably brief. We return to this problem and study it in detail in Chapter 6. Here we have three objectives in mind:

1. The introduction of time-varying linear filters and simple minimization techniques.
2. The development of a specific result to be used in subsequent chapters; specifically, the integral equation whose solution is the optimum linear filter.
3. The illustration of how the orthogonal expansion techniques we have just developed will enable us to obtain a formal solution to an integral equation.

The system of interest is shown in Fig. 3.14. The message $a(t)$ is a sample function from a zero-mean random process with a finite mean-square value and a covariance function $K_a(t, u)$. It is corrupted by an uncorrelated

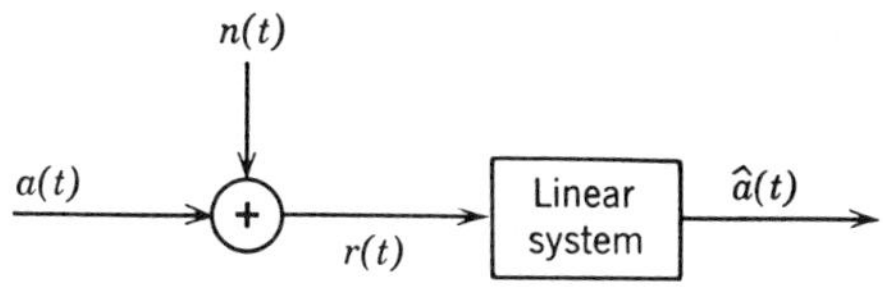

Fig. 3.14 Linear filter problem.

additive zero-mean noise $n(t)$ with covariance function $K_n(t, u)$. We observe the sum of these two processes,

$$r(t) = a(t) + n(t), \qquad 0 \le t \le T. \tag{130}$$

We pass $r(t)$ through a linear filter to obtain an estimate of $a(t)$ denoted by $\hat{a}(t)$.

Because $a(t)$ is not necessarily stationary and the observation interval is finite, we anticipate that to obtain the best estimate we may require a time-varying filter. We characterize the filter by its impulse response $h(t, u)$, which is the value of the output at time t when the input is an impulse at time u. If the system is physically realizable, then

$$h(t, u) = 0, \qquad t < u,$$

for the output cannot precede the input. If the system is time-invariant, then $h(t, u)$ depends only on the difference $(t - u)$. We assume that $r(t)$ equals zero for $t < 0$ and $t > T$. Because the system is linear, the output due to $r(t)$, $0 \le t \le T$, can be written as

$$\hat{a}(t) = \int_0^T h(t, u)\, r(u)\, du, \tag{131}$$

which is an obvious generalization of the convolution integral.

We want to choose $h(t, u)$ to minimize the mean of the squared error integrated over the interval $[0, T]$. In other words, we want to choose $h(t, u)$ to minimize the quantity

$$\begin{aligned} \xi_I &\triangleq E\left\{\frac{1}{T}\int_0^T [a(t) - \hat{a}(t)]^2\, dt\right\} \\ &= E\left\{\frac{1}{T}\int_0^T \left[a(t) - \int_0^T h(t, u)\, r(u)\, du\right]^2 dt\right\}. \end{aligned} \tag{132}$$

Thus we are minimizing the mean-square error integrated over the interval. We refer to ξ_I as the *interval estimation error.*

Similarly, we can define a *point estimation error*:

$$\xi_P(t) = E\left\{\left[a(t) - \int_0^T h(t, u)\, r(u)\, du\right]^2\right\}, \qquad 0 \le t \le T. \tag{133}$$

Clearly, if we minimize the point error at each time, the total interval error will be minimized. One way to solve this minimization problem is to use standard variational techniques (e.g., [31], Chapter 2). Our approach is less formal and leads directly to a necessary and sufficient condition. We require the filter $h(t, u)$ to be a continuous function in both variables over the area $0 \le t, u \le T$ and denote the $h(t, u)$ that minimizes $\xi_P(t)$ as $h_o(t, u)$. Any other filter function $h(t, u)$ in the allowed class can be written as

$$h(t, u) = h_o(t, u) + \epsilon h_\epsilon(t, u), \qquad 0 \le t, u \le T, \tag{134}$$

where ϵ is a real parameter and $h_\epsilon(t, u)$ is in the allowable class of filters. Taking the expectation of (133), substituting (134) into the result, and grouping terms according to the power of ϵ, we obtain

$$\begin{aligned}\xi_P(t:\epsilon) = K_a(t, t) &- 2\int_0^T h(t, u)\, K_a(t, u)\, du \\ &+ \int_0^T dv \int_0^T du\, h(t, v)\, h(t, u)\, K_r(u, v)\end{aligned} \tag{135}$$

or

$$\begin{aligned}\xi_P(t:\epsilon) = K_a(t, t) &- 2\int_0^T h_o(t, u)\, K_a(t, u)\, du \\ &+ \int_0^T dv \int_0^T du\, h_o(t, u)\, h_o(t, v)\, K_r(u, v) \\ &- 2\epsilon \int_0^T du\, h_\epsilon(t, u)\left[K_a(t, u) - \int_0^T h_o(t, v)\, K_r(u, v)\, dv\right] \\ &+ \epsilon^2 \iint_0^T h_\epsilon(t, v)\, h_\epsilon(t, u)\, K_r(u, v)\, du\, dv.\end{aligned} \tag{136}$$

If we denote the first three terms as $\xi_{P_o}(t)$ and the last two terms as $\Delta\xi(t:\epsilon)$, then (136) becomes

$$\xi_P(t:\epsilon) = \xi_{P_o}(t) + \Delta\xi(t:\epsilon). \tag{137}$$

Now, if $h_o(t, u)$ is the optimum filter, then $\Delta\xi(t:\epsilon)$ must be greater than or equal to zero for all allowable $h_\epsilon(t, u)$ and all $\epsilon \ne 0$. We show that a necessary and sufficient condition for this to be true is that

$$\boxed{K_a(t, u) - \int_0^T h_o(t, v)\, K_r(u, v)\, dv = 0, \qquad \begin{array}{l} 0 \le t \le T \\ 0 < u < T. \end{array}} \tag{138}$$

The equation for $\Delta\xi(t:\epsilon)$ is

$$\begin{aligned}\Delta\xi(t:\epsilon) = &-2\epsilon \int_0^T du\, h_\epsilon(t, u)\left[K_a(t, u) - \int_0^T h_o(t, v)\, K_r(u, v)\, dv\right] \\ &+ \epsilon^2 \iint_0^T h_\epsilon(t, v)\, h_\epsilon(t, u)\, K_r(u, v)\, du\, dv.\end{aligned} \tag{139}$$

Three observations are needed:

1. The second term is nonnegative for any choice of $h_\epsilon(t, v)$ and ϵ because $K_r(t, u)$ is nonnegative definite.
2. Unless

$$\int_0^T h_\epsilon(t, u)\left[K_a(t, u) - \int_0^T h_o(t, v)\, K_r(u, v)\, dv\right] du = 0, \tag{140}$$

there exists for every continuous $h_\epsilon(t, u)$ a range of values of ϵ that will cause $\Delta\xi(t:\epsilon)$ to be negative. Specifically, $\Delta\xi(t:\epsilon) < 0$ for all

$$0 < \epsilon < \frac{2\int_0^T h_\epsilon(t, u)\left[K_a(t, u) - \int_0^T h_o(t, v)\, K_r(u, v)\, dv\right] du}{\iint_0^T h_\epsilon(t, v)\, h_\epsilon(t, u)\, K_r(u, v)\, du\, dv} \tag{141}$$

if the numerator on the right side of (141) is positive. $\Delta\xi(t:\epsilon)$ is negative for all negative ϵ greater than the right side of (141) if the numerator is negative.

3. In order that (140) may hold, it is necessary and sufficient that the term in the bracket be identically zero for all $0 < u < T$. Thus

$$K_a(t, u) - \int_0^T h_o(t, v)\, K_r(u, v)\, dv = 0, \qquad \begin{array}{l} 0 \le t \le T \\ 0 < u < T. \end{array} \tag{142}$$

The inequality on u is strict if there is a white noise component in $r(t)$ because the second term is discontinuous at $u = 0$ and $u = T$. If (142) is not true, we can make the left side of (140) positive by choosing $h_\epsilon(t, u) > 0$ for those values of u in which the left side of (142) is greater than zero and $h_\epsilon(t, u) < 0$ elsewhere. These three observations complete the proof of (138).

The result in (138) is fundamental to many of our later problems. For the case of current interest we assume that the additive noise is white. Then

$$K_r(t, u) = \frac{N_0}{2}\, \delta(t - u) + K_a(t, u). \tag{143}$$

Substituting (143) into (138), we obtain

$$\boxed{\frac{N_0}{2}\, h_o(t, u) + \int_0^T h_o(t, v)\, K_a(u, v)\, dv = K_a(t, u), \qquad \begin{array}{l} 0 \le t \le T \\ 0 < u < T. \end{array}} \tag{144}$$

Observe that $h_o(t, 0)$ and $h_o(t, T)$ are uniquely specified by the continuity requirement

$$h_o(t, 0) = \lim_{u \to 0^+} h_o(t, u) \tag{145a}$$

$$h_o(t, T) = \lim_{u \to T^-} h_o(t, u). \tag{145b}$$

Because $a(t)$ has a finite mean-square value, (145a) and (145b) imply that (144) is also valid for $u = 0$ and $u = T$.

The resulting error for the optimum processor follows easily. It is simply the first term in (137).

$$\xi_{P_o}(t) = K_a(t, t) - 2 \int_0^T h_o(t, u)\, K_a(t, u)\, du + \iint_0^T h_o(t, u)\, h_o(t, v)\, K_r(u, v)\, du\, dv \tag{146}$$

or

$$\xi_{P_o}(t) = K_a(t, t) - \int_0^T h_o(t, u)\, K_a(t, u)\, du - \int_0^T h_o(t, u)\left[K_a(t, u) - \int_0^T h_o(t, v)\, K_r(u, v)\, dv\right] du. \tag{147}$$

But (138) implies that the term in brackets is zero. Therefore

$$\xi_{P_o}(t) = K_a(t, t) - \int_0^T h_o(t, u)\, K_a(t, u)\, du. \tag{148}$$

For the white noise case, substitution of (144) into (148) gives

$$\boxed{\xi_{P_o}(t) = \frac{N_0}{2}\, h_o(t, t).} \tag{149}$$

As a final result in our present discussion of optimum linear filters, we demonstrate how to obtain a solution to (144) in terms of the eigenvalues and eigenfunctions of $K_a(t, u)$. We begin by expanding the message covariance function in a series,

$$K_a(t, u) = \sum_{i=1}^{\infty} \lambda_i \phi_i(t)\, \phi_i(u), \tag{150}$$

where λ_i and $\phi_i(t)$ are solutions to (46) when the kernel is $K_a(t, u)$. Using (127), we can expand the white noise component in (143)

$$K_w(t, u) = \frac{N_0}{2}\, \delta(t - u) = \sum_{i=1}^{\infty} \frac{N_0}{2}\, \phi_i(t)\, \phi_i(u). \tag{151}$$

To expand the white noise we need a CON set. If $K_a(t, u)$ is not positive definite we augment its eigenfunctions to obtain a CON set. (See Property 9 on p. 181).

Then

$$K_r(t, u) = \sum_{i=1}^{\infty} \left(\lambda_i + \frac{N_0}{2}\right) \phi_i(t)\, \phi_i(u). \tag{152}$$

Because the $\phi_i(t)$ are a CON set, we try a solution of the form

$$h_o(t, u) = \sum_{i=1}^{\infty} h_i \, \phi_i(t) \, \phi_i(u). \tag{153}$$

Substituting (150), (152), and (153) into (144), we find

$$h_o(t, u) = \sum_{i=1}^{\infty} \frac{\lambda_i}{\lambda_i + N_0/2} \phi_i(t) \, \phi_i(u). \tag{154}$$

Thus the optimum linear filter can be expressed in terms of the eigenfunctions and eigenvalues of the message covariance function. A K-term approximation is shown in Fig. 3.15.

The nonrealizability could be eliminated by a T-second delay in the second multiplication. Observe that (154) represents a practical solution only when the number of significant eigenvalues is small. In most cases the solution in terms of eigenfunctions will be useful only for theoretical purposes. When we study filtering and estimation in detail in later chapters, we shall find more practical solutions.

The error can also be expressed easily in terms of eigenvalues and eigenfunctions. Substitution of (154) into (149) gives

$$\boxed{\xi_{P_o}(t) = \frac{N_0}{2} \sum_{i=1}^{\infty} \frac{\lambda_i}{\lambda_i + N_0/2} \phi_i^2(t), \qquad 0 \le t \le T} \tag{155}$$

and

$$\boxed{\xi_I = \frac{1}{T} \int_0^T \xi_{P_o}(t) \, dt = \frac{N_0}{2T} \sum_{i=1}^{\infty} \frac{\lambda_i}{\lambda_i + N_0/2}.} \tag{156}$$

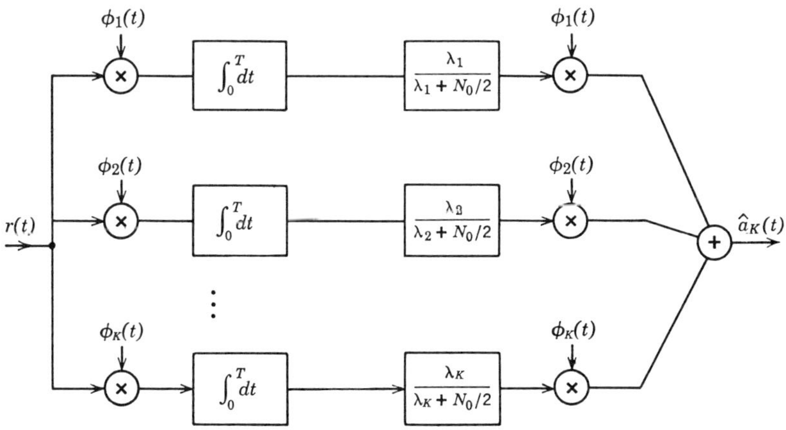

Fig. 3.15 Optimum filter.

In addition to the development of a useful result, this problem has provided an excellent example of the use of eigenfunctions and eigenvalues in finding a series solution to an integral equation. It is worthwhile to re-emphasize that all the results in this section were based on the original constraint of a linear processor and that a Gaussian assumption was not needed. We now return to the general discussion and develop several properties of interest.

3.4.6 Properties of Eigenfunctions and Eigenvalues

In this section we derive two interesting properties that will be useful in the sequel.

***Monotonic Property.*†** Consider the integral equation

$$\lambda_i(T)\,\phi_i(t:T) = \int_0^T K_x(t, u)\,\phi_i(u:T)\,du, \qquad 0 \le t \le T, \tag{157}$$

where $K_x(t, u)$ is a square-integrable covariance function. [This is just (46) rewritten to emphasize the dependence of the solution on T.] Every eigenvalue $\lambda_i(T)$ is a monotone-increasing function of the length of the interval T.

Proof. Multiplying both sides by $\phi_i(t:T)$ and integrating with respect to T over the interval $[0, T]$, we have,

$$\lambda_i(T) = \iint_0^T \phi_i(t:T)\,K_x(t, u)\,\phi_i(u:T)\,dt\,du. \tag{158}$$

Differentiating with respect to T we have,

$$\begin{aligned}\frac{\partial\lambda_i(T)}{\partial T} &= 2\int_0^T \frac{\partial\phi_i(t:T)}{\partial T}\,dt \int_0^T K_x(t, u)\,\phi_i(u:T)\,du \\ &\quad + 2\phi_i(T:T)\int_0^T K_x(T, u)\,\phi_i(u:T)\,du.\end{aligned} \tag{159}$$

Using (157), we obtain

$$\frac{\partial\lambda_i(T)}{\partial T} = 2\lambda_i(T)\int_0^T \frac{\partial\phi_i(t:T)}{\partial T}\,\phi_i(t:T)\,dt + 2\lambda_i(T)\,\phi_i^2(T:T). \quad (160$$

To reduce this equation, recall that

$$\int_0^T \phi_i^2(t:T)\,dt = 1. \tag{161}$$

† This result is due to R. Huang [23].

Differentiation of (161) gives

$$2\int_0^T \frac{\partial\phi_i(t:T)}{\partial T}\,\phi_i(t:T)\,dt + \phi_i^2(T:T) = 0. \tag{162}$$

By substituting (162) into (160), we obtain

$$\frac{\partial\lambda_i(T)}{\partial T} = \lambda_i(T)\,\phi_i^2(T:T) \geq 0, \tag{163}$$

which is the desired result.

The second property of interest is the behavior of the eigenfunctions and eigenvalues of stationary processes for *large T*.

Asymptotic Behavior Properties. In many cases we are dealing with stationary processes and are interested in characterizing them over an infinite interval. To study the behavior of the eigenfunctions and eigenvalues we return to (46); we assume that the process is stationary and that the observation interval is infinite. Then (46) becomes

$$\lambda\,\phi(t) = \int_{-\infty}^{\infty} K_x(t-u)\,\phi(u)\,du, \qquad -\infty < t < \infty. \tag{164}$$

In order to complete the solution by inspection, we recall the simple linear filtering problem shown in Figure 3.16. The input is $y(t)$, the impulse response is $h(\tau)$, and the output is $z(t)$. They are related by the convolution integral:

$$z(t) = \int_{-\infty}^{\infty} h(t-u)\,y(u)\,du, \qquad -\infty < t < \infty. \tag{165}$$

In a comparison of (164) and (165) we see that the solution to (164) is simply a function that, when put into a linear system with impulse response $K_x(\tau)$, will come out of the system unaltered except for a gain change. It is well known from elementary linear circuit theory that complex exponentials meet this requirement. Thus

$$\phi(t) = e^{j\omega t}, \qquad -\infty < \omega < \infty, \tag{166†}$$

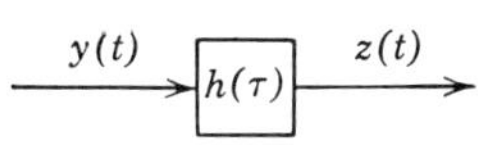

Fig. 3.16 Linear filter.

† The function $e^{(\sigma + j\omega)t}$ also satisfies (165) for values of σ where the exponential transform of $h(\tau)$ exists. The family of exponentials with $\sigma = 0$ is adequate for our purposes. This is our first use of a complex eigenfunction. As indicated at the beginning of Section 3.2, the modifications should be clear. (See Problem 3.4.11)

is an eigenfunction for any ω on the real line. Substituting into (164), we have

$$\lambda e^{+j\omega t} = \int_{-\infty}^{\infty} K_x(t-u)e^{j\omega u}\,du \tag{167}$$

or

$$\lambda = \int_{-\infty}^{\infty} K_x(t-u)e^{-j\omega(t-u)}\,du = S_x(\omega). \tag{168}$$

Thus the eigenvalue for a particular ω is the value of the power density spectrum of the process at that ω.

Now the only difficulty with this discussion is that we no longer have a countable set of eigenvalues and eigenfunctions to deal with and the idea of a series expansion of the sample function loses its meaning. There are two possible ways to get out of this difficulty.

1. Instead of trying to use a series representation of the sample functions, we could try to find some integral representation. The transition would be analogous to the Fourier series–Fourier integral transition for deterministic functions.
2. Instead of starting with the infinite interval, we could consider a finite interval and investigate the behavior as the length increases. This might lead to some simple approximate expressions for large T.

In Sections 3.5 and 3.6 we develop the first approach. It is an approach for dealing with the infinite interval that can be made rigorous. The second, which we now demonstrate, is definitely heuristic but leads to the correct results and is easy to apply.

We start with (46) and assume that the limits are $-T/2$ and $T/2$:

$$\lambda\,\phi(t) = \int_{-T/2}^{+T/2} K_x(t-u)\,\phi(u)\,du, \qquad -\frac{T}{2} \le t \le \frac{T}{2}. \tag{169}$$

We define

$$f_0 = \frac{1}{T}, \tag{170}$$

and try a solution of the form,

$$\phi_n(u) = e^{+j2\pi f_0 nu}, \qquad -\frac{T}{2} \le u \le \frac{T}{2}, \tag{171}$$

where $n = 0, \pm 1, \pm 2, \ldots$. (We index over both positive and negative integers for convenience).

Define

$$f_n = nf_0. \tag{172}$$

Substituting (171) into (169), we have

$$\lambda_n\,\phi_n(t) = \int_{-T/2}^{T/2} K_x(t-u)e^{+j2\pi f_n u}\,du. \tag{173}$$

Now,

$$K_x(t-u) = \int_{-\infty}^{\infty} S_x(f)e^{+j2\pi f(t-u)}\,df. \tag{174}$$

Substituting (174) into (173) and integrating with respect to u, we obtain

$$\lambda_n\,\phi_n(t) = \int_{-\infty}^{\infty} S_x(f)e^{+j2\pi ft}\left[\frac{\sin \pi T(f_n - f)}{\pi(f_n - f)}\right] df. \tag{175}$$

The function in the bracket, shown in Fig. 3.17, is centered at $f = f_n$ where its height is T. Its width is inversely proportional to T and its area equals one for all values of T. We see that for large T the function in the bracket is approximately an impulse at f_n. Thus

$$\lambda_n\,\phi_n(t) \simeq \int_{-\infty}^{\infty} S_x(f)e^{+j2\pi ft}\,\delta(f - f_n)\,df = S_x(f_n)e^{+j2\pi f_n t}. \tag{176}$$

Therefore

$$\lambda_n \simeq S_x(f_n) = S_x(nf_0) \tag{177}$$

and

$$\phi_n(t) \simeq \frac{1}{\sqrt{T}}e^{+j2\pi f_n t}, \qquad -\frac{T}{2} \le t \le \frac{T}{2}, \tag{178}$$

for large T.

From (175) we see that the magnitude of T needed for the approximation to be valid depends on how quickly $S_x(f)$ varies near f_n.

In (156) we encountered the infinite sum of a function of the eigenvalues

$$\xi_I = \frac{N_0}{2T}\sum_{i=1}^{\infty}\frac{\lambda_i}{\lambda_i + N_0/2}. \tag{179}$$

More generally we encounter sums of the form

$$g_\lambda \triangleq \sum_{i=1}^{\infty} g(\lambda_i). \tag{180}$$

An approximate expression for g_λ useful for large T follows directly from the above results. In Fig. 3.18 we sketch a typical spectrum and the approximate eigenvalues based on (177). We see that

$$g_\lambda \simeq \sum_{n=-\infty}^{+\infty} g(S_x(nf_0)) = T\sum_{n=-\infty}^{+\infty} g(S_x(nf_0))f_0, \tag{181}$$

where the second equality follows from the definition in (170). Now, for large T we can approximate the sum by an integral,

$$\boxed{g_\lambda \cong T\int_{-\infty}^{\infty} g(S_x(f))\,df,} \tag{182}$$

which is the desired result.

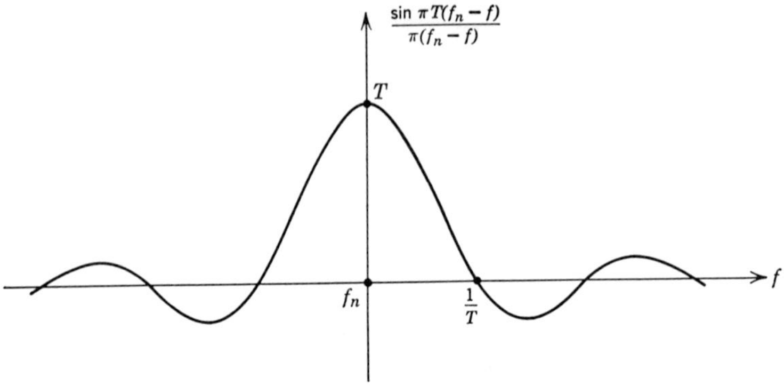

Fig. 3.17 Weighting function in (175).

The next properties concern the size of the largest eigenvalue.

Maximum and Minimum Properties. Let $x(t)$ be a stationary random process represented over an interval of length T. The largest eigenvalue $\lambda_{\max}(T)$ satisfies the inequality

$$\lambda_{\max}(T) \le \max_f S_x(f)$$

for any interval T. This result is obtained by combining (177) with the monotonicity property.

Another bound on the maximum eigenvalue follows directly from Property 10 on p. 181.

$$\lambda_{\max}(T) \le \int_{-T/2}^{T/2} K_x(t, t)\, dt = T \int_{-\infty}^{\infty} S_x(f)\, df.$$

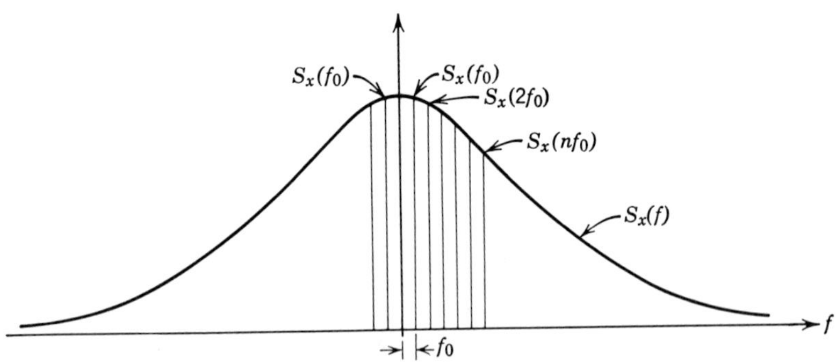

Fig. 3.18 Approximate eigenvalues; large T.

A lower bound is derived in Problem 3.4.4,

$$\lambda_{\max}(T) \geq \iint\limits_{-T/2}^{T/2} f(t)\, K_x(t, u) f(u)\, dt\, du,$$

where $f(t)$ is any function with unit energy in the interval $(-T/2, T/2)$.

The asymptotic properties developed on pp. 205–207 are adequate for most of our work. In many cases, however, we shall want a less heuristic method of dealing with stationary processes and an infinite interval.

In Section 3.6 we develop a method of characterization suitable to stationary processes over an infinite interval. A convenient way to approach this problem is as a limiting case of a periodic process. Therefore in Section 3.5 we digress briefly and develop representations for periodic processes.

3.5 PERIODIC PROCESSES†

Up to this point we have emphasized the representation of processes over a finite time interval. It is often convenient, however, to consider the infinite time interval. We begin our discussion with the definition of a periodic process.

Definition. A periodic process is a stationary random process whose correlation function $R_x(\tau)$ is periodic with period T:

$$R_x(\tau) = R_x(\tau + T), \qquad \text{for all } \tau.$$

It is easy to show that this definition implies that almost every sample function is periodic [i.e., $x(t) = x(t + T)$]. The expectation of the difference is

$$E[(x(t) - x(t + T))^2] = 2R_x(0) - 2R_x(T) = 0.$$

Therefore the probability that $x(t) = x(t + T)$ is one.

We want to represent $x(t)$ in terms of a conventional Fourier series with random coefficients. We first consider a cosine-sine series and assume that the process is zero-mean for notational simplicity.

Cosine-Sine Expansion. The series expansion for the process is

$$x(t) = \underset{N\to\infty}{\text{l.i.m.}} \sum_{i=1}^{N} \left[x_{ci} \cos\left(\frac{2\pi}{T} it\right) + x_{si} \sin\left(\frac{2\pi}{T} it\right)\right], \quad -\infty < t < \infty, \quad (183)$$

† Sections 3.5 and 3.6 are not essential to most of the discussions in the sequel and may be omitted in the first reading.

where

$$x_{ci} = \frac{2}{T}\int_{-T/2}^{T/2} x(t)\cos\left(\frac{2\pi}{T}it\right)dt \tag{184}$$

and

$$x_{si} = \frac{2}{T}\int_{-T/2}^{T/2} x(t)\sin\left(\frac{2\pi}{T}it\right)dt. \tag{185}$$

The covariance function can be expanded as

$$K_x(\tau) = \sum_{i=1}^{\infty} p_i \cos\left(\frac{2\pi}{T}i\tau\right), \tag{186}$$

where

$$p_i = \frac{2}{T}\int_{-T/2}^{T/2} K_x(\tau)\cos\left(\frac{2\pi}{T}i\tau\right)d\tau. \tag{187}$$

It follows easily that

$$\begin{aligned} E(x_{ci}x_{cj}) = E(x_{si}x_{sj}) &= 0, \qquad i \neq j, \\ E(x_{ci}x_{sj}) &= 0, \qquad \text{all } i, j. \end{aligned} \tag{188}$$

Thus the coefficients in the series expansion are uncorrelated random variables. (This means that the eigenfunctions of any periodic process for the interval $(-T/2, T/2)$ are harmonically related cosines and sines.) Similarly,

$$E(x_{ci}^2) = E(x_{si}^2) = p_i. \tag{189}$$

Observe that we have not normalized the coordinate functions. The motivation for this is based on the fact that the *power*, not the energy, at a given frequency is the quantity of interest. By omitting the $\sqrt{T}$ in the coordinate functions the value $(x_{ci}^2 + x_{si}^2)$ represents the power and not the energy. The *expected value* of the power at frequency $\omega_i \triangleq 2\pi i/T \triangleq i\omega_0$ is p_i.

Complex Exponential Expansion. Alternately, we could expand the process by using complex exponentials:

$$K_x(\tau) = \sum_{\substack{i=-\infty \\ i\neq 0}}^{\infty} \frac{p_i}{2}\exp\left(j\frac{2\pi}{T}it\right), \tag{190}$$

and

$$x(t) = \underset{N\to\infty}{\text{l.i.m.}} \sum_{\substack{i=-N \\ i\neq 0}}^{i=N} x_i \exp\left(j\frac{2\pi}{T}it\right), \qquad -\infty < t < \infty. \tag{191}$$

For positive i,

$$x_i = \tfrac{1}{2}(x_{ci} - jx_{si}), \qquad i \geq 1. \tag{192}$$

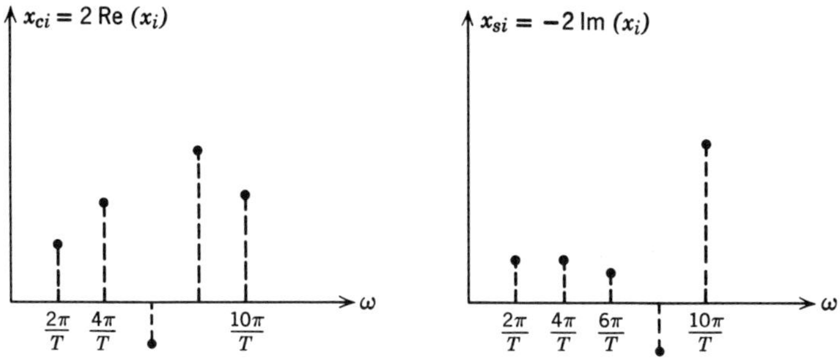

(*a*) (*b*)

Fig. 3.19 Coefficients for a typical sample function.

The values for negative indices are conjugates of the values for positive indices:

$$x_i = x^*_{-i}, \tag{193}$$

$$E(x_i x_k^*) = \frac{p_i}{2}\,\delta_{ik}, \tag{194}$$

and

$$x_i = \frac{1}{T}\int_{-T/2}^{T/2} x(t)\exp\left(-j\frac{2\pi}{T}it\right)dt. \tag{195}$$

We see that the coefficients are uncorrelated.

Just as in the finite interval case, every sample function is determined in the mean-square sense by its coefficients. We can conveniently catalog these coefficients as a function of ω. In Fig. 3.19 we show them for a *typical* sample function. In Fig. 3.20 we show the *statistical average* of the square of the coefficients (the variance).

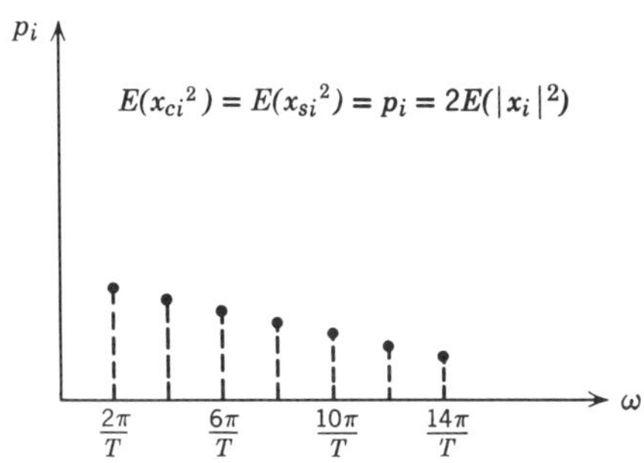

Fig. 3.20 Variance of coefficients: periodic process.

3.6 INFINITE TIME INTERVAL: SPECTRAL DECOMPOSITION

3.6.1 Spectral Decomposition

We now consider the effect of letting T, the period of the process, approach infinity.

Cosine-Sine Representation. Looking at Fig. 3.19, we see that the lines come closer together as T increases. In anticipation of this behavior, a more convenient sketch might be the cumulative amplitude plot shown in Fig. 3.21 for a *typical* sample function. The function $Z_c(\omega_n)$ is the sum of the cosine coefficients from 1 through $\omega_n(\triangleq n\omega_0)$.

$$Z_c(\omega_n) = \sum_{i=1}^{n} x_{ci}. \tag{196}$$

Similarly,

$$Z_s(\omega_n) = \sum_{i=1}^{n} x_{si}. \tag{197}$$

We see that because of the zero-mean assumption,

$$Z_c(0) = Z_s(0) = 0, \tag{198}$$

and

$$Z_c(\omega_n) = \sum_{i=1}^{n} x_{ci} = \sum_{i=1}^{n} \frac{2}{T} \int_{-T/2}^{T/2} x(t) \cos(i\omega_0 t)\, dt. \tag{199}$$

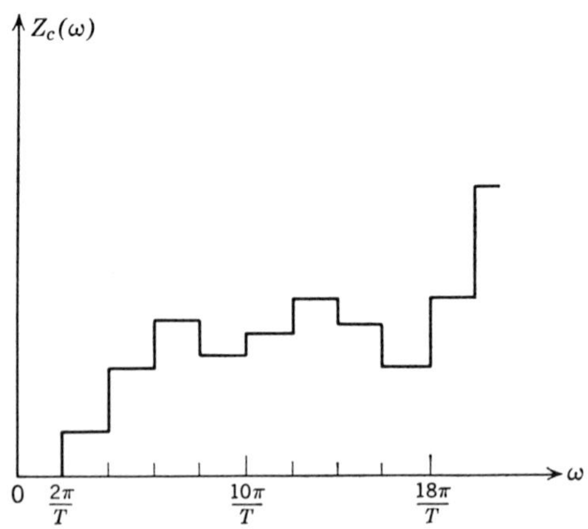

Fig. 3.21 Cumulative voltage function: periodic process.

Looking at (183), we see that we can write

$$x(t) = \underset{N\to\infty}{\text{l.i.m.}} \sum_{n=1}^{N} [Z_c(\omega_n) - Z_c(\omega_{n-1})] \cos \omega_n t$$
$$+ \underset{N\to\infty}{\text{l.i.m.}} \sum_{n=1}^{N} [Z_s(\omega_n) - Z_s(\omega_{n-1})] \sin \omega_n t. \tag{200}$$

From the way it is defined we see that

$$E\{[Z_c(\omega_n) - Z_c(\omega_{n-1})]^2\} = E\{[Z_s(\omega_n) - Z_s(\omega_{n-1})]^2\} = p_n. \tag{201}$$

We can indicate the cumulative mean power by the function $G_c(\omega_n)$, where

$$G_c(\omega_n) = \sum_{i=1}^{n} p_i = G_s(\omega_n), \qquad \omega_n \geq 0. \tag{202}$$

A typical function is shown in Fig. 3.22.

The covariance function can be expressed in terms of $G_c(\omega_n)$ by use of (186) and (202).

$$K_x(\tau) = \sum_{n=1}^{\infty} [G_c(\omega_n) - G_c(\omega_{n-1})] \cos \omega_n \tau. \tag{203}$$

Complex Exponential Representation. Alternately, in complex notation,

$$Z(\omega_n) - Z(\omega_m) = \sum_{i=m+1}^{n} x_i = \sum_{i=m+1}^{n} \frac{1}{T} \int_{-T/2}^{T/2} x(t) e^{-j(i\omega_0)t}\, dt$$

and

$$x(t) = \underset{N\to\infty}{\text{l.i.m.}} \sum_{n=-N}^{N} [Z(\omega_n) - Z(\omega_{n-1})] e^{j\omega_n t}. \tag{204}$$

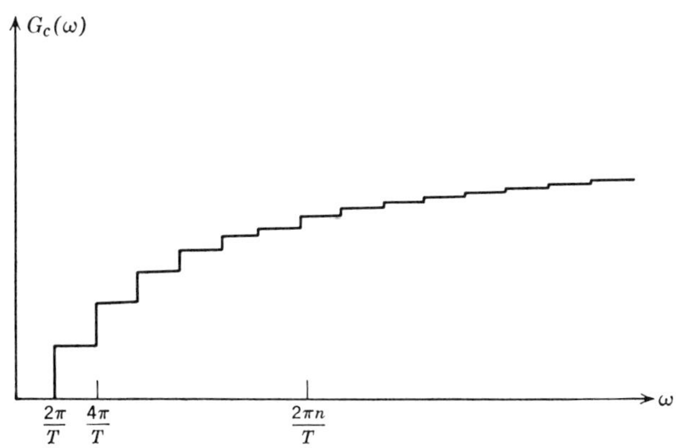

Fig. 3.22 Cumulative power spectrum: periodic process.

The mean-square value of the nth coefficient is,

$$E[|Z(\omega_n) - Z(\omega_{n-1})|^2] = \frac{p_n}{2}, \qquad n = -\infty, \ldots, -1, 1, \ldots, \infty, \tag{205}$$

The cumulative mean power is,

$$G(\omega_n) = \sum_{i=-\infty}^{n} \frac{p_i}{2}, \qquad \omega_n > -\infty, \tag{206}$$

and

$$K_x(\tau) = \sum_{n=-\infty}^{\infty} [G(\omega_n) - G(\omega_{n-1})]e^{j\omega_n\tau}. \tag{207}$$

Cosine-Sine Representation. We now return to the cosine-sine representation and look at the effect of letting $T \to \infty$. First reverse the order of summation and integration in (199). This gives

$$Z_c(\omega_n) = \int_{-T/2}^{T/2} 2x(t) \sum_{i=1}^{n} \cos(i\omega_0 t) \frac{1}{T}\, dt. \tag{208}$$

Now let

$$\frac{\Delta\omega}{2\pi} = \Delta f = \frac{1}{T},$$

and

$$\omega = n\omega_0 = \omega_n.$$

Holding $n\omega_0$ constant and letting $T \to \infty$, we obtain

$$Z_c(\omega) = \int_{-\infty}^{\infty} 2x(t)\, dt \int_0^{\omega} \cos \omega t \frac{d\omega}{2\pi} \tag{209}$$

or

$$Z_c(\omega) = \frac{1}{2\pi} \int_{-\infty}^{\infty} 2\frac{\sin \omega t}{t} x(t)\, dt. \tag{210}$$

Similarly,

$$Z_s(\omega) = \frac{1}{2\pi} \int_{-\infty}^{\infty} 2\frac{1 - \cos \omega t}{t} x(t)\, dt. \tag{211}$$

The sum representing $x(t)$ also becomes an integral,

$$x(t) \triangleq \int_0^{\infty} dZ_c(\omega) \cos \omega t + \int_0^{\infty} dZ_s(\omega) \sin \omega t. \tag{212}$$

We have written these integrals as Stieltjes integrals. They are defined as the limit of the sum in (200) as $T \to \infty$. It is worthwhile to note that we shall never be interested in evaluating a Stieltjes integral. Typical plots of $Z_c(\omega)$ and $Z_s(\omega)$ are shown in Fig. 3.23. They are zero-mean processes with the following useful properties:

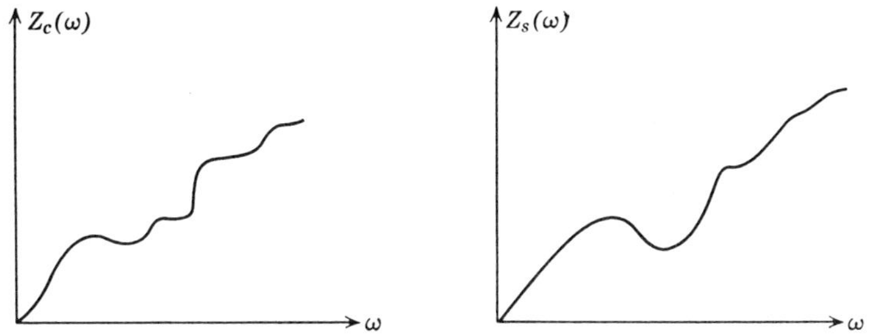

Fig. 3.23 Typical integrated voltage spectrum.

1. The increments in disjoint intervals are uncorrelated; that is,

$$E\{[Z_c(\omega_1) - Z_c(\omega_1 - \Delta\omega_1)][Z_c(\omega_2) - Z_c(\omega_2 - \Delta\omega_2)]\} = 0 \quad (213)$$

and

$$E\{[Z_s(\omega_1) - Z_s(\omega_1 - \Delta\omega_1)][Z_s(\omega_2) - Z_s(\omega_2 - \Delta\omega_2)]\} = 0 \quad (214)$$

if $(\omega_1 - \Delta\omega_1, \omega_1]$ and $(\omega_2 - \Delta\omega_2, \omega_2]$ are disjoint. This result is directly analogous to the coefficients in a series expansion being uncorrelated.

2. The quadrature components are uncorrelated even in the same interval; that is,

$$E\{[Z_c(\omega_1) - Z_c(\omega_1 - \Delta\omega_1)][Z_s(\omega_2) - Z_s(\omega_2 - \Delta\omega_2)]\} = 0 \quad (215)$$

for all ω_1 and ω_2.

3. The mean-square value of the increment variable has a simple physical interpretation,

$$E\{[Z_c(\omega_1) - Z_c(\omega_1 - \Delta\omega)]^2\} = G_c(\omega_1) - G_c(\omega_1 - \Delta\omega). \quad (216)$$

The quantity on the right represents the mean power contained in the frequency interval $(\omega_1 - \Delta\omega, \omega_1]$.

4. In many cases of interest the function $G_c(\omega)$ is differentiable.

$$G_c(\omega_2) - G_c(\omega_1) = \int_{\omega_1}^{\omega_2} 2S_x(\omega)\,\frac{d\omega}{2\pi}. \quad (217)$$

(The "2" inside the integral is present because $S_x(\omega)$ is a double-sided spectrum.)

$$\frac{dG_c(\omega)}{d\omega} \triangleq \frac{2S_x(\omega)}{2\pi}. \quad (218)$$

5. If $x(t)$ contains a periodic component of frequency ω_c, $G_c(\omega)$ will have a step discontinuity at ω_c and $S_x(\omega)$ will contain an impulse at ω_c.

The functions $Z_c(\omega)$ and $Z_s(\omega)$ are referred to as the *integrated Fourier transforms* of $x(t)$. The function $G_c(\omega)$ is the *integrated spectrum* of $x(t)$.

A logical question is: why did we use $Z_c(\omega)$ instead of the usual Fourier transform of $x(t)$?

The difficulty with the ordinary Fourier transform can be shown. We define

$$X_{c,T}(\omega) = \int_{-T/2}^{T/2} x(t) \cos \omega t \, dt \tag{219}$$

and examine the behavior as $T \to \infty$. Assuming $E[x(t)] = 0$,

$$E[X_{c,T}(\omega)] = 0, \tag{220}$$

and

$$E[|X_{c,T}(\omega)|^2] = \int_{-T/2}^{T/2} dt \int_{-T/2}^{T/2} du \, R_x(t-u) \cos \omega t \cos \omega u. \tag{221}$$

It is easy to demonstrate that the right side of (221) will become arbitrarily large as $T \to \infty$. Thus, for every ω, the usual Fourier transform is a random variable with an unbounded variance.

Complex Exponential Representation. A similar result to that in (210) can be obtained for the complex representation.

$$Z(\omega_n) - Z(\omega_m) = \frac{1}{2\pi} \int_{-\infty}^{\infty} \left(\frac{e^{-j\omega_n t} - e^{-j\omega_m t}}{-jt} \right) x(t) \, dt \tag{222}$$

and

$$x(t) = \int_{-\infty}^{\infty} dZ(\omega) e^{j\omega t}. \tag{223}$$

The expression in (222) has a simple physical interpretation. Consider the complex bandpass filter and its transfer function shown in Fig. 3.24. The impulse response is complex:

$$h_g(t) = \frac{1}{2\pi} \frac{e^{j\omega_n t} - e^{j\omega_m t}}{jt}. \tag{224}$$

The output at $t = 0$ is

$$y(0) = \int_{-\infty}^{\infty} \frac{1}{2\pi} \left(\frac{e^{-j\omega_n \tau} - e^{-j\omega_m \tau}}{-j\tau} \right) x(\tau) \, d\tau = Z(\omega_m) - Z(\omega_n). \tag{225}$$

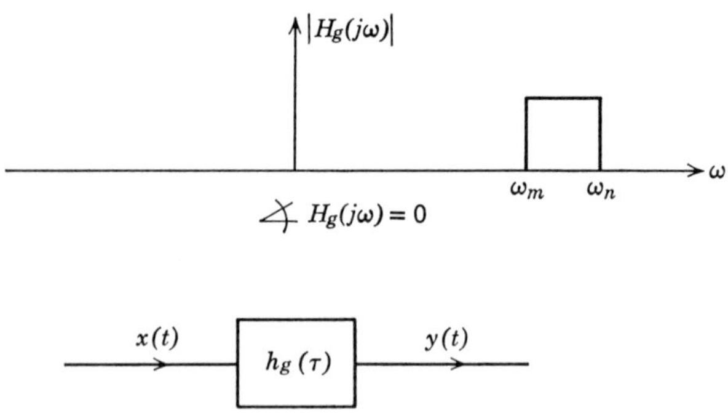

Fig. 3.24 A complex filter.

Thus the increment variables in the $Z(\omega)$ process correspond to the output of a complex linear filter when its input is $x(t)$.

The properties of interest are directly analogous to (213)–(218) and are listed below:

$$E[|Z(\omega) - Z(\omega - \Delta\omega)|^2] = G(\omega) - G(\omega - \Delta\omega). \tag{226}$$

If $G(\omega)$ is differentiable, then

$$\frac{dG(\omega)}{d\omega} \triangleq \frac{S_x(\omega)}{2\pi}. \tag{227}$$

A typical case is shown in Fig. 3.25.

$$E\{|Z(\omega) - Z(\omega - \Delta\omega)|^2\} = \frac{1}{2\pi}\int_{\omega-\Delta\omega}^{\omega} S_x(\omega)\, d\omega. \tag{228}$$

If $\omega_3 > \omega_2 > \omega_1$, then

$$E\{[Z(\omega_3) - Z(\omega_2)][Z^*(\omega_2) - Z^*(\omega_1)]\} = 0. \tag{229}$$

In other words, the increment variables are uncorrelated. These properties can be obtained as limiting relations from the exponential series or directly from (225) using the second-moment relations for a linear system.

Several observations will be useful in the sequel:

1. The quantity $dZ(\omega)$ plays exactly the same role as the Fourier transform of a finite energy signal.

For example, consider the linear system shown in Fig. 3.26. Now,

$$y(t) = \int_{-\infty}^{\infty} h(\tau)\, x(t - \tau)\, d\tau \tag{230}$$

or

$$\int_{-\infty}^{\infty} dZ_y(\omega)e^{j\omega t} = \int_{-\infty}^{\infty} d\tau\, h(\tau) \int_{-\infty}^{\infty} dZ_x(\omega)e^{j\omega(t-\tau)} \tag{231}$$

$$= \int_{-\infty}^{\infty} H(j\omega)\, dZ_x(\omega)e^{j\omega t}. \tag{232}$$

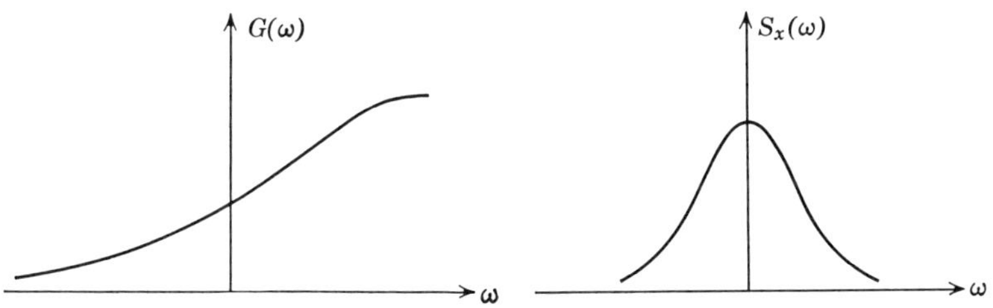

Fig. 3.25 An integrated power spectrum and a power spectrum.

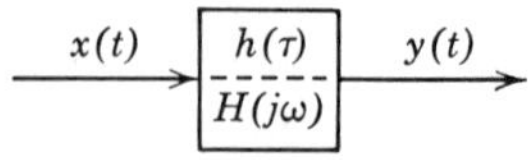

Fig. 3.26 Linear filter.

Thus

$$dZ_y(\omega) = H(j\omega)\, dZ_x(\omega) \tag{233}$$

and

$$S_y(\omega) = |H(j\omega)|^2 S_x(j\omega). \tag{234}$$

2. If the process is Gaussian, the random variables $[Z(\omega_1) - Z(\omega_1 - \Delta\omega)]$ and $[Z(\omega_2) - Z(\omega_2 - \Delta\omega)]$ are *statistically independent* whenever the intervals are disjoint.

We see that the spectral decomposition† of the process accomplishes the same result for *stationary processes* over the *infinite interval* that the Karhunen-Loève decomposition did for the finite interval. It provides us with a function $Z(\omega)$ associated with each sample function. Moreover, we can divide the ω axis into arbitrary nonoverlapping frequency intervals; the resulting increment random variables are uncorrelated (or statistically independent in the Gaussian case).

To illustrate the application of these notions we consider a simple estimation problem.

3.6.2 An Application of Spectral Decomposition: MAP Estimation of a Gaussian Process

Consider the simple system shown in Fig. 3.27:

$$r(t) = a(t) + n(t), \qquad -\infty < t < \infty. \tag{235}$$

We assume that $a(t)$ is a message that we want to estimate. In terms of the integrated transform,

$$Z_r(\omega) = Z_a(\omega) + Z_n(\omega), \qquad -\infty < \omega < \infty. \tag{236}$$

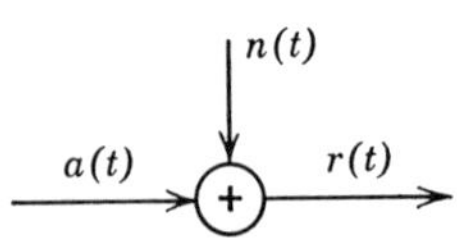

Fig. 3.27 System for estimation example.

† Further discussion of spectral decomposition is available in Gnedenko [27] or Bartlett ([28], Section 6-2).

Assume that $a(t)$ and $n(t)$ are sample functions from uncorrelated zero-mean Gaussian random processes with spectral densities $S_a(\omega)$ and $S_n(\omega)$, respectively. Because $Z_a(\omega)$ and $Z_n(\omega)$ are linear functionals of a Gaussian process, they also are Gaussian processes.

If we divide the frequency axis into a set of disjoint intervals, the increment variables will be independent. (See Fig. 3.28.) Now consider a particular interval $(\omega - d\omega, \omega]$ whose length is $d\omega$. Denote the increment variables for this interval as $dZ_r(\omega)$ and $dZ_n(\omega)$. Because of the statistical independence, we can estimate each increment variable, $dZ_a(\omega)$, separately, and because MAP and MMSE estimation commute over linear transformations it is equivalent to estimating $a(t)$.

The a posteriori probability of $dZ_a(\omega)$, given that $dZ_r(\omega)$ was received, is just

$$p_{dZ_a(\omega)|dZ_r(\omega)}[dZ_a(\omega)|dZ_r(\omega)]$$
$$= k \exp\left(-\frac{1}{2}\frac{|dZ_r(\omega) - dZ_a(\omega)|^2}{S_n(\omega)\, d\omega/2\pi} - \frac{1}{2}\frac{|dZ_a(\omega)|^2}{S_a(\omega)\, d\omega/2\pi}\right). \quad (237)$$

[This is simply (2-141) with $N = 2$ because $dZ_r(\omega)$ is complex.]

Because the a posteriori density is Gaussian, the MAP and MMSE estimates coincide. The solution is easily found by completing the square and recognizing the conditional mean. This gives

$$dZ_{\hat{a}}(\omega) = \widehat{dZ_a(\omega)} = \frac{S_a(\omega)}{S_a(\omega) + S_n(\omega)}\, dZ_r(\omega). \quad (238)$$

Therefore the minimum-mean square error estimate is obtained by passing $r(t)$ through a *linear filter*,

$$H_o(j\omega) = \frac{S_a(\omega)}{S_a(\omega) + S_n(\omega)}. \quad (239)$$

We see that the Gaussian assumption and MMSE criterion lead to a *linear* filter. In the model in Section 3.4.5 we required linearity but did not assume Gaussianness. Clearly, the two filters should be identical. To verify this, we take the limit of the finite time interval result. For the special case of white noise we can modify the result in (154) to take in account the complex eigenfunctions and the doubly infinitc sum. The result is,

$$h_o(t, u) = \sum_{i=-\infty}^{\infty} \frac{\lambda_i}{\lambda_i + N_0/2}\, \phi_i(t)\, \phi_i^*(u). \quad (240)$$

Using (177) and (178), we have

$$\lim_{T\to\infty} h_o(t, u) = 2\int_0^{\infty} \frac{S_a(\omega)}{S_a(\omega) + N_0/2} \cos \omega(t - u)\, \frac{d\omega}{2\pi}, \quad (241)$$

which corresponds to (239).

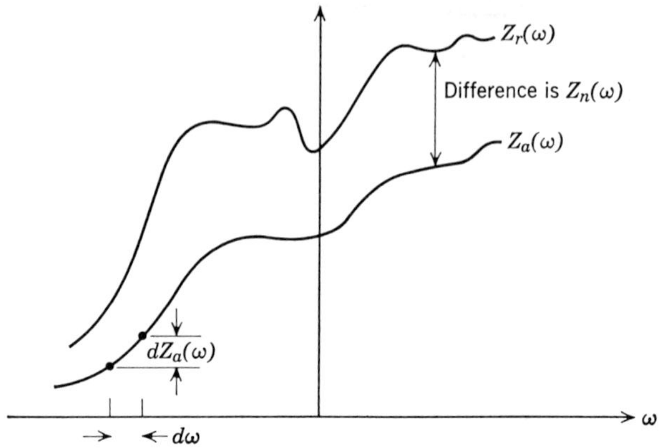

Fig. 3.28 Integrated transforms of $a(t)$ and $r(t)$.

In most of our developments we consider a finite time interval and use the orthogonal series expansion of Section 3.3. Then, to include the infinite interval-stationary process case we use the asymptotic results of Section 3.4.6. This leads us heuristically to the correct answer for infinite time. A rigorous approach for the infinite interval would require the use of the integrated transform technique we have just developed.

Before summarizing the results in this chapter, we discuss briefly how the results of Section 3.3 can be extended to vector random processes.

3.7 VECTOR RANDOM PROCESSES

In many cases of practical importance we are concerned with more than one random process at the same time; for example, in the phased arrays used in radar systems the input at each element must be considered. Analogous problems are present in sonar arrays and seismic arrays in which the received signal has a number of components. In telemetry systems a number of messages are sent simultaneously.

In all of these cases it is convenient to work with a single vector random process $\mathbf{x}(t)$ whose components are the processes of interest. If there are N processes, $x_1(t), x_2(t) \cdots x_N(t)$, we define $\mathbf{x}(t)$ as a column matrix,

$$\mathbf{x}(t) \triangleq \begin{bmatrix} x_1(t) \\ x_2(t) \\ \vdots \\ x_N(t) \end{bmatrix}. \tag{242}$$

The dimension N may be finite or countably infinite. Just as in the single process case, the second moment properties are described by the process means and covariance functions. In addition, the cross-covariance functions between the various processes must be known. The mean value function is a vector

$$\mathbf{m_x}(t) \triangleq E\begin{bmatrix} x_1(t) \\ x_2(t) \\ \vdots \\ x_N(t) \end{bmatrix} = \begin{bmatrix} m_1(t) \\ m_2(t) \\ \vdots \\ m_N(t) \end{bmatrix}, \tag{243}$$

and the covariances may be described by an $N \times N$ matrix, $\mathbf{K_x}(t, u)$, whose elements are

$$K_{ij}(t, u) \triangleq E\{[x_i(t) - m_i(t)][x_j(u) - m_j(u)]\}. \tag{244}$$

We want to derive a series expansion for the vector random process $\mathbf{x}(t)$. There are several possible representations, but two seem particularly efficient. In the first method we use a set of vector functions as coordinate functions and have scalar coefficients. In the second method we use a set of scalar functions as coordinate functions and have vector coefficients. For the first method and finite N, the modification of the properties on pp. 180–181 is straightforward. For infinite N we must be more careful. A detailed derivation that is valid for infinite N is given in [24]. In the text we go through some of the details for finite N. In Chapter II-5 we use the infinite N result without proof. For the second method, additional restrictions are needed. Once again we consider zero-mean processes.

Method 1. Vector Eigenfunctions, Scalar Eigenvalues. Let

$$\mathbf{x}(t) = \underset{N \to \infty}{\text{l.i.m.}} \sum_{i=1}^{N} x_i \boldsymbol{\phi}_i(t), \tag{245}$$

where

$$x_i = \int_0^T \boldsymbol{\phi}_i^T(t)\, \mathbf{x}(t)\, dt = \int_0^T \mathbf{x}^T(t)\, \boldsymbol{\phi}_i(t)\, dt = \sum_{k=1}^{N} \int_0^T x_k(t)\, \phi_i^k(t)\, dt, \tag{246}$$

and

$$\boldsymbol{\phi}_i(t) \triangleq \begin{bmatrix} \phi_i^1(t) \\ \phi_i^2(t) \\ \vdots \\ \phi_i^N(t) \end{bmatrix} \tag{247}$$

is chosen to satisfy

$$\lambda_i \boldsymbol{\phi}_i(t) = \int_0^T \mathbf{K_x}(t, u)\, \boldsymbol{\phi}_i(u)\, du, \qquad 0 \le t \le T. \tag{248}$$

Observe that the eigenfunctions are *vectors* but that the eigenvalues are still scalars.

Equation 248 can also be written as,

$$\sum_{j=1}^{N} \int_0^T K_{kj}(t, u)\, \phi_i^{\,j}(u)\, du = \lambda_i \phi_i^{\,k}(t), \qquad k = 1, \ldots, N, \qquad 0 \le t \le T. \tag{249}$$

The scalar properties carry over directly. In particular,

$$E(x_i x_j) = \lambda_i \delta_{ij}, \tag{250}$$

and the coordinate functions are orthonormal; that is,

$$\int_0^T \boldsymbol{\phi}_i^T(t)\, \boldsymbol{\phi}_j(t)\, dt = \delta_{ij}, \tag{251}$$

or

$$\sum_{k=1}^{N} \int_0^T \phi_i^{\,k}(t)\, \phi_j^{\,k}(t)\, dt = \delta_{ij}. \tag{252}$$

The matrix

$$\begin{aligned}\mathbf{K_x}(t, u) &= E[\mathbf{x}(t)\, \mathbf{x}^T(u)] - \mathbf{m_x}(t)\, \mathbf{m_x}^T(u)\\ &= \sum_{i=1}^{\infty} \sum_{j=1}^{\infty} \operatorname{Cov}[x_i x_j]\, \boldsymbol{\phi}_i(t)\, \boldsymbol{\phi}_j^T(u)\\ &= \sum_{i=1}^{\infty} \lambda_i \boldsymbol{\phi}_i(t) \boldsymbol{\phi}_i^T(u)\end{aligned} \tag{253}$$

or

$$\mathbf{K}_{\mathbf{x},kj}(t, u) = \sum_{i=1}^{\infty} \lambda_i \phi_i^{\,k}(t)\, \phi_i^{\,j}(u), \qquad k, j = 1, \ldots, N. \tag{254}$$

This is the multidimensional analog of (50).

One property that makes the expansion useful is that the coefficient is a scalar variable and not a vector. This point is perhaps intuitively troublesome. A trivial example shows how it comes about.

Example. Let

$$\begin{aligned} x_1(t) &= a\, s_1(t), \qquad 0 \le t \le T,\\ x_2(t) &= b\, s_2(t), \qquad 0 \le t \le T,\end{aligned} \tag{255}$$

where a and b are independent, zero-mean random variables and $s_1(t)$ and $s_2(t)$ are orthonormal functions

$$\int_0^T s_i(t)\, s_j(t)\, dt = \delta_{ij}, \qquad i, j = 1, 2, \tag{256}$$

and

$$\begin{aligned}\operatorname{Var}(a) &= \sigma_a^2,\\ \operatorname{Var}(b) &= \sigma_b^2.\end{aligned} \tag{257}$$

Then

$$\mathbf{K_x}(t, u) = \begin{bmatrix} \sigma_a^2\, s_1(t)\, s_1(u) & 0 \\ 0 & \sigma_b^2\, s_2(t)\, s_2(u) \end{bmatrix}. \tag{258}$$

We can verify that there are two vector eigenfunctions:

$$\boldsymbol{\phi}_1(t) = \begin{bmatrix} s_1(t) \\ 0 \end{bmatrix}; \qquad \lambda_1 = \sigma_a^2, \tag{259}$$

and

$$\boldsymbol{\phi}_2(t) = \begin{bmatrix} 0 \\ s_2(t) \end{bmatrix}; \qquad \lambda_2 = \sigma_b^2. \tag{260}$$

Thus we see that in this degenerate case† we can achieve simplicity in the coefficients by increasing the number of vector eigenfunctions. Clearly, when there is an infinite number of eigenfunctions, this is unimportant.

A second method of representation is obtained by incorporating the complexity into the eigenvalues.

Method 2. Matrix Eigenvalues, Scalar Eigenfunctions

In this approach we let

$$\mathbf{x}(t) = \sum_{i=1}^{\infty} \mathbf{x}_i \psi_i(t), \qquad 0 \le t \le T, \tag{261}$$

and

$$\mathbf{x}_i = \int_0^T \mathbf{x}(t)\, \psi_i(t)\, dt. \tag{262}$$

We would like to find a set of $\boldsymbol{\Lambda}_i$ and $\psi_i(t)$ such that

$$E[\mathbf{x}_i \mathbf{x}_j^T] = \boldsymbol{\Lambda}_i\, \delta_{ij} \tag{263}$$

and

$$\int_0^T \psi_i(t) \psi_j(t)\, dt = \delta_{ij}. \tag{264}$$

These requirements lead to the equation

$$\boldsymbol{\Lambda}_i\, \psi_i(t) = \int_0^T \mathbf{K_x}(t, u) \psi_i(u)\, du, \qquad 0 \le t \le T. \tag{265}$$

For arbitrary time intervals (265) does *not* have a solution except for a few trivial cases. However, if we restrict our attention to stationary processes and *large* time intervals then certain asymptotic results may be obtained. Defining

$$\mathbf{S_x}(\omega) \triangleq \int_{-\infty}^{\infty} \mathbf{K_x}(\tau) e^{j\omega\tau}\, d\tau, \tag{266}$$

† It is important not to be misled by this degenerate example. The useful application is in the case of correlated processes. Here the algebra of calculating the actual eigenfunctions is tedious but the representation is still simple.

and assuming the interval is large, we find

$$\psi_i(t) \simeq \frac{1}{\sqrt{T}} e^{j\omega_i t} \tag{267}$$

and

$$\boldsymbol{\Lambda}_i \simeq \mathbf{S_x}(\omega_i). \tag{268}$$

As before, to treat the infinite time case rigorously we must use the integrated transform

$$\mathbf{Z_x}(\omega_n) - \mathbf{Z_x}(\omega_m) \triangleq \int_{-\infty}^{\infty} \frac{e^{-j\omega_n t} - e^{-j\omega_m t}}{-jt} \mathbf{x}(t)\, dt \tag{269}$$

and

$$\mathbf{x}(t) = \int_{-\infty}^{\infty} d\mathbf{Z}_x(\omega) e^{j\omega t}. \tag{270}$$

The second method of representation has a great deal of intuitive appeal in the large time interval case where it is valid, but the first method enables us to treat a more general class of problems. For this reason we shall utilize the first representation in the text and relegate the second to the problems.

It is difficult to appreciate the importance of the first expansion until we get to some applications. We shall then find that it enables us to obtain results for multidimensional problems almost by inspection. The key to the simplicity is that we can still deal with *scalar* statistically independent random variables.

It is worthwhile to re-emphasize that we did not *prove* that the expansions had the desired properties. Specifically, we did not demonstrate that solutions to (248) existed and had the desired properties, that the multidimensional analog for Mercer's theorem was valid, or that the expansion converged in the mean-square sense ([24] does this for the first expansion).

3.8 SUMMARY

In this chapter we developed means of characterizing random processes. The emphasis was on a method of representation that was particularly well suited to solving detection and estimation problems in which the random processes were Gaussian. For non-Gaussian processes the representation provides an adequate second-moment characterization but may not be particularly useful as a complete characterization method.

For finite time intervals the desired representation was a series of orthonormal functions whose coefficients were uncorrelated random variables.

The choice of coordinate functions depended on the covariance function of the process through the integral equation

$$\lambda\, \phi(t) = \int_{T_i}^{T_f} K(t, u)\, \phi(u)\, du, \qquad T_i \le t \le T_f. \tag{271}$$

The eigenvalues λ corresponded physically to the expected value of the energy along a particular coordinate function $\phi(t)$. We indicated that this representation was useful for both theoretical and practical purposes. Several classes of processes for which solutions to (271) could be obtained were discussed in detail. One example, the simple Wiener process, led us logically to the idea of a white noise process. As we proceed, we shall find that this process is a useful tool in many of our studies.

To illustrate a possible application of the expansion techniques we solved the optimum linear filtering problem for a finite interval. The optimum filter for the additive white noise case was the solution to the integral equation

$$\frac{N_0}{2}\, h_o(t, u) + \int_{T_i}^{T_f} K_a(t, z)\, h_o(z, u)\, dz = K_a(t, u), \qquad T_i \le t, u \le T_f. \tag{272}$$

The solution could be expressed in terms of the eigenfunctions and eigenvalues.

For large time intervals we found that the eigenvalues of a stationary process approached the power spectrum of the process and the eigenfunctions became sinusoids. Thus for this class of problem the expansion could be interpreted in terms of familiar quantities.

For infinite time intervals and stationary processes the eigenvalues were not countable and no longer served a useful purpose. In this case, by starting with a periodic process and letting the period go to infinity, we developed a useful representation. Instead of a series representation for each sample function, there was an integral representation,

$$x(t) = \int_{-\infty}^{\infty} dZ_x(\omega) e^{j\omega t}. \tag{273}$$

The function $Z_x(\omega)$ was the *integrated transform* of $x(t)$. It is a sample function of a random process with *uncorrelated increments* in frequency. For the Gaussian process the increments were statistically independent. Thus the increment variables for the infinite interval played exactly the same role as the series coefficients in the finite interval. A simple example showed one of the possible applications of this property.

Finally, we extended these ideas to vector random processes. The significant result here was the ability to describe the process in terms of *scalar* coefficients.

In Chapter 4 we apply these representation techniques to solve the detection and estimation problem.

3.9 PROBLEMS

Many of the problems in Section P3.3 are of a review nature and may be omitted by the reader with an adequate random process background. Problems 3.3.19–23 present an approach to the continuous problem which is different from that in the text.

Section P3.3 Random Process Characterizations

SECOND MOMENT CHARACTERIZATIONS

Problem 3.3.1. In chapter 1 we formulated the problem of choosing a linear filter to maximize the output signal-to-noise ratio.

$$\left(\frac{S}{N}\right)_o \triangleq \frac{\left[\int_0^T h(T-\tau)\, s(\tau)\, d\tau\right]^2}{N_0/2 \int_0^T h^2(\tau)\, d\tau}.$$

1. Use the Schwarz inequality to find the $h(\tau)$ which maximizes $(S/N)_o$.
2. Sketch $h(\tau)$ for some typical $s(t)$.

Comment. The resulting filter is called a matched filter and was first derived by North [34].

Problem 3.3.2. Verify the result in (3-26).

Problem 3.3.3. [1]. The input to a stable linear system with a transfer function $H(j\omega)$ is a zero-mean process $x(t)$ whose correlation function is

$$R_x(\tau) = \frac{N_0}{2}\, \delta(\tau).$$

1. Find an expression for the variance of the output $y(t)$.
2. The noise bandwidth of a network is defined as

$$B_N \triangleq \frac{\int_{-\infty}^{\infty} |H(j\omega)|^2\, d\omega/2\pi}{|H_{\text{max}}|^2}, \qquad \text{(double-sided in cps).}$$

Verify that

$$\sigma_y^2 = \frac{N_0 B_N |H_{\text{max}}|^2}{2}.$$

Problem 3.3.4. [1]. Consider the fixed-parameter linear system defined by the equation

$$v(t) = x(t-\delta) - x(t)$$

and

$$y(t) = \int_{-\infty}^{t} v(u)\, du.$$

1. Determine the impulse response relating the input $x(t)$ and output $y(t)$.
2. Determine the system function.
3. Determine whether the system is stable.
4. Find B_N.

Problem 3.3.5. [1]. The transfer function of an *RC* network is

$$\frac{Y(j\omega)}{X(j\omega)} = \frac{1}{1 + RCj\omega}.$$

The input consists of a noise which is a sample function of a stationary random process with a flat spectral density of height $N_0/2$, plus a signal which is a sequence of constant-amplitude rectangular pulses. The pulse duration is δ and the minimum interval between pulses is T, where $\delta \ll T$.

A signal-to-noise ratio at the system output is defined here as the ratio of the maximum amplitude of the output signal with no noise at the input to the rms value of the output noise.

1. Derive an expression relating the output signal-to-noise ratio as defined above to the input pulse duration and the effective noise bandwidth of the network.
2. Determine what relation should exist between the input pulse duration and the effective noise bandwidth of the network to obtain the maximum output signal-to-noise.

Alternate Representations and Non-Gaussian Processes

Problem 3.3.6. (sampling representation). When the observation interval is infinite and the processes of concern are bandlimited, it is sometimes convenient to use a sampled representation of the process. Consider the stationary process $x(t)$ with the spectrum shown in Fig. P3.1. Assume that $x(t)$ is sampled every $1/2W$ seconds. Denote the samples as $x(i/2W)$, $i = -\infty, \ldots, 0, \ldots$.

1. Prove

$$x(t) = \underset{K\to\infty}{\text{l.i.m.}} \sum_{-K}^{K} x\left(\frac{i}{2W}\right) \frac{\sin 2\pi W(t - i/2W)}{2\pi W(t - i/2W)}.$$

2. Find $E[x(i/2W)x(j/2W)]$.

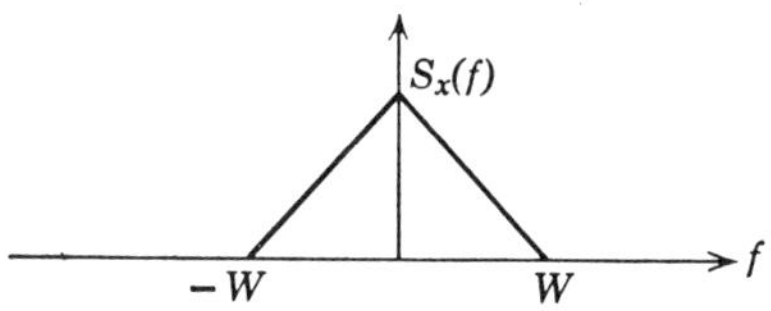

Fig. P3.1

Problem 3.3.7 (continuation). Let

$$\phi_i(t) = \sqrt{2W}\,\frac{\sin 2\pi W(t - i/2W)}{2\pi W(t - i/2W)}, \qquad -\infty < t < \infty.$$

Define

$$x(t) = \underset{K\to\infty}{\text{l.i.m.}} \sum_{i=-K}^{K} x_i\, \phi_i(t).$$

Prove that if

$$E(x_i x_j) = P\delta_{ij} \qquad \text{for all } i, j,$$

then

$$S_x(f) = \frac{P}{2W}, \qquad |f| \le W.$$

Problem 3.3.8. Let $x(t)$ be a bandpass process "centered" around f_c.

$$S_x(f) = 0, \qquad \begin{array}{ll} |f - f_c| > W, & f > 0, \\ |f + f_c| > W, & f < 0. \end{array}$$

We want to represent $x(t)$ in terms of two low-pass processes $x_c(t)$ and $x_s(t)$. Define

$$\hat{x}(t) = \sqrt{2}\, x_c(t) \cos(2\pi f_c t) + \sqrt{2}\, x_s(t) \sin(2\pi f_c t),$$

where $x_c(t)$ and $x_s(t)$ are obtained physically as shown in Fig. P3.2.

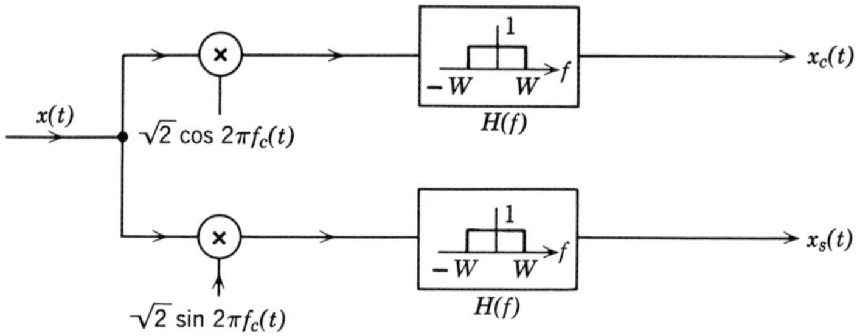

Fig. P3.2

1. Prove
$$E\{[x(t) - \hat{x}(t)]^2\} = 0.$$
2. Find $S_{x_c}(f)$, $S_{x_s}(f)$, and $S_{x_c x_s}(f)$.
3. What is a necessary and sufficient condition for $S_{x_c x_s}(f) = 0$?

Observe that this enables us to replace any bandpass process by *two* low-pass processes or a vector low-pass process.

$$\mathbf{x}(t) = \begin{bmatrix} x_c(t) \\ x_s(t) \end{bmatrix}$$

Problem 3.3.9. Show that the n-dimensional probability density of a Markov process can be expressed as

$$p_{x_{t_1}\cdots x_{t_n}}(X_{t_1}, \cdots, X_{t_n}) = \frac{\prod_{k=2}^{n} p_{x_{t_{k-1}} x_{t_k}}(X_{t_{k-1}}, X_{t_k})}{\prod_{k=2}^{n-1} p_{x_{t_k}}(X_{t_k})}, \qquad n \geq 3.$$

Problem 3.3.10. Consider a Markov process at three ordered time instants, $t_1 < t_2 < t_3$. Show that the conditional density relating the first and third time must satisfy the following equation:

$$p_{x_{t_3}|x_{t_1}}(X_{t_3}|X_{t_1}) = \int dX_{t_2} p_{x_{t_3}|x_{t_2}}(X_{t_3}|X_{t_2}) p_{x_{t_2}|x_{t_1}}(X_{t_2}|X_{t_1}).$$

Problem 3.3.11. A continuous-parameter random process is said to have independent increments if, for all choices of indices $t_0 < t_1 \cdots < t_n$, the n random variables

$$x(t_1) - x(t_0), \ldots, x(t_n) - x(t_{n-1})$$

are independent. Assuming that $x(t_0) = 0$, show that

$$M_{x_{t_1}x_{t_2}\cdots x_{t_n}}(jv_1, \ldots, jv_n) = M_{x_{t_1}}(jv_1 + jv_2 + \cdots + jv_n) \prod_{k=0}^{n} M_{x_{t_k} - x_{t_{k-1}}}(jv_k + \cdots + jv_n).$$

Gaussian Processes

Problem 3.3.12. ***(Factoring of higher order moments).*** Let $x(t)$, $t \in T$ be a Gaussian process with zero mean value function

$$E[x(t)] = 0.$$

1. Show that all odd-order moments of $x(t)$ vanish and that the even-order moments may be expressed in terms of the second-order moments by the following formula:

Let n be an even integer and let $t_1, \ldots, t_n$ be points in T, some of which may coincide. Then

$$E[x(t_1)\cdots x(t_n)] = \sum E[x(t_{i_1})\, x(t_{i_2})]\, E[x(t_{i_3})\, x(t_{i_4})] \cdots E[x(t_{i_{n-1}})\, x(t_{i_n})],$$

in which the sum is taken over all possible ways of dividing the n points into $n/2$ combinations of pairs. The number of terms in the sum is equal to

$$1\cdot 3\cdot 5\cdots(n-3)(n-1);$$

for example,

$$E[x(t_1)\, x(t_2)\, x(t_3)\, x(t_4)] = E[x(t_1)\, x(t_2)]E[x(t_3)\, x(t_4)] + E[x(t_1)\, x(t_3)]E[x(t_2)\, x(t_4)] + E[x(t_1)\, x(t_4)]E[x(t_2)\, x(t_3)].$$

Hint. Differentiate the characteristic function.

2. Use your result to find the fourth-order correlation function

$$R_x(t_1, t_2, t_3, t_4) = E[x(t_1)\, x(t_2)\, x(t_3)\, x(t_4)]$$

of a stationary Gaussian process whose spectral density is

$$S_x(f) = \frac{N_0}{2}, \qquad |f| \le W,$$
$$= 0, \qquad \text{elsewhere}.$$

What is $\lim_{W\to\infty} R_x(t_1, t_2, t_3, t_4)$?

Problem 3.3.13. Let $x(t)$ be a sample function of a stationary real Gaussian random process with a zero mean and finite mean-square value. Let a new random process be defined with the sample functions

$$y(t) = x^2(t).$$

Show that

$$R_y(\tau) = R_x^2(0) + 2R_x^2(\tau).$$

Problem 3.3.14. [1]. Consider the system shown in Fig. P3.3. Let the input $e_0(t)$ be a sample function of stationary real Gaussian process with zero mean and *flat* spectral density at all frequencies of interest; that is, we may assume that

$$S_{e_0}(f) = N_0/2$$

1. Determine the autocorrelation function or the spectral density of $e_2(t)$.
2. Sketch the autocorrelation function or the spectral density of $e_0(t)$, $e_1(t)$, and $e_2(t)$.

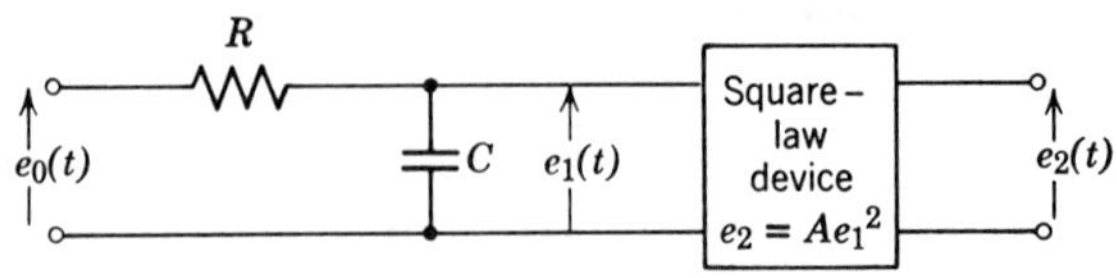

Fig. P3.3

Problem 3.3.15. The system of interest is shown in Fig. P3.4, in which $x(t)$ is a sample function from an ergodic Gaussian random process.

$$R_x(\tau) = \frac{N_0}{2}\,\delta(\tau).$$

The transfer function of the linear system is

$$\begin{aligned} H(f) &= e^{2jf} \qquad |f| \le W, \\ &= 0, \qquad \text{elsewhere.} \end{aligned}$$

1. Find the dc power in $z(t)$.
2. Find the ac power in $z(t)$.

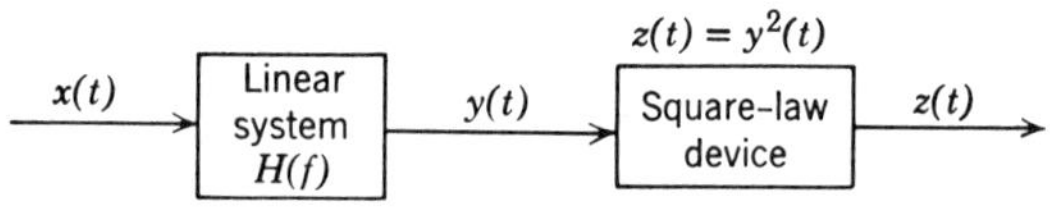

Fig. P3.4

Problem 3.3.16. The output of a linear system is $y(t)$, where

$$y(t) = \int_0^{\infty} h(\tau)\, x(t - \tau)\, d\tau.$$

The input $x(t)$ is a sample function from a *stationary* Gaussian process with correlation function

$$R_x(\tau) = \delta(\tau).$$

We should like the output at a particular time t_1 to be statistically independent of the input at that time. Find a necessary and sufficient condition on $h(\tau)$ for $x(t_1)$ and $y(t_1)$ to be statistically independent.

Problem 3.3.17. Let $x(t)$ be a real, wide-sense stationary, Gaussian random process with zero mean. The process $x(t)$ is passed through an ideal limiter. The output of the limiter is the process $y(t)$,

$$y(t) = L[x(t)],$$

where

$$L(u) = \begin{cases} +1 & u \ge 0, \\ -1 & u < 0. \end{cases}$$

Show that the autocorrelation functions of the two processes are related by the formula

$$R_y(\tau) = \frac{2}{\pi}\sin^{-1}\left[\frac{R_x(\tau)}{R_x(0)}\right].$$

Problem 3.3.18. Consider the bandlimited Gaussian process whose spectrum is shown in Fig. P3.5.

Write

$$x(t) = v(t) \cos\,[2\pi f_c t + \theta(t)].$$

Find

$$p_{v(t)}(V) \quad \text{and} \quad p_{\theta(t)}(\theta).$$

Are $v(t)$ and $\theta(t)$ independent random variables?

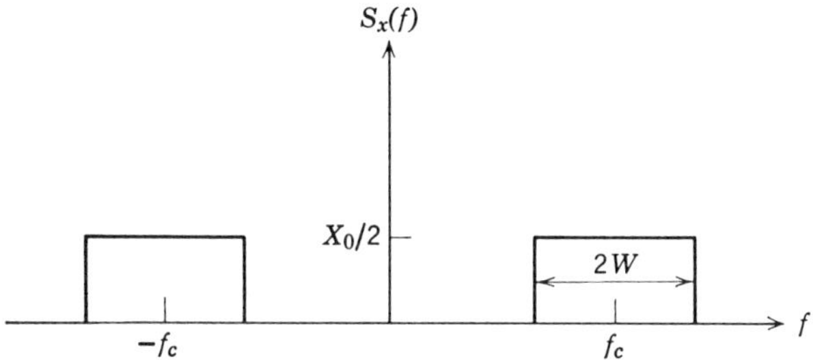

Fig. P3.5

Sampling Approach to Continuous Gaussian Processes†

In Chapters 4, 5, and II-3 we extend the classical results to the waveform case by using the Karhunen-Loève expansion. If, however, we are willing to use a heuristic argument, most of the results that we obtained in Section 2.6 for the general Gaussian problem can be extended easily to the waveform case in the following manner.

The processes and signals are sampled every ϵ seconds as shown in Fig. P3.6*a*. The gain in the sampling device is chosen so that

$$\int_0^T m^2(t)\,dt = \lim_{\epsilon \to 0} \sum_{i=1}^{T/\epsilon} m_i^2. \tag{1}$$

This requires

$$m_i = \sqrt{\epsilon}\, m(t_i). \tag{2}$$

Similarly, for a random process,

$$n_i = \sqrt{\epsilon}\, n(t_i) \tag{3}$$

and

$$E[n_i n_j] = \epsilon E[n(t_i)\, n(t_j)] = \epsilon K_n(t_i, t_j). \tag{4}$$

To illustrate the procedure consider the simple model shown in Fig. P3.6*b*. The continuous waveforms are

$$\begin{aligned} r(t) &= m(t) + n(t), \qquad 0 \le t \le T : H_1, \\ r(t) &= n(t), \qquad 0 \le t \le T : H_0, \end{aligned} \tag{5}$$

where $m(t)$ is a known function and $n(t)$ is a sample function from a Gaussian random process.

The corresponding sampled problem is

$$\begin{aligned} \mathbf{r} &= \mathbf{m} + \mathbf{n} : H_1, \\ \mathbf{r} &= \mathbf{n} : H_0, \end{aligned} \tag{6}$$

† We introduce the sampling approach here because of its widespread use in the literature and the feeling of some instructors that it is easier to understand. The results of these five problems are derived and discussed thoroughly later in the text.

where

$$\mathbf{r} \triangleq \sqrt{\epsilon}\begin{bmatrix} r(t_1) \\ r(t_2) \\ \vdots \\ r(t_N) \end{bmatrix}$$

and $N = T/\epsilon$. Assuming that the noise $n(t)$ is band-limited to $1/\epsilon$ (double-sided, cps) and has a flat spectrum of $N_0/2$, the samples are statistically independent Gaussian variables (Problem 3.3.7).

$$E[\mathbf{n}\mathbf{n}^T] = E[n^2(t)]\mathbf{I} = \frac{N_0}{2}\mathbf{I} \triangleq \sigma^2\mathbf{I}. \tag{7}$$

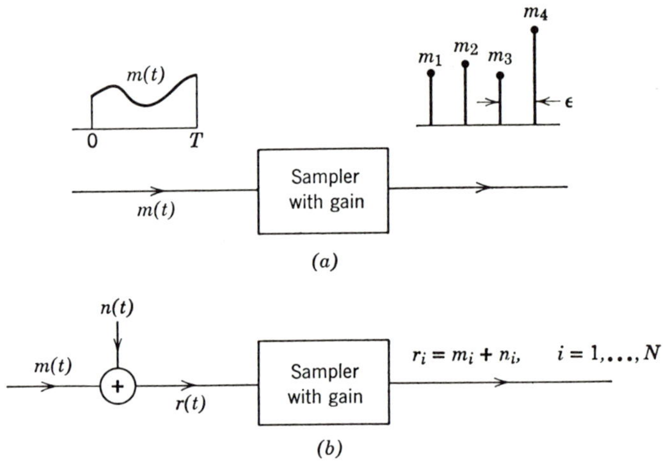

Fig. P3.6

The vector problem in (6) is familiar from Section 2.6 of Chapter 2. From (2.350) the sufficient statistic is

$$l(\mathbf{R}) = \frac{1}{\sigma^2}\sum_{i=1}^{N} m_i R_i. \tag{8}$$

Using (2) and (3) in (8), we have

$$l(\mathbf{R}) = \frac{2}{N_0}\sum_{i=1}^{T/\epsilon} \sqrt{\epsilon}\, m(t_i)\cdot\sqrt{\epsilon}\, r(t_i).$$

As $\epsilon \rightarrow 0$, we have (letting $dt = \epsilon$)

$$\underset{\epsilon\rightarrow 0}{\text{l.i.m.}}\; l(\mathbf{R}) = \frac{2}{N_0}\int_0^T m(t)\, r(t)\, dt \triangleq l(r(t))$$

which is the desired result. Some typical problems of interest are developed below.

Problem 3.3.19. Consider the simple example described in the introduction.

1. Show that

$$d^2 = \frac{2E}{N_0},$$

where E is the energy in $m(t)$.

2. Draw a block diagram of the receiver and compare it with the result in Problem 3.3.1.

Problem 3.3.20. Consider the discrete case defined by (2.328). Here

$$E[\mathbf{n}\mathbf{n}^T] = \mathbf{K}$$

and

$$\mathbf{Q} = \mathbf{K}^{-1}.$$

1. Sample the bandlimited noise process $n(t)$ every ϵ seconds to obtain $n(t_1)$, $n(t_2), \ldots, n(t_K)$. Verify that in the limit $\mathbf{Q}$ becomes a function with two arguments defined by the equation

$$\int_0^T Q(t, u)\, K(u, z)\, du = \delta(t - z).$$

Hint: Define a function $Q(t_i, t_j) = (1/\epsilon)Q_{ij}$.

2. Use this result to show that

$$l = \iint_0^T m_\Delta(t)\, Q(t, u)\, r(u)\, dt\, du$$

in the limit.

3. What is d^2?

Problem 3.3.21. In the example defined in (2.387) the means are equal but the covariance matrices are different. Consider the continuous waveform analog to this and show that

$$l = \iint_0^T r(t)\, h_\Delta(t, u)\, r(u)\, dt\, du,$$

where

$$h_\Delta(t, u) = Q_0(t, u) - Q_1(t, u).$$

Problem 3.3.22. In the linear estimation problem defined in Problem 2.6.8 the received vector was

$$\mathbf{r} = \mathbf{a} + \mathbf{n}$$

and the MAP estimate was

$$\mathbf{K}_r\hat{\mathbf{a}} = \mathbf{K}_\mathbf{a}\mathbf{R}.$$

Verify that the continuous analog to this result is

$$\int_0^T K_r(t, u)\, \hat{a}(u)\, du = \int_0^T K_a(t, u)\, r(u)\, du, \qquad 0 \le t \le T.$$

Problem 3.3.23. Let

$$r(t) = a(t) + n(t), \qquad 0 \le t \le T,$$

where $a(t)$ and $n(t)$ are independent zero-mean Gaussian processes with covariance functions $K_a(t, u)$ and $K_n(t, u)$, respectively. Consider a specific time t_1 in the interval. Find

$$p_{a(t_1)}|r(t), \quad 0 \le t \le T \qquad [A_{t_1}|r(t), \quad 0 \le t \le T].$$

Hint. Sample $r(t)$ every ϵ seconds and then let $\epsilon \to 0$.

Section P3.4 Integral equations

Problem 3.4.1. Consider the integral equation

$$\int_{-T}^{T} du\, P \exp(-\alpha|t - u|)\phi_i(u) = \lambda_i\phi_i(t), \qquad -T \le t \le T.$$

1. Prove that $\lambda = 0$ and $\lambda = 2P/\alpha$ are not eigenvalues.
2. Prove that all values of $\lambda > 2P/\alpha$ cannot be eigenvalues of the above integral equation.

Problem 3.4.2. Plot the behavior of the largest eigenvalue of the integral equation in Problem 3.4.1 as a function of αT.

Problem 3.4.3. Consider the integral equation (114).

$$\lambda\, \phi(t) = \sigma^2 \int_0^t u\, \phi(u)\, du + \sigma^2 t \int_t^T \phi(u)\, du, \qquad 0 \le t \le T.$$

Prove that values of $\lambda \le 0$ are not eigenvalues of the equation.

Problem 3.4.4. Prove that the largest eigenvalue of the integral equation

$$\lambda \phi(t) = \int_{-T}^{T} K_n(t, u)\, \phi(u)\, du, \qquad -T \le t \le T,$$

satisfies the inequality.

$$\lambda_1 \ge \iint_{-T}^{T} f(t)\, K_n(t, u)\, f(u)\, dt\, du,$$

where $f(t)$ is *any* function with unit energy in $[-T, T]$.

Problem 3.4.5. Compare the bound in Problem 3.4.4, using the function

$$f(t) = \frac{1}{\sqrt{2T}}, \qquad -T \le t \le T,$$

with the actual value found in Problem 3.4.2.

Problem 3.4.6. [15]. Consider a function whose total energy in the interval $-\infty < t < \infty$ is E.

$$E = \int_{-\infty}^{\infty} |f(t)|^2\, dt.$$

Now, time-limit $f(t)$, $-T/2 \le t \le T/2$ and then band-limit the result to $(-W, W$ cps). Call this resulting function $f_{DB}(t)$. Denote the energy in $f_{DB}(t)$ as E_{DB}.

$$E_{DB} = \int_{-\infty}^{\infty} |f_{DB}(t)|^2\, dt.$$

1. Choose $f(t)$ to maximize

$$\gamma \triangleq \frac{E_{DB}}{E}.$$

2. What is the resulting value of γ when $WT = 10.2$?

Problem 3.4.7. [15]. Assume that $f(t)$ is first band-limited.

$$f_B(t) = \int_{-2\pi W}^{2\pi W} F(\omega)\, e^{j\omega t}\, \frac{d\omega}{2\pi}.$$

Now, time-limit $f_B(t)$, $-T/2 \le t \le T/2$ and band-limit the result to $(-W, W)$ to obtain $f_{BDB}(t)$. Repeat Problem 3.4.6 with *BDB* replacing *DB*.

Problem 3.4.8 [35]. Consider the triangular correlation function

$$\begin{aligned} K_n(t - u) &= 1 - |t - u|, \qquad && |t - u| \le 1, \\ &= 0, && \text{elsewhere.} \end{aligned}$$

Find the eigenfunctions and eigenvalues over the interval $(0, T)$ when $T < 1$.

Problem 3.4.9. Consider the integral equation

$$\lambda\,\phi(t) = \int_{T_i}^{T_f} K_n(t, u)\,\phi(u)\,du, \qquad T_i \le t \le T_f,$$

where

$$K_n(t, u) = \sum_{i=1}^{6} \sigma_i^2 \cos\left(\frac{2\pi i t}{T}\right) \cos\left(\frac{2\pi i u}{T}\right),$$

and

$$T \triangleq T_f - T_i.$$

Find the eigenfunctions and eigenvalues of this equation.

Problem 3.4.10. The input to an unrealizable linear time-invariant system is $x(t)$ and the output is $y(t)$. Thus, we can write

$$y(t) = \int_{-\infty}^{\infty} h(\tau)\,x(t - \tau)\,d\tau.$$

We assume that

(i) $$\int_{-\infty}^{\infty} x^2(t)\,dt = 1.$$

(ii) $$h(\tau) = \frac{1}{\sqrt{2\pi}\,\sigma_c} \exp\left(-\frac{\tau^2}{2\sigma_c^2}\right), \qquad -\infty < \tau < \infty.$$

(iii) $$E_y \triangleq \int_{-\infty}^{\infty} y^2(t)\,dt.$$

1. What is the maximum value of E_y that can be obtained by using an $x(t)$ that satisfies the above constraints?
2. Find an $x(t)$ that gives an E_y arbitrarily close to the maximum E_y.
3. Generalize your answers to (1) and (2) to include an arbitrary $H(j\omega)$.

Problem 3.4.11. All of our basic derivations assumed that the coefficients and the coordinate functions in the series expansions of signals and processes were real. In this problem we want to derive the analogous relations for complex coefficients and coordinate functions. We still assume that the signals and processes are real.

Derive the analogous results to those obtained in (12), (15), (18), (20), (40), (44), (46), (50), (128), and (154).

Problem 3.4.12. In (180) we considered a function

$$g_\lambda \triangleq \sum_{i=1}^{\infty} g(\lambda_i)$$

and derived its asymptotic value (181). Now consider the finite energy signal $s(t)$ and define

$$s_i \triangleq \int_{-T/2}^{T/2} s(t)\,\phi_i(t)\,dt,$$

where the $\phi_i(t)$ are the same eigenfunctions used to define (180). The function of interest is

$$g'_\lambda \triangleq \sum_{i=1}^{\infty} s_i^2\,g(\lambda_i).$$

Show that

$$g'_\lambda \simeq \int_{-\infty}^{\infty} |S(j\omega)|^2\,g(S_x(\omega))\,\frac{d\omega}{2\pi}$$

for large T, where the function $S(j\omega)$ is the Fourier transform of $s(t)$.

Section P3.5 Periodic Processes

Problem 3.5.1. Prove that if $R_x(\tau)$ is periodic then almost every sample function is periodic.

Problem 3.5.2. Show that if the autocorrelation $R_x(\tau)$ of a random process is such that

$$R_x(\tau_1) = R_x(0), \qquad \text{for some } \tau_1 \neq 0,$$

then $R_x(\tau)$ is periodic.

Problem 3.5.3. Consider the random process

$$x(t) = \sum_{n=1}^{N} a_n \cos(n\omega t + \theta_n).$$

The a_n ($n = 1, 2, \ldots, N$) are independent random variables:

$$E(a_1) = E(a_2) = \cdots E(a_N) = 0.$$

Var (a_1), Var (a_2), ..., Var (a_n) are different. The θ_n ($n = 1, 2, \ldots, N$) are identically distributed, independent random variables.

$$p_{\theta_n}(\theta) = \frac{1}{2\pi}, \qquad 0 \le \theta \le 2\pi.$$

The θ_n and a_n are independent.

1. Find $R_x(t_1, t_2)$.
2. Is the process wide-sense stationary?
3. Can you make any statements regarding the structure of particular sample functions?

Section P3.6 Integrated Transforms

Problem 3.6.1. Consider the feedback system in Fig. P3.7. The random processes $a(t)$ and $n(t)$ are statistically independent and stationary. The spectra $S_a(\omega)$ and $S_n(\omega)$ are known.

1. Find an expression for $Z_x(\omega)$, the integrated Fourier transform of $x(t)$.
2. Express $S_x(\omega)$ in terms of $S_a(\omega)$, $S_n(\omega)$, $G_1(j\omega)$, and $G_2(j\omega)$.

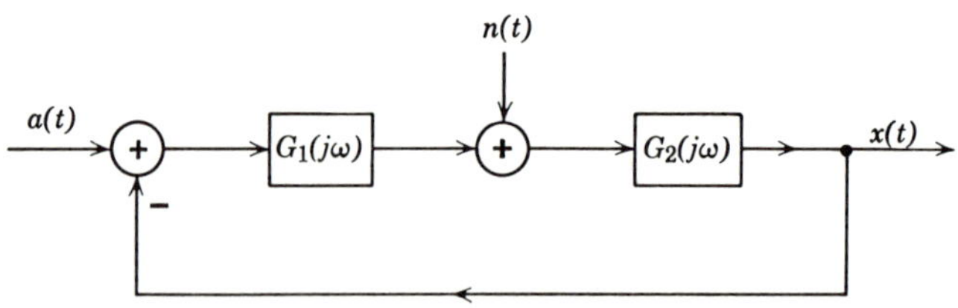

Fig. P3.7

Problem 3.6.2. From (221)

$$E[|X_{c,T}(\omega)|^2] = \int_{-T/2}^{T/2} dt \int_{-T/2}^{T/2} du\, R_x(t-u) \cos \omega t \cos \omega u.$$

Prove that the right side becomes arbitrarily large as $T \to \infty$.

Section P3.7 Vector Random Processes

Problem 3.7.1. Consider the spectral matrix

$$\mathbf{S}(\omega) = \begin{bmatrix} 2 & \dfrac{\sqrt{2k}}{j\omega + k} \\ \dfrac{\sqrt{2k}}{-j\omega + k} & \dfrac{2k}{\omega^2 + k^2} \end{bmatrix}.$$

Extend the techniques in Section 3.4 to find the vector eigenfunctions and eigenvalues for Method 1.

Problem 3.7.2. Investigate the asymptotic behavior (i.e., as T becomes large) of the eigenvalues and eigenfunctions in Method 1 for an arbitrary stationary matrix kernel.

Problem 3.7.3. Let $x_1(t)$ and $x_2(t)$ be statistically independent zero-mean random processes with covariance functions $K_{x_1}(t, u)$ and $K_{x_2}(t, u)$, respectively. The eigenfunctions and eigenvalues are

$$K_{x_1}(t, u): \lambda_i, \ \phi_i(t), \qquad i = 1, 2, \ldots,$$
$$K_{x_2}(t, u): \mu_i, \ \psi_i(t), \qquad i = 1, 2, \ldots.$$

Prove that the vector eigenfunctions and scalar eigenvalues can always be written as

$$\lambda_1, \begin{bmatrix} \phi_1(t) \\ 0 \end{bmatrix} : \mu_1, \begin{bmatrix} 0 \\ \psi_1(t) \end{bmatrix} : \lambda_2, \begin{bmatrix} \phi_2(t) \\ 0 \end{bmatrix} : \cdots.$$

Problem 3.7.4. Consider the vector process $\mathbf{r}(t)$ in which

$$\mathbf{r}(t) = \mathbf{a}(t) + \mathbf{n}(t), \qquad -\infty < t < \infty.$$

The processes $\mathbf{a}(t)$ and $\mathbf{n}(t)$ are statistically independent, with spectral matrices $\mathbf{S_a}(\omega)$ and $\mathbf{S_n}(\omega)$, respectively. Extend the idea of the integrated transform to the vector case. Use the approach in Section 3.6.2 and the differential operator introduced in Section 2.4.3 (2-239) to find the MAP estimate of $\mathbf{a}(t)$.

REFERENCES

[1] W. B. Davenport, Jr., and W. L. Root, *An Introduction to the Theory of Random Signals and Noise*, McGraw-Hill, New York, 1958.

[2] D. Middleton, *Introduction to Statistical Communication Theory*, McGraw-Hill, New York, 1960.

[3] R. Courant and D. Hilbert, *Methods of Mathematical Physics*, Interscience Publishers, New York, 1953.

[4] F. Riesz and B. Sz.-Nagy, *Functional Analysis*, Ungar, New York, 1955.

[5] W. V. Lovitt, *Linear Integral Equations*, McGraw-Hill, New York, 1924.

[6] F. G. Tricomi, *Integral Equations*, Interscience Publishers, New York, 1957.

[7] S. O. Rice, "Mathematical Analysis of Random Noise," *Bell System Tech. J.*, **23**, 282–332 (1944).

[8] A. Van der Ziel, *Noise*, Prentice-Hall, Englewood Cliffs, New Jersey, 1954.

[9] D. Slepian, "Estimation of Signal Parameters in the Presence of Noise," *Trans. IRE*, **PGIT-3**, 68 (March 1954).

[10] D. Youla, "The Solution of a Homogeneous Wiener-Hopf Integral Equation Occurring in the Expansion of Second-Order Stationary Random Functions," *Trans. IRE*, **IT-3**, 187–193 (September 1957).

[11] J. H. Laning and R. H. Battin, *Random Processes in Automatic Control,* McGraw-Hill, New York, 1956.
[12] S. Darlington, "Linear Least-Squares Smoothing and Predictions with Applications," *Bell System Tech. J.*, **37**, 1221–1294 (September 1958).
[13] C. W. Helstrom, *Statistical Theory of Signal Detection*, Pergamon, London, 1960.
[14] C. W. Helstrom, "Solution of the Detection Integral Equation for Stationary Filtered White Noise," *IEEE Trans. Inform. Theory*, **IT-11**, 335–339 (July 1965).
[15] D. Slepian and H. O. Pollak, "Prolate Spheroidal Wave Functions, Fourier Analysis and Uncertainty—I," *Bell System Tech. J.*, **40**, 43–64 (1961).
[16] H. J. Landau and H. O. Pollak, "Prolate Spheroidal Wave Functions, Fourier Analysis and Uncertainty—II," *Bell System Tech. J.*, **40**, 65–84 (1961).
[17] H. J. Landau and H. O. Pollak, "Prolate Spheroidal Wave Functions, Fourier Analysis and Uncertainty—III: The Dimension of the Space of Essentially Time- and Band-Limited Signals," *Bell System Tech J.*, **41**, 1295, 1962.
[18] C. Flammer, *Spheroidal Wave Functions*, Stanford University Press, Stanford, 1957.
[19] J. A. Stratton, P. M. Morse, L. J. Chu, J. D. C. Little, and F. J. Corbato, *Spheroidal Wave Functions*, MIT Press and Wiley, New York, 1956.
[20] E. Parzen, *Stochastic Processes*, Holden-Day, San Francisco, 1962.
[21] M. Rosenblatt, *Random Processes*, Oxford University Press, New York, 1962.
[22] L. A. Zadeh and J. R. Ragazzini, "Optimum Filters for the Detection of Signals in Noise," *Proc. I.R.E.*, **40**, 1223, 1952.
[23] R. Y. Huang and R. A. Johnson, "Information Transmission with Time-Continuous Random Processes," *IEEE Trans. Inform. Theory*, **IT-9**, No. 2, 84–95 (April 1963).
[24] E. J. Kelly and W. L. Root, "A Representation of Vector-Valued Random Processes," *Group Rept. 55–21*, revised, MIT, Lincoln Laboratory, April 22, 1960.
[25] K. Karhunen, "Über Linearen Methoden in der Wahrscheinlichkeitsrechnung," *Ann. Acad. Sci. Fennical*, Ser. A, **1**, No. 2.
[26] M. Loéve, *Probability Theory*, Van Nostrand, Princeton, New Jersey, 1955.
[27] B. V. Gnedenko, *The Theory of Probability*, Chelsea, New York, 1962.
[28] M. S. Bartlett, *An Introduction to Stochastic Processes*, Cambridge University Press, Cambridge, 1961.
[29] A. Papoulis, *Probability, Random Variables, and Stochastic Processes*, McGraw-Hill, New York, 1965.
[30] M. Loeve, "Sur les Fonctions Aleatoires Stationnaires de Second Order," *Rev. Sci.*, **83**, 297–310 (1945).
[31] F. B. Hildebrand, *Methods of Applied Mathematics*, Prentice-Hall, Inc., Englewood Cliffs, N.J., 1952.
[32] A. B. Baggeroer, "A State-Variable Technique for the Solution of Fredholm Integral Equations," 1967 IEEE International Conference on Communication, Minneapolis, Minnesota, June 12–14, 1967.
[33] A. B. Baggeroer, "A State-Variable Technique for the Solution of Fredholm Integral Equations," R.L.E., Technical Report No. 459, M.I.T. July 15, 1967.
[34] D. O. North, "Analysis of the Factors Which Determine Signal/Noise Discrimination in Radar," RCA Tech. Rep. PTR-6-C, June 1943; reprinted in *Proc. IRE*, Vol. 51, pp. 1016–1028 (July 1963).
[35] T. Kailath, "Some Integral Equations with 'Nonrational' Kernels," *IEEE Transactions on Information Theory*, Vol. IT-12, No. 4, pp. 442–447 (October 1966).

4

Detection of Signals —Estimation of Signal Parameters

4.1 INTRODUCTION

In Chapter 2 we formulated the detection and estimation problems in the classical context. In order to provide background for several areas, we first examined a reasonably general problem. Then, in Section 2.6 of Chapter 2, we investigated the more precise results that were available in the general Gaussian case.

In Chapter 3 we developed techniques for representing continuous processes by sets of numbers. The particular representation that we considered in detail was appropriate primarily for Gaussian processes.

We now want to use these representations to extend the results of the classical theory to the case in which the observations consist of *continuous waveforms*.

4.1.1 Models

The problems of interest to us in this chapter may be divided into two categories. The first is the detection problem which arises in three broad areas: digital communications, radar/sonar, and pattern recognition and classification. The second is the signal parameter estimation problem which also arises in these three areas.

Detection. The conventional model of a simple digital communication system is shown in Fig. 4.1. The source puts out a binary digit (either 0 or 1) every T seconds. The most straightforward system would transmit either $\sqrt{E_{t_0}}\, s_0(t)$ or $\sqrt{E_{t_1}}\, s_1(t)$ during each interval. In a typical space communication system an attenuated version of the transmitted signal would be received with negligible distortion. The received signal consists of $\sqrt{E_0}\, s_0(t)$ or $\sqrt{E_1}\, s_1(t)$ plus an additive noise component.

The characterization of the noise depends on the particular application. One source, always present, is thermal noise in the receiver front end. This

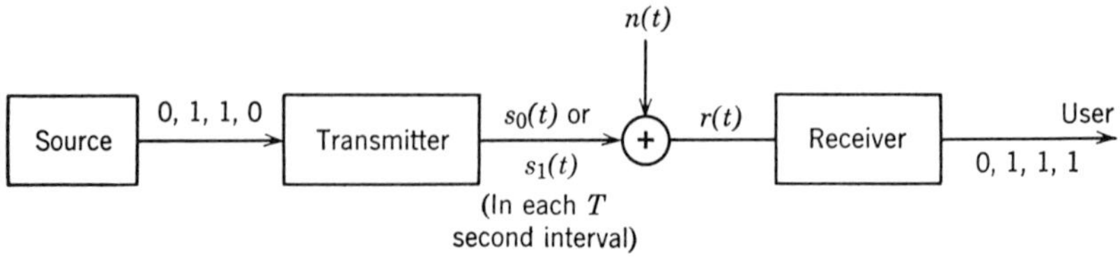

Fig. 4.1 A digital communication system.

noise can be modeled as a sample function from a Gaussian random process. As we proceed to more complicated models, we shall encounter other sources of interference that may turn out to be more important than the thermal noise. In many cases we can redesign the system to eliminate these other interference effects almost entirely. Then the thermal noise will be the disturbance that limits the system performance. In most systems the spectrum of the thermal noise is flat over the frequency range of interest, and we may characterize it in terms of a spectral height of $N_0/2$ joules. An alternate characterization commonly used is effective noise temperature T_e (e.g., Valley and Wallman [1] or Davenport and Root [2], Chapter 10). The two are related simply by

$$N_0 = kT_e, \tag{1}$$

where k is Boltzmann's constant, 1.38×10^{-23} joule/°K and T_e is the effective noise temperature, °K.

Thus in this particular case we could categorize the receiver design as a problem of detecting one of two *known signals in the presence of additive white Gaussian noise.*

If we look into a possible system in more detail, a typical transmitter could be as shown in Fig. 4.2. The transmitter has an oscillator with nominal center frequency of ω_c. It is biphase modulated according to whether the source output is 1 (0°) or 0 (180°). The oscillator's instantaneous phase varies slowly, and the receiver must include some auxiliary equipment to measure the oscillator phase. If the phase varies slowly enough, we shall see that accurate measurement is possible. If this is true, the problem may be modeled as above. If the measurement is not accurate, however, we must incorporate the phase uncertainty in our model.

A second type of communication system is the point-to-point ionospheric scatter system shown in Fig. 4.3 in which the transmitted signal is scattered by the layers in the ionosphere. In a typical system we can transmit a "one" by sending a sine wave of a given frequency and a "zero" by a sine wave of another frequency. The receiver signal may vary as shown in Fig. 4.4. Now, the receiver has a signal that fluctuates in amplitude and phase.

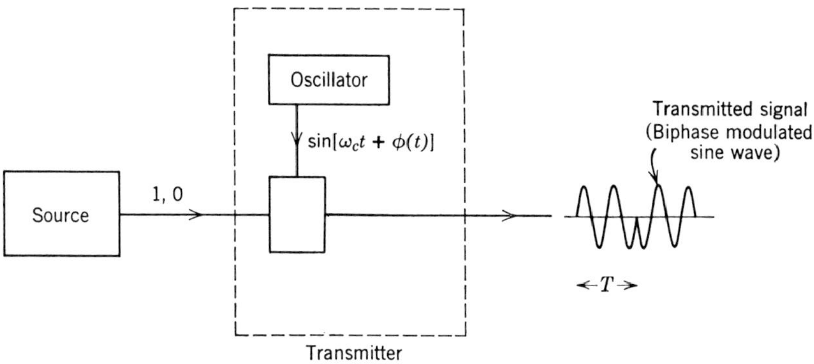

Fig. 4.2 Details of typical system.

In the commonly-used frequency range most of the additive noise is Gaussian.

Corresponding problems are present in the radar context. A conventional pulsed radar transmits a signal as shown in Fig. 4.5. If a target is present, the sequence of pulses is reflected. As the target fluctuates, the amplitude and phase of the reflected pulses change. The returned signal consists of a sequence of pulses whose amplitude and phase are unknown. The problem

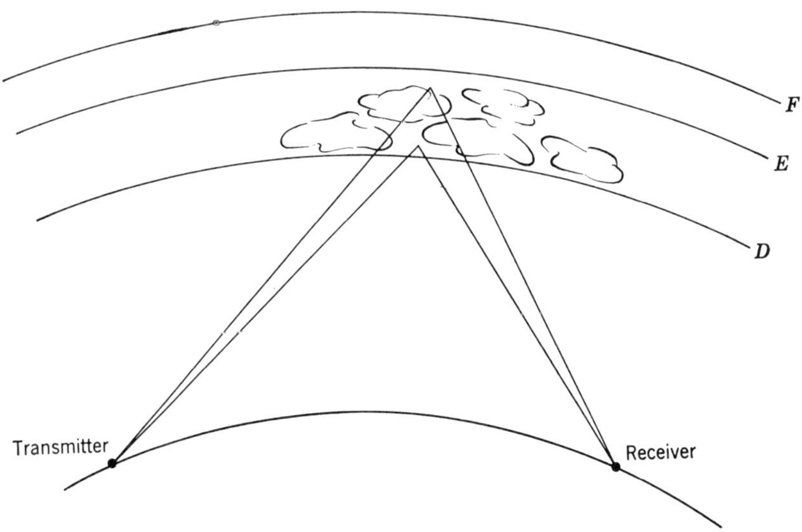

Fig. 4.3 Ionospheric scatter link.

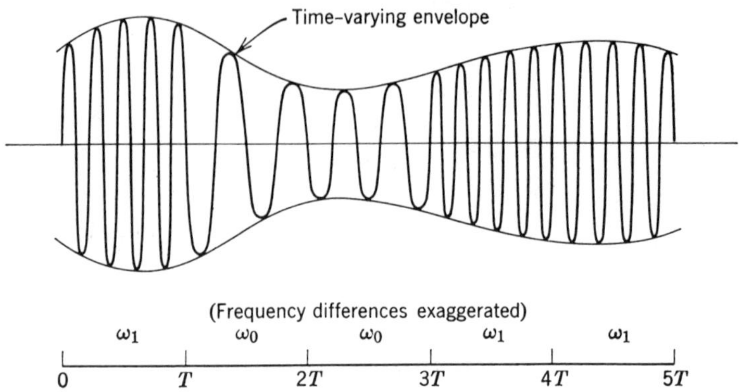

Fig. 4.4 Signal component in time-varying channel.

is to examine this sequence in the presence of receiver noise and decide whether a target is present.

There are obvious similarities between the two areas, but there are also some differences:

1. In a digital communication system the two types of error (say 1, when 0 was sent, and vice versa) are usually of equal importance. Furthermore, a signal may be present on both hypotheses. This gives a symmetry to the problem that can be exploited. In a radar/sonar system the two types of

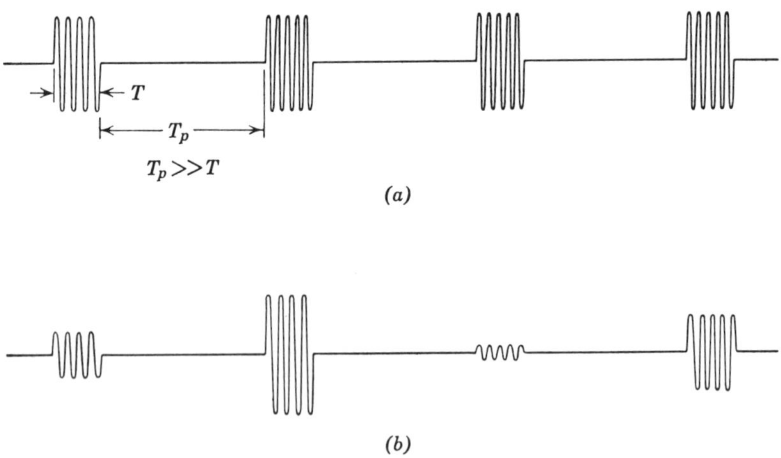

Fig. 4.5 Signals in radar model: (*a*) transmitted sequence of rf pulses; (*b*) received sequence [amplified (time-shift not shown)].

error are almost always of unequal importance. In addition, a signal is present only on one hypothesis. This means that the problem is generally nonsymmetric.

2. In a digital communication system the probability of error is usually an adequate measure of system performance. Normally, in radar/sonar a reasonably complete ROC is needed.

3. In a digital system we are sending a sequence of digits. Thus we can correct digit errors by putting some structure into the sequence. In the radar/sonar case this is not an available alternative.

In spite of these differences, a great many of the basic results will be useful for both areas.

Estimation. The second problem of interest is the estimation of signal parameters, which is encountered in both the communications and radar/sonar areas. We discuss a communication problem first.

Consider the analog message source shown in Fig. 4.6*a*. For simplicity we assume that it is a sample function from a bandlimited random process ($2W$ cps: double-sided). We could then sample it every $1/2W$ seconds without losing any information. In other words, given these samples at the

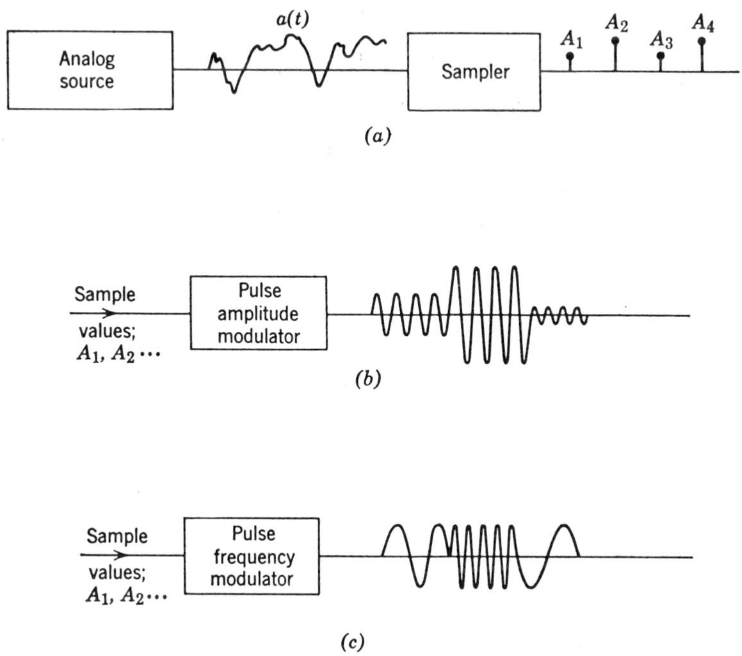

Fig. 4.6 Analog message transmission.

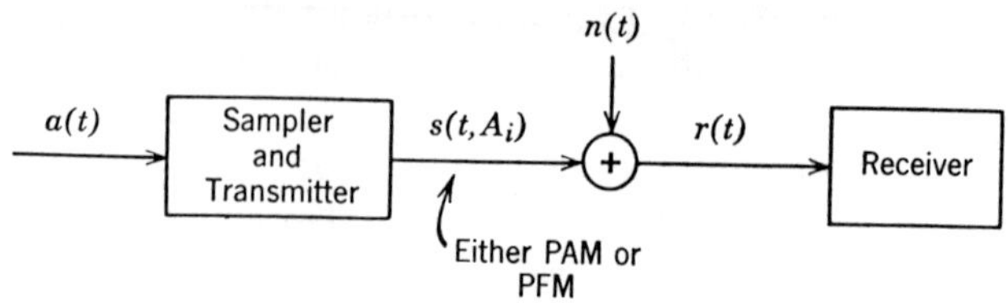

Fig. 4.7 A parameter transmission system.

receiver, we could reconstruct the message exactly (e.g. Nyquist, [4] or Problem 3.3.6). Every T seconds ($T = 1/2W$) we transmit a signal that depends on the particular value A_i at the last sampling time. In the system in Fig. 4.6*b* the amplitude of a sinusoid depends on the value of A_i. This system is referred to as a pulse-amplitude modulation system (PAM). In the system in Fig. 4.6*c* the frequency of the sinusoid depends on the sample value. This system is referred to as a pulse frequency modulation system (PFM). The signal is transmitted over a channel and is corrupted by noise (Fig. 4.7). The received signal in the ith interval is:

$$r(t) = s(t, A_i) + n(t), \qquad T_i \le t \le T_{i+1}. \tag{2}$$

The purpose of the receiver is to estimate the values of the successive A_i and use these estimates to reconstruct the message.

A typical radar system is shown in Fig. 4.8. In a conventional pulsed radar the transmitted signal is a sinusoid with a rectangular envelope.

$$\begin{aligned} s_t(t) &= \sqrt{2E_t} \sin \omega_c t, \qquad && 0 \le t \le T, \\ &= 0, && \text{elsewhere.} \end{aligned} \tag{3a}$$

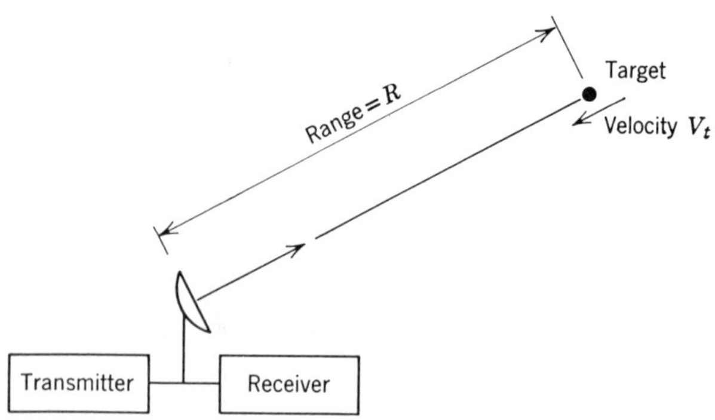

Fig. 4.8 Radar system diagram.

The returned signal is delayed by the round-trip time to the target. If the target is moving, there is Doppler shift. Finally there is a random amplitude and phase due to the target fluctuation. The received signal in the absence of noise is

$$\begin{aligned} s_r(t) &= v\sqrt{2E_t}\sin[(\omega_c + \omega_D)(t - \tau) + \phi], &\quad \tau \le t \le \tau + T. \\ &= 0, &\quad \text{elsewhere.} \end{aligned} \tag{3b}$$

Here we estimate τ and ω_D (or, equivalently, target range and velocity). Once again, there are obvious similarities between the communication and radar/sonar problem. The basic differences are the following:

1. In the communications context A_i is a random variable with a probability density that is usually known. In radar the range or velocity limits of interest are known. The parameters, however, are best treated as nonrandom variables (e.g., discussion in Section 2.4).
2. In the radar case the difficulty may be compounded by a lack of knowledge regarding the target's presence. Thus the detection and estimation problem may have to be combined.
3. In almost all radar problems a phase reference is not available.

Other models of interest will appear naturally in the course of our discussion.

4.1.2 Format

Because this chapter is long, it is important to understand the over-all structure. The basic approach has three steps:

1. The observation consists of a waveform $r(t)$. Thus the observation space may be infinite dimensional. Our first step is to map the received signal into some convenient decision or estimation space. This will reduce the problem to one studied in Chapter 2.
2. In the detection problem we then select decision regions and compute the ROC or $\Pr(\epsilon)$. In the estimation problem we evaluate the variance or mean-square error.
3. We examine the results to see what they imply about system design and performance.

We carry out these steps for a sequence of models of increasing complexity (Fig. 4.9) and develop the detection and estimation problem in parallel. By emphasizing their parallel nature for the simple cases, we can save appreciable effort in the more complex cases by considering only one problem in the text and leaving the other as an exercise. We start with the simple models and then proceed to the more involved.

Channel	Signal detection	Signal parameter estimation
Additive white noise	Simple binary	Single parameter, linear
Additive colored noise	General binary	Single parameter, nonlinear
Simple random	M-ary	Multiple parameter
Multiple channels		

Fig. 4.9 Sequence of models.

A logical question is: if the problem is so simple, why is the chapter so long? This is a result of our efforts to determine how the model and its parameters affect the design and performance of the system. We feel that only by examining some representative problems in detail can we acquire an appreciation for the implications of the theory.

Before proceeding to the solution, a brief historical comment is in order. The mathematical groundwork for our approach to this problem was developed by Grenander [5]. The detection problem relating to optimum radar systems was developed at the M.I.T. Radiation Laboratory, (e.g., Lawson and Uhlenbeck [6]) in the early 1940's. Somewhat later Woodward and Davies [7, 8] approached the radar problem in a different way. The detection problem was formulated at about the same time in a manner similar to ours by both Peterson, Birdsall, and Fox [9] and Middleton and Van Meter [10], whereas the estimation problem was first done by Slepian [11]. Parallel results with a communications emphasis were developed by Kotelnikov [12, 13] in Russia. Books that deal almost exclusively with radar include Helstrom [14] and Wainstein and Zubakov [15]. Books that deal almost exclusively with communication include Kotelnikov [13], Harman [16], Baghdady (ed.) [17], Wozencraft and Jacobs [18], and Golomb et al. [19]. The last two parts of Middleton [47] cover a number of topics in both areas. By presenting the problems side by side we hope to emphasize their inherent similarities and contrast their differences.

4.2 DETECTION AND ESTIMATION IN WHITE GAUSSIAN NOISE

In this section we formulate and solve the detection and estimation problems for the case in which the interference is additive white Gaussian noise.

We consider the detection problem first: the simple binary case, the general binary case, and the M-ary case are discussed in that order. By using the concept of a sufficient statistic the optimum receiver structures are simply derived and the performances for a number of important cases are evaluated. Finally, we study the sensitivity of the optimum receiver to the detailed assumptions of our model.

As we have seen in the classical context, the decision and estimation problems are closely related; linear estimation will turn out to be essentially the same as simple binary detection. When we proceed to the nonlinear estimation problem, new issues will develop, both in specifying the estimator structure and in evaluating its performance.

4.2.1 Detection of Signals in Additive White Gaussian Noise

Simple Binary Detection. In the simplest binary decision problem the received signal under one hypothesis consists of a completely known signal, $\sqrt{E}\, s(t)$, corrupted by an additive zero-mean white Gaussian noise $w(t)$ with spectral height $N_0/2$; the received signal under the other hypothesis consists of the noise $w(t)$ alone. Thus

$$\begin{aligned} r(t) &= \sqrt{E}\, s(t) + w(t), \qquad && 0 \le t \le T: H_1, \\ &= w(t), && 0 \le t \le T: H_0. \end{aligned} \tag{4}$$

For convenience we assume that

$$\int_0^T s^2(t)\, dt = 1, \tag{5}$$

so that E represents the received signal energy. The problem is to observe $r(t)$ over the interval $[0, T]$ and decide whether H_0 or H_1 is true. The criterion may be either Bayes or Neyman-Pearson.

The following ideas will enable us to solve this problem easily:

1. Our observation is a time-continuous random waveform. The first step is to reduce it to a set of random variables (possibly a countably infinite set).
2. One method is the series expansion of Chapter 3:

$$r(t) = \underset{K \to \infty}{\text{l.i.m.}} \sum_{i=1}^{K} r_i\, \phi_i(t); \qquad 0 \le t \le T. \tag{6}$$

When $K = K'$, there are K' coefficients in the series, $r_1, \ldots, r_{K'}$ which we could denote by the vector $\mathbf{r}_{K'}$. In our subsequent discussion we suppress the K' subscript and denote the coefficients by $\mathbf{r}$.

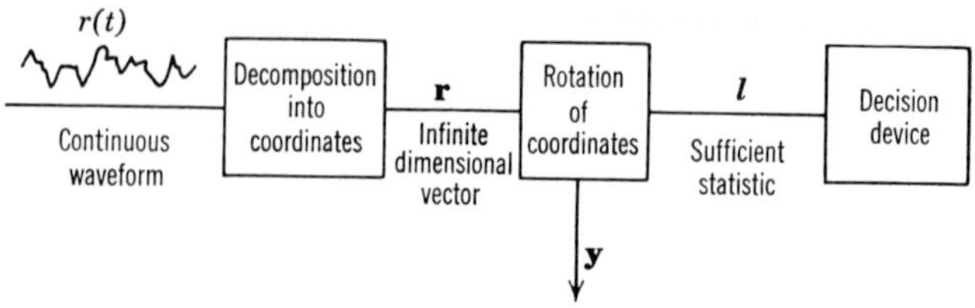

Fig. 4.10 Generation of sufficient statistics.

3. In Chapter 2 we saw that if we transformed **r** into two independent vectors, $\boldsymbol{l}$ (the sufficient statistic) and **y**, as shown in Fig. 4.10, our decision could be based only on $\boldsymbol{l}$, because the values of **y** did not depend on the hypothesis. The advantage of this technique was that it reduced the dimension of the decision space to that of $\boldsymbol{l}$. Because this is a binary problem we know that $\boldsymbol{l}$ will be one-dimensional.

Here the method is straightforward. If we choose the first orthonormal function to be $s(t)$, the first coefficient in the decomposition is the Gaussian random variable,

$$r_1 = \begin{cases} \displaystyle\int_0^T s(t)\, w(t)\, dt \triangleq w_1 : H_0, \\ \displaystyle\int_0^T s(t)[\sqrt{E}\, s(t) + w(t)]\, dt = \sqrt{E} + w_1 : H_1. \end{cases} \tag{7}$$

The remaining r_i $(i > 1)$ are Gaussian random variables which can be generated by using some arbitrary orthonormal set whose members are orthogonal to $s(t)$.

$$r_i = \begin{cases} \displaystyle\int_0^T \phi_i(t)\, w(t)\, dt \triangleq w_i : H_0, \\ \displaystyle\int_0^T \phi_i(t)[\sqrt{E}\, s(t) + w(t)]\, dt = w_i : H_1, \end{cases} \qquad i \neq 1. \tag{8}$$

From Chapter 3 (44) we know that

$$E(w_i w_j) = 0; \qquad i \neq j.$$

Because w_i and w_j are jointly Gaussian, they are statistically independent (see Property 3 on p. 184).

We see that *only* r_1 depends on which hypothesis is true. Further, all r_i $(i > 1)$ are statistically independent of r_1. Thus r_1 is a sufficient statistic $(r_1 = l)$. The other r_i correspond to **y**. Because they will not affect the decision, there is no need to compute them.

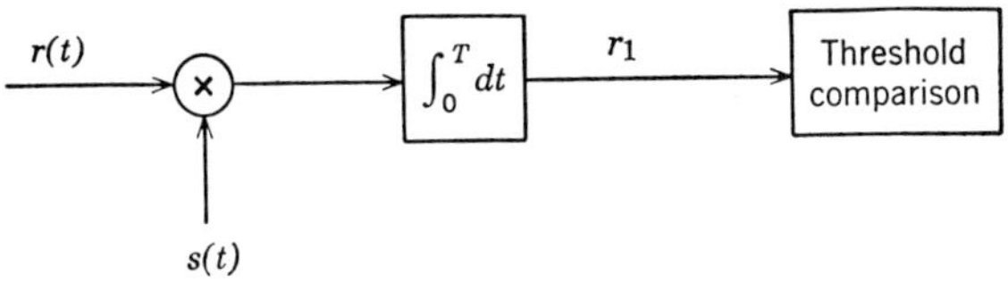

Fig. 4.11 Correlation receiver.

Several equivalent receiver structures follow immediately. The structure in Fig. 4.11 is called a *correlation receiver*. It correlates the input $r(t)$ with a stored replica of the signal $s(t)$. The output is r_1, which is a sufficient statistic ($r_1 = l$) and is a Gaussian random variable. Once we have obtained r_1, the decision problem will be identical to the classical problem in Chapter 2 (specifically, Example 1 on pp. 27–28). We compare l to a threshold in order to make a decision.

An equivalent realization is shown in Fig. 4.12. The impulse response of the linear system is simply the signal reversed in time and shifted,

$$h(\tau) = s(T - \tau). \tag{9}$$

The output at time T is the desired statistic l. This receiver is called a *matched filter receiver*. (It was first derived by North [20].) The two structures are mathematically identical; the choice of which structure to use depends solely on ease of realization.

Just as in Example 1 of Chapter 2, the sufficient statistic l is Gaussian under either hypothesis. Its mean and variance follow easily:

$$\begin{aligned} E(l|\,H_1) &= E(r_1|\,H_1) = \sqrt{E}, \\ E(l|\,H_0) &= E(r_1|\,H_0) = 0, \\ \text{Var}\,(l|\,H_0) &= \text{Var}\,(l|\,H_1) = \frac{N_0}{2}. \end{aligned} \tag{10}$$

Thus we can use the results of Chapter 2, (64)–(68), with

$$d = \left(\frac{2E}{N_0}\right)^{1/2}. \tag{11}$$

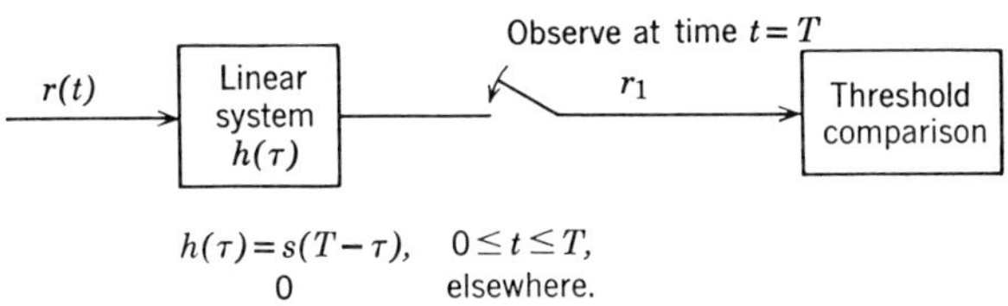

Fig. 4.12 Matched filter receiver.

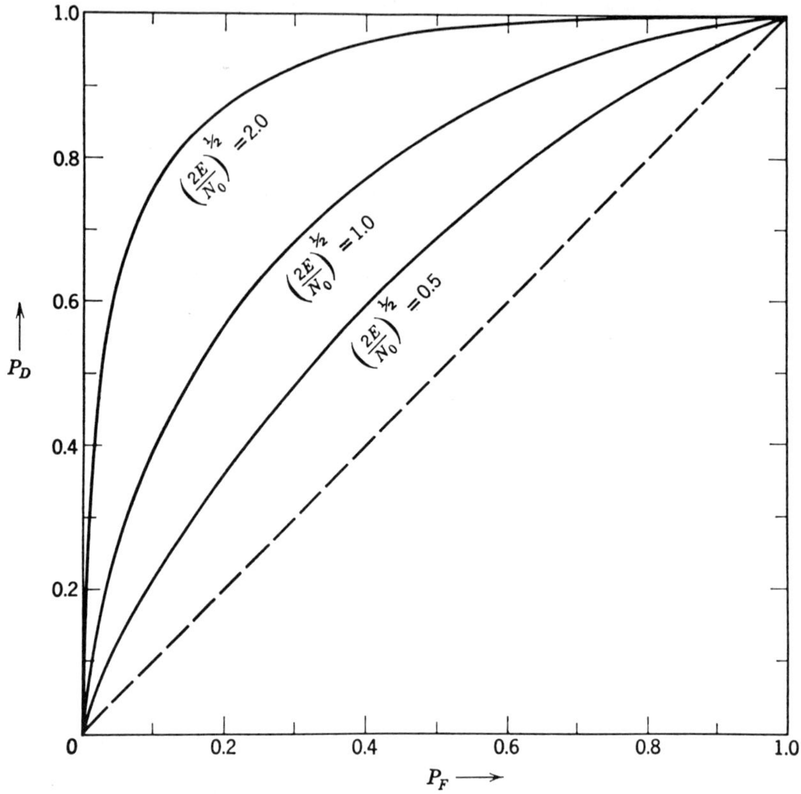

Fig. 4.13 Receiver operating characteristic: known signal in additive white Gaussian noise.

The curves in Figs. 2.9*a* and 2.9*b* of Chapter 2 are directly applicable and are reproduced as Figs. 4.13 and 4.14. We see that the performance depends only on the received signal energy E and the noise spectral height N_0—the signal shape is not important. This is intuitively logical because the noise is the same along any coordinate.

The key to the simplicity in the solution was our ability to reduce an infinite dimensional observation space to a one-dimensional decision space by exploiting the idea of a sufficient statistic. Clearly, we should end up with the same receiver even if we do not recognize that a sufficient statistic is available. To demonstrate this we construct the likelihood ratio directly. Three observations lead us easily to the solution.

1. If we approximate $r(t)$ in terms of some finite set of numbers,

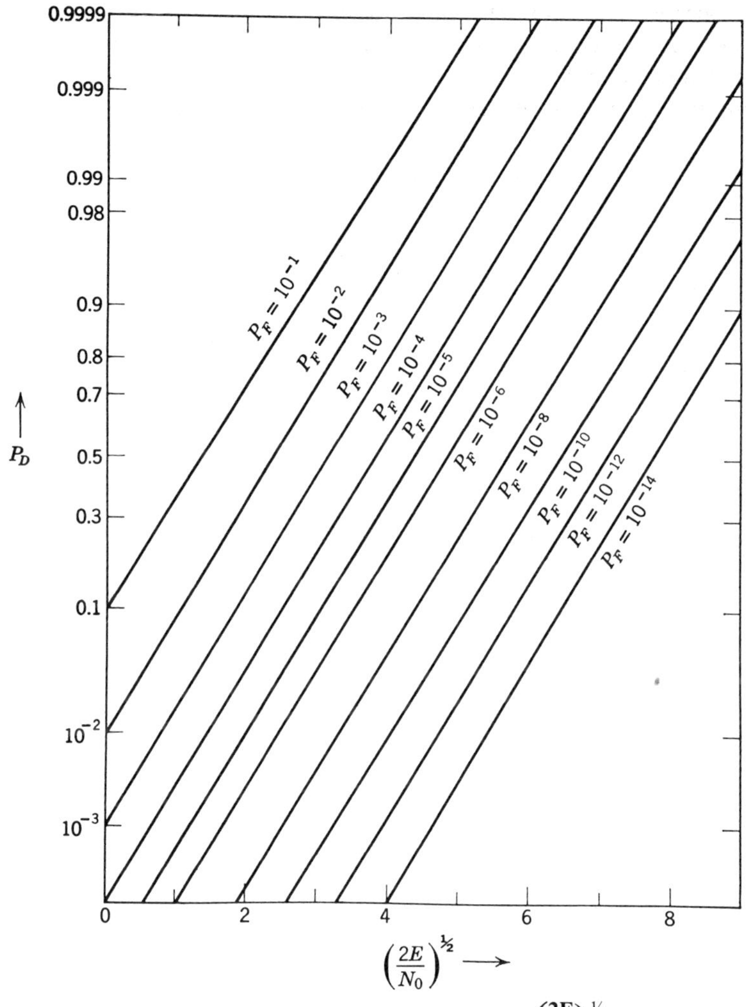

Fig. 4.14 Probability of detection vs $\left(\frac{2E}{N_0}\right)^{1/2}$.

$r_1, \ldots, r_K$, $(\mathbf{r})$, we have a problem in classical detection theory that we can solve.

2. If we choose the set $r_1, r_2, \ldots, r_K$ so that

$$p_{r_1, r_2, \ldots, r_K | H_i}(R_1, R_2, \ldots, R_K | H_i) = \prod_{j=1}^{K} p_{r_j | H_i}(R_j | H_i), \qquad i = 0, 1, \quad (12)$$

that is, the observations are conditionally independent, we have an *easy* problem to solve.

3. Because we know that it requires an infinite set of numbers to represent $r(t)$ completely, we want to get the solution in a convenient form so that we can let $K \to \infty$.

We denote the approximation that uses K coefficients as $r_K(t)$. Thus

$$r_K(t) = \sum_{i=1}^{K} r_i \, \phi_i(t), \qquad 0 \le t \le T, \tag{13}$$

where

$$r_i = \int_0^T r(t) \, \phi_i(t) \, dt, \qquad i = 1,2,\ldots K, \tag{14}$$

and the $\phi_i(t)$ belong to an *arbitrary* complete orthonormal set of functions. Using (14), we see that under H_0

$$r_i = \int_0^T w(t) \, \phi_i(t) \, dt = w_i, \tag{15}$$

and under H_1

$$r_i = \int_0^T \sqrt{E} s(t) \, \phi_i(t) \, dt + \int_0^T w(t) \, \phi_i(t) \, dt = s_i + w_i. \tag{16}$$

The coefficients s_i correspond to an expansion of the signal

$$s_K(t) \triangleq \sum_{i=1}^{K} s_i \, \phi_i(t), \qquad 0 \le t \le T, \tag{17}$$

and

$$\sqrt{E} \, s(t) = \lim_{K \to \infty} s_K(t). \tag{18}$$

The r_i's are Gaussian with known statistics:

$$\begin{aligned} E(r_i|H_0) &= 0, \\ E(r_i|H_1) &= s_i, \\ \operatorname{Var}(r_i|H_0) &= \operatorname{Var}(r_i|H_1) = \frac{N_0}{2}. \end{aligned} \tag{19}$$

Because the noise is "white," these coefficients are independent along any set of coordinates. The likelihood ratio is

$$\Lambda[r_K(t)] = \frac{p_{\mathbf{r}|H_1}(\mathbf{R}|H_1)}{p_{\mathbf{r}|H_0}(\mathbf{R}|H_0)} = \frac{\prod_{i=1}^{K} \frac{1}{\sqrt{\pi N_0}} \exp\left(-\frac{1}{2}\frac{(R_i - s_i)^2}{N_0/2}\right)}{\prod_{i=1}^{K} \frac{1}{\sqrt{\pi N_0}} \exp\left(-\frac{1}{2}\frac{R_i^2}{N_0/2}\right)}. \tag{20}$$

Taking the logarithm and canceling common terms, we have

$$\ln \Lambda[r_K(t)] = \frac{2}{N_0} \sum_{i=1}^{K} R_i s_i - \frac{1}{N_0} \sum_{i=1}^{K} s_i^2. \tag{21}$$

The two sums are easily expressed as integrals. From Parseval's theorem,

$$\sum_{i=1}^{K} R_i s_i = \int_0^T r_K(t)\, s_K(t)\, dt$$

and

$$\sum_{i=1}^{K} s_i^2 = \int_0^T s_K^2(t)\, dt. \tag{22}$$

We now have the log likelihood ratio in a form in which it is convenient to pass to the limit:

$$\underset{K\to\infty}{\text{l.i.m.}}\ \ln \Lambda[r_K(t)] \triangleq \ln \Lambda[r(t)] = \frac{2}{N_0}\int_0^T r(t)\, s(t)\, dt - \frac{E}{N_0}. \tag{23}$$

The first term is just the sufficient statistic we obtained before. The second term is a bias. The resulting likelihood ratio test is

$$\frac{2}{N_0}\int_0^T r(t)\, s(t)\, dt \underset{H_0}{\overset{H_1}{\gtrless}} \ln \eta + \frac{E}{N_0}. \tag{24}$$

(Recall from Chapter 2 that η is a constant which depends on the costs and a priori probabilities in a Bayes test and the desired P_F in a Neyman-Pearson test.) It is important to observe that even though the probability density $p_{r(t)|H_i}(r(t)|H_i)$ is not well defined for either hypothesis, the likelihood ratio is.

Before going on to more general problems it is important to emphasize the two separate features of the signal detection problem:

1. First we reduce the received waveform to a single number which is a point in a decision space. This operation is performed physically by a correlation operation and is invariant to the decision criterion that we plan to use. This invariance is important because it enables us to construct the waveform processor without committing ourselves to a particular criterion.

2. Once we have transformed the received waveform into the decision space we have only the essential features of the problem left to consider. Once we get to the decision space the problem is the same as that studied in Chapter 2. The actual received waveform is no longer important and all physical situations that lead to the same picture in a decision space are identical for our purposes. In our simple example we saw that all signals of equal energy map into the same point in the decision space. It is therefore obvious that the signal shape is unimportant.

The separation of these two parts of the problem leads to a clearer understanding of the fundamental issues.

General Binary Detection in White Gaussian Noise. The results for the simply binary problem extend easily to the general binary problem. Let

$$\begin{aligned} r(t) &= \sqrt{E_1}\, s_1(t) + w(t), \qquad 0 \le t \le T{:}H_1, \\ &= \sqrt{E_0}\, s_0(t) + w(t), \qquad 0 \le t \le T{:}H_0, \end{aligned} \tag{25}$$

where $s_0(t)$ and $s_1(t)$ are normalized but are *not* necessarily orthogonal. We denote the correlation between the two signals as

$$\rho \triangleq \int_0^T s_0(t)\, s_1(t)\, dt.$$

(Note that $|\rho| \le 1$ because the signals are normalized.)

We choose our first two orthogonal functions as follows:

$$\phi_1(t) = s_1(t), \qquad 0 \le t \le T, \tag{26}$$

$$\phi_2(t) = \frac{1}{\sqrt{1-\rho^2}}\,[s_0(t) - \rho s_1(t)], \qquad 0 \le t \le T. \tag{27}$$

We see that $\phi_2(t)$ is obtained by subtracting out the component of $s_0(t)$ that is correlated with $\phi_1(t)$ and normalizing the result. The remaining $\phi_i(t)$ consist of an arbitrary orthonormal set whose members are orthogonal to $\phi_1(t)$ and $\phi_2(t)$ and are chosen so that the entire set is complete. The coefficients are

$$r_i = \int_0^T r(t)\, \phi_i(t)\, dt; \qquad i = 1, 2, \ldots. \tag{28}$$

All of the r_i except r_1 and r_2 do not depend on which hypothesis is true and are statistically independent of r_1 and r_2. Thus a two-dimensional decision region, shown in Fig. 4.15*a*, is adequate. The mean value of r_i along each coordinate is

$$E[r_i|H_0] = \sqrt{E_0}\int_0^T s_0(t)\, \phi_i(t)\, dt, \triangleq s_{0i}, \qquad i = 1, 2, \quad {:}H_0, \tag{29}$$

and

$$E[r_i|H_1] = \sqrt{E_1}\int_0^T s_1(t)\, \phi_i(t)\, dt, \triangleq s_{1i}, \qquad i = 1, 2, \quad {:}H_1. \tag{30}$$

The likelihood ratio test follows directly from Section 2.6 (2.327)

$$\ln \Lambda = -\frac{1}{N_0}\sum_{i=1}^{2} (R_i - s_{1i})^2 + \frac{1}{N_0}\sum_{i=1}^{2} (R_i - s_{0i})^2 \underset{H_0}{\overset{H_1}{\gtrless}} \ln \eta, \tag{31a}$$

$$\ln \Lambda = -\frac{1}{N_0}|\mathbf{R} - \mathbf{s}_1|^2 + \frac{1}{N_0}|\mathbf{R} - \mathbf{s}_0|^2 \underset{H_0}{\overset{H_1}{\gtrless}} \ln \eta, \tag{31b}$$

or, canceling common terms and rearranging the result,

$$\mathbf{R}^T(\mathbf{s}_1 - \mathbf{s}_0) \underset{H_0}{\overset{H_1}{\gtrless}} \frac{N_0}{2}\ln \eta + \frac{1}{2}(|\mathbf{s}_1|^2 - |\mathbf{s}_0|^2). \tag{31c}$$

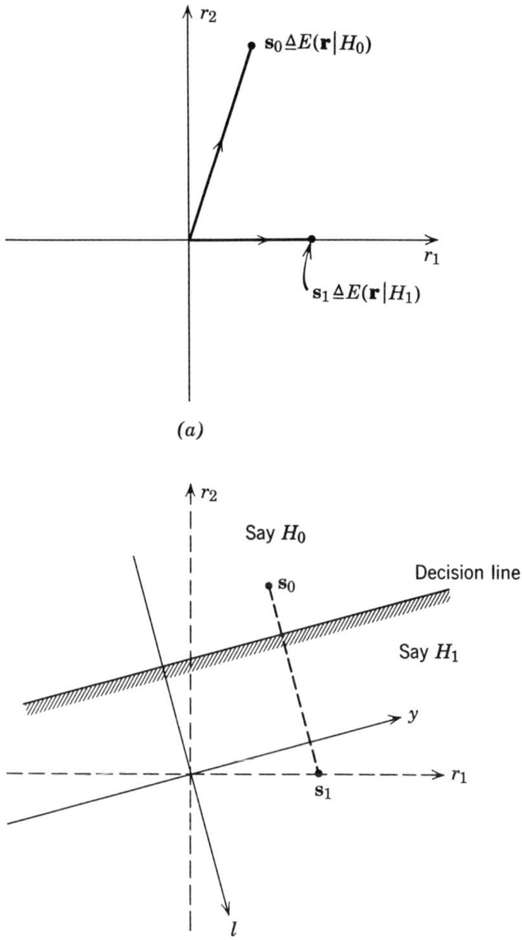

(*a*)

(*b*)

Fig. 4.15 Decision spaces.

Thus only the product of $\mathbf{R}^T$ with the difference vector $\mathbf{s}_1 - \mathbf{s}_0$ is used to make a decision. Therefore the decision space is divided into two parts by a line perpendicular to $\mathbf{s}_1 - \mathbf{s}_0$ as shown in Fig. 4.15*b*. The noise components along the r_1 and r_2 axes are independent and identically distributed.

Now observe that we can transform the coordinates as shown in Fig. 4.15*b*. The noises along the new coordinates are still independent, but only the coefficient along the l coordinate depends on the hypothesis and the y coefficient may be disregarded. Therefore we can simplify our receiver by generating l instead of r_1 and r_2. The function needed to generate the statistic is just the normalized version of the difference signal. Denote the difference signal by $s_\Delta(t)$:

$$s_\Delta(t) \triangleq \sqrt{E_1}\, s_1(t) - \sqrt{E_0}\, s_0(t). \tag{32}$$

The normalized function is

$$f_\Delta(t) = \frac{\sqrt{E_1}\, s_1(t) - \sqrt{E_0}\, s_0(t)}{(E_1 - 2\rho\sqrt{E_0 E_1} + E_0)^{1/2}}. \tag{33}$$

The receiver is shown in Fig. 4.16. (Note that this result could have been obtained directly by choosing $f_\Delta(t)$ as the first orthonormal function.)

Thus once again the binary problem reduces to a one-dimensional decision space. The statistic l is Gaussian:

$$E(l|H_1) = \frac{E_1 - \sqrt{E_0 E_1}\,\rho}{(E_1 - 2\rho\sqrt{E_0 E_1} + E_0)^{1/2}}, \tag{34}$$

$$E(l|H_0) = \frac{\sqrt{E_0 E_1}\,\rho - E_0}{(E_1 - 2\rho\sqrt{E_0 E_1} + E_0)^{1/2}}. \tag{35}$$

The variance is $N_0/2$ as before. Thus

$$\boxed{d^2 = \frac{2}{N_0}(E_1 + E_0 - 2\rho\sqrt{E_0 E_1}).} \tag{36}$$

Observe that if we normalized our coordinate system so that noise variance was unity then d would be the distance between the two signals. The resulting probabilities are

$$P_F = \operatorname{erfc}_*\left(\frac{\ln \eta}{d} + \frac{d}{2}\right), \tag{37}$$

$$P_D = \operatorname{erfc}_*\left(\frac{\ln \eta}{d} - \frac{d}{2}\right). \tag{38}$$

[These equations are just (2.67) and (2.68).]

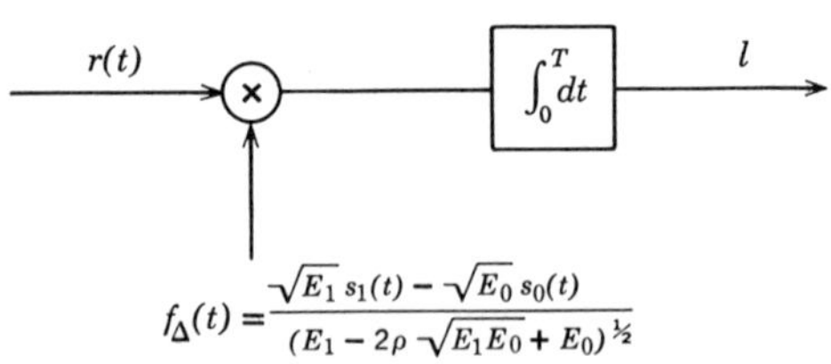

Fig. 4.16 Optimum correlation receiver, general binary problem.

The best choice of signals follows easily. The performance index d is monotonically related to the distance between the two signals in the decision space. For fixed energies the best performance is obtained by making $\rho = -1$. In other words,

$$s_0(t) = -s_1(t). \tag{39}$$

Once again the signal shape is not important.

When the criterion is minimum probability of error (as would be the logical choice in a binary communication system) *and* the a priori probabilities of the two hypotheses are equal, the decision region boundary has a simple interpretation. It is the perpendicular bisector of the line connecting the signal points (Fig. 4.17). Thus the receiver under these circumstances can be interpreted as a *minimum-distance* receiver and the error probability is

$$\Pr(\epsilon) = \int_{d/2}^{\infty} \frac{1}{\sqrt{2\pi}} \exp\left(-\frac{x^2}{2}\right) dx = \operatorname{erfc}_*\left(\frac{d}{2}\right). \tag{40}$$

If, *in addition*, the signals have *equal energy*, the bisector goes through the origin and we are simply choosing the signal that is most correlated with $r(t)$. This can be referred to as a "largest-of" receiver (Fig. 4.18).

The discussion can be extended to the M-ary problem in a straightforward manner.

***M*-ary Detection in White Gaussian Noise.** Assume that there are M-hypotheses:

$$r(t) = \sqrt{E_i}\, s_i(t) + w(t); \qquad 0 \le t \le T: H_i. \tag{41}$$

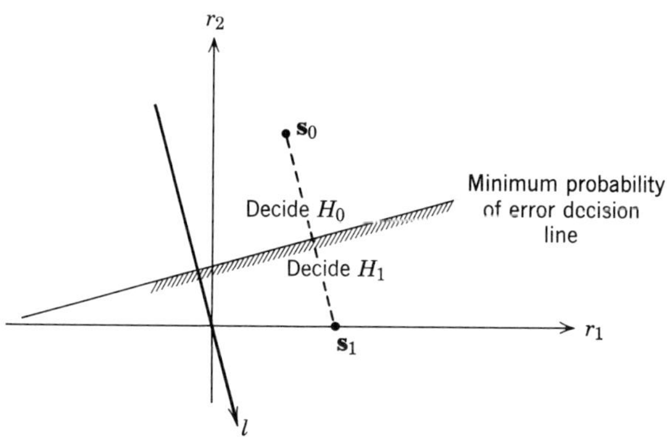

Fig. 4.17 Decision space.

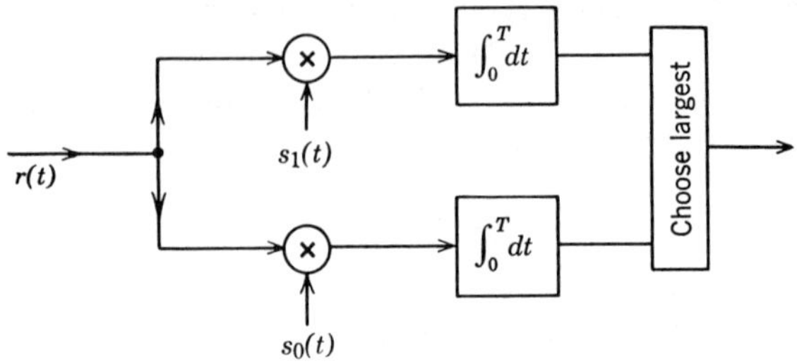

Fig. 4.18 "Largest of" receiver.

The $s_i(t)$ all have unit energy but may be correlated:

$$\int_0^T s_i(t)\, s_j(t)\, dt = \rho_{ij}. \qquad i, j = 1, 2, \ldots, M. \tag{42}$$

This problem is analogous to the M-hypothesis problem in Chapter 2. We saw that the main difficulty for a likelihood ratio test with arbitrary costs was the specification of the boundaries of the decision regions. We shall devote our efforts to finding a suitable set of sufficient statistics and evaluating the minimum probability of error for some interesting cases.

First we construct a suitable coordinate system to find a decision space with the minimum possible dimensionality. The procedure is a simple extension of the method used for two dimensions. The first coordinate function is just the first signal. The second coordinate function is that component of the second signal which is linearly independent of the first and so on. We let

$$\phi_1(t) = s_1(t), \tag{43a}$$

$$\phi_2(t) = (1 - \rho_{12}^{\,2})^{-1/2}[s_2(t) - \rho_{12}\, s_1(t)]. \tag{43b}$$

To construct the third coordinate function we write

$$\phi_3(t) = c_3[s_3(t) - c_1\phi_1(t) - c_2\phi_2(t)], \tag{43c}$$

and find c_1 and c_2 by requiring orthogonality and c_3 by requiring $\phi_3(t)$ to be normalized. (This is called the Gram-Schmidt procedure and is developed in detail in Problem 4.2.7.) We proceed until one of two things happens:

1. M orthonormal functions are obtained.
2. N $(< M)$ orthonormal functions are obtained and the remaining signals can be represented by linear combinations of these orthonormal functions. Thus the decision space will consist of *at most M* dimensions and fewer if the signals are linearly dependent.†

† Observe that we are talking about algebraic dependence.

We then use this set of orthonormal functions to generate N coefficients ($N \leq M$)

$$r_i \triangleq \int_0^T r(t)\, \phi_i(t)\, dt, \qquad i = 1, 2, \ldots, N. \tag{44a}$$

These are statistically independent Gaussian random variables with variance $N_0/2$ whose means depend on which hypothesis is true.

$$E[r_i|H_j] \triangleq m_{ij}, \qquad \begin{aligned} i &= 1, \ldots, N, \\ j &= 1, \ldots, M. \end{aligned} \tag{44b}$$

The likelihood ratio test follows directly from our results in Chapter 2 (Problem No. 2.6.1). When the criterion is minimum Pr (ϵ), we compute

$$l_j = \ln P_j - \frac{1}{N_0} \sum_{i=1}^{N} (R_i - m_{ij})^2, \qquad j = 1, \ldots, M, \tag{45}$$

and choose the largest. (The modification for other cost assignments is given in Problem No. 2.3.2.)

Two examples illustrate these ideas.

Example 1. Let

$$s_i(t) = \left(\frac{2}{T}\right)^{1/2} \sin\left[\omega_c t + (i-1)\frac{\pi}{2}\right], \qquad 0 \leq t \leq T, \qquad i = 1, 2, 3, 4,$$

$$E_i = E, \qquad i = 1, 2, 3, 4, \tag{46}$$

and

$$\omega_c = \frac{2\pi n}{T}$$

(n is an arbitrary integer). We see that

$$\phi_1(t) = \left(\frac{2}{T}\right)^{1/2} \sin \omega_c t, \qquad 0 \leq t \leq T,$$

and

$$\phi_2(t) = \left(\frac{2}{T}\right)^{1/2} \cos \omega_c t, \qquad 0 \leq t \leq T. \tag{47}$$

We see $s_3(t)$ and $s_4(t)$ are $-\phi_1(t)$ and $-\phi_2(t)$ respectively. Thus, in this case, $M = 4$ and $N = 2$. The decision space is shown in Fig. 4.19*a*. The decision regions follow easily when the criterion is minimum probability of error and the a priori probabilities are equal. Using the result in (45), we obtain the decision regions in Fig. 4.19*b*.

Example 2. Let

$$s_1(t) = \left(\frac{2}{T}\right)^{1/2} \sin \frac{2\pi n}{T} t, \qquad 0 \leq t \leq T,$$

$$s_2(t) = \left(\frac{2}{T}\right)^{1/2} \sin \frac{4\pi n}{T} t, \qquad 0 \leq t \leq T, \tag{48}$$

$$s_3(t) = \left(\frac{2}{T}\right)^{1/2} \sin \frac{6\pi n}{T} t, \qquad 0 \leq t \leq T,$$

(n is an arbitrary integer) and

$$E_i = E, \qquad i = 1, 2, 3.$$

Now

$$\phi_i(t) = s_i(t). \tag{49}$$

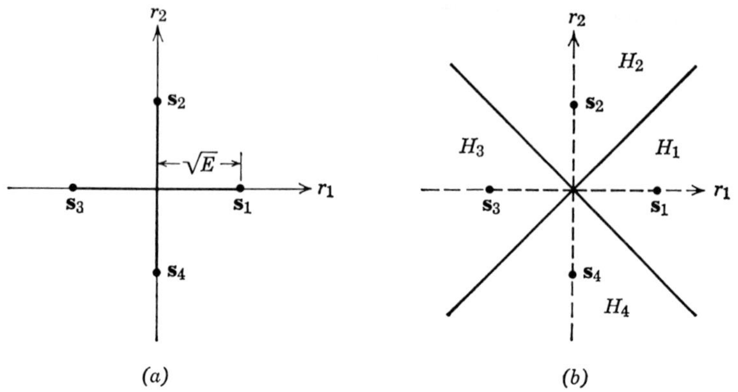

Fig. 4.19 Decision space.

In this case, $M = N = 3$ and the decision space is three-dimensional, as shown in Fig. 4.20*a*. For min Pr (ϵ) and equal a priori probabilities the decision regions follow easily from (45). The boundaries are planes perpendicular to the plane through $\mathbf{s}_1$, $\mathbf{s}_2$, and $\mathbf{s}_3$. Thus it is only the projection of $\mathbf{R}$ on this plane that is used to make a decision, and we can reduce the decision space to two dimensions as shown in Fig. 4.20*b*. (The coefficients r_1' and r_2' are along the two orthonormal coordinate functions used to define the plane.)

Note that in Examples 1 and 2 the signal sets were so simple that the Gram-Schmidt procedure was not needed.

It is clear that these results are directly analogous to the M hypothesis case in Chapter 2. As we have already seen, the calculation of the errors is conceptually simple but usually tedious for $M > 2$. To illustrate the procedure, we compute the Pr (ϵ) for Example 1.

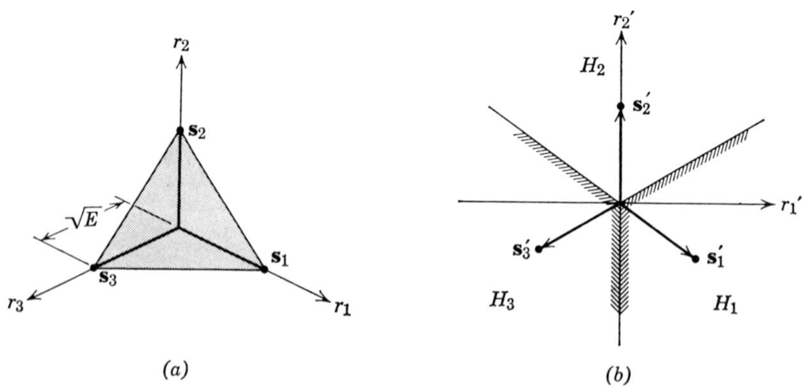

Fig. 4.20 Decision space: orthogonal signals.

Example 1 (*continued*). We assume that the hypotheses are equally likely. Now the problem is symmetrical. Thus it is sufficient to assume that $s_1(t)$ was transmitted and compute the resulting $\Pr(\epsilon)$. (Clearly, $\Pr(\epsilon) = \Pr(\epsilon|H_i)$, $i = 1, \ldots, 4$.) We also observe that the answer would be invariant to a 45° rotation of the signal set because the noise is circularly symmetric.

Thus the problem of interest reduces to the simple diagram shown in Fig. 4.21.

The $\Pr(\epsilon)$ is simply the probability that $\mathbf{r}$ lies outside the first quadrant when H_1 is true.

Now r_1 and r_2 are independent Gaussian variables with identical means and variances:

$$E(r_1|H_1) = E(r_2|H_1) = \left(\frac{E}{2}\right)^{1/2}$$

and

$$\operatorname{Var}(r_1|H_1) = \operatorname{Var}(r_2|H_1) = \frac{N_0}{2}. \tag{50}$$

The $\Pr(\epsilon)$ can be obtained by integrating $p_{r_1,r_2|H_1}(R_1, R_2|H_1)$ over the area outside the first quadrant. Equivalently, $\Pr(\epsilon)$ is the integral over the first quadrant subtracted from unity.

$$\Pr(\epsilon) = 1 - \left[\int_0^\infty \left(2\pi\frac{N_0}{2}\right)^{-1/2} \exp\left(-\frac{(R_1 - \sqrt{E/2})^2}{N_0}\right) dR_1\right]^2 \tag{51}$$

Changing variables, we have

$$\Pr(\epsilon) = 1 - \left(\int_{-\sqrt{E/N_0}}^\infty \frac{1}{\sqrt{2\pi}} \exp\left(-\frac{x^2}{2}\right) dx\right)^2 = 1 - \left(\operatorname{erfc}_*\left[-\left(\frac{E}{N_0}\right)^{1/2}\right]\right)^2, \tag{52}$$

which is the desired result.

Another example of interest is a generalization of Example 2.

Example 3. Let us assume that

$$r(t) = \sqrt{E}\, s_i(t) + w(t), \qquad 0 \le t \le T, \qquad H_i, \qquad i = 1, 2, \ldots, M \tag{53}$$

and

$$\rho_{ij} = 0, \tag{54}$$

and the hypotheses are equally likely. Because the energies are equal, it is convenient to

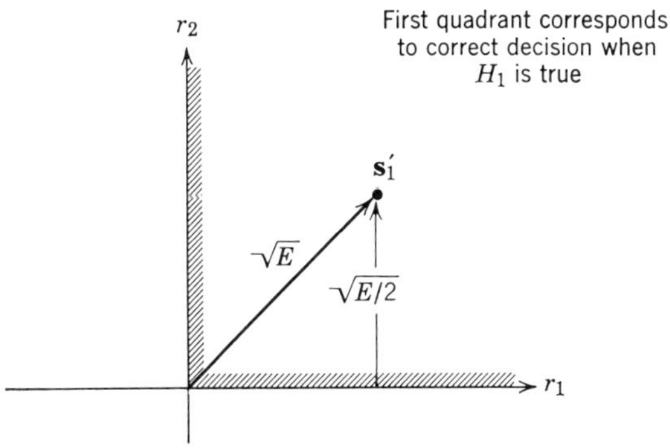

Fig. 4.21 Rotation of signal.

implement the LRT as a "greatest of" receiver as shown in Fig. 4.22. Once again the problem is symmetric, so we may assume H_1 is true. Then an error occurs if any $l_j > l_1 : j \neq 1$, where

$$l_j \triangleq \int_0^T r(t)s_j(t)\,dt, \qquad j = 1, 2, \ldots, M.$$

Thus

$$\Pr(\epsilon) = \Pr(\epsilon|H_1) = 1 - \Pr(\text{all } l_j < l_1 : j \neq 1) \tag{55}$$

or, noting that the $l_j\,(j \neq 1)$ have the same density on H_1,

$$\Pr(\epsilon) = 1 - \int_{-\infty}^{\infty} p_{l_1|H_1}(L_1|H_1)\left[\int_{-\infty}^{L_1} p_{l_2|H_1}(L_2|H_1)\,dL_2\right]^{M-1} dL_1. \tag{56}$$

In this particular case the densities are

$$p_{l_1|H_1}(L_1|H_1) = \frac{1}{\sqrt{\pi N_0}} \exp\left\{-\frac{1}{2}\frac{(L_1 - \sqrt{E})^2}{N_0/2}\right\} \tag{57}$$

and

$$p_{l_j|H_1}(L_j|H_1) = \frac{1}{\sqrt{\pi N_0}} \exp\left\{-\frac{1}{2}\frac{L_j^2}{N_0/2}\right\}, \qquad j \neq 1. \tag{58}$$

Substituting these densities into (56) and normalizing the variables, we obtain

$$\Pr(\epsilon) = 1 - \int_{-\infty}^{\infty} dx \frac{1}{\sqrt{2\pi}} \exp\left\{-\frac{[x - (2E/N_0)^{1/2}]^2}{2}\right\}\left(\int_{-\infty}^{x} \frac{1}{\sqrt{2\pi}} \exp\left[-\frac{y^2}{2}\right] dy\right)^{M-1}. \tag{59}$$

Unfortunately, we cannot integrate this analytically. Numerical results for certain

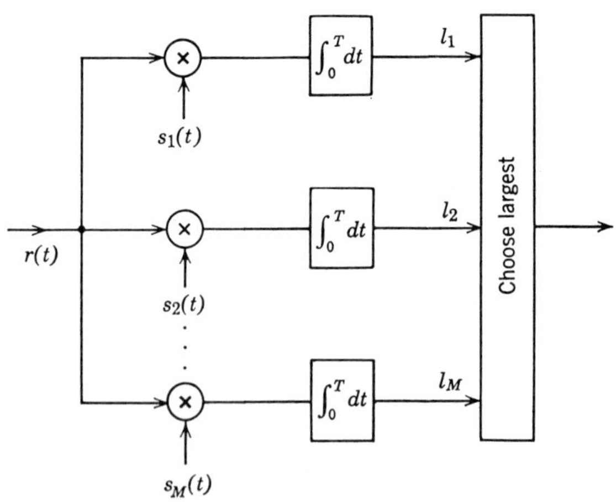

Fig. 4.22 "Largest of" receiver.

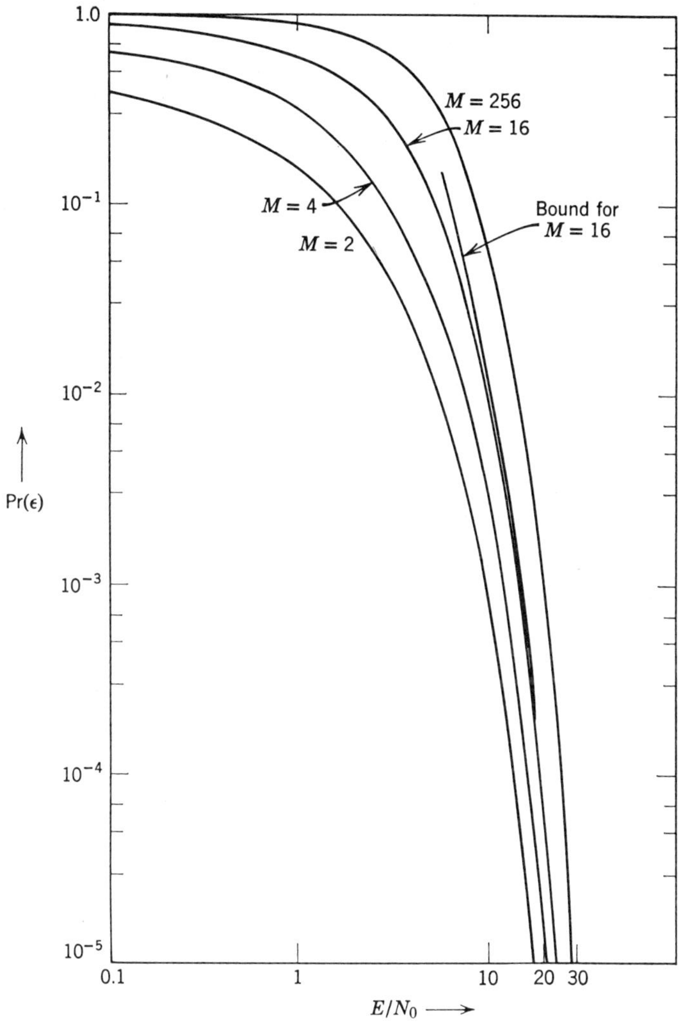

Fig. 4.23 Error probability: M orthogonal signals.

values of M and E/N_0 are tabulated in [21] and shown in Fig. 4.23. For some of our purposes an *approximate* analytic expression is more interesting. We derive a very simple bound. Some other useful bounds are derived in the problems. Looking at (55), we see that we could rewrite the Pr (ε) as

$$\Pr(\epsilon) = \Pr(\textit{any } l_j > l_1 : j \neq 1), \tag{60}$$

$$\Pr(\epsilon) = \Pr(l_2 > l_1 \quad \text{or} \quad l_3 > l_1 \quad \text{or} \quad \cdots \quad \text{or} \quad l_M > l_1). \tag{61}$$

Now, several l_j can be greater than l_1. (The events are not mutually exclusive.) Thus

$$\Pr(\epsilon) \leq \Pr(l_2 > l_1) + \Pr(l_3 > l_1) + \cdots + \Pr(l_M > l_1), \tag{62}$$

$$\Pr(\epsilon) \leq (M-1)\left\{\int_{-\infty}^{\infty} \frac{1}{\sqrt{2\pi}} \exp\left[-\frac{(x-\sqrt{2E/N_0})^2}{2}\right]\left(\int_{x}^{\infty} \frac{1}{\sqrt{2\pi}} \exp\left(-\frac{y^2}{2}\right) dy\right) dx\right\}; \tag{63}$$

but the term in the bracket is just the expression of the probability of error for two orthogonal signals. Using (36) with $\rho = 0$ and $E_1 = E_0 = E$ in (40), we have

$$\boxed{\Pr(\epsilon) \leq (M-1) \int_{\sqrt{E/N_0}}^{\infty} \frac{1}{\sqrt{2\pi}} \exp\left(-\frac{y^2}{2}\right) dy.} \tag{64}$$

(Equation 64 also follows directly from (63) by a change of variables.) We can further simplify this equation by using (2.71):

$$\Pr(\epsilon) \leq \frac{(M-1)}{\sqrt{2\pi}\sqrt{E/N_0}} \exp\left(-\frac{E}{2N_0}\right). \tag{65}$$

We observe that the upper bound increases linearly with M. The bound on the $\Pr(\epsilon)$ given by this expression is plotted in Fig. 4.23 for $M = 16$.

A related problem in which M orthogonal signals arise is that of transmitting a sequence of binary digits.

Example 4. Sequence of Digits. Consider the simple digital system shown in Fig. 4.24, in which the source puts out a binary digit every T seconds. The outputs 0 and 1 are equally likely. The available transmitter power is P. For simplicity we assume that we are using orthogonal signals. The following choices are available:

1. Transmit one of two orthogonal signals every T seconds. The energy per signal is PT.
2. Transmit one of four orthogonal signals every $2T$ seconds. The energy per signal is $2PT$. For example, the encoder could use the mapping,

$$\begin{aligned} 00 &\to s_0(t), \\ 01 &\to s_1(t), \\ 10 &\to s_2(t), \\ 11 &\to s_3(t). \end{aligned}$$

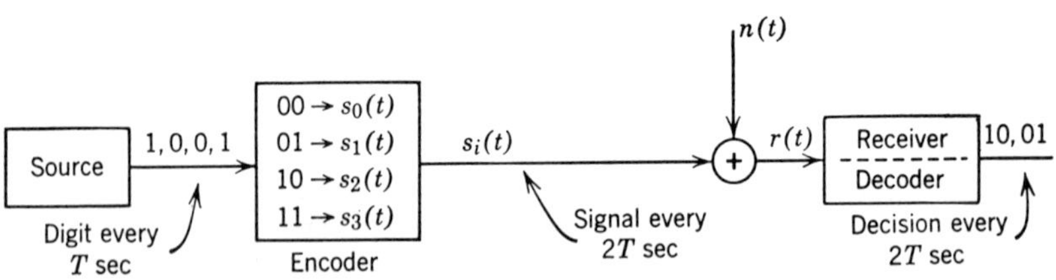

$$\int_0^{2T} s_i(t)\, s_j(t)\, dt = \delta_{ij}$$

Fig. 4.24 Digital communication system.

3. In general, we could transmit one of M orthogonal signals every $T \log_2 M$ seconds. The energy per signal is $TP \log_2 M$. To compute the probability of error we use (59):

$$\Pr(\epsilon) = 1 - \int_{-\infty}^{\infty} dx \frac{1}{\sqrt{2\pi}} \exp\left\{-\frac{1}{2}\left[x - \left(\frac{2PT \log_2 M}{N_0}\right)^{1/2}\right]^2\right\} \times \left[\int_{-\infty}^{x} \frac{1}{\sqrt{2\pi}} \exp\left(-\frac{y^2}{2}\right) dy\right]^{M-1}. \tag{66}$$

The results have been calculated numerically [19] and are plotted in Fig. 4.25. The behavior is quite interesting. Above a certain value of PT/N_0 the error probability decreases with increased M. Below this value the converse is true. It is instructive to investigate the behavior as $M \to \infty$. We obtain from (66), by a simple change of variables,

$$\lim_{M \to \infty} (1 - \Pr(\epsilon)) = \int_{-\infty}^{\infty} dy \frac{e^{-y^2/2}}{\sqrt{2\pi}} \lim_{M \to \infty} \left\{\operatorname{erf}_*^{M-1}\left[y + \left(\frac{2PT \log_2 M}{N_0}\right)^{1/2}\right]\right\}. \tag{67}$$

Now consider the limit of the logarithm of the expression in the brace:

$$\lim_{M \to \infty} \frac{\ln \operatorname{erf}_*\left[y + \left(\frac{2PT \log_2 M}{N_0}\right)^{1/2}\right]}{(M-1)^{-1}}. \tag{68}$$

Evaluating the limit by treating M as a continuous variable and using L'Hospital's rule, we find that (see Problem 4.2.15)

$$\lim_{M \to \infty} \ln \{\sim\} = \begin{cases} -\infty, & \dfrac{PT}{N_0} < \ln 2, \\ 0, & \dfrac{PT}{N_0} > \ln 2. \end{cases} \tag{69}$$

Thus, from the continuity of logarithm,

$$\lim_{M \to \infty} \Pr(\epsilon) = \begin{cases} 0, & \dfrac{PT}{N_0} > \ln 2, \\ 1, & \dfrac{PT}{N_0} < \ln 2. \end{cases} \tag{70}$$

Thus we see that there is a definite threshold effect. The value of T is determined by how fast the source produces digits. Specifically, the rate in binary digits per second is

$$R \triangleq \frac{1}{T} \text{ binary digits/sec.} \tag{71}$$

Using orthogonal signals, we see that if

$$R < \frac{1}{\ln 2} \frac{P}{N_0} \tag{72}$$

the probability of error will go to zero. The obvious disadvantage is the bandwidth requirement. As $M \to \infty$, the transmitted bandwidth goes to ∞.

The result in (72) was derived for a particular set of signals. Shannon has shown (e.g., [22] or [23]) that the right-hand side is the bound on the

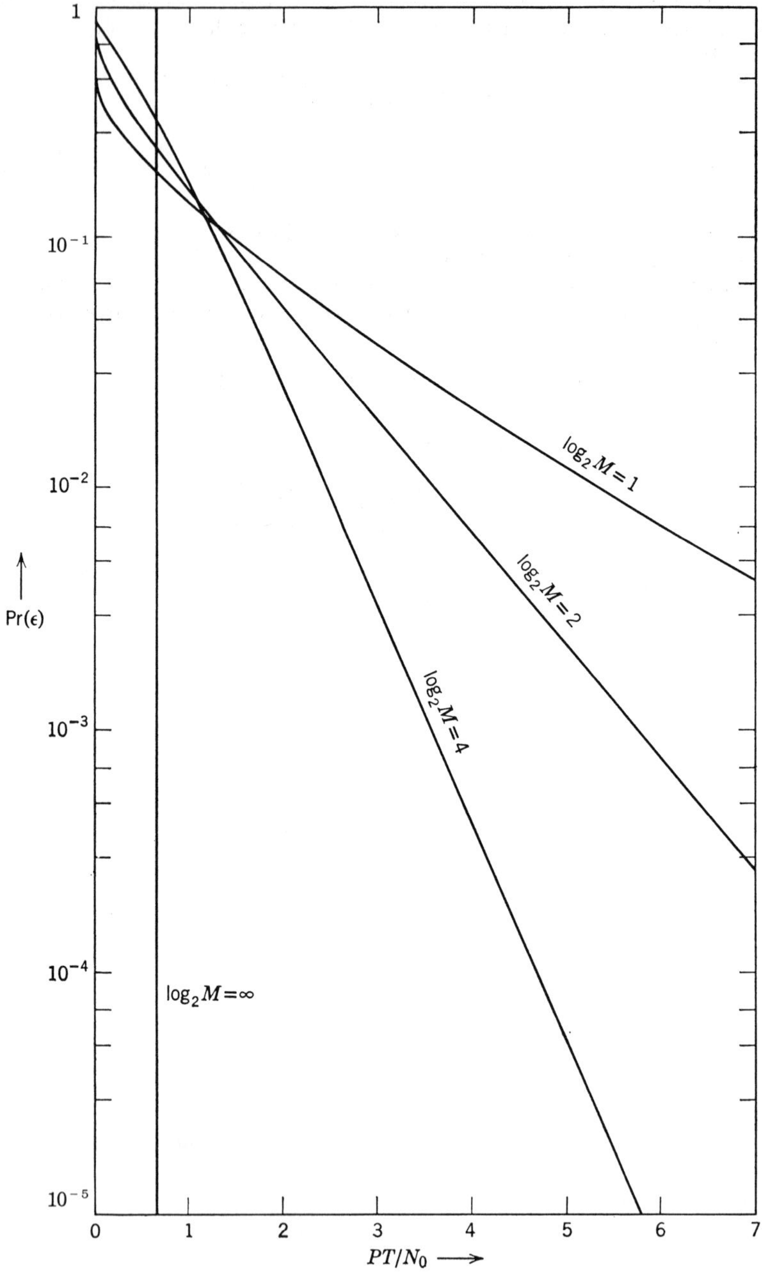

Fig. 4.25 Probability of decision error: M orthogonal signals, power constraint.

rate for error-free transmission for any communication scheme. This rate is referred to as the capacity of an infinite bandwidth, additive white Gaussian noise channel,

$$C_\infty = \frac{1}{\ln 2}\frac{P}{N_0} \text{ bits/sec.} \tag{73}$$

Shannon has also derived an expression for a bandlimited channel (W_{ch}: single-sided):

$$C = W_{ch}\log_2\left(1 + \frac{P}{W_{ch}N_0}\right). \tag{74}$$

These capacity expressions are fundamental to the problem of sending sequences of digits. We shall not consider this problem, for an adequate discussion would take us too far afield. Suitable references are [18] and [66].

In this section we have derived the canonic receiver structures for the M-ary hypothesis problem in which the received signal under each hypothesis is a known signal plus additive white Gaussian noise. The simplicity resulted because we were always able to reduce an infinite dimensional observation space to a finite ($\leq M$) dimensional decision space.

In the problems we consider some of the implications of these results. Specific results derived in the problems include the following:

1. The probability of error for any set of M equally correlated signals can be expressed in terms of an equivalent set of M orthogonal signals (Problem 4.2.9).
2. The lowest value of uniform correlation is $-(M-1)^{-1}$. Signals with this property are optimum when there is no bandwidth restriction (Problems 4.2.9–4.2.12). They are referred to as Simplex signals.
3. For large M, orthogonal signals are essentially optimum.

Sensitivity. Before leaving the problem of detection in the presence of white noise we shall discuss an important issue that is frequently overlooked. We have been studying the mathematical model of a physical system and have assumed that we know the quantities of interest such as $s(t)$, E, and N_0 exactly. In an actual system these quantities will vary from their nominal values. It is important to determine how the performance of the optimum receiver will vary when the nominal values are perturbed. If the performance is highly sensitive to small perturbations, the validity of the nominal performance calculation is questionable. We shall discuss sensitivity in the context of the simple binary detection problem.

The model for this problem is

$$\begin{aligned} r(t) &= \sqrt{E}\, s(t) + w(t), && 0 \leq t \leq T: H_1 \\ r(t) &= w(t), && 0 \leq t \leq T: H_0. \end{aligned} \tag{75}$$

The receiver consists of a matched filter followed by a decision device. The impulse response of the matched filter depends on the shape of $s(t)$. The energy and noise levels affect the decision level in the general Bayes case. (In the Neyman–Pearson case only the noise level affects the threshold setting). There are several possible sensitivity analyses. Two of these are the following:

1. Assume that the actual signal energy and signal shape are identical to those in the model. Calculate the change in P_D and P_F due to a change in the white noise level.
2. Assume that the signal energy and the noise level are identical to those in the model. Calculate the change in P_D and P_F due to a change in the signal.

In both cases we can approach the problem by first finding the change in d due to the changes in the model and then seeing how P_D and P_F are affected by a change in d. In this section we shall investigate the effect of an inaccurate knowledge of signal shape on the value of d. The other questions mentioned above are left as an exercise. We assume that we have designed a filter that is matched to the assumed signal $s(t)$,

$$h(T - t) = s(t), \qquad 0 \le t \le T, \tag{76}$$

and that the received waveform on H_1 is

$$r(t) = s_a(t) + w(t), \qquad 0 \le t \le T, \tag{77}$$

where $s_a(t)$ is the actual signal received. There are two general methods of relating $s_a(t)$ to $s(t)$. We call the first the function-variation method.

Function-Variation Method. Let

$$s_a(t) = \sqrt{E}\, s(t) + \sqrt{E_\epsilon}\, s_\epsilon(t), \qquad 0 \le t \le T, \tag{78}$$

where $s_\epsilon(t)$ is a normalized waveform representing the inaccuracy. The energy in the error signal is constrained to equal E_ϵ.

The effect can be most easily studied by examining the decision space (more precisely an augmented decision space). To include all of $s_\epsilon(t)$ in the decision space we *think* of adding another matched filter,

$$h_2(T - t) = \phi_2(t) = \frac{s_\epsilon(t) - \rho_\epsilon\, s(t)}{\sqrt{1 - \rho_\epsilon^{\,2}}}, \qquad 0 \le t \le T, \tag{79}$$

where ρ_ϵ is the correlation between $s_\epsilon(t)$ and $s(t)$. (Observe that we do not do this physically.) We now have a two-dimensional space. The effect of the constraint is clear. Any $s_a(t)$ will lead to a point on the circle surrounding $\mathbf{s}$, as shown in Fig. 4.26. Observe that the decision still uses only the

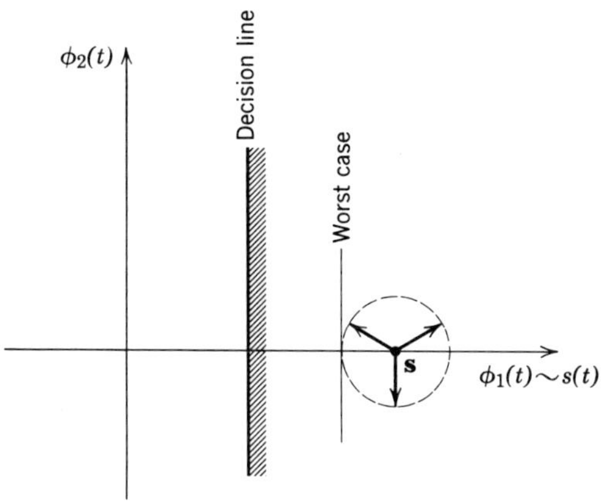

Fig. 4.26 Signal locus: fixed energy in error signal.

coordinate along $s(t)$. The effect is obvious. The error signal that causes the largest performance degradation is

$$s_\epsilon(t) = -s(t). \tag{80}$$

Then

$$d_a{}^2 = \frac{2}{N_0}(\sqrt{E} - \sqrt{E_\epsilon})^2. \tag{81}$$

To state the result another way,

$$\frac{\Delta d}{d} = -\frac{\sqrt{2E_\epsilon/N_0}}{\sqrt{2E/N_0}} = -\left(\frac{E_\epsilon}{E}\right)^{1/2} \tag{82}$$

where

$$\Delta d \triangleq d_a - d. \tag{83}$$

We see that small energy in the error signal implies a small change in performance. Thus the test is insensitive to small perturbations. The second method is called the parameter-variation method.

Parameter-Variation Method. This method can best be explained by an example. Let

$$s(t) = \left(\frac{2}{T}\right)^{1/2} \sin \omega_c t, \qquad 0 \le t \le T, \tag{84}$$

be the nominal signal. The actual signal is

$$s_a(t) = \left(\frac{2}{T}\right)^{1/2} \sin (\omega_c t + \theta), \qquad 0 \le t \le T, \tag{85}$$

which, for $\theta = 0$, corresponds to the nominal signal. The augmented

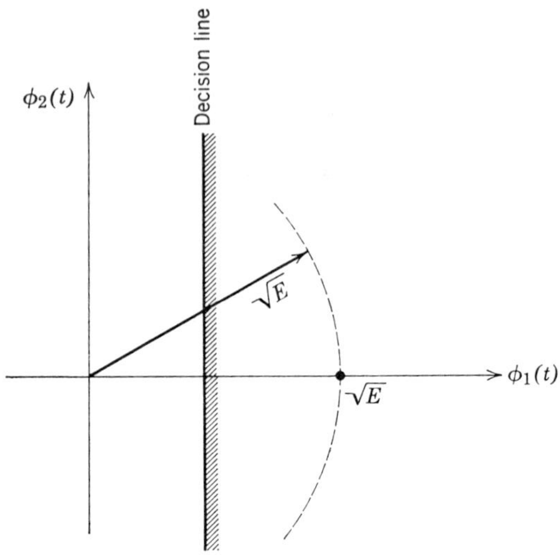

Fig. 4.27 Signal locus: fixed energy in total signal.

decision space is shown in Fig. 4.27. The vector corresponding to the actual signal moves on a circle around the origin.

$$d_a = \left(\frac{2E}{N_0}\right)^{1/2} \cos \theta \tag{86}$$

and

$$\frac{\Delta d}{d} = -(1 - \cos \theta). \tag{87}$$

Once again we see that the test is insensitive to small perturbations.

The general conclusion that we can infer from these two methods is that the results of detection in the presence of white noise are insensitive to the detailed assumptions. In other words, small perturbations from the design assumptions lead to small perturbations in performance. In almost all cases this type of insensitivity is necessary if the mathematical model is going to predict the actual system performance accurately.

Many statistical analyses tend to ignore this issue. The underlying reason is probably psychological. After we have gone through an involved mathematical optimization, it would be pleasant to demonstrate an order-of-magnitude improvement over the system designed by using an intuitive approach. Unfortunately, this does not always happen. When it does, we must determine whether the mathematical result is sensitive to some detailed assumption. In the sequel we shall encounter several examples of this sensitivity.

We now turn to the problem of linear estimation.

4.2.2 Linear Estimation

In Section 4.1.1 we formulated the problem of estimating signal parameters in the presence of additive noise. For the case of additive white noise the received waveform is

$$r(t) = s(t, A) + w(t), \qquad 0 \le t \le T, \tag{88a}$$

where $w(t)$ is a sample function from a white Gaussian noise process with spectral height $N_0/2$. The parameter A is the quantity we wish to estimate. If it is a random parameter we will assume that the a priori density is known and use a Bayes estimation procedure. If it is a nonrandom variable we will use ML estimates. The function $s(t, A)$ is a deterministic mapping of A into a time function. If $s(t, A)$ is a linear mapping (in other words, superposition holds), we refer to the system using the signal as a *linear signaling* (or *linear modulation*) system. Furthermore, for the criterion of interest the estimator will turn out to be linear so we refer to the problem as a *linear estimation* problem. In this section we study linear estimation and in Section 4.2.3, nonlinear estimation. For linear modulation (88*a*) can always be written as

$$r(t) = A\sqrt{E}\, s(t) + w(t), \qquad 0 \le t \le T, \tag{88b}$$

where $s(t)$ has unit energy.

We can solve the linear estimation problem easily by exploiting its similarity to the detection problem that we just solved. From Section 2.4 we know that the likelihood function is needed. We recall, however, that the problem is greatly simplified if we can find a sufficient statistic and work with it instead of the received waveform. If we compare (88*b*) and (4)–(7), it is clear that a sufficient statistic is r_1, where

$$r_1 = \int_0^T r(t)\, s(t)\, dt. \tag{89}$$

Just as in Section 4.2.1, the probability density of r_1, given $a = A$, is Gaussian:

$$\begin{aligned} E(r_1|A) &= A\sqrt{E}, \\ \operatorname{Var}(r_1|A) &= \frac{N_0}{2}. \end{aligned} \tag{90}$$

It is easy to verify that the coefficients along the other orthogonal functions [see (8)] are independent of a. Thus the waveform problem reduces to the classical estimation problem (see pp. 58–59).

The logarithm of the likelihood function is

$$l(A) = -\frac{1}{2}\frac{(R_1 - A\sqrt{E})^2}{N_0/2}. \tag{91}$$

If A is a nonrandom variable, the ML estimate is the value of A at which this function is a maximum. Thus

$$\hat{a}_{ml}(R_1) = \frac{R_1}{\sqrt{E}}. \tag{92}$$

The receiver is shown in Fig. 4.28*a*. We see the estimate is unbiased.

If a is a random variable with a probability density $p_a(A)$, then the MAP estimate is the value of A where

$$l_p(A) = -\frac{1}{2}\frac{(R_1 - A\sqrt{E})^2}{N_0/2} + \ln p_a(A), \tag{93}$$

is a maximum. For the special case in which a is Gaussian, $N(0, \sigma_a)$, the

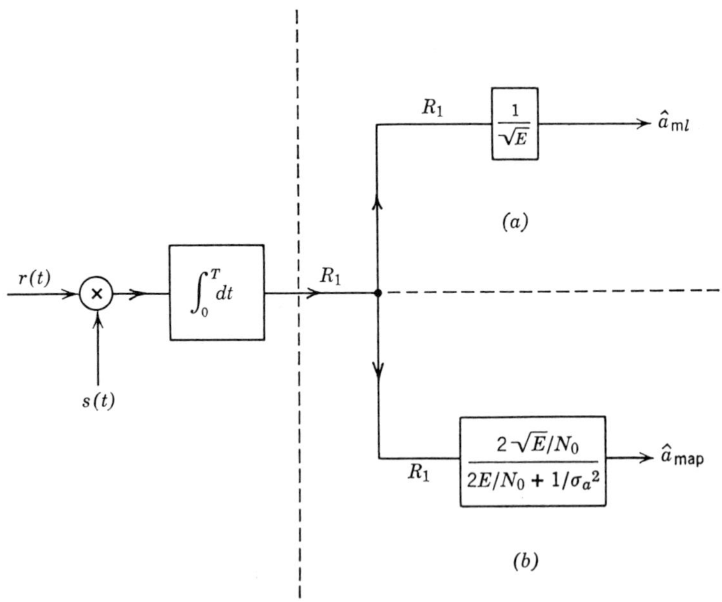

Fig. 4.28 Optimum receiver, linear estimation: (*a*) ML receiver; (*b*) MAP receiver.

MAP estimate is easily obtained by differentiating $l_p(A)$ and equating the result to zero:

$$\frac{\partial l_p(A)}{\partial A} = \frac{R_1 - A\sqrt{E}}{N_0/2}\sqrt{E} - \frac{A}{\sigma_a^2} \tag{94}$$

and

$$\hat{a}_{\text{map}}(R_1) = \frac{2E/N_0}{2E/N_0 + 1/\sigma_a^2}\frac{R_1}{\sqrt{E}}. \tag{95}$$

In both the ML and MAP cases it is easy to show that the result is the absolute maximum.

The MAP receiver is shown in Fig. 4.28*b*. Observe that the only difference between the two receivers is a gain. The normalized error variances follow easily: for MAP

$$\frac{E[a_\epsilon^2]}{\sigma_a^2} = \sigma_{a_\epsilon n}^2 \triangleq \frac{\sigma_{a_\epsilon}^2}{\sigma_a^2} = \left(1 + \frac{2\sigma_a^2 E}{N_0}\right)^{-1} \qquad \text{(MAP)}. \tag{96}$$

The quantity $\sigma_a^2 E$ is the expected value of the received energy. For ML

$$\sigma_{a_\epsilon n}^2 \triangleq \frac{\sigma_{a_\epsilon}^2}{A^2} = \left(\frac{2A^2E}{N_0}\right)^{-1} \qquad \text{(ML)}. \tag{97}$$

Here A^2E is the actual value of the received energy. We see that the variance of the maximum likelihood estimate is the reciprocal of d^2, the performance index of the simple binary problem. In both cases we see that the only way to decrease the mean-square error is to increase the energy-to-noise ratio. In many situations the available energy-to-noise ratio is not adequate to provide the desired accuracy. In these situations we try a nonlinear signaling scheme in an effort to achieve the desired accuracy. In the next section we discuss the nonlinear estimation.

Before leaving linear estimation, we should point out that the MAP estimate is also the Bayes estimate for a large class of criteria. Whenever a is Gaussian the a posteriori density is Gaussian and Properties 1 and 2 on pp. 60–61 are applicable. This invariance to criterion depends directly on the linear signaling model.

4.2.3 Nonlinear Estimation

The system in Fig. 4.7 illustrates a typical nonlinear estimation problem. The received signal is

$$r(t) = s(t, A) + w(t), \qquad 0 \le t \le T. \tag{98}$$

From our results in the classical case we know that a sufficient statistic does not exist in general. As before, we can construct the likelihood function. We approach the problem by making a K-coefficient approximation to $r(t)$. By proceeding as on p. 252 with obvious notation we have

$$\Lambda[r_K(t), A] = p_{\mathbf{r}|a}(\mathbf{R}|A) = \prod_{i=1}^{K} \frac{1}{\sqrt{\pi N_0}} \exp\left(-\frac{1}{2}\frac{[R_i - s_i(A)]^2}{N_0/2}\right). \tag{99}$$

where

$$s_i(A) \triangleq \int_0^T s(t, A)\, \phi_i(t)\, dt.$$

Now, if we let $K \to \infty$, $\Lambda[r_K(t), A]$ is not well defined. We recall from Chapter 2 that we can divide a likelihood function by anything that does not depend on A and still have a likelihood function. On p. 252 we avoided the convergence problem by dividing by

$$p_{r_K(t)|H_0}[r_K(t)|H_0] = \prod_{i=1}^{K} \frac{1}{\sqrt{\pi N_0}} \exp\left(-\frac{1}{2}\frac{R_i^2}{N_0/2}\right),$$

before letting $K \to \infty$. Because this function does not depend on A, it is legitimate to divide by it here. Define

$$\Lambda_1[r_K(t), A] = \frac{\Lambda[r_K(t), A]}{p_{r_K(t)|H_0}[r_K(t)|H_0]}. \tag{100}$$

Substituting into this expression, canceling common terms, letting $K \to \infty$, and taking the logarithm we obtain

$$\ln \Lambda_1[r(t), A] = \frac{2}{N_0}\int_0^T r(t)\, s(t, A)\, dt - \frac{1}{N_0}\int_0^T s^2(t, A)\, dt. \tag{101}$$

To find $\hat{a}_{ml}$ we must find the absolute maximum of this function. To find $\hat{a}_{map}$ we add $\ln p_a(A)$ to (101) and find the absolute maximum. The basic operation on the received data consists of generating the first term in (101) as a function of A. The physical device that we actually use to accomplish it will depend on the functional form of $s(t, A)$. We shall consider some specific cases and find the actual structure.

Before doing so we shall derive a result for the general case that will be useful in the sequel. Observe that if the maximum is interior and $\ln \Lambda_1(A)$ is differentiable at the maximum, then a necessary, but not sufficient, condition is obtained by first differentiating (101):

$$\frac{\partial \ln \Lambda_1(A)}{\partial A} = \frac{2}{N_0}\int_0^T [r(t) - s(t, A)]\, \frac{\partial s(t, A)}{\partial A}\, dt \tag{102}$$

(assuming that $s(t, A)$ is differentiable with respect to A). For $\hat{a}_{ml}$, a necessary condition is obtained by setting the right-hand side of (102) equal to zero. For $\hat{a}_{map}$ we add $d \ln p_a(A)/dA$ to the right-hand side of (102) and set the sum equal to zero. In the special case in which $p_a(A)$ is Gaussian, $N(0, \sigma_a)$, we obtain

$$\hat{a}_{\text{map}} = \frac{2\sigma_a^2}{N_0} \int_0^T [r(t) - s(t, A)] \frac{\partial s(t, A)}{\partial A} \, dt \Big|_{A=\hat{a}_{\text{map}}}. \tag{103}$$

In the linear case (103) reduces to (95) and gives a unique solution. A number of solutions may exist in the nonlinear case and we must examine the sum of (101) and $\ln p_a(A)$ to guarantee an absolute maximum.

However, just as in Chapter 2, (102) enables us to find a bound on the variance of any unbiased estimate of a nonrandom variable and the addition of $d^2 \ln p_a(A)/dA^2$ leads to a bound on the mean-square error in estimating a random variable. For nonrandom variables we differentiate (102) and take the expectation

$$E\left[\frac{\partial^2 \ln \Lambda_1(A)}{\partial A^2}\right] = \frac{2}{N_0} \left\{ E \int_0^T [r(t) - s(t, A)] \frac{\partial^2 s(t, A)}{\partial A^2} \, dt - E \int_0^T \left[\frac{\partial s(t, A)}{\partial A}\right]^2 dt \right\}, \tag{104}$$

where we assume the derivatives exist. In the first term we observe that

$$E[r(t) - s(t, A)] = E[w(t)] = 0. \tag{105}$$

In the second term there are no random quantities; therefore the expectation operation gives the integral itself.

Substituting into (2.179), we have

$$\text{Var}(\hat{a} - A) \geq \frac{N_0}{2 \displaystyle\int_0^T \left[\frac{\partial s(t, A)}{\partial A}\right]^2 dt} \tag{106}$$

for any unbiased estimate $\hat{a}$. Equality holds in (106) if and only if

$$\frac{\partial \ln \Lambda_1(A)}{\partial A} = k(A)\{\hat{a}[r(t)] - A\} \tag{107}$$

for all A and $r(t)$. Comparing (102) and (107), we see that this will hold only for linear modulation. Then $\hat{a}_{ml}$ is the minimum variance estimate.

Similarly, for random variables

$$E[\hat{a} - a]^2 \geq \left(\frac{2}{N_0} E_a\left\{\int_0^T \left[\frac{\partial s(t, A)}{\partial A}\right]^2 dt - \frac{d^2 \ln p_a(A)}{dA^2}\right\}\right)^{-1}, \tag{108}$$

where E_a denotes an expectation over the random variable a. Defining

$$\gamma_a^2 \triangleq E_a \int_0^T \left[\frac{\partial s(t, A)}{\partial A}\right]^2 dt, \tag{109}$$

we have

$$E(\hat{a} - a)^2 \geq \left(\frac{2}{N_0} \left\{\gamma_a^2 - E_a\left[\frac{d^2 \ln p_a(A)}{dA^2}\right]\right\}\right)^{-1}. \tag{110}$$

Equality will hold if and only if (see 2.226)

$$\frac{\partial^2 \ln \Lambda_1(A)}{\partial A^2} + \frac{d^2 \ln p_a(A)}{dA^2} = \text{constant}. \tag{111}$$

Just as in Chapter 2 (p. 73), in order for (111) to hold it is necessary and sufficient that $p_{a|r(t)}[A|r(t) : 0 \le t \le T]$ be a Gaussian probability density. This requires a linear signaling scheme and a Gaussian a priori density.

What is the value of the bound if it is satisfied only for linear signaling schemes?

As in the classical case, it has two principal uses:

1. It always provides a lower bound.
2. In many cases, the actual variance (or mean-square error) in a nonlinear signaling scheme will approach this bound under certain conditions. These cases are the analogs to the *asymptotically efficient* estimates in the classical problem. We shall see that they correspond to large E/N_0 values.

To illustrate some of the concepts in the nonlinear case we consider two simple examples.

Example 1. Let $s(t)$ be the pulse shown in Fig. 4.29*a*. The parameter a is the arrival

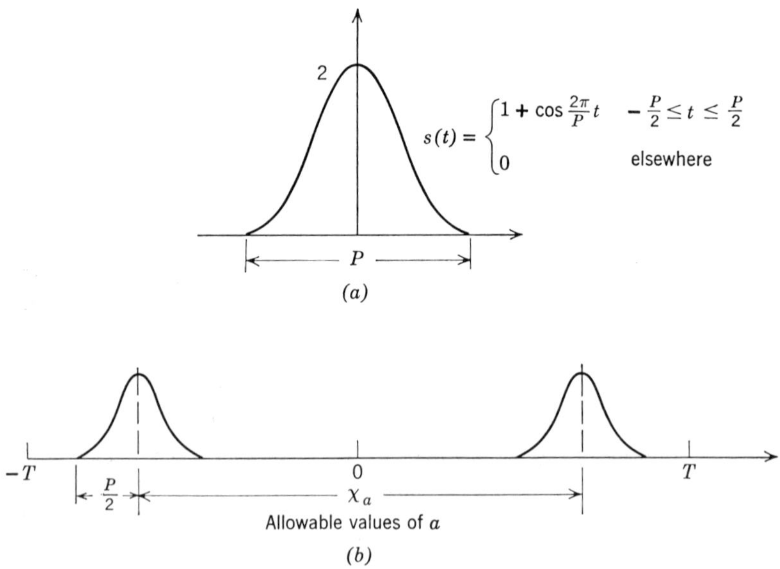

Fig. 4.29 (*a*) **Pulse shape;** (*b*) **Allowable parameter range.**

time of the pulse. We want to find a MAP estimate of a. We know a range of values χ_a that a may assume (Fig. 4.29*b*). Inside this range the probability density is uniform. For simplicity we let the observation interval $[-T, T]$ be long enough to completely contain the pulse.

From (101) we know that the operation on the received waveform consists of finding $\ln \Lambda_1[r(t), A]$. Here

$$\ln \Lambda_1[r(t), A] = \frac{2}{N_0}\int_{-T}^{T} r(u)\, s(u - A)\, du - \frac{1}{N_0}\int_{-T}^{T} s^2(u - A)\, du. \tag{112}$$

For this particular case the second term does not depend on A, for the entire pulse is always in the interval. The first term is a convolution operation. The output of a linear filter with impulse response $h(\tau)$ and input $r(u)$ over the interval $[-T, T]$ is

$$y(t) = \int_{-T}^{T} r(u)\, h(t - u)\, du, \qquad -T \le t \le T \tag{113}$$

Clearly, if we let

$$h(\tau) = s(-\tau), \tag{114}$$

the output as a function of time over the range x_a will be identical to the likelihood function as a function of A. We simply pick the peak of the filter output as a function of time. The time at which the peak occurs is $\hat{a}_{\text{map}}$ .The filter is the matched filter that we have already encountered in the detection problem.

In Fig. 4.30 we indicate the receiver structure. The output due to the signal component is shown in line (*a*). Typical total outputs for three noise levels are shown in lines (*b*), (*c*), and (*d*). In line (*b*) we see that the peak of $\ln \Lambda(A)$ is large compared to

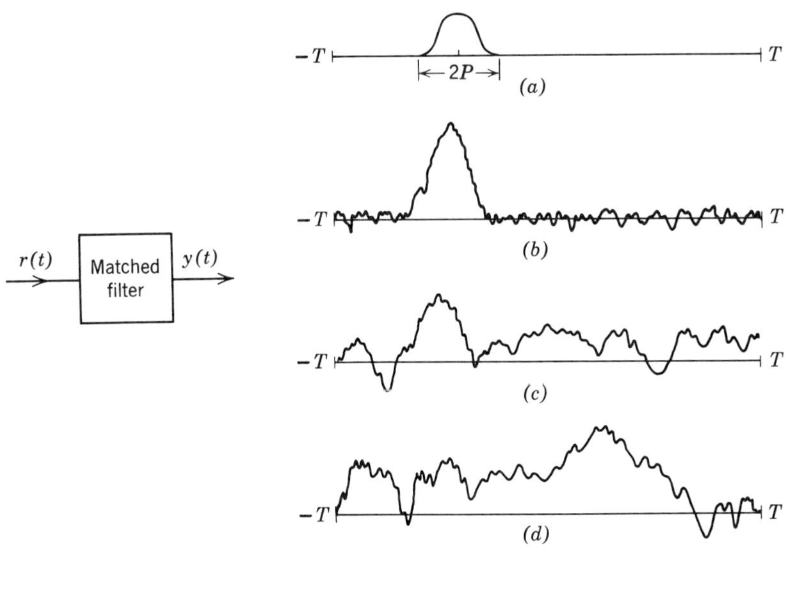

Fig. 4.30 Receiver outputs [arrival time estimation]: (*a*) signal component; (*b*) low noise level; (*c*) moderate noise level; (*d*) high noise level.

the noise background. The actual peak is near the correct peak, and we can expect that the error will be accurately predicted by using the expression in (110). In line (*c*) the noise has increased and large subsidiary peaks which have no relation to the correct value of A are starting to appear. Finally, in line (*d*) the noise has reached the point at which the maximum bears no relation to the correct value. Thus two questions are posed:

1. Under what conditions does the lower bound given by (110) accurately predict the error?
2. How can one predict performance when (110) is not useful?

Before answering these questions, we consider a second example to see if similar questions arise.

Example 2. Another common example of a nonlinear signaling technique is discrete frequency modulation (also referred to as pulse frequency modulation, PFM). Every T seconds the source generates a new value of the parameter a. The transmitted signal is $s(t, A)$, where

$$s(t, A) = \left(\frac{2E}{T}\right)^{1/2} \sin(\omega_c + \beta A)t, \qquad -\frac{T}{2} \leq t \leq \frac{T}{2}. \tag{115}$$

Here ω_c is a known carrier frequency, β is a known constant, and E is the transmitted energy (also the received signal energy). We assume that $p_a(A)$ is a uniform variable over the interval $(-\sqrt{3}\,\sigma_a, \sqrt{3}\,\sigma_a)$.

To find $\hat{a}_{map}$ we construct the function indicated by the first term in (101): (The second term in (101) and the a priori density are constant and may be discarded.)

$$\begin{aligned} l_1(A) &= \int_{-T/2}^{T/2} r(t) \sin(\omega_c t + \beta A t)\, dt, &\quad -\sqrt{3}\,\sigma_a \leq A \leq \sqrt{3}\,\sigma_a \\ &= 0, &\quad \text{elsewhere.} \end{aligned} \tag{116}$$

One way to construct $l_1(A)$ would be to record $r(t)$ and perform the multiplication and integration indicated by (116) for successive values of A over the range. This is obviously a time-consuming process. An alternate approach† is to divide the range into increments of length Δ and perform the parallel processing operation shown in Fig. 4.31 for discrete values of A:

$$\begin{aligned} A_1 &= -\sqrt{3}\,\sigma_a + \frac{\Delta}{2}, \\ A_2 &= -\sqrt{3}\,\sigma_a + \frac{3\Delta}{2}, \\ &\;\;\vdots \\ A_M &= -\sqrt{3}\,\sigma_a + (M + \tfrac{1}{2})\,\Delta, \end{aligned} \qquad M = \left\lfloor \frac{2\sqrt{3}\,\sigma_a}{\Delta} + \frac{1}{2} \right\rfloor, \tag{117}$$

† This particular type of approach and the resulting analysis were first done by Woodward (radar range measurement [8]) and Kotelnikov (PPM and PFM, [13]). Subsequently, they have been used with various modifications by a number of authors (e.g., Darlington [87], Akima [88], Wozencraft and Jacobs [18], Wainstein and Zubakov [15]). Our approach is similar to [18]. A third way to estimate A is discussed in [87]. (See also Chapter II.4.)

where [·] denotes the largest integer smaller than or equal to the argument. The output of this preliminary processing is M numbers. We choose the largest and assume that the correct value of A is in that region.

To get the final estimate we conduct a local maximization by using the condition

$$\int_{-T/2}^{T/2} [r(t) - s(t, A)] \frac{\partial s(t, A)}{\partial A} \, dt \Big|_{A=\hat{a}_{\text{map}}} = 0. \tag{118}$$

(This assumes the maximum is interior.)

A possible way to accomplish this maximization is given in the block diagram in Fig. 4.32. We expect that if we chose the correct interval in our preliminary processing the final accuracy would be closely approximated by the bound in (108). This bound can be evaluated easily. The partial derivative of the signal is

$$\frac{\partial s(t, A)}{\partial A} = \left(\frac{2E}{T}\right)^{1/2} \beta t \cos(\omega_c t + \beta A t), \qquad -T/2 \le t \le T/2, \tag{119}$$

and

$$\gamma_a{}^2 = \frac{2E}{T} \beta^2 \int_{-T/2}^{T/2} \overline{t^2 \cos^2(\omega_c t + \beta A t)} \, dt \simeq \frac{ET^2}{12} \beta^2, \tag{120}$$

when

$$T \gg \frac{1}{\omega_c}.$$

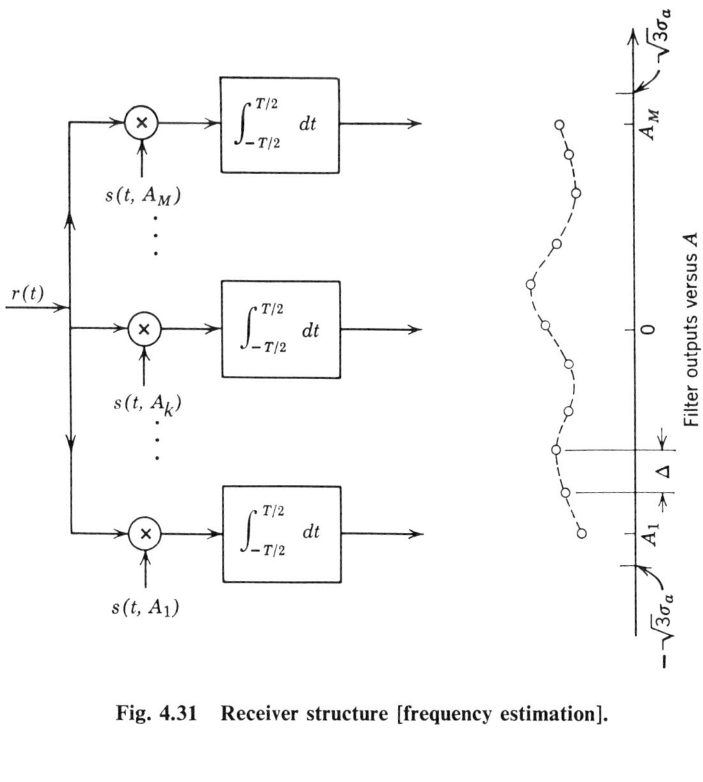

Fig. 4.31 Receiver structure [frequency estimation].

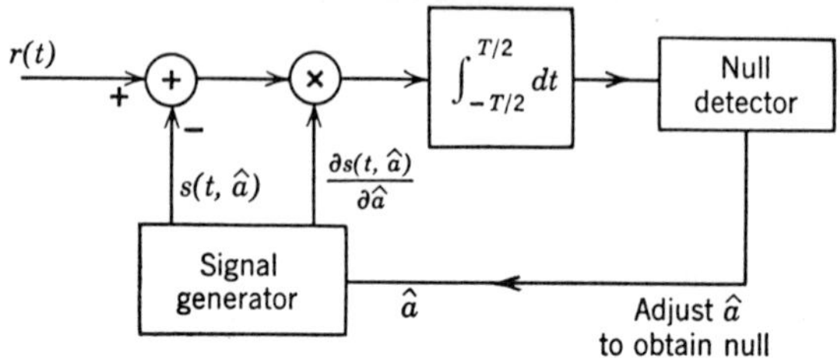

Fig. 4.32 Local estimator.

Then the normalized mean-square error of any estimate is bounded by

$$\sigma_{a_{\epsilon n}}^2 \triangleq \frac{\sigma_\epsilon^2}{\sigma_a^2} \geq \frac{N_0}{2\gamma_a^2} = \frac{12}{T^2} \frac{N_0}{2E} \frac{1}{\beta^2} \frac{1}{\sigma_a^2}, \tag{121}$$

which seems to indicate that, regardless of how small E/N_0 is, we can make the mean-square error arbitrarily small by increasing β. Unfortunately, this method neglects an important part of the problem. How is the probability of an initial interval error affected by the value of β?

With a few simplifying assumptions we can obtain an approximate expression for this probability. We denote the actual value of A as A_a. (This subscript is necessary because A is the argument in our likelihood function.) A plot of

$$\frac{1}{E} \int_{-T/2}^{T/2} s(t, A_a)\, s(t, A)\, dt$$

for the signal in (115) as a function of $A_x \triangleq A - A_a$ is given in Fig. 4.33. (The double frequency term is neglected.) We see that the signal component of $l_1(A)$ passes through zero every $2\pi/\beta T$ units. This suggests that a logical value of Δ is $2\pi/\beta T$.

To calculate the probability of choosing the wrong interval we use the approximation that we can replace all A in the first interval by A_1 and so forth. We denote the probability of choosing the wrong interval as $\Pr(\epsilon_I)$. With this approximation the problem is reduced to detecting which of M orthogonal, equal energy signals is present. For large M we neglect the residual at the end of the interval and let

$$M \cong \sqrt{3}\, \sigma_a \beta \frac{T}{\pi}; \tag{122}$$

but this is a problem we have already solved (64). Because large βT is the case of interest, we may use the approximate expression in (65):

$$\Pr(\epsilon_I) \leq \frac{(\sqrt{3}\, \sigma_a \beta T/\pi - 1)}{\sqrt{2\pi E/N_0}} \exp\left(-\frac{E}{2N_0}\right). \tag{123}$$

We see that as $\sigma_a \beta T$ increases the probability that we will choose the wrong interval also increases.† The conclusion that can be inferred from this result is of fundamental importance in the nonlinear estimation problem.

For a fixed E/N_0 and T we can increase β so that the local error will be arbitrarily small *if* the receiver has chosen the correct interval. As β increases, however, the

† This probability is sometimes referred to as the probability of anomaly or ambiguity.

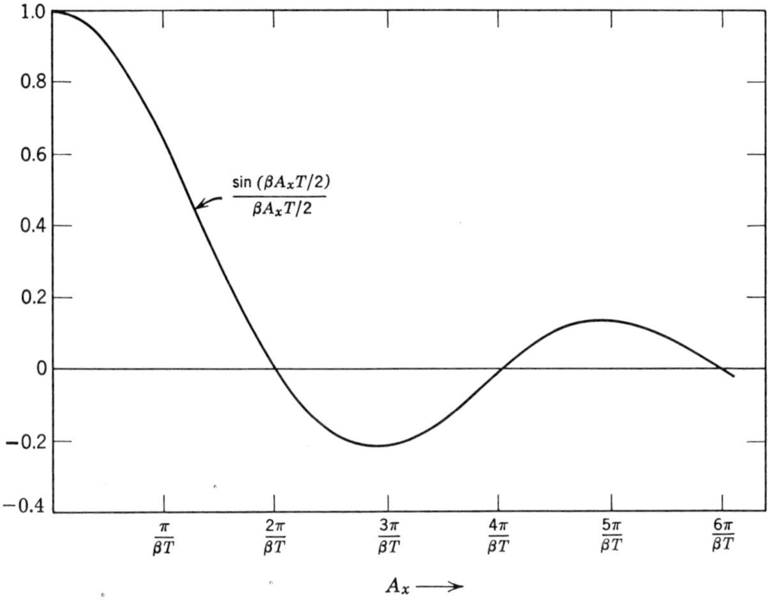

Fig. 4.33 Signal component vs. A_x.

probability that we will be in the correct interval goes to zero. Thus, for a particular β, we must have some minimum E/N_0 to ensure that the probability of being in the wrong interval is adequately small.

The expression in (123) suggests the following design procedure. We decide that a certain Pr (ϵ_I) (say p_0) is acceptable. In order to minimize the mean-square error subject to this constraint we choose β such that (123) is satisfied with equality. Substituting p_0 into the left side of (123), solving for $\sigma_a \beta T$, and substituting the result in (121), we obtain

$$\frac{E[a_\epsilon^2]}{\sigma_a^2} = \sigma_{a_\epsilon n}^2 \simeq \frac{1}{{p_0}^2} \frac{9}{\pi^3} \left(\frac{N_0}{E}\right)^2 e^{-E/N_0}. \tag{124†}$$

The reciprocal of the normalized mean-square error as a function of E/N_0 for typical values of p_0 is shown in Fig. 4.34. For reasons that will become obvious shortly, we refer to the constraint imposed by (123) as a threshold constraint.

The result in (123) indicates one effect of increasing β. A second effect can be seen directly from (115). Each value of A shifts the frequency of the transmitted signal from ω_c to $\omega_c + \beta A$. Therefore we must have enough bandwidth available in the channel to accommodate the maximum possible frequency excursion. The pulse

† Equation 124 is an approximation, for (123) is a bound and we neglected the 1 in the parentheses because large βT is the case of interest.

bandwidth is approximately $2\pi/T$ rad/sec. The maximum frequency shift is $\pm\sqrt{3}\,\beta\sigma_a$. Therefore the required channel bandwidth centered at ω_c is approximately

$$2\pi W_{\text{ch}} \cong 2\sqrt{3}\,\beta\sigma_a + \frac{2\pi}{T} = \frac{1}{T}(2\sqrt{3}\,\beta\sigma_a T + 2\pi) \tag{125a}$$

When $\sigma_a\beta T$ is large we can neglect the 2π and use the approximation

$$2\pi W_{\text{ch}} \simeq 2\sqrt{3}\,\beta\sigma_a. \tag{125b}$$

In many systems of interest we have only a certain bandwidth available. (This bandwidth limitation may be a legal restriction or may be caused by the physical nature of the channel.) If we assume that E/N_0 is large enough to guarantee an acceptable $\Pr(\epsilon_I)$, then (125*b*) provides the constraint of the system design. We simply increase β until the available bandwidth is occupied. To find the mean-square error using this design procedure we substitute the expression for $\beta\sigma_a$ in (125*b*) into (121) and obtain

$$\frac{E[a^2{}_\epsilon]}{\sigma_a{}^2} = \sigma^2_{a_\epsilon n} = \frac{18}{\pi^2}\frac{N_0}{E}\frac{1}{(W_{\text{ch}}T)^2} \qquad \text{(bandwidth constraint)}. \tag{126}$$

We see that the two quantities that determine the mean-square error are E/N_0, the energy-to-noise ratio, and $W_{\text{ch}}T$, which is proportional to the time-bandwidth product of the transmitted pulse. The reciprocal of the normalized mean-square error is plotted in Fig. 4.34 for typical values of $W_{\text{ch}}T$.

The two families of constraint lines provide us with a complete design procedure for a PFM system. For low values of E/N_0 the threshold constraint dominates. As E/N_0 increases, the MMSE moves along a fixed p_0 line until it reaches a point where the available bandwidth is a constraint. Any further increase in E/N_0 moves the MMSE along a fixed β line.

The approach in which we consider two types of error separately is useful and contributes to our understanding of the problem. To compare the results with other systems it is frequently convenient to express them as a single number, the over-all mean-square error.

We can write the mean-square error as

$$\begin{aligned} E(a_\epsilon{}^2) = \sigma_{\epsilon_T}{}^2 = {} & E[a_\epsilon{}^2|\text{interval error}] \Pr[\text{interval error}] \\ & + E[a_\epsilon{}^2|\text{no interval error}] \Pr[\text{no interval error}] \end{aligned} \tag{127}$$

We obtained an approximation to $\Pr(\epsilon_I)$ by collecting each incremental range of A at a single value A_i. With this approximation there is no signal component at the other correlator outputs in Fig. 4.31. Thus, if an interval error is made, it is equally likely to occur in any one of the wrong intervals. Therefore the resulting estimate $\hat{a}$ will be uncorrelated with a.

$$\begin{aligned} E[a_\epsilon{}^2|\text{interval error}] &= E[(\hat{a} - a)^2|\text{interval error}] \\ &= E[\hat{a}^2|\text{interval error}] + E[a^2|\text{interval error}] \\ &\quad - 2E[\hat{a}a|\text{interval error}]. \end{aligned} \tag{128}$$

Our approximation makes the last term zero. The first two terms both equal $\sigma_a{}^2$. Therefore

$$E[a_\epsilon{}^2|\text{interval error}] = 2\sigma_a{}^2. \tag{129}$$

If we assume that p_0 is fixed, we then obtain by using (124) and (129) in (127)

$$\sigma^2_{\epsilon_T n} = \frac{E(a_\epsilon{}^2)}{\sigma_a{}^2} = 2p_0 + (1 - p_0)\frac{9}{\pi^3 p_0^2}\left(\frac{N_0}{E}\right)^2 e^{-E/N_0}. \tag{130}$$

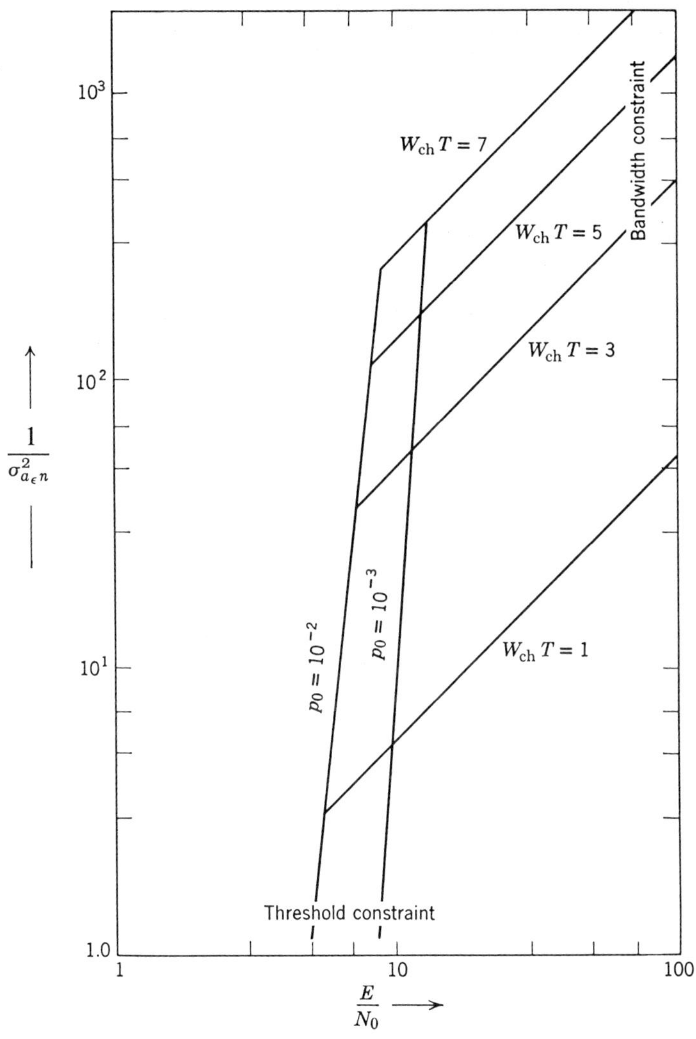

Fig. 4.34 Reciprocal of the mean-square error under threshold and bandwidth constraints.

In this case the modulation index β must be changed as E/N_0 is changed. For a fixed β we use (121) and (123) to obtain

$$\sigma^2_{\epsilon_T n} = \frac{12}{(\sigma_a \beta T)^2} \frac{N_0}{2E}\left[1 - \frac{\sqrt{3}\,\sigma_a \beta T/\pi}{\sqrt{2\pi E/N_0}}\, e^{-E/2N_0}\right] + 2\frac{\sqrt{3}\,\sigma_a \beta T/\pi}{\sqrt{2\pi E/N_0}}\, e^{-E/2N_0}. \tag{131}$$

The result in (131) is plotted in Fig. 4.35, and we see that the mean-square error exhibits a definite threshold. The reciprocal of the normalized mean-square error for a PAM system is also shown in Fig. 4.35 (from 96). The magnitude of this improvement can be obtained by dividing (121) by (96).

$$\frac{\sigma^2_{a_\epsilon n|\text{PFM}}}{\sigma^2_{a_\epsilon n|\text{PAM}}} \simeq \frac{12}{\beta^2 T^2}, \qquad \frac{2\sigma_a{}^2 E}{N_0} \gg 1.$$

Thus the improvement obtained from PFM is proportional to the square of βT. It is important to re-emphasize that this result assumes E/N_0 is such that the system is above threshold. If the noise level should increase, the performance of the PFM system can decrease drastically.

Our approach in this particular example is certainly plausible. We see, however, that it relies on a two-step estimation procedure. In discrete frequency modulation this procedure was a natural choice because it was also a logical practical implementation. In the first example there was no need for the two-step procedure. However, in order to obtain a parallel set of results for Example 1 we can carry out an analogous two-step analysis and similar results. Experimental studies of both types of systems indicate that the analytic results correctly describe system performance. It would still be desirable to have a more rigorous analysis.

We shall discuss briefly in the context of Example 2 an alternate approach in which we bound the mean-square error directly. From (115) we see that $s(t, A)$ is an analytic function with respect to the parameter A. Thus all derivatives will exist and can be expressed in a simple form:

$$\frac{\partial^n s(t, A)}{\partial A^n} = \begin{cases} \left(\dfrac{2E}{T}\right)^{1/2} (\beta t)^n (-1)^{(n-1)/2} \cos(\omega_c t + \beta A t), & n \text{ odd}, \\ \left(\dfrac{2E}{T}\right)^{1/2} (\beta t)^n (-1)^{n/2} \sin(\omega_c t + \beta A t), & n \text{ even}. \end{cases} \tag{132}$$

This suggests that a generalization of the Bhattacharyya bound that we developed in the problem section of Chapter 2 would enable us to get as good an estimate of the error as desired. This extension is conceptually straightforward (Van Trees [24]). For $n = 2$ the answer is still simple. For $n \geq 3$, however, the required matrix inversion is tedious and it is easier to proceed numerically. The detailed calculations have been carried out [29]. In this particular case the series does not converge fast enough to give a good approximation to the actual error in the high noise region.

One final comment is necessary. There are some cases of interest in which the signal is not differentiable with respect to the parameter. A simple example of this type of signal arises when we approximate the transmitted signal in a radar system by an ideal rectangular pulse and want to estimate the time of arrival of the returning pulse. When the noise is weak, formulas for these cases can be developed easily (e.g., Mallinckrodt

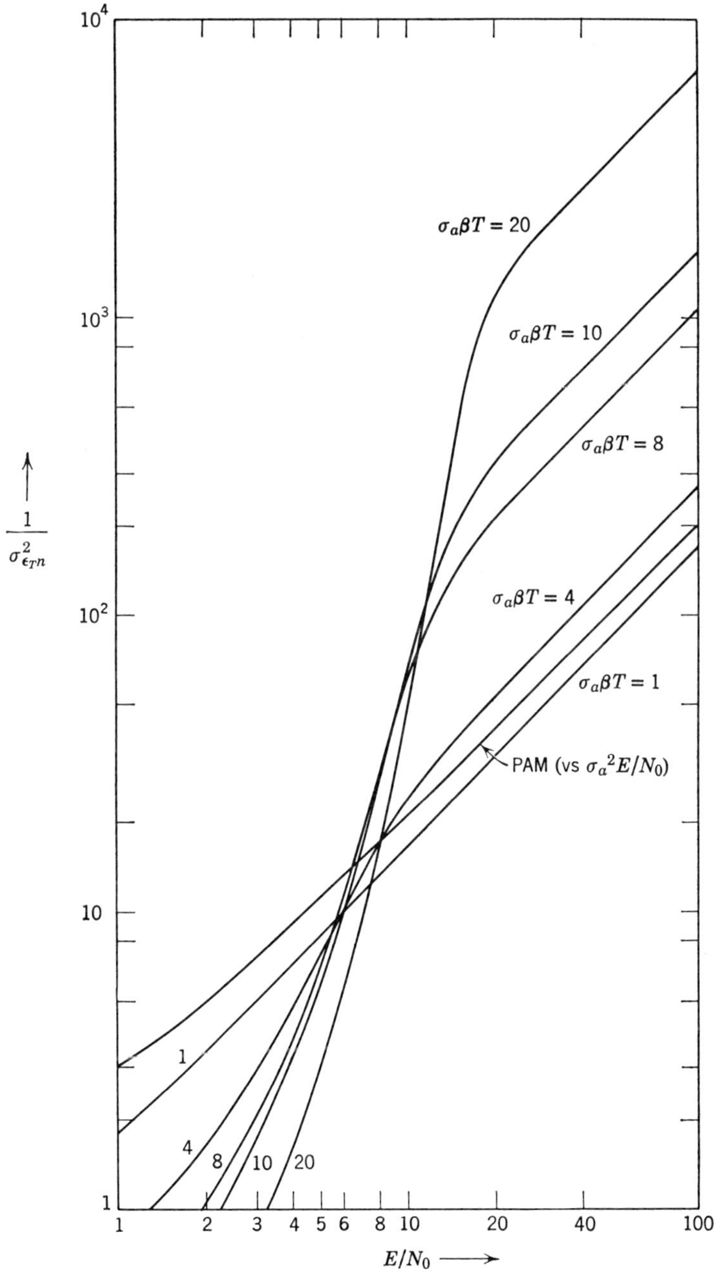

Fig. 4.35 Reciprocal of the total mean-square error.

and Sollenberger [25], Kotelnikov [13], Manasse [26], Skolnik [27]). For arbitrary noise levels we can use the approach in Example 2 or the Barankin bound (e.g., Swerling [28]) which does not require differentiability.

4.2.4 Summary: Known Signals in White Gaussian Noise

It is appropriate to summarize some of the important results that have been derived for the problem of detection and estimation in the presence of additive white Gaussian noise.

Detection. 1. For the simple binary case the optimum receiver can be realized as a matched filter or a correlation receiver, as shown in Figs. 4.11 and 4.12, respectively.

2. For the general binary case the optimum receiver can be realized by using a single matched filter or a pair of filters.

3. In both cases the performance is determined completely by the normalized distance d between the signal points in the decision space,

$$d^2 = \frac{2}{N_0}(E_1 + E_0 - 2\rho\sqrt{E_1 E_0}). \tag{133}$$

The resulting errors are

$$P_F = \text{erfc}_*\left(\frac{\ln \eta}{d} + \frac{d}{2}\right), \tag{134}$$

$$P_M = \text{erf}_*\left(\frac{\ln \eta}{d} - \frac{d}{2}\right). \tag{135}$$

For equally-likely hypotheses and a minimum $\Pr(\epsilon)$ criterion, the total error probability is

$$\Pr(\epsilon) = \text{erfc}_*\left(\frac{d}{2}\right) \leq \left(\frac{2}{\pi d^2}\right)^{1/2} e^{-d^2/8}. \tag{136}$$

4. The performance of the optimum receiver is insensitive to small signal variations.

5. For the M-ary case the optimum receiver requires at most $M - 1$ matched filters, although frequently M matched filters give a simpler implementation. For M orthogonal signals a simple bound on the error probability is

$$\Pr(\epsilon) \leq \frac{M-1}{\sqrt{2\pi(E/N_0)}} \exp\left(-\frac{E}{2N_0}\right). \tag{137}$$

6. A simple example of transmitting a sequence of digits illustrated the idea of a channel capacity. At transmission rates below this capacity the

Pr (ϵ) approaches zero as the length of encoded sequence approaches infinity. Because of the bandwidth requirement, the orthogonal signal technique is not efficient.

Estimation. 1. Linear estimation is a trivial modification of the detection problem. The optimum estimator is a simple correlator or matched filter followed by a gain.

2. The nonlinear estimation problem introduced several new ideas. The optimum receiver is sometimes difficult to realize exactly and an approximation is necessary. Above a certain energy-to-noise level we found that we could make the estimation error appreciably smaller than in the linear estimation case which used the same amount of energy. Specifically,

$$\text{Var}\,[\hat{a} - A] \approx \frac{N_0/2}{\displaystyle\int_0^T \left[\frac{\partial s(t, A)}{\partial A}\right]^2 dt}. \tag{138}$$

As the noise level increased however, the receiver exhibited a *threshold* phenomenon and the error variance increased rapidly. Above the threshold we found that we had to consider the problem of a bandwidth constraint when we designed the system.

We now want to extend our model to a more general case. The next step in the direction of generality is to consider known signals in the presence of nonwhite additive Gaussian noise.

4.3 DETECTION AND ESTIMATION IN NONWHITE GAUSSIAN NOISE

Several situations in which nonwhite Gaussian interference can occur are of interest:

1. Between the actual noise source and the data-processing part of the receiver are elements (such as an antenna and RF filters) which shape the noise spectrum.
2. In addition to the desired signal at the receiver, there may be an interfering signal that can be characterized as a Gaussian process. In radar/sonar it is frequently an interfering target.

With this motivation we now formulate and solve the detection and estimation problem. As we have seen in the preceding section, a close coupling exists between detection and estimation. In fact, the development through construction of the likelihood ratio (or function) is identical. We derive the simple binary case in detail and then indicate how the results extend to other cases of interest. The first step is to specify the model.

When colored noise is present, we have to be more careful about our model. We assume that the transmitted signal on hypothesis 1 is

$$\sqrt{E}\,s(t) \triangleq \begin{cases} \sqrt{E}\,s_T(t), & 0 \le t \le T, \\ 0 \quad , & \text{elsewhere.} \end{cases} \tag{139}$$

Observe that $s(t)$ is defined for all time. Before reception the signal is corrupted by additive Gaussian noise $n(t)$. The received waveform $r(t)$ is observed over the interval $T_i \le t \le T_f$. Thus

$$\begin{aligned} r(t) &= \sqrt{E}\,s(t) + n(t), \qquad & T_i \le t \le T_f : H_1 \\ &= n(t), & T_i \le t \le T_f : H_0. \end{aligned} \tag{140}$$

Sometimes T_i will equal zero and T_f will equal T. In general, however, we shall let T_i (≤ 0) and T_f $(\ge T)$ remain arbitrary. Specifically, we shall frequently examine the problem in which $T_i = -\infty$ and $T_f = +\infty$. A logical question is; why should we observe the received waveform when the signal component is zero? The reason is that the noise outside the interval is correlated with the noise inside the interval, and presumably the more knowledge available about the noise inside the interval the better we can combat it and improve our system performance. A trivial example can be used to illustrate this point.

Example. Let

$$\begin{aligned} \sqrt{E}\,s(t) &= 1, \qquad & 0 \le t \le 1 \\ &= 0, & \text{elsewhere.} \end{aligned} \tag{141}$$

Let

$$n(t) = n, \qquad 0 \le t \le 2, \tag{142}$$

where n is a Gaussian random variable. We can decide which hypothesis is true in the following way:

$$l = \int_0^1 r(t)\,dt - \int_1^2 r(t)\,dt. \tag{143}$$

If

$$\begin{aligned} l &= 0, \qquad & \text{say } H_0 \\ &\ne 0 & \text{say } H_1. \end{aligned}$$

Clearly, we can make error-free decisions. Here we used the extended interval to estimate the noise inside the interval where the signal was nonzero. Unfortunately, the actual situation is not so simple, but the idea of using an extended observation interval carries over to more realistic problems.

Initially, we shall find it useful to assume that the noise always contains an *independent* white component. Thus

$$n(t) = w(t) + n_c(t) \tag{144}$$

where $n_c(t)$ is the *colored* noise component. Then,

$$K_n(t, u) = \frac{N_0}{2}\,\delta(t - u) + K_c(t, u). \tag{145}$$

We assume the $n_c(t)$ has a finite mean-square value $[E(n_c^2(t)) < \infty$ for all $T_i \leq t \leq T_f]$ so $K_c(t, u)$ is a square-integrable function over $[T_i, T_f]$.

The white noise assumption is included for two reasons:

1. The physical reason is that regardless of the region of the spectrum used there will be a nonzero noise level. Extension of this level to infinity is just a convenience.
2. The mathematical reason will appear logically in our development. The white noise component enables us to guarantee that our operations will be meaningful. There are other ways to accomplish this objective but the white noise approach is the simplest.

Three logical approaches to the solution of the nonwhite noise problem are the following:

1. We choose the coordinates for the orthonormal expansion of $r(t)$ so that the coefficients are statistically independent. This will make the construction of the likelihood ratio straightforward. From our discussion in Chapter 3 we know how to carry out this procedure.
2. We operate on $r(t)$ to obtain a sufficient statistic and then use it to perform the detection.
3. We perform preliminary processing on $r(t)$ to transform the problem into a white Gaussian noise problem and then use the white Gaussian noise solution obtained in the preceding section. It is intuitively clear that if the preliminary processing is reversible it can have no effect on the performance of the system. Because we use the idea of reversibility repeatedly, however, it is worthwhile to provide a simple proof.

Reversibility. It is easy to demonstrate the desired result in a general setting. In Fig. 4.36*a* we show a system that operates on $r(u)$ to give an output that is optimum according to some desired criterion. (The problem of interest may be detection or estimation.) In system 2, shown in Fig. 4.36*b*, we first operate on $r(u)$ with a reversible operation $k[t, r(u)]$ to obtain $z(t)$. We then design a system that will perform an operation on $z(t)$ to obtain an output that is optimum according to the same criterion as in system 1. We now claim that the performances of the two systems are identical. Clearly, system 2 cannot perform better than system 1 or this would contradict our statement that system 1 is the optimum operation on $r(u)$. We now show that system 2 cannot be worse than system 1. Suppose that system 2 were worse than system 1. If this were true, we could design the system shown in Fig. 4.36*c*, which operates on $z(t)$ with the inverse of $k[t, r(u)]$ to give $r(u)$ and then passes it through system 1. This over-all system will work as well as system 1 (they are identical from the input-output standpoint). Because the result in Fig. 4.36*c* is obtained by operating on $z(t)$, it cannot be better than system 2 or it will contradict the statement that the second operation in system 2 is optimum. Thus system 2 cannot be worse than system 1.

Therefore *any* reversible operation can be included to facilitate the solution. We observe that linearity is not an issue, only the existence of an inverse. Reversibility is only *sufficient*, not *necessary*. (This is obvious from our discussion of sufficient statistics in Chapter 2.)

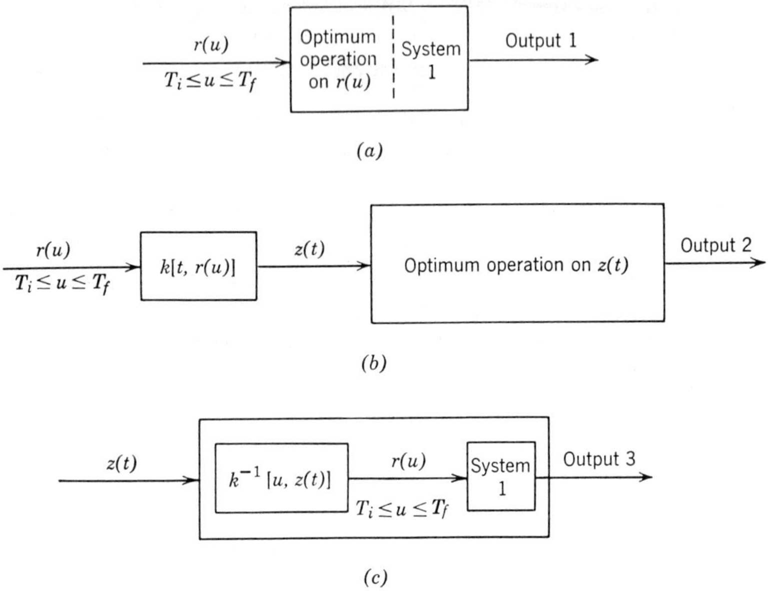

Fig. 4.36 Reversibility proof: (*a*) system 1; (*b*) system 2; (*c*) system 3.

We now return to the problem of interest. The first two of these approaches involve much less work and also extend in an easy fashion to more general cases. The third approach however, using reversibility, seems to have more intuitive appeal, so we shall do it first.

4.3.1 "Whitening" Approach

First we shall derive the structures of the optimum detector and estimator. In this section we require a nonzero white noise level.

Structures. As a preliminary operation, we shall pass $r(t)$ through a linear time-varying filter whose impulse response is $h_w(t, u)$ (Fig. 4.37). The impulse response is assumed to be zero for either t or u outside the interval $[T_i, T_f]$. For the moment, we shall not worry about realizability

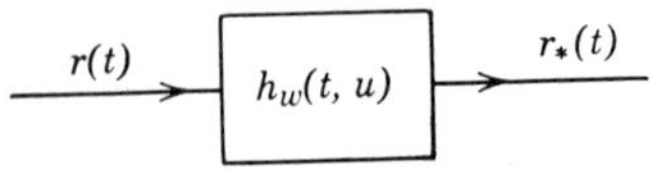

Fig. 4.37 "Whitening" filter.

and shall allow $h_w(t, u)$ to be nonzero for $u > t$. Later, in specific examples, we also look for realizable whitening filters. The output is

$$\begin{aligned} r_*(t) &\triangleq \int_{T_i}^{T_f} h_w(t, u)\, r(u)\, du \\ &= \int_{T_i}^{T_f} h_w(t, u)\sqrt{E}\, s(u)\, du + \int_{T_i}^{T_f} h_w(t, u)\, n(u)\, du \\ &\triangleq s_*(t) + n_*(t), \qquad T_i \le t \le T_f, \end{aligned} \tag{146}$$

when H_1 is true and

$$r_*(t) = n_*(t), \qquad T_i \le t \le T_f, \tag{147}$$

when H_0 is true. We want to choose $h_w(t, u)$ so that

$$K_{n_*}(t, u) = E[n_*(t)\, n_*(u)] = \delta(t - u), \qquad T_i \le t, u \le T_f. \tag{148}$$

Observe that we have arbitrarily specified a unity spectral height for the noise level at the output of the whitening filter. This is merely a convenient normalization.

The following logical question arises:

What conditions on $K_n(t, u)$ will guarantee that a reversible whitening filter exists? Because the whitening filter is linear, we can show reversibility by finding a filter $h_w^{-1}(t, u)$ such that

$$\int_{T_i}^{T_f} h_w^{-1}(t, z)\, h_w(z, u)\, dz = \delta(t - u), \qquad T_i \le t, u \le T_f. \tag{149}$$

For the moment we shall assume that we can find a suitable set of conditions and proceed with the development.

Because $n_*(t)$ is "white," we may use (22) and (23) directly ($N_0 = 2$):

$$\ln \Lambda[r_*(t)] = \int_{T_i}^{T_f} r_*(u)\, s_*(u)\, du - \tfrac{1}{2}\int_{T_i}^{T_f} s_*^2(u)\, du. \tag{150}$$

We can also write this directly in terms of the original waveforms and $h_w(t, u)$:

$$\begin{aligned} \ln \Lambda[r(t)] &= \int_{T_i}^{T_f} du \int_{T_i}^{T_f} h_w(u, z)\, r(z)\, dz \int_{T_i}^{T_f} h_w(u, v)\sqrt{E}\, s(v)\, dv \\ &\quad - \tfrac{1}{2}\int_{T_i}^{T_f} du \int_{T_i}^{T_f} h_w(u, z)\sqrt{E}\, s(z)\, dz \int_{T_i}^{T_f} h_w(u, v)\sqrt{E}\, s(v)\, dv. \end{aligned} \tag{151}$$

This expression can be formally simplified by defining a new function:

$$Q_n(z, v) = \int_{T_i}^{T_f} h_w(u, z)\, h_w(u, v)\, du, \qquad T_i < z, v < T_f. \tag{152†}$$

For the moment we can regard it as a function that we accidentally stumbled on in an effort to simplify an equation. Later we shall see that it plays a fundamental role in many of our discussions. Rewriting (151), we have

$$\ln \Lambda[r(t)] = \int_{T_i}^{T_f} r(z)\, dz \int_{T_i}^{T_f} Q_n(z, v)\sqrt{E}\, s(v)\, dv - \frac{E}{2}\int_{T_i}^{T_f} s(z)\, dz \int_{T_i}^{T_f} Q_n(z, v)\, s(v)\, dv. \tag{153}$$

We can simplify (153) by writing

$$g(z) = \int_{T_i}^{T_f} Q_n(z, v)\sqrt{E}\, s(v)\, dv, \qquad T_i < z < T_f. \tag{154}$$

We have used a strict inequality in (154). Looking at (153), we see that $g(z)$ only appears inside an integral. Therefore, if $g(z)$ *does not* contain singularities at the endpoints, we can assign $g(z)$ any finite value at the endpoint and $\ln \Lambda[r(t)]$ will be unchanged. Whenever there is a white noise component, we can show that $g(z)$ is square-integrable (and therefore contains no singularities). For convenience we make $g(z)$ continuous at the endpoints.

$$g(T_f) = \lim_{z \to T_f^-} g(z),$$

$$g(T_i) = \lim_{z \to T_i^+} g(z).$$

We see that the construction of the likelihood function involves a correlation operation between the actual received waveform and a function $g(z)$. Thus, from the standpoint of constructing the receiver, the function $g(z)$ is the only one needed. Observe that the correlation of $r(t)$ with $g(t)$ is simply the reduction of the observation space to a single sufficient statistic.

Three canonical receiver structures for simple binary detection are

† Throughout this section we must be careful about the endpoints of the interval. The difficulty is with factors of 2 which arise because of the delta function in the noise covariance. We avoid this by using an open interval and then show that endpoints are not important in this problem. We suggest that the reader ignore the comments regarding endpoints until he has read through Section 4.3.3. This strategy will make these sections more readable.

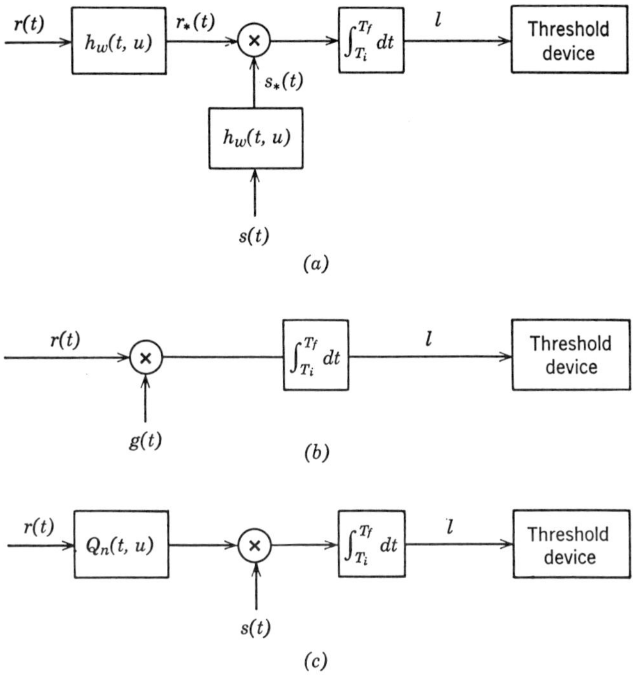

Fig. 4.38 Alternate structures for colored noise problem.

shown in Fig. 4.38. We shall see that the first two are practical implementations, whereas the third affords an interesting interpretation. The modification of Fig. 4.38*b* to obtain a matched filter realization is obvious. To implement the receivers we must solve (149), (152), or (154). Rather than finding closed-form solutions to these equations we shall content ourselves in this section with series solutions in terms of the eigenfunctions and eigenvalues of $K_c(t, u)$. These series solutions have two purposes:

1. They demonstrate that solutions exist.
2. They are useful in certain optimization problems.

After deriving these solutions, we shall look at the receiver performance and extensions to general binary detection, M-ary detection, and estimation problems. We shall then return to the issue of closed-form solutions. The advantage of this approach is that it enables us to obtain an integrated picture of the colored noise problem and many of its important features without getting lost in the tedious details of solving integral equations.

Construction of $Q_n(t, u)$ and $g(t)$. The first step is to express $Q_n(t, u)$ directly in terms of $K_n(t, u)$. We recall our definition of $h_w(t, u)$. It is a time-varying linear filter chosen so that when the input is $n(t)$ the output will be $n_*(t)$, a sample function from a white Gaussian process. Thus

$$n_*(t) = \int_{T_i}^{T_f} h_w(t, x)\, n(x)\, dx \qquad T_i \le t \le T_f \tag{155}$$

and

$$E[n_*(t) n_*(u)] = K_{n_*}(t, u) = \delta(t - u). \qquad T_i \le t \le T_f. \tag{156}$$

Substituting (155) into (156), we have

$$\delta(t - u) = E \iint_{T_i}^{T_f} h_w(t, x)\, h_w(u, z)\, n(x)\, n(z)\, dx\, dz. \tag{157}$$

By bringing the expectation inside the integrals, we have

$$\delta(t - u) = \iint_{T_i}^{T_f} h_w(t, x)\, h_w(u, z)\, K_n(x, z)\, dx\, dz, \qquad T_i < t, u < T_f. \tag{158}$$

In order to get (158) into a form such that we can introduce $Q_n(t, u)$, we multiply both sides by $h_w(t, v)$ and integrate with respect to t. This gives

$$h_w(u, v) = \int_{T_i}^{T_f} dz\, h_w(u, z) \int_{T_i}^{T_f} K_n(x, z)\, dx \int_{T_i}^{T_f} h_w(t, v)\, h_w(t, x)\, dt. \tag{159}$$

Looking at (152), we see that the last integral is just $Q_n(v, x)$. Therefore

$$h_w(u, v) = \int_{T_i}^{T_f} dz\, h_w(u, z) \int_{T_i}^{T_f} K_n(x, z)\, Q_n(v, x)\, dx. \tag{160}$$

This implies that the inner integral must be an impulse over the open interval,

$$\boxed{\delta(z - v) = \int_{T_i}^{T_f} K_n(x, z)\, Q_n(v, x)\, dx, \qquad T_i < z, v < T_f.} \tag{161}$$

This is the desired result that relates $Q_n(v, x)$ directly to the original covariance function. Because $K_n(x, z)$ is the kernel of many of the integral equations of interest to us, $Q_n(v, x)$ is frequently called the *inverse kernel.*

From (145) we know that $K_n(x, z)$ consists of an impulse and a well-behaved term. A logical approach is to try and express $Q_n(v, x)$ in a similar manner. We try a solution to (161) of the form

$$Q_n(v, x) = \frac{2}{N_0}[\delta(v - x) - h_o(v, x)] \qquad T_i < v, x < T_f. \tag{162}$$

Substituting (145) and (162) into (161) and rearranging terms, we obtain an equation that $h_0(v, x)$ must satisfy:

$$\frac{N_0}{2} h_o(v, z) + \int_{T_i}^{T_f} h_o(v, x)\, K_c(x, z)\, dx = K_c(v, z), \qquad T_i < z, v < T_f \quad (163)$$

This equation is familiar to us from the section on optimum linear filters in Chapter 3 [Section 3.4.5; particularly, (3-144)]. The significance of this similarity is seen by re-drawing the system in Fig. 4.38*c* as shown in Fig. 4.39. The function $Q_n(t, u)$ is divided into two parts. We see that the output of the filter in the bottom path is precisely the minimum mean-square error estimate of the colored noise component, assuming that H_0 is true. If we knew $n_c(t)$, it is clear that the optimum processing would consist of subtracting it from $r(t)$ and passing the result into a matched filter or correlation receiver. The optimum receiver does exactly that, except that it does not know $n_c(t)$; therefore it makes a MMSE estimate $\hat{n}_c(t)$ and uses it. This is an intuitively pleasing result of a type that we shall encounter frequently.†

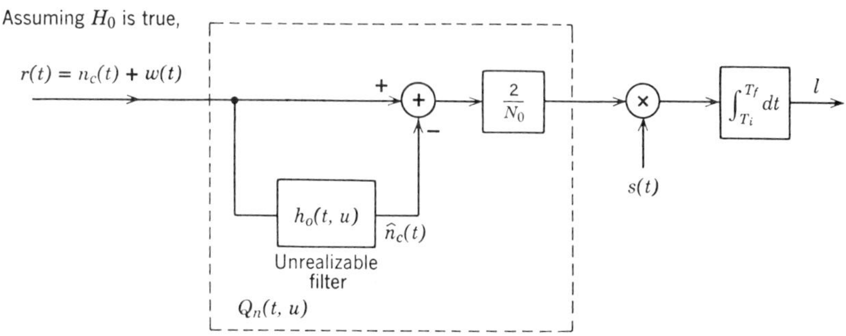

Fig. 4.39 Realization of detector using an optimum linear filter.

† The reader may wonder why we care whether a result is intuitively pleasing, if we know it is optimum. There are two reasons for this interest: (a) It is a crude error-checking device. For the type of problems of interest to us, when we obtain a mathematical result that is unintuitive it is usually necessary to go back over the model formulation and the subsequent derivation and satisfy ourselves that either the model omits some necessary feature of the problem or that our intuition is wrong. (b) In many cases the solution for the optimum receiver may be mathematically intractable. Having an intuitive interpretation for the solutions to the various Gaussian problems equips us to obtain a good receiver by using intuitive reasoning when we cannot get a mathematical solution.

From our results in Chapter 3 (3.154) we can write a formal solution for $h_o(t, u)$ in terms of the eigenvalues of $K_c(t, u)$. Using (3.154),

$$h_o(t, u) = \sum_{i=1}^{\infty} \frac{\lambda_i^c}{\lambda_i^c + N_0/2} \phi_i(t) \phi_i(u), \qquad T_i < t, u < T_f, \tag{164}$$

where λ_i^c and $\phi_i(t)$ are the eigenvalues and eigenfunctions, respectively, of $K_c(t, u)$. We can write the entire inverse kernel as

$$Q_n(t, u) = \frac{2}{N_0}\left[\delta(t - u) - \sum_{i=1}^{\infty} \frac{\lambda_i^c}{\lambda_i^c + N_0/2} \phi_i(t) \phi_i(u)\right]. \tag{165}$$

It is important to re-emphasize that our ability to write $Q_n(t, u)$ as an impulse function and a well-behaved function rests heavily on our assumption that there is a nonzero white noise level. This is the mathematical reason for the assumption.

We can also write $Q_n(t, u)$ as a single series. We express the impulse in terms of a series by using (3.128) and then combine the series to obtain

$$Q_n(t, u) = \sum_{i=1}^{\infty} \left(\frac{N_0}{2} + \lambda_i^c\right)^{-1} \phi_i(t) \phi_i(u) = \sum_{i=1}^{\infty} \frac{1}{\lambda_i^T} \phi_i(t) \phi_i(u), \tag{166}$$

where

$$\lambda_i^T \triangleq \frac{N_0}{2} + \lambda_i^c. \tag{167}$$

(T denotes total). The series in (166) does not converge. However, in most cases $Q_n(t, u)$ is inside an integral and the overall expression will converge.

As a final result, we want to find an equation that will specify $g(t)$ directly in terms of $K_n(t, z)$. We start with (154):

$$g(z) = \int_{T_i}^{T_f} Q_n(z, v)\sqrt{E}\, s(v)\, dv, \qquad T_i < z < T_f. \tag{168}$$

The technique that we use is based on the inverse relation between $K_n(t, z)$ and $Q_n(t, z)$, expressed by (161). To get rid of $Q_n(z, v)$ we simply multiply (168) by $K_n(t, z)$, integrate with respect to z, and use (161). The result is

$$\boxed{\int_{T_i}^{T_f} K_n(t, z)\, g(z)\, dz = \sqrt{E}\, s(t), \qquad T_i < t < T_f.} \tag{169a}$$

Substituting (145) into (169a), we obtain an equation for the open interval (T_i, T_f). Our continuity assumption after (154) extends the range to the closed interval $[T_i, T_f]$. The result is

$$\boxed{\frac{N_0}{2} g(t) + \int_{T_i}^{T_f} K_c(t, z)\, g(z)\, dz = \sqrt{E}\, s(t), \qquad T_i \le t \le T_f.} \tag{169b}$$

To implement the receiver, as shown in Fig. 4.38b, we would solve (169b)

directly. We shall develop techniques for obtaining closed-form solutions in 4.3.6. A series solution can be written easily by using (168) and (165):

$$g(z) = \frac{2}{N_0}\sqrt{E}\,s(z) - \frac{2}{N_0}\sum_{i=1}^{\infty}\frac{\lambda_i^c s_i}{\lambda_i^c + N_0/2}\,\phi_i(z), \tag{170}$$

where

$$s_i = \int_{T_i}^{T_f}\sqrt{E}\,s(t)\,\phi_i(t)\,dt. \tag{171}$$

The first term is familiar from the white noise case. The second term indicates the effect of nonwhite noise. Observe that $g(t)$ is *always* a square-integrable function over (T_i, T_f) when a white noise component is present. We defer checking the endpoint behavior until 4.3.3.

Summary

In this section we have derived the solution for the optimum receiver for the simple binary detection problem of a known signal in nonwhite Gaussian noise. Three realizations were the following:

1. Whitening realization (Fig. 4.38*a*).
2. Correlator realization (Fig. 4.38*b*).
3. Estimator-subtractor realization (Fig. 4.38*c*).

Coupled with each of these realizations was an integral equation that must be solved to build the receiver: 1. (158). 2. (169). 3. (163).

We demonstrated that series solutions could be obtained in terms of eigenvalues and eigenfunctions, but we postponed the problem of actually finding a closed-form solution. The concept of an "inverse kernel" was introduced and a simple application shown. The following questions remain:

1. How well does the system perform?
2. How do we find closed-form solutions to the integral equations of interest?
3. What are the analogous results for the estimation problem?

Before answering these questions we digress briefly and rederive the results without using the idea of whitening. In view of these alternate derivations, we leave the proof that $h_w(t, u)$ is a reversible operator as an exercise for the reader (Problem 4.3.1).

4.3.2 A Direct Derivation Using the Karhunen-Loève Expansion†

In this section we consider a more fundamental approach. It is not only

† This approach to the problem is due to Grenander [30]. (See also: Kelly, Reed, and Root [31].)

more direct for this particular problem but extends easily to the general case. The derivation is analogous to the one on pp. 250–253.

The reason that the solution to the white noise detection problem in Section 4.2 was so straightforward was that regardless of the orthonormal set we chose, the resulting observables $r_1, r_2, \ldots, r_K$ were conditionally independent.

From our work in Chapter 3 we know that we can achieve the same simplicity if we choose an orthogonal set in a particular manner. Specifically, we want the orthogonal functions to be the eigenfunctions of the integral equation (3-46)

$$\lambda_i^c \phi_i(t) = \int_{T_i}^{T_f} K_c(t, u)\, \phi_i(u)\, du, \qquad T_i \le t \le T_f. \tag{172}$$

Observe that the λ_i^c are the eigenvalues of the colored noise process only. (If $K_c(t, u)$ is not positive-definite, we augment the set to make it complete.) Then we expand $r(t)$ in this coordinate system:

$$r(t) = \underset{K\to\infty}{\text{l.i.m.}} \sum_{i=1}^{K} r_i\, \phi_i(t) = \underset{K\to\infty}{\text{l.i.m.}} \sum_{i=1}^{K} s_i\, \phi_i(t) + \underset{K\to\infty}{\text{l.i.m.}} \sum_{i=1}^{K} n_i\, \phi_i(t), \qquad T_i \le t \le T_f, \tag{173}$$

where

$$r_i = \int_{T_i}^{T_f} r(t)\, \phi_i(t)\, dt, \tag{174}$$

$$s_i = \int_{T_i}^{T_f} \sqrt{E}\, s(t)\, \phi_i(t)\, dt, \tag{175}$$

and

$$n_i = \int_{T_i}^{T_f} n(t)\, \phi_i(t)\, dt. \tag{176}$$

From (3.42) we know

$$E(n_i) = 0, \qquad E(n_i n_j) = \lambda_i^T \delta_{ij}, \tag{177}$$

where

$$\lambda_i^T \triangleq \frac{N_0}{2} + \lambda_i^c. \tag{178}$$

Just as on p. 252 (20) we consider the first K coordinates. The likelihood ratio is

$$\Lambda\, [r_K(t)] = \frac{\prod_{i=1}^{K} \frac{1}{\sqrt{2\pi\lambda_i^T}} \exp\left[-\frac{1}{2}\frac{(R_i - s_i)^2}{\lambda_i^T}\right]}{\prod_{i=1}^{K} \frac{1}{\sqrt{2\pi\lambda_i^T}} \exp\left[-\frac{1}{2}\frac{R_i^2}{\lambda_i^T}\right]}. \tag{179}$$

Canceling common terms, letting $K \to \infty$, and taking the logarithm, we obtain

$$\ln \Lambda[r(t)] = \sum_{i=1}^{\infty} \frac{R_i s_i}{\lambda_i^T} - \frac{1}{2} \sum_{i=1}^{\infty} \frac{s_i^2}{\lambda_i^T}. \tag{180}$$

Using (174) and (175), we have

$$\ln \Lambda[r(t)] = \int_{T_i}^{T_f} dt \int_{T_i}^{T_f} du\, r(t) \sum_{i=1}^{\infty} \frac{\phi_i(t)\,\phi_i(u)}{\lambda_i^T} \sqrt{E}\, s(u)$$
$$- \frac{E}{2} \int_{T_i}^{T_f} dt \int_{T_i}^{T_f} du\, s(t) \sum_{i=1}^{\infty} \frac{\phi_i(t)\,\phi_i(u)}{\lambda_i^T} s(u). \tag{181}$$

From (166) we recognize the sum as $Q_n(t, u)$. Thus

$$\ln \Lambda[r(t)] = \int_{T_i}^{T_f} dt \int_{T_i}^{T_f} du\, r(t)\, Q_n(t, u) \sqrt{E}\, s(u)$$
$$- \frac{E}{2} \int_{T_i}^{T_f} dt \int_{T_i}^{T_f} du\, s(t)\, Q_n(t, u)\, s(u). \tag{182†}$$

This expression is identical to (153).

Observe that if we had not gone through the whitening approach we would have simply defined $Q_n(t, u)$ to fit our needs when we arrived at this point in the derivation. When we consider more general detection problems later in the text (specifically Chapter II.3), the direct derivation can easily be extended.

4.3.3 A Direct Derivation with a Sufficient Statistic‡

For convenience we rewrite the detection problem of interest (140):

$$\begin{aligned} r(t) &= \sqrt{E}\, s(t) + n(t), \qquad & T_i \le t \le T_f : H_1 \\ &= n(t), & T_i \le t \le T_f : H_0. \end{aligned} \tag{183}$$

In this section we will not require that the noise contain a white component.

From our work in Chapter 2 and Section 4.2 we know that if we can write

$$r(t) = r_1\, s(t) + y(t), \qquad T_i \le t \le T_f. \tag{184}$$

† To proceed rigorously from (181) to (182) we require $\sum_{i=1}^{\infty} (s_i^2/\lambda_i^{T^2}) < \infty$ (Grenander [30]; Kelly, Reed, and Root [31]). This is always true when white noise is present. Later, when we look at the effect of removing the white noise assumption, we shall see that the divergence of this series leads to an unstable test.

‡ This particular approach to the colored noise problem seems to have been developed independently by several people (Kailath [32]; Yudkin [39]). Although the two derivations are essentially the same, we follow the second.

where r_1 is a random variable obtained by operating on $r(t)$ and demonstrate that:

(a) r_1 and $y(t)$ are statistically independent on both hypotheses,
(b) the statistics of $y(t)$ do not depend on which hypothesis is true,

then r_1 is a sufficient statistic. We can then base our decision solely on r_1 and disregard $y(t)$. [Note that conditions (a) and (b) are sufficient, but not necessary, for r_1 to be a sufficient statistic (see pp. 35–36).]

To do this we hypothesize that r_1 can be obtained by the operation

$$r_1 = \int_{T_i}^{T_f} r(u)\, g(u)\, du \tag{185}$$

and try to find a $g(u)$ that will lead to the desired properties. Using (185), we can rewrite (184) as

$$\begin{aligned} r(t) &= (s_1 + n_1)\, s(t) + y(t) \qquad &:H_1\\ &= n_1\, s(t) + y(t) &:H_0. \end{aligned} \tag{186}$$

where

$$s_1 \triangleq \int_{T_i}^{T_f} \sqrt{E}\, s(u)\, g(u)\, du \tag{187}$$

and

$$n_1 \triangleq \int_{T_i}^{T_f} n(u)\, g(u)\, du. \tag{188}$$

Because a sufficient statistic can be multiplied by any nonzero constant and remain a sufficient statistic we can introduce a constraint,

$$\int_{T_i}^{T_f} s(u) g(u)\, du = 1. \tag{189a}$$

Using (189a) in (187), we have

$$s_1 = \sqrt{E}. \tag{189b}$$

Clearly, n_1 is a zero-mean random variable and

$$n(t) = n_1\, s(t) + y(t), \qquad T_i \le t \le T_f. \tag{190}$$

This puts the problem in a convenient form and it remains only to find a condition on $g(u)$ such that

$$E[n_1\, y(t)] = 0, \qquad T_i \le t \le T_f, \tag{191}$$

or, equivalently,

$$E\{n_1[n(t) - n_1\, s(t)]\} = 0, \qquad T_i \le t \le T_f, \tag{192}$$

or

$$E[n_1 \cdot n(t)] = E[n_1^2]\, s(t), \qquad T_i \le t \le T_f. \tag{193}$$

Using (188)

$$\int_{T_i}^{T_f} K_n(t, u)\, g(u)\, du = s(t) \iint_{T_i}^{T_f} g(\sigma)\, K_n(\sigma, \beta)\, g(\beta)\, d\sigma\, d\beta, \qquad T_i \leq t \leq T_f. \tag{194}$$

Equations 189*a* and 194 will both be satisfied if

$$\int_{T_i}^{T_f} K_n(t, u)\, g(u)\, du = \sqrt{E}\, s(t), \qquad T_i \leq t \leq T_f. \tag{195}$$

[Substitute (195) into the right side of (194) and use (189*a*).] Our sufficient statistic r_1 is obtained by correlating $r(u)$ with $g(u)$. After obtaining r_1 we use it to construct a likelihood ratio test in order to decide which hypothesis is true.

We observe that (195) is over the closed interval $[T_i, T_f]$, whereas (169*a*) was over the open interval (T_i, T_f). The reason for this difference is that in the absence of white noise $g(u)$ may contain singularities at the endpoints. These singularities change the likelihood ratio so we can no longer arbitrarily choose the endpoint values. An advantage of our last derivation is that the correct endpoint conditions are included. We should also observe that if there is a white noise component (195) and (169*a*) will give different values for $g(T_i)$ and $g(T_f)$. However, because both sets of values are finite they lead to the same likelihood ratio.

In the last two sections we have developed two alternate derivations of the optimum receiver. Other derivations are available (a mathematically inclined reader might read Parzen [40], Hajek [41], Galtieri [43], or Kadota [45]). We now return to the questions posed on p. 297.

4.3.4 Detection Performance

The next question is: "How does the presence of colored noise affect performance?" In the course of answering it a number of interesting issues appear. We consider the simple binary detection case first.

Performance: Simple Binary Detection Problem. Looking at the receiver structure in Fig. 4.38*a*, we see that the performance is identical to that of a receiver in which the input signal is $s_*(t)$ and the noise is white with a spectral height of 2. Using (10) and (11), we have

$$d^2 = \int_{T_i}^{T_f} [s_*(t)]^2\, dt. \tag{196}$$

Thus the performance index d^2 is simply equal to the energy in the whitened signal. We can also express d^2 in terms of the original signal.

$$d^2 = \int_{T_i}^{T_f} dt \left[\int_{T_i}^{T_f} h_w(t, u)\sqrt{E}\, s(u)\, du\right]\left[\int_{T_i}^{T_f} h_w(t, z)\sqrt{E}\, s(z)\, dz\right]. \tag{197}$$

We use the definition of $Q_n(u, z)$ to perform the integration with respect to t. This gives

$$\boxed{\begin{aligned} d^2 &= E \iint_{T_i}^{T_f} du\, dz\, s(u)\, Q_n(u, z)\, s(z) \\ d^2 &= \sqrt{E} \int_{T_i}^{T_f} du\, s(u)\, g(u). \end{aligned}} \tag{198}$$

It is clear that the performance is no longer independent of the signal shape. The next logical step is to find the best possible signal shape. There are three cases of interest:

1. $T_i = 0$, $T_f = T$: the signal interval and observation interval coincide.
2. $T_i < 0$, $T_f > T$: the observation interval extends beyond the signal interval in one or both directions but is still finite.
3. $T_i = -\infty$, $T_f = \infty$: the observation interval is doubly infinite.

We consider only the first case.

Optimum Signal Design: Coincident Intervals. The problem is to constrain the signal energy E and determine how the detailed shape of $s(t)$ affects performance. The answer follows directly. Write

$$Q_n(t, u) = \sum_{i=1}^{\infty} \left(\frac{N_0}{2} + \lambda_i^c\right)^{-1} \phi_i(t)\, \phi_i(u). \tag{199}$$

Then

$$d^2 = \sum_{i=1}^{\infty} \frac{s_i^2}{N_0/2 + \lambda_i^c}, \tag{200}$$

where

$$s_i = \int_0^T \sqrt{E}\, s(t)\, \phi_i(t)\, dt. \tag{201}$$

Observe that

$$\sum_{i=1}^{\infty} s_i^2 = E, \tag{202}$$

because the functions are normalized.

Looking at (200), we see that d^2 is just a weighted sum of the s_i^2. Because (202) constrains the sum of the s_i^2, we want to distribute the energy so that those s_i with large weighting are large. If there exists a smallest eigenvalue, say $\lambda_j^c = \lambda_{\min}^c$, then d^2 will be maximized by letting $s_j = \sqrt{E}$ and all other $s_i = 0$. There are two cases of interest:

1. If $K_c(t, u)$ is positive-definite, the number of eigenvalues is infinite. There is no smallest eigenvalue. We let $s_j = \sqrt{E}$ and all other $s_i = 0$. Then, assuming the eigenvalues are ordered according to decreasing size,

$$d^2 \to \frac{2E}{N_0}$$

as we increase j. For many of the colored noises that we encounter in practice (e.g., the one-pole spectrum shown in Fig. 3.9), the frequency of the eigenfunction increases as the eigenvalues decrease. In other words, we increase the frequency of the signal until the colored noise becomes negligible. In these cases we obtain a more realistic signal design problem by including a bandwidth constraint.

2. If $K_c(t, u)$ is only nonnegative definite, there will be zero eigenvalues. If $s(t)$ is the eigenfunction corresponding to any one of these eigenvalues, then

$$d^2 = \frac{2E}{N_0}.$$

We see that the performance of the best signal is limited by the white noise.

Singularity. It is easy to see the effect of removing the white noise by setting N_0 equal to zero in (200). When the colored noise is positive-definite (Case 1), all eigenvalues are nonzero. We can achieve *perfect detection* ($d^2 = \infty$) if and only if the sum

$$d^2 = \sum_{i=1}^{\infty} \frac{s_i^2}{\lambda_i^c} \tag{203}$$

diverges.

It can be accomplished by choosing $s(t)$ so that s_i^2 is proportional to λ_i^c. We recall that

$$\sum_{i=1}^{\infty} \lambda_i^c = \int_{T_i}^{T_f} K_c(t, t)\, dt < M.$$

The right side is finite by our assumption below (145). Thus the energy in the signal ($E = \sum_{i=1}^{\infty} s_i^2$) will be finite. If there were a white noise component, we could not achieve this proportionality for all i with a finite energy signal. In (Case 2) there are zero eigenvalues. Thus we achieve $d^2 = \infty$ by choosing $s(t) = \phi_i(t)$ for any i that has a zero eigenvalue.

These two cases are referred to as *singular* detection. For arbitrarily small time intervals and arbitrarily small energy levels we achieve perfect detection. We know that this kind of performance cannot be obtained in an actual physical situation. Because the purpose of our mathematical model is to predict performance of an actual system, it is important that we make it realistic enough to eliminate singular detection. We have eliminated the possibility of singular detection by insisting on a nonzero white noise component. This accounts for the thermal noise in the receiver. Often it will appear to be insignificant. If, however, we design the signal to eliminate the effect of all other noises, it becomes the quantity that limits the performance and keeps our mathematical model from predicting results that would not occur in practice.

From (196) we know that d^2 is the energy in the whitened signal. Therefore, if the whitened signal has finite energy, the test is not singular. When the observation interval is infinite and the noise process is stationary with a rational spectrum, it is easy to check the finiteness of the energy of $s_*(t)$. We first find the transfer function of the whitening filter. Recall that

$$n_*(t) = \int_{-\infty}^{\infty} h_w(u)\, n(t-u)\, du. \tag{204}$$

We require that $n_*(t)$ be white with unity spectral height. This implies that

$$\iint_{-\infty}^{\infty} du\, dz\; h_w(u)\, h_w(z)\, K_n(t-u+z-v) = \delta(t-v),$$
$$-\infty < t, v < \infty. \tag{205}$$

Transforming, we obtain

$$|H_w(j\omega)|^2 S_n(\omega) = 1 \tag{206a}$$

or

$$|H_w(j\omega)|^2 = \frac{1}{S_n(\omega)}. \tag{206b}$$

Now assume that $S_n(\omega)$ has a rational spectrum

$$S_n(\omega) = \frac{c_q\omega^{2q} + c_{q-1}\omega^{2q-2} + \cdots + c_0}{d_p\omega^{2p} + d_{p-1}\omega^{2p-2} + \cdots + d_0}. \tag{207a}$$

We define the difference between the order of denominator and numerator (as a function of ω^2) as r.

$$r \triangleq p - q \tag{207b}$$

If $n(t)$ has finite power then $r \geq 1$. However, if the noise consists of white noise plus colored noise with finite power, then $r = 0$. Using (207a)

in (206*b*), we see that we can write $H_w(j\omega)$ as a ratio of two polynomials in $j\omega$.

$$H_w(j\omega) = \frac{a_p(j\omega)^p + a_{p-1}(j\omega)^{p-1} + \cdots + a_0}{b_q(j\omega)^q + b_{q-1}(j\omega)^{q-1} + \cdots + b_0}. \tag{208a}$$

In Chapter 6 we develop an algorithm for finding the coefficients. For the moment their actual values are unimportant. Dividing the numerator by the denominator, we obtain

$$H_w(j\omega) = f_r(j\omega)^r + f_{r-1}(j\omega)^{r-1} + \cdots + f_0 + \frac{R(j\omega)}{b_q(j\omega)^q + \cdots + b_0}, \tag{208b}$$

where $f_r, \ldots, f_0$ are constants and $R(j\omega)$ is the remainder polynomial of order less than q. Recall that $(j\omega)^r$ in the frequency domain corresponds to taking the rth derivative in the time domain. Therefore, in order for the test to be nonsingular, the rth derivative must have finite energy. In other words, if

$$\int_{-\infty}^{\infty} \left(\frac{d^r s(t)}{dt^r}\right)^2 dt < M \tag{209}$$

the test is nonsingular; for example, if

$$S_n(\omega) = \frac{2\alpha\sigma_n^{\;2}}{\omega^2 + \alpha^2} \tag{210a}$$

then

$$p - q = r = 1 \tag{210b}$$

and $s'(t)$ must have finite energy. If we had modeled the signal as an ideal rectangular pulse, then our model would indicate perfect detectability. We know that this perfect detectability will not occur in practice, so we must modify our model to accurately predict system performance. In this case we can eliminate the singular result by giving the pulse a finite rise time or by adding a white component to the noise. Clearly, whenever there is finite-power colored noise plus an independent white noise component, the integral in (209) is just the energy in the signal and singularity is never an issue.

Our discussion has assumed an infinite observation interval. Clearly, if the test is nonsingular on the infinite interval, it is nonsingular on the finite interval because the performance is related monotonically to the length of the observation interval. The converse is not true. Singularity on the infinite interval does not imply singularity on the finite interval. In this case we must check (203) or look at the finite-time whitening operation.

Throughout most of our work we retain the white noise assumption so singular tests never arise. Whenever the assumption is removed, it is necessary to check the model to ensure that it does not correspond to a singular test.

General Binary Receivers. Our discussion up to this point has considered only the simple binary detection problem. The extension to general binary receivers is straightforward. Let

$$\begin{aligned} r(t) &= \sqrt{E_1}\, s_1(t) + n(t), \qquad T_i \le t \le T_f : H_1, \\ r(t) &= \sqrt{E_0}\, s_0(t) + n(t), \qquad T_i \le t \le T_f : H_0, \end{aligned} \tag{211}$$

where $s_0(t)$ and $s_1(t)$ are normalized over the interval $(0, T)$ and are zero elsewhere. Proceeding in exactly the same manner as in the simple binary case, we obtain the following results. One receiver configuration is shown in Fig. 4.40*a*. The function $g_\Delta(t)$ satisfies

$$\begin{aligned} s_\Delta(t) &\triangleq \sqrt{E_1}\, s_1(t) - \sqrt{E_0}\, s_0(t) \\ &= \int_{T_i}^{T_f} g_\Delta(u)\, K_n(t, u)\, du, \qquad T_i \le t \le T_f. \end{aligned} \tag{212}$$

The performance is characterized by d^2:

$$d^2 = \iint_{T_i}^{T_f} s_\Delta(t)\, Q_n(t, u)\, s_\Delta(u)\, dt\, du. \tag{213}$$

The functions $K_n(t, u)$ and $Q_n(t, u)$ were defined in (145) and (161), respectively. As an alternative, we can use the whitening realization shown in Fig. 4.40*b*. Here $h_w(t, u)$ satisfies (158) and

$$s_{\Delta_*}(t) \triangleq \int_{T_i}^{T_f} h_w(t, u)\, s_\Delta(u)\, du, \qquad T_i \le t \le T_f. \tag{214}$$

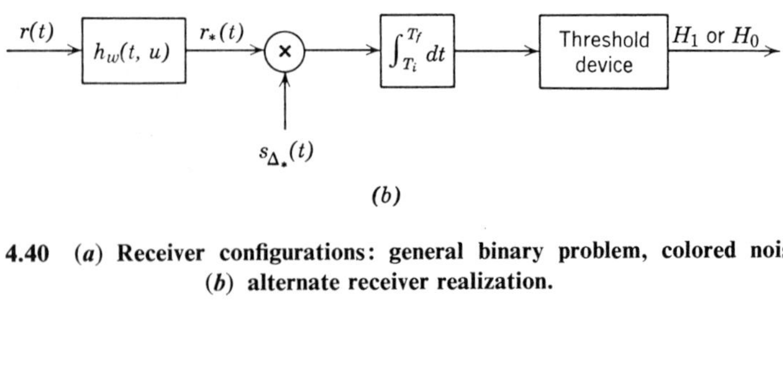

Fig. 4.40 (*a*) Receiver configurations: general binary problem, colored noise; (*b*) alternate receiver realization.

The performance is characterized by the energy in the whitened difference signal:

$$d^2 = \int_{T_i}^{T_f} s_{\Delta *}^2(t)\,dt. \tag{215}$$

The M-ary detection case is also a straightforward extension (see Problem 4.3.5). From our discussion of white noise we would expect that the estimation case would also follow easily. We discuss it briefly in the next section.

4.3.5 Estimation

The model for the received waveform in the parameter estimation problem is

$$r(t) = s(t, A) + n(t), \qquad T_i \le t \le T_f. \tag{216}$$

The basic operation on the received waveform consists of constructing the likelihood function, for which it is straightforward to derive an expression. If, however, we look at (98–101), and (146–153), it is clear that the answer will be:

$$\ln \Lambda_1[r(t), A] = \int_{T_i}^{T_f} r(z)\,dz \int_{T_i}^{T_f} Q_n(z, v)\, s(v, A)\,dv - \frac{1}{2}\int_{T_i}^{T_f} dz\, s(z, A) \int_{T_i}^{T_f} Q_n(z, v)\, s(v, A)\,dv. \tag{217}$$

This result is analogous to (153) in the detection problem. If we define

$$g(z, A) = \int_{T_i}^{T_f} Q_n(z, v)\, s(v, A)\,dv, \qquad T_i < z < T_f, \tag{218}$$

or, equivalently,

$$s(v, A) = \int_{T_i}^{T_f} K_n(v, z)\, g(z, A)\,dz, \qquad T_i < v < T_f, \tag{219}$$

(217) reduces to

$$\ln \Lambda_1[r(t), A] = \int_{T_i}^{T_f} r(z)\, g(z, A)\,dz - \frac{1}{2}\int_{T_i}^{T_f} s(z, A)\, g(z, A)\,dz. \tag{220}$$

The discussions in Sections 4.2.2 and 4.2.3 carry over to the colored noise case in an obvious manner. We summarize some of the important results for the linear and nonlinear estimation problems.

Linear Estimation. The received waveform is

$$r(t) = A\sqrt{E}\, s(t) + n(t), \qquad T_i \le t \le T_f, \tag{221}$$

where $s(t)$ is normalized [0, T] and zero elsewhere. Substituting into (218), we see that

$$g(t, A) = A\, g(t), \tag{222}$$

where $g(t)$ is the function obtained in the simple binary detection case by solving (169).

Thus the linear estimation problem is essentially equivalent to simple binary detection. The estimator structure is shown in Fig. 4.41, and the estimator is completely specified by finding $g(t)$. If A is a nonrandom variable, the normalized error variance is

$$\sigma^2_{a_\epsilon n} = (A^2 d^2)^{-1}, \tag{223}$$

where d^2 is given by (198). If A is a value of a random variable a with a Gaussian a priori density, $N(0, \sigma_a)$, the minimum mean-square error is

$$\sigma^2_{a_\epsilon n} = (1 + \sigma_a{}^2 d^2)^{-1}. \tag{224}$$

(These results correspond to (96) and (97) in the white noise case) All discussion regarding singular tests and optimum signals carries over directly.

Nonlinear Estimation. In nonlinear estimation, in the presence of colored noise, we encounter all the difficulties that occur in the white noise case. In addition, we must find either $Q_n(t, u)$ or $g(t, A)$. Because all of the results are obvious modifications of those in 4.2.3, we simply summarize the results:

1. A necessary, but not sufficient, condition on $\hat{a}_{ml}$:

$$0 = \int_{T_i}^{T_f} dt \int_{T_i}^{T_f} du\, [r(t) - s(t, A)]\, Q_n(t, u) \frac{\partial s(u, A)}{\partial A}\bigg|_{A = \hat{a}_{ml}}. \tag{225}$$

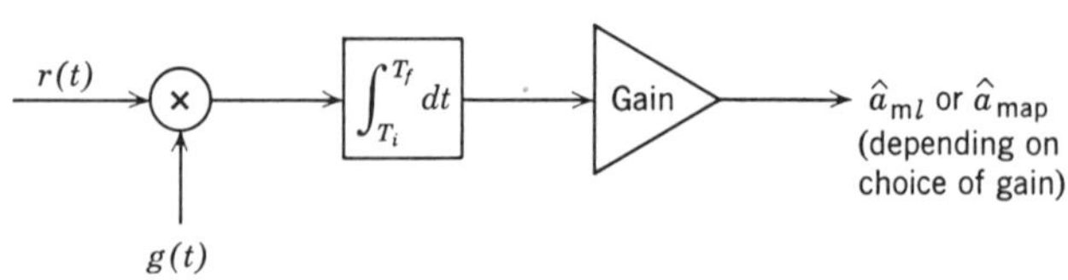

Fig. 4.41 Linear estimation, colored noise.

2. A necessary, but not sufficient, condition on $\hat{a}_{\text{map}}$ (assuming that a has a Gaussian a priori density):

$$\hat{a}_{\text{map}} = \sigma_a^2 \int_{T_i}^{T_f} dt\, [r(t) - s(t, A)] \int_{T_i}^{T_f} du\, Q_n(t, u) \left.\frac{\partial s(u, A)}{\partial A}\right|_{A=\hat{a}_{\text{map}}}. \tag{226}$$

3. A lower bound on the variance of any *unbiased* estimate of the non-random variable A:

$$\operatorname{Var}(\hat{a} - A) \geq \left[\iint_{T_i}^{T_f} \frac{\partial s(t, A)}{\partial A} Q_n(t, u) \frac{\partial s(u, A)}{\partial A}\, dt\, du\right]^{-1}, \tag{227a}$$

or, equivalently,

$$\operatorname{Var}(\hat{a} - A) \geq \left[\int_{T_i}^{T_f} \frac{\partial s(t, A)}{\partial A} \frac{\partial g(t, A)}{\partial A}\, dt\right]^{-1}. \tag{227b}$$

4. A lower bound on the mean-square error in the estimate of a zero-mean Gaussian random variable a:

$$E[(\hat{a} - a)^2] \geq \left[\frac{1}{\sigma_a^2} + E_a\left(\int_{T_i}^{T_f} dt \int_{T_i}^{T_f} du\, \frac{s\partial(t, A)}{\partial A} Q_n(t, u) \frac{\partial s(u, A)}{\partial A}\right)\right]^{-1}. \tag{228}$$

5. A lower bound on the variance of any *unbiased* estimate of a non-random variable for the special case of an infinite observation interval and a stationary noise process:

$$\operatorname{Var}(\hat{a} - A) \geq \left[\int_{-\infty}^{\infty} \frac{\partial S^*(j\omega, A)}{\partial A} S_n^{-1}(\omega) \frac{\partial S(j\omega, A)}{\partial A} \frac{d\omega}{2\pi}\right]^{-1}, \tag{229}$$

where

$$S(j\omega, A) \triangleq \int_{-\infty}^{\infty} s(t, A) e^{-j\omega t}\, dt.$$

As we discussed in 4.2.3, results of this type are always valid, but we must always look at the over-all likelihood function to investigate their usefulness. In other words, we must not ignore the threshold problem.

The only remaining issue in the matter of colored noise is a closed form solution for $Q_n(t, u)$ or $g(t)$. We consider this problem in the next section.

4.3.6 Solution Techniques for Integral Equations

As we have seen above, to specify the receiver structure completely we must solve the integral equation for $g(t)$ or $Q_n(t, u)$.

In this section we consider three cases of interest:

1. Infinite observation interval; stationary noise process.
2. Finite observation interval; separable kernel.
3. Finite observation interval; stationary noise process.

Infinite Observation Interval; Stationary Noise. In this particular case $T_i = -\infty, T_f = \infty$, and the covariance function of the noise is a function only of the difference in the arguments. Then (161) becomes

$$\delta(z - v) = \int_{-\infty}^{\infty} Q_n(x - z)\, K_n(v - x)\, dx, \qquad -\infty < v, z < \infty, \tag{230}$$

where we assume that we can find a $Q_n(x, z)$ of this form. By denoting the Fourier transform of $K_n(\tau)$ by $S_n(\omega)$ and the Fourier transform of $Q_n(\tau)$ by $S_Q(\omega)$ and transforming both sides of (230) with respect to $\tau = z - v$, we obtain

$$\boxed{S_Q(\omega) = \frac{1}{S_n(\omega)}.} \tag{231}$$

We see that $S_Q(\omega)$ is just the inverse of the noise spectrum. Further, in the stationary case (152) can be written as

$$Q_n(z - v) = \int_{-\infty}^{\infty} h_w(u - z)\, h_w(u - v)\, du. \tag{232}$$

By denoting the Fourier transform of $h_w(\tau)$ by $H_w(j\omega)$, we find that (232) implies

$$\frac{1}{S_n(\omega)} = S_Q(\omega) = |H_w(j\omega)|^2. \tag{233}$$

Finally, for the detection and linear estimation cases (154) is useful. Transforming, we have

$$\boxed{G_\infty(j\omega) = \sqrt{E}\, S_Q(\omega)\, S(j\omega) = \frac{S(j\omega)\sqrt{E}}{S_n(\omega)},} \tag{234}$$

where the subscript ∞ indicates that we are dealing with an infinite interval.

To illustrate the various results, we consider some particular examples.

Example 1. We assume that the colored noise component has a rational spectrum. A typical case is

$$S_c(\omega) = \frac{2k\sigma_n^2}{\omega^2 + k^2}, \tag{235}$$

and

$$S_n(\omega) = \frac{N_0}{2} + \frac{2k\sigma_n^2}{\omega^2 + k^2}. \tag{236}$$

Then

$$S_Q(\omega) = \frac{\omega^2 + k^2}{\frac{N_0}{2}\,[\omega^2 + k^2(1 + \Lambda)]}, \tag{237}$$

where $\Lambda = 4\sigma_n^2/kN_0$. Writing

$$S_Q(\omega) = \frac{(j\omega + k)(-j\omega + k)}{(N_0/2)(j\omega + k\sqrt{1 + \Lambda})(-j\omega + k\sqrt{1 + \Lambda})}, \tag{238}$$

we want to choose an $H_w(j\omega)$ so that (233) will be satisfied. To obtain a realizable whitening filter we assign the term $(j\omega + k(1 + \Lambda)^{1/2})$ to $H_w(j\omega)$ and its conjugate to $H_w^*(j\omega)$. The term $(j\omega + k)$ in the numerator can be assigned to $H_w(j\omega)$ or $H_w^*(j\omega)$. Thus there are two equally good choices† for the whitening filter:

$$H_{w_1}(j\omega) = \left(\frac{2}{N_0}\right)^{1/2} \frac{j\omega + k}{j\omega + k(1 + \Lambda)^{1/2}} = \left(\frac{2}{N_0}\right)^{1/2}\left[1 - \frac{k(\sqrt{1 + \Lambda} - 1)}{j\omega + k\sqrt{1 + \Lambda}}\right] \tag{239}$$

and

$$H_{w_2}(j\omega) = \left(\frac{2}{N_0}\right)^{1/2} \frac{-j\omega + k}{j\omega + k(1 + \Lambda)^{1/2}} = \left(\frac{2}{N_0}\right)^{1/2}\left[-1 + \frac{k(\sqrt{1 + \Lambda} + 1)}{j\omega + k\sqrt{1 + \Lambda}}\right]. \tag{240}$$

Thus the optimum receiver (detector) can be realized in the whitening forms shown in Fig. 4.42. A sketch of the waveforms for the case in which $s(t)$ is a rectangular pulse is also shown. Three observations follow:

1. The whitening filter has an infinite memory. Thus it uses the entire past of $r(t)$ to generate the input to the correlator.
2. The signal input to the multiplier will start at $t = 0$, but even after time $t = T$ the input will continue.
3. The actual integration limits are $(0, \infty)$, because one multiplier input is zero before $t = 0$.

It is easy to verify that these observations are true whenever the noise consists of white noise plus an independent colored noise with a rational spectrum. It is also true, but less easy to verify directly, when the colored noise has a nonrational spectrum. Thus we conclude that under the above conditions an increase in observation interval will always improve the performance. It is worthwhile to observe that if we use $H_{w_1}(j\omega)$ as the whitening filter the output of the filter in the bottom path will be $\hat{n}_{c_r}(t)$, the minimum mean-square error *realizable* point estimate of $n_c(t)$. We shall verify that this result is always true when we study realizable estimators in Chapter 6.

Observe that we can just as easily (conceptually, at least) operate with $S_Q(\omega)$ directly. In this particular case it is *not* practical, but it does lead to an interesting interpretation of the optimum receiver. Notice that $S_Q(\omega)$ corresponds to an unrealizable filter. We see that we could pass $r(t)$ through this filter and then cross-correlate it with $s(t)$, as shown in Figure 4.43*a*. Observe that the integration is just over $[0, T]$ because $s(t)$ is zero elsewhere; $r_{**}(t)$, $0 \le t \le T$, however, is affected by $r(t)$, $-\infty < t < \infty$. We see that the receiver structure in Fig. 4.43*b* is the estimator-subtractor configuration shown in Fig. 4.39. Therefore the signal at the output of the bottom path must be $\hat{n}_{c_u}(t)$, the minimum mean-square error *unrealizable* estimate of

† There are actually an infinite number, for we can cascade $H_{w_1}(j\omega)$ with any filter whose transfer function has unity magnitude. Observe that we choose a realizable filter so that we can build it. Nothing in our mathematical model requires realizability.

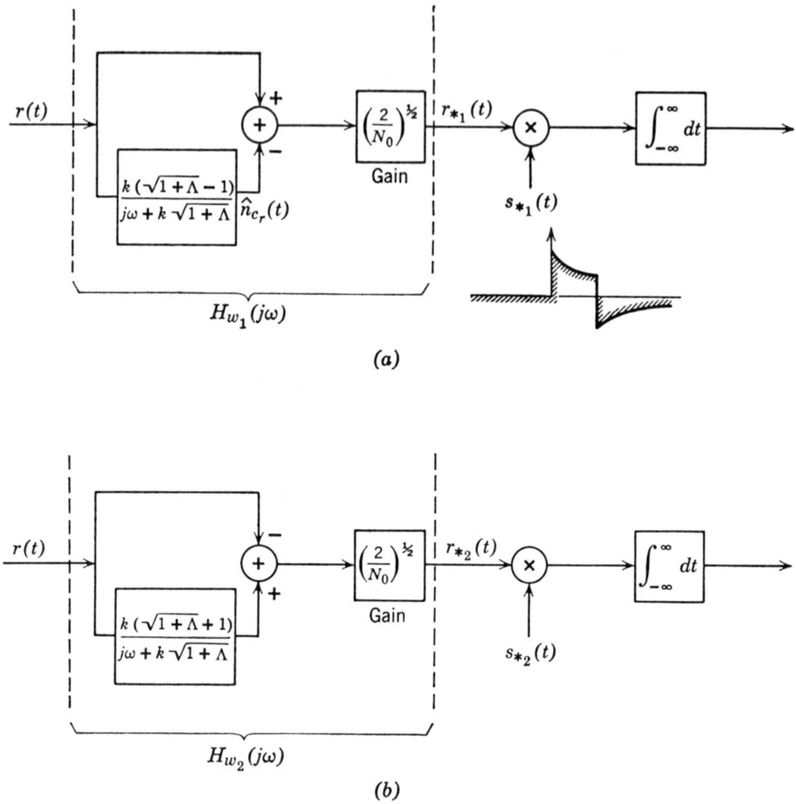

Fig. 4.42 "Optimum receiver": "whitening" realizations: (*a*) configuration 1; (*b*) configuration 2.

$n_c(t)$. This can be verified directly by substituting (235) and (236) into (3.239). We shall see that exactly the same result occurs in the general colored noise detection problem. Comparing Figs. 4.42 and 4.43, we see that they both contain estimates of colored noise but use them differently.

As a second example we investigate what happens when we *remove* the white noise component.

Example 2.

$$S_n(\omega) = \frac{2k\sigma_n^2}{\omega^2 + k^2}. \tag{241}$$

Then

$$S_Q(\omega) = \frac{\omega^2 + k^2}{2k\sigma_n^2}. \tag{242}$$

If we use a whitening realization, then one choice for the whitening filter is

$$H_w(j\omega) = \frac{1}{\sqrt{2k\sigma_n^2}}(j\omega + k). \tag{243}$$

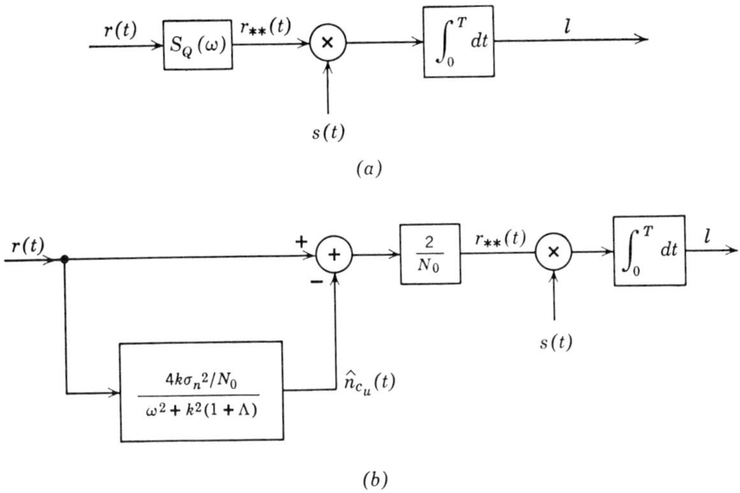

Fig. 4.43 Optimum receiver: estimator-subtractor interpretation.

Thus the whitening filter is a differentiator and gain in parallel (Fig. 4.44*a*). Alternately, using (234), we see that $G_\infty(j\omega)$ is,

$$G_\infty(j\omega) = \frac{\sqrt{E}\, S(j\omega)}{S_n(\omega)} = \frac{\sqrt{E}}{2k\sigma_n^2}\,(\omega^2 + k^2)\, S(j\omega). \tag{244}$$

Remembering that $j\omega$ in the frequency domain corresponds to differentiation in the time domain, we obtain

$$g_\infty(t) = \frac{\sqrt{E}}{2k\sigma_n^2}\,[-s''(t) + k^2\, s(t)], \tag{245}$$

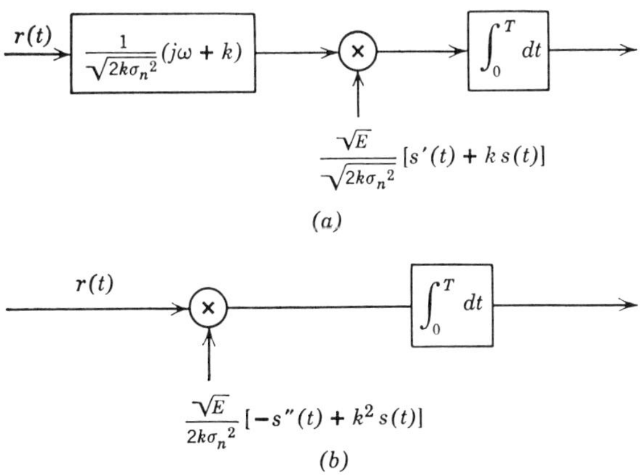

Fig. 4.44 Optimum receiver: no white noise component.

as shown in Fig. 4.44*b*. Observe that $s(t)$ must be differentiable everywhere in the interval $-\infty < t < \infty$; but we assume that $s(t) = 0$, $t < 0$, and $t > T$. Therefore $s(0)$ and $s(T)$ must also be zero. This restriction is intuitively logical. Recall the loose argument we made previously: if there *were* a step in the signal and it was differentiated formally, the result would be an impulse plus a white noise and lead to perfect detection. This is obviously not the actual physical case. By giving the pulse a finite rise time or including some white noise we avoid this condition.

We see that the receiver does not use any of the received waveform outside the interval $0 \le t \le T$, even though it is available. Thus we should expect the solution for $T_i = 0$ and $T_f = T$ to be identical. We shall see shortly that it is.

Clearly, this result will hold whenever the noise spectrum has *only poles*, because the whitening filter is a weighted sum of derivative operators. When the total noise spectrum has zeros, a longer observation time will help the detectability. Observe that when independent white noise is present the total noise spectrum will always have zeros.

Before leaving the section, it is worthwhile to summarize some of the important results.

1. For rational colored noise spectra and nonzero independent white noise, the infinite interval performance is better than any finite observation interval. Thus, the infinite interval performance which is characterized by $d_\infty^{\,2}$ provides a simple bound on the finite interval performance. For the particular one-pole spectrum in Example 1 a realizable, stable whitening filter can be found. This filter is *not* unique. In Chapter 6 we shall again encounter whitening filters for rational spectra. At that time we demonstrate how to find whitening filters for arbitrary rational spectra.

2. For rational colored noise spectra with no zeros and no white noise the interval in which the signal is nonzero is the only region of importance. In this case the whitening filter is realizable but not stable (it contains differentiators).

We now consider stationary noise processes and a finite observation interval.

***Finite Observation Interval; Rational Spectra*†.** In this section we consider some of the properties of integral equations over a finite interval. Most of the properties have been proved in standard texts on integral equations (e.g., [33] and [34]). They have also been discussed in a clear manner in the detection theory context by Helstrom [14]. We now state some simple properties that are useful and work some typical examples.

The first equation of interest is (195),

$$\sqrt{E}\, s(t) = \int_{T_i}^{T_f} g(u)\, K_n(t, u)\, du; \qquad T_i \le t \le T_f, \tag{246}$$

† The integral equations in Section 3.4 are special cases of the equations studied in this section. Conversely, if the equation specifying the eigenfunctions and eigenvalues has already been solved, then the solutions to the equations in the section follow easily.

where $s(t)$ and $K_n(t, u)$ are known. We want to solve for $g(t)$. Two special cases should be considered separately.

Case 1. The kernel $K_n(t, u)$ *does not* contain singularities. Physically, this means that there is *no white noise* present. Here (246) is a *Fredholm* equation of the *first kind*, and we can show (see [33]) that if the range (T_i, T_f) is finite a continuous square-integrable solution will *not* exist in general. We shall find that we can always obtain a solution if we allow singularity functions (impulses and their derivatives) in $g(u)$ at the end points of the observation interval.

In Section 4.3.7 we show that whenever $g(t)$ is *not* square-integrable the test is unstable with respect to small perturbations in the model assumptions.

We have purposely excluded Case No. 1 from most of our discussion on physical grounds. In this section we shall do a simple exercise to show the result of letting the white noise level go to zero. We shall find that in the absence of white noise we must put additional restrictions on $s(t)$ to get physically meaningful results.

Case 2. The noise contains a *nonzero* white-noise term. We may then write

$$K_n(t, u) = \frac{N_0}{2}\,\delta(t - u) + K_c(t, u), \tag{247}$$

where $K_c(t, u)$ is a continuous square-integrable function. Then (169*b*) is the equation of interest,

$$\sqrt{E}\, s(t) = \frac{N_0}{2}\, g(t) + \int_{T_i}^{T_f} K_c(t, u)\, g(u)\, du, \qquad T_i \le t \le T_f. \tag{248}$$

This equation is called a *Fredholm* equation of the *second kind.* A continuous, square-integrable solution for $g(t)$ will always exist when $K_c(t, u)$ is a continuous square-integrable function.

We now discuss two types of kernels in which straightforward procedures for solving (246) and (248) are available.

Type A (Rational Kernels). The noise $n_c(t)$ is the steady-state response of a lumped, linear passive network excited with white Gaussian noise. Here the covariance function depends only on $(t - u)$ and we may write

$$K_c(t, u) = K_c(t - u) = K_c(\tau). \tag{249}$$

The transform is

$$S_c(\omega) = \int_{-\infty}^{\infty} K_c(\tau) e^{-j\omega\tau}\, d\tau \triangleq \frac{N(\omega^2)}{D(\omega^2)} \tag{250}$$

and is a ratio of two polynomials in ω^2. The numerator is of order q in ω^2 and the denominator is of order p in ω^2. We assume that $n_c(t)$ has finite power so $p - q \geq 1$. Kernels whose transforms satisfy (250) are called *rational* kernels.

Integral equations with this type of kernel have been studied in detail in [35–37], [47, pp. 1082–1102] [54, pp. 309–329], and [62]. We shall discuss a simple example that illustrates the techniques and problems involved.

Type B (Separable Kernels). The covariance function of the noise can be written as

$$K_c(t, u) = \sum_{i=1}^{K} \lambda_i \, \phi_i(t) \, \phi_i(u), \qquad T_i \leq t, u \leq T_f, \tag{251}$$

where K is *finite*. This type of kernel is frequently present in radar problems when there are multiple targets. As we shall see in a later section, the solution to (246) is straightforward. We refer to this type of kernel as *separable*. Observe that if we *had* allowed $K = \infty$ all kernels would be considered separable, for we can always write

$$K_c(t, u) = \sum_{i=1}^{\infty} \lambda_i \, \phi_i(t) \, \phi_i(u), \qquad T_i \leq t, u \leq T_f, \tag{252}$$

where the λ_i and $\phi_i(t)$ are the eigenvalues and eigenfunctions. Clearly, this is *not* a practical solution technique because we have to solve another integral equation to find the $\phi_i(t)$.

We consider rational kernels in this section and separable kernels in the next.

Fredholm Equations of the First Kind: Rational Kernels. The basic technique is to find a differential equation corresponding to the integral equation. Because of the form of the kernel, this will be a differential equation with constant coefficients whose solution can be readily obtained. In fact, the particular solution of the differential equation is precisely the $g_\infty(t)$ that we derived in the last section (234). An integral equation with a rational kernel corresponds to a differential equation *plus* a set of boundary conditions. To incorporate the boundary conditions, we substitute the particular solution plus a weighted sum of the homogeneous solutions back into the integral equation and try to adjust the weightings so that the equation will be satisfied. It is at this point that we may have difficulty. To illustrate the technique and the possible difficulties we may meet, we consider a simple example. The first step is to show how $g_\infty(t)$ enters the picture. Assume that

$$S_n(\omega) = \frac{N(\omega^2)}{D(\omega^2)} \tag{253}$$

and recall that

$$\delta(t-u) = \int_{-\infty}^{\infty} e^{j\omega(t-u)} \frac{d\omega}{2\pi}. \tag{254}$$

Differentiation with respect to t gives

$$p\,\delta(t-u) = \int_{-\infty}^{\infty} j\omega\, e^{j\omega(t-u)} \frac{d\omega}{2\pi}, \tag{255}$$

where $p \triangleq d/dt$. More generally,

$$N(-p^2)\,\delta(t-u) = \int_{-\infty}^{\infty} N(\omega^2)\, e^{j\omega(t-u)} \frac{d\omega}{2\pi}. \tag{256}$$

In an analogous fashion

$$D(-p^2)\,K_n(t-u) = \int_{-\infty}^{\infty} D(\omega^2)\,S_n(\omega)\, e^{j\omega(t-u)} \frac{d\omega}{2\pi}. \tag{257}$$

From (253) we see that the right sides of (256) and (257) are identical. Therefore the kernel satisfies the differential equation obtained by equating the left sides of (256) and (257):

$$N(-p^2)\,\delta(t-u) = D(-p^2)\,K_n(t-u). \tag{258}$$

Now the integral equation of interest is

$$\sqrt{E}\,s(t) = \int_{T_i}^{T_f} K_n(t-u)\,g(u)\,du, \qquad T_i \le t \le T_f. \tag{259}$$

Operating on both sides of this equation with $D(-p^2)$, we obtain

$$D(-p^2)\sqrt{E}\,s(t) = \int_{T_i}^{T_f} D(-p^2)\,K_n(t-u)\,g(u)\,du, \qquad T_i \le t \le T_f. \tag{260}$$

Using (258) on the right-hand side, we have

$$D(-p^2)\sqrt{E}\,s(t) = N(-p^2)\,g(t), \qquad T_i \le t \le T_f, \tag{261}$$

but from our previous results (234) we know that if the observation interval were infinite,

$$D(\omega^2)\sqrt{E}\,S(j\omega) = N(\omega^2)\,G_\infty(j\omega), \tag{262}$$

or

$$D(-p^2)\sqrt{E}\,s(t) = N(-p^2)\,g_\infty(t), \qquad -\infty < t < \infty. \tag{263}$$

Thus $g_\infty(t)$ corresponds to the *particular* solution of (261). There are also *homogeneous* solutions to (261):

$$0 = N(-p^2)\, g_{h_i}(t), \qquad i = 1, 2, \dots, 2q. \tag{264}$$

We now add the particular solution $g_\infty(t)$ to a weighted sum of the $2q$ homogeneous solutions $g_{h_i}(t)$, substitute the result back into the integral equation, and adjust the weightings to satisfy the equation. At this point the discussion will be clearer if we consider a specific example.

Example. We consider (246) and use limits $[0, T]$ for algebraic simplicity.

$$K_n(t-u) = K_n(\tau) = \sigma_n^2 e^{-k|\tau|}, \qquad -\infty < \tau < \infty \tag{265}$$

or

$$S_n(\omega) = \frac{2k\sigma_n^2}{\omega^2 + k^2}. \tag{266}$$

Thus

$$N(\omega^2) = 2k\sigma_n^2 \tag{267}$$

and

$$D(\omega^2) = \omega^2 + k^2. \tag{268}$$

The differential equation (261) is

$$\sqrt{E}(-s''(t) + k^2 s(t)) = 2k\sigma_n^2\, g(t). \tag{269}$$

The particular solution is

$$g_\infty(t) = \frac{\sqrt{E}}{2k\sigma_n^2}[-s''(t) + k^2 s(t)] \tag{270}$$

and there is no homogeneous solution as

$$q = 0. \tag{271}$$

Substituting back into the integral equation, we obtain

$$\sqrt{E}\, s(t) = \sigma_n^2 \int_0^T \exp(-k|t-u|)\, g(u)\, du, \qquad 0 \le t \le T, \tag{272}$$

For $g(t)$ to be a solution, we require,

$$s(t) = \sigma_n^2 \Bigg\{ e^{-kt} \int_0^t e^{+ku} \left[\frac{-s''(u) + k^2 s(u)}{2k\sigma_n^2}\right] du + e^{+kt} \int_t^T e^{-ku} \left[\frac{-s''(u) + k^2 s(u)}{2k\sigma_n^2}\right] du \Bigg\}, \qquad 0 \le t \le T. \tag{273}$$

Because there are no homogeneous solutions, there are no weightings to adjust. Integrating by parts we obtain the equivalent requirement,

$$0 = e^{-kt}\left\{\frac{1}{2k}[s'(0) - ks(0)]\right\} - e^{+k(t-T)}\left\{\frac{1}{2k}[s'(T) + ks(T)]\right\}, \qquad 0 \le t \le T. \tag{274}$$

Clearly, the two terms in brackets must vanish independently in order for $g_\infty(t)$ to satisfy the integral equation. *If* they do, then our solution is complete. Unfortunately, the signal behavior at the end points often will cause the terms in the brackets to be nonzero. We must add something to $g_\infty(t)$ to cancel the e^{-kt} and $e^{k(t-T)}$ terms. We denote this additional term by $g_\delta(t)$ and choose it so that

$$\sigma_n^2 \int_0^T \exp(-k|t-u|)g_\delta(u)\,du$$
$$= -\frac{1}{2k}[s'(0) - ks(0)]\,e^{-kt} + \frac{1}{2k}[s'(T) + ks(T)]\,e^{+k(t-T)}, \qquad 0 \le t \le T. \quad (275)$$

To generate an e^{-kt} term $g_\delta(u)$ must contain an impulse $c_1\,\delta(u)$. To generate an $e^{+k(t-T)}$ term $g_\delta(u)$ must contain an impulse $c_2\,\delta(u-T)$. Thus

$$g_\delta(u) = c_1\,\delta(u) + c_2\,\delta(u-T), \quad (276)$$

where

$$c_1 = \frac{k\,s(0) - s'(0)}{k\sigma_n^2},$$
$$c_2 = \frac{k\,s(T) + s'(T)}{k\sigma_n^2}, \quad (277)$$

to satisfy (274).† Thus the complete solution to the integral equation is

$$g(t) = g_\infty(t) + g_\delta(t), \qquad 0 \le t \le T. \quad (278)$$

From (153) and (154) we see that the output of the processor is

$$l = \int_0^T r(t)\,g(t)\,dt$$
$$= \frac{c_1}{2}r(0) + \frac{c_2}{2}r(T) + \int_0^T r(t)\left\{\sqrt{E}\left[\frac{k^2\,s(t) - s''(t)}{2k\sigma_n^2}\right]\right\}dt. \quad (279)$$

Thus the optimum processor consists of a *filter* and a *sampler*.

Observe that $g(t)$ will be square-integrable only when c_1 and c_2 are zero. We discuss the significance of this point in Section 4.3.7.

When the spectrum has more poles, higher order singularities must be added at the end points. When the spectrum has zeros, there will be homogeneous solutions, which we denote as $g_{h_i}(t)$. Then we can show that the general solution is of the form

$$g(t) = g_\infty(t) + \sum_{i=1}^{2q} a_i\,g_{h_i}(t) + \sum_{k=0}^{p-q-1}[b_k\,\delta^{(k)}(t) + c_k\,\delta^{(k)}(t-T)], \quad (280)$$

where $2p$ is the order of $D(\omega^2)$ as a function of ω and $2q$ is the order of $N(\omega^2)$ as a function of ω (e.g., [35]). The function $\delta^{(k)}(t)$ is the kth derivative of $\delta(t)$. A great deal of effort has been devoted to finding efficient methods of evaluating the coefficients in (280) (e.g., [63], [3]).

As we have pointed out, whenever we assume that white noise is present, the resulting integral equation will be a Fredholm equation of the second kind. For rational spectra the solution techniques are similar but the character of the solution is appreciably different.

† We assume that the impulse is symmetric. Thus only one half its area is in the interval.

Fredholm Equations of the Second Kind: Rational Kernels. The equation of interest is (248):

$$\sqrt{E}\, s(t) = \frac{N_0}{2} g(t) + \int_{T_i}^{T_f} K_c(t, u)\, g(u)\, du, \qquad T_i \le t \le T_f. \tag{281}$$

We assume that the noise is stationary with spectrum $S_n(\omega)$,

$$S_n(\omega) = \frac{N_0}{2} + S_c(\omega) \triangleq \frac{N(\omega^2)}{D(\omega^2)}. \tag{282}$$

[Observe that $N(\omega^2)$ and $D(\omega^2)$ are of the same order. (This is because $S_c(\omega)$ has finite power.)] Proceeding in a manner identical to the preceding section, we obtain a differential equation that has a particular solution, $g_\infty(t)$, and homogeneous solutions, $g_{h_i}(t)$. Substituting

$$g(t) = g_\infty(t) + \sum_{i=1}^{2q} a_i g_{h_i}(t), \tag{283}$$

into the integral equation, we find that by suitably choosing the a_i we can always obtain a solution to the integral equation. (No $g_\delta(t)$ is necessary because we have enough weightings (or degrees of freedom) to satisfy the boundary conditions.) A simple example illustrates the technique.

Example. Let

$$K_c(t, u) = \sigma_c^2 \exp(-k|t-u|); \tag{284}$$

the corresponding spectrum is

$$S_c(\omega) = \frac{\sigma_c^2 2k}{\omega^2 + k^2}. \tag{285}$$

Then

$$S_n(\omega) = \frac{N_0}{2} + \frac{\sigma_c^2 2k}{\omega^2 + k^2} = \frac{(N_0/2)[\omega^2 + k^2(1 + 4\sigma_c^2/kN_0)]}{\omega^2 + k^2}$$

$$\triangleq \frac{N(\omega^2)}{D(\omega^2)}. \tag{286}$$

The integral equation is (using the interval (0, T) for simplicity)

$$\sqrt{E}\, s(t) = \frac{N_0}{2} g(t) + \sigma_c^2 \int_0^T e^{-k|t-u|} g(u)\, du, \qquad 0 \le t \le T. \tag{287}$$

The corresponding differential equation follows easily from (286),

$$\sqrt{E}(-s''(t) + k^2 s(t)) = \frac{N_0}{2}[-g''(t) + \gamma^2 g(t)], \tag{288}$$

where $\gamma^2 \triangleq k^2(1 + 4\sigma_c^2/kN_0)$. The particular solution is just $g_\infty(t)$. This can be obtained by solving the differential equation directly or by transform methods.

$$g_\infty(t) = \int_{-\infty}^{\infty} e^{+j\omega t} G_\infty(j\omega) \frac{d\omega}{2\pi}, \qquad 0 \le t \le T, \tag{289}$$

$$g_\infty(t) = \frac{2\sqrt{E}}{N_0} \int_{-\infty}^{\infty} e^{+j\omega t} \left(\frac{\omega^2 + k^2}{\omega^2 + \gamma^2}\right) S(j\omega) \frac{d\omega}{2\pi}, \qquad 0 \le t \le T. \tag{290}$$

The homogeneous solutions are

$$\begin{aligned} g_{h_1}(t) &= e^{\gamma t}, \\ g_{h_2}(t) &= e^{-\gamma t}. \end{aligned} \tag{291}$$

Then

$$g(t) = g_\infty(t) + a_1 e^{+\gamma t} + a_2 e^{-\gamma t}, \qquad 0 \le t \le T. \tag{292}$$

Substitution of (292) into (287) will lead to two simultaneous equations that a_1 and a_2 must satisfy. Solving for a_1 and a_2 explicitly gives the complete solution. Several typical cases are contained in the problems.

The particular property of interest is that a solution can always be found without having to add singularity functions. Thus the white noise assumption guarantees a square-integrable solution. (The convergence of the series in (164) and (170) implies that the solution is square-integrable.)

The final integral equation of interest is the one that specifies $h_o(t, u)$, (163). Rewriting it for the interval $[0, T]$, we have

$$h_o(t, z) + \frac{2}{N_0}\int_0^T K_c(t, u)\, h_o(u, z)\, du = \frac{2}{N_0} K_c(t, z), \qquad 0 \le t, z \le T. \tag{293}$$

We observe that this is identical to (281) in the preceding problem, except that there is an extra variable in each expression. Thus we can think of t as a fixed parameter and z as a variable or vice versa. In either case we have a Fredholm equation of the second kind.

For rational kernels the procedure is identical. We illustrate this with a simple example.

Example.

$$K_c(u, z) = \sigma_s^2 \exp(-k|u - z|), \tag{294}$$

$$h_o(t, z) + \frac{2}{N_0}\int_0^T h_o(t, u)\sigma_s^2 \exp(-k|u - z|)\, du = \frac{2}{N_0}\sigma_s^2 \exp(-k|t - z|), \qquad 0 \le t, z \le T. \tag{295}$$

Using the operator $k^2 - p^2$ and the results of (258) and (286), we have

$$\left(p \triangleq \frac{d}{dz}\right),$$

$$(k^2 - p^2)\, h_o(t, z) + \frac{2\sigma_s^2}{N_0}\cdot 2k\, h_o(t, z) = \frac{2\sigma_s^2}{N_0}\, 2k\, \delta(t - z), \tag{296}$$

or

$$(1 + \Lambda)h_o(t, z) - \frac{p^2}{k^2}\, h_o(t, z) = \Lambda\, \delta(t - z), \tag{297}$$

where

$$\Lambda = \frac{4\sigma_s^2}{kN_0}. \tag{298}$$

Let $\beta^2 = k^2(1 + \Lambda)$. The particular solution is

$$h_{op}(t, z) = \frac{2\sigma_s^2}{N_0\sqrt{1 + \Lambda}} \exp(-k\sqrt{1 + \Lambda}\,|t - z|), \qquad 0 \le t, z \le T. \tag{299}$$

Now add homogeneous solutions $a_1(t)e^{+\beta z}$ and $a_2(t)e^{-\beta z}$ to the particular solution in (299) and substitute the result into (295). We find that we require

$$a_1(t) = \frac{2k\sigma_s^2(\beta - k)[(\beta + k)e^{+\beta t} + (\beta - k)e^{-\beta t}]e^{-\beta T}}{N_0\beta[(\beta + k)^2e^{\beta T} - (\beta - k)^2e^{-\beta T}]} \tag{300}$$

and

$$a_2(t) = \frac{2k\sigma_s^2(\beta - k)[(\beta + k)e^{+\beta(T-t)} + (\beta - k)e^{-\beta(T-t)}]}{N_0\beta[(\beta + k)^2e^{\beta T} - (\beta - k)^2e^{-\beta T}]}. \tag{301}$$

The entire solution is

$$h_o(z, t) = \frac{2k\sigma_s^2[(\beta + k)e^{+\beta z} + (\beta - k)e^{-\beta z}][(\beta + k)e^{+\beta(T-t)} + (\beta - k)e^{-\beta(T-t)}]}{N_0\beta[(\beta + k)^2e^{+\beta T} - (\beta - k)^2e^{-\beta T}]},$$
$$0 \le z \le t \le T. \tag{302}$$

The solution is symmetric in z and t. This is clearly not a very appealing function to mechanize. An important special case that we will encounter later is the one in which the colored noise component is small. Then $\beta \simeq k$ and

$$h_o(z, t) \simeq \frac{2\sigma_s^2}{N_0} \exp - \beta|t - z|, \qquad 0 \le z, t \le T. \tag{303}$$

The important property to observe about (293) is that the extra variable complicates the algebra but the basic technique is still applicable.

This completes our discussion of integral equations with rational kernels and finite time intervals.

Several observations may be made:

1. The procedure is straightforward but tedious.
2. When there is no white noise, certain restrictions must be placed on $s(t)$ to guarantee that $g(t)$ will be square-integrable.
3. When white noise is present, increasing the observation interval always improves the performance.
4. The solution for $h_o(t, u)$ for arbitrary colored noise levels appears to be too complex to implement. We can use the d^2 derived from it (198) as a basis of comparison for simpler mechanizations. [In Section 6.7 we discuss an easier implementation of $h_o(t, u)$.]

Finite Observation Time: Separable Kernels. As a final category, we consider integral equations with separable kernels. By contrast with the tedium of the preceding section, the solution for separable kernels follows almost by inspection. In this case

$$K_c(t, u) = \sum_{i=1}^{K} \lambda_i\, \phi_i(t)\, \phi_i(u), \qquad T_i \le t, u \le T_f, \tag{304}$$

where λ_i and $\phi_i(t)$ are the eigenvalues and eigenfunctions of $K_c(t, u)$. Observe that (304) says that the noise has only K nonzero eigenvalues. Thus, unless we include a white noise component, we may have a singular problem. We include the white noise component and then observe that the solution for $h_o(t, u)$ is just a truncated version of the infinite series in (164). Thus

$$h_o(t, u) = \sum_{i=1}^{K} \frac{\lambda_i}{N_0/2 + \lambda_i} \phi_i(t)\, \phi_i(u), \qquad T_i \le t, u \le T_f, \tag{305}$$

The solution to (154) follows easily. Using (305) in (162) and the result in (154), we obtain

$$g(t) = \int_{T_i}^{T_f} du \sqrt{E}\, s(u) \frac{2}{N_0} \left[\delta(t-u) - \sum_{i=1}^{K} \frac{\lambda_i}{N_0/2 + \lambda_i} \phi_i(t)\, \phi_i(u)\right] \qquad T_i < t < T_f. \tag{306}$$

Recalling the definition of s_i in (201) and recalling that $g(t)$ is continuous at the end-points, we have

$$\begin{aligned} g(t) &= \frac{2}{N_0}\left[\sqrt{E}\, s(t) - \sum_{i=1}^{K} \frac{s_i \lambda_i}{N_0/2 + \lambda_i} \phi_i(t)\right], && T_i \le t \le T_f, \\ g(t) &= 0, && \text{elsewhere.} \end{aligned} \tag{307}$$

This receiver structure is shown in Fig. 4.45. Fortunately, in addition to having a simple solution, the separable kernel problem occurs frequently in practice.

A typical case is shown in Fig. 4.46. Here we are trying to detect a target in the presence of an interfering target and white noise (Siebert [38]).

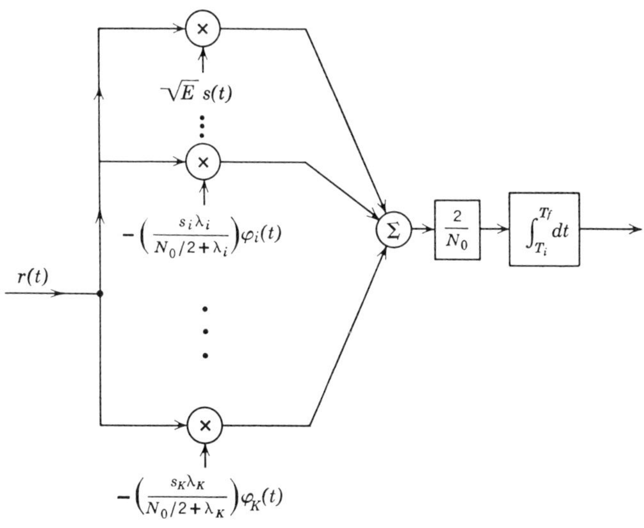

Fig. 4.45 Optimum receiver: separable noise process.

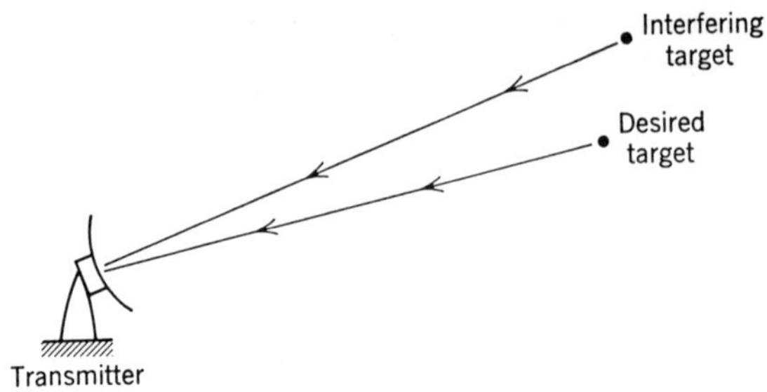

Fig. 4.46 Detection in presence of interfering target.

Let

$$r(t) = \sqrt{E}\, s(t) + a_I\, s_I(t) + w(t) \qquad T_i \le t \le T_f : H_1$$
$$= a_I\, s_I(t) + w(t) \qquad T_i \le t \le T_f : H_0. \tag{308}$$

If we assume that a_I and $s_I(t)$ are known, the problem is trivial. The simplest nontrivial model is to assume that $s_I(t)$ is a known normalized waveform but a_I is a zero-mean Gaussian random variable, $N(0, \sigma_I)$. Then

$$K_n(t, u) = \sigma_I^2 s_I(t)\, s_I(u) + \frac{N_0}{2}\, \delta(t - u), \qquad T_i \le t, u \le T_f. \tag{309}$$

This is a special case of the problem we have just solved. The receiver is shown in Fig. 4.47. The function $g(t)$ is obtained from (307). It can be redrawn, as shown in Fig. 4.47*b*, to illustrate the estimator-subtractor interpretation (this is obviously not an efficient realization). The performance index is obtained from (198),

$$d^2 = \frac{2E}{N_0}\left(1 - \frac{2\sigma_I^2/N_0}{1 + 2\sigma_I^2/N_0}\,\rho_I^2\right), \tag{310}$$

where

$$\rho_I \triangleq \int_{T_i}^{T_f} s(t)\, s_I(t)\, dt. \tag{311}$$

Rewriting (310), we have

$$d^2 = \frac{2E}{N_0}\left[\frac{1 + 2\sigma_I^2/N_0(1 - \rho_I^2)}{1 + 2\sigma_I^2/N_0}\right] \tag{312a}$$

as $\rho_I \to 0$, $d^2 \to 2E/N_0$. This result is intuitively logical. If the interfering signal is orthogonal to $s(t)$, then, regardless of its strength, it should not degrade the performance. On the other hand, as $\rho_I \to 1$,

$$d^2 \to \frac{2E/N_0}{1 + 2\sigma_I^2/N_0}. \tag{312b}$$

Now the signals on the two hypotheses are equal and the difference in their amplitudes is the only basis for making a decision.

We have introduced this example for two reasons:

1. It demonstrates an important case of nonwhite noise in which the inverse kernel is particularly simple to calculate.
2. It shows all of the concepts (but not the detail) that is necessary to solve the problem of detection (or estimation) in the presence of clutter (radar) or reverberation (sonar). In Chapter II-4, after we have developed a detailed model for the reverberation problem, we shall see how these results can be extended to handle the actual problem.

Summary of Integral Equations. In this section we have developed techniques for solving the types of integral equation encountered in the detection and estimation problems in the presence of nonwhite noise. The character of the solution was determined by the presence or absence of a white noise component. The simplicity of the solution in the infinite-interval, stationary process case should be emphasized. Because the performance in this case always bounds the finite interval, stationary process case, it is a useful preliminary calculation.

As a final topic for the colored noise problem, we consider the sensitivity of the result to perturbations in the initial assumptions.

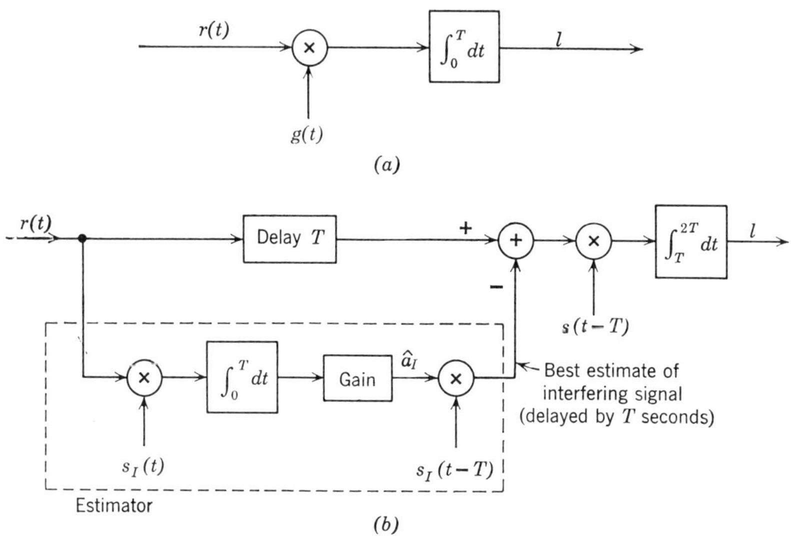

Fig. 4.47 Optimum receiver: interfering targets.

4.3.7 Sensitivity

Up to this point in our discussion we have assumed that all the quantities needed to design the optimum receiver were known exactly. We want to investigate the effects of imperfect knowledge of these quantities. In order to obtain some explicit results we shall discuss the sensitivity issue in the context of the simple binary decision problem developed in Section 4.3.1. Specifically, the model assumed is

$$\begin{aligned} r(t) &= \sqrt{E}\, s(t) + n(t), \qquad && T_i \le t \le T_f : H_1, \\ r(t) &= n(t), && T_i \le t \le T_f : H_0, \end{aligned} \tag{313}$$

where $s(t)$, the signal, and $K_n(t, u)$, the noise covariance function, are assumed known. Just as in the white noise case, there are two methods of sensitivity analysis: the parameter variation approach and the functional variation approach. In the white noise case we varied the signal. Now the variations can include both the signal and the noise.

Typical parameter variation examples are formulated below:

1. Let the assumed signal be

$$s(t) = \begin{cases} \left(\dfrac{2}{T}\right)^{1/2} \sin \omega_c t, & 0 \le t \le T, \\ 0, & \text{elsewhere,} \end{cases} \tag{314}$$

and the actual signal be

$$s_a(t) = \begin{cases} \left(\dfrac{2}{T}\right)^{1/2} \sin (\omega_c + \Delta\omega) t, & 0 \le t \le T, \\ 0, & \text{elsewhere.} \end{cases} \tag{315}$$

Find $\Delta d/d$ as a function of $\Delta\omega$.

2. Let the assumed noise covariance be

$$K_n(t, u) = \frac{N_0}{2}\, \delta(t - u) + K_c(t, u) \tag{316}$$

and the actual covariance be

$$K_{na}(t, u) = \left(\frac{N_0 + \Delta N_0}{2}\right) \delta(t - u) + K_c(t, u). \tag{317}$$

Find $\Delta d/d$ as a function of ΔN_0.

3. In the interfering target example of the last section (308) let the assumed interference signal be

$$s_I(t) = s(t - \tau). \tag{318}$$

In other words, it is a delayed version of the desired signal. Let the actual interference signal be

$$s_{Ia}(t) = s(t - \tau - \Delta\tau). \tag{319}$$

Find $\Delta d/d$ as a function of $\Delta\tau$.

These examples illustrate typical parameter variation problems. Clearly, the appropriate variations depend on the physical problem of interest. In almost all of them the succeeding calculations are straightforward. Some typical cases are included in the problems.

The functional variation approach is more interesting. As before, we do a "worst-case" analysis. Two examples are the following:

1. Let the actual signal be

$$s_a(t) = \sqrt{E}\, s(t) + \sqrt{E_\epsilon}\, s_\epsilon(t), \qquad T_i \le t \le T_f, \tag{320}$$

where

$$\int_{T_i}^{T_f} s_\epsilon^2(t)\, dt = 1. \tag{321}$$

To find the worst case we choose $s_\epsilon(t)$ to make Δd as negative as possible.

2. Let the actual noise be

$$n_a(t) = n(t) + n_\epsilon(t) \tag{322a}$$

whose covariance function is

$$K_{na}(t, u) = K_n(t, u) + K_{n\epsilon}(t, u), \tag{322b}$$

We assume that $n_\epsilon(t)$ has finite energy in the interval

$$E \int_{T_i}^{T_f} n_\epsilon^2(t)\, dt \le \Delta_n. \tag{323a}$$

This implies that

$$\iint_{T_i}^{T_f} K_{n\epsilon}^2(t, u)\, dt\, du \le \Delta_n. \tag{323b}$$

To find the worst case we choose $K_{n\epsilon}(t, u)$ to make Δd as negative as possible.

Various other perturbations and constraints are also possible. We now consider a simple version of the first problem. The second problem is developed in detail in [42].

We assume that the noise process is stationary with a spectrum $S_n(\omega)$ and that the observation interval is infinite. The optimum receiver, using a whitening realization (see Fig. 4.38*a*), is shown in Fig. 4.48*a*. The

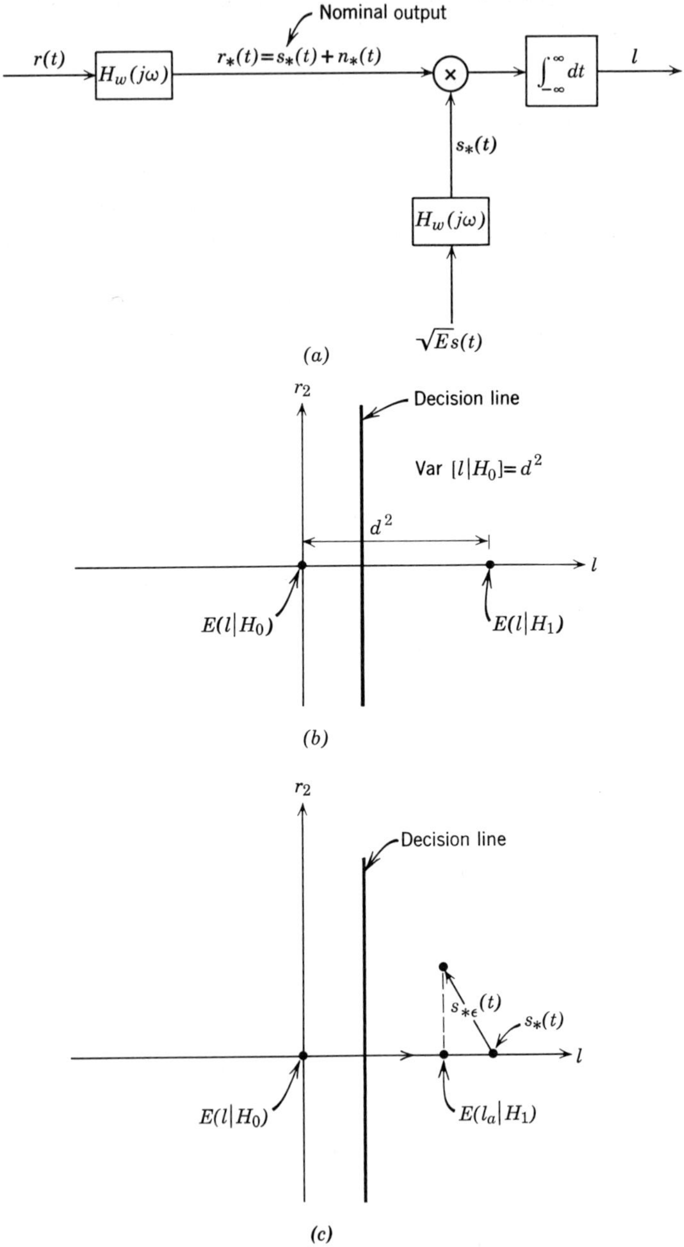

Fig. 4.48 Sensitivity analysis: (*a*) filter with nominal input; (*b*) nominal decision space; (*c*) actual design space.

corresponding decision space is shown in Fig. 4.48*b*. The nominal performance is

$$d = \frac{E(l|H_1) - E(l|H_0)}{[\text{Var}\,(l|H_0)]^{1/2}} = \frac{\int_{-\infty}^{\infty} s_*^2(t)\,dt}{\left[\int_{-\infty}^{\infty} s_*^2(t)\,dt\right]^{1/2}}, \tag{324}$$

or

$$d = \left[\int_{-\infty}^{\infty} s_*^2(t)\,dt\right]^{1/2}. \tag{325}$$

We let the actual signal be

$$s_a(t) = \sqrt{E}\,s(t) + \sqrt{E_\epsilon}\,s_\epsilon(t), \qquad -\infty < t < \infty, \tag{326}$$

where $s(t)$ and $s_\epsilon(t)$ have unit energy. The output of the whitening filter will be

$$r_{*a}(t) \triangleq s_*(t) + s_{*\epsilon}(t) + n_*(t), \qquad -\infty < t < \infty, \tag{327}$$

and the decision space will be as shown in Fig. 4.48*c*. The only quantity that changes is $E(l_a|H_1)$. The variance is still the same because the noise covariance is unchanged. Thus

$$\Delta d = \frac{1}{d}\int_{-\infty}^{\infty} s_{*\epsilon}(t)\,s_*(t)\,dt. \tag{328}$$

To examine the sensitivity we want to make Δd as negative as possible. If we can make $\Delta d = -d$, then the actual operating characteristic will be the $P_D = P_F$ line which is equivalent to a random test. If $\Delta d < -d$, the actual test will be worse than a random test (see Fig. 2.9*a*). It is important to note that the constraint is on the energy in $s_\epsilon(t)$, *not* $s_{*\epsilon}(t)$. Using Parseval's theorem, we can write (328) as

$$\Delta d = \frac{1}{d}\int_{-\infty}^{\infty} S_{*\epsilon}(j\omega)\,S_*^*(j\omega)\,\frac{d\omega}{2\pi}. \tag{329}$$

This equation can be written in terms of the original quantities by observing that

$$S_{*\epsilon}(j\omega) = \sqrt{E_\epsilon}\,H_w(j\omega)\,S_\epsilon(j\omega) \tag{330}$$

and

$$S_*(j\omega) = \sqrt{E}\,H_w(j\omega)\,S(j\omega). \tag{331}$$

Thus

$$\begin{aligned}\Delta d &= \frac{\sqrt{EE_\epsilon}}{d}\int_{-\infty}^{\infty} \frac{d\omega}{2\pi}\,S_\epsilon(j\omega)|H_w(j\omega)|^2\,S^*(j\omega)\\ &= \frac{\sqrt{EE_\epsilon}}{d}\int_{-\infty}^{\infty} S_\epsilon(j\omega)\,\frac{S^*(j\omega)}{S_n(\omega)}\,\frac{d\omega}{2\pi}.\end{aligned} \tag{332}$$

The constraint in (321) can be written as

$$\int_{-\infty}^{\infty} |S_\epsilon(j\omega)|^2 \frac{d\omega}{2\pi} = 1. \tag{333}$$

To perform a worst-case analysis we minimize Δd subject to the constraint in (333) by using Lagrange multipliers. Let

$$F = \Delta d + \lambda \left[\int_{-\infty}^{\infty} |S_\epsilon(j\omega)|^2 \frac{d\omega}{2\pi} - 1\right]. \tag{334}$$

Minimizing with respect to $S_\epsilon(j\omega)$, we obtain

$$S_{\epsilon_o}(j\omega) = -\frac{\sqrt{EE_\epsilon}}{2\lambda d}\frac{S(j\omega)}{S_n(\omega)}, \tag{335}$$

(the subscript o denotes optimum). To evaluate λ we substitute into the constraint equation (333) and obtain

$$\frac{EE_\epsilon}{4\lambda^2 d^2}\int_{-\infty}^{\infty} \frac{|S(j\omega)|^2}{S_n^2(\omega)}\frac{d\omega}{2\pi} = 1. \tag{336}$$

If the integral exists, then

$$2\lambda = \frac{\sqrt{EE_\epsilon}}{d}\left[\int_{-\infty}^{\infty} \frac{|S(j\omega)|^2}{S_n^2(\omega)}\frac{d\omega}{2\pi}\right]^{1/2}. \tag{337}$$

Substituting into (335) and then (332), we have

$$\Delta d = -\left(\frac{EE_\epsilon}{d^2}\right)^{1/2}\left[\int_{-\infty}^{\infty} \frac{|S(j\omega)|^2}{S_n^2(\omega)}\frac{d\omega}{2\pi}\right]^{1/2}. \tag{338}$$

(Observe that we could also obtain (338) by using the Schwarz inequality in (332).) Using the frequency domain equivalent of (325), we have

$$\frac{\Delta d}{d} = -\left(\frac{E_\epsilon}{E}\right)^{1/2}\left\{\frac{\left[\int_{-\infty}^{\infty} \frac{|S(j\omega)|^2}{S_n^2(\omega)}\frac{d\omega}{2\pi}\right]^{1/2}}{\left[\int_{-\infty}^{\infty} \frac{|S(j\omega)|^2}{S_n(\omega)}\frac{d\omega}{2\pi}\right]}\right\}. \tag{339}$$

In the white noise case the term in the brace reduces to one and we obtain the same result as in (82). When the noise is not white, several observations are important:

1. If there is a white noise component, both integrals exist and the term in the braces is greater than or equal to one. (Use the Schwarz inequality on the denominator.) Thus in the colored noise case a small signal perturbation may cause a large change in performance.
2. If there is *no* white noise component *and* the nominal test is *not singular*, the integral in the denominator exists. Without further restrictions

on $S(j\omega)$ and $S_n(\omega)$ the integral in the numerator may not exist. If it does not exist, the above derivation is not valid. In this case we can find an $S_\epsilon(j\omega)$ so that Δd will be less than any desired Δd_x. Choose

$$S_\epsilon(j\omega) = \begin{cases} k\dfrac{S(j\omega)}{S_n(\omega)}, & \omega \text{ in } \Omega, \\ 0, & \omega \text{ not in } \Omega, \end{cases} \tag{340}$$

where Ω is a region such that

$$k\frac{\sqrt{EE_\epsilon}}{d}\left(\int_\Omega \frac{|S(j\omega)|^2}{S_n(\omega)}\frac{d\omega}{2\pi}\right)^{1/2} = \Delta d_x \tag{341}$$

and k is chosen to satisfy the energy constraint on $s_\epsilon(t)$. We see that in the absence of white noise a signal perturbation exists that will make the test performance arbitrarily bad. Such tests are referred to as *unstable* (or infinitely sensitive) tests. We see that stability is a stronger requirement than nonsingularity and that the white noise assumption guarantees a nonsingular, stable test. Clearly, even though a test is stable, it may be extremely sensitive.

3. Similar results can be obtained for a finite interval and nonstationary processes in terms of the eigenvalues. Specifically, we can show (e.g., [42]) that the condition

$$\sum_{i=1}^{\infty} \frac{s_i^2}{\lambda_i^2} < \infty$$

is necessary and sufficient for stability. This is identical to the condition for $g(t)$ to be square-integrable.

In this section we have illustrated some of the ideas involved in a sensitivity analysis of an optimum detection procedure. Although we have eliminated unstable tests by the white noise assumption, it is still possible to encounter sensitive tests. In any practical problem it is essential to check the test sensitivity against possible parameter and function variations. We can find cases in which the test is too sensitive to be of any practical value. In these cases we try to design a test that is nominally suboptimum but less sensitive. Techniques for finding this test depend on the problem of interest.

Beforc leaving the colored noise problem we consider briefly a closely related problem.

4.3.8 Known Linear Channels

There is an almost complete duality between the colored additive noise problem and the problem of transmitting through a known linear channel with memory. The latter is shown in Fig. 4.49*a*.

The received waveform on H_1 in the simple binary problem is

$$r(t) = \int_{T_i}^{T_f} h_{ch}(t, u)\sqrt{E}\, s(u)du + w(t), \qquad T_i \le t \le T_f. \tag{342}$$

This is identical in form to (146). Thus $h_{ch}(t, u)$ plays an analogous role to the whitening filter. The optimum receiver is shown in Fig. 4.49*b*. The performance index is

$$\begin{aligned} d^2 &= \frac{2}{N_0}\int_{T_i}^{T_f} s_*^2(t)\, dt \\ &= \frac{2E}{N_0}\int_{T_i}^{T_f} dt \left[\int_a^b h_{ch}(t, u)\, s(u)\, du \int_a^b h_{ch}(t, v)\, s(v)\, dv\right], \end{aligned} \tag{343}$$

where the limits (a, b) depend on the channel's impulse response and the input signal duration. We assume that $T_i \le a \le b \le T_f$. We can write this in a familiar quadratic form:

$$d^2 = \frac{2E}{N_0}\iint_a^b du\, dv\, s(u)\, Q_{ch}(u, v)\, s(v) \tag{344}$$

by defining

$$Q_{ch}(u, v) = \int_{T_i}^{T_f} h_{ch}(t, u)\, h_{ch}(t, v)\, dt, \qquad a \le u, v \le b. \tag{345}$$

The only difference is that now $Q_{ch}(u, v)$ has the properties of a covariance function rather than an inverse kernel. A problem of interest is to choose

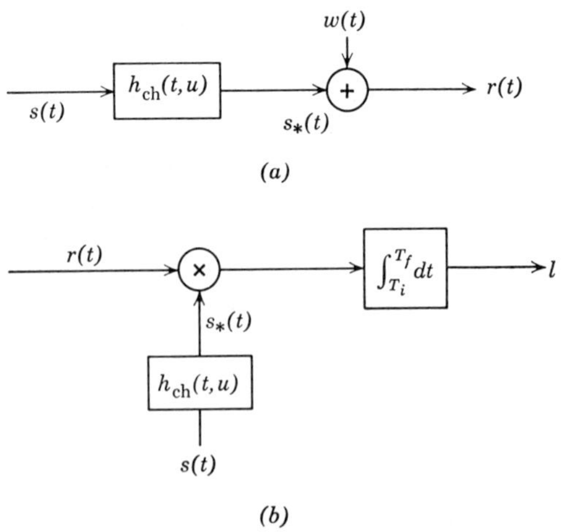

Fig. 4.49 Known dispersive channel.

$s(t)$ to maximize d^2. The solution follows directly from our earlier signal design results (p. 302). We can express d^2 in terms of the channel eigenvalues and eigenfunctions

$$d^2 = \frac{2}{N_0} \sum_{i=1}^{\infty} \lambda_i^{\text{ch}} s_i^2, \tag{346}$$

where

$$s_i \triangleq \int_a^b \sqrt{E}\, s(u)\, \phi_i(u)\, du \tag{347}$$

and λ_i^{ch} and $\phi_i(u)$ correspond to the kernel $Q_{\text{ch}}(u, v)$. To maximize d^2 we choose

$$s_1 = \sqrt{E},$$

and

$$s_i = 0, \qquad i \neq 1, \tag{348}$$

because λ_1^{ch} is defined as the largest eigenvalue of the channel kernel $Q_{\text{ch}}(u, v)$. Some typical channels and their optimum signals are developed in the problems.

When we try to communicate sequences of signals over channels with memory, another problem arises. Looking at the basic communications system in Fig. 4.1, we see that inside the basic interval $0 \leq t \leq T$ there is interference due to noise and the sequence of signals corresponding to previous data. This second interference is referred to as the intersymbol interference and it turns out to be the major disturbance in many systems of interest. We shall study effective methods of combatting intersymbol interference in Chapter II.4.

4.4 SIGNALS WITH UNWANTED PARAMETERS: THE COMPOSITE HYPOTHESIS PROBLEM

Up to this point in Chapter 4 we have assumed that the signals of concern were completely known. The only uncertainty was caused by the additive noise. As we pointed out at the beginning of this chapter, in many physical problems of interest this assumption is not realistic. One example occurs in the radar problem. The transmitted signal is a high frequency pulse that acquires a random phase angle (and perhaps a random amplitude) when it is reflected from the target. Another example arises in the communications problem in which there is an uncertainty in the oscillator phase. Both problems are characterized by the presence of an unwanted parameter.

Unwanted parameters appear in both detection and estimation problems. Because of the inherent similarities, it is adequate to confine our present discussion to the detection problem. In particular, we shall discuss general binary detection. In this case the received signals under the two hypotheses are

$$\begin{aligned} r(t) &= s_1(t, \boldsymbol{\theta}) + n(t), \qquad T_i \le t \le T_f : H_1, \\ r(t) &= s_0(t, \boldsymbol{\theta}) + n(t), \qquad T_i \le t \le T_f : H_0. \end{aligned} \tag{349}$$

The vector $\boldsymbol{\theta}$ denotes an unwanted vector parameter. The functions $s_0(t, \boldsymbol{\theta})$ and $s_1(t, \boldsymbol{\theta})$ are conditionally deterministic (i.e., if the value of $\boldsymbol{\theta}$ were known, the values of $s_0(t, \boldsymbol{\theta})$ and $s_1(t, \boldsymbol{\theta})$ would be known for all t in the observation interval). We see that this problem is just the waveform counterpart to the classical composite hypothesis testing problem discussed in Section 2.5. As we pointed out in that section, three types of situations can develop:

1. $\boldsymbol{\theta}$ is a random variable with a known a priori density;
2. $\boldsymbol{\theta}$ is a random variable with an unknown a priori density;
3. $\boldsymbol{\theta}$ is a nonrandom variable.

We shall confine our discussion here to the first situation. At the end of the section we comment briefly on the other two. The reason for this choice is that the two physical problems encountered most frequently in practice can be modeled by the first case. We discuss them in detail in Sections 4.4.1 and 4.4.2, respectively.

The technique for solving problems in the first category is straightforward. We choose a finite set of observables and denote them by the K-dimensional vector $\mathbf{r}$. We construct the likelihood ratio and then let $K \to \infty$.

$$\Lambda[r(t)] \triangleq \lim_{K \to \infty} \frac{p_{\mathbf{r}|H_1}(\mathbf{R}|H_1)}{p_{\mathbf{r}|H_0}(\mathbf{R}|H_0)}. \tag{350}$$

The only new feature is finding $p_{\mathbf{r}|H_1}(\mathbf{R}|H_1)$ and $p_{\mathbf{r}|H_0}(\mathbf{R}|H_0)$ in the presence of $\boldsymbol{\theta}$. If $\boldsymbol{\theta}$ were known, we should then have a familiar problem. Thus an obvious approach is to write

$$p_{\mathbf{r}|H_1}(\mathbf{R}|H_1) = \int_{\chi_{\boldsymbol{\theta}}} p_{\mathbf{r}|\boldsymbol{\theta},H_1}(\mathbf{R}|\boldsymbol{\theta}, H_1)\, p_{\boldsymbol{\theta}|H_1}(\boldsymbol{\theta}|H_1)\, d\boldsymbol{\theta}, \tag{351}$$

and

$$p_{\mathbf{r}|H_0}(\mathbf{R}|H_0) = \int_{\chi_{\boldsymbol{\theta}}} p_{\mathbf{r}|\boldsymbol{\theta},H_0}(\mathbf{R}|\boldsymbol{\theta}, H_0)\, p_{\boldsymbol{\theta}|H_0}(\boldsymbol{\theta}|H_0)\, d\boldsymbol{\theta}. \tag{352}$$

Substituting (351) and (352) into (350) gives the likelihood ratio. The tractability of the procedure depends on the form of the functions to be integrated. In the next two sections we consider two physical problems in which the procedure leads to easily interpretable results.

4.4.1 Random Phase Angles

In this section we look at several physical problems in which the uncertainty in the received signal is due to a random phase angle. The first problem of interest is a radar problem. The transmitted signal is a bandpass waveform which may be both amplitude- and phase-modulated. We can write the transmitted waveform as

$$s_t(t) = \begin{cases} \sqrt{2E_t}\, f(t) \cos\,[\omega_c t + \phi(t)], & 0 \le t \le T, \\ 0, & \text{elsewhere.} \end{cases} \tag{353}$$

Two typical waveforms are shown in Fig. 4.50. The function $f(t)$ corresponds to the envelope and is normalized so that the transmitted energy is E_t. The function $\phi(t)$ corresponds to a phase modulation. Both functions are low frequency in comparison to ω_c.

For the present we assume that we simply want to decide whether a target is present at a particular range. If a target is present, the signal will be reflected. In the simplest case of a fixed point target, the received waveform will be an attenuated version of the transmitted waveform with a random phase angle added to the carrier. In addition, there is an additive white noise component $w(t)$ at the receiver whether the target is present or

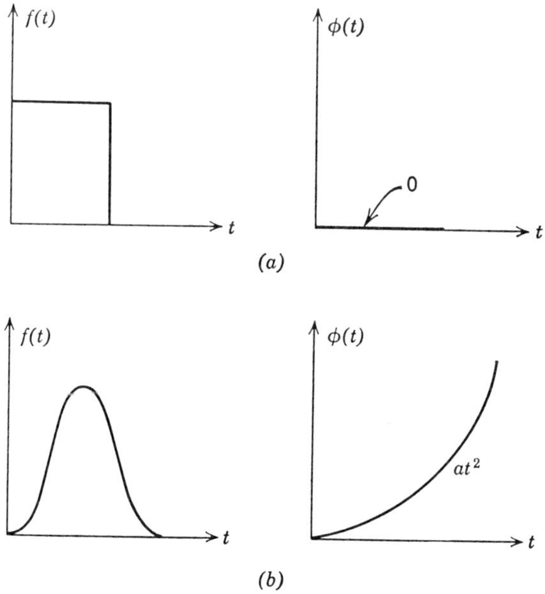

Fig. 4.50 Typical envelope and phase functions.

not. If we define H_1 as the hypothesis that the target is present and H_0 as the hypothesis that the target is absent, the following detection problem results:

$$\begin{aligned} H_1: r(t) &= \sqrt{2E_r}\, f(t-\tau)\cos(\omega_c(t-\tau)+\phi(t-\tau)+\theta) \\ &\qquad\qquad + w(t), \qquad \tau \le t \le \tau + T, \\ &= \qquad\qquad w(t), \qquad T_i \le t < \tau,\ \tau + T < t \le T_f, \\ &\triangleq s_r(t-\tau, \theta) + w(t), \qquad T_i \le t \le T_f. \end{aligned} \tag{354a}$$

$$H_0: r(t) = w(t); \qquad T_i \le t \le T_f. \tag{354b}$$

Because the noise is white, we need only observe over the interval $\tau \le t \le \tau + T$. Under the assumption that we are interested only in a particular τ, the model is the same if we let $\tau = 0$. Thus we need only consider the problem

$$H_1: r(t) = s_r(t, \theta) + w(t), \qquad 0 \le t \le T, \tag{355a}$$

$$H_0: r(t) = \qquad w(t), \qquad 0 \le t \le T. \tag{355b}$$

Here we have a simple binary detection problem in which the unknown parameter occurs only on one hypothesis. Before solving it we indicate how a similar problem can arise in the communications context.

In a simple on-off communication system we send a signal when the source output is "one" and nothing when the source output is "zero". The transmitted signals on the two hypotheses are

$$\begin{aligned} H_1: s_t(t) &= \sqrt{2E_t}\, f(t)\cos(\omega_c t + \phi(t) + \theta_a), \qquad 0 \le t \le T, \\ H_0: s_t(t) &= 0, \qquad\qquad 0 \le t \le T. \end{aligned} \tag{356}$$

Frequently, we try to indicate to the receiver what θ_a is. One method of doing this is to send an auxiliary signal that contains information about θ_a. If this signal were transmitted through a noise-free channel, the receiver would know θ_a exactly and the problem would reduce to the known signal problem. More frequently the auxiliary signal is corrupted by noise and the receiver operates on the noise-corrupted auxiliary signal and tries to estimate θ_a. We denote this estimate by $\hat{\theta}_a$. A block diagram is shown in Fig. 4.51. We discuss the detailed operation of the lower box in Chapter II.2. Now, if the estimate $\hat{\theta}_a$ equals θ_a, the problem is familiar. If they are unequal, the uncertainty is contained in the difference $\theta = \theta_a - \hat{\theta}_a$, which is a random variable. Therefore we may consider the problem:

$$H_1: r(t) = \sqrt{2E_r}\, f(t)\cos(\omega_c t + \phi(t) + \theta) + w(t), \qquad 0 \le t \le T, \tag{357}$$

$$H_0: r(t) = w(t), \qquad 0 \le t \le T, \tag{358}$$

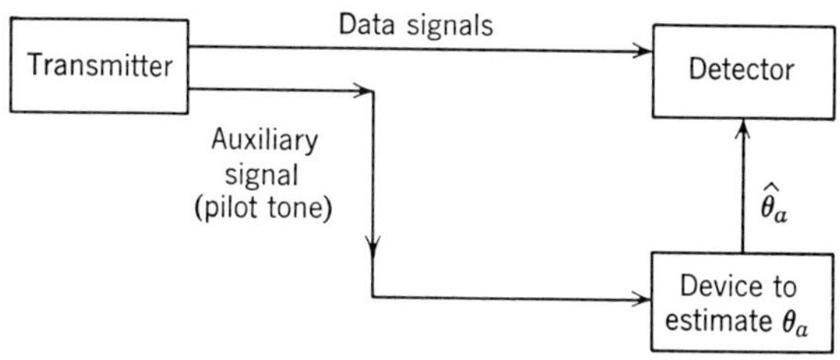

Fig. 4.51 A phase estimation system.

where E_r is the actual received signal energy and θ is the phase measurement error. We see that the radar and communication problems lead to the same mathematical model.

The procedure for finding the likelihood ratio was indicated at the beginning of Section 4.4. In this particular case the model is so familiar (see (23)) that we can write down the form for $K \to \infty$ immediately. The resulting likelihood ratio is

$$\Lambda[r(t)] = \int_{-\pi}^{\pi} p_\theta(\theta)\, d\theta \exp\left[+\frac{2}{N_0}\int_0^T r(t)\, s_r(t,\theta)\, dt - \frac{1}{N_0}\int_0^T s_r^2(t,\theta)\, dt\right], \tag{359}$$

where we assume the range of θ is $[-\pi, \pi]$. The last integral corresponds to the received energy. In most cases of interest it will not be a function of the phase so we incorporate it in the threshold. To evaluate the other integral, we expand the cosine term in (357),

$$\cos[\omega_c t + \phi(t) + \theta] = \cos[\omega_c t + \phi(t)]\cos\theta - \sin[\omega_c t + \phi(t)]\sin\theta, \tag{360}$$

and define

$$L_c \triangleq \int_0^T \sqrt{2}\, r(t)\, f(t) \cos[\omega_c t + \phi(t)]\, dt, \tag{361}$$

and

$$L_s \triangleq \int_0^T \sqrt{2}\, r(t)\, f(t) \sin[\omega_c t + \phi(t)]\, dt. \tag{362}$$

Thus the integral of interest is

$$\Lambda'[r(t)] = \int_{-\pi}^{\pi} p_\theta(\theta)\, d\theta \exp\left[\frac{2\sqrt{E_r}}{N_0}(L_c\cos\theta - L_s\sin\theta)\right]. \tag{363}$$

To proceed we must specify $p_\theta(\theta)$. Instead of choosing a particular density, we specify a family of densities indexed by a single parameter. We want to choose a family that will enable us to model as many cases of interest as

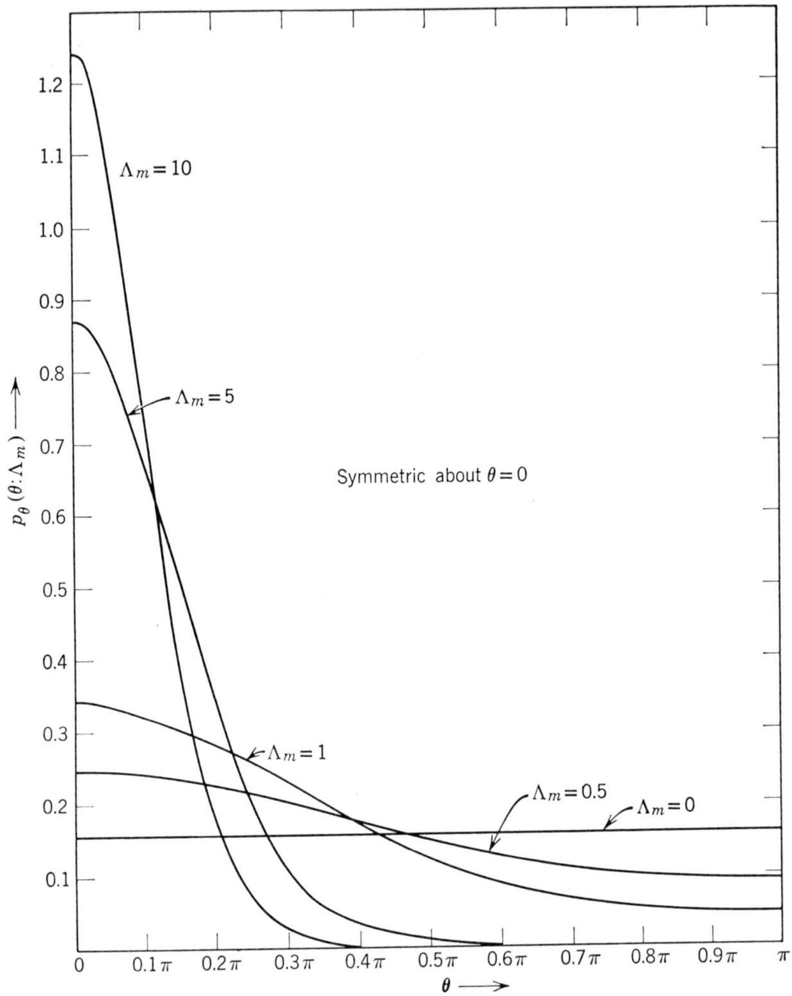

Fig. 4.52 Family of probability densities for the phase angle.

possible. A family that will turn out to be useful is given in (364) and shown in Fig. 4.52†:

$$p_\theta(\theta:\Lambda_m) = \frac{\exp[\Lambda_m \cos\theta]}{2\pi\, I_0(\Lambda_m)}; \qquad -\pi \le \theta \le \pi. \tag{364}$$

The function $I_0(\Lambda_m)$ is a modified Bessel function of the first kind which is included so that the density will integrate to unity. For the present Λ_m

† This density was first used for this application by Viterbi [44].

can be regarded simply as a parameter that controls the spread of the density. When we study phase estimators in Chapter II.2, we shall find that it has an important physical significance.

Looking at Fig. 4.52, we see that for $\Lambda_m = 0$

$$p_\theta(\theta) = \frac{1}{2\pi}, \qquad -\pi \leq \theta \leq \pi. \tag{365}$$

This is the logical density for the radar problem. As Λ_m increases, the density becomes more peaked. Finally, as $\Lambda_m \to \infty$, we approach the known signal case. Thus by varying Λ_m we can move continuously from the known signal problem through the intermediate case, in which there is some information about the phase, to the other extreme, the uniform phase problem.

Substituting (364) into (363), we have

$$\Lambda'[r(t)] = \int_{-\pi}^{\pi} \frac{1}{2\pi\, I_0(\Lambda_m)} \exp\left[\left(\Lambda_m + \frac{2\sqrt{E_r}}{N_0} L_c\right)\cos\theta - \frac{2\sqrt{E_r}}{N_0} L_s \sin\theta\right] d\theta. \tag{366}$$

This is a standard integral (e.g., [45]). Thus

$$\Lambda'[r(t)] = \frac{1}{I_0(\Lambda_m)} I_0\left\{\left[\left(\Lambda_m + \frac{2\sqrt{E_r}}{N_0} L_c\right)^2 + \left(\frac{2\sqrt{E_r}}{N_0} L_s\right)^2\right]^{1/2}\right\}. \tag{367}$$

Substituting (367) into (359), incorporating the threshold, and taking the logarithm, we obtain

$$\ln I_0\left\{\left[\left(\Lambda_m + \frac{2\sqrt{E_r}\, L_c}{N_0}\right)^2 + \left(\frac{2\sqrt{E_r}\, L_s}{N_0}\right)^2\right]^{1/2}\right\} \underset{H_0}{\overset{H_1}{\gtrless}} \ln\eta + \frac{E_r}{N_0} + \ln I_0(\Lambda_m). \tag{368}$$

The formation of the test statistic is straightforward (Fig. 4.53). The function $I_0(\cdot)$ is shown in Fig. 4.54. For large x

$$I_0(x) \simeq \frac{e^x}{\sqrt{2\pi x}}, \qquad x \gg 1, \tag{369}$$

whereas for small x

$$I_0(x) \simeq 1 + \frac{x^2}{4}, \qquad x \ll 1, \tag{370a}$$

and

$$\ln I_0(x) \simeq \frac{x^2}{4}, \qquad x \ll 1. \tag{370b}$$

Observe that because $\ln I_0(x)$ is monotone we can remove it by modifying

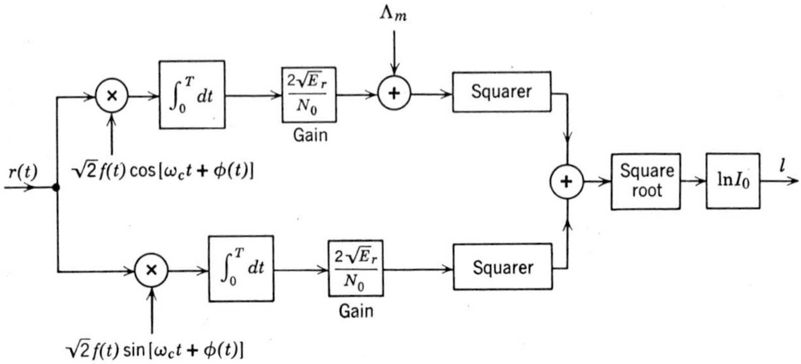

Fig. 4.53 Optimum receiver: random phase angle

the threshold. Thus two tests equivalent to (368) are

$$\left(L_c + \frac{N_0 \Lambda_m}{2\sqrt{E_r}}\right)^2 + L_s^2 \underset{H_0}{\overset{H_1}{\gtrless}} \gamma \tag{371a}$$

and

$$\left(\frac{2\sqrt{E_r}}{N_0}\right)^2 (L_c^2 + L_s^2) + 2\Lambda_m \frac{2\sqrt{E_r}}{N_0} L_c \underset{H_0}{\overset{H_1}{\gtrless}} \gamma'. \tag{371b}$$

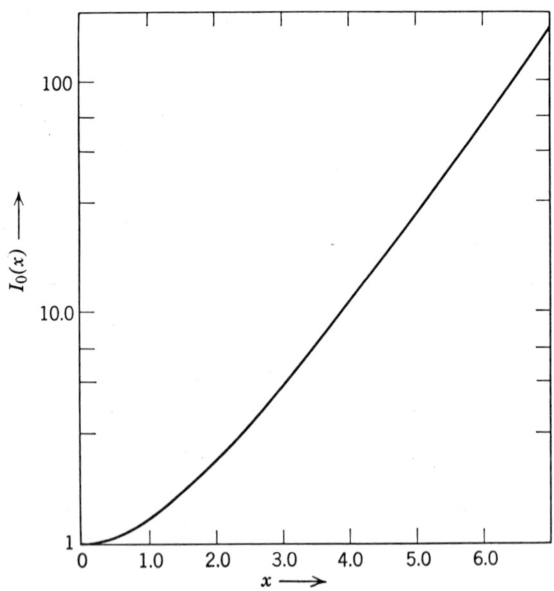

Fig. 4.54 Plot of $I_0(x)$.

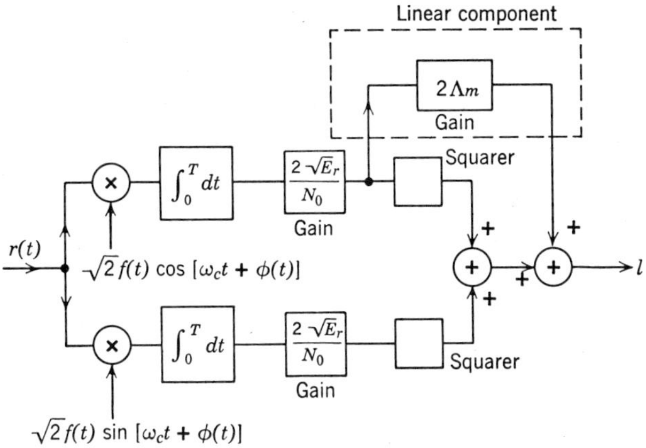

Fig. 4.55 Alternate realization of optimum receiver.

Redrawing the receiver structure as shown in Fig. 4.55, we see that the optimum receiver consists of a linear component and a square-law component.

Looking at (371*a*), we see that the region in the L_c, L_s plane corresponding to the decision H_0 is the interior of a circle centered at $(-N_0\Lambda_m/2\sqrt{E_r}, 0)$ with radius $\gamma^{1/2}$. We denote this region as Ω_0. The probability density of L_c and L_s under H_0 is a circularly symmetric Gaussian density centered at the origin. Therefore, if γ is fixed and Λ_m is allowed to increase, Ω_0 will move to the left and the probability of being in it on H_0 will decrease. Thus, to maintain a constant P_F we increase γ as Λ_m is increased. Several decision regions are shown in Fig. 4.56. In the limit, as $\Lambda_m \to \infty$, the decision boundary approaches a straight line and we have the familiar known signal problem of Section 4.2. The probability density on H_1 depends on θ. A typical case is shown in the figure. We evaluate P_F and P_D for some interesting special cases on p. 344 and in the problems. Before doing this it will be worthwhile to develop an alternate receiver realization for the case in which $\Lambda_m = 0$. In many cases this alternate realization will be more convenient to implement.

Matched Filter-Envelope Detector Realization. When $\Lambda_m = 0$, we must find $\sqrt{L_c^2 + L_s^2}$. We can do so by using a bandpass filter followed by an envelope detector, as shown in Fig. 4.57. Because $h(t)$ is the impulse response of a bandpass filter, it is convenient to write it as

$$h(t) = h_L(t) \cos [\omega_c t + \psi_L(t)], \tag{372}$$

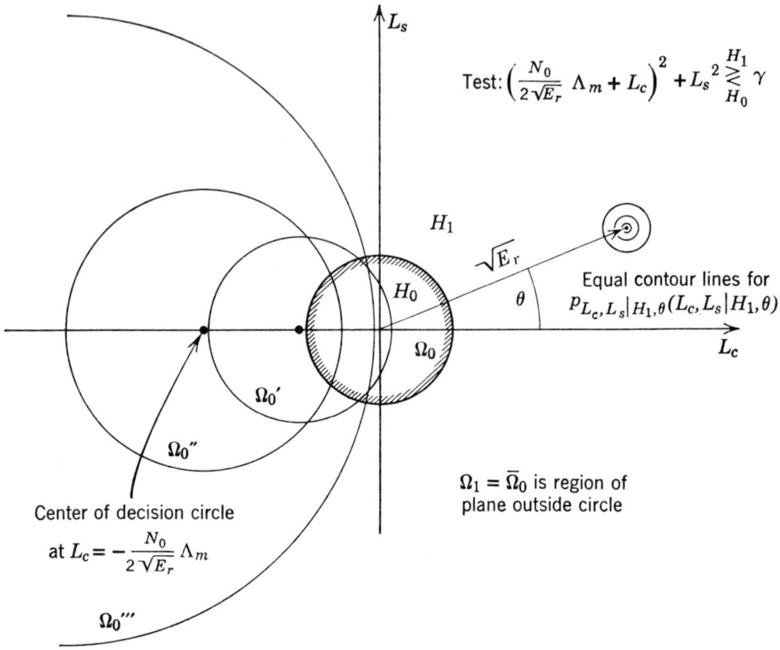

Fig. 4.56 Decision regions, partially coherent case.

where $h_L(t)$ and $\psi_L(t)$ are low-pass functions. The output at time T is

$$y(T) = \int_0^T h(T - \tau)\, r(\tau)\, d\tau. \tag{373}$$

Using (372), we can write this equation as

$$\begin{aligned} y(T) &= \int_0^T r(\tau)\, h_L(T - \tau) \cos\,[\omega_c(T - \tau) + \psi_L(T - \tau)]\, d\tau \\ &= \cos \omega_c T \int_0^T r(\tau)\, h_L(T - \tau) \cos\,[\omega_c \tau - \psi_L(T - \tau)]\, d\tau \\ &\quad + \sin \omega_c T \int_0^T r(\tau)\, h_L(T - \tau) \sin\,[\omega_c \tau - \psi_L(T - \tau)]\, d\tau. \end{aligned} \tag{374}$$

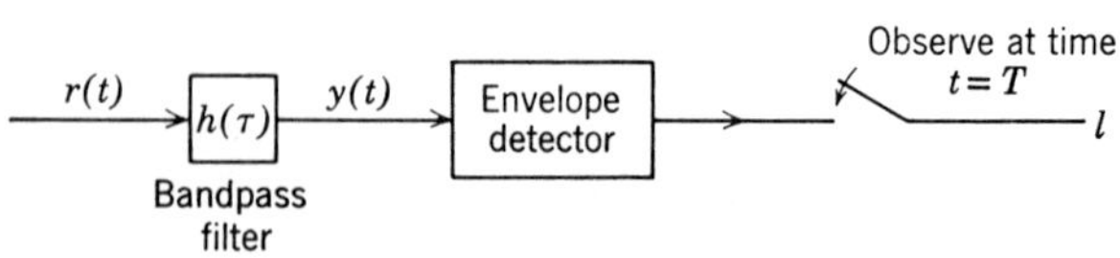

Fig. 4.57 Matched filter-envelope detector for uniform phase case.

This can be written as

$$y(T) \triangleq y_c(T) \cos \omega_c T + y_s(T) \sin \omega_c T$$
$$= \sqrt{y_c^2(T) + y_s^2(T)} \cos \left[\omega_c T - \tan^{-1} \frac{y_s(T)}{y_c(T)}\right]. \tag{375}$$

Observing that

$$y_c(T) = \text{Re} \int_0^T r(\tau)\, h_L(T - \tau) \exp\,[+j\omega_c \tau - j\psi_L(T - \tau)]\, d\tau \tag{376a}$$

and

$$y_s(T) = \text{Im} \int_0^T r(\tau)\, h_L(T - \tau) \exp\,[+j\omega_c \tau - j\psi_L(T - \tau)]\, d\tau, \tag{376b}$$

we see that the output of the envelope detector is

$$\sqrt{y_c^2(T) + y_s^2(T)} = \left| \int_0^T r(\tau)\, h_L(T - \tau) \exp\,[-j\psi_L(T - \tau) + j\omega_c \tau]\, d\tau \right|. \tag{377}$$

From (361) and (362) we see that the desired test statistic is

$$\sqrt{L_c^2 + L_s^2} = \left| \int_0^T r(\tau) \sqrt{2}\, f(\tau) e^{+j\phi(\tau)} e^{+j\omega_c \tau}\, d\tau \right|. \tag{378}$$

We see the two expressions will be identical if

$$h_L(T - \tau) = \sqrt{2}\, f(\tau) \tag{379}$$

and

$$\psi_L(T - \tau) = -\phi(\tau). \tag{380}$$

This bandpass matched filter provides a simpler realization for the *uniform* phase case.

The receiver in the uniform phase case is frequently called an incoherent receiver, but the terminology tends to be misleading. We see that the matched filter utilizes all the *internal* phase structure of the signal. The only thing missing is an *absolute* phase reference. The receiver for the known signal case is called a coherent receiver because it requires an oscillator at the receiver that is coherent with the transmitter oscillator. The general case developed in this section may be termed the partially coherent case.

To complete our discussion we consider the performance for some simple cases. There is no conceptual difficulty in evaluating the error probabilities but the resulting integrals often cannot be evaluated analytically. Because various modifications of this particular problem are frequently encountered in both radar and communications, a great deal of effort has been expended in finding convenient closed-form expressions and in numerical evaluations. We have chosen two typical examples to illustrate the techniques employed.

First we consider the radar problem defined at the beginning of this section (354–355).

Example 1 **(*Uniform Phase*).** Because this model corresponds to a radar problem, the uniform phase assumption is most realistic. To construct the ROC we must compute P_F and P_D. (Recall that P_F and P_D are the probabilities that we will exceed the threshold γ when noise only and signal plus noise are present, respectively.)

Looking at Fig. 4.55, we see that the test statistic is

$$l = L_c^2 + L_s^2, \tag{381}$$

where L_c and L_s are Gaussian random variables. The decision region is shown in Fig. 4.56. We can easily verify that

$$H_0: E(L_c) = E(L_s) = 0; \quad \text{Var}(L_c) = \text{Var}(L_s) = \frac{N_0}{2},$$

$$H_1: E(L_c|\theta) = \sqrt{E_r}\cos\theta; \quad E(L_s|\theta) = \sqrt{E_r}\sin\theta; \quad \text{Var}(L_c) = \text{Var}(L_s) = \frac{N_0}{2}. \tag{382}$$

Then

$$P_F \triangleq \Pr[l > \gamma|H_0] = \iint_{\bar{\Omega}_0} \left(2\pi\frac{N_0}{2}\right)^{-1} \exp\left(-\frac{L_c^2 + L_s^2}{N_0}\right) dL_c\, dL_s. \tag{383}$$

Changing to polar coordinates and evaluating, we have

$$P_F = \exp\left(-\frac{\gamma}{N_0}\right). \tag{384}$$

Similarly, the probability of detection for a particular θ is

$$P_D(\theta) = \iint_{\bar{\Omega}_0} \left(2\pi\frac{N_0}{2}\right)^{-1} \exp\left(-\frac{(L_c - \sqrt{E_r}\cos\theta)^2 + (L_s - \sqrt{E_r}\sin\theta)^2}{N_0}\right) dL_c\, dL_s. \tag{385}$$

Letting $L_c = R\cos\beta$, $L_s = R\sin\beta$, and performing the integration with respect to β, we obtain

$$P_D(\theta) = P_D = \int_{\gamma}^{\infty} \frac{2}{N_0} R \exp\left(-\frac{R^2 + E_r}{N_0}\right) I_0\left(\frac{2R\sqrt{E_r}}{N_0}\right) dR. \tag{386}$$

As we expected, P_D does not depend on θ. We can normalize this expression by letting $z = \sqrt{2/N_0}\, R$. This gives

$$P_D = \int_{\gamma\sqrt{2/N_0}}^{\infty} z \exp\left(-\frac{z^2 + d^2}{2}\right) I_0(zd)\, dz, \tag{387}$$

where $d^2 \triangleq 2E_r/N_0$.

This integral cannot be evaluated analytically. It was first tabulated by Marcum [46, 48] in terms of a function commonly called Marcum's Q function:

$$Q(\alpha, \beta) \triangleq \int_{\beta}^{\infty} z \exp\left(-\frac{z^2 + \alpha^2}{2}\right) I_0(\alpha z)\, dz. \tag{388}$$

This function has been studied extensively and tabulated for various values of α, β (e.g., [48], [49], and [50]). Thus

$$P_D = Q\left(d, \left(\frac{2}{N_0}\right)^{1/2} \gamma\right). \tag{389}$$

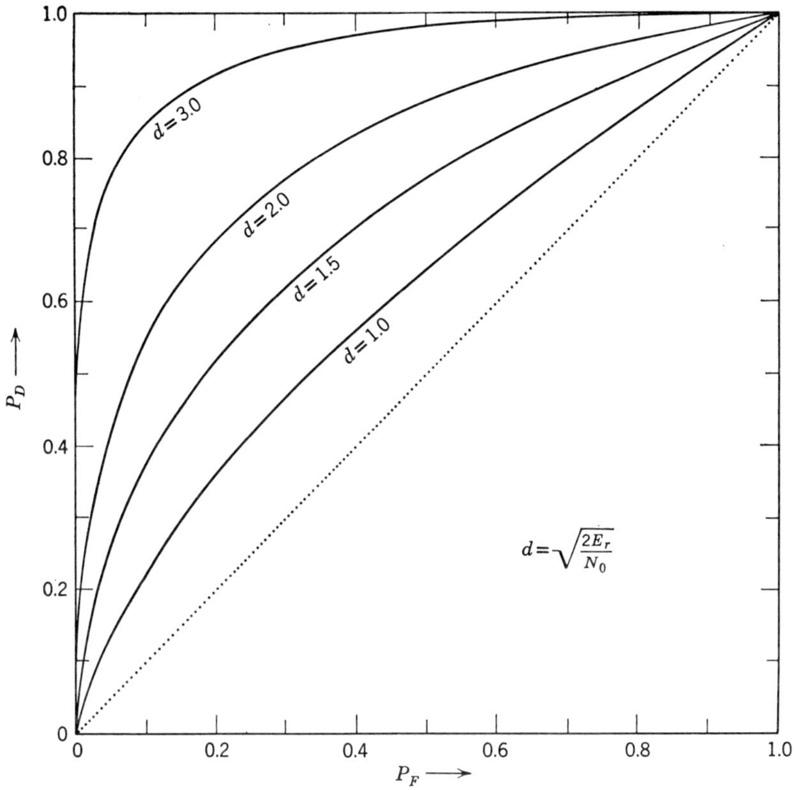

Fig. 4.58 Receiver operating characteristic, random phase with uniform density.

This can be written in terms of P_F. Using (384), we have

$$P_D = Q(d, \sqrt{-2 \ln P_F}). \tag{390}$$

The ROC is shown in Fig. 4.58. The results can also be plotted in the form of P_D versus d with P_F as a parameter. This is done in Fig. 4.59. Comparing Figs. 4.14 and 4.59, we see that a negligible increase of d is required to maintain the same P_D for a fixed P_F when we go from the known signal model to the uniform phase model for the parameter ranges shown in Fig. 4.59.

The second example of interest is a binary communication system in which some phase information is available.

Example 2. *Partially Coherent Binary Communication.* The criterion is minimum probability of error and the hypotheses are equally likely. We assume that the signals under the two hypotheses are

$$\begin{aligned} H_1 : r(t) &= \sqrt{2E_r}\, f_1(t) \cos(\omega_c t + \theta) + w(t), \qquad 0 \le t \le T, \\ H_0 : r(t) &= \sqrt{2E_r}\, f_0(t) \cos(\omega_c t + \theta) + w(t), \qquad 0 \le t \le T, \end{aligned} \tag{391}$$

where $f_0(t)$ and $f_1(t)$ are normalized and

$$\int_0^T f_0(t)\, f_1(t)\, dt = \rho; \qquad -1 \le \rho \le 1. \tag{392}$$

The noise spectral height is $N_0/2$ and $p_\theta(\theta)$ is given by (364). The likelihood ratio test is obtained by an obvious modification of the simple binary problem and the receiver structure is shown in Fig. 4.60.

We now look at Pr (ϵ) as a function of ρ, d^2, and Λ_m. Intuitively, we expect that as $\Lambda_m \to \infty$ we would approach the known signal problem, and $\rho = -1$ (the equal and opposite signals of (39)) would give the best result. On the other hand, as $\Lambda_m \to 0$,

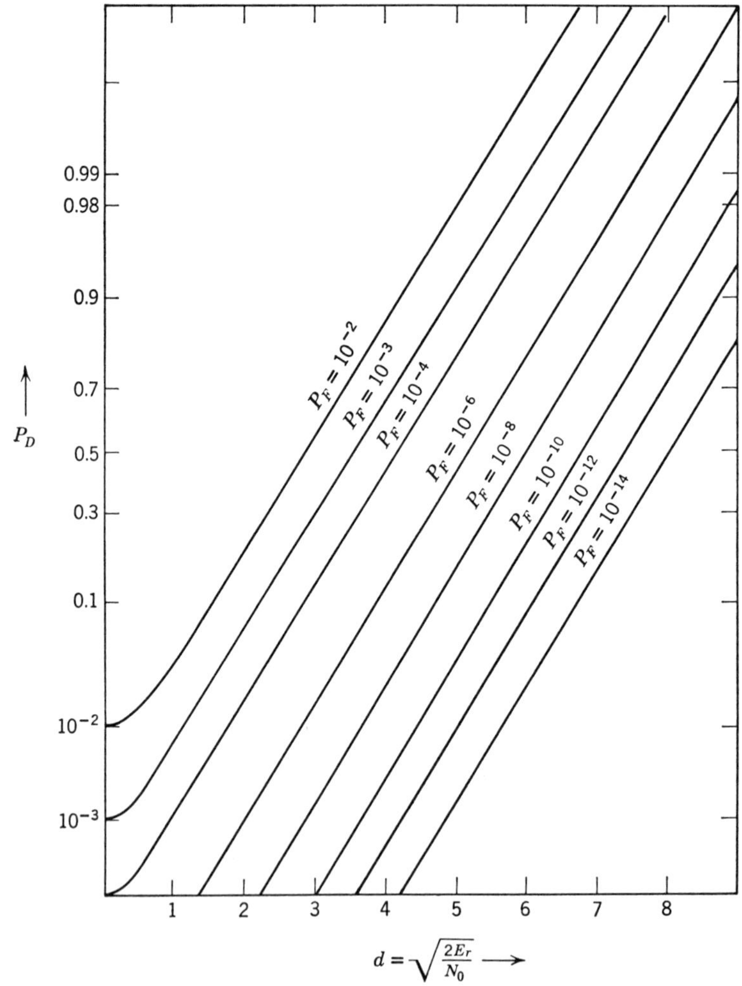

Fig. 4.59 Probability of detection vs d, uniform phase.

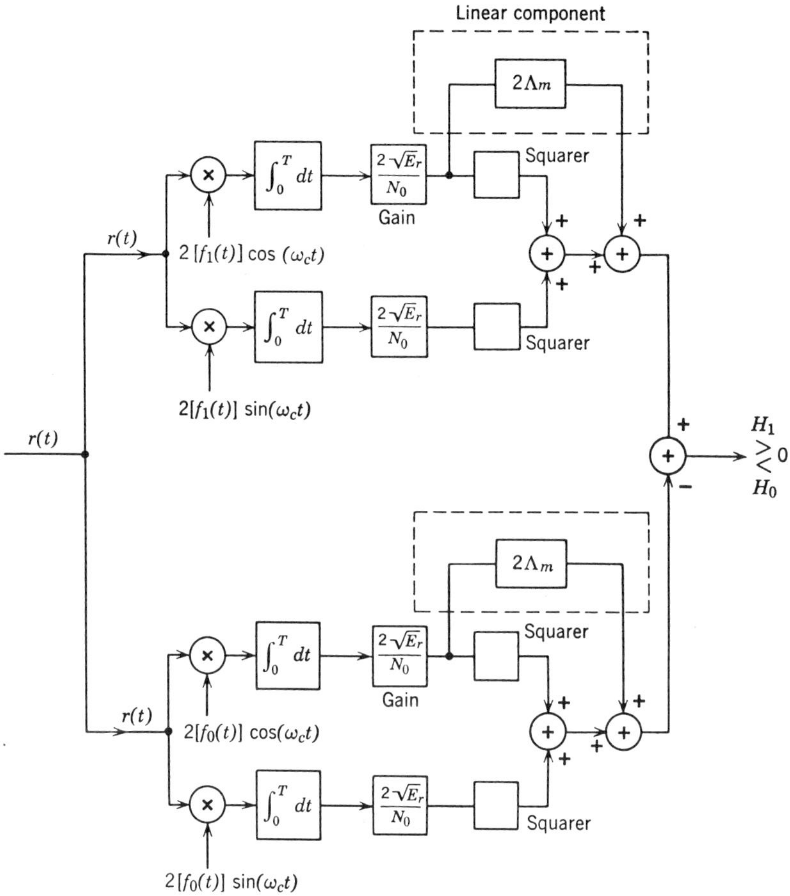

Fig. 4.60 Receiver: binary communication system.

the phase becomes uniform. Now, any correlation (+ or −) would move the signal points closer together. Thus, we expect that $\rho = 0$ would give the best performance. As we go from the first extreme to the second, the best value of ρ should move from -1 to 0.

We shall do only the details for the easy case in which $\rho = -1$; $\rho = 0$ is done in Problem 4.4.9. The error calculation for arbitrary ρ is done in [44].

When $\rho = -1$, we observe that the output of the square-law section is identical on both hypotheses. Thus the receiver is linear. The effect of the phase error is to rotate the signal points in the decision space as shown in Fig. 4.61.

Using the results of Section 4.2.1 (p. 257),

$$\Pr(\epsilon|\theta) = \int_{-\infty}^{0} \left(2\pi \frac{N_0}{2}\right)^{-1/2} \exp\left[-\frac{(x - \sqrt{E_r}\cos\theta)^2}{N_0}\right] dx \tag{393}$$

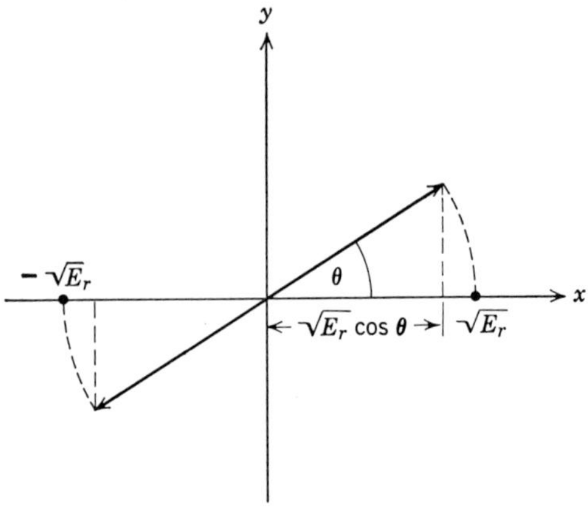

Fig. 4.61 Effect of phase errors in decision space.

or

$$\Pr(\epsilon|\theta) = \int_{-\infty}^{-\sqrt{2E_r/N_0}\cos\theta} \frac{1}{\sqrt{2\pi}} \exp\left(-\frac{z^2}{2}\right) dz. \tag{394}$$

Using (364),

$$\Pr(\epsilon) = \int_{-\pi}^{+\pi} \frac{\exp(\Lambda_m \cos\theta)}{2\pi I_0(\Lambda_m)} \Pr(\epsilon|\theta)\, d\theta. \tag{395}$$

This can be integrated numerically. The results for two particular values of d^2 are shown in Figs. 4.62 and 4.63.† The results for other ρ were also evaluated in [44] and are given in these figures. We see that for Λ_m greater than about 2 the negatively correlated signals become more efficient than orthogonal signals. For $\Lambda_m \geq 10$ the difference is significant. The physical significance of Λ_m will become clearer when we study phase estimating systems in Chapter II.2.

In this section we have studied a particular case of an unwanted parameter, a random phase angle. By using a family of densities we were able to demonstrate how to progress smoothly from the known signal case to the uniform phase case. The receiver consisted of a weighted sum of a linear operation and a quadratic operation. We observe that the specific receiver structure is due to the precise form of the density chosen. In many cases the probability density for the phase angle would not correspond to any of these densities. Intuitively we expect that the receiver developed here should be "almost" optimum for *any* single-peaked density with the same variance as the member of the family for which it was designed.

We now turn to a case of equal (or perhaps greater) importance in which both the amplitude and phase of the received signal vary.

† The values of d^2 were chosen to give a $\Pr(\epsilon) = 10^{-3}$ and 10^{-5}, respectively, at $\Lambda_m = \infty$.

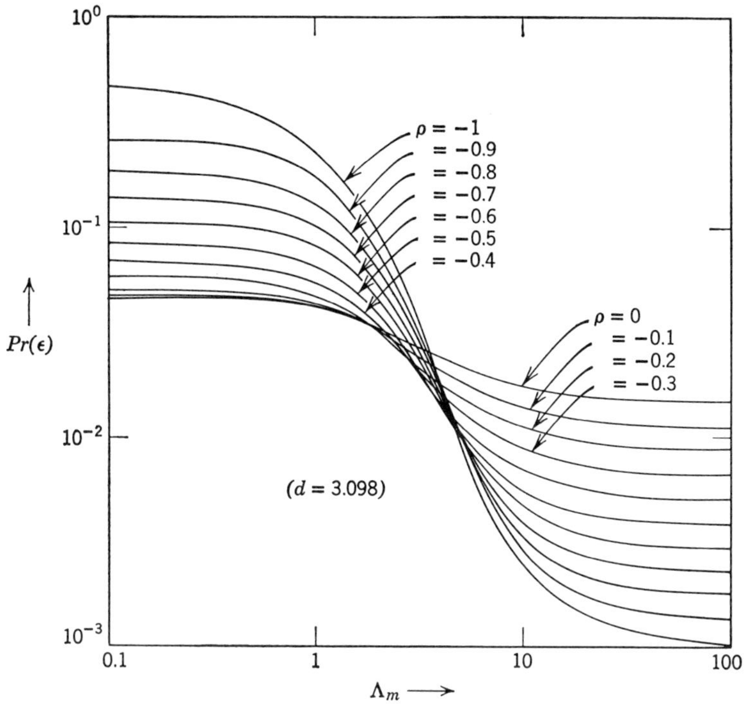

Fig. 4.62 Pr (ϵ), **partially coherent binary system** (10^{-3} **asymptote**) **[44].**

4.4.2 Random Amplitude and Phase

As we discussed in Section 4.1, there are cases in which both the amplitude and phase of the received signal vary. In the communication context this situation is encountered in ionospheric links operating above the maximum usable frequency (e.g., [51]) and in some tropospheric links (e.g., [52]). In the radar context it is encountered when the target's aspect or effective radar cross section changes from pulse to pulse (e.g., Swerling [53]).

Experimental results for a number of physical problems indicate that when the input is a sine wave, $\sqrt{2} \sin \omega_c t$, the output (in the absence of additive noise) is

$$r(t) = v_{\text{ch}}(t) \sin [\omega_c t + \theta_{\text{ch}}(t)]. \tag{396}$$

An exaggerated sketch is shown in Fig. 4.64*a*. The envelope and phase vary continuously. The envelope $v_{\text{ch}}(t)$ has the Rayleigh probability density shown in Fig. 4.64*b* and that the phase angle $\theta_{\text{ch}}(t)$ has a uniform density.

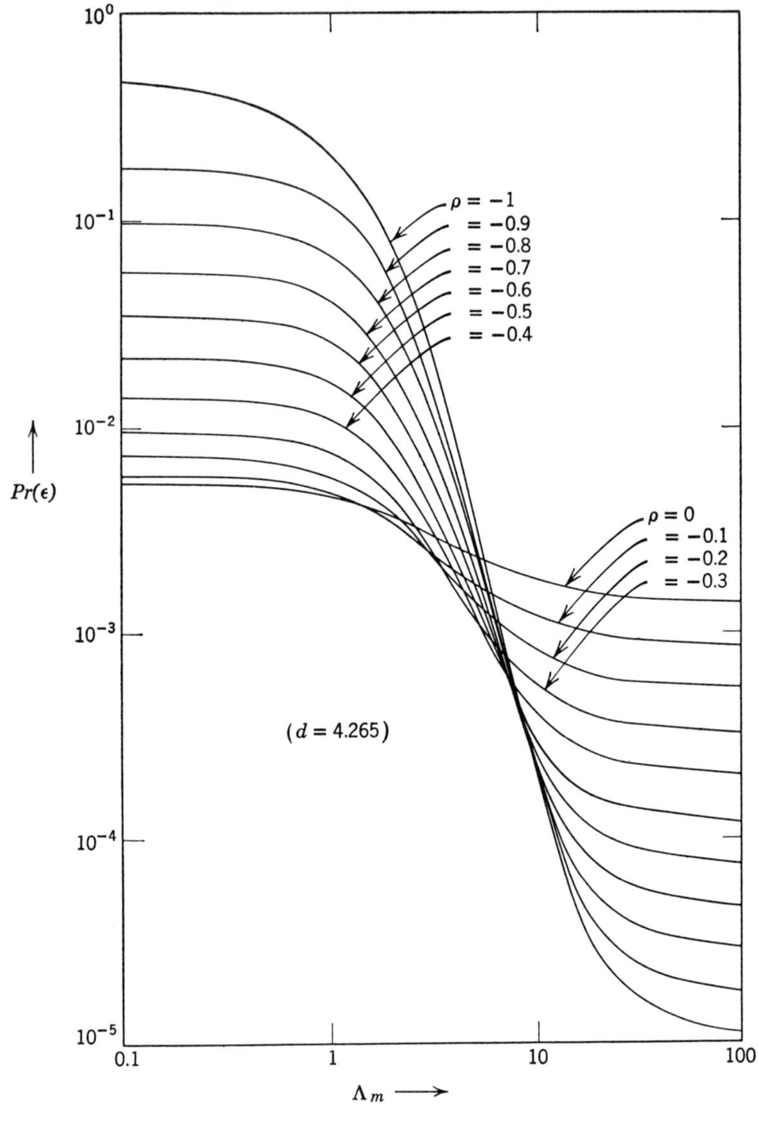

Fig. 4.63 Pr (ϵ), **partially coherent binary system** (10^{-5} **asymptote**) **[44].**

There are several ways to model this channel. The simplest technique is to replace the actual channel functions by piecewise constant functions (Fig. 4.65). This would be valid when the channel does not vary significantly

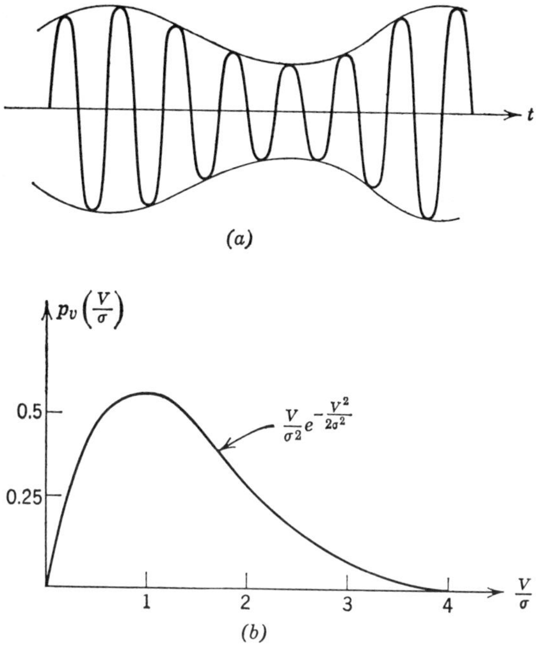

Fig. 4.64 Narrow-band process at output of channel, and the probability of its envelope.

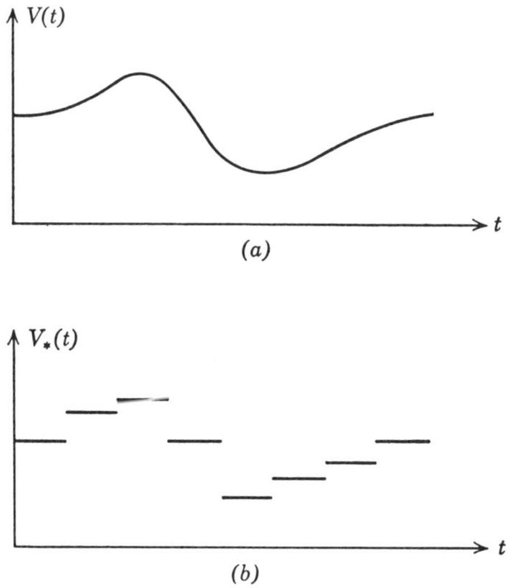

Fig. 4.65 Piecewise constant approximation: (*a*) actual envelope; (*b*) piecewise constant model.

in a T second interval. Given this "slow-fading" model, two choices are available. We can process each signaling interval independently or exploit the channel continuity by measuring the channel and using the measurements in the receiver. We now explore the first alternative.

For the simple binary detection problem in additive white Gaussian noise we may write the received signal under the two hypotheses as†

$$\begin{aligned} H_1 &: r(t) = v\sqrt{2}\, f(t) \cos [\omega_c t + \phi(t) + \theta] + w(t), \qquad 0 \le t \le T, \\ H_0 &: r(t) = w(t), \qquad 0 \le t \le T, \end{aligned} \tag{397}$$

where v is a Rayleigh random variable and θ is a uniform random variable.

We can write the signal component equally well in terms of its quadrature components:

$$\begin{aligned} \sqrt{2}\, v\, f(t) \cos [\omega_c t + \phi(t) + \theta] = a_1 \sqrt{2}\, f(t) \cos [\omega_c t + \phi(t)] \\ + a_2 \sqrt{2}\, f(t) \sin [\omega_c t + \phi(t)], \\ 0 \le t \le T, \end{aligned} \tag{398}$$

where a_1 and a_2 are independent zero-mean Gaussian random variables with variance σ^2 (where $E[v^2] = 2\sigma^2$; see pp. 158–161 of Davenport and Root [2]). We also observe that the two terms are *orthogonal.* Thus the signal out of a Rayleigh fading channel can be viewed as the sum of two orthogonal signals, each multiplied by an independent Gaussian random variable. This seems to be an easier way to look at the problem. As a matter of fact, it is just as easy to solve the more general problem in which the received waveform on H_1 is,

$$r(t) = \sum_{i=1}^{M} a_i\, s_i(t) + w(t), \qquad 0 \le t \le T, \tag{399}$$

where the a_i are independent, zero-mean Gaussian variables $N(0, \sigma_{a_i})$ and

$$\int_0^T s_i(t)\, s_j(t)\, dt = \delta_{ij}. \tag{400}$$

The likelihood ratio is

$$\begin{aligned} \Lambda[r(t)] = \\ \underbrace{\int_{-\infty}^{\infty} \cdots \int_{-\infty}^{\infty}}_{M} p_{a_1}(A_1)\, p_{a_2}(A_2) \cdots p_{a_M}(A_M) \exp \left[+\frac{2}{N_0} \int_0^T r(t) \sum_{i=1}^{M} A_i s_i(t)\, dt \right. \\ \left. - \frac{1}{N_0} \int_0^T \sum_{i=1}^{M} \sum_{j=1}^{M} A_i A_j\, s_i(t)\, s_j(t)\, dt \right] dA_1 \cdots dA_M. \end{aligned} \tag{401}$$

† For simplicity we assume that the transmitted signal has unit energy and adjust the received energy by changing the characteristics of v.

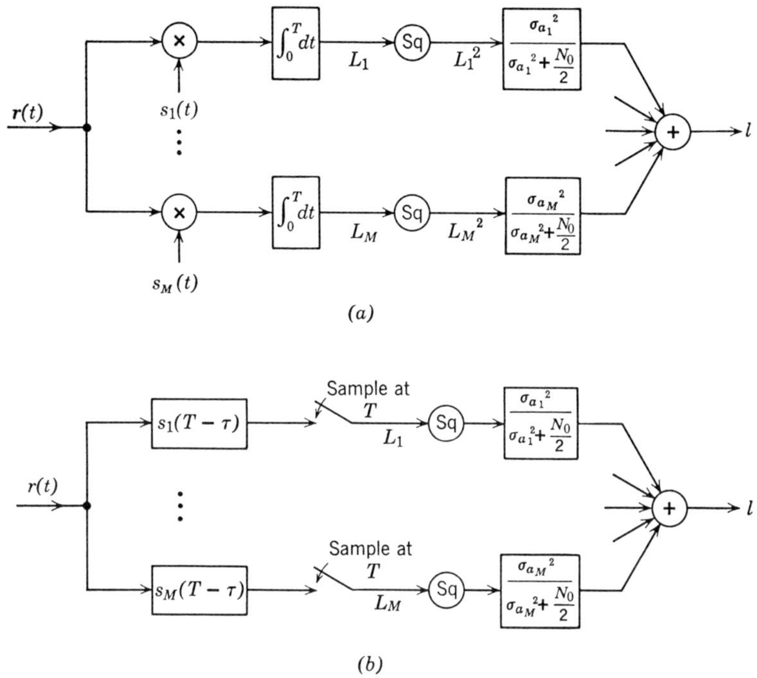

Fig. 4.66 Receivers for Gaussian amplitude signals: (*a*) correlation-squarer receiver; (*b*) filter-squarer receiver.

Defining

$$L_i = \int_0^T r(t)\, s_i(t)\, dt, \tag{402}$$

using the orthogonality of the $s_i(t)$, and completing the square in each of the M integrals, we find the test reduces to

$$l \triangleq \sum_{i=1}^{M} L_i^2 \left(\frac{\sigma_{a_i}^{\ 2}}{\sigma_{a_i}^{\ 2} + N_0/2} \right) \underset{H_0}{\overset{H_1}{\gtrless}} \gamma. \tag{403}†$$

Two receivers corresponding to (403) and shown in Fig. 4.66 are commonly called a correlator-squarer receiver and a filter-squarer receiver, respectively. Equation 403 can be rewritten as

$$l = \sum_{i=1}^{M} L_i \left(\frac{\sigma_{a_i}^{\ 2} L_i}{\sigma_{a_i}^{\ 2} + N_0/2} \right) = \sum_{i=1}^{M} L_i \hat{a}_i. \tag{404}$$

† Note that we could have also obtained (403) by observing that the L_i are jointly Gaussian on both hypotheses and are sufficient statistics. Thus the results of Section 2.6 [specifically (2.326)] are directly applicable. Whenever the L_i have nonzero means or are correlated, the use of 2.326 is the simplest method (e.g., Problem 4.4.21).

This structure, shown in Fig. 4.67, can be interpreted as an estimator–correlator receiver (i.e., we are correlating $r(t)$ with our estimate of the received signal.) The identification of the term in braces as $\hat{a}_i$ follows from our estimation discussion in Section 4.2. It is both a minimum mean-square error estimate and a maximum a posteriori probability estimate. In Fig. 4.67*a* we show a practical implementation. The realization in Fig. 4.67*b* shows that we could actually obtain the estimate of the signal component as a waveform in the optimum receiver. This interpretation is quite important in later applications.

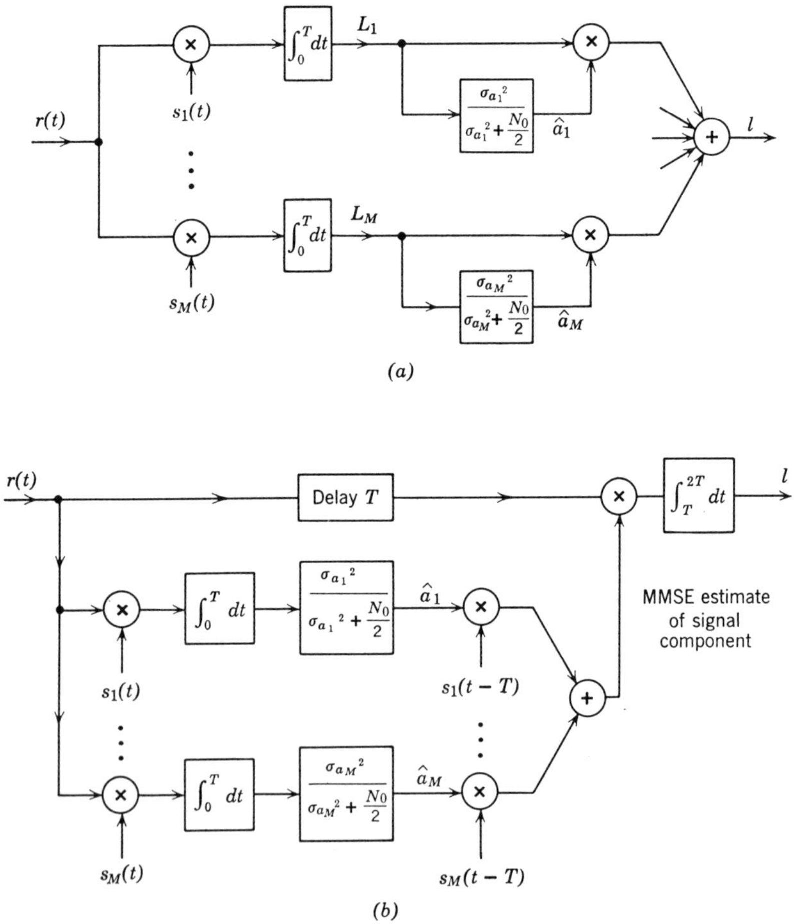

Fig. 4.67 Estimator-correlator receiver.

We now apply these results to the original problem in (397). If we relate L_1 to L_c and L_2 to L_s, we see that the receiver is

$$L_c^2 + L_s^2 \underset{H_0}{\overset{H_1}{\gtrless}} \gamma \tag{405}$$

(where L_c and L_s were defined in the random phase example). This can be realized, as shown in the preceding section, by a bandpass matched filter followed by an envelope detector. The two alternate receiver realizations are shown in Figs. 4.68*a* and *b*.

The next step is to evaluate the performance. We observe that L_c and L_s are Gaussian random variables with identical distributions. Thus the Rayleigh channel corresponds *exactly* to Example 2 on p. 41 of Chapter 2. In Equation (2.80), we showed that

$$P_D = (P_F)^{\sigma_0^2/\sigma_1^2}, \tag{2.80}$$

where σ_0^2 is the variance of L_c on H_0 and σ_1^2 is the variance of L_c on H_1. Looking at Fig. 4.58*a*, we see that

$$\sigma_0^2 = \frac{N_0}{2} \tag{406}$$

and

$$\sigma_1^2 = \frac{N_0}{2} + E(v^2) \triangleq \frac{N_0}{2} + \bar{E}_r, \tag{407}$$

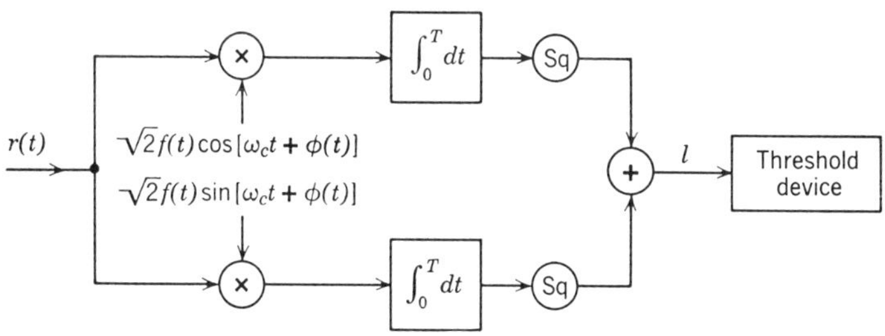

(*a*)

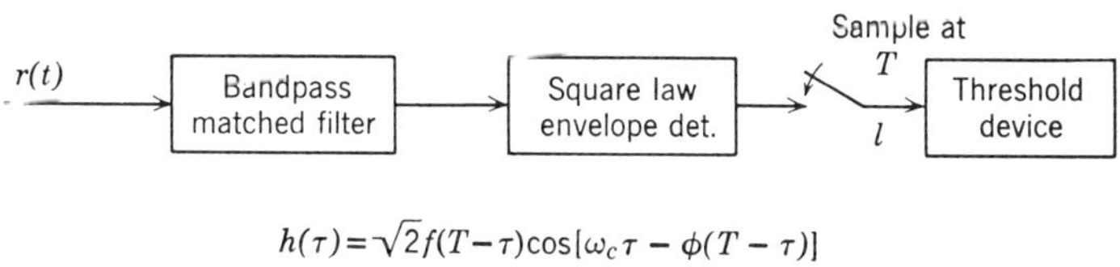

(*b*)

Fig. 4.68 Optimum receivers, Rayleigh channel: (*a*) squarer realization; (*b*) matched filter-envelope detector realization.

because v^2 is the received signal energy. Substituting (406) and (407) into (2.80), we obtain

$$P_F = (P_D)^{1+2\bar{E}_r/N_0} \tag{408}$$

The ROC is plotted in Fig. 4.69.

The solution to the analogous binary communication problem for arbitrary signals follows in a similar fashion (e.g., Masonson [55] and

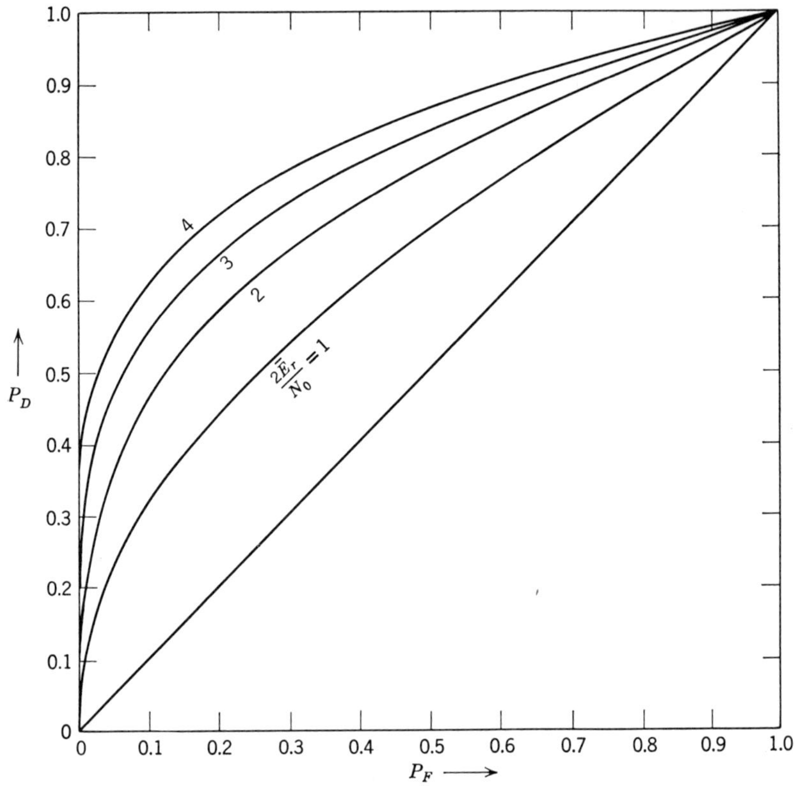

Fig. 4.69 (*a*) Receiver operating characteristic, Rayleigh channel.

Turin [56]). We discuss a typical system briefly. Recall that the phase angle θ has a uniform density. From our results in the preceding section (Figs. 4.62 and 4.63) we would expect that orthogonal signals would be optimum. We discuss briefly a simple FSK system using orthogonal signals. The received signals under the two hypotheses are

$$\begin{aligned} H_1: r(t) &= \sqrt{2}\, v\, f(t) \cos[\omega_1 t + \phi(t) + \theta] + w(t), \qquad 0 \le t \le T, \\ H_0: r(t) &= \sqrt{2}\, v\, f(t) \cos[\omega_0 t + \phi(t) + \theta] + w(t), \qquad 0 \le t \le T. \end{aligned} \tag{409}$$

The frequencies are separated enough to guarantee orthogonality. Assuming equal a priori probabilities and a minimum probability of error criterion, $\eta = 1$. The likelihood ratio test follows directly (see Problem 4.4.24).

$$L_{c1}^2 + L_{s1}^2 \underset{H_0}{\overset{H_1}{\gtrless}} L_{c0}^2 + L_{s0}^2. \tag{410}$$

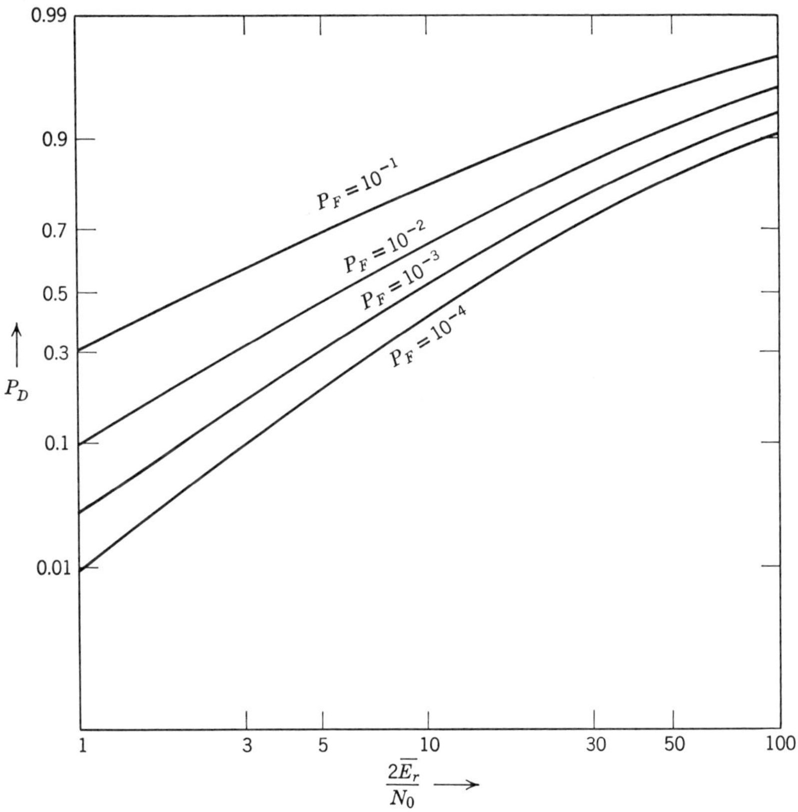

Fig. 4.69 **(*b*) probability of detection vs. $2\bar{E}_r/N_0$.**

The receiver structure is shown in Fig. 4.70. The probability of error can be evaluated analytically:

$$\Pr(\epsilon) = \frac{1}{2}\left[1 + \frac{1}{2}\frac{\bar{E}_r}{N_0}\right]^{-1}. \tag{411}$$

(See Problem 4.4.24.) In Fig. 4.71 we have plotted the $\Pr(\epsilon)$ as a function of $\bar{E}_r/N_0$. For purposes of comparison we have also shown the $\Pr(\epsilon)$ for

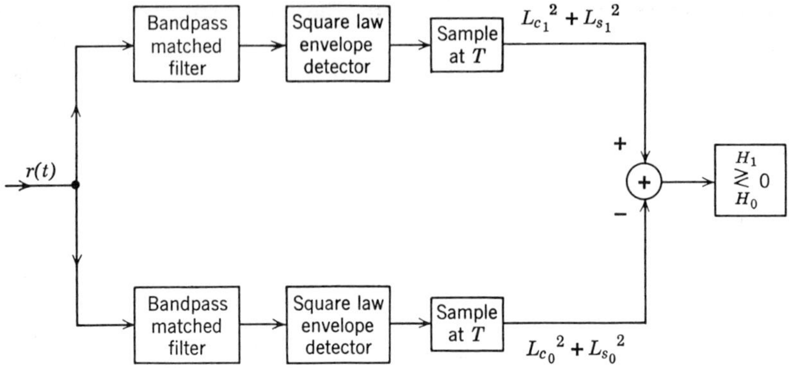

Fig. 4.70 Optimum receiver: binary communication system with orthogonal signals.

the known signal case and the uniform random phase case. We see that for both nonfading cases the probability of error decreases exponentially for large $\bar{E}_r/N_0$, whereas the fading case decreases only linearly. This is intuitively logical. Regardless of how large the average received energy becomes, during a deep signal fade the probability of error is equal or nearly equal to $\frac{1}{2}$. Even though this does not occur often, its occurrence keeps the probability of error from improving exponentially. In Chapter II.3 we shall find that by using diversity (for example, sending the signal over several independent Rayleigh channels in parallel) we can achieve an exponential decrease.

As we have already pointed out, an alternate approach is to measure the channel characteristics and use this measurement in the receiver structure. We can easily obtain an estimate of the possible improvement available by assuming that the channel measurement is *perfect.* If the measurement is perfect, we can use a coherent receiver. The resulting Pr (ϵ) is easy to evaluate. First we write the error probability conditioned on the channel variable v being equal to V. We then average over the Rayleigh density in Fig. 4.64*b*. Using coherent or known signal reception and orthogonal signals the probability of error for a given value V is given by (36) and (40),

$$\Pr(\epsilon|V) = \int_{V\sqrt{1/N_0}}^{\infty} \frac{1}{\sqrt{2\pi}} e^{-y^2/2}\, dy, \qquad (V \geq 0) \tag{412}$$

and

$$p_v(V) = \begin{cases} \dfrac{V}{\sigma^2} e^{-V^2/2\sigma^2}, & V \geq 0, \\ 0, & V < 0. \end{cases} \tag{413}$$

Thus

$$\Pr(\epsilon) = \int_0^\infty dV \frac{V}{\sigma^2} e^{-V^2/2\sigma^2} \int_{V\sqrt{1/N_0}}^\infty \frac{1}{\sqrt{2\pi}} e^{-y^2/2}\, dy. \tag{414}$$

Changing to polar coordinates and integrating, we obtain

$$\Pr(\epsilon) = \frac{1}{2}\left[1 - \left(\frac{\bar{E}_r/N_0}{1 + \bar{E}_r/N_0}\right)^{1/2}\right]. \tag{415}$$

The result is shown in Fig. 4.72.

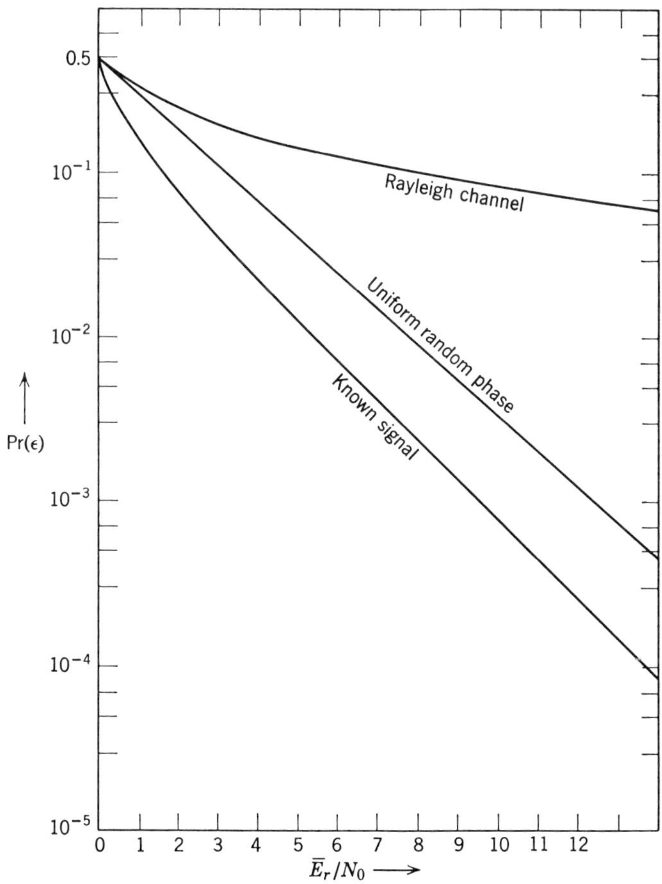

Fig. 4.71 Probability of error, binary orthogonal signals, Rayleigh channel.

Comparing (415) and (411) (or looking at Fig. 4.72), we see that perfect measurement gives about a 3-db improvement for high $\bar{E}_r/N_0$ values and orthogonal signals. In addition, if we measured the channel, we could use equal-and-opposite signals to obtain another 3 db.

Rician Channel In many physical channels there is a fixed (or "specular") component in addition to the Rayleigh component. A typical example is an ionospheric radio link operated below the maximum usable frequency (e.g., [57], [58], or [59]). Such channels are called Rician channels. We now illustrate the behavior of this type of channel for a

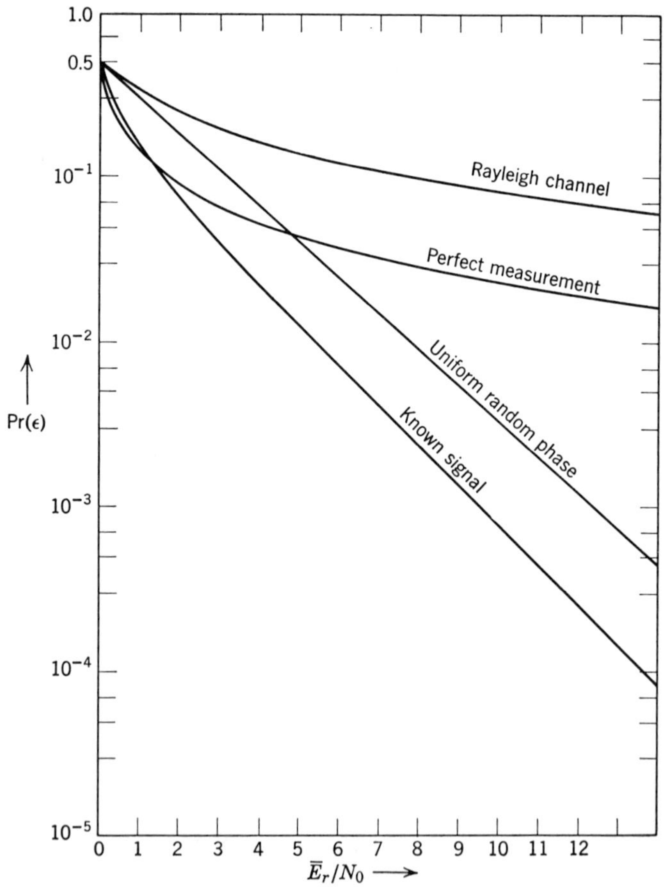

Fig. 4.72 Probability of error, Rayleigh channel with perfect measurement.

binary communication system, using orthogonal signals. The received signals on the two hypotheses are

$$\begin{aligned} H_1\!:r(t) &= \sqrt{2}\,\alpha\, f_1(t) \cos\left[\omega_c t + \phi_1(t) + \delta\right] \\ &\quad + \sqrt{2}\, v\, f_1(t) \cos\left[\omega_c t + \phi_1(t) + \theta\right] + w(t), \\ &\qquad\qquad 0 \le t \le T, \\ H_0\!:r(t) &= \sqrt{2}\,\alpha\, f_0(t) \cos\left[\omega_c t + \phi_0(t) + \delta\right] \\ &\quad + \sqrt{2}\, v\, f_0(t) \cos\left[\omega_c t + \phi_0(t) + \theta\right] + w(t), \\ &\qquad\qquad 0 \le t \le T, \end{aligned} \tag{416}$$

where α and δ are the amplitude and phase of the specular component. The transmitted signals are orthonormal. In the simplest case α and δ are assumed to be known (see Problem 4.5.26 for unknown δ). Under this assumption, with no loss in generality, we can let $\delta = 0$. We may now write the signal component on H_i as

$$a_1\{\sqrt{2}\, f_i(t) \cos\left[\omega_c t + \phi_i(t)\right]\} + a_2\{\sqrt{2}\, f_i(t) \sin\left[\omega_c t + \phi_i(t)\right]\}, \qquad (i = 0, 1). \tag{417}$$

Once again, a_1 and a_2 are independent Gaussian random variables:

$$\begin{aligned} E(a_1) &= \alpha, \qquad E(a_2) = 0, \\ \text{Var}\,(a_1) &= \text{Var}\,(a_2) = \sigma^2. \end{aligned} \tag{418}$$

The expected value of the received energy in the signal component on either hypothesis is

$$E(E_r) = 2\sigma^2 + \alpha^2 \triangleq \sigma^2(2 + \gamma^2). \tag{419}$$

where γ^2 is twice the ratio of the energy in the specular component to the average energy in the random component.

If we denote the total received amplitude and phase angle as

$$v' = \sqrt{a_1^2 + a_2^2}, \qquad \theta' = \tan^{-1} \frac{a_2}{a_1}. \tag{420}$$

The density of the normalized envelope $(V_n' = V/\sigma)p_{v_n'}(X)$ and the density of the phase angle $p_{\theta'}(\theta')$ are shown in Fig. 4.73 ([60] and [56]). As we would expect, the phase angle probability density becomes quite peaked as γ increases.

The receiver structure is obtained by a straightforward modification of (398) to (405). The likelihood ratio test is

$$\left(\frac{\alpha}{2\sigma^2} + \frac{1}{N_0} L_{c1}\right)^2 + \left(\frac{1}{N_0} L_{s1}\right)^2 \underset{H_0}{\overset{H_1}{\gtrless}} \left(\frac{\alpha}{2\sigma^2} + \frac{1}{N_0} L_{c0}\right)^2 + \left(\frac{1}{N_0} L_{s0}\right)^2. \tag{421}$$

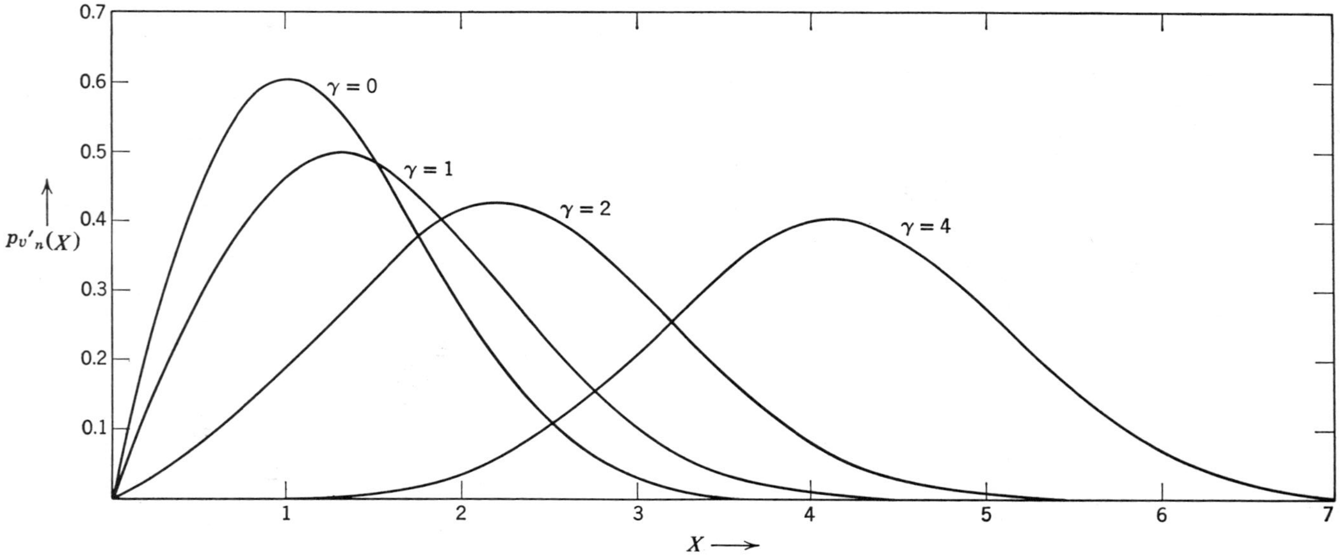

Fig. 4.73 **(*a*) Probability density for envelope, Rician channel.**

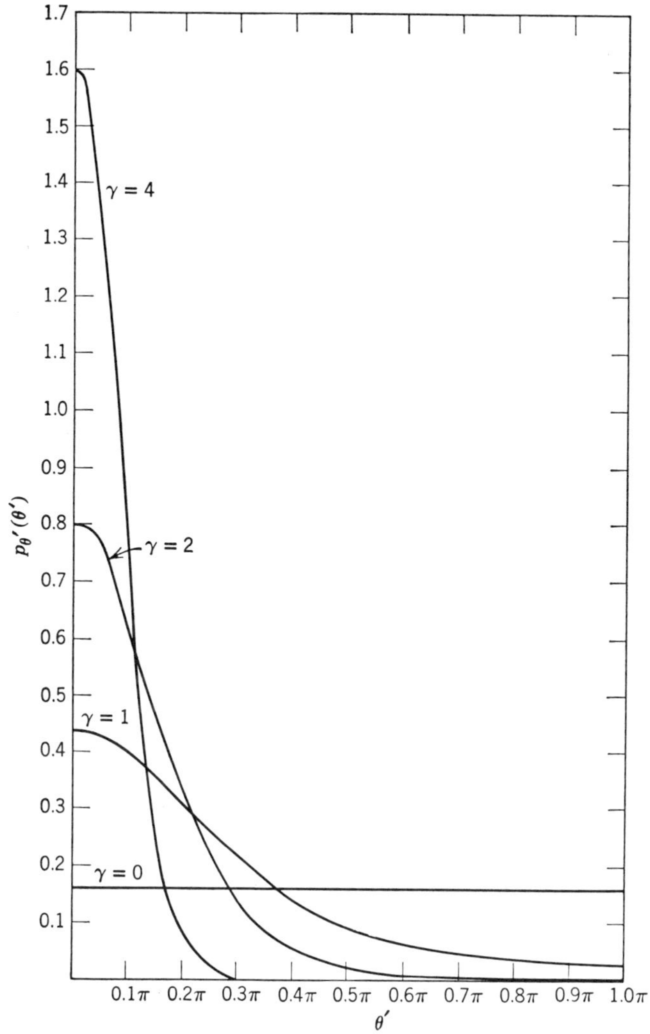

Fig. 4.73 (*b*) probability density for phase angle, Rician channel.

The receiver structure is shown in Fig. 4.74. The calculation of the error probability is tedious (e.g., [56]), but the result is

$$\Pr(\epsilon) = Q\left[\frac{\gamma}{(\beta + \frac{1}{2})^{1/2}\beta^{1/2}}, \frac{\gamma(\beta + 1)}{(\beta + 2)^{1/2}\beta^{1/2}}\right] - \left(\frac{\beta + 1}{\beta + 2}\right) \exp\left[-\frac{\gamma^2}{2}\left(\frac{\beta^2 + 2\beta + 2}{\beta^2 + 2\beta}\right)\right] I_0\left[\gamma^2 \frac{\beta + 1}{\beta(\beta + 2)}\right] \quad (422)$$

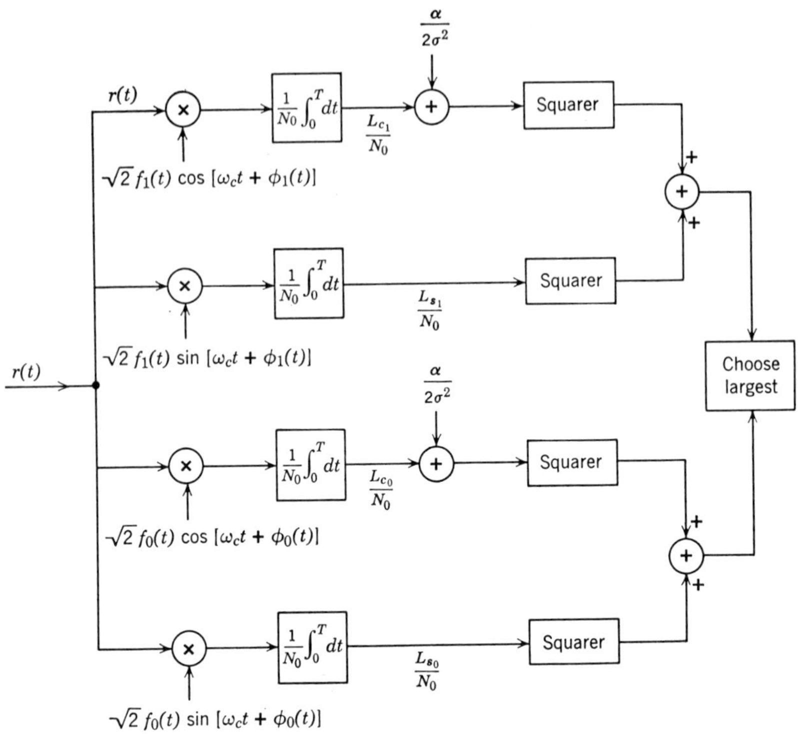

Fig. 4.74 Optimum receiver for binary communication over a Rician channel.

where $\beta \triangleq 2\sigma^2/N_0$ is the expected value of the received signal energy in the random component divided by N_0. The probability of error is plotted for typical values of γ in Fig. 4.75. Observe that $\gamma = 0$ is the Rayleigh channel and $\gamma = \infty$ is the completely known signal. We see that even when the power in the specular component is twice that of the random component the performance lies close to the Rayleigh channel performance. Once again, because the Rician channel is a channel of practical importance, considerable effort has been devoted to studying its error behavior under various conditions (e.g., [56]).

Summary As we would expect, the formulation for the *M*-ary signaling problem is straightforward. Probability of error calculations are once again involved (e.g., [61] or [15]). In Chapter II.3 we shall see that both the Rayleigh and Rician channels are special cases of the *general Gaussian* problem.

In this section we have studied in detail two important cases in which unwanted random parameters are contained in the signal components.

Because the probability density was known, the optimum test procedure followed directly from our general likelihood ratio formulation. The particular examples of densities we considered gave integrals that could be evaluated analytically and consequently led to explicit receiver structures. Even when we could not evaluate the integrals, the method of setting up the likelihood ratio was clear.

When the probability density of $\boldsymbol{\theta}$ is unknown, the best procedure is not obvious. There are two logical possibilities:

1. We can hypothesize a density and use it as if it were correct. We can investigate the dependence of the performance on the assumed density by using sensitivity analysis techniques analogous to those we have demonstrated for other problems.

2. We can use a minimax procedure. This is conceptually straightforward. For example, in a binary communication problem we find the

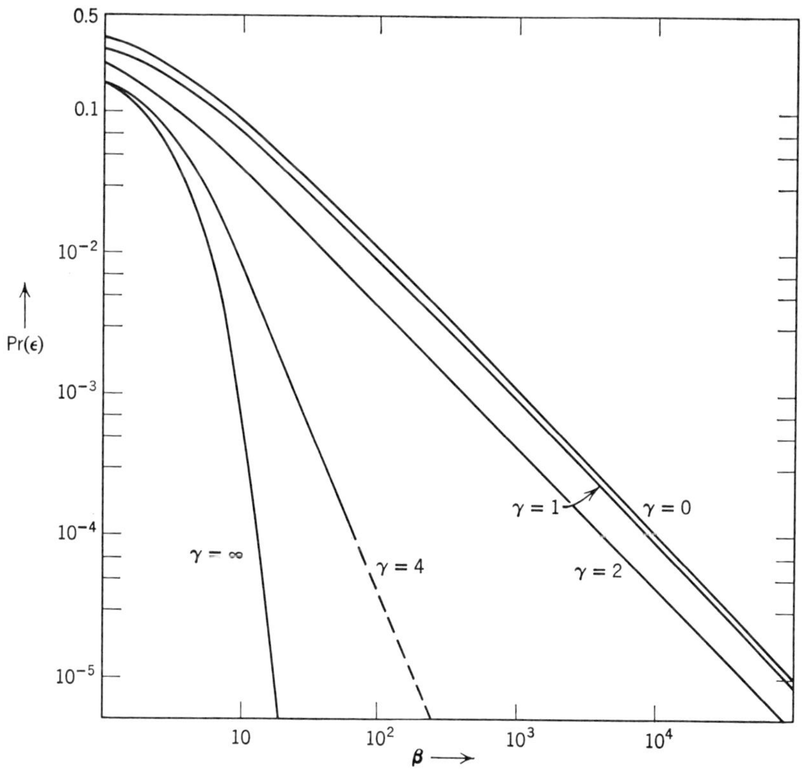

Fig. 4.75 Probability of error for binary orthogonal signals, Rician channel.

Pr (ϵ) as a function of $p_{\boldsymbol{\theta}}(\boldsymbol{\theta})$ and then choose the $p_{\boldsymbol{\theta}}(\boldsymbol{\theta})$ that maximizes Pr (ϵ) and design for this case. The two objections to this procedure are its difficulty and the conservative result.

The final possibility is for $\boldsymbol{\theta}$ to be a nonrandom variable. To deal with this problem we simply extend the composite hypothesis testing techniques that we developed in Section 2.5 to include waveform observations. The techniques are straightforward. Fortunately, in many cases of practical importance either a UMP test will exist or a generalized likelihood ratio test will give satisfactory results. Some interesting examples are discussed in the problems. Helstrom [14] discusses the application of generalized likelihood ratio tests to the radar problem of detecting signals of unknown arrival time.

4.5 MULTIPLE CHANNELS

In Chapter 3 we introduced the idea of a vector random process. We now want to solve the detection and estimation problems for the case in which the received waveform is a sample function from a vector random process.

In the simple binary detection problem, the received waveforms are

$$\begin{aligned} H_1&:\mathbf{r}(t) = \mathbf{s}(t) + \mathbf{n}(t), \qquad T_i \le t \le T_f, \\ H_0&:\mathbf{r}(t) = \mathbf{n}(t), \qquad\qquad\quad T_i \le t \le T_f. \end{aligned} \tag{423†}$$

In the estimation case, the received waveform is

$$\mathbf{r}(t) = \mathbf{s}(t, A) + \mathbf{n}(t), \qquad T_i \le t \le T_f. \tag{424}$$

Two issues are involved in the vector case:

1. The first is a compact formulation of the problem. By using the vector Karhunen-Loéve expansion with scalar coefficients introduced in Chapter 3 we show that the construction of the likelihood ratio is a trivial extension of the scalar case. (This problem has been discussed in great detail by Wolf [63] and Thomas and Wong [64].)

2. The second is to study the performance of the resulting receiver structures to see whether problems appear that did *not* occur in the scalar case. We discuss only a few simple examples in this section. In Chapter II.5 we return to the multidimensional problem and investigate some of the interesting phenomena.

† In the scalar case we wrote the signal energy separately and worked with normalized waveforms. In the vector case this complicates the notation needlessly, and we use unnormalized waveforms.

4.5.1 Formulation

We assume that $\mathbf{s}(t)$ is a known vector signal. The additive noise $\mathbf{n}(t)$ is a sample function from an M-dimensional Gaussian random process. We assume that it contains a white noise component:

$$\mathbf{n}(t) = \mathbf{w}(t) + \mathbf{n}_c(t), \tag{425}$$

where

$$E[\mathbf{w}(t)\mathbf{w}^T(u)] = \frac{N_0}{2}\mathbf{I}\,\delta(t-u). \tag{426a}$$

a more general case is,

$$E[\mathbf{w}(t)\,\mathbf{w}^T(u)] = \mathbf{N}\,\delta(t-u). \tag{426b}$$

The matrix $\mathbf{N}$ contains only numbers. We assume that it is *positive-definite*. Physically this means that all components of $\mathbf{r}(t)$ or any linear transformation of $\mathbf{r}(t)$ will contain a white noise component. The general case is done in Problem 4.5.2. We consider the case described by (426*a*) in the text. The covariance function matrix of the colored noise is

$$E[\mathbf{n}_c(t)\,\mathbf{n}_c^T(u)] \triangleq \mathbf{K}_c(t, u). \tag{427}$$

We assume that each element in $\mathbf{K}_c(t, u)$ is square-integrable and that the white and colored components are independent. Using (425–427), we have

$$\mathbf{K}_\mathbf{n}(t, u) = \frac{N_0}{2}\mathbf{I}\,\delta(t-u) + \mathbf{K}_c(t, u). \tag{428}$$

To construct the likelihood ratio we proceed as in the scalar case. Under hypothesis H_1

$$\begin{aligned} r_i &\triangleq \int_{T_i}^{T_f} \mathbf{r}^T(t)\,\boldsymbol{\phi}_i(t)\,dt \\ &= \int_{T_i}^{T_f} \mathbf{s}^T(t)\,\boldsymbol{\phi}_i(t)\,dt + \int_{T_i}^{T_f} \mathbf{n}^T(t)\,\boldsymbol{\phi}_i(t)\,dt \\ &= s_i + n_i. \end{aligned} \tag{429}$$

Notice that all of the coefficients are scalars. Thus (180) is directly applicable:

$$\ln \Lambda[\mathbf{r}(t)] = \sum_{i=1}^{\infty} \frac{R_i s_i}{\lambda_i} - \frac{1}{2}\sum_{i=1}^{\infty} \frac{s_i^2}{\lambda_i}. \tag{430}$$

Substituting (429) into (430), we have

$$\begin{aligned} \ln \Lambda[\mathbf{r}(t)] &= \iint_{T_i}^{T_f} \mathbf{r}^T(t) \sum_{i=1}^{\infty} \frac{\boldsymbol{\phi}_i(t)\,\boldsymbol{\phi}_i^T(u)}{\lambda_i}\,\mathbf{s}(u)\,dt\,du \\ &\quad - \tfrac{1}{2}\iint_{T_i}^{T_f} \mathbf{s}^T(t) \sum_{i=1}^{\infty} \frac{\boldsymbol{\phi}_i(t)\,\boldsymbol{\phi}_i^T(u)}{\lambda_i}\,\mathbf{s}(u)\,dt\,du. \end{aligned} \tag{431}$$

Defining

$$\mathbf{Q_n}(t, u) = \sum_{i=1}^{\infty} \frac{\boldsymbol{\phi}_i(t)\, \boldsymbol{\phi}_i^T(u)}{\lambda_i}, \qquad T_i < t, u < T_f, \tag{432}$$

we have

$$\ln \Lambda[\mathbf{r}(t)] = \iint_{T_i}^{T_f} \mathbf{r}^T(t)\, \mathbf{Q_n}(t, u)\, \mathbf{s}(u)\, dt\, du$$

$$- \frac{1}{2} \iint_{T_i}^{T_f} \mathbf{s}^T(t)\, \mathbf{Q_n}(t, u)\, \mathbf{s}(u)\, dt\, du. \tag{433}$$

Using the vector form of Mercer's theorem (2.253) and (432), we observe that

$$\boxed{\int_{T_i}^{T_f} \mathbf{K_n}(t, u)\, \mathbf{Q_n}(u, z)\, du = \delta(t - z)\mathbf{I}, \qquad T_i < t, z < T_f.} \tag{434}$$

By analogy with the scalar case we write

$$\mathbf{Q_n}(t, u) = \frac{2}{N_0} \mathbf{I}[\delta(t - u) - \mathbf{h}_o(t, u)] \tag{435}$$

and show that $\mathbf{h}_o(t, u)$ can be represented by a convergent series. The details are in Problems 4.5.1. As in the scalar case, we simplify (433) by defining,

$$\boxed{\mathbf{g}(t) = \int_{T_i}^{T_f} \mathbf{Q_n}(t, u)\, \mathbf{s}(u)\, du, \qquad T_i < t < T_f.} \tag{436}$$

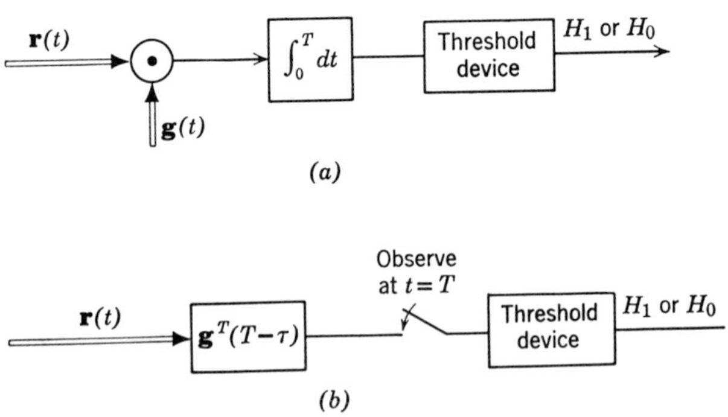

Fig. 4.76 Vector receivers: (*a*) matrix correlator; (*b*) matrix matched filter.

The optimum receiver is just a vector correlator or vector matched filter, as shown in Fig. 4.76. The double lines indicate vector operations and the symbol $\odot$ denotes the dot product of the two input vectors. We can show that the performance index is

$$d^2 = \iint_{T_i}^{T_f} \mathbf{s}^T(t)\, \mathbf{Q_n}(t, u)\, \mathbf{s}(u)\, dt\, du$$

$$= \iint_{T_i}^{T_f} \mathbf{s}^T(t)\, \mathbf{g}(t)\, dt. \tag{437}$$

4.5.2 Application

Consider a simple example.

Example.

$$\mathbf{s}(t) = \begin{bmatrix} \sqrt{E_1}\, s_1(t) \\ \sqrt{E_2}\, s_2(t) \\ \vdots \\ \sqrt{E_M}\, s_M(t) \end{bmatrix}, \qquad 0 \le t \le T, \tag{438}$$

where the $s_i(t)$ are orthonormal.

Assume that the channel noises are independent and white:

$$E[\mathbf{w}(t)\mathbf{w}^T(u)] = \begin{bmatrix} \frac{N_0}{2} & & & 0 \\ & \frac{N_0}{2} & & \\ & & \ddots & \\ 0 & & & \frac{N_0}{2} \end{bmatrix} \delta(t-u). \tag{439}$$

Then

$$\mathbf{g}(t) = \begin{bmatrix} \dfrac{2\sqrt{E_1}}{N_1}\, s_1(t) \\ \vdots \\ \dfrac{2\sqrt{E_M}}{N_M}\, s_M(t) \end{bmatrix}. \tag{440}$$

The resulting receiver is the vector correlator shown in Fig. 4.77 and the performance index is

$$d^2 = \sum_{i=1}^{M} \frac{2E_i}{N_0}. \tag{441}$$

This receiver is commonly called a maximal ratio combiner [65] because the inputs are weighted to maximize the output signal-to-noise ratio. The appropriate combiners for colored noise are developed in the problems.

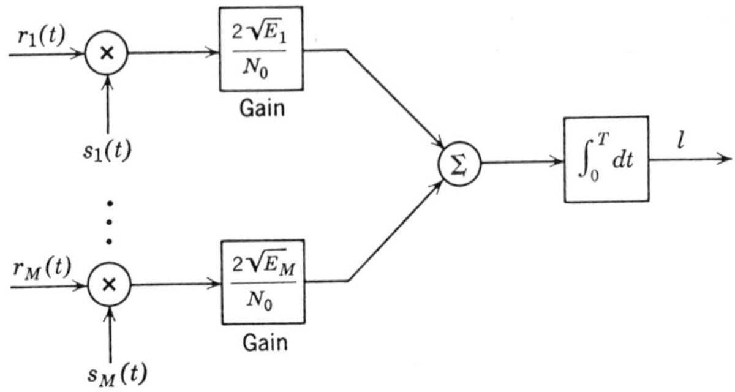

Fig. 4.77 Maximal ratio combiner.

Most of the techniques of the scalar case carry over directly to the vector case at the expense of algebraic complexity. Some of them are illustrated in the problems and a more detailed discussion is contained in Chapter II.5. The modifications for linear and nonlinear estimation are straightforward (see Problems 4.5.4 and 4.5.5). The modifications for unwanted parameters can also be extended to the vector case. The formulation for M channels of the random phase, Rayleigh, and Rician types is carried out in the problems.

4.6 MULTIPLE PARAMETER ESTIMATION

In this section we consider the problem of estimating a finite set of parameters, $a_1, a_2, \ldots, a_m$. We denote the parameters by the vector $\mathbf{a}$. We will consider only the additive white noise channel. The results are obtained by combining the classical multiple parameter estimation result of Chapter 2 with those of Section 4.2.

Our motivation for studying this problem is twofold:

1. One obvious reason is that multiple parameter problems are present in many physical situations of interest. A common example in radar is finding the range and velocity of a target by estimating the delay and Doppler shift of the returned pulse.

2. The second reason is less obvious. In Chapter 5 we shall consider the estimation of a continuous waveform, and we shall see that by expanding the waveform in a series we can estimate the coefficients of the series and use them to construct a waveform estimate. Thus the multiple parameter problem serves as a method of transition from single parameter estimation to waveform estimation.

4.6.1 Additive White Gaussian Noise Channel

Joint MAP Estimates. We assume that the signal depends on the parameter values $A_1, A_2, \ldots, A_M$. Then, for the additive channel, we may write the received signal as

$$r(t) = s(t, \mathbf{A}) + w(t), \qquad T_i \le t \le T_f, \tag{442}$$

where the $\mathbf{A}$ is a column matrix:

$$\mathbf{A} = \begin{bmatrix} A_1 \\ \vdots \\ A_M \end{bmatrix}. \tag{443}$$

We want to form the a posteriori density in terms of a suitable set of observables which we denote by the K-dimensional vector $\mathbf{r}$. We then find the estimate $\hat{\mathbf{a}}$ that maximizes the a posteriori density and let $K \to \infty$ to get the desired result.

The parameters $a_1, a_2, \ldots, a_M$ can be coupled either in the signal structure or because of an a priori statistical dependence. We can categorize this statistical dependence in the following way:

1. The $a_1, a_2, \ldots, a_M$ are *jointly Gaussian.*
2. The $a_1, a_2, \ldots, a_M$ are statistically independent and Gaussian.
3. The $a_1, a_2, \ldots, a_M$ are statistically independent but *not* Gaussian.
4. The $a_1, a_2, \ldots, a_M$ are not statistically independent and are *not* jointly Gaussian.

Our first observation is that Case 1 can be transformed into Case 2. The following property was proved in Chapter 2 (2.237).

Property. If $\mathbf{b}$ is a nonsingular linear transformation on $\mathbf{a}$ (i.e., $\mathbf{b} = \mathbf{La}$), then

$$\hat{\mathbf{b}}_{\text{map}} = \mathbf{L}\hat{\mathbf{a}}_{\text{map}},$$

and

$$\hat{\mathbf{a}}_{\text{map}} = \mathbf{L}^{-1}\hat{\mathbf{b}}_{\text{map}}. \tag{444}$$

We know that there is a nonsingular linear transformation that transforms any set of dependent Gaussian variables into a set of independent Gaussian variables (Chapter 2, pp. 101–105). Thus, if the a_i are dependent, we can estimate the b_i instead. Therefore the assumption

$$p_{\mathbf{a}}(\mathbf{A}) \triangleq p_{a_1 a_2 \cdots a_M}(A_1, \ldots, A_M) = \prod_{i=1}^{M} p_{a_i}(A_i) \tag{445}$$

effectively includes Cases 1, 2, and 3. Case 4 is much more involved in detail (but not in concept) and is of no importance in the sequel. We do not consider it here.

Assuming that the noise is white and that (445) holds, it follows in the same manner as the scalar case that the MAP estimates are solutions to the following set of M simultaneous equations:

$$0 = \left\{\frac{2}{N_0}\int_{T_i}^{T_f} \frac{\partial s(t, \mathbf{A})}{\partial A_i}[r(t) - s(t, \mathbf{A})]\,dt + \frac{\partial \ln p_{a_i}(A_i)}{\partial A_i}\right\}\Bigg|_{\mathbf{A}=\hat{\mathbf{a}}_{map}}, \quad (i = 1, 2, \ldots, M). \tag{446}$$

If the a_i are Gaussian with zero-mean and variances $\sigma_{a_i}^2$, the equations reduce to a simple form:

$$\hat{a}_i = \frac{2\sigma_{a_i}^2}{N_0}\int_{T_i}^{T_f} \frac{\partial s(t, \mathbf{A})}{\partial A_i}[r(t) - s(t, \mathbf{A})]\,dt\Bigg|_{\mathbf{A}=\hat{\mathbf{a}}_{map}}, \qquad (i = 1, 2, \ldots, M). \tag{447}$$

This set of simultaneous equations imposes a set of *necessary* conditions on the MAP estimates. (We assume that the maximum is interior to the allowed region of $\mathbf{A}$ and that the indicated derivatives exist at the maximum.)

The second result of interest is the bound matrix. From Section 2.4.3 we know that the first step is to find the information matrix. From Equation 2.289

$$\mathbf{J}_T = \mathbf{J}_D + \mathbf{J}_P, \tag{448}$$

$$\mathbf{J}_{D_{ij}} = -E\left(\frac{\partial^2 \ln \Lambda(\mathbf{A})}{\partial A_i\,\partial A_j}\right), \tag{449}$$

and for a Gaussian a priori density

$$\mathbf{J}_P = \mathbf{\Lambda_a}^{-1}, \tag{450}$$

where $\mathbf{\Lambda_a}$ is the covariance matrix. The term in (449) is analogous to (104) in Section 4.2. Thus it follows easily that (449) reduces to,

$$\mathbf{J}_{D_{ij}} = \frac{2}{N_0}E_{\mathbf{a}}\left[\int_{T_i}^{T_f} \frac{\partial s(t, \mathbf{A})}{\partial A_i}\frac{\partial s(t, \mathbf{A})}{\partial A_j}\,dt\right]. \tag{451}$$

We recall that this is a bound with respect to the correlation matrix $\mathbf{R}_\epsilon$ in the sense that

$$\mathbf{J}_T - \mathbf{R}_\epsilon^{-1} \tag{452}$$

is nonnegative definite. If the a posteriori density is Gaussian, $\mathbf{R}_\epsilon^{-1} = \mathbf{J}_T$.

A similar result is obtained for unbiased estimates of nonrandom variables by letting $\mathbf{J}_P = \mathbf{0}$. The conditions for the existence of an efficient estimate carry over directly. Equality will hold for the ith parameter if and only if

$$\hat{a}_i[r(t)] - A_i = \sum_{j=1}^{K} k_j(\mathbf{A})\int_{T_i}^{T_f} [r(t) - s(t, \mathbf{A})]\frac{\partial s(t, \mathbf{A})}{\partial A_j}\,dt. \tag{453}$$

To illustrate the application of this result we consider a simple example.

Example. Suppose we simultaneously amplitude- and frequency-modulate a sinusoid with two independent Gaussian parameters a, $N(0, \sigma_a)$, and b, $N(0, \sigma_b)$. Then

$$r(t) = s(t, A, B) + w(t) = \left(\frac{2E}{T}\right)^{1/2} B \sin(\omega_c t + \beta A t) + w(t); \qquad -\frac{T}{2} \le t \le \frac{T}{2}. \tag{454}$$

The likelihood function is

$$\ln \Lambda[r(t)|A, B] = \frac{1}{N_0} \int_{-T/2}^{T/2} \left[2r(t) - \left(\frac{2E}{T}\right)^{1/2} B \sin(\omega_c t + \beta A t)\right] \times \left(\frac{2E}{T}\right)^{1/2} B \sin(\omega_c t + \beta A t)\, dt. \tag{455}$$

Then

$$\frac{\partial s(t, A, B)}{\partial A} = \left(\frac{2E}{T}\right)^{1/2} B\, \beta t \cos(\omega_c t + \beta A t) \tag{456}$$

and

$$\frac{\partial s(t, A, B)}{\partial B} = \left(\frac{2E}{T}\right)^{1/2} \sin(\omega_c t + \beta A t). \tag{457}$$

Because the variables are independent, $\mathbf{J}_P$ is diagonal.

The elements of $\mathbf{J}_T$ are

$$\begin{aligned} J_{11} &= \frac{2}{N_0} E_{a,b} \int_{-T/2}^{T/2} \frac{2E}{T} B^2 \beta^2 t^2 \cos^2(\omega_c t + \beta A t)\, dt + \frac{1}{\sigma_a^2} \\ &\cong \sigma_b^2 \frac{T^2}{12} \frac{2E}{N_0} \beta^2 + \frac{1}{\sigma_a^2}, \end{aligned} \tag{458}$$

$$J_{22} = \frac{2}{N_0} E_{a,b} \int_{-T/2}^{T/2} \frac{2E}{T} \sin^2(\omega_c t + \beta A t) + \frac{1}{\sigma_b^2} \cong \frac{2E}{N_0} + \frac{1}{\sigma_b^2}, \tag{459}$$

and

$$\begin{aligned} J_{12} &= \frac{2}{N_0} E_{a,b} \left[\int_{-T/2}^{T/2} \frac{\partial s(t, A, B)}{\partial A} \cdot \frac{\partial s(t, A, B)}{\partial B}\, dt\right] \\ &= \frac{2}{N_0} E_{a,b} \left[\int_{-T/2}^{T/2} \frac{2E}{T} B \beta t \sin(\omega_c t + \beta A t) \cos(\omega_c t + \beta A t)\, dt\right] \cong 0. \end{aligned} \tag{460}$$

Thus the $\mathbf{J}$ matrix is diagonal. This means that

$$E[(\hat{a} - a)^2] \ge \left(\frac{1}{\sigma_a^2} + \sigma_b^2 \frac{T^2}{12} \frac{2E}{N_0} \beta^2\right)^{-1} \tag{461}$$

and

$$E[(\hat{b} - b)^2] \ge \left(\frac{1}{\sigma_b^2} + \frac{2E}{N_0}\right)^{-1}. \tag{462}$$

Thus we observe that the *bounds* on the estimates of a and b are uncorrelated. We can show that for large E/N_0 the actual variances approach these bounds.

We can interpret this result in the following way. If, each time the experiment was conducted, the receiver were *given* the value of b, the performance in estimating a would not be improved over the case in which the receiver was required to estimate b (assuming large E/N_0).

We observe that there are two ways in which $\mathbf{J}_{12}$ can be zero. If

$$\int_{-T/2}^{T/2} \frac{\partial s(t, A, B)}{\partial A} \frac{\partial s(t, A, B)}{\partial B}\, dt = 0 \tag{463}$$

before the expectation is taken, it means that for any value of A or B the partial derivatives are orthogonal. This is required for ML estimates to be uncoupled.

Even if the left side of (463) were not zero, however, the value *after* taking the expectation might be zero, which gives uncoupled MAP estimates.

Several interesting examples of multiple parameter estimation are included in the problems.

4.6.2 Extensions

The results can be modified in a straightforward manner to include other cases of interest.

1. Nonrandom variables, ML estimation.
2. Additive colored noise.
3. Random phase channels.
4. Rayleigh and Rician channels.
5. Multiple received signals.

Some of these cases are considered in the problems. One that will be used in the sequel is the additive colored noise case, discussed in Problem 4.6.7. The results are obtained by an obvious modification of (447) which is suggested by (226).

$$\hat{a}_i = \sigma_{a_i}^{\ 2} \int_{T_i}^{T_f} \frac{\partial s(z, \mathbf{A})}{\partial A_i}\bigg|_{\mathbf{A}=\hat{\mathbf{a}}_{\text{map}}} [r_g(z) - g(z)]\, dz, \qquad i = 1, 2, \dots, M, \quad (464)$$

where

$$r_g(z) - g(z) \triangleq \int_{T_i}^{T_f} Q_n(z, u)[r(u) - s(u, \hat{\mathbf{a}}_{\text{map}})]\, du, \qquad T_i \le z \le T_f. \quad (465)$$

4.7 SUMMARY AND OMISSIONS

4.7.1 Summary

In this chapter we have covered a wide range of problems. The central theme that related them was an additive Gaussian noise component. Using this theme as a starting point, we examined different types of problems and studied their solutions and the implications of these solutions. It turned out that the formal solution was the easiest part of the problem and that investigating the implications consumed most of our efforts. It is worthwhile to summarize some of the more general results.

The simplest detection problem was binary detection in the presence of white Gaussian noise. The optimum receiver could be realized as a matched filter or a correlation receiver. The performance depended only on the normalized distance between the two signal points in the decision space. This distance was characterized by the signal energies, their correlation coefficient, and the spectral height of the additive noise. For equal energy signals, a correlation coefficient of -1 was optimum. In all cases the signal shape was unimportant. The performance was insensitive to the detailed assumptions of the model.

The solution for the M signal problem followed easily. The receiver structure consisted of at most $M - 1$ matched filters or correlators. Except for a few special cases, performance calculations for arbitrary cost assignments and a priori probabilities were unwieldy. Therefore we devoted our attention to minimum probability of error decisions. For *arbitrary* signal sets the calculation of the probability of error was still tedious. For orthogonal and nonorthogonal equally-correlated signals simple expressions could be found and evaluated numerically. Simple bounds on the error probability were derived that were useful for certain ranges of parameter values. The question of the optimum signal set was discussed briefly in the text and in more detail in the problems. We found that for large M, orthogonal signals were essentially optimum.

The simple detection problem was then generalized by allowing a nonwhite additive Gaussian noise component. This generalization also included known linear channels. The formal extension by means of the whitening approach or a suitable set of observable coordinates was easy. As we examined the result, some issues developed that we had not encountered before. By including a nonzero white noise component we guaranteed that the matched filter would have a square-integrable impulse response and that perfect (or singular) detection would be impossible. The resulting test was stable, but its sensitivity depended on the white noise level. In the presence of a white noise component the performance could always be improved by extending the observation interval. In radar this was easy because of the relatively long time between successive pulses. Next we studied the effect of removing the white noise component. We saw that unless we put additional "smoothness" restrictions on the signal shape our mathematical model could lead us to singular and/or unstable tests.

The next degree of generalization was to allow for uncertainties in the signal even in the absence of noise. For the case in which these uncertainties could be parameterized by random variables with known densities, the desired procedure was clear. We considered in detail the random phase case and the random amplitude and phase case. In the random phase problem, we introduced the idea of a simple estimation system that measured the phase angle and used the measurement in the detector. This gave us a method of transition from the known signal case to situations, such as the radar problem, in which the phase is uniformly distributed. For binary signals we found that the optimum signal set depended on the quality of the phase measurement. As we expected, the optimum correlation coefficient ranged from $\rho = -1$ for perfect measurement to $\rho = 0$ for the uniform density.

The random amplitude and phase case enabled us to model a number of communication links that exhibited Rayleigh and Rician fading. Here we examined no-measurement receivers and perfect measurement receivers. We found that perfect measurement offered a 6-db improvement. However, even with perfect measurement, the channel fading caused the error probability to decrease linearly with $\bar{E}_r/N_0$ instead of exponentially as in a nonfading channel.

We next considered the problem of multiple channel systems. The vector Karhunen-Loève expansion enabled us to derive the likelihood ratio test easily. Except for a simple example, we postponed our discussion of vector systems to later chapters.

The *basic* ideas in the *estimation* problem were similar, and the entire formulation up through the likelihood function was identical. For linear estimation, the resulting receiver structures were identical to those obtained in the simple binary problem. The mean-square estimation error in white noise depended only on E/N_0.

The nonlinear estimation problem gave rise to a number of issues. The first difficulty was that a sufficient statistic did not exist, which meant that the mapping from the observation space to the estimation space depended on the parameter we were trying to estimate. In some cases this could be accommodated easily. In others approximate techniques were necessary. The resulting function in the estimation space had a number of local maxima and we had to choose the absolute maximum. Given that we were near the correct maximum, the mean-square error could be computed easily. The error could be reduced significantly over the linear estimation error by choosing a suitable signaling scheme. If we tried to reduce the error too far, however, a new phenomenon developed, which we termed threshold. In the cascade approximation to the optimum estimator the physical mechanism for the occurrence of a threshold was clear. The first stage chose the wrong interval in which to make its local estimate. In the continuous realization (such as range estimation) the occurrence was clear but a quantitative description was more difficult. Because the actual threshold level will depend on the signal structure, the quantitative results for the particular example discussed are less important than the realization that whenever we obtain an error decrease without an increase in signal energy or a decrease in noise level a threshold effect will occur at some signal-to-noise level.

The final problem of interest was multiple-parameter estimation. This served both to complete our discussion and as a starting point for the problem of waveform estimation. Here the useful results were relations that showed how estimation errors were coupled by the signal structure and the a priori densities.

In addition to summarizing what we have covered, it is equally important to point out some related issues that we have not.

4.7.2 TOPICS OMITTED

Digital Communications. We have done a great deal of the groundwork necessary for the study of modern digital systems. Except for a few cases, however, we have considered only single-digit transmission. (This is frequently referred to as the one-shot problem in the literature.) From the simple example in Section 4.2 it is clear that performance can be improved by transmitting and detecting blocks of digits. The study of efficient methods is one of the central problems of coding theory. Suitable references are given in [66] and [18]. This comment does not imply that all digital communication systems should employ coding, but it does imply that coding should always be considered as one of the possible alternatives in the over-all system design.

Non-Gaussian Interference. It is clear that in many applications the prime source of interference is non-Gaussian. Simple examples are man-made interference at lower frequencies, impulse noise, and galactic, solar, and atmospheric noise.

Our reason for the omission of non-Gaussian interferences is not because of a lack of interest in or appreciation of their importance. Neither is it because of our inability to solve a *particular* non-Gaussian problem. It is probable that if we can model or measure the pertinent statistics adequately a close-to-optimum receiver can be derived (e.g., [67], [68]). The reason is that it is too difficult to derive useful but general results.

Our goal with respect to the non-Gaussian problem is modest. First, it is to leave the user with an awareness that in any given situation we must verify that the Gaussian model is either valid or an adequate approximation to obtain useful results. Second, if the Gaussian model does not hold, we should be willing to try to solve the actual problem (even approximately) and not to retain the Gaussian solution because of its neatness.

In this chapter we have developed solutions for the problems of detection and finite parameter estimation. We now turn to waveform estimation.

4.8 PROBLEMS

The problems are divided according to sections in the text. Unless otherwise stated, all problems use the model from the corresponding section of the text; for example, the received signals are corrupted by additive zero-mean Gaussian noise which is independent of the hypotheses.

Section P4.2 Additive White Gaussian Noise

Binary Detection

Problem 4.2.1. Derive an expression for the probability of detection P_D, in terms of d and P_F, for the known signal in the additive white Gaussian noise detection problem. [see (37) and (38)].

Problem 4.2.2. In a binary FSK system one of two sinusoids of different frequencies is transmitted; for example,

$$\begin{aligned} s_1(t) &= f(t)\cos 2\pi f_c t, & 0 \le t \le T,\\ s_2(t) &= f(t)\cos 2\pi(f_c + \Delta f)t, & 0 \le t \le T, \end{aligned}$$

where $f_c \gg 1/T$ and Δf. The correlation coefficient is

$$\rho = \frac{\int_0^T f^2(t)\cos(2\pi\Delta f t)\,dt}{\int_0^T f^2(t)\,dt}.$$

The transmitted signal is corrupted by additive white Gaussian noise ($N_0/2$).

1. Evaluate ρ for a rectangular pulse; that is,

$$\begin{aligned} f(t) &= \left(\frac{2E}{T}\right)^{1/2}, & 0 \le t \le T,\\ &= 0, & \text{elsewhere.} \end{aligned}$$

Sketch the result as a function of $\Delta f T$.

2. Assume that we require $\Pr(\epsilon) = 0.01$. What value of E/N_0 is necessary to achieve this if $\Delta f = \infty$? Plot the *increase* in E/N_0 over this asymptotic value that is necessary to achieve the same $\Pr(\epsilon)$ as a function of $\Delta f T$.

Problem 4.2.3. The risk involved in an experiment is

$$\mathcal{R} = C_F P_F P_0 + C_M P_M P_1.$$

The applicable ROC is Fig. 2.9. You are given (a) $C_M = 2$; (b) $C_F = 1$; (c) P_1 may vary between 0 and 1. Sketch the line on the ROC that will minimize your maximum possible risk (i.e., assume P_1 is chosen to make $\mathcal{R}$ as *large* as possible. Your line should be a locus of the thresholds that will cause the maximum to be as small as possible).

Problem 4.2.4. Consider the linear feedback system shown below

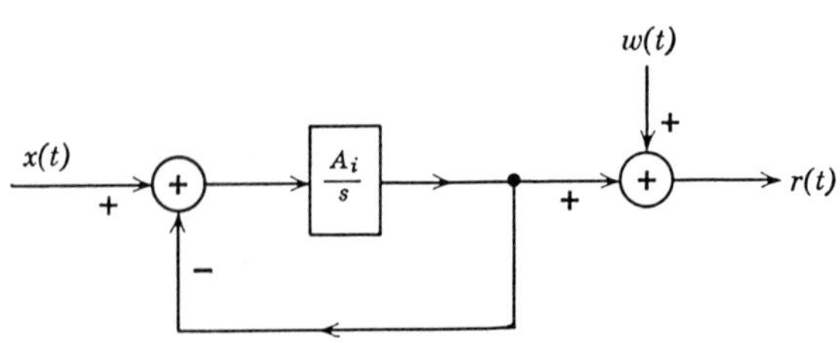

Fig. P4.1

The function $x(t)$ is a known deterministic function that is zero for $t < 0$. Under H_1, $A_i = A_1$. Under H_0, $A_i = A_0$. The noise $w(t)$ is a sample function from a white Gaussian process of spectral height $N_0/2$. We observe $r(t)$ over the interval $(0, T)$. All initial conditions in the feedback system are zero.

1. Find the likelihood ratio test.
2. Find an expression for P_D and P_F for the special case in which $x(t) = \delta(t)$ (an impulse) and $T = \infty$.

Problem 4.2.5. Three commonly used methods for transmitting binary signals over an additive Gaussian noise channel are on-off keying (ASK), frequency-shift keying (FSK), and phase-shift keying (PSK):

$$H_0: r(t) = s_0(t) + w(t), \qquad 0 \le t \le T,$$
$$H_1: r(t) = s_1(t) + w(t), \qquad 0 \le t \le T,$$

where $w(t)$ is a sample function from a white Gaussian process of spectral height $N_0/2$. The signals for the three cases are as follows:

	ASK	FSK	PSK
$s_0(t)$	0	$\sqrt{2E/T}\sin\omega_1 t$	$\sqrt{2E/T}\sin\omega_0 t$
$s_1(t)$	$\sqrt{2E/T}\sin\omega_1 t$	$\sqrt{2E/T}\sin\omega_0 t$	$\sqrt{2E/T}\sin(\omega_0 t + \pi)$

where $\omega_0 - \omega_1 = 2\pi n/T$ for some nonzero integer n and $w_0 = 2\pi m T$ for some nonzero integer m.

1. Draw appropriate signal spaces for the three techniques.
2. Find d^2 and the resulting probability of error for the three schemes (assume that the two hypotheses are equally likely).
3. Comment on the relative efficiency of the three schemes (a) with regard to utilization of transmitter energy, (b) with regard to ease of implementation.
4. Give an example in which the model of this problem does not accurately describe the actual physical situation.

Problem 4.2.6. Suboptimum Receivers. In this problem we investigate the degradation in performance that results from using a filter other than the optimum receiver filter. A reasonable performance comparison is the increase in transmitted energy required to overcome the decrease in d^2 that results from the mismatching. We would hope that for many practical cases the equipment simplification that results from using other than the matched filter is well worth the required increase in transmitted energy. The system of interest is shown in Fig. P4.2, in which

$$\int_0^T s^2(t)\,dt = 1 \quad \text{and} \quad E[w(t)\,w(\tau)] = \frac{N_0}{2}\,\delta(t - \tau).$$

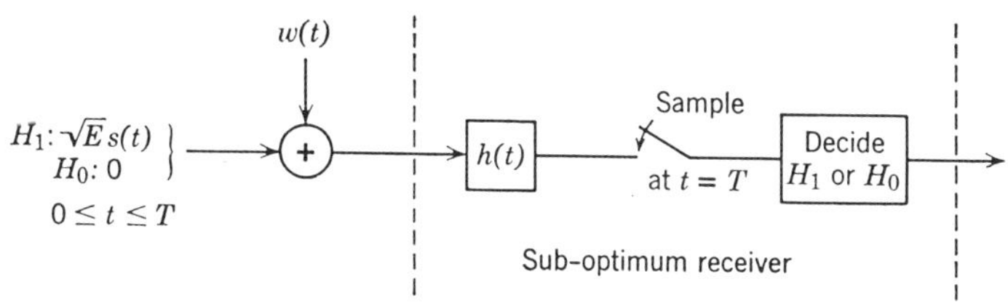

Fig. P4.2

The received waveform is

$$H_1: r(t) = \sqrt{E}\, s(t) + w(t), \qquad -\infty < t < \infty,$$
$$H_0: r(t) = \qquad\qquad w(t), \qquad -\infty < t < \infty.$$

We know that

$$h_{\text{opt}}(t) = \begin{cases} s(T-t), & 0 \le t \le T, \\ 0, & \text{elsewhere,} \end{cases}$$

and

$$d_{\text{opt}}^2 = \frac{2E}{N_0}.$$

Suppose that

$$h(t) = e^{-at} u_{-1}(t), \qquad -\infty < t < \infty,$$

$$s(t) = \begin{cases} \left(\dfrac{1}{T}\right)^{1/2}, & 0 \le t \le T, \\ 0, & \text{elsewhere.} \end{cases}$$

1. Choose the parameter a to maximize the output signal-to-noise ratio d^2.
2. Compute the resulting d^2 and compare with d_{opt}^2. How many decibels must the transmitter energy be increased to obtain the same performance?

M-ARY SIGNALS

Problem 4.2.7. Gram-Schmidt. In this problem we go through the details of the geometric representation of a set of M waveforms in terms of $N (N \le M)$ orthogonal signals.

Consider the M signals $s_1(t), \dots, s_M(t)$ which are either linearly independent or linearly dependent. If they are linearly dependent, we can write (by definition)

$$\sum_{i=1}^{M} a_i s_i(t) = 0.$$

1. Show that if M signals are linearly dependent, then $s_M(t)$ can be expressed in terms of $s_i(t): i = 1, \dots, M-1$.
2. Continue this procedure until you obtain N-linearly independent signals and M-N signals expressed in terms of them. N is called the *dimension* of the signal set.
3. Carry out the details of the Gram-Schmidt procedure described on p. 258.

Problem 4.2.8. Translation/Simplex Signals [18]. For maximum a posteriori reception the probability of error is not affected by a linear translation of the signals in the decision space; for example, the two decision spaces in Figs. P4.3*a* and P4.3*b* have the same Pr (ϵ). Clearly, the sets do not require the same energy. Denote the average energy in a signal set as

$$\bar{E} \triangleq \sum_{i=1}^{M} \Pr(H_i)\, |\mathbf{s}_i|^2 = \sum_{i=1}^{M} \Pr(H_i) E_i \int_0^T s_i^2(t)\, dt.$$

1. Find the linear translation that minimizes the average energy of the translated signal set; that is, minimize

$$\bar{E} \triangleq \sum_{i=1}^{M} \Pr(H_i) |\mathbf{s}_i - \mathbf{m}|^2.$$

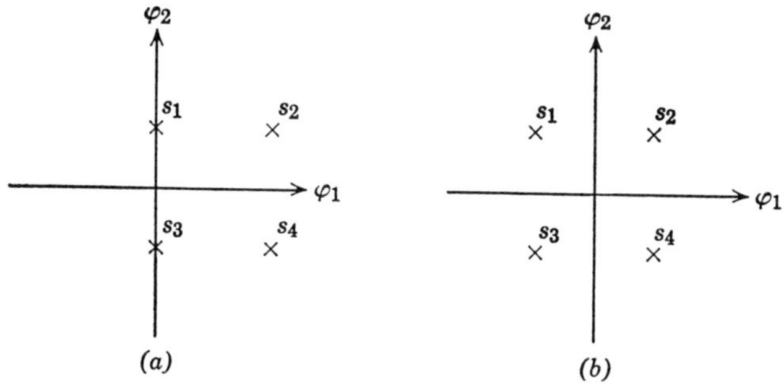

Fig. P4.3

2. Explain the geometric meaning of the result in part 1.

3. Apply the result in part 1 to the case of M orthogonal equal-energy signals representing equally likely hypotheses. The resulting signals are called *Simplex* signals. Sketch the signal vectors for $M = 2, 3, 4$.

3. What is the energy required to transmit each signal in the Simplex set?

4. Discuss the energy reduction obtained in going from the orthogonal set to the Simplex set while keeping the same Pr (ϵ).

Problem 4.2.9. Equally correlated signals. Consider M equally correlated signals

$$E\int_0^T s_i(t)s_j(t)\,dt = \begin{cases} E, & i = j, \\ \rho E, & i \neq j. \end{cases}$$

1. Prove

$$-\frac{1}{M-1} \leq \rho \leq 1.$$

2. Verify that the left inequality is given by a Simplex set.

3. Prove that an equally-correlated set with energy E has the same Pr (ϵ) as an orthogonal set with energy $E_{\text{orth}} = E(1 - \rho)$.

4. Express the Pr (ϵ) of the Simplex set in terms of the Pr (ϵ) for the orthogonal set and M.

Problem 4.2.10. M Signals, Arbitrary Correlation. Consider an M-ary system used to transmit equally likely messages. The signals have equal energy and may be correlated:

$$\rho_{ij} = \int_0^T s_i(t)\, s_j(t)\, dt, \qquad i, j = 1, 2, \ldots, M.$$

The channel adds white Gaussian noise with spectral height $N_0/2$. Thus

$$r(t) = \sqrt{E}\, s_i(t) + w(t), \qquad 0 \leq t \leq T : H_i, \qquad i = 1, \ldots, M.$$

1. Draw a block diagram of an optimum receiver containing M matched filters. What is the minimum number of matched filters that can be used?

2. Let $\boldsymbol{\rho}$ be the signal correlation matrix. The ij element is ρ_{ij}. If $\boldsymbol{\rho}$ is nonsingular, what is the dimension of the signal space?

3. Find an expression for Pr ($\epsilon|H_1$), the probability of error, assuming H_1 is true. Assume that $\boldsymbol{\rho}$ is nonsingular.

4. Find an expression for Pr (ϵ).

5. Is this error expression valid for Simplex signals? (Is $\boldsymbol{\rho}$ singular?)

Problem 4.2.11 (continuation). Error Probability [69]. In this problem we derive an alternate expression for the Pr (ϵ) for the system in Problem 4.2.10. The desired expression is

$$1 - \Pr(\epsilon) = \frac{1}{M} \exp\left(-\frac{E}{N_0}\right) \int_{-\infty}^{\infty} \exp\left[\left(\frac{2E}{N_0}\right)^{1/2} x\right] \times \left[\frac{d}{dx} \int_{-\infty}^{x} \cdots \int_{-\infty}^{x} \frac{\exp(-\frac{1}{2}\mathbf{y}^T \boldsymbol{\rho}^{-1} \mathbf{y})}{(2\pi)^{M/2} |\boldsymbol{\rho}|^{1/2}} d\mathbf{y}\right] dx. \quad \text{(P.1)}$$

Develop the following steps:

1. Rewrite the receiver in terms of M orthonormal functions $\phi_i(t)$. Define

$$s_i(t) = \sum_{k=1}^{M} s_{ik}\, \phi_k(t), \qquad i = 1, 2, \ldots, M,$$

$$r(t) = \sum_{k=1}^{\infty} r_k\, \phi_k(t).$$

Verify that the optimum receiver forms the statistics

$$l_i = \int_0^T r(t)\, s_i(t)\, dt = \sum_{k=1}^{M} s_{ik} R_k$$

and chooses the greatest.

2. Assume that $s_m(t)$ is transmitted. Show

$$\Pr(\bar{\epsilon}|m) \triangleq \Pr(\mathbf{R} \text{ in } Z_m) = \Pr\left(\sum_{k=1}^{M} s_{mk} R_k = \max_j \sum_{k=1}^{M} s_{jk} R_k\right).$$

3. Verify that

$$\Pr(\bar{\epsilon}) = \frac{1}{M} \exp\left(-\frac{E}{N_0}\right) \sum_{m=1}^{M} \int_{-\infty}^{\infty} \cdots \int_{-\infty}^{\infty} \frac{\exp[-(1/N_0) \sum_{k=1}^{M} R_k^2]}{(\pi N_0)^{M/2}} \times \exp\left(\frac{2}{N_0} \max_j \sum_{k=1}^{M} R_k s_{jk}\right). \quad \text{(P.2)}$$

4. Define

$$f(\mathbf{R}) = \exp\left\{\max_j \left[\left(\frac{2}{EN_0}\right)^{1/2} \sum_{k=1}^{M} s_{jk} R_k\right]\right\}$$

and observe that (P.2) can be viewed as the expectation of $f(\mathbf{R})$ over a set of statistically independent zero-mean Gaussian variables, R_k, with variance $N_0/2$. To evaluate this expectation, define

$$z_j \triangleq \left(\frac{2}{EN_0}\right)^{1/2} \sum_{k=1}^{M} s_{jk} R_k, \qquad j = 1, 2, \ldots, M,$$

and

$$\mathbf{z} = \begin{bmatrix} z_1 \\ z_2 \\ \vdots \\ z_M \end{bmatrix}.$$

Find $p_{\mathbf{z}}(\mathbf{Z})$. Define

$$x = \max_j z_j.$$

Find $p_x(x)$.

5. Using the results in (4), we have

$$1 - \Pr(\epsilon) = \frac{1}{M} \exp\left(-\frac{E}{N_0}\right) \int_{-\infty}^{\infty} \exp\left[\left(\frac{2E}{N_0}\right)^{1/2} X\right] p_x(X)\, dX.$$

Use $p_x(X)$ from (4) to obtain the desired result.

Problem 4.2.12 (continuation).

1. Using the expression in (P.1) of Problem 4.2.11, show that $\partial \Pr(\epsilon)/\partial \rho_{12} > 0$. Does your derivation still hold if $1 \to i$ and $2 \to j$?
2. Use the results of part 1 and Problem 4.2.9 to develop an intuitive argument that the Simplex set is locally optimum.

Comment. The proof of local optimality is contained in [70]. The proof of global optimality is contained in [71].

Problem 4.2.13. Consider the system in Problem 4.2.10. Define

$$\rho_{\max} = \max_{i \neq j} \rho_{ij}.$$

1. Prove that $\Pr(\epsilon)$ on any signal set is less than the $\Pr(\epsilon)$ for a set of equally correlated signals with correlation equal to $\rho_{\max}$.
2. Express this in terms of the error probability for a set of orthogonal signals.
3. Show that the $\Pr(\epsilon)$ is upper bounded by

$$\Pr(\epsilon) \leq (M-1)\left\{\operatorname{erfc}_*\left(\left[\frac{E}{N_0}(1-\rho_{\max})\right]^{1/2}\right)\right\}.$$

Problem 4.2.14 [72]. Consider the system in Problem 4.2.10. Define

d_i: distance between the ith message point and the nearest neighbor.

Observe

$$d_i = \min_j 2\sqrt{(1-\rho_{ij})E/N_0}$$

$$\bar{d} = \frac{1}{M}\sum_{i=1}^{M} d_i,$$

$$d_{\min} = \min_i d_i.$$

Prove

$$\operatorname{erfc}_*(\bar{d}) \leq \Pr(\epsilon) \leq (M-1)\operatorname{erfc}_*(d_{\min})$$

Note that this result extends to signals with unequal energies in an obvious manner.

Problem 4.2.15. In (68) of the text we used the limit

$$\lim_{M\to\infty} \frac{\ln \operatorname{erf}_*\left[y + \left(\frac{2PT\log_2 M}{N_0}\right)^{1/2}\right]}{1/(M-1)}.$$

Use l'Hospital's rule to verify the limits asserted in (69) and (70).

Problem 4.2.16. The error probability in (66) is the probability of error in deciding which signal was sent. Each signal corresponds to a sequence of digits; for example, if $M = 8$,

$000 \to s_0(t)$	$100 \to s_5(t)$
$001 \to s_1(t)$	$101 \to s_6(t)$
$010 \to s_2(t)$	$110 \to s_7(t)$
$011 \to s_3(t)$	$111 \to s_8(t)$.

Therefore an error in the signal decision does not necessarily mean that all digits will be in error. Frequently the digit (or bit) error probability [$\Pr_B(\epsilon)$] is the error of interest.

1. Verify that if an error is made any of the other $M - 1$ signals are equally likely to be chosen.
2. Verify that the expected number of bits in error, given a signal error is made, is

$$\left[\frac{\sum_{i=1}^{\log_2 M} i\binom{\log_2 M}{i}}{\sum_{i=1}^{\log_2 M}\binom{\log_2 M}{i}}\right] = \frac{(\log_2 M)M}{2(M-1)}.$$

3. Verify that the bit error probability is

$$\Pr_B(\epsilon) = \frac{M}{2(M-1)} \Pr(\epsilon).$$

4. Sketch the behavior of the bit error probability for $M = 2$, 4, and 8 (use Fig. 4.25).

Problem 4.2.17. Bi-orthogonal Signals. Prove that for a set of M bi-orthogonal signals with energy E and equally likely hypotheses the $\Pr(\epsilon)$ is

$$\Pr(\epsilon) = 1 - \int_0^{\infty} \frac{1}{\sqrt{\pi N_0}} \exp\left[-\frac{1}{N_0}(x - \sqrt{E})^2\right]\left[\int_{-x}^{x} \frac{1}{\sqrt{\pi N_0}} \exp\left(-\frac{y^2}{N_0}\right) dy\right]^{M/2-1} dx.$$

Verify that this $\Pr(\epsilon)$ approaches the error probability for orthogonal signals for large M and d^2. What is the advantage of the bi-orthogonal set?

Problem 4.2.18. Consider the following digital communication system. There are four equally probable hypotheses. The signals transmitted under the hypotheses are

$$\begin{aligned}
&H_1: \left(\frac{2}{T}\right)^{1/2} A \sin \omega_c t, && 0 \le t \le T,\\
&H_2: \frac{1}{3}\left(\frac{2}{T}\right)^{1/2} A \sin \omega_c t, && 0 \le t \le T, \qquad \omega_c = \frac{2\pi n}{T}\\
&H_3: -\frac{1}{3}\left(\frac{2}{T}\right)^{1/2} A \sin \omega_c t, && 0 \le t \le T,\\
&H_4: -\left(\frac{2}{T}\right)^{1/2} A \sin \omega_c t, && 0 \le t \le T.
\end{aligned}$$

The signal is corrupted by additive Gaussian white noise $w(t)$, $(N_0/2)$.

1. Draw a block diagram of the minimum probability of error receiver and the decision space and *compute* the resulting probability of error.
2. How does the probability of error behave for large A^2/N_0?

Problem 4.2.19. M-ary ASK [72]. An ASK system is used to transmit equally likely messages

$$s_i(t) = \sqrt{E_i}\,\phi(t), \qquad i = 1, 2, \ldots, M,$$

where

$$\sqrt{E_i} = (i-1)\Delta, \qquad \int_0^T \phi^2(t)\, dt = 1.$$

The received signal under the ith hypothesis is

$$r(t) = s_i(t) + w(t), \qquad 0 \le t \le T : H_i, \qquad i = 1, 2, \ldots, M,$$

where $w(t)$ is a white noise with spectral height $N_0/2$.

1. Draw a block diagram of the optimum receiver.
2. Draw the decision space and compute the Pr (ϵ).
3. What is the average transmitted energy?

Note. $\displaystyle\sum_{j=1}^{n-1} j^2 = \frac{(n-1)\, n(2n-1)}{6}.$

4. What translation of the signal set in the decision space would maintain the Pr (ϵ) while minimizing the average transmitted energy?

Problem 4.2.20 (continuation). Use the sequence transmission model on pp. 264–265 with the ASK system in part 4 of Problem 4.2.19. Consider specifically the case in which $M = 4$. How should the digit sequence be mapped into signals to minimize the bit error probability? Compute the signal error probability and the bit error probability.

Problem 4.2.21. M-ary PSK [72]. A communication system transmitter sends one of M messages over an additive white Gaussian noise channel (spectral height $N_0/2$) using the signals

$$s_i(t) = \begin{cases} \left(\dfrac{2E}{T}\right)^{1/2} \cos\left(2\pi \dfrac{n}{T} t + \dfrac{2\pi i}{M}\right), & 0 \le t \le T, \\ 0, & \text{elsewhere}, \qquad i = 0, 1, 2, \ldots, M-1, \end{cases}$$

where n is an integer. The messages are equally likely. This type of system is called an M-ary phase-shift-keyed (PSK) system.

1. Draw a block diagram of the optimum receiver. Use the minimum number of filters.
2. Draw the decision-space and decision lines for various M.
3. Prove

$$\alpha \le \Pr(\epsilon) \le 2\alpha,$$

where

$$\alpha = \operatorname{erfc}_* \left(\left(\frac{2E}{N_0}\right)^{1/2} \sin \frac{\pi}{M}\right).$$

Problem 4.2.22 (continuation). Optimum PSK [73]. The basic system is shown in Fig. 4.24. The possible signaling strategies are the following:

1. Use a binary PSK set with the energy in each signal equal to PT.
2. Use an M-ary PSK set with the energy in each signal equal to PT $\log_2$ M.

Discuss how you would choose M to minimize the digit error probability. Compare bi-phase and four phase PSK on this basis.

Problem 4.2.23 (continuation). In the context of an M-ary PSK system discuss qualitatively the effect of an incorrect phase reference. In other words, the nominal signal set is given in Problem 4.2.22 and the receiver is designed on that basis. The actual signal set, however, is

$$s_i(t) = \begin{cases} \left(\dfrac{2E}{T}\right)^{1/2} \cos\left(\dfrac{2\pi n}{T}t + \dfrac{2\pi i}{M} + \theta\right), & 0 \le t \le T, \quad i = 1, 2, \dots, M, \\ & n \text{ is an integer,} \\ 0, & \text{elsewhere,} \end{cases}$$

where θ is a random phase angle. How does the importance of a phase error change as M increases?

ESTIMATION

Problem 4.2.24. Bhattacharyya Bound. Let

$$r(t) = s(t, A) + w(t), \qquad 0 \le t \le T,$$

where $s(t, A)$ is differentiable k times with respect to A. The noise has spectral height $N_0/2$.

1. Extend the Bhattacharyya bound technique developed in Problem 2.4.23 to the waveform for the $n = 2$ case. Assume that A is nonrandom variable.
2. Repeat for the case in which A is a Gaussian random variable; $N(0, \sigma_a)$.
3. Extend the results in parts 1 and 2 to the case in which $n = 3$.

Problem 4.2.25. Consider the problem in Example 1 on p. 276. In addition to the unknown time of arrival, the pulse has an unknown amplitude. Thus

$$r(t) = b\, s(t - a) + w(t), \qquad -T \le t \le T,$$

where a is a uniformly distributed random variable (see Fig. 4.29*b*) and b is Gaussian, $N(0, \sigma_b)$.

Draw a block diagram of a receiver to generate the joint MAP estimates, $\hat{a}_{\text{map}}$ and $\hat{b}_{\text{map}}$.

Problem 4.2.26. The known signal $s(t)$, $0 \le t \le T$, is transmitted over a channel with unknown *nonnegative* gain A and additive Gaussian noise $n(t)$:

$$\int_0^T s^2(t)\, dt = E,$$

$$K_n(t, \tau) = \frac{N_0}{2}\, \delta(t - \tau).$$

1. What is the maximum likelihood estimate of A?
2. What is the bias in the estimate?
3. Is the estimate asymptotically unbiased?

Problem 4.2.27. Consider the stationary Poisson random process $x(t)$. A typical sample function is shown in Fig. P4.4.

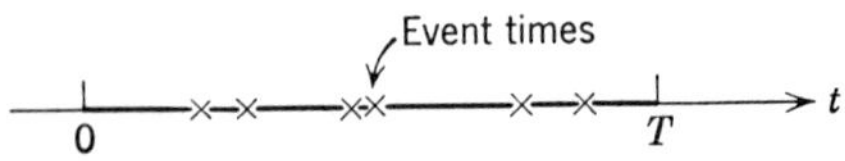

Fig. P4.4

The probability of n events in any interval τ is

$$\Pr(n, \tau) = \frac{(k\tau)^n}{n!} e^{-k\tau}.$$

The parameter k of the process is an unknown nonrandom variable which we want to estimate. We observe $x(t)$ over an interval $(0, T)$.

1. Is it necessary to record the event times or is it adequate to count the number of events that occur in the interval? Prove that n^*, the number of events that occur in the interval $(0, T)$ is a sufficient statistic.
2. Find the Cramér-Rao inequality for any unbiased estimate of k.
3. Find the maximum-likelihood estimate of k. Call this estimate $\hat{k}$.
4. Prove that $\hat{k}$ is unbiased.
5. Find

$$\text{Var}(\hat{k} - k).$$

6. Is the maximum-likelihood estimate efficient?

Problem 4.2.28. When a signal is transmitted through a particular medium, the amplitude of the output is inversely proportional to the murkiness of the medium. Before observation the output of the medium is corrupted by additive, white Gaussian noise. (Spectral height $N_0/2$, double-sided.) Thus

$$r(t) = \frac{1}{M} f(t) + w(t), \qquad 0 \leq t \leq T,$$

where $f(t)$ is a known signal and

$$\int_0^T f^2(t)\, dt = E.$$

We want to design an optimum Murky-Meter.

1. Assume that M is a nonrandom variable. *Derive* the block diagram of a system whose output is the maximum-likelihood estimate of M (denoted by $\hat{m}_{ml}$).
2. Now assume that M is a Gaussian random variable with zero mean and variance σ_M^2. Find the *equation* that specifies the maximum a posteriori estimate of M (denoted by $\hat{m}_{\text{map}}$).
3. Show that

$$\hat{m}_{\text{map}} \to \hat{m}_{ml}$$

as

$$\sigma_M^2 \to \infty.$$

Section 4.3 Nonwhite Additive Gaussian Noise

Mathematical Preliminaries

Problem 4.3.1. Reversibility. Prove that $h_w(t, u)$ [defined in (157)] is a reversible operation by demonstrating an $h_w^{-1}(t, u)$ such that

$$\int_{T_i}^{T_f} h_w(t, u)\, h_w^{-1}(u, z)\, du = \delta(t - z).$$

What restrictions on the noise are needed?

Problem 4.3.2. We saw in (163) that the integral equation

$$\frac{N_0}{2} h_o(z, v) + \int_{T_i}^{T_f} h_o(v, x) K_c(x, z)\, dx = K_c(z, v), \qquad T_i \le z, v \le T_f,$$

specifies the inverse kernel

$$Q_n(t, \tau) = \frac{2}{N_0} [\delta(t - \tau) - h_o(t, \tau)].$$

Show that an equivalent equation is

$$\frac{N_0}{2} h_o(z, v) + \int_{T_i}^{T_f} h_o(z, v) K_c(x, v)\, dx = K_c(z, v), \qquad T_i \le z, v \le T_f.$$

Problem 4.3.3 [74] We saw in Problem 4.3.2 that the inverse kernel $Q_n(t, \tau)$ can be obtained from the solution to an integral equation:

$$\frac{N_0}{2} h_o(t, \tau) + \int_{T_i}^{T_f} h_o(t, u) K_c(u, \tau)\, du = K_c(t, \tau), \qquad T_i \le t, \tau \le T_f,$$

where

$$Q_n(t, \tau) = \frac{2}{N_0} [\delta(t - \tau) - h_o(t, \tau)].$$

Suppose we let T_f, the end point of the interval, be a variable. We indicate this by writing $h_o(t, \tau{:}T_f)$ instead of $h_o(t, \tau)$:

$$\frac{N_0}{2} h_o(t, \tau{:}T_f) + \int_{T_i}^{T_f} h_o(t, u{:}T_f) K_c(u, \tau)\, du = K_c(t, \tau), \qquad T_i \le t, \tau \le T_f.$$

Now differentiate this equation with respect to T_f and show that

$$\frac{\partial h_o(t, \tau{:}T_f)}{\partial T_f} = -h_o(t, T_f{:}T_f)\, h_o(T_f, \tau{:}T_f).$$

Hint.

$$\int_{T_i}^{T_f} f(\tau) K_c(t, \tau) d\tau = \lambda f(t), \qquad T_i \le t \le T_f$$

has no solution for $\lambda < 0$.

Problem 4.3.4. Realizable Whitening Filters [91] In the text, two equivalent realizations of the optimum receiver for the colored noise problem were given in Figs. 4.38*a* and *b*. We also saw that $Q_n(t, u)$ was an unrealizable filter specified by (162) and (163). Furthermore, we found one solution for $h_w(t, \tau)$, the whitening filter, in terms of eigenfunctions that was an unrealizable filter. We want to investigate the possibility of finding a *realizable* whitening filter. Recall that we were able to do so in the simple example on p. 311.

1. Write down the log-likelihood ratio in terms of $h_o(t, \tau) = h_o(t, \tau{:}T_f)$ (see Problem 4.3.3).
2. Write

$$\ln \Lambda(r(t)) = \int_{T_i}^{T_f} dt \left[\int_{T_i}^{T_f} h_{wr}(t, u)\, s(u)\, du\right]^2 \triangleq L(T_f) = \int_{T_i}^{T_f} \frac{\partial L(t)}{\partial t}\, dt.$$

The additional subscript r denotes realizable.

3. Use the result from Problem 4.3.3 that

$$\frac{\partial h_o(u, v:t)}{\partial t} = -h_o(t, u:t)\, h_o(t, v:t)$$

to show that

$$h_{wr}(t, u) = \left(\frac{2}{N_0}\right)^{1/2} [\delta(t - \tau) - h_o(t, \tau:t)].$$

Observe that $h_o(t, \tau:t)$ is a *realizable* filter.

4. Write down the integral equation satisfied by $h_o(t, \tau:t)$. In Chapter 6 we discuss techniques for solving this equation.

Problem 4.3.5. M-ary Signals, Colored Noise. Let the received signal on the ith hypothesis be

$$r(t) = \sqrt{E_i}\, s_i(t) + n_c(t) + w(t), \qquad T_i \le t \le T_f : H_i, \qquad i = 1, 2, \dots, M,$$

where $w(t)$ is zero-mean white Gaussian noise with spectral height $N_0/2$ and $n_c(t)$ is independent zero-mean colored noise with covariance function $K_c(t, u)$. The signals $s_i(t)$ are normalized over $(0, T)$ and are zero outside that interval. Assume that the hypotheses are equally likely and that the criterion is minimum Pr (ϵ). Draw a block diagram of the optimum receiver.

Estimation

Problem 4.3.6. Consider the following estimation problem:

$$r(t) = A\, s(t) + \sum_{i=1}^{3} b_i s_i(t) + w(t), \qquad 0 \le t \le T,$$

where A is a nonrandom variable, b_i are independent, zero-mean, Gaussian random variables $[E(b_i^2) = \sigma_i^2]$, $w(t)$ is white noise $(N_0/2)$, $s(t) = \sum_{i=1}^{3} c_i s_i(t)$,

$$\int_0^T s_i(t)\, s_j(t)\, dt = \delta_{ij}, \quad \text{and} \quad \int_0^T = s^2(t)\, dt = 1.$$

1. Draw a block diagram of the maximum-likelihood estimator of A, $\hat{a}_{ml}$.
2. Choose c_1, c_2, c_3 to minimize the variance of the estimate.

Integral Equation Solutions

Problem 4.3.7. In this problem we solve a simple Fredholm equation of the second kind,

$$\sqrt{E}\, s(t) = \frac{N_0}{2} g(t) + \int_{T_i}^{T_f} K_c(t, u)\, g(u)\, du, \qquad T_i \le t \le T_f,$$

where

$$K_c(t, u) = \sigma_c^2 \exp[-k|t - u|],$$

$$s(t) = \frac{1}{\sqrt{T}}, \qquad 0 \le t \le T,$$

$$T_i = 0,$$
$$T_f = T.$$

1. Find $g(t)$.
2. Evaluate the performance index d^2.

Problem 4.3.8 (continuation). Solve Problem 4.3.7 for the case in which $T_i = -\infty$ and $T_f = \infty$. Compare the value of d^2 that you obtain with the value obtained in that problem.

Problem 4.3.9. Solve the Fredholm equation of the first kind,

$$\int_0^T K(t, u)\, g(u) = s(t), \qquad 0 \le t \le T,$$

for the triangular kernel

$$K(t, u) = \begin{cases} 1 - |t - u| & \text{for } |t - u| < 1, \\ 0, & \text{elsewhere.} \end{cases}$$

Assume that $s(t)$ is twice differentiable and that $T < 1$.

Now apply this result to the problem of detecting a known signal $s(t)$, $0 \le t \le T$, which is observed in additive Gaussian noise with covariance

$$K_n(t, u) = \begin{cases} 1 - |t - u| & \text{for } |t - u| \le 1, \\ 0, & \text{elsewhere.} \end{cases}$$

1. What is the optimum receiver? Note that we cannot physically generate impulses so that correlation with $g(t)$ is not a satisfactory answer.
2. Calculate d^2.

What is a necessary and sufficient condition for singular detection in this problem if $s(t)$ is bounded?

Problem 4.3.10.

1. Evaluate d^2 for the example given on p. 318.
2. Provided that $s(t)$ is bounded and has finite energy, what is a necessary and sufficient condition on $s(t)$ for a nonsingular test?

Problem 4.3.11 (continuation). The optimum receiver for Problem 4.3.10 includes a matched filter plus a sampler. Find d^2 for the suboptimum receiver that has the matched filter but not the sampler.

Problem 4.3.12. The opposition is using a binary communication system to transmit data. The two signals used are the following:

$$s_1(t) = \sin^2 \frac{2\pi}{T}\, t, \qquad 0 \le t \le T,$$

$$s_0(t) = -\sin^2 \frac{2\pi}{T}, \qquad 0 \le t \le T.$$

The received signal is either

$$\begin{aligned} H_1\!: r(t) &= s_1(t) + n(t), \qquad 0 \le t \le T, \\ H_0\!: r(t) &= s_0(t) + n(t), \qquad 0 \le t \le T, \end{aligned}$$

where $n(t)$ is a sample function from a zero-mean Gaussian random process with covariance function

$$K_n(\tau) = e^{-\alpha|\tau|}.$$

Assume that he knows α and builds a min Pr (ϵ) receiver. Choose α to minimize his performance.

SENSITIVITY AND SINGULARITY

Problem 4.3.13. Singularity [2]. Consider the simple binary detection problem shown in Fig. P4.5. On H_1 the transmitted signal is a finite energy signal $x(t)$. On H_0 there is no transmitted signal. The additive noise $w(t)$ is a sample function from a white process unity. The received waveform $r(t)$ is passed through a filter whose transfer function is $H(j\omega)$. The output $y(t)$, $0 \le t \le T$ is the signal available for processing. Let λ_k and $\phi_k(t)$ be the eigenvalues and eigenfunctions, respectively, of $n(t)$, $0 \le t \le T$. To have singular detection, we require

$$\sum_{k=1}^{\infty} \frac{s_k{}^2}{\lambda_k} = \infty.$$

We want to prove that this cannot happen in this case.

1. From

$$s_k = \int_0^T \phi_k(t)\, s(t)\, dt$$

show that

$$s_k = \int_{-\infty}^{\infty} X(f)\, H(j2\pi f)\, \Phi_k^*(f)\, df,$$

where

$$X(f) = \int_{-\infty}^{\infty} x(t) e^{-j2\pi f t}\, dt$$

and

$$\Phi_k(f) = \int_{-\infty}^{\infty} \phi_k(t) e^{-j2\pi f t}\, dt = \int_0^T \phi_k(t) e^{-j2\pi f t}\, dt.$$

2. Show that

$$\int_{-\infty}^{\infty} [H^*(j2\pi f)\, \Phi_m(f)][H(j2\pi f)\, \Phi_k^*(f)]\, df = \begin{cases} \lambda_k & \text{if } m = k, \\ 0 & \text{if } m \neq k. \end{cases}$$

3. Observe from part 2 that for some set of numbers c_k

$$X(f) = \sum_{k=1}^{\infty} \frac{c_k}{\lambda_k} H^*(j2\pi f)\, \Phi_k(f) + U(f),$$

where

$$\int_{-\infty}^{\infty} U(f)\, H(j2\pi f)\, \Phi_k^*(f)\, df = 0.$$

4. Using (1), (2), and (3), show that

$$\sum_{k=1}^{\infty} \frac{s_k{}^2}{\lambda_k} \le \int_{-\infty}^{\infty} x^2(t)\, dt,$$

hence that perfect detection is impossible in this situation.

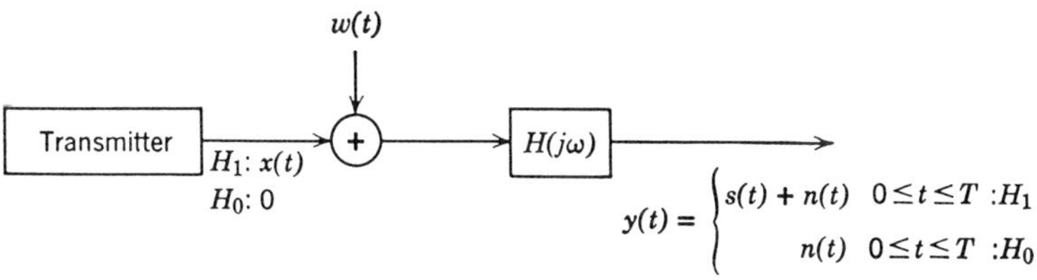

Fig. P4.5

Problem 4.3.14. Consider the following system:

$$H_1: r(t) = s_1(t) + n(t), \qquad 0 \le t \le T,$$
$$H_0: r(t) = n(t), \qquad 0 \le t \le T.$$

It is given that

$$n(t) = \sum_{n=1}^{6} a_n \cos n \frac{2\pi}{T} t,$$

where a_n are zero-mean random variables. The signal energy is

$$\int_0^T s_1^2(t)\, dt = E.$$

Choose $s_1(t)$ and the corresponding receiver so that perfect decisions can be made with probability 1.

Problem 4.3.15. Because white noise is a mathematical fiction (it has infinite energy, which is physically impossible), we sometimes talk about band-limited white noise; that is,

$$S_n(\omega) = \begin{cases} \dfrac{N_0}{2}, & \text{for } \omega_1 \le |\omega| \le \omega_2, \\ 0, & \text{otherwise.} \end{cases}$$

Now suppose that we wish to detect a strictly time-limited signal

$$s(t) = \begin{matrix} \left(\dfrac{E}{T}\right)^{1/2}, & 0 \le t \le T, \\ 0, & \text{otherwise.} \end{matrix}$$

Is this a good mathematical model for a physical problem? Justify your answer.

Problem 4.3.16. Sensitivity to White Noise Level. The received waveforms under the two hypotheses are

$$r(t) = s(t) + n_c(t) + w(t), \qquad -\infty < t < \infty : H_1,$$
$$r(t) = n_c(t) + w(t), \qquad -\infty < t < \infty : H_0.$$

The signal waveform $s(t)$ and the colored noise spectrum $S_n(\omega)$ are known exactly. The white noise level is

$$\frac{N_a}{2} = \frac{N_0}{2}(1 + x),$$

where $N_0/2$ is the nominal value and x is a small variation. Assume that the receiver is designed on the basis of the nominal white noise level.

1. Find an expression for $\left.\dfrac{\partial d/\partial x}{d}\right|_{x=0} = \left.\dfrac{\partial \ln d}{\partial x}\right|_{x=0} \triangleq \Delta.$

2. Assume that

$$s(t) = \begin{matrix} \sqrt{2kP}\, e^{-kt}, & t \ge 0, \\ 0, & t < 0 \end{matrix}$$

and

$$S_{n_c}(\omega) = \frac{2k\sigma_c^2}{\omega^2 + k^2}.$$

Evaluate Δ as a function of $\Lambda \triangleq 4\sigma_c^2/kN_0$.

Problem 4.3.17. Sensitivity to Noise Spectrum. Assume the same nominal model as in Problem 4.3.16.

1. Now let

$$\frac{N_a}{2} = \frac{N_0}{2}$$

and

$$S_{n_c}(\omega) = \frac{2k_a\sigma_a^2}{\omega^2 + k_a^2},$$

where

$$k_a = k\,(1 + y)$$
$$\sigma_a^2 = \sigma_c^2\,(1 + z)$$

Find

$$\left.\frac{\partial d/\partial y}{d}\right|_{\substack{z=0\\y=0}} \triangleq \Delta_y \quad \text{and} \quad \left.\frac{\partial d/\partial z}{d}\right|_{\substack{z=0\\y=0}} \triangleq \Delta_z.$$

2. Evaluate Δ_y and Δ_z for the signal shape in Problem 4.3.16.

Problem 4.3.18. Sensitivity to Delay and Gain. The received waveforms under the two hypotheses are

$$r(t) = \sqrt{E}\,s(t) + b_I s(t - \tau) + w(t), \qquad -\infty < t < \infty : H_1,$$
$$r(t) = b_I s(t - \tau) + w(t), \qquad -\infty < t < \infty : H_0,$$

where b_I is $N(0, \sigma_I)$ and $w(t)$ is white with spectral height $N_0/2$. The signal is

$$s(t) = \left(\frac{1}{T}\right)^{1/2}, \qquad 0 \le t \le T,$$
$$= 0, \qquad \text{elsewhere.}$$

1. Design the optimum receiver, assuming that τ is known.
2. Evaluate d^2 as a function of τ and σ_I.
3. Now assume

$$\tau_a = \tau\,(1 + x).$$

Find an expression for d^2 of the nominal receiver as a function of x. Discuss the implications of your results.

4. Now we want to study the effect of changing σ_I. Let

$$\sigma_{Ia}^2 = \sigma_I^2\,(1 + y)$$

and find an expression for d^2 as a function of y.

Linear Channels

Problem 4.3.19. Optimum Signals. Consider the system shown in Fig. 4.49*a*. Assume that the channel is time-invariant with impulse response $h(\tau)$ [or transfer function $H(f)$]. Let

$$H(f) = 1, \qquad |f| < W,$$
$$= 0, \qquad \text{otherwise.}$$

The output observation interval is infinite. The signal input is $s(t)$, $0 \le t \le T$ and is normalized to have unity energy. The additive white Gaussian noise has spectral height $N_0/2$.

1. Find the optimum receiver.
2. Choose $s(t)$, $0 \le t \le T$, to maximize d^2.

Problem 4.3.20. Repeat Problem 4.3.19 for the $h(t, \tau)$ given below:

$$h(t, \tau) = \delta(t - \tau), \qquad 0 \le \tau \le \frac{T}{4}, \frac{T}{2} \le \tau \le \frac{3T}{4}, \ -\infty < t < \infty,$$
$$= 0, \qquad \text{elsewhere.}$$

Problem 4.3.21. The system of interest is shown in Fig. P4.6. Design an optimum *binary* signaling system subject to the constraints:

1. $\int_0^T s^2(t)\, dt = E_t.$

2. $s(t) = 0, \qquad t < 0,$
   $\qquad t < T.$

3. $h(\tau) = e^{-k\tau}, \qquad \tau \ge 0,$
   $= 0, \qquad \tau < 0.$

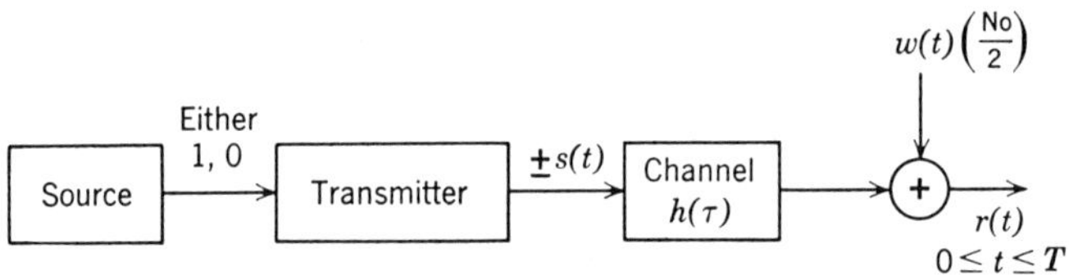

Fig. P4.6

Section P.4.4 Signals with Unwanted Parameters

MATHEMATICAL PRELIMINARIES

Formulas. Some of the problems in this section require the manipulation of Bessel functions and Q functions. A few convenient formulas are listed below. Other relations can be found in [75] and the appendices of [47] and [92]

I. ***Modified Bessel Functions***

$$I_n(z) \triangleq \frac{1}{2\pi} \int_0^{2\pi} \exp(\pm jn\theta) \exp(z \cos\theta)\, d\theta, \tag{F.1.1}$$

$$I_n(z) = I_{-n}(z), \tag{F.1.2}$$

$$I_v(z) \simeq \frac{(\frac{1}{2}z)^v}{\Gamma(v + 1)}, \qquad v \ne -1, -2, \ldots, z \ll 1, \tag{F.1.3}$$

$$I_v(z) \simeq \frac{e^z}{\sqrt{2\pi z}} \left[1 - \frac{4v^2 - 1}{8z}\right], \qquad z \gg 1, \tag{F.1.4}$$

$$\frac{1}{z^k} \frac{d^k}{dz^k} (z^{-v} I_v(z)) = z^{-v-k} I_{v+k}(z), \tag{F.1.5}$$

$$\frac{1}{z^k} \frac{d^k}{dz^k} (z^{v} I_v(z)) = z^{v-k} I_{v-k}(z). \tag{F.1.6}$$

II. ***Marcum's Q-function*** [92]

$$Q(\sqrt{2a}, \sqrt{2b}) = \int_b^\infty \exp(-a + x)\, I_0(2\sqrt{ax})\, dx, \tag{F.2.1}$$

$$Q(a, a) = \tfrac{1}{2}[1 + I_0(a^2) \exp(-a^2)], \tag{F.2.2}$$

$$1 + Q(a, b) - Q(b, a)$$
$$= \frac{b^2 - a^2}{b^2 + a^2} \int_{a^2+b^2/2}^{\infty} \exp(-x) I_0\left(\frac{2abx}{a^2+b^2}\right) dx, \qquad b > a > 0, \tag{F.2.3}$$

$$\int_0^\infty Q\left(\frac{\alpha_2}{\sigma_2}, \frac{R_1}{\sigma_2}\right) \frac{R_1}{\sigma_1^2} \exp\left[-\frac{\alpha_1^2 + R_1^2}{2\sigma_1^2}\right] I_0\left(\frac{\alpha_1 R_1}{\sigma_1^2}\right) dR_1$$
$$= \frac{\sigma_1^2}{\sigma_1^2 + \sigma_2^2}\left[1 - Q\left(\sqrt{\frac{\alpha_1^2}{\sigma_1^2 + \sigma_2^2}}, \sqrt{\frac{\alpha_2^2}{\sigma_1^2 + \sigma_2^2}}\right)\right]$$
$$+ \frac{\sigma_2^2}{\sigma_1^2 + \sigma_2^2} Q\left(\sqrt{\frac{\alpha_2^2}{\sigma_1^2 + \sigma_2^2}}, \sqrt{\frac{\alpha_1^2}{\sigma_1^2 + \sigma_2^2}}\right), \tag{F.2.4}$$

$$Q(a, b) \simeq \operatorname{erfc}_*(b - a), \qquad b \gg 1, b \gg b - a. \tag{F.2.5}$$

III. ***Rician variables*** [76]

Consider the two statistically independent Rician variables, x_1 and x_2 with probability densities,

$$p_{x_k}(X_k) = \frac{X_k}{\sigma_k^2} \exp\left(-\frac{a_k^2 + X_k^2}{2\sigma_k^2}\right) I_0\left(\frac{a_k X_k}{\sigma_k^2}\right), \qquad \begin{array}{l} 0 < a_k < \infty, \\ 0 < X_k < \infty, \\ k = 1, 2. \end{array} \tag{F.3.1}$$

The probability of interest is

$$P_* = \Pr[x_2 > x_1].$$

Define the constants

$$a = \frac{a_2^2}{\sigma_1^2 + \sigma_2^2}, \qquad b = \frac{a_1^2}{\sigma_1^2 + \sigma_2^2}, \qquad c = \frac{\sigma_1}{\sigma_2}.$$

Then

$$P_* = Q(\sqrt{a}, \sqrt{b}) - \frac{c^2}{1 + c^2} \exp\left(-\frac{a+b}{2}\right) I_0(\sqrt{ab}), \tag{F.3.2}$$

or

$$P_* = \frac{c^2}{1 + c^2}[1 - Q(\sqrt{b}, \sqrt{a})] + \frac{1}{1 + c^2} Q(\sqrt{a}, \sqrt{b}), \tag{F.3.3}$$

or

$$P_* = \tfrac{1}{2}[1 - Q(\sqrt{b}, \sqrt{a}) + Q(\sqrt{a}, \sqrt{b})] - \frac{1}{2}\frac{c^2 - 1}{c^2 + 1} \exp\left(-\frac{a+b}{2}\right) I_0(\sqrt{ab}). \tag{F.3.4}$$

Problem 4.4.1. Q-function Properties. Marcum's Q-function appears frequently in the calculation of error probabilities:

$$Q(\alpha, \beta) = \int_\beta^\infty x \exp\left[-\tfrac{1}{2}(x^2 + \alpha^2)\right] I_0(\alpha x)\, dx.$$

Verify the following properties:

1. $Q(\alpha, 0) = 1,$
2. $Q(0, \beta) = e^{-\beta^2/2}.$
3. $Q(\alpha, \beta) = e^{-(\alpha^2+\beta^2)/2} \sum_{n=0}^{\infty} \left(\frac{\alpha}{\beta}\right)^n I_n(\alpha\beta), \qquad \alpha < \beta,$

$$= 1 - e^{-(\alpha^2+\beta^2)/2} \sum_{n=1}^{\infty} \left(\frac{\beta}{\alpha}\right)^n I_n(\alpha\beta), \qquad \beta < \alpha.$$

4. $Q(\alpha, \beta) + Q(\beta, \alpha) = 1 + (e^{-(\alpha^2+\beta^2)/2})\, I_0(\alpha\beta)$.

5. $Q(\alpha, \beta) \simeq 1 - \dfrac{1}{\alpha - \beta}\left(\dfrac{\beta}{2\pi\alpha}\right)^{1/2}(e^{-(\alpha-\beta)^2/2}), \qquad \alpha \gg \beta \gg 1.$

6. $Q(\alpha, \beta) \simeq \dfrac{1}{\beta - \alpha}\left(\dfrac{\beta}{2\pi\alpha}\right)^{1/2}(e^{-(\beta-\alpha)^2/2}), \qquad \beta \gg \alpha \gg 1.$

Problem 4.4.2. Let x be a Gaussian random variable $N(m_x, \sigma_x)$.

1. Prove that

$$M_{x^2}(jv) \triangleq E[\exp(+jvx^2)] = \frac{\exp[jvm_x^2/(1 - 2jv\sigma_x^2)]}{(1 - 2jv\sigma_x^2)^{1/2}}.$$

Hint.

$$M_{x^2}(jv) = [M_{x^2}(jv)\, M_{y^2}(jv)]^{1/2},$$

where y is an independent Gaussian random variable with identical statistics.

2. Let z be a complex number. Modify the derivation in part 1 to show that

$$E[\exp(+zx^2)] = \frac{\exp[zm_x^2/(1 - 2z\sigma_x^2)]}{(1 - 2z\sigma_x^2)^{1/2}}, \qquad \text{Re}(z) < \frac{1}{2\sigma_x^2}.$$

3. Let

$$y^2 = \sum_{i=1}^{2M} \lambda_i x_i^2,$$

where the x_i are statistically independent Gaussian variables, $N(m_i, \sigma_i)$.

Find $M_{y^2}(jv)$ and $E[\exp(+zy^2)]$. What condition must be imposed on Re (z) in order for the latter expectation to exist.

4. Consider the special case in which $\lambda_i = 1$ and $\sigma_i^2 = \sigma^2$. Verify that the probability density of y^2 is

$$p_{y^2}(Y) = \frac{1}{2\sigma^2}\left(\frac{Y}{S\sigma^2}\right)^{\frac{M-1}{2}} \exp\left(-\frac{Y + S\sigma^2}{2\sigma^2}\right) I_{M-1}\left[\left(\frac{YS}{\sigma^2}\right)^{1/2}\right], \qquad Y \geq 0,$$
$$= 0, \qquad \text{elsewhere},$$

where $S = \sum_{i=1}^{2M} m_i^2$. (See Erdelyi [75], p. 197, eq. 18.)

Problem 4.4.3. Let $Q(\mathbf{x})$ be a quadratic form of correlated Gaussian random variables,

$$Q(\mathbf{x}) \triangleq \mathbf{x}^T\mathbf{A}\mathbf{x}.$$

1. Show that the characteristic function of Q is

$$M_Q(jv) \triangleq E(e^{jvQ}) = \frac{\exp\{-\frac{1}{2}\mathbf{m_x}^T\mathbf{\Lambda}^{-1}[\mathbf{I} - (\mathbf{I} - 2jv\mathbf{\Lambda}\mathbf{A})^{-1}]\mathbf{m_x}\}}{|\mathbf{I} - 2jv\mathbf{\Lambda}\mathbf{A}|^{1/2}}.$$

2. Consider the special case in which $\mathbf{\Lambda}^{-1} = \mathbf{A}$ and $\mathbf{m_x} = \mathbf{0}$. What is the resulting density?

3. Extend the result in part 1 to find $E(e^{zQ})$, where z is a complex number. What restrictions must be put on Re (z)?

Problem 4.4.4. [76] Let x_1, x_2, x_3, x_4 be statistically independent, identically distributed Gaussian random variables. Prove

$$\Pr(x_1^2 + x_2^2 \geq x_3^2 + x_4^2) = \tfrac{1}{2}[1 - Q(\beta, \alpha) + Q(\alpha, \beta)],$$

where

$$\alpha = \left(\frac{\overline{x_1}^2 + \overline{x_2}^2}{2\sigma^2}\right)^{1/2}, \quad \beta = \left(\frac{\overline{x_3}^2 + \overline{x_4}^2}{2\sigma^2}\right)^{1/2}.$$

RANDOM PHASE CHANNELS

Problem. 4.4.5. On-Off Signaling: Partially Coherent Channel. Consider the hypothesis testing problem stated in (357) and (358) with the probability density given by (364). From (371) we see that an equivalent test statistic is

$$(\beta + L_c)^2 + L_s^2 \underset{H_0}{\overset{H_1}{\gtrless}} \gamma,$$

where

$$\beta \triangleq \frac{N_0}{2} \frac{\Lambda_m}{\sqrt{E_r}}.$$

1. Express P_F as a Q-function.
2. Express P_D as an integral of a Q-function.

Problem 4.4.6. M-orthogonal Signals: Partially Coherent Channel. Assume that each of the M hypotheses are equally likely. The received signals at the output of a random phase channel are

$$r(t) = \sqrt{2E_r}\, f_i(t) \cos[\omega_c t + \phi_i(t) + \theta] + w(t), \quad 0 \le t \le T: H_i, \quad i = 1, 2, \ldots, M,$$

where $p_\theta(\theta)$ satisfies (364) and $w(t)$ is white with spectral height $N_0/2$. Find the LRT and draw a block diagram of the minimum probability of error receiver.

Problem 4.4.7 (continuation). Error Probability; Uniform Phase. [18] Consider the special case of the above model in which the signals are orthogonal and θ has a uniform density.

1. Show that

$$\Pr(\epsilon|\theta) = 1 = E\left\{\left[1 - \exp\left(-\frac{x^2 + y^2}{2}\right)\right]^{M-1}\right\},$$

where x and y are statistically independent Gaussian random variables with unit variance.

$$E[x|\theta] = \sqrt{2E_r/N_0}\cos\theta,$$
$$E[y|\theta] = \sqrt{2E_r/N_0}\sin\theta.$$

The expectation is over x and y, given θ.

2. Show that

$$\Pr(\epsilon) = \sum_{k=1}^{M-1} \binom{M-1}{k} (-1)^{k+1} \left(\frac{\exp[-(E_r/N_0)k/(k+1)]}{k+1}\right).$$

Problem 4.4.8. In the binary communication problem on pp. 345–348 we assumed that the signals on the two hypotheses were *not* phase-modulated. The general binary problem in white noise is

$$r(t) = \sqrt{2E_r}\, f_1(t) \cos[\omega_c t + \phi_1(t) + \theta] + w(t), \qquad 0 \le t \le T: H_1,$$
$$r(t) = \sqrt{2E_r}\, f_0(t) \cos[\omega_c t + \phi_0(t) + \theta] + w(t), \qquad 0 \le t \le T: H_0,$$

where E_r is the energy received in the signal component. The noise is white with spectral height $N_0/2$, and $p_\theta(\theta)$ satisfies (364). Verify that the optimum receiver structure is as shown in Fig. P4.7 for $i = 0, 1$, and that the minimum probability of error test is

$$z_1^2 \underset{H_0}{\overset{H_1}{\gtrless}} z_0^2.$$

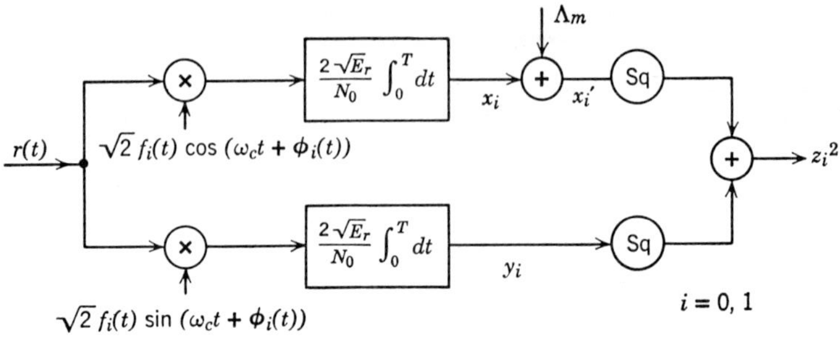

Fig. P4.7

Problem 4.4.9 (continuation) [44]. Assume that the signal components on the two hypotheses are orthogonal.

1. Assuming that H_0 is true, verify that

$$\begin{aligned} E(x_0') &= \Lambda_m + d^2 \cos\theta, \\ E(y_0) &= d^2 \sin\theta, \\ E(x_1') &= \Lambda_m, \\ E(y_1) &= 0, \end{aligned}$$

where $d^2 \triangleq 2E_r/N_0$ and

$$\text{Var}(x_0') = \text{Var}(y_0) = \text{Var}(x_1) = \text{Var}(y_1) = d^2.$$

2. Prove that

$$p_{z_0|H_0,\theta}(Z_0|H_0, \theta) = \frac{Z_0}{d^2}\exp\left(-\frac{Z_0^2 + \Lambda_m^2 + d^4 + 2\Lambda_m d^2 \cos\theta}{2d^2}\right) \times \left\{I_0\left(\frac{[(\Lambda_m^2 + d^4 + 2\Lambda_m d^2 \cos\theta)^{1/2} Z_0]}{d^2}\right)\right\}$$

and

$$p_{z_1|H_0,\theta}(Z_1|H_0, \theta) = \frac{Z_1}{d^2}\exp\left(-\frac{z_1^2 + \Lambda_m^2}{2d^2}\right) I_0\left(\frac{\Lambda_m}{d^2} Z_1\right).$$

3. Show that

$$\Pr(\epsilon) = \Pr(\epsilon|H_0) = \Pr(z_0 < z_1|H_0) = \int_{-\pi}^{\pi} p_\theta(\theta)\, d\theta \int_0^\infty p_{z_0 H_0,\theta}(Z_0|H_0, \theta)\, dZ_0 \times \int_{z_0}^\infty p_{z_1|H_1,\theta}(Z_1|H_1, \theta)\, dZ_1,$$

4. Prove that the inner two integrals can be rewritten as

$$\Pr(\epsilon|\theta) = Q(a, b) - \tfrac{1}{2}\exp\left(\frac{a^2 + b^2}{2}\right) I_0(ab),$$

where

$$a = \frac{\sqrt{2}\,\Lambda_m}{d},$$

$$b = \frac{[2(\Lambda_m^2 + d^4 + 2\Lambda_m d^2 \cos\theta)]^{1/2}}{d}.$$

5. Check your result for the two special cases in which $\Lambda_m \to 0$ and $\Lambda_m \to \infty$. Compare the resulting $\Pr(\epsilon)$ for these two cases in the region where d is large.

Problem 4.4.10 (continuation). Error Probability, Binary Nonorthogonal Signals [77]. When bandpass signals are not orthogonal, it is conventional to define their correlation in the following manner:

$$\tilde{f}_1(t) \triangleq f_1(t)e^{j\phi_1(t)}$$

$$\tilde{f}_1(t) \triangleq f_0(t)e^{j\phi_0(t)}$$

$$\tilde{\rho} \triangleq \int_0^T \tilde{f}_0(t)\,\tilde{f}_1^*(t)\,dt,$$

which is a complex number.

1. Express the actual signals in terms of $\tilde{f}_i(t)$.
2. Express the actual correlation coefficient of two signals in terms of $\tilde{\rho}$.
3. Assume $\Lambda_m = 0$ (this corresponds to a uniform density) and define the quantity

$$\lambda = (1 - |\tilde{\rho}|^2)^{1/2}.$$

Show that

$$\Pr(\epsilon) = Q\left(\frac{d}{2}\sqrt{1-\lambda}, \frac{d}{2}\sqrt{1+\lambda}\right) - \tfrac{1}{2}\exp\left(-\frac{d^2}{2}\right) I_0\left(\frac{d^2}{4}|\tilde{\rho}|\right).$$

Problem 4.4.11 (continuation). When $p_\theta(\theta)$ is nonuniform and the signals are nonorthogonal, the calculations are much more tedious. Set up the problem and then refer to [44] for the detailed manipulations.

Problem 4.4.12. M-ary PSK. Consider the M-ary PSK communication system in Problem 4.2.21. Assume that

$$p_\theta(\theta) = \frac{\exp(\Lambda_m \cos\theta)}{2\pi\, I_0(\Lambda_m)}, \qquad -\pi \le \theta \le \pi.$$

1. Find the optimum receiver.
2. Write an expression for the $\Pr(\epsilon)$.

Problem 4.4.13. ASK: Incoherent Channel [72]. An ASK system transmits equally likely messages

$$s_i(t) = \sqrt{2E_i}\, f(t)\cos\omega_c t, \qquad i = 1, 2, \ldots, M, \qquad 0 \le t \le T,$$

where

$$\sqrt{E_i} = (i-1)\,\Delta,$$

$$\int_0^T f^2(t)\,dt = 1,$$

and

$$(M-1)\Delta \triangleq E.$$

The received signal under the ith hypothesis is

$$r(t) = \sqrt{2E_i}\, f(t)\cos(\omega_c t + \theta) + w(t), \qquad 0 \le t \le T : H_i, \qquad i = 1, 2, \ldots, M,$$

where $w(t)$ is white noise ($N_0/2$). The phase θ is a random variable with a uniform density $(0, 2\pi)$.

1. Find the minimum $\Pr(\epsilon)$ receiver.
2. Draw the decision space and compute the $\Pr(\epsilon)$.

Problem 4.4.14. Asymptotic Behavior of Incoherent M-ary Systems [78]. In the text we saw that the probability of error in a communication system using M orthogonal signals approached zero as $M \to \infty$ as long as the rate in digits per second was less than $P/N_0 \ln 2$ (the channel capacity) (see pp. 264–267). Use exactly the same model as in Example 4 on pp. 264–267. Assume, however, that the channel adds a random phase angle. Prove that exactly the same results hold for this case. (*Comment*. The derivation is somewhat involved. The result is due originally to Turin [78]. A detailed derivation is given in Section 8.10 of [69].)

Problem 4.4.15 [79]. Calculate the moment generating function, mean, and variance of the test statistic $G = L_c^2 + L_s^2$ for the random phase problem of Section 4.4.1 under the hypothesis H_1.

Problem 4.4.16 (continuation) [79]. We can show that for $d \gtrsim 3$ the equivalent test statistic

$$R = \sqrt{L_c^2 + L_s^2}$$

is approximately Gaussian (see Fig. 4.73*a*). Assuming that this is true, find the mean and variance of R.

Problem 4.4.17 (continuation) [79]. Now use the result of Problem 4.4.16 to derive an approximate expression for the probability of detection. Express the result in terms of d and P_F and show that P_D is approximately a straight line when plotted versus d on probability paper. Compare a few points with Fig. 4.59. Evaluate the increase in d over the known signal case that is necessary to achieve the same performance.

Problem 4.4.18. Amplitude Estimation. We consider a simple estimation problem in which an unwanted parameter is present. The received signal is

$$r(t) = A\sqrt{E}\, s(t, \theta) + w(t),$$

where A is a nonrandom parameter we want to estimate (assume that it is nonnegative):

$$s(t, \theta) = f(t) \cos\,[\omega_c t + \phi(t) + \theta],$$

where $f(t)$ and $\phi(t)$ are slowly varying known functions of time and θ is a random variable whose probability density is,

$$p_\theta(\theta) = \begin{cases} \dfrac{1}{2\pi}, & -\pi < \theta < \pi, \\ 0, & \text{otherwise}, \end{cases}$$

$w(t)$ is white Gaussian noise of spectral height $N_0/2$.

Find the transcendental equation satisfied by the maximum-likelihood estimate of A.

Problem 4.4.19. Frequency Estimation: Random Phase Channel. The received signal is

$$r(t) = \sqrt{2E}\, f(t) \cos\,(\omega_c t + \phi(t) + \omega t + \theta) + w(t), \qquad 0 \le t \le T,$$

where $\int_0^T f^2(t)\, dt = 1$ and $f(t)$, $\phi(t)$, and E are known. The noise $w(t)$ is a sample function from a white Gaussian noise process ($N_0/2$). The frequency shift ω is an unknown nonrandom variable.

1. Find $\Lambda[r(t)|\omega]$.
2. Find the likelihood equation.
3. Draw a receiver whose output is a good approximation to $\hat{\omega}_{ml}$.

Problem 4.4.20 (continuation). Estimation Errors.

1. Compute a bound on the variance of any unbiased estimate of ω.
2. Compare the variance of $\hat{\omega}_{ml}$ under the assumption of a small error. Compare the result with the bound in part 1.
3. Compare the result of this problem with example 2 on p. 278.

RANDOM AMPLITUDE AND PHASE

Problem 4.4.21. Consider the detection problem in which

$$r(t) = \sum_{i=1}^{M} a_i s_i(t) + n(t), \qquad 0 \le t \le T : H_1,$$

$$= n(t), \qquad 0 \le t \le T : H_0.$$

The a_i are jointly Gaussian variables which we denote by the vector $\mathbf{a}$. The signals are denoted by the vector $s(t)$.

$$E(\mathbf{a}) \triangleq \mathbf{m_a},$$

$$E[(\mathbf{a} - \mathbf{m_a})(\mathbf{a}^T - \mathbf{m_a}^T)] \triangleq \mathbf{\Lambda}_a,$$

and

$$\boldsymbol{\rho} = \int_0^T \mathbf{s}(t)\, \mathbf{s}^T(t)\, dt,$$

$$E[n(t)\, n(u)] = \frac{N_0}{2}\, \delta(t - u)$$

1. Find the optimum receiver structure. Draw the various interpretations analogous to Figs. 4.66 and 4.67. *Hint.* Find a set of sufficient statistics and use (2.326).
2. Find $\mu(s)$ for this system. (See Section 2.7 of Chapter 2.)

Problem 4.4.22 (continuation). Extend the preceding problem to the case in which

$$E[n(t)\, n(u)] = \frac{N_0}{2}\, \delta(t - u) + K_c(t, u).$$

Problem 4.4.23. Consider the ASK system in Problem 4.4.13, operating over a Rayleigh channel. The received signal under the kth hypothesis is

$$r(t) = \sqrt{2E_k}\, v\, f(t) \cos(\omega_c t + \theta) + w(t), \qquad 0 \le t \le T : H_k, \qquad k = 1, 2, \ldots, M.$$

All quantities are described in Problem 4.4.13 except v:

$$p_v(V) = \begin{cases} V \exp\left(-\dfrac{V^2}{2}\right), & V \ge 0, \\ 0, & \text{elsewhere}, \end{cases}$$

and is independent of H_k. The hypotheses are equally likely.

1. Find the minimum Pr (ϵ) receiver.
2. Find the Pr (ϵ).

Problem 4.4.24. M-Orthogonal Signals: Rayleigh Channel [80]. One of M-orthogonal signals is used to transmit equally likely hypotheses over a Rayleigh channel. The received signal under the ith hypothesis is

$$r(t) = \sqrt{2}\, v\, f(t) \cos(\omega_i t + \theta) + w(t), \qquad 0 \le t \le T : H_i, \qquad i = 1, 2, \ldots, M,$$

where v is Rayleigh with variance $\bar{E}_r$, θ is uniform, $w(t)$ is white $(N_0/2)$, and $f(t)$ is normalized.

1. Draw a block diagram of the minimum probability of error receiver.
2. Show that

$$\Pr(\epsilon) = \sum_{n=1}^{M-1} \binom{M-1}{n} \frac{(-1)^{n+1}}{n+1+n\beta},$$

where

$$\beta \triangleq \frac{\bar{E}_r}{N_0}.$$

Problem 4.4.25 [90]. In this problem we investigate the improvement obtained by using M orthogonal signals instead of two orthogonal signals to transmit information over a Rayleigh channel.

1. Show that

$$\Pr(\epsilon) = 1 - \frac{\Gamma[1/(\beta+1)+1]\Gamma(M)}{\Gamma[1/(\beta+1)+M]}.$$

Hint. Use the familiar expression

$$\frac{\Gamma(z)\,\Gamma(a+1)}{\Gamma(z+a)} = \sum_{n=0}^{a-1} (-1)^n \frac{a(a-1)\cdots(a-n)}{n!} \frac{1}{z+n}.$$

2. Consider the case in which $\beta \gg 1$. Use a Taylor series expansion and the properties of $\psi(x) \triangleq \dfrac{\Gamma'(x)}{\Gamma(x)}$ to obtain the approximate expression

$$\Pr(\epsilon) \simeq \frac{1}{\beta}\left(\ln M - \frac{1}{2M} + 0.577\right).$$

Recall that

$$\psi(1) = 0.577,$$
$$\psi(z) = \ln z - \frac{1}{2z} + o(z).$$

3. Now assume that the M hypotheses arise from the simple coding system in Fig. 4.24. Verify that the bit error probability is

$$P_B(\epsilon) = \frac{1}{2}\frac{M}{M-1}\Pr(\epsilon).$$

4. Find an expression for the ratio of the $\Pr_B(\epsilon)$ in a binary system to the $\Pr_B(\epsilon)$ in an M-ary system.

5. Show that $M \to \infty$, the power saving resulting from using M orthogonal signals, approaches $2/\ln 2 = 4.6$ db.

Problem 4.4.26. M Orthogonal Signals: Rician Channel. Consider the same system as in Problem 4.4.24, but assume that v is Rician.

1. Draw a block diagram of the minimum $\Pr(\epsilon)$ receiver.
2. Find the $\Pr(\epsilon)$.

Problem 4.4.27. Binary Orthogonal Signals: Square-Law Receiver [18]. Consider the problem of transmitting two equally likely bandpass orthogonal signals with energy E_t over the Rician channel defined in (416). Instead of using the optimum receiver shown in Fig. 4.74, we use the receiver for the Rayleigh channel (i.e., let $\alpha = 0$ in Fig. 4.74). Show that

$$\Pr(\epsilon) = \left(2 + \frac{\sigma^2 E_t}{N_0}\right)^{-1} \exp\left[\frac{-\alpha^2 E_t}{N_0(2 + \sigma^2 E_t/N_0)}\right]$$

Problem 4.4.28. Repeat Problem 4.4.27 for the case of M orthogonal signals.

Composite Signal Hypotheses

Problem 4.4.29. Detecting One of M Orthogonal Signals. Consider the following binary hypothesis testing problem. Under H_1 the signal is one of M orthogonal signals $\sqrt{E_1}\, s_1(t), \sqrt{E_2}\, s_2(t), \ldots, \sqrt{E_M}\, s_M(t)$:

$$\int_0^T s_i(t)\, s_j(t)\, dt = \delta_{ij}, \qquad i, j = 1, 2, \ldots, M.$$

Under H_1 the ith signal occurs with probability p_i ($\sum_{i=1}^M p_i = 1$). Under H_0 there is no signal component. Under both hypotheses there is additive white Gaussian noise with spectral height $N_0/2$:

$$r(t) = \sqrt{E_i}\, s_i(t) + w(t), \qquad 0 \le t \le T \text{ with probability } p_i: H_1,$$

$$r(t) = w(t), \qquad 0 \le t \le T: H_0.$$

1. Find the likelihood ratio test.
2. Draw a block diagram of the optimum receiver.

Problem 4.4.30 (continuation). Now assume that

$$p_i = \frac{1}{M}, \qquad i = 1, 2, \ldots, M$$

and

$$E_i = E.$$

One method of approximating the performance of the receiver was developed in Problem 2.2.14. Recall that we computed the variance of Λ (*not* $\ln \Lambda$) on H_0 and used the equation

$$d^2 = \ln(1 + \text{Var}[\Lambda|H_0]). \tag{P.1}$$

We then used these values of d on the ROC of the known signal problem to find P_F and P_D.

1. Find $\text{Var}[\Lambda|H_0]$.
2. Using (P.1), verify that

$$\frac{2E}{N_0} = \ln(1 - M + Me^{d^2}). \tag{P.2}$$

3. For $2E/N_0 \gtrsim 3$ verify that we may approximate (P.2) by

$$\frac{2E}{N_0} \simeq \ln M + \ln(e^{d^2} - 1). \tag{P.3}$$

The significance of (P.3) is that if we have a certain performance level (P_F, P_D) for a single known signal then to maintain the performance level when the signal is equally likely to be any one of M orthogonal signals requires an increase in the energy-to-noise ratio of $\ln M$. This can be considered as the cost of signal uncertainty.

4. Now remove the equal probability restriction. Show that (P.3) becomes

$$\frac{2E}{N_0} \simeq -\ln\left(\sum_{i=1}^{M} p_i^2\right) + \ln\,(e^{d^2} - 1).$$

What probability assignment maximizes the first term? Is this result intuitively logical?

Problem 4.4.31 (alternate continuation). Consider the special case of Problem 4.4.29 in which $M = 2$, $E_1 = E_2 = E$, and $p_1 = p_2 = \frac{1}{2}$. Define

$$l_i[r(t)] = \left[\frac{2\sqrt{E}}{N_0}\int_0^T dt\, r(t)\, s_i(t) - \frac{E}{N_0}\right], \qquad i = 1, 2. \tag{P.4}$$

1. Sketch the optimum decision boundary in l_1, l_2-plane for various values of η.
2. Verify that the decision boundary approaches the asymptotes $l_1 = 2\eta$ and $l_2 = 2\eta$.
3. Under what conditions would the following test be close to optimum.

Test. If either l_1 *or* $l_2 \geq 2\eta$, say H_1 is true. Otherwise say H_0 is true.

4. Find P_D and P_F for the suboptimum test in Part 3.

Problem 4.4.32 (continuation). Consider the special case of Problem 4.4.29 in which $E_i = E$, $i = 1, 2, \ldots, M$ and $p_i = 1/M$, $i = 1, 2, \ldots, M$. Extending the definition of $l_i[r(t)]$ in (P.4) to $i = 1, 2, \ldots, M$, we consider the suboptimum test.

Test. If one or more $l_i \geq \ln M\eta$, say H_1. Otherwise say H_0.

1. Define

$$\alpha = \Pr\,[l_1 > \ln M\eta \,|\, s_1(t) \text{ is not present}],$$
$$\beta = \Pr\,[l_1 < \ln M\eta \,|\, s_1(t) \text{ is present}].$$

Show

$$P_F = 1 - (1 - \alpha)^M$$

and

$$P_D = 1 - \beta(1 - \alpha)^{M-1}.$$

2. Verify that

$$P_F \leq M\alpha$$

and

$$P_D \leq \beta.$$

When are these bounds most accurate?

3. Find α and β.
4. Assume that $M = 1$ and E/N_0 gives a certain P_F, P_D performance. How must E/N_0 increase to maintain the same performance at M increases? (Assume that the relations in part 2 are exact.) Compare these results with those in Problem 4.4.30.

Problem 4.4.33. A similar problem is encountered when each of the M orthogonal signals has a random phase.

Under H_1:

$$r(t) = \sqrt{2E}\, f_i(t) \cos\,[\omega_c t + \phi_i(t) + \theta_i] + w(t), \qquad 0 \leq t \leq T \quad \text{(with probability } p_i\text{)}.$$

Under H_0:

$$r(t) = w(t), \qquad 0 \leq t \leq T.$$

The signal components are orthogonal. The white noise has spectral height $N_0/2$. The probabilities, p_i, equal $1/M$, $i = 1, 2, \ldots, M$. The phase term in each signal θ_i is an independent, uniformly distributed random variable $(0, 2\pi)$.

1. Find the likelihood ratio test and draw a block diagram of the optimum receiver.
2. Find Var $(\Lambda | H_0)$.
3. Using the same approximation techniques as in Problem 4.4.30, show that the correct value of d to use on the known signal ROC is

$$d \triangleq \ln [1 + \text{Var}(\Lambda | H_0)] = \ln \left[1 - \frac{1}{M} + \frac{1}{M} I_0 \left(\frac{2E}{N_0}\right)\right].$$

Problem 4.4.34 (continuation). Use the same reasoning as in Problem 4.4.31 to derive a suboptimum test and find an expression for its performance.

Problem 4.4.35. Repeat Problem 4.4.33(1) and 4.4.34 for the case in which each of M orthogonal signals is received over a Rayleigh channel.

Problem 4.4.36. In Problem 4.4.30 we saw in the "one-of-M" orthogonal signal problem that to maintain the same performance we had to increase $2E/N_0$ by $\ln M$. Now suppose that under H_1 one of $N(N > M)$ equal-energy signals occurs with equal probability. The N signals, however, lie in an M-dimensional space. Thus, if we let $\phi_j(t)$, $j = 1, 2, \ldots, M$, be a set of orthonormal functions $(0, T)$, then

$$s_i(t) = \sum_{j=1}^{M} a_{ij}\phi_j(t), \qquad i = 1, 2, \ldots, N,$$

where

$$\sum_{j=1}^{M} a_{ij}^2 = 1, \qquad i = 1, 2, \ldots, N.$$

The other assumptions in Problem 4.4.29 remain the same.

1. Find the likelihood ratio test.
2. Discuss qualitatively (or quantitatively, if you wish) the cost of uncertainty in this problem.

Channel Measurement Receivers

Problem 4.4.37. Channel Measurement [18]. Consider the following approach to exploiting the phase stability in the channel. Use the first half of the signaling interval to transmit a channel measuring signal $\sqrt{2}\, s_m(t) \cos \omega_c t$ with energy E_m. Use the other half to send one of two equally likely signals $\pm \sqrt{2}\, s_d(t) \cos \omega_c t$ with energy E_d. Thus

$$r(t) = [s_m(t) + s_d(t)]\sqrt{2} \cos(\omega_c t + \theta) + w(t): H_1, \qquad 0 \le t \le T,$$

$$r(t) = [s_m(t) - s_d(t)]\sqrt{2} \cos(\omega_c t + \theta) + w(t): H_0, \qquad 0 \le t \le T,$$

and

$$p_\theta(\theta) = \frac{1}{2\pi}, \qquad 0 \le \theta \le 2\pi.$$

1. Draw the optimum receiver and decision rule for the case in which $E_m = E_d$.
2. Find the optimum receiver and decision rule for the case in part 1.
3. Prove that the optimum receiver can also be implemented as shown in Fig. P4.8.
4. What is the Pr (ϵ) of the optimum system?

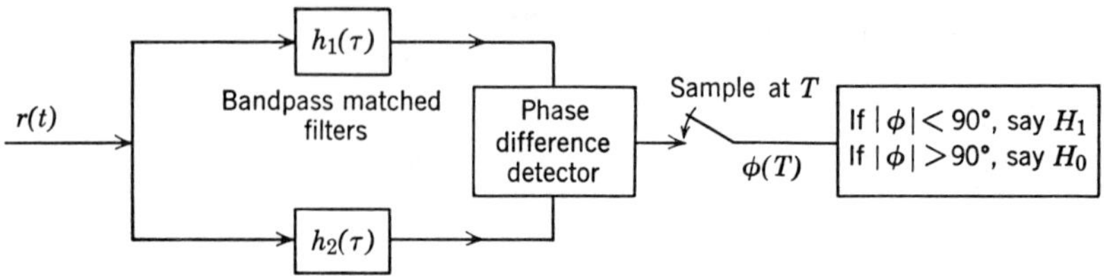

$$h_1(\tau) = \sqrt{2}\, s_m(T-\tau) \cos \omega_c \tau$$
$$h_2(\tau) = \sqrt{2}\, s_d(T-\tau) \cos \omega_c \tau$$

Fig. P4.8

Problem 4.4.38 (continuation). Kineplex [81]. A clever way to take advantage of the result in Problem 4.4.37 is employed in the Kineplex system. The information is transmitted by the phase relationship between successive bauds. If $s_d(t)$ is transmitted in one interval, then to send H_1 in the next interval we transmit $+s_d(t)$; and to send H_0 we transmit $-s_d(t)$. A typical sequence is shown in Fig. P4.9.

Source		1	1	0	0	0	1	1
Trans. sequence	$+s_m(t)$ (Initial reference)	$+s_m(t)$	+	−	+	−	−	−

Fig. P4.9

1. Assuming that there is no phase change from baud-to-baud, adapt the receiver in Fig. P4.8 to this problem. Show that the resulting Pr (ϵ) is

$$\Pr(\epsilon) = \tfrac{1}{2} \exp\left(-\frac{E}{N_0}\right),$$

(where E is the energy per baud, $E = E_d = E_m$).

2. Compare the performance of this system with the optimum coherent system in the text for large E/N_0. Are decision errors in the Kineplex system independent from baud to baud?

3. Compare the performance of Kineplex to the partially coherent system performance shown in Figs. 4.62 and 4.63.

Problem 4.4.39 (continuation). Consider the signal system in Problem 4.4.37 and assume that $E_m \neq E_d$.

1. Is the phase-comparison receiver of Fig. P4.8 optimum?
2. Compute the Pr (ϵ) of the optimum receiver.

Comment. It is clear that the ideas of phase-comparison can be extended to M-ary systems. [72], [82], and [83] discuss systems of this type.

MISCELLANEOUS

Problem 4.4.40. Consider the communication system described below. A known signal $s(t)$ is transmitted. It arrives at the receiver through *one* of two possible channels. The output is corrupted by additive white Gaussian noise $w(t)$. *If* the signal passes through channel 1, the input to the receiver is

$$r(t) = a\,s(t) + w(t), \qquad 0 \le t \le T,$$

where a is constant over the interval. It is the value of a Gaussian random variable $N(0, \sigma_a)$. If the signal passes through channel 2, the input to the receiver is

$$r(t) = s(t) + w(t), \qquad 0 \le t \le T.$$

It is given that

$$\int_0^T s^2(t)\,dt = E.$$

The probability of passing through channel 1 is equal to the probability of passing through channel 2 (i.e., $P_1 = P_2 = \frac{1}{2}$).

1. Find a receiver that decides which channel the signal passed through with minimum probability of error.
2. Compute the $\Pr(\epsilon)$.

Problem 4.4.41. A new engineering graduate is told to design an optimum detection system for the following problem:

$$H_1: r(t) = s(t) + w(t), \qquad T_i \le t \le T_f,$$

$$H_0: r(t) = n(t), \qquad T_i \le t \le T_f.$$

The signal $s(t)$ is known. To find a suitable covariance function $K_n(t, u)$ for the noise, he asks several engineers for an opinion.

Engineer A says

$$K_n(t, u) = \frac{N_0}{2}\,\delta(t - u).$$

Engineer B says

$$K_n(t, u) = \frac{N_0}{2}\,\delta(t - u) + K_c(t, u),$$

where $K_c(t, u)$ is a known, square-integrable, positive-definite function.

He must now reconcile these different opinions in order to design a signal detection system.

1. He decides to combine their opinions probabilistically. Specifically,

$$\Pr(\text{Engineer A is correct}) = P_A,$$
$$\Pr(\text{Engineer B is correct}) = P_B,$$

where $P_A + P_B = 1$.
   (a) Construct an optimum Bayes test (threshold η) to decide whether H_1 or H_0 is true.
   (b) Draw a block diagram of the receiver.
   (c) Check your answer for $P_A = 0$ and $P_B = 0$.
2. Discuss some other possible ways you might reconcile these different opinions.

Problem 4.4.42. Resolution. The following detection problem is a crude model of a simple radar resolution problem:

$$H_1: r(t) = b_d\, s_d(t) + b_I\, s_I(t) + w(t), \qquad T_i \le t \le T_f$$
$$H_0: r(t) = b_I\, s_I(t) + w(t), \qquad T_i \le t \le T_f.$$

1. $\int_{T_i}^{T_f} s_d(t)\, s_I(t)\, dt = \rho.$
2. $s_d(t)$ and $s_I(t)$ are normalized to unit energy.
3. The multipliers b_d and b_I are *independent* zero-mean Gaussian variables with variances σ_d^2 and σ_I^2, respectively.
4. The noise $w(t)$ is white Gaussian with spectral height $N_0/2$ and is independent of the multipliers.

Find an explicit solution for the optimum likelihood ratio receiver. You do *not* need to specify the threshold.

Section P.4.5. Multiple Channels.

Mathematical Derivations

Problem 4.5.1. The definition of a matrix inverse kernel given in (4.434) is

$$\int_{T_i}^{T_f} \mathbf{K_n}(t, u)\, \mathbf{Q_n}(u, z)\, du = \mathbf{I}\, \delta(t - z).$$

1. Assume that

$$\mathbf{K_n}(t, u) = \frac{N_0}{2}\mathbf{I}\, \delta(t - u) + \mathbf{K}_c(t, u).$$

Show that we can write

$$\mathbf{Q_n}(t, u) = \frac{2}{N_0}\, [\mathbf{I}\, \delta(t - u) - \mathbf{h}_o(t, u)],$$

where $\mathbf{h}_o(t, u)$ is a square-integrable function. Find the matrix integral equation that $\mathbf{h}_o(t, u)$ must satisfy.

2. Consider the problem of a matrix linear filter operating on $\mathbf{n}(t)$.

$$\mathbf{d}(t) = \int_{T_i}^{T_f} \mathbf{h}(t, u)\, \mathbf{n}(u)\, du,$$

where

$$\mathbf{n}(t) = \mathbf{n}_c(t) + \mathbf{w}(t)$$

has the covariance function given in part 1. We want to choose $\mathbf{h}(t, u)$ so that

$$\xi_I \triangleq E \int_{T_i}^{T_f} [\mathbf{n}_c(t) - \mathbf{d}(t)]^T [\mathbf{n}_c(t) - \mathbf{d}(t)]\, dt$$

is minimized. Show that the linear matrix filter that does this is the $\mathbf{h}_o(t, u)$ found in part 1.

Problem 4.5.2 (continuation). In this problem we extend the derivation in Section 4.5 to include the case in which

$$\mathbf{K_n}(t, u) = \mathbf{N}\, \delta(t - u) + \mathbf{K}_c(t, u), \qquad T_i \le t, u \le T_f,$$

where $\mathbf{N}$ is a positive-definite matrix of numbers. We denote the eigenvalues of $\mathbf{N}$ as $\lambda_1, \lambda_2, \ldots, \lambda_M$ and define a diagonal matrix,

$$\mathbf{I}_\lambda{}^k = \begin{bmatrix} \lambda_1{}^k & & & & 0 \\ & \lambda_2{}^k & & & \\ & & \cdot & & \\ & & & \cdot & \\ & & & & \cdot \\ 0 & & & & \lambda_M{}^k \end{bmatrix}.$$

To find the LRT we first perform two preliminary transformations on $\mathbf{r}$ as shown in Fig. P4.10.

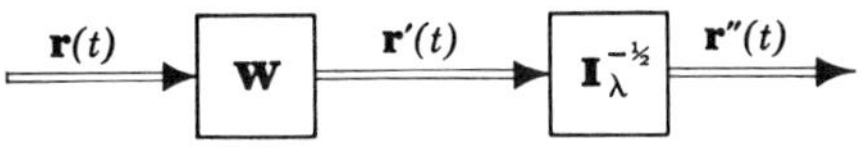

Fig. P4.10

The matrix $\mathbf{W}$ is an orthogonal matrix defined in (2.369) and has the properties

$$\mathbf{W}^T = \mathbf{W}^{-1},$$
$$\mathbf{N} = \mathbf{W}^{-1}\mathbf{I}_\lambda\mathbf{W}.$$

1. Verify that $\mathbf{r}''(t)$ has a covariance function matrix which satisfies (428).
2. Express l in terms of $\mathbf{r}''(t)$, $\mathbf{Q}_\mathbf{n}''(t, u)$, and $\mathbf{s}''(t)$.
3. Prove that

$$l = \iint_{T_i}^{T_f} \mathbf{r}^T(t)\, \mathbf{Q}_\mathbf{n}(t, u)\, \mathbf{s}(u)\, dt\, du,$$

where

$$\mathbf{Q}_\mathbf{n}(t, u) \triangleq \mathbf{N}^{-1}[\delta(t-u) - \mathbf{h}_o(t, u)],$$

and $\mathbf{h}_o(t, u)$ satisfies the equation

$$\mathbf{K}_c(t, u) = \mathbf{h}_o(t, u)\mathbf{N} + \int_{T_i}^{T_f} \mathbf{h}_o(t, z)\, \mathbf{K}_c(z, u)\, dz, \qquad T_i \le t, u \le T_f.$$

4. Repeat part (2) of Problem 4.5.1.

Problem 4.5.3. Consider the vector detection problem defined in (4.423). Assume that $\mathbf{K}_c(t, u) = 0$ and that $\mathbf{N}$ is not positive-definite. Find a signal vector $\mathbf{s}(t)$ with total energy E and a receiver that leads to perfect detectability.

Problem 4.5.4. Let

$$\mathbf{r}(t) = \mathbf{s}(t, A) + \mathbf{n}(t), \qquad T_i \le t \le T_f,$$

where the covariance of $\mathbf{n}(t)$ is given by (425) to (428) and A is a nonrandom parameter.

1. Find the equation the maximum-likelihood estimate of A must satisfy.
2. Find the Cramér-Rao inequality for an unbiased estimate $\hat{a}$.
3. Now assume that a is Gaussian, $N(0, \sigma_a)$. Find the MAP equation and the lower bound on the mean-square error.

Problem 4.5.5 (continuation). Let $\mathbf{L}$ denote a nonsingular linear transformation on a, where a is a zero-mean Gaussian random variable.

1. Show that an efficient estimate of A will exist if

$$\mathbf{s}(t, A) = \mathbf{L}A\,\mathbf{s}(t).$$

2. Find an explicit solution for $\hat{a}_{\text{map}}$ and an expression for the resulting mean-square error.

Problem 4.5.6. Let

$$r_i(t) = \sum_{j=1}^{K} a_{ij}\,s_{ij}(t) + w_i(t), \qquad i = 1, 2, \ldots, M: H_1,$$

$$r_i(t) = w_i(t), \qquad i = 1, 2, \ldots, M: H_0.$$

The noise in each channel is a sample function from a zero-mean white Gaussian random process

$$E[\mathbf{w}(t)\,\mathbf{w}^T(u)] = \frac{N_0}{2}\,\mathbf{I}\,\delta(t - u).$$

The a_{ij} are jointly Gaussian and zero-mean. The $s_{ij}(t)$ are orthogonal. Find an expression for the optimum Bayes receiver.

Problem 4.5.7. Consider the binary detection problem in which the received signal is an M-dimensional vector:

$$\mathbf{r}(t) = \mathbf{s}(t) + \mathbf{n}_c(t) + \mathbf{w}(t), \qquad -\infty < t < \infty: H_1,$$

$$= \mathbf{n}_c(t) + \mathbf{w}(t), \qquad -\infty < t < \infty: H_0.$$

The total signal energy is ME:

$$\int_0^T \mathbf{s}^T(t)\,\mathbf{s}(t)\,dt = ME.$$

The signals are zero outside the interval $(0, T)$.

1. Draw a block diagram of the optimum receiver.
2. Verify that

$$d^2 = \int_{-\infty}^{\infty} \mathbf{S}^T(j\omega)\,\mathbf{S}_n^{-1}(\omega)\,\mathbf{S}(j\omega)\,\frac{d\omega}{2\pi}.$$

Problem 4.5.8. Maximal-Ratio Combiners. Let

$$\mathbf{r}(t) = \mathbf{s}(t) + \mathbf{w}(t), \qquad 0 \le t \le T.$$

The received signal $\mathbf{r}(t)$ is passed into a time-invariant matrix filter with M inputs and one output $y(t)$:

$$y(t) = \int_0^T \mathbf{h}(t - \tau)\,\mathbf{r}(\tau)\,d\tau.$$

The subscript s denotes the output due to the signal. The subscript n denotes the output due to the noise. Define

$$\left(\frac{S}{N}\right)_{\text{out}} \triangleq \frac{y_s^2(T)}{E[y_n^2(T)]}.$$

1. Assume that the covariance matrix of $\mathbf{w}(t)$ satisfies (439). Find the matrix filter $\mathbf{h}(\tau)$ that maximizes $(S/N)_{\text{out}}$. Compare your answer with (440).
2. Repeat part 1 for a noise vector with an arbitrary covariance matrix $\mathbf{K}_c(t, u)$.

RANDOM PHASE CHANNELS

Problem 4.5.9 [14]. Let

$$x = \sum_{i=1}^{M} a_i^2,$$

where each a_i is an independent random variable with the probability density

$$p_{a_i}(A) = \frac{A}{\sigma^2} \exp\left(-\frac{A^2 + \alpha_i^2}{2\sigma^2}\right) I_0\left(\frac{\alpha_i A}{\sigma^2}\right), \qquad 0 \le A < \infty,$$
$$= 0, \qquad \text{elsewhere.}$$

Show that

$$p_x(X) = \frac{1}{2\sigma^2}\left(\frac{X}{P}\right)^{\frac{M-1}{2}} \exp\left(-\frac{X + P}{2\sigma^2}\right) I_{M-1}\left(\frac{\sqrt{PX}}{\sigma^2}\right), \qquad 0 \le X < \infty,$$
$$= 0, \qquad \text{elsewhere,}$$

where

$$P = \sigma^2 \sum_{i=1}^{M} \alpha_i^2.$$

Problem 4.5.10. Generalized Q-Function.

The generalization of the Q-function to M channels is

$$Q_M(\alpha, \beta) = \int_\beta^\infty x\left(\frac{x}{\alpha}\right)^{M-1} \exp\left(-\frac{x^2 + \alpha^2}{2}\right) I_{M-1}(\alpha x)\, dx.$$

1. Verify the relation

$$Q_M(\alpha, \beta) = Q(\alpha, \beta) + \exp\left(-\frac{\alpha^2 + \beta^2}{2}\right) \sum_{k=1}^{M-1} \left(\frac{\beta}{\alpha}\right)^k I_k(\alpha\beta).$$

2. Find $Q_M(\alpha, 0)$.
3. Find $Q_M(0, \beta)$.

Problem 4.5.11. On-Off Signaling: N Incoherent Channels. Consider an on-off communication system that transmits over N fixed-amplitude random-phase channels. When H_1 is true, a bandpass signal is transmitted over each channel. When H_0 is true, no signal is transmitted. The received waveforms under the two hypotheses are

$$r_i(t) = \sqrt{2E_i}\, f_i(t) \cos(\omega_i t + \phi_i(t) + \theta_i) + w(t), \qquad 0 \le t \le T: H_1,$$
$$r_i(t) = w(t), \qquad 0 \le t \le T: H_0,$$
$$i = 1, 2, \ldots, N.$$

The carrier frequencies are separated enough so that the signals are in disjoint frequency bands. The $f_i(t)$ and $\phi_i(t)$ are known low-frequency functions. The amplitudes $\sqrt{E_i}$ are known. The θ_i are statistically independent phase angles with a uniform distribution. The additive noise $w(t)$ is a sample function from a white Gaussian random process ($N_0/2$) which is independent of the θ_i.

1. Show that the likelihood ratio test is

$$\Lambda = \prod_{i=1}^{N} \exp\left(-\frac{E_i}{N_0}\right) I_0\left[\frac{2E_i^{1/2}}{N_0}(L_{c_i}^2 + L_{s_i}^2)^{1/2}\right] \underset{H_0}{\overset{H_1}{\gtrless}} \eta,$$

where L_{c_i} and L_{s_i} are defined as in (361) and (362).

2. Draw a block diagram of the optimum receiver based on ln Λ.
3. Using (371), find a good approximation to the optimum receiver for the case in which the argument of $I_0(\cdot)$ is small.
4. Repeat for the case in which the argument is large.
5. If the E_i are unknown nonrandom variables, does a UMP test exist?

Problem 4.5.12 (continuation). In this problem we analyze the performance of the suboptimum receiver developed in part 3 of the preceding problem. The test statistic is

$$l = \sum_{i=1}^{N} (L_{c_i}^2 + L_{s_i}^2) \underset{H0}{\overset{H1}{\gtrless}} \gamma.$$

1. Find $E[L_{c_i}|H_0]$, $E[L_{s_i}|H_0]$, Var $[L_{c_i}|H_0]$, Var $[L_{s_i}|H_0]$, $E[L_{c_i}|H_1, \theta]$, $E[L_{s_i}|H_1, \theta]$, Var $[L_{c_i}|H_1, \theta]$, Var $[L_{s_i}|H_1, \theta]$.
2. Use the result in Problem 2.6.4 to show that

$$M_{l|H_1}(jv) = (1 - jvN_0)^{-N} \exp\left(\frac{jv \sum_{i=1}^{N} E_i}{1 - jvN_0}\right)$$

and

$$M_{l|H_0}(jv) = (1 - jvN_0)^{-N}.$$

3. What is $p_{l|H_0}(X|H_0)$? Write an expression for P_F. The probability density of H_1 can be obtained from Fourier transform tables (e.g., [75], p. 197), It is

$$\begin{aligned} p_{l|H_1}(X|H_1) &= \frac{1}{N_0}\left(\frac{X}{E_T}\right)^{\frac{N-1}{2}} \exp\left(-\frac{X + E_T}{N_0}\right) I_{N-1}\left(\frac{2\sqrt{XE_T}}{N_0}\right), & X \geq 0, \\ &= 0, & \text{elsewhere,} \end{aligned}$$

where

$$E_T \triangleq \sum_{i=1}^{N} E_i.$$

4. Express P_D in terms of the generalized Q-function.

Comment. This problem was first studied by Marcum [46].

Problem 4.5.13 (continuation). Use the bounding and approximation techniques of Section 2.7 to evaluate the performance of the square-law receiver in Problem 4.5.11. Observe that the test statistic l is *not* equal to ln Λ, so that the results in Section 2.7 must be modified.

Problem 4.5.14. N Pulse Radar: Nonfluctuating Target. In a conventional pulse radar the target is illuminated by a sequence of pulses, as shown in Fig. 4.5. If the target strength is constant during the period of illumination, the return signal will be

$$r(t) = \sqrt{2E} \sum_{k=1}^{M} f(t - \tau - kT_p) \cos(\omega_c t + \theta_i) + w(t), \qquad -\infty < t < \infty : H_1,$$

where τ is the round-trip time to the target, which is assumed known, and T_p is the interpulse time which is much larger than the pulse length T $[f(t) = 0 : t < 0, t > T]$. The phase angles of the received pulses are statistically independent random variables with uniform densities. The noise $w(t)$ is a sample function of a zero-mean white Gaussian process ($N_0/2$). Under H_0 no target is present and

$$r(t) = w(t), \qquad -\infty < t < \infty : H_0.$$

1. Show that the LRT for this problem is identical to that in Problem 4.5.11 (except for notation). This implies that the results of Problems 4.5.11 to 13 apply to this model also.
2. Draw a block diagram of the optimum receiver. Do not use more than one bandpass filter.

Problem 4.5.15. Orthogonal Signals: N Incoherent Channels. An alternate communication system to the one described in Problem 4.5.11 would transmit a signal on both hypotheses. Thus

$$r_i(t) = \sqrt{2E_{1i}}\, f_{1i}(t) \cos [\omega_i t + \phi_{1i}(t) + \theta_i] + w(t), \qquad 0 \le t \le T: H_1, \quad i = 1, 2, \ldots, N,$$

$$r_i(t) = \sqrt{2E_{0i}}\, f_{0i}(t) \cos [\omega_i t + \phi_{0i}(t) + \theta_i] + w(t), \qquad 0 \le t \le T: H_0, \quad i = 1, 2, \ldots, N.$$

All of the assumptions in 4.5.11 are valid. In addition, the signals on the two hypotheses are orthogonal.

1. Find the likelihood ratio test under the assumption of equally likely hypotheses and minimum Pr (ϵ) criterion.
2. Draw a block diagram of the suboptimum square-law receiver.
3. Assume that $E_i = E$. Find an expression for the probability of error in the square-law receiver.
4. Use the techniques of Section 2.7 to find a bound on the probability of error and an approximate expression for Pr (ϵ).

Problem 4.5.16 (continuation). N Partially Coherent Channels.

1. Consider the model in Problem 4.5.11. Now assume that the phase angles are independent random variables with probability density

$$p_{\theta_i}(\theta) = \frac{\exp(\Lambda_m \cos \theta)}{2\pi I_0(\Lambda_m)}, \qquad -\pi < \theta < \pi, \qquad i = 1, 2, \ldots, N.$$

Do parts 1, 2, and 3 of Problem 4.5.11, using this assumption.

2. Repeat part 1 for the model in Problem 4.5.15.

Random Amplitude and Phase Channels

Problem 4.5.17. Density of Rician Envelope and Phase. If a narrow-band signal is transmitted over a Rician channel, the output contains a specular component and a random component. Frequently it is convenient to use complex notation. Let

$$s_t(t) \triangleq \sqrt{2}\, \text{Re}\, [f(t) e^{j\phi(t)} e^{j\omega_c t}]$$

denote the transmitted signal. Then, using (416), the received signal (without the additive noise) is

$$s_r(t) \triangleq \sqrt{2}\, \text{Re}\, \{v' f(t) \exp [j\phi(t) + j\theta' + j\omega_c t]\},$$

where

$$v' e^{j\theta'} \triangleq \alpha e^{j\delta} + v e^{j\theta}$$

in order to agree with (416).

1. Show that

$$p_{v',\theta'}(V', \theta') = \begin{cases} \dfrac{V'}{2\pi\sigma^2} \exp\left(-\dfrac{V'^2 + \alpha^2 - 2V'\alpha\cos(\theta' - \delta)}{2\sigma^2}\right), & 0 \le V' < \infty, \quad 0 \le \theta' - \delta \le 2\pi, \\ 0, & \text{elsewhere.} \end{cases}$$

2. Show that

$$p_{v'}(V') = \begin{cases} \dfrac{V'}{\sigma^2} \exp\left(-\dfrac{V'^2 + \alpha^2}{2\sigma^2}\right) I_0\left(\dfrac{\alpha V'}{\sigma^2}\right), & 0 \le V' < \infty, \\ 0, & \text{elsewhere.} \end{cases}$$

3. Find $E(v')$ and $E(v'^2)$.
4. Find $p_{\theta'}(\theta')$, the probability density of θ'.

The probability densities in parts 2 and 4 are plotted in Fig. 4.73.

Problem 4.5.18. On-off Signaling: N Rayleigh Channels. In an on-off communication system a signal is transmitted over each of N Rayleigh channels when H_1 is true. The received signals are

$$H_1 : r_i(t) = v_i \sqrt{2}\, f_i(t) \cos[\omega_i t + \phi_i(t) + \theta_i] + w_i(t), \qquad 0 \le t \le T, \quad i = 1, 2, \dots, N,$$

$$H_0 : r_i(t) = w_i(t), \qquad 0 \le t \le T, \quad i = 1, 2, \dots, N,$$

where $f_i(t)$ and $\phi_i(t)$ are known waveforms, the v_i are statistically independent Rayleigh random variables with variance E_i, the θ_i are statistically independent random variables uniformly distributed $0 \le \theta \le 2\pi$, and $w_i(t)$ are independent white Gaussian noises $(N_0/2)$.

1. Find the LRT.
2. Draw a block diagram of the optimum receiver. Indicate both a bandpass filter realization and a filter-squarer realization.

Problem 4.5.19 (continuation). Optimum Diversity.
Now assume that $E_i = E$, $(i = 1, 2, \dots, N)$.

1. Verify that this problem is mathematically identical to Case 1A on p. 108 in Section 2.6. Find the relationships between the parameters in the two problems.
2. Use the identity in part 1 and the results in Example 2 on pp. 127–129 to find $\mu(s)$ and $\dot{\mu}(s)$ for this problem.
3. Assume that the hypotheses are equally likely and that minimum $\Pr(\epsilon)$ is the criterion. Find an upper bound on the $\Pr(\epsilon)$ and an approximate expression for the $\Pr(\epsilon)$.
4. Constrain $NE = E_T$. Use an approximate expression of the type given in (2.508) to find the optimum number of diversity channels.

Problem 4.5.20. N Pulse Radar: Fluctuating Target. Consider the pulsed model developed in Problem 4.5.14. If the target fluctuates, the amplitude of the reflected signal will change from pulse to pulse. A good model for this fluctuation is the Rayleigh model. Under H_1 the received signal is

$$r(t) = \sqrt{2} \sum_{i=1}^{N} v_i\, f(t - \tau - kT_p) \cos(\omega_c t + \theta_i) + w(t), \qquad -\infty < t < \infty,$$

where v_i, θ_i, and $w(t)$ are specified in Problem 4.5.18.

1. Verify that this problem is mathematically identical to Problem 4.5.18.
2. Draw a block diagram of the optimum receiver.
3. Verify that the results in Figs. 2.35 and 2.42 are immediately applicable to this problem.

4. If the required $P_F = 10^{-4}$ and the total average received energy is constrained $E[Nv_i^2] = 64$, what is the optimum number of pulses to transmit in order to maximize P_D?

Problem 4.5.21. Binary Orthogonal Signals: N Rayleigh Channels. Consider a binary communication system using orthogonal signals and operating over N Rayleigh channels. The hypotheses are equally likely and the criterion is minimum Pr (ϵ). The received waveforms are

$$\begin{aligned} r_i(t) &= \sqrt{2}\, v_i\, f_1(t) \cos [\omega_{1i}t + \phi_1(t) + \theta_i] + w_i(t), \qquad 0 \le t \le T, \\ &\qquad\qquad\qquad\qquad\qquad\qquad\qquad\qquad\qquad i = 1, 2, \dots, N: H_1 \\ &= \sqrt{2}\, v_i\, f_0(t) \cos [\omega_{0i}t + \phi_0(t) + \theta_i] + w_i(t), \qquad 0 < t < T, \\ &\qquad\qquad\qquad\qquad\qquad\qquad\qquad\qquad\qquad i = 1, 2, \dots, N: H_0. \end{aligned}$$

The signals are orthogonal. The quantities v_i, θ_i, and $w_i(t)$ are described in Problem 4.5.18. The system is an FSK system with diversity.

1. Draw a block diagram of the optimum receiver.
2. Assume $E_i = E$, $i = 1, 2, \dots, N$. Verify that this model is mathematically identical to Case 2A on p. 115. The resulting Pr (ϵ) is given in (2.434). Express this result in terms of E and N_0.

Problem 4.5.22 (continuation). Error Bounds: Optimal Diversity. Now assume the E_i may be different.

1. Compute $\mu(s)$. (Use the result in Example 3A on p. 130.)
2. Find the value of s which corresponds to the threshold $\gamma = \dot{\mu}(s)$ and evaluate $\mu(s)$ for this value.
3. Evaluate the upper bound on the Pr (ϵ) that is given by the Chernoff bound.
4. Express the result in terms of the probability of error in the individual channels:

$$P_i \triangleq \text{Pr } (\epsilon \text{ on the } i\text{th diversity channel})$$

$$P_i = \frac{1}{2}\left[\left(1 + \frac{E_i}{2N_0}\right)^{-1}\right].$$

5. Find an approximate expression for Pr(ϵ) using a Central Limit Theorem argument.
6. Now assume that $E_i = E$, $i = 1, 2, \dots, N$, and $NE = E_T$. Using an approximation of the type given in (2.473), find the optimum number of diversity channels.

Problem 4.5.23. M-ary Orthogonal Signals: N Rayleigh Channels. A generalization of the binary diversity system is an M-ary diversity system. The M hypotheses are equally likely. The received waveforms are

$$\begin{aligned} r_i(t) = \sqrt{2}\, v_i\, f_k(t) \cos [\omega_{ki}t + \phi_k(t) + \theta_i] + w_i(t), \qquad & 0 \le t \le T: H_k, \\ & i = 1, 2, \dots, N, \\ & k = 1, 2, \dots, M. \end{aligned}$$

The signals are orthogonal. The quantities v_i, θ_i, and $w_i(t)$ are described in Problem 4.5.18. This type of system is usually referred to as multiple FSK (MFSK) with diversity.

1. Draw a block diagram of the optimum receiver.
2. Find an expression for the probability of error in deciding which hypothesis is true.

Comment. This problem is discussed in detail in Hahn [84] and results for various M and N are plotted.

Problem 4.5.24 (continuation). Bounds.

1. Combine the bounding techniques of Section 2.7 with the simple bounds in (4.63) through (4.65) to obtain a bound on the probability of error in the preceding problem.
2. Use a Central Limit Theorem argument to obtain an approximate expression.

Problem 4.5.25. M Orthogonal Signals: N Rician Channels. Consider the M-ary system in Problem 4.5.23. All the assumptions remain the same except now we assume that the channels are independent Rician instead of Rayleigh. (See Problem 4.5.17.) The amplitude and phase of the specular component are known.

1. Find the LRT and draw a block diagram of the optimum receiver.
2. What are some of the difficulties involved in implementing the optimum receiver?

Problem 4.5.26 (continuation). Frequently the phase of specular component is not accurately known. Consider the model in Problem 4.5.25 and assume that

$$p_{\delta_i}(X) = \frac{\exp(\Lambda_m \cos X)}{2\pi\, I_0(\Lambda_m)}, \qquad \pi \leq X \leq \pi,$$

and that the δ_i are independent of each other and all the other random quantities in the model.

1. Find the LRT and draw a block diagram of the optimum receiver.
2. Consider the special case where $\Lambda_m = 0$. Draw a block diagram of the optimum receiver.

Commentary. The preceding problems show the computational difficulties that are encountered in evaluating error probabilities for multiple-channel systems. There are two general approaches to the problem. The direct procedure is to set up the necessary integrals and attempt to express them in terms of Q-functions, confluent hypergeometric functions, Bessel functions, or some other tabulated function. Over the years a large number of results have been obtained. A summary of solved problems and an extensive list of references are given in [89]. A second approach is to try to find analytically tractable bounds to the error probability. The bounding technique derived in Section 2.7 is usually the most fruitful. The next two problems consider some useful examples.

Problem 4.5.27. Rician Channels: Optimum Diversity [86].

1. Using the approximation techniques of Section 2.7, find $\Pr(\epsilon)$ expressions for binary orthogonal signals in N Rician channels.
2. Conduct the same type of analysis for a suboptimum receiver using square-law combining.
3. The question of optimum diversity is also appropriate in this case. Check your expressions in parts 1 and 2 with [86] and verify the optimum diversity results.

Problem 4.5.28. In part 3 of Problem 4.5.27 it was shown that if the ratio of the energy in the specular component to the energy in the random component exceeded a certain value, then infinite diversity was optimum. This result is not practical because it assumes perfect knowledge of the phase of the specular component. As N increases, the effect of small phase errors will become more important and should always lead to a finite optimum number of channels. Use the phase probability density in Problem 4.5.26 and investigate the effects of imperfect phase knowledge.

Section P4.6 Multiple Parameter Estimation

Problem 4.6.1. The received signal is

$$r(t) = s(t, \mathbf{A}) + w(t), \qquad 0 \le t \le T.$$

The parameter $\mathbf{a}$ is a Gaussian random vector with probability density

$$p_{\mathbf{a}}(\mathbf{A}) = [(2\pi)^{M/2}|\mathbf{\Lambda_a}|^{1/2}]^{-1} \exp\left(-\tfrac{1}{2}\mathbf{A}^T\mathbf{\Lambda_a}^{-1}\mathbf{A}\right).$$

1. Using the derivative matrix notation of Chapter 2 (p. 76), derive an integral equation for the MAP estimate of $\mathbf{a}$.
2. Use the property in (444) and the result in (447) to find the $\hat{\mathbf{a}}_{\text{map}}$.
3. Verify that the two results are identical.

Problem 4.6.2. Modify the result in Problem 4.6.1 to include the case in which $\mathbf{\Lambda_a}$ is singular.

Problem 4.6.3. Modify the result in part 1 of Problem 4.6.1 to include the case in which $E(\mathbf{a}) = \mathbf{m_a}$.

Problem 4.6.4. Consider the example on p. 372. Show that the actual mean-square errors approach the bound as E/N_0 increases.

Problem 4.6.5. Let

$$r(t) = s(t, a(t)) + n(t), \qquad 0 \le t \le T.$$

Assume that $a(t)$ is a zero-mean Gaussian random process with covariance function $K_a(t, u)$. Consider the function $a^*(t)$ obtained by sampling $a(t)$ every T/M seconds and reconstructing a waveform from the samples.

$$a^*(t) = \sum_{i=1}^{M} a(t_i) \frac{\sin[(\pi M/T)(t - t_i)]}{(\pi M/T)(t - t_i)}, \qquad t_i = 0, \frac{T}{M}, \frac{2T}{M}, \ldots.$$

1. Define

$$\hat{a}^*(t) = \sum_{i=1}^{M} \hat{a}(t_i) \frac{\sin[(\pi M/T)(t - t_i)]}{(\pi M/T)(t - t_i)}.$$

Find an equation for $\hat{a}^*(t)$.

2. Proceeding formally, show that as $M \to \infty$ the equation for the MAP estimate of $a(t)$ is

$$\hat{a}(t) = \frac{2}{N_0}\int_0^T [r(u) - s(u, \hat{a}(u))] \frac{\partial s(u, \hat{a}(u))}{\partial \hat{a}(u)} K_a(t, u)\, du, \qquad 0 \le t \le T.$$

Problem 4.6.6. Let

$$r(t) = s(t, \mathbf{A}) + n(t), \qquad 0 \le t \le T,$$

where $\mathbf{a}$ is a zero-mean Gaussian vector with a diagonal covariance matrix and $n(t)$ is a sample function from a zero-mean Gaussian random process with covariance function $K_n(t, u)$. Find the MAP estimate of $\mathbf{a}$.

Problem 4.6.7. The multiple channel estimation problem is

$$\mathbf{r}(t) = \mathbf{s}(t, \mathbf{A}) + \mathbf{n}(t), \qquad 0 \le t \le T,$$

where $\mathbf{r}(t)$ is an N-dimensional vector and $\mathbf{a}$ is an M-dimensional parameter. Assume that $\mathbf{a}$ is a zero-mean Gaussian vector with a diagonal covariance matrix. Let

$$E[\mathbf{n}(t)\,\mathbf{n}^T(u)] = \mathbf{K_n}(t, u).$$

Find an equation that specifies the MAP estimate of $\mathbf{a}$.

Problem 4.6.8. Let

$$r(t) = \sqrt{2}\, v\, f(t, \mathbf{A}) \cos [\omega_c t + \phi(t, \mathbf{A}) + \theta] + w(t), \qquad 0 \le t \le T,$$

where v is a Rayleigh variable and θ is a uniform variable. The additive noise $w(t)$ is a sample function from a white Gaussian process with spectral height $N_0/2$. The parameter $\mathbf{a}$ is a zero-mean Gaussian vector with a diagonal covariance matrix; $\mathbf{a}$, v, θ, and $w(t)$ are statistically independent. Find the likelihood function as a function of $\mathbf{a}$.

Problem 4.6.9. Let

$$r(t) = \sqrt{2}\, v\, f(t - \tau) \cos [\omega_c t + \phi(t - \tau) + \omega t + \theta] + w(t), \qquad -\infty < t < \infty,$$

where $w(t)$ is a sample function from a zero-mean white Gaussian noise process with spectral height $N_0/2$. The functions $f(t)$ and $\phi(t)$ are deterministic functions that are low-pass compared with ω_c. The random variable v is Rayleigh and the random variable θ is uniform. The parameters τ and ω are nonrandom.

1. Find the likelihood function as a function of τ and ω.
2. Draw the block diagram of a receiver that provides an approximate implementation of the maximum-likelihood estimator.

Problem 4.6.10. A sequence of amplitude modulated signals is transmitted. The signal transmitted in the kth interval is

$$s_k(t, A) = A_k\, s(t), \qquad (k - 1)T \le t \le kT, \qquad k = 1, 2, \ldots.$$

The sequence of random variables is zero-mean Gaussian; the variables are related in the following manner:

$$a_1 \text{ is } N(0, \sigma_a)$$
$$a_2 = \Phi a_1 + u_1$$
$$\vdots$$
$$a_k = \Phi a_{k-1} + u_{k-1}.$$

The multiplier Φ is fixed. The u_i are independent, zero-mean Gaussian random variables, $N(0, \sigma_u)$. The received signal in the kth interval is

$$r(t) = s_k(t, A) + w(t), \qquad (k - 1)T \le t \le kT, \qquad k = 1, 2, \ldots.$$

Find the MAP estimate of a_k, $k = 1, 2, \ldots$. (Note the similarity to Problem 2.6.15.)

REFERENCES

[1] G. E. Valley and H. Wallman, *Vacuum Tube Amplifiers*, McGraw-Hill, New York, 1948.

[2] W. B. Davenport and W. L. Root, *An Introduction to the Theory of Random Signals and Noise*, McGraw-Hill, New York, 1958.

[3] C. Helstrom, "Solution of the Detection Integral Equation for Stationary Filtered White Noise," *IEEE Trans. Inform. Theory*, **IT-11**, 335–339 (July 1965).

[4] H. Nyquist, "Certain Factors Affecting Telegraph Speed," *Bell System Tech. J.*, **3**, No. 2, 324–346 (1924).

[5] U. Grenander, "Stochastic Processes and Statistical Inference," *Arkiv Mat.*, **1**, 295 (1950).

[6] J. L. Lawson and G. E. Uhlenbeck, *Threshold Signals*, McGraw-Hill, New York, 1950.

[7] P. M. Woodward and I. L. Davies, "Information Theory and Inverse Probability in Telecommunication," *Proc. Inst. Elec. Engrs.*, **99**, Pt. III, 37 (1952).

[8] P. M. Woodward, *Probability and Information Theory, with Application to Radar*, McGraw-Hill, New York, 1955.

[9] W. W. Peterson, T. G. Birdsall, and W. C. Fox, "The Theory of Signal Detectability," *IRE Trans. Inform. Theory*, **PGIT-4**, 171 (September 1954).

[10] D. Middleton and D. Van Meter, "Detection and Extraction of Signals in Noise from the Point of View of Statistical Decision Theory," *J. Soc. Ind. Appl. Math.*, **3**, 192 (1955); **4**, 86 (1956).

[11] D. Slepian, "Estimation of Signal Parameters in the Presence of Noise," *IRE Trans. on Information Theory*, **PGIT-3**, 68 (March 1954).

[12] V. A. Kotelnikov, "The Theory of Optimum Noise Immunity," Doctoral Dissertation, Molotov Institute, Moscow, 1947.

[13] V. A. Kotelnikov, *The Theory of Optimum Noise Immunity*, McGraw-Hill, New York, 1959.

[14] C. W. Helstrom, *Statistical Theory of Signal Detection*, Pergamon, New York, 1960.

[15] L. A. Wainstein and V. D. Zubakov, *Extraction of Signals from Noise*, Prentice-Hall, Englewood Cliffs, New Jersey, 1962.

[16] W. W. Harman, *Principles of the Statistical Theory of Communication*, McGraw-Hill, New York, 1963.

[17] E. J. Baghdady, ed., *Lectures on Communication System Theory*, McGraw-Hill, New York, 1961.

[18] J. M. Wozencraft and I. M. Jacobs, *Principles of Communication Engineering*, Wiley, New York, 1965.

[19] S. W. Golomb, L. D. Baumert, M. F. Easterling, J. J. Stiffer, and A. J. Viterbi, *Digital Communications*, Prentice-Hall, Englewood Cliffs, New Jersey, 1964.

[20] D. O. North, "Analysis of the Factors which Determine Signal/Noise Discrimination in Radar," RCA Laboratories, Princeton, New Jersey, Report PTR-6C, June 1943; "Analysis of the Factors which Determine Signal/Noise Discrimination in Radar," *Proc. IEEE*, **51**, July 1963.

[21] R. H. Urbano, "Analysis and Tabulation of the M Positions Experiment Integral and Related Error Function Integrals," Tech. Rep. No. AFCRC TR-55-100, April 1955, AFCRC, Bedford, Massachusetts.

[22] C. E. Shannon and W. Weaver, *The Mathematical Theory of Communication*, University of Illinois Press, Urbana, Illinois, 1949.

[23] C. E. Shannon, "Communication in the Presence of Noise," *Proc. IRE*, **37**, 10–21 (1949).

[24] H. L. Van Trees, "A Generalized Bhattacharyya Bound," Internal Memo, IM-VT-6, M.I.T., January 15, 1966.

[25] A. J. Mallinckrodt and T. E. Sollenberger, "Optimum-Pulse-Time Determination," *IRE Trans. Inform. Theory*, PGIT-3, 151–159 (March 1954).

[26] R. Manasse, "Range and Velocity Accuracy from Radar Measurements," M.I.T. Lincoln Laboratory (February 1955).

[27] M. I. Skolnik, *Introduction to Radar Systems*, McGraw-Hill, New York, 1962.

[28] P. Swerling, "Parameter Estimation Accuracy Formulas," *IEEE Trans. Inform. Theory*, IT-10, No. 1, 302–313 (October 1964).

[29] M. Mohajeri, "Application of Bhattacharyya Bounds to Discrete Angle Modulation," 6.681 Project, Dept. of Elec. Engr., MIT.; May 1966.

[30] U. Grenander, "Stochastic Processes and Statistical Inference," *Arkiv Mat.*, **1**, 195–277 (1950).

[31] E. J. Kelly, I. S. Reed, and W. L. Root, "The Detection of Radar Echoes in Noise, Pt. I," *J. SIAM*, **8**, 309–341 (September 1960).

[32] T. Kailath, "The Detection of Known Signals in Colored Gaussian Noise," *Proc. National Electronics Conf.*, **21** (1965).

[33] R. Courant and D. Hilbert, *Methoden der Mathematischen Physik*, Springer-Verlag, Berlin. English translation: Interscience, New York, 1953.

[34] W. V. Lovitt, *Linear Integral Equations*, McGraw-Hill, New York.

[35] L. A. Zadeh and J. R. Ragazzini, "Optimum Filters for the Detection of Signals in Noise," *Proc. IRE*, **40**, 1223 (1952).

[36] K. S. Miller and L. A. Zadeh, "Solution of an Integral Equation Occurring in the Expansion of Second-Order Stationary Random Functions," *IRE Trans. Inform. Theory*, **IT-3**, 187, **IT-2**, No. 2, pp. 72–75, June 1956.

[37] D. Youla, "The Solution of a Homogeneous Wiener-Hopf Integral Equation Occurring in the Expansion of Second-Order Stationary Random Functions," *IRE Trans. Inform. Theory*, **IT-3**, 187 (1957).

[38] W. M. Siebert, *Elements of High-Powered Radar Design*, J. Friedman and L. Smullin, eds., to be published.

[39] H. Yudkin, "A Derivation of the Likelihood Ratio for Additive Colored Noise," unpublished note, 1966.

[40] E. Parzen, "Statistical Inference on Time Series by Hilbert Space Methods," Tech. Rept. No. 23, Appl. Math. and Stat. Lab., Stanford University, 1959; "An Approach to Time Series Analysis," *Ann. Math. Stat.*, **32**, 951–990 (December 1961).

[41] J. Hajek, "On a Simple Linear Model in Gaussian Processes," *Trans. 2nd Prague Conf., Inform. Theory*, 185–197 (1960).

[42] W. L. Root, "Stability in Signal Detection Problems," Stochastic Processes in Mathematical Physics and Engineering, *Proc. Symposia Appl. Math.*, **16** (1964).

[43] C. A. Galtieri, "On the Characterization of Continuous Gaussian Processes," Electronics Research Laboratory, University of California, Berkeley, Internal Tech. Memo, No. M-17, July 1963.

[44] A. J. Viterbi, "Optimum Detection and Signal Selection for Partially Coherent Binary Communication," *IEEE Trans. Inform. Theory*, **IT-11**, No. 2, 239–246 (April 1965).

[45] T. T. Kadota, "Optimum Reception of *M*-ary Gaussian Signals in Gaussian Noise," *Bell System Tech. J.*, 2187–2197 (November 1965).

[46] J. Marcum, "A Statistical Theory of Target Detection by Pulsed Radar," Math. Appendix, Report RM-753, Rand Corporation, July 1, 1948. Reprinted in *IEEE Trans. Inform. Theory*, **IT-6**, No. 2, 59–267 (April 1960).

[47] D. Middleton, *An Introduction to Statistical Communication Theory*, McGraw-Hill, New York, 1960.

[48] J. I. Marcum, "Table of *Q*-Functions," Rand Corporation Rpt. RM-339, January 1, 1950.

[49] A. R. Di Donato and M. P. Jarnagin, "A Method for Computing the Circular Coverage Function," *Math. Comp.*, **16**, 347–355 (July 1962).

[50] D. E. Johansen, "New Techniques for Machine Computation of the *Q*-functions, Truncated Normal Deviates, and Matrix Eigenvalues," Sylvania ARL, Waltham, Mass., Sci. Rept. No. 2, July 1961.

[51] R. A. Silverman and M. Balser, "Statistics of Electromagnetic Radiation Scattered by a Turbulent Medium," *Phys. Rev.*, **96**, 560–563 (1954).

[52] K. Bullington, W. J. Inkster, and A. L. Durkee, "Results of Propagation Tests at 505 Mc and 4090 Mc on Beyond-Horizon Paths," *Proc. IRE*, **43**, 1306–1316 (1955).

[53] P. Swerling, "Probability of Detection for Fluctuating Targets," *IRE Trans. Inform. Theory*, **IT-6**, No. 2, 269–308 (April 1960). Reprinted from Rand Corp. Report RM-1217, March 17, 1954.

[54] J. H. Laning and R. H. Battin, *Random Processes in Automatic Control*, McGraw-Hill, New York, 1956.

[55] M. Masonson, "Binary Transmission Through Noise and Fading," 1956 *IRE Natl. Conv. Record*, Pt. 2, 69–82.

[56] G. L. Turin, "Error Probabilities for Binary Symmetric Ideal Reception through Non-Selective Slow Fading and Noise," *Proc. IRE*, **46**, 1603–1619 (1958).

[57] R. W. E. McNicol, "The Fading of Radio Waves on Medium and High Frequencies," *Proc. Inst. Elec. Engrs.*, **96**, Pt. 3, 517–524 (1949).

[58] D. G. Brennan and M. L. Philips, "Phase and Amplitude Variability in Medium-Frequency Ionospheric Transmission," Tech. Rep. 93, M.I.T., Lincoln Lab., September 16, 1957.

[59] R. Price, "The Autocorrelogram of a Complete Carrier Wave Received over the Ionosphere at Oblique Incidence," *Proc. IRE*, **45**, 879–880 (1957).

[60] S. O. Rice, "Mathematical Analysis of Random Noise," *Bell System Tech. J.*, **23**, 283–332 (1944); **24**, 46–156 (1945).

[61] J. G. Proakis, "Optimum Pulse Transmissions for Multipath Channels," Group Report 64-G-3, M.I.T. Lincoln Lab., August 16, 1963.

[62] S. Darlington, "Linear Least-Squares Smoothing and Prediction, with Applications," *Bell System Tech. J.*, **37**, 1221–1294 (1958).

[63] J. K. Wolf, "On the Detection and Estimation Problem for Multiple Non-stationary Random Processes," Ph.D. Thesis, Princeton University, Princeton, New Jersey, 1959.

[64] J. B. Thomas and E. Wong, "On the Statistical Theory of Optimum Demodulation," *IRE Trans. Inform. Theory* **IT-6**, 420–425 (September 1960).

[65] D. G. Brennan, "On the Maximum Signal-to-Noise Ratio Realizable from Several Noisy Signals," *Proc. IRE*, **43**, 1530 (1955).

[66] R. M. Fano, *The Transmission of Information*, M.I.T. Press and Wiley, New York, 1961.

[67] V. R. Algazi and R. M. Lerner, "Optimum Binary Detection in Non-Gaussian Noise," *IEEE Trans. Inform. Theory*, **IT-12**, No. 2, p. 269; also Lincoln Laboratory Preprint, DS-2138, M.I.T., 1965.

[68] H. Sherman and B. Reiffen, "An Optimum Demodulator for Poisson Processes: Photon Source Detectors," *Proc. IEEE*, **51**, 1316–1320 (October 1963).

[69] A. J. Viterbi, *Principles of Coherent Communication*, McGraw-Hill, New York, 1966.

[70] A. V. Balakrishnan, "A Contribution to the Sphere Packing Problem of Communication Theory," *J. Math. Anal. Appl.*, **3**, 485–506 (December 1961).

[71] H. J. Landau and D. Slepian, "On the Optimality of the Regular Simplex Code," *Bell System Tech. J.*; to be published.

[72] E. Arthurs and H. Dym, "On the Optimum Detection of Digital Signals in the Presence of White Gaussian Noise—a Geometric Interpretation and a Study of Three Basic Data Transmission Systems," *IRE Trans. Commun. Systems*, **CS-10**, 336–372 (December 1962).

[73] C. R. Cahn, "Performance of Digital Phase-Modulation Systems," *IRE Trans. Commun. Systems*, **CS-7**, 3–6 (May 1959).

[74] L. Collins, "An expression for $\frac{\partial h_o(s, \tau:t)}{\partial t}$," Detection and Estimation Theory Group Internal Memo., IM-LDC-6, M.I.T. (April 1966).

[75] A. Erdelyi et al., *Higher Transcendental Functions*, McGraw-Hill, New York, (1954).

[76] S. Stein, "Unified Analysis of Certain Coherent and Non-Coherent Binary Communication Systems," *IEEE Trans. Inform. Theory*, **IT-10**, 43–51 (January 1964).

[77] C. W. Helstrom, "The Resolution of Signals in White Gaussian Noise," *Proc. IRE*, **46**, 1603–1619 (September 1958).

[78] G. L. Turin, "The Asymptotic Behavior of Ideal *M*-ary Systems," *Proc. IRE* (correspondence), **47**, 93–94 (1959).

[79] L. Collins, "An Asymptotic Expression for the Error Probability for Signals with Random Phase," Detection and Estimation Theory Group Internal Memo, IM-LDC-20, M.I.T. (February 3, 1967).

[80] J. N. Pierce, "Theoretical Diversity Improvement in Frequency-Shift Keying," *Proc. IRE*, **46**, 903–910 (May 1958).

[81] M. Doelz, E. Heald, and D. Martin, "Binary Data Transmission Techniques for Linear Systems," *Proc. IRE*, **45**, 656–661 (May 1957).

[82] J. J. Bussgang and M. Leiter, "Error Rate Approximation for Differential Phase-Shift Keying," *IEEE Trans. Commun. Systems*, **CS-12, 18**–**27** (March 1964).

[83] M. Leiter and J. Proakis, "The Equivalence of Two Test Statistics for an Optimum Differential Phase-Shift Keying Receiver," *IEEE Trans. Commun. Tech.*, **COM-12**, No. 4, 209–210 (December 1964).

[84] P. Hahn, "Theoretical Diversity Improvement in Multiple FSK," *IRE Trans. Commun. Systems*. **CS-10**, 177–184 (June 1962).

[85] J. C. Hancock and W. C. Lindsey, "Optimum Performance of Self-Adaptive Systems Operating Through a Rayleigh-Fading Medium," *IEEE Trans. Commun. Systems*, **CS-11**, 443–453 (December 1963).

[86] Irwin Jacobs, "Probability-of-Error Bounds for Binary Transmission on the Slowly Fading Rician Channel," *IEEE Trans. Inform. Theory*, **IT-12**, No. 4, 431–441 (October 1966).

[87] S. Darlington, "Demodulation of Wideband, Low-Power FM Signals," *Bell System Tech. J., 43,* No. 1., Pt. 2, 339–374 (1964).

[88] H. Akima, "Theoretical Studies on Signal-to-Noise Characteristics of an FM System, *IEEE Trans. on Space Electronics and Telemetry,* SET-9, 101–108 (1963).

[89] W. C. Lindsey, "Error Probabilities for Incoherent Diversity Reception," *IEEE Trans. Inform. Theory*, **IT-11**, No. 4, 491–499 (October 1965).

[90] S. M. Sussman, "Simplified Relations for Bit and Character Error Probabilities for *M*-ary Transmission over Rayleigh Fading Channels," *IEEE Trans. Commun. Tech.*, **COM-12**, No. 4 (December 1964).

[91] L. Collins, "Realizable Likelihood Function Computers for Known Signals in Colored Noise," Detection and Estimation Theory Group Internal Memo, IM-LDC-7, M.I.T. (April 1966).

[92] M. Schwartz, W. R. Bennett, and S. Stein, *Communication Systems and Techniques*, McGraw-Hill, New York, 1966.

5

Estimation of Continuous Waveforms

5.1 INTRODUCTION

Up to this point we have considered the problems of detection and parameter estimation. We now consider the problem of estimating a *continuous* waveform. Just as in the parameter estimation problem, we shall find it convenient to discuss both nonrandom waveforms and waveforms that are sample functions from a random process. We shall find that the estimation procedure for nonrandom waveforms is straightforward. By contrast, when the waveform is a sample function from a random process, the formulation is straightforward but the solution is more complex.

Before solving the estimation problem it will be worthwhile to investigate some of the physical problems in which we want to estimate a continuous waveform. We consider the random waveform case first.

An important situation in which we want to estimate a random waveform is in analog modulation systems. In the simplest case the message $a(t)$ is the input to a no-memory modulator whose output is $s(t, a(t))$ which is then transmitted as shown in Fig. 5.1. The transmitted signal is deterministic in the sense that a given sample function $a(t)$ causes a unique output $s(t, a(t))$. Some common examples are the following:

$$s(t, a(t)) = \sqrt{2P}\, a(t) \sin \omega_c t. \tag{1}$$

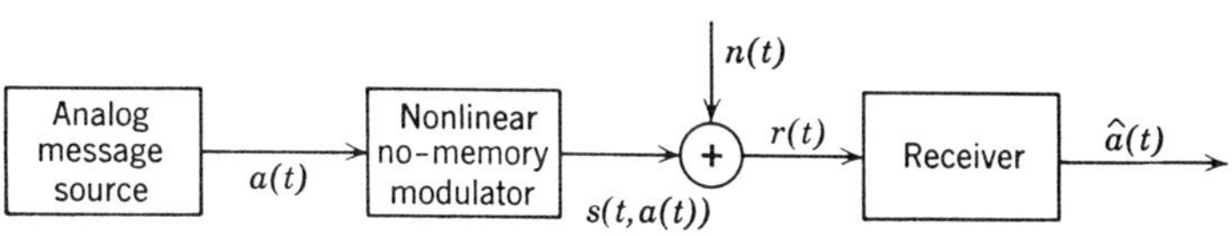

Fig. 5.1 A continuous no-memory modulation system.

This is double sideband, suppressed carrier, amplitude modulation (DSB-SC-AM).

$$s(t, a(t)) = \sqrt{2P}\,[1 + ma(t)] \sin \omega_c t. \tag{2}$$

This is conventional DSB-AM with a residual carrier component.

$$s(t, a(t)) = \sqrt{2P} \sin [\omega_c t + \beta a(t)]. \tag{3}$$

This is phase modulation (PM).

The transmitted waveform is corrupted by a sample function of zero-mean Gaussian noise process which is independent of the message process. The noise is completely characterized by its covariance function $K_n(t, u)$. Thus, for the system shown in Fig.5.1, the received signal is

$$r(t) = s(t, a(t)) + n(t), \qquad T_i \le t \le T_f. \tag{4}$$

The simple system illustrated is not adequate to describe many problems of interest. The first step is to remove the no-memory restriction. A modulation system with memory is shown in Fig. 5.2. Here, $h(t, u)$ represents the impulse response of a linear, not necessarily time-invariant, filter. Examples are the following:

1. The linear system is an integrator and the no-memory device is a phase modulator. In this case the transmitted signal is

$$s(t, x(t)) = \sqrt{2P} \sin \left[\omega_c t + \int_{T_i}^{t} a(u)\, du\right]. \tag{5}$$

This is frequency modulation (FM).

2. The linear system is a realizable time-invariant network and the no-memory device is a phase modulator. The transmitted signal is

$$s(t, x(t)) = \sqrt{2P} \sin \left[\omega_c t + \int_{T_i}^{t} h(t - u)\, a(u)\, du\right]. \tag{6}$$

This is pre-emphasized angle modulation.

Figures 5.1 and 5.2 describe a broad class of analog modulation systems which we shall study in some detail. We denote the waveform of interest, $a(t)$, as the *message*. The message may come from a variety of sources. In

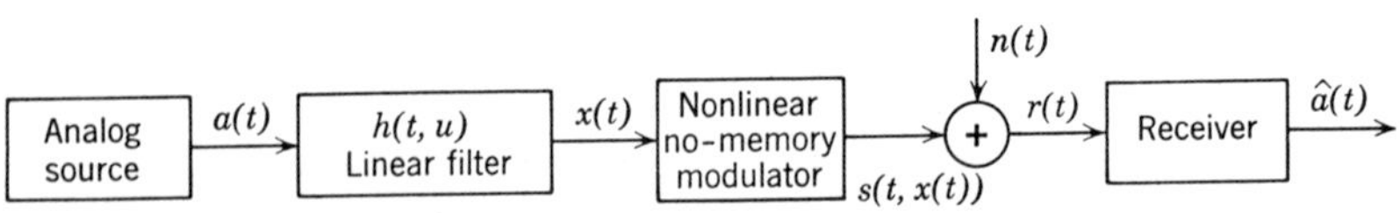

Fig. 5.2 A modulation system with memory.

commercial FM it corresponds to music or speech. In a satellite telemetry system it might correspond to analog data from a sensor (e.g., temperature or attitude).

The waveform estimation problem occurs in a number of other areas. If we remove the modulator in Fig. 5.1,

$$r(t) = a(t) + n(t), \qquad T_i \leq t \leq T_f. \tag{7}$$

If $a(t)$ represents the position of some object we are trying to track in the presence of measurement noise, we have the simplest form of the control problem.

Many more complicated systems also fit into the model. Three are shown in Fig. 5.3. The system in Fig. 5.3*a* is an FM/FM system. This type of system is commonly used when we have a number of messages to transmit. Each message is modulated onto a subcarrier at a different frequency, the modulated subcarriers are summed, and modulated onto the main carrier. In the Fig. 5.3*a* we show the operations for a single message. The system in Fig. 5.3*b* represents an FM modulation system

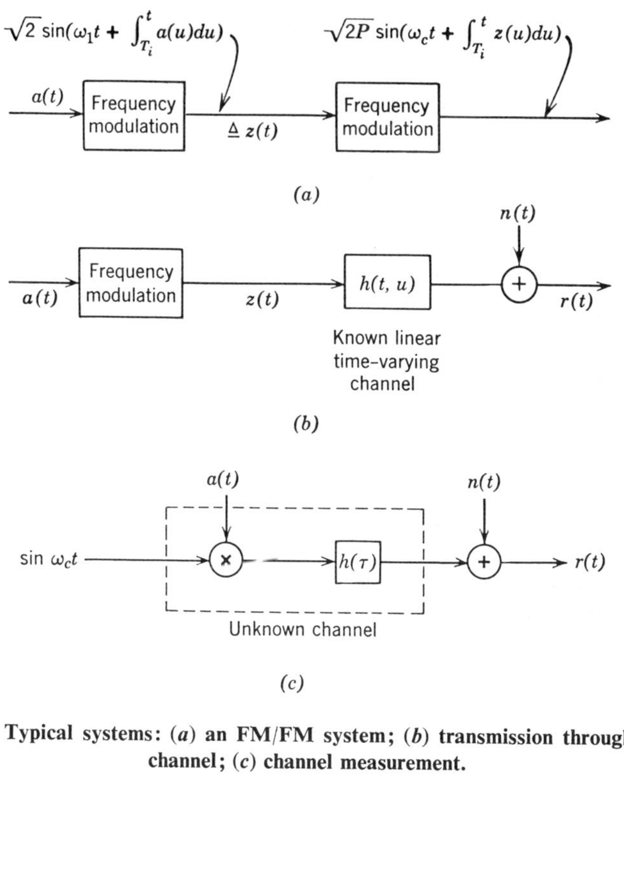

Fig. 5.3 Typical systems: (*a*) an FM/FM system; (*b*) transmission through varying channel; (*c*) channel measurement.

transmitting through a *known* linear time-varying channel. In Fig. 5.3c the channel has an impulse response that depends on the random process $a(t)$. The input is a deterministic signal and we want to estimate the channel impulse response. Measurement problems of this type arise frequently in digital communication systems. A simple example was encountered when we studied the Rayleigh channel in Section 4.4. Other examples will arise in Chapters II.2 and II.3. Note that the channel process is the "message" in this class of problems.

We see that all the problems we have described correspond to the first level in the hierarchy described in Chapter 1. We referred to it as the known signal-in-noise problem. It is important to understand the meaning of this description in the context of continuous waveform estimation. *If* $a(t)$ were known, then $s(t, a(t))$ would be known. In other words, *except* for the additive noise, the mapping from $a(t)$ to $r(t)$ is deterministic.

We shall find that in order to proceed it is expedient to assume that $a(t)$ is a sample function from a Gaussian random process. In many cases this is a valid assumption. In others, such as music or speech, it is *not*. Fortunately, we shall find experimentally that if we use the Gaussian assumption in system design, the system will work well for many non-Gaussian inputs.

The chapter proceeds in the following manner. In Section 5.2 we derive the equations that specify the optimum estimate $\hat{a}(t)$. In Section 5.3 we derive bounds on the mean-square estimation error. In Section 5.4 we extend the results to vector messages and vector received signals. In Section 5.5 we solve the nonrandom waveform estimation problem.

The purpose of the chapter is to develop the necessary equations and to look at some of the properties that can be deduced without solving them. A far more useful end result is the solutions of these equations and the resulting receiver structures. In Chapter 6 we shall study the linear modulation problem in detail. In Chapter II.2 we shall study the nonlinear modulation problem.

5.2 DERIVATION OF ESTIMATOR EQUATIONS

In this section we want to solve the estimation problem for the type of system shown in Fig. 5.1. The general category of interest is defined by the property that the mapping from $a(t)$ to $s(t, a(t))$ is a no-memory transformation.

The received signal is

$$r(t) = s(t, a(t)) + n(t), \qquad T_i \leq t \leq T_f. \tag{8}$$

By a no-memory transformation we mean that the transmitted signal at some time t_0 depends only on $a(t_0)$ and not on the past of $a(t)$.

5.2.1 No-Memory Modulation Systems. Our specific assumptions are the following:

1. The message $a(t)$ and the noise $n(t)$ are sample functions from independent, continuous, zero-mean Gaussian processes with covariance functions $K_a(t, u)$ and $K_n(t, u)$, respectively.
2. The signal $s(t, a(t))$ has a derivative with respect to $a(t)$. As an example, for the DSB-SC-AM signal in (1) the derivative is

$$\frac{\partial s(t, a(t))}{\partial a(t)} = \sqrt{2P} \sin \omega_c t. \tag{9}$$

Clearly, whenever the transformation $s(t, a(t))$ is a linear transformation, the derivative will not be a function of $a(t)$. We refer to these cases as linear modulation schemes. For PM

$$\frac{\partial s(t, a(t))}{\partial a(t)} = \sqrt{2P}\, \beta \cos (\omega_c t + \beta a(t)). \tag{10}$$

The derivative is a function of $a(t)$. This is an example of a nonlinear modulation scheme. These ideas are directly analogous to the linear signaling and nonlinear signaling schemes in the parameter estimation problem.

As in the parameter estimation case, we must select a suitable criterion. The mean-square error criterion and the maximum a posteriori probability criterion are the two logical choices. Both are conceptually straightforward and lead to identical answers for linear modulation schemes.

For nonlinear modulation schemes both criteria have advantages and disadvantages. In the minimum mean-square error case, if we formulate the a posteriori probability density of $a(t)$ over the interval $[T_i, T_f]$ as a Gaussian-type quadratic form, it is difficult to find an explicit expression for the conditional mean. On the other hand, if we model $a(t)$ as a component of a vector Markov process, we shall see that we can find a differential equation for the conditional mean that represents a formal explicit solution to the problem. This particular approach requircs background we have not developed, and we defer it until Chapter II.2. In the maximum a posteriori probability criterion case we are led to an integral equation whose solution is the MAP estimate. This equation provides a simple physical interpretation of the receiver. The MAP estimate will turn out to be asymptotically efficient. Because the MAP formulation is more closely related to our previous work, we shall emphasize it.†

† After we have studied the problem in detail we shall find that in the region in which we get an estimator of practical value the MMSE and MAP estimates coincide.

To help us in solving the waveform estimation problem let us recall some useful facts from Chapter 4 regarding parameter estimation.

In (4.464) and (4.465) we obtained the integral equations that specified the optimum estimates of a set of parameters. We repeat the result. If $a_1, a_2, \ldots, a_K$ are independent zero-mean Gaussian random variables, which we denote by the vector $\mathbf{a}$, the MAP estimates $\hat{a}_i$ are given by the simultaneous solution of the equations,

$$\hat{a}_i = \sigma_i^2 \int_{T_i}^{T_f} \frac{\partial s(z, \mathbf{A})}{\partial A_i}\bigg|_{\mathbf{A}=\hat{\mathbf{a}}} [r_g(z) - g(z)]\, dz, \qquad i = 1, 2, \ldots, K, \tag{11}$$

where

$$\sigma_i^2 \triangleq \mathrm{Var}\,(a_i), \tag{12}$$

$$r_g(z) \triangleq \int_{T_i}^{T_f} Q_n(z, u)\, r(u)\, du, \qquad T_i \leq z \leq T_f, \qquad (13)\dagger$$

$$g(z) \triangleq \int_{T_i}^{T_f} Q_n(z, u)\, s(u, \hat{\mathbf{a}})\, du, \qquad T_i \leq z \leq T_f, \tag{14}$$

and the received waveform is

$$r(t) = s(t, \mathbf{A}) + n(t). \qquad T_i \leq t \leq T_f. \tag{15}$$

Now we want to apply this result to our problem. From our work in Chapter 3 we know that we can represent the message, $a(t)$, in terms of an orthonormal expansion:

$$a(t) = \underset{K\to\infty}{\mathrm{l.i.m.}} \sum_{i=1}^{K} a_i\, \psi_i(t), \qquad T_i \leq t \leq T_f, \tag{16}$$

where the $\psi_i(t)$ are solutions to the integral equation

$$\mu_i\, \psi_i(t) = \int_{T_i}^{T_f} K_a(t, u)\, \psi_i(u)\, du, \qquad T_i \leq t \leq T_f \tag{17}$$

and

$$a_i = \int_{T_i}^{T_f} a(t)\, \psi_i(t)\, dt. \tag{18}$$

The a_i are independent Gaussian variables:

$$E(a_i) = 0 \tag{19}$$

and

$$\mathrm{E}\,(a_i a_j) = \mu_i\, \delta_{ij}. \tag{20}$$

Now we consider a subclass of processes, those that can be represented by the first K terms in the orthonormal expansion. Thus

$$a_K(t) = \sum_{i=1}^{K} a_i\, \psi_i(t), \qquad T_i \leq t \leq T_f. \tag{21}$$

† Just as in the colored noise discussions of Chapter 4, the end-points must be treated carefully. Throughout Chapter 5 we shall include the end-points in the interval.

Our logic is to show how the problem of estimating $a_K(t)$ in (21) is identical to the problem we have already solved of estimating a set of K independent parameters. We then let $K \to \infty$ to obtain the desired result. An easy way to see this problem is identical is given in Fig. 5.4*a*. If we look only at the modulator, we may logically write the transmitted signal as

$$s(t, a_K(t)) = s\left(t, \sum_{i=1}^{K} A_i\, \psi_i(t)\right). \tag{22}$$

By grouping the elements as shown in Fig. 5.4*b*, however, we may logically write the output as $s(t, \mathbf{A})$. Clearly the two forms are equivalent:

$$s(t, \mathbf{A}) = s\left(t, \sum_{i=1}^{K} A_i\, \psi_i(t)\right). \tag{23}$$

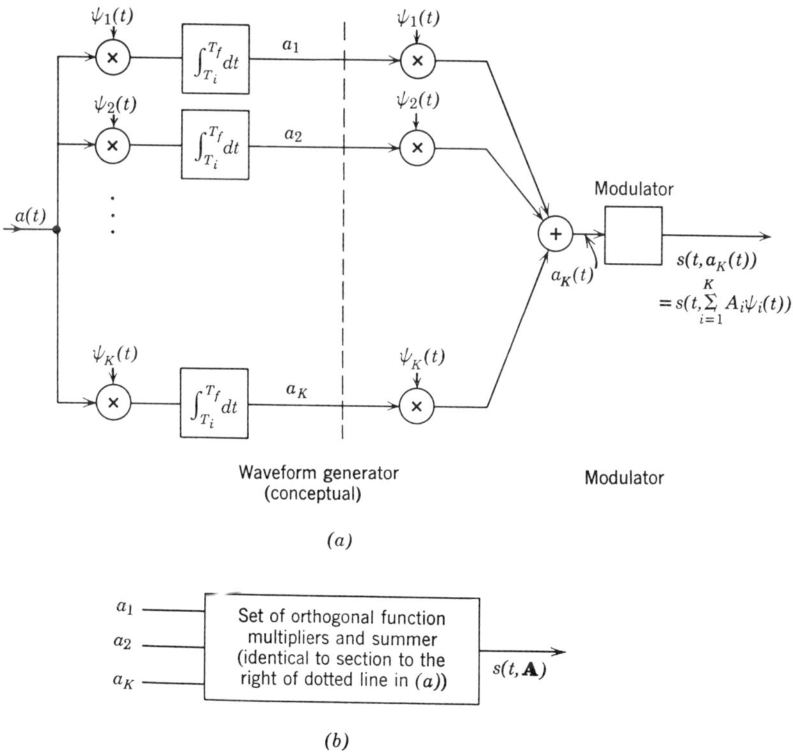

Fig. 5.4 Equivalence of waveform representation and parameter representation.

We *define* the MAP estimate of $a_K(t)$ as

$$\hat{a}_K(t) = \sum_{r=1}^{K} \hat{a}_r \, \psi_r(t), \qquad T_i \le t \le T_f. \tag{24}$$

We see that $\hat{a}_K(t)$ is an *interval estimate*. In other words, we are estimating the *waveform* $a_K(t)$ over the entire interval $T_i \le t \le T_f$ rather than the value at a single instant of time in the interval. To find the estimates of the coefficients we can use (11).

Looking at (22) and (23), we see that

$$\begin{aligned} \frac{\partial s(z, \mathbf{A})}{\partial A_r} &= \frac{\partial s(z, a_K(z))}{\partial A_r} \\ &= \frac{\partial s(z, a_K(z)}{\partial a_K(z)} \cdot \frac{\partial a_K(z)}{\partial A_r} \\ &= \left[\frac{\partial s(z, a_K(z))}{\partial a_K(z)}\right] \psi_r(z), \end{aligned} \tag{25}$$

where the last equality follows from (21). From (11)

$$\hat{a}_r = \mu_r \int_{T_i}^{T_f} \frac{\partial s(z, a_K(z))}{\partial a_K(z)}\bigg|_{a_K(z) = \hat{a}_K(z)} \psi_r(z)[r_g(z) - g(z)] \, dz, \qquad r = 1, 2, \ldots, K. \tag{26}$$

Substituting (26) into (24) we see that

$$\hat{a}_K(t) = \sum_{r=1}^{K} \psi_r(t)\mu_r \int_{T_i}^{T_f} \frac{\partial s(z, a_K(z))}{\partial a_K(z)}\bigg|_{a_K(z) = \hat{a}_K(z)} \psi_r(z)[r_g(z) - g(z)] \, dz \tag{27}$$

or

$$\hat{a}_K(t) = \int_{T_i}^{T_f} \frac{\partial s(z, a_K(z))}{\partial a_K(z)}\bigg|_{a_K(z) = \hat{a}_K(z)} \left[\sum_{r=1}^{K} \mu_r \, \psi_r(t) \, \psi_r(z)\right] [r_g(z) - g(z)] \, dz. \tag{28}$$

In this form it is now easy to let $K \to \infty$. From Mercer's theorem in Chapter 3

$$\lim_{K \to \infty} \sum_{r=1}^{K} \mu_r \, \psi_r(t) \, \psi_r(z) = K_a(t, z). \tag{29}$$

Now define

$$\hat{a}(t) = \underset{K \to \infty}{\text{l.i.m.}} \; \hat{a}_K(t), \qquad T_i \le t \le T_f. \tag{30}$$

The resulting equation is

$$\hat{a}(t) = \int_{T_i}^{T_f} \frac{\partial s(z, \hat{a}(z))}{\partial \hat{a}(z)} K_a(t, z)[r_g(z) - g(z)]\, dz, \qquad T_i \le t \le T_f, \quad (31)\dagger$$

where

$$r_g(z) = \int_{T_i}^{T_f} Q_n(z, u)\, r(u)\, du, \qquad T_i \le z \le T_f, \quad (32)$$

and

$$g(z) = \int_{T_i}^{T_f} Q_n(z, u)\, s(u, \hat{a}(u))\, du, \qquad T_i \le z \le T_f. \quad (33)$$

Equations 31, 32, and 33 specify the MAP estimate of the waveform $a(t)$. These equations (and their generalizations) form the basis of our study of analog modulation theory. For the special case in which the additive noise is white, a much simpler result is obtained. If

$$K_n(t, u) = \frac{N_0}{2} \delta(t - u), \tag{34}$$

then

$$Q_n(t, u) = \frac{2}{N_0} \delta(t - u). \tag{35}$$

Substituting (35) into (32) and (33), we obtain

$$r_g(z) = \frac{2}{N_0} r(z) \tag{36}$$

and

$$g(z) = \frac{2}{N_0} s(z, \hat{a}(z)). \tag{37}$$

Substituting (36) and (37) into (31), we have

$$\hat{a}(t) = \frac{2}{N_0} \int_{T_i}^{T_f} K_a(t, z) \frac{\partial s(z, \hat{a}(z))}{\partial \hat{a}(z)} [r(z) - s(z, \hat{a}(z))]\, dz, \qquad T_i \le t \le T_f. \tag{38}$$

Now the estimate is specified by a single nonlinear integral equation.

In the parameter estimation case we saw that it was useful to interpret the integral equation specifying the MAP estimate as a block diagram. This interpretation is even more valuable here. As an illustration, we consider two simple examples.

† The results in (31)–(33) were first obtained by Youla [1]. In order to simplify the notation we have made the substitution

$$\frac{\partial s(z, \hat{a}(z))}{\partial \hat{a}(z)} \triangleq \left.\frac{\partial s(z, a(z))}{\partial a(z)}\right|_{a(z) = \hat{a}(z)}.$$

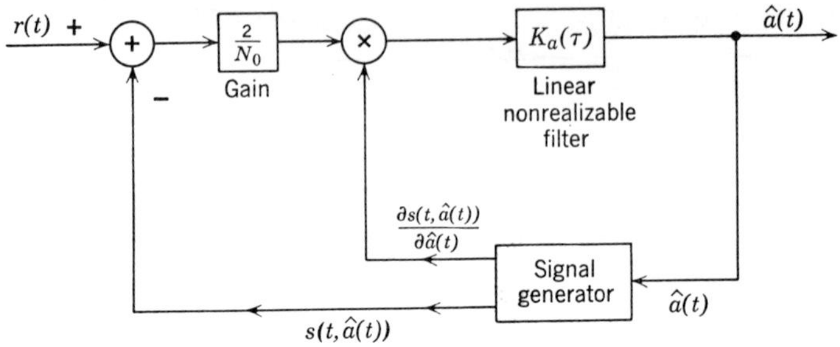

Fig. 5.5 A block diagram of an unrealizable system: white noise.

Example 1. Assume that

$$T_i = -\infty,\; T_f = \infty, \tag{39}$$

$$K_a(t, u) = K_a(t - u), \tag{40}$$

$$K_n(t, u) = \frac{N_0}{2}\,\delta(t - u). \tag{41}$$

In this case (38) is appropriate.† Substituting into (38), we have

$$\hat{a}(t) = \frac{2}{N_0}\int_{-\infty}^{\infty} K_a(t - z)\left\{\frac{\partial s(z, \hat{a}(z))}{\partial \hat{a}(z)}\,[r(z) - s(z, \hat{a}(z))]\right\} dz, \qquad -\infty < t < \infty. \tag{42}$$

We observe that this is simply a convolution of the term inside the braces with a linear filter whose impulse response is $K_a(\tau)$. Thus we can visualize (42) as the block diagram in Fig. 5.5. Observe that the linear filter is *unrealizable*. It is important to emphasize that the block diagram is only a conceptual aid in interpreting (42). It is clearly not a practical solution (in its present form) to the nonlinear integral equation because we cannot build the unrealizable filter. One of the problems to which we shall devote our attention in succeeding chapters is finding a practical approximation to the block diagram.

A second easy example is the nonwhite noise case.

Example 2. Assume that

$$T_i = -\infty,\; T_f = \infty, \tag{43}$$

$$K_a(t, u) = K_a(t - u), \tag{44}$$

$$K_n(t, u) = K_n(t - u). \tag{45}$$

Now (43) and (45) imply that

$$Q_n(t, u) = Q_n(t - u) \tag{46}$$

As in Example 1, we can interpret the integrals (31), (32), and (33) as the block diagram shown in Fig. 5.6. Here, $Q_n(\tau)$ is an unrealizable time-invariant filter.

† For this case we should derive the integral equation by using a spectral representation based on the integrated transform of the process instead of a Karhunen-Loève expansion representation. The modifications in the derivation are straightforward and the result is identical; therefore we relegate the derivation to the problems (see Problem 5.2.6).

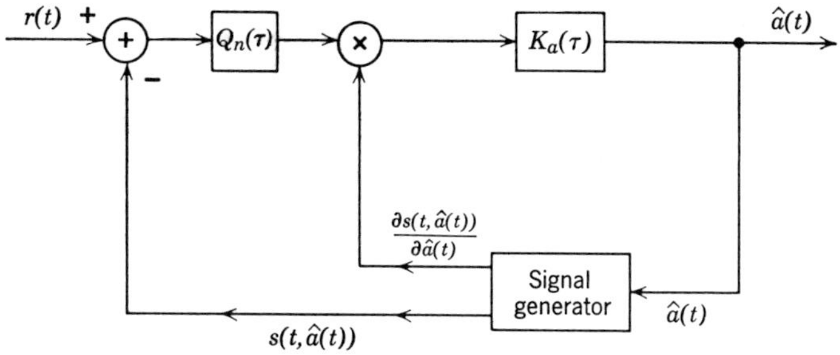

Fig. 5.6 A block diagram of an unrealizable system: colored noise.

Before proceeding we recall that we assumed that the modulator was a no-memory device. This assumption is too restrictive. As we pointed out in Section 1, this assumption excludes such common schemes as FM. While the derivation is still fresh in our minds we can modify it to eliminate this restriction.

5.2.2 Modulation Systems with Memory†

A more general modulation system is shown in Fig. 5.7. The linear system is described by a deterministic impulse response $h(t, u)$. It may be time-varying or unrealizable. Thus we may write

$$x(t) = \int_{T_i}^{T_f} h(t, u)\, a(u)\, du, \qquad T_i \le t \le T_f. \tag{47}$$

The modulator performs a no-memory operation on $x(t)$,

$$s(t, x(t)) = s\left(t, \int_{T_i}^{T_f} h(t, u)\, a(u)\, du\right). \tag{48}$$

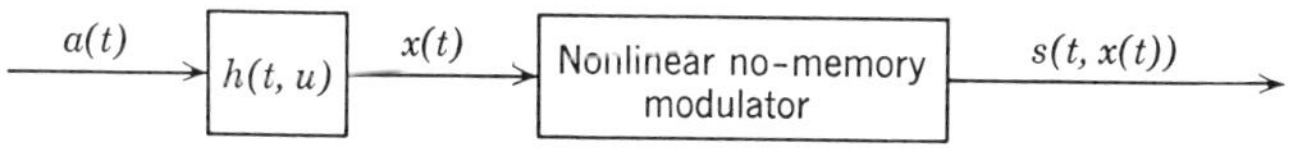

Fig. 5.7 Modulator with memory.

† The extension to include FM is due to Lawton [2], [3]. The extension to arbitrary linear operations is due to Van Trees [4]. Similar results were derived independently by Rauch in two unpublished papers [5], [6].

As an example, for FM,

$$x(t) = d_f \int_{T_i}^{t} a(u)\, du, \tag{49}$$

where d_f is the frequency deviation. The transmitted signal is

$$s(t, x(t)) = \sqrt{2P} \sin\left(\omega_c t + d_f \int_{T_i}^{t} a(u)\, du\right). \tag{50}$$

Looking back at our derivation we see that everything proceeds identically until we want to take the partial derivative with respect to A_r in (25). Picking up the derivation at this point, we have

$$\begin{aligned} \frac{\partial s(z, \mathbf{A})}{\partial A_r} &= \frac{\partial s(z, x_K(z))}{\partial A_r} \\ &= \frac{\partial s(z, x_K(z))}{\partial x_K(z)} \frac{\partial x_K(z)}{\partial A_r}; \end{aligned} \tag{51}$$

but

$$\begin{aligned} \frac{\partial x_K(z)}{\partial A_r} &= \frac{\partial}{\partial A_r} \int_{T_i}^{T_f} h(z, y)\, a_K(y)\, dy \\ &= \frac{\partial}{\partial A_r} \int_{T_i}^{T_f} h(z, y) \sum_{i=1}^{K} A_i\, \psi_i(y)\, dy \\ &= \int_{T_i}^{T_f} h(z, y)\, \psi_r(y)\, dy. \end{aligned} \tag{52}$$

It is convenient to give a label to the output of the linear operation when the input is $\hat{a}(t)$. We define

$$\tilde{x}_K(t) = \int_{T_i}^{T_f} h(t, y)\, \hat{a}_K(y)\, dy. \tag{53}$$

It should be observed that $\tilde{x}_K(t)$ is not defined to be the MAP estimate of $x_K(t)$. In view of our results with finite sets of parameters, we suspect that it is. For the present, however, it is simply a function defined by (53) From (11),

$$\hat{a}_r = \mu_r \int_{T_i}^{T_f} \frac{\partial s(z, \tilde{x}_K(z))}{\partial \tilde{x}_K(z)} \left(\int_{T_i}^{T_f} h(z, y)\, \psi_r(y)\, dy\right)(r_g(z) - g(z))\, dz. \tag{54}$$

As before,

$$\hat{a}_K(t) = \sum_{r=1}^{K} \hat{a}_r\, \psi_r(t), \tag{55}$$

and from (54),

$$\hat{a}_K(t) = \int_{T_i}^{T_f} \frac{\partial s(z, \tilde{x}_K(z))}{\partial \tilde{x}_K(z)} \left\{ \int_{T_i}^{T_f} h(z, y) \left[\sum_{r=1}^{K} \mu_r \, \psi_r(t) \, \psi_r(y) \right] dy \right\} \times [r_g(z) - g(z)] \, dz. \tag{56}$$

Letting $K \rightarrow \infty$, we obtain

$$\boxed{\hat{a}(t) = \iint_{T_i}^{T_f} dy \, dz \, \frac{\partial s(z, \tilde{x}(z))}{\partial \tilde{x}(z)} \, h(z, y) \, K_a(t, y)[r_g(z) - g(z)], \qquad T_i \le t \le T_f,} \tag{57}$$

where $r_g(z)$ and $g(z)$ were defined in (32) and (33) [replace $\hat{a}(u)$ by $\tilde{x}(u)$]. Equation 57 is similar in form to the no-memory equation (31). If we care to, we can make it identical by performing the integration with respect to y in (57)

$$\int_{T_i}^{T_f} h(z, y) \, K_a(t, y) \, dy \triangleq h_a(z, t) \tag{58}$$

so that

$$\hat{a}(t) = \int_{T_i}^{T_f} dz \, \frac{\partial s(z, \tilde{x}(z))}{\partial \tilde{x}(z)} \, h_a(z, t)(r_g(z) - g(z)) \, dz, \qquad T_i \le t \le T_f. \tag{59}$$

Thus the block diagram we use to represent the equation is identical in structure to the no-memory diagram given in Fig. 5.8. This similarity in structure will prove to be useful as we proceed in our study of modulation systems.

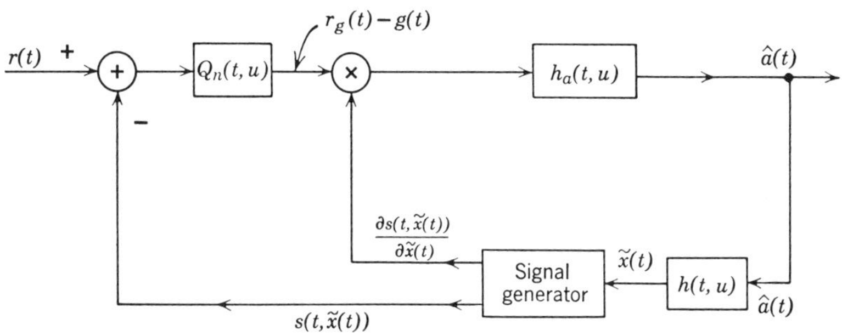

Fig. 5.8 A block diagram.

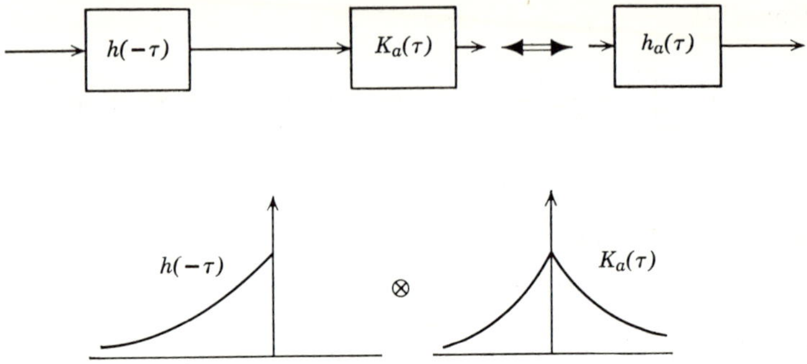

Fig. 5.9 Filter interpretation.

An interesting interpretation of the filter $h_a(z, t)$ can be made for the case in which $T_i = -\infty$, $T_f = \infty$, $h(z, y)$ is time-invariant, and $a(t)$ is stationary. Then

$$h_a(z, t) = \int_{-\infty}^{\infty} h(z - y)\, K_a(t - y)\, dy = h_a(z - t). \tag{60}$$

We see that $h_a(z - t)$ is a cascade of two filters, as shown in Fig. 5.9. The first has an impulse response corresponding to that of the filter in the modulator *reversed in time*. This is familiar in the context of a matched filter. The second filter is the correlation function.

A final question of interest with respect to modulation systems with memory is: Does

$$\tilde{x}(t) = \hat{x}(t)? \tag{61}$$

In other words, is a linear operation on a MAP estimate of a continuous random process equal to the MAP estimate of the output of the linear operation? We shall prove that (61) is true for the case we have just studied. More general cases follow easily. From (53) we have

$$\tilde{x}(\tau) = \int_{T_i}^{T_f} h(\tau, t)\, \hat{a}(t)\, dt. \tag{62}$$

Substituting (57) into (62), we obtain

$$\tilde{x}(\tau) = \int_{T_i}^{T_f} \frac{\partial s(z, \tilde{x}(z))}{\partial \tilde{x}(z)} [r_g(z) - g(z)] \times \left[\iint_{T_i}^{T_f} h(\tau, t)\, h(z, y)\, K_a(t, y)\, dt\, dy\right] dz. \tag{63}$$

We now want to write the integral equation that specifies $\hat{x}(\tau)$ and compare it with the right-hand side of (63). The desired equation is identical to (31), with $a(t)$ replaced by $x(t)$. Thus

$$\hat{x}(\tau) = \int_{T_i}^{T_f} \frac{\partial s(z, \hat{x}(z))}{\partial \hat{x}(z)} K_x(\tau, z)[r_g(z) - g(z)]\, dz. \tag{64}$$

We see that $\tilde{x}(\tau) = \hat{x}(\tau)$ if

$$\iint_{T_i}^{T_f} h(\tau, t)\, h(z, y)\, K_a(t, y)\, dt\, dy = K_x(\tau, z), \qquad T_i \leq \tau, z \leq T_f; \tag{65}$$

but

$$K_x(\tau, z) \triangleq E[x(\tau)\, x(z)] = E\left[\int_{T_i}^{T_f} h(\tau, t)\, a(t)\, dt \int_{T_i}^{T_f} h(z, y)\, a(y)\, dy\right]$$

$$= \iint_{T_i}^{T_f} h(\tau, t)\, h(z, y)\, K_a(t, y)\, dt\, dy, \qquad T_i \leq \tau, z \leq T_f, \tag{66}$$

which is the desired result. Thus we see that the operations of maximum a posteriori interval estimation and linear filtering commute. This result is one that we might have anticipated from the analogous results in Chapter 4 for parameter estimation.

We can now proceed with our study of the characteristics of MAP estimates of waveforms and the structure of the optimum estimators.

The important results of this section are contained in (31), (38), and (57). An alternate derivation of these equations with a variational approach is given in Problem 5.2.1.

5.3 A LOWER BOUND ON THE MEAN-SQUARE ESTIMATION ERROR†

In our work with estimating finite sets of variables we found that an extremely useful result was the lower bound on the mean-square error that any estimate could have. We shall see that in waveform estimation such a bound is equally useful. In this section we derive a lower bound on the mean-square error that any estimate of a random process can have.

First, define the error waveform

$$a_\epsilon(t) = a(t) - \hat{a}(t) \tag{67}$$

and

$$e_I = \frac{1}{T_f - T_i} \int_{T_i}^{T_f} [a(t) - \hat{a}(t)]^2\, dt = \frac{1}{T} \int_{T_i}^{T_f} a_\epsilon^2(t)\, dt, \tag{68}$$

† This section is based on Van Trees [7].

where $T \triangleq T_f - T_i$. The subscript I emphasizes that we are making an interval estimate. Now e_I is a random variable. We are concerned with its expectation,

$$T\xi_I \triangleq TE(e_I) = E\left\{\int_{T_i}^{T_f} \sum_{i=1}^{\infty} (a_i - \hat{a}_i)\,\psi_i(t) \sum_{j=1}^{\infty} (a_j - \hat{a}_j)\,\psi_j(t)\,dt\right\}. \quad (69)$$

Using the orthogonality of the eigenfunctions, we have

$$\xi_I T = \sum_{i=1}^{\infty} E(a_i - \hat{a}_i)^2. \quad (70)$$

We want to find a lower bound on the sum on the right-hand side. We first consider the sum $\sum_{i=1}^{K} E(a_i - \hat{a}_i)^2$ and then let $K \to \infty$. The problem of bounding the mean-square error in estimating K random variables is familiar to us from Chapter 4.

From Chapter 4 we know that the first step is to find the information matrix $\mathbf{J}_T$, where

$$\mathbf{J}_T = \mathbf{J}_D + \mathbf{J}_P, \quad (71a)$$

$$J_{D_{ij}} = -E\left[\frac{\partial^2 \ln \Lambda(\mathbf{A})}{\partial A_i\,\partial A_j}\right], \quad (71b)$$

and

$$J_{P_{ij}} = -E\left[\frac{\partial^2 \ln p_{\mathbf{a}}(\mathbf{A})}{\partial A_i\,\partial A_j}\right]. \quad (71c)$$

After finding $\mathbf{J}_T$, we invert it to obtain $\mathbf{J}_T^{-1}$. Throughout the rest of this chapter we shall always be interested in $\mathbf{J}_T$ so we suppress the subscript T for convenience. The expression for $\ln \Lambda(\mathbf{A})$ is the vector analog to (4.217)

$$\ln \Lambda(\mathbf{A}) = \iint_{T_i}^{T_f} [r(t) - \tfrac{1}{2} s(t, \mathbf{A})]\, Q_n(t, u)\, s(u, \mathbf{A})\, dt\, du \quad (72a)$$

or, in terms of $a_K(t)$,

$$\ln \Lambda(a_K(t)) = \iint_{T_i}^{T_f} [r(t) - \tfrac{1}{2} s(t, a_K(t))]\, Q_n(t, u)\, s(u, a_K(u))\, dt\, du. \quad (72b)$$

From (19) and (20)

$$\ln p_{\mathbf{a}}(\mathbf{A}) = \sum_{i=1}^{K}\left[-\frac{A_i^2}{2\mu_i} - \frac{1}{2}\ln(2\pi)\right]. \quad (72c)$$

Adding (72b) and (72c) and differentiating with respect to A_i, we obtain

$$\frac{\partial[\ln p_{\mathbf{a}}(\mathbf{A}) + \ln \Lambda(a_K(t))]}{\partial A_i} =$$

$$-\frac{A_i}{\mu_i} + \int_{T_i}^{T_f} dt\, \psi_i(t)\, \frac{\partial s(t, a_K(t))}{\partial a_K(t)} \int_{T_i}^{T_f} Q_n(t, u)[r(u) - s(u, a_K(u))]\, du. \quad (72d)$$

Differentiating with respect to A_j and including the minus signs, we have

$$J_{ij} = \frac{\delta_{ij}}{\mu_i} + E\int\!\!\int_{T_i}^{T_f} dt\,du\,\psi_i(t)\,\psi_j(u)\,\frac{\partial s(t, a_K(t))}{\partial a_K(t)}\,Q_n(t, u)\,\frac{\partial s(u, a_K(u))}{\partial a_K(u)}$$

$$+ \text{ terms with zero expectation.} \tag{73}$$

Looking at (73), we see that an efficient estimate will exist only when the modulation is linear (see p. 84).

To interpret the first term recall that

$$K_a(t, u) = \sum_{i=1}^{\infty} \mu_i\,\psi_i(t)\,\psi_i(u), \qquad T_i \le t, u \le T_f. \tag{74}$$

Because we are using only K terms, we define

$$K_{a_K}(t, u) \triangleq \sum_{i=1}^{K} \mu_i\,\psi_i(t)\,\psi_i(u), \qquad T_i \le t, u \le T_f. \tag{75}$$

The form of the first term in (73) suggests defining

$$Q_{a_K}(t, u) = \sum_{i=1}^{K} \frac{1}{\mu_i}\,\psi_i(t)\,\psi_i(u), \qquad T_i \le t, u \le T_f. \tag{76}$$

We observe that

$$\int_{T_i}^{T_f} Q_{a_K}(t, u)\,K_{a_K}(u, z)\,du = \sum_{i=1}^{K} \psi_i(t)\,\psi_i(z), \qquad T_i \le t, z \le T_f. \tag{77}$$

Once again, $Q_{a_K}(t, u)$ is an inverse kernel, but because the message $a(t)$ does not contain a white noise component, the limit of the sum in (76) as $K \to \infty$ will not exist in general. Thus we must eliminate $Q_{a_K}(t, u)$ from our solution before letting $K \to \infty$. Observe that we may write the first term as

$$\frac{\delta_{ij}}{\mu_i} = \int\!\!\int_{T_i}^{T_f} Q_{a_K}(t, u)\,\psi_i(t)\,\psi_j(u)\,dt\,du, \tag{78}$$

so that if we define

$$J_K(t, u) \triangleq Q_{a_K}(t, u) + E\left[\frac{\partial s(t, a_K(t))}{\partial a_K(t)}\,Q_n(t, u)\,\frac{\partial s(u, a_K(u))}{\partial a_K(u)}\right] \tag{79}$$

we can write the elements of the **J** matrix as

$$J_{ij} = \int\!\!\int_{T_i}^{T_f} J_K(t, u)\,\psi_i(t)\,\psi_j(u)\,dt\,du, \qquad i, j = 1, 2, \ldots, K. \tag{80}$$

Now we find the inverse matrix $\mathbf{J}^{-1}$. We can show (see Problem 5.3.6) that

$$J^{ij} = \iint_{T_i}^{T_f} J_K^{-1}(t, u)\, \psi_i(t)\, \psi_j(u)\, dt\, du \tag{81}$$

where the function $J_K^{-1}(t, u)$ satisfies the equation

$$\int_{T_i}^{T_f} J_K^{-1}(t, u) J_K(u, z)\, du = \sum_{i=1}^{K} \psi_i(t)\, \psi_i(z). \tag{82}$$

(Recall that the superscript ij denotes an element in $\mathbf{J}^{-1}$.) We now want to put (82) into a more usable form.

If we denote the derivative of $s(t, a_K(t))$ with respect to $a_K(t)$ as $d_s(t, a_K(t))$, then

$$E\left[\frac{\partial s(t, a_K(t))}{\partial a_K(t)} \frac{\partial s(u, a_K(u))}{\partial a_K(u)}\right] = E[d_s(t, a_K(t))\, d_s(u, a_K(u))]$$
$$\triangleq R_{d_s K}(t, u). \tag{83a}$$

Similarly

$$E\left[\frac{\partial s(t, a(t))}{\partial a(t)} \frac{\partial s(u, a(u))}{\partial a(u)}\right] = E[d_s(t, a(t))\, d_s(u, a(u))] \triangleq R_{d_s}(t, u). \tag{83b}$$

Therefore

$$J_K(u, z) \triangleq Q_{a_K}(u, z) + R_{d_s K}(u, z)\, Q_n(u, z). \tag{84}$$

Substituting (84) into (82), multiplying by $K_{a_K}(z, x)$, integrating with respect to z, and letting $K \to \infty$, we obtain the following integral equation for $J^{-1}(t, x)$,

$$\boxed{\begin{aligned} J^{-1}(t, x) + \int_{T_i}^{T_f} du \int_{T_i}^{T_f} dz\, J^{-1}(t, u)\, R_{d_s}(u, z)\, Q_n(u, z)\, K_a(z, x) \\ = K_a(t, x), \qquad T_i \le t, x \le T_f. \end{aligned}} \tag{85}$$

From (2.292) we know that the diagonal elements of $\mathbf{J}^{-1}$ are lower bounds on the mean-square errors. Thus

$$E[(a_i - \hat{a}_i)^2] \ge J^{ii}. \tag{86a}$$

Using (81) in (86*a*) and the result in (70), we have

$$T\xi_I \ge \lim_{K \to \infty} \sum_{i=1}^{K} \iint_{T_i}^{T_f} J_K^{-1}(t, u)\, \psi_i(t)\, \psi_i(u)\, dt\, du \tag{86b}$$

or, using (3.128),

$$\boxed{\xi_I \ge \frac{1}{T} \int_{T_i}^{T_f} J^{-1}(t, t)\, dt.} \tag{87}$$

Therefore to evaluate the lower bound we must solve (85) for $J^{-1}(t, x)$ and evaluate its trace. By analogy with the classical case, we refer to $J(t, x)$ as the *information kernel.*

We now want to interpret (85). First consider the case in which there is only a white noise component so that,

$$Q_n(t, u) = \frac{2}{N_0}\,\delta(t - u). \tag{88}$$

Then (85) becomes

$$J^{-1}(t, x) + \int_{T_i}^{T_f} du\,\frac{2}{N_0}\,J^{-1}(t, u)\,R_{d_s}(u, u)\,K_a(u, x) = K_a(t, x),$$
$$T_i \le t, x \le T_f. \tag{89}$$

The succeeding work will be simplified if $R_{d_s}(t, t)$ is a constant. A sufficient, but not necessary, condition for this to be true is that $d_s(t, a(t))$ be a sample function from a stationary process. We frequently encounter estimation problems in which we can approximate $R_{d_s}(t, t)$ with a constant without requiring $d_s(t, a(t))$ to be stationary. A case of this type arises when the transmitted signal is a bandpass waveform having a spectrum centered around a carrier frequency ω_c; for example, in PM,

$$s(t, a(t)) = \sqrt{2P}\sin\,[\omega_c t + \beta a(t)]. \tag{90}$$

Then

$$d_s(t, a(t)) = \frac{\partial s(t, a(t))}{\partial a(t)} = \sqrt{2P}\,\beta\cos\,[\omega_c t + \beta\,a(t)] \tag{91}$$

and

$$R_{d_s}(t, u) = \beta^2 P E_a\{\cos\,[\omega_c(t - u) + \beta\,a(t) - \beta\,a(u)] + \cos\,[\omega_c(t + u) + \beta\,a(t) + \beta\,a(u)]\}. \tag{92}$$

Letting $u = t$, we observe that

$$R_{d_s}(t, t) = \beta^2 P(1 + E_a\{\cos\,[2\omega_c t + 2\beta\,a(t)]\}). \tag{93}$$

We assume that the frequencies contained in $a(t)$ are low relative to ω_c. To develop the approximation we fix t in (89). Then (89) can be represented as the linear time-varying system shown in Fig. 5.10. The input is a function of u, $J^{-1}(t, u)$. Because $K_a(u, x)$ corresponds to a low-pass filter and $K_a(t, x)$, a low-pass function, we see that $J^{-1}(t, x)$ must be low-pass and the double-frequency term in $R_{d_s}(u, u)$ may be neglected. Thus we can make the approximation

$$R_{d_s}(t, t) \simeq \beta^2 P \simeq R_{d_s}^*(0) \tag{94}$$

to solve the integral equation. The function $R_{d_s}^*(0)$ is simply the stationary component of $R_{d_s}(t, t)$. In this example it is the low-frequency component.

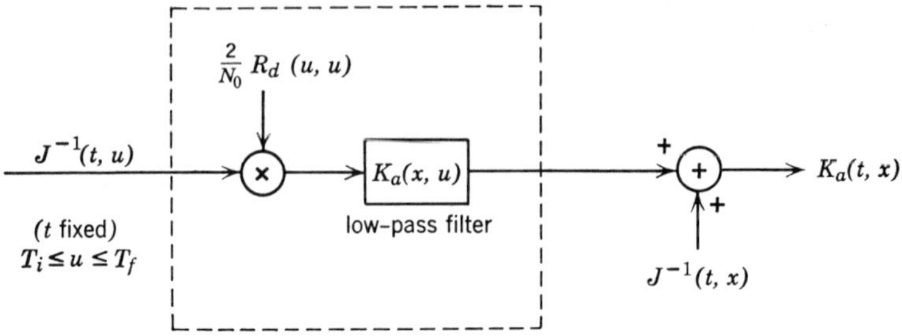

Fig. 5.10 Linear system interpretation.

(Note that $R_{d_s}^*(0) = R_{d_s}(0)$ when $d_s(t, a(t))$ is stationary.) For the cases in which (94) is valid, (89) becomes

$$J^{-1}(t, x) + \frac{2R_{d_s}^*(0)}{N_0} \int_{T_i}^{T_f} du\, J^{-1}(t, u)\, K_a(u, x) = K_a(t, x), \qquad T_i \leq t, x \leq T_f, \tag{95}$$

which is an integral equation whose solution is the desired function. From the mathematical viewpoint, (95) is an adequate final result.

We can, however, obtain a very useful physical interpretation of our result by observing that (95) is familiar in a different context. Recall the following *linear* filter problem (see Chapter 3, p. 198).

$$r(t) = a(t) + n_1(t), \qquad T_i \leq t \leq T_f, \tag{96}$$

where $a(t)$ is the same as our message and $n_1(t)$ is a sample function from a white noise process ($N_1/2$, double-sided). We want to design a linear filter $h(t, x)$ whose output is the estimate of $a(t)$ which minimizes the mean-square error. This is the problem that we solved in Chapter 3. The equation that specifies the filter $h_o(t, x)$ is

$$\frac{N_1}{2} h_o(t, x) + \int_{T_i}^{T_f} h_o(t, u)\, K_a(u, x)\, du = K_a(t, x), \qquad T_i \leq t, x \leq T_f, \tag{97}$$

where $N_1/2$ is the height of the white noise. We see that if we let

$$N_1 = \frac{N_0}{R_{d_s}^*(0)} \tag{98}$$

then

$$J^{-1}(t, x) = \frac{N_0}{2R_{d_s}^*(0)} h_o(t, x). \tag{99}$$

The error in the linear filtering problem is

$$\xi_I = \frac{1}{T}\int_{T_i}^{T_f} \frac{N_1}{2}\, h_o(x, x)\, dx = \frac{1}{T}\int_{T_i}^{T_f} J^{-1}(x, x)\, dx, \tag{100}$$

Our bound for the nonlinear modulation problem corresponds to the mean-square error in the linear filter problem except that the noise level is *reduced* by a factor $R_{d_s}^*(0)$.

The quantity $R_{d_s}^*(0)$ may be greater or less than one. We shall see in Examples 1 and 2 that in the case of linear modulation we can increase $R_{d_s}(0)$ only by *increasing* the transmitted power. In Example 3 we shall see that in a nonlinear modulation scheme such as phase modulation we can increase $R_{d_s}^*(0)$ by increasing the modulation index. This result corresponds to the familiar PM improvement.

It is easy to show a similar interpretation for colored noise. First define an effective noise whose inverse kernel is,

$$Q_{ne}(t, u) = R_{d_s}(t, u)\, Q_n(t, u). \tag{101}$$

Its covariance function satisfies the equation

$$\int_{T_i}^{T_f} K_{ne}(t, u)\, Q_{ne}(u, z)\, du = \delta(t - z), \qquad T_i < t, z < T_f. \tag{102}$$

Then we can show that

$$J^{-1}(u, z) = \int_{T_i}^{T_f} K_{ne}(u, x)\, h_o(x, z)\, dx, \qquad T_i \le u, z \le T_f \tag{103}$$

where $h_o(x, z)$ is the solution to,

$$\int_{T_i}^{T_f} [K_a(x, t) + K_{ne}(x, t)]\, h_o(t, z)\, dt = K_a(x, z), \quad T_i \le x, z \le T_f. \tag{104}$$

This is the colored noise analog to (95).

Two special but important cases lead to simpler expressions.

Case 1. $\mathbf{J^{-1}(t, u) = J^{-1}(t - u)}$. Observe that when $J^{-1}(t, u)$ is a function only of the difference of its two arguments

$$J^{-1}(t, t) = J^{-1}(t - t) = J^{-1}(0). \tag{105}$$

Then (87) becomes,

$$\xi_I \ge J^{-1}(0). \tag{106}$$

If we define

$$\mathfrak{J}^{-1}(\omega) = \int_{-\infty}^{\infty} J^{-1}(\tau) e^{-j\omega\tau}\, d\tau, \tag{107}$$

then

$$\xi_I = \int_{-\infty}^{\infty} \mathfrak{J}^{-1}(\omega)\, \frac{d\omega}{2\pi}. \tag{108}$$

A further simplification develops when the observation interval includes the infinite past and future.

Case 2. Stationary Processes, Infinite Interval.† Here, we assume

$$T_i = -\infty, \tag{109}$$

$$T_f = \infty, \tag{110}$$

$$K_a(t, u) = K_a(t - u), \tag{111}$$

$$K_n(t, u) = K_n(t - u), \tag{112}$$

$$R_{d_s}(t, u) = R_{d_s}(t - u). \tag{113}$$

Then

$$J^{-1}(t, u) = J^{-1}(t - u). \tag{114}$$

The transform of $J(\tau)$ is

$$\mathfrak{J}(\omega) = \int_{-\infty}^{\infty} J(\tau) e^{-j\omega\tau}\, d\tau. \tag{115}$$

Then, from (82) and (85),

$$\mathfrak{J}^{-1}(\omega) = \frac{1}{\mathfrak{J}(\omega)} = \left[\frac{1}{S_a(\omega)} + S_{d_s}(\omega) \otimes \frac{1}{S_n(\omega)}\right]^{-1} \tag{116}$$

(where $\otimes$ denotes convolution‡) and the resulting error is

$$\xi_I \geq \int_{-\infty}^{\infty} \left[\frac{1}{S_a(\omega)} + S_{d_s}(\omega) \otimes \frac{1}{S_n(\omega)}\right]^{-1} \frac{d\omega}{2\pi}. \tag{117}$$

Several simple examples illustrate the application of the bound.

Example 1. We assume that Case 2 applies. In addition, we assume that

$$s(t, a(t)) = a(t). \tag{118}$$

Because the modulation is linear, an efficient estimate exists. There is no carrier so $\partial s(t, a(t))/\partial a(t) = 1$ and

$$S_{d_s}(\omega) = 2\pi\delta(\omega). \tag{119}$$

Substituting into (117), we obtain

$$\xi_I = \int_{-\infty}^{\infty} \frac{S_a(\omega)\, S_n(\omega)}{S_a(\omega) + S_n(\omega)} \frac{d\omega}{2\pi}. \tag{120}$$

The expression on the right-hand side of (120) will turn out to be the minimum mean-square error with an unrealizable linear filter (Chapter 6). Thus, as we would expect, the *efficient* estimate is obtained by processing $r(t)$ with a linear filter.

A second example is linear modulation onto a sinusoid.

Example 2. We assume that Case 2 applies and that the carrier is amplitude-modulated by the message,

$$s(t, a(t)) = \sqrt{2P}\, a(t) \sin \omega_c t, \tag{121}$$

† See footnote on p. 432 and Problem 5.3.3.
‡ We include $1/2\pi$ in the convolution operation when ω is the variable.

where $a(t)$ is low-pass compared with ω_c. The derivative is,

$$\frac{\partial s(t, a(t))}{\partial a(t)} = \sqrt{2P} \sin \omega_c t. \tag{122}$$

For simplicity, we assume that the noise has a flat spectrum whose bandwidth is much larger than that of $a(t)$. It follows easily that

$$\xi_I = \int_{-\infty}^{\infty} \frac{S_a(\omega)}{1 + S_a(\omega)(2P/N_0)} \frac{d\omega}{2\pi}. \tag{123}$$

We can verify that an estimate with this error can be obtained by multiplying $r(t)$ by $\sqrt{2/P} \sin \omega_c t$ and passing the output through the same linear filter as in Example 1. Thus once again an *efficient* estimate exists and is obtained by using a linear system at the receiver.

Example 3. Consider a phase-modulated sine wave in additive white noise. Assume that $a(t)$ is stationary. Thus

$$s(t, a(t)) = \sqrt{2P} \sin [\omega_c t + \beta\, a(t)], \tag{124}$$

$$\frac{\partial s(t, a(t))}{\partial a(t)} = \sqrt{2P}\, \beta \cos [\omega_c t + \beta\, a(t)] \tag{125}$$

and

$$K_n(t, u) = \frac{N_0}{2} \delta(t-u). \tag{126}$$

Then, using (92), we see that

$$R_{d_s}^*(0) = P\beta^2. \tag{127}$$

* By analogy with Example 2 we have

$$\xi_I \geq \int_{-\infty}^{\infty} \frac{S_a(\omega)}{1 + S_a(\omega)(2P\beta^2/N_0)} \frac{d\omega}{2\pi}. \tag{128}$$

In linear modulation the error was only a function of the spectrum of the message, the transmitted power and the white noise level. For a given spectrum and noise level the only way to decrease the mean-square error is to increase the transmitted power. In the nonlinear case we see that by increasing β, the modulation index, we can decrease the bound on the mean-square error. We shall show that as P/N_0 is increased the mean-square error of a MAP estimate approaches the bound given by (128). Thus the MAP estimate is asymptotically efficient. On the other hand, if β is large and P/N_0 is decreased, any estimation scheme will exhibit a "threshold." At this point the estimation error will increase rapidly and the bound will no longer be useful. This result is directly analogous to that obtained for parameter estimation (Example 2, Section 4.2.3). We recall that if we tried to make β too large the result obtained by considering the local estimation problem was meaningless. In Chapter II.2, in which we discuss nonlinear modulation in more detail, we shall see that an analogous phenomenon occurs. We shall also see that for large signal-to-noise ratios the mean-square error approaches the value given by the bound.

The principal results of this section are (85), (95), and (97). The first equation specifies $J^{-1}(t, x)$, the inverse of the information kernel. The trace of this inverse kernel provides a lower bound on the mean-square interval error in continuous waveform estimation. This is a generalization of the classical Cramér-Rao inequality to random processes. The second equation is a special case of (85) which is valid when the additive noise is white and the component of $d_s(t, a(t))$ which affects the integral equation is stationary. The third equation (97) shows how the bound on the mean-square interval estimation error in a nonlinear system is identical to the actual mean-square interval estimation error in a linear system whose white noise level is divided by $R^*_{d_s}(0)$.

In our discussion of detection and estimation we saw that the receiver often had to process multiple inputs. Similar situations arise in the waveform estimation problem.

5.4 MULTIDIMENSIONAL WAVEFORM ESTIMATION

In Section 4.5 we extended the detection problem to *M* received signals. In Problem 4.5.4 of Chapter 4 it was demonstrated that an analogous extension could be obtained for linear and nonlinear estimation of a single parameter. In Problem 4.6.7 of Chapter 4 a similar extension was obtained for multiple parameters. In this section we shall estimate *N* continuous messages by using *M* received waveforms. As we would expect, the derivation is a simple combination of those in Problems 4.6.7 and Section 5.2.

It is worthwhile to point out that all one-dimensional *concepts* carry over directly to the multidimensional case. We can almost guess the form of the particular results. Thus most of the interest in the multidimensional case is based on the solution of these equations for actual physical problems. It turns out that many issues not encountered in the scalar case must be examined. We shall study these issues and their implications in detail in Chapter II.5. For the present we simply derive the equations that specify the MAP estimates and indicate a bound on the mean-square errors.

Before deriving these equations, we shall find it useful to discuss several physical situations in which this kind of problem occurs.

5.4.1 Examples of Multidimensional Problems

Case 1. Multilevel Modulation Systems. In many communication systems a number of messages must be transmitted simultaneously. In one common method we perform the modulation in two steps. First, each of the

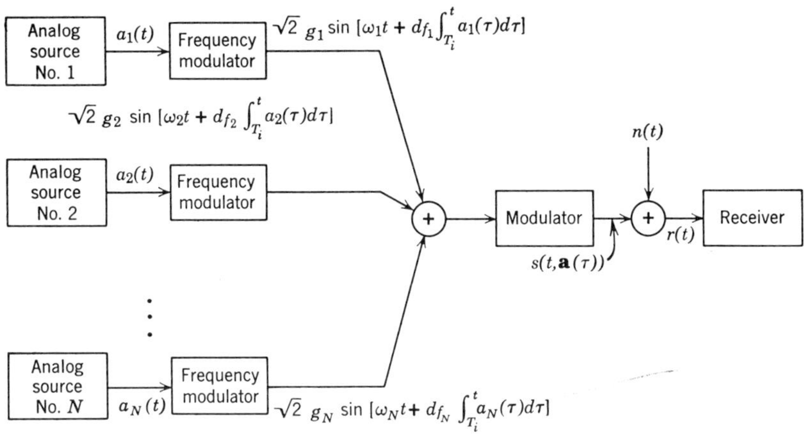

Fig. 5.11 An FM/FM system.

messages is modulated onto individual subcarriers. The modulated subcarriers are then summed and the result is modulated onto a main carrier and transmitted. A typical system is the FM/FM system shown in Fig. 5.11, in which each message $a_i(t)$, $(i = 1, 2, \dots, N)$, is frequency-modulated onto a sine wave of frequency ω_i. The ω_i are chosen so that the modulated subcarriers are in disjoint frequency bands. The modulated subcarriers are amplified, summed and the result is frequency-modulated onto a main carrier and transmitted.

Notationally, it is convenient to denote the N messages by a column matrix,

$$\mathbf{a}(\tau) \triangleq \begin{bmatrix} a_1(\tau) \\ a_2(\tau) \\ \vdots \\ a_N(\tau) \end{bmatrix}. \tag{129}$$

Using this notation, the transmitted signal is

$$s(t, \mathbf{a}(\tau)) = \sqrt{2P} \sin \left[\omega_c t + d_{f_c} \int_{T_i}^{t} z(u)\, du\right], \tag{130†}$$

where

$$z(u) = \sum_{j=1}^{N} \sqrt{2}\, g_j \sin \left[\omega_j u + d_{f_j} \int_{T_i}^{u} a_j(\tau)\, d\tau\right]. \tag{131}$$

† The notation $s(t, \mathbf{a}(\tau))$ is an abbreviation for $s(t; \mathbf{a}(\tau), T_i \le \tau \le t)$. The second variable emphasizes that the modulation process has memory.

The channel adds noise to the transmitted signal so that the received waveform is

$$r(t) = s(t, \mathbf{a}(\tau)) + n(t). \tag{132}$$

Here we want to estimate the N messages simultaneously. Because there are N messages and one received waveform, we refer to this as an $N \times 1$-dimensional problem.

FM/FM is typical of many possible multilevel modulation systems such as SSB/FM, AM/FM, and PM/PM. The possible combinations are essentially unlimited. A discussion of schemes currently in use is available in [8].

Case 2. Multiple-Channel Systems. In Section 4.5 we discussed the use of diversity systems for digital communication systems. Similar systems can be used for analog communication. Figure 5.12 in which the message $a(t)$ is frequency-modulated onto a set of carriers at different frequencies is typical. The modulated signals are transmitted over separate channels, each of which attenuates the signal and adds noise. We see that there are M received waveforms,

$$r_i(t) = s_i(t, a(\tau)) + n_i(t), \qquad (i = 1, 2, \ldots, M), \tag{133}$$

where

$$s_i(t, a(\tau)) = g_i \sqrt{2P_i} \sin\left(\omega_c t + d_{f_i} \int_{T_i}^{t} a(u)\, du\right). \tag{134}$$

Once again matrix notation is convenient. We define

$$\mathbf{s}(t, a(\tau)) = \begin{bmatrix} s_1(t, a(\tau)) \\ s_2(t, a(\tau)) \\ \vdots \\ s_M(t, a(\tau)) \end{bmatrix} \tag{135}$$

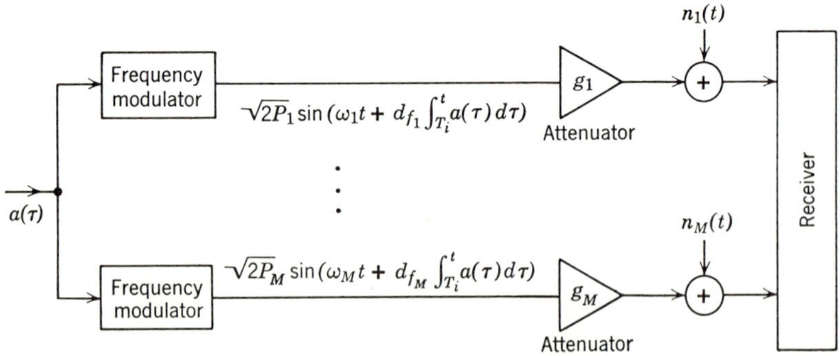

Fig. 5.12 Multiple channel system.

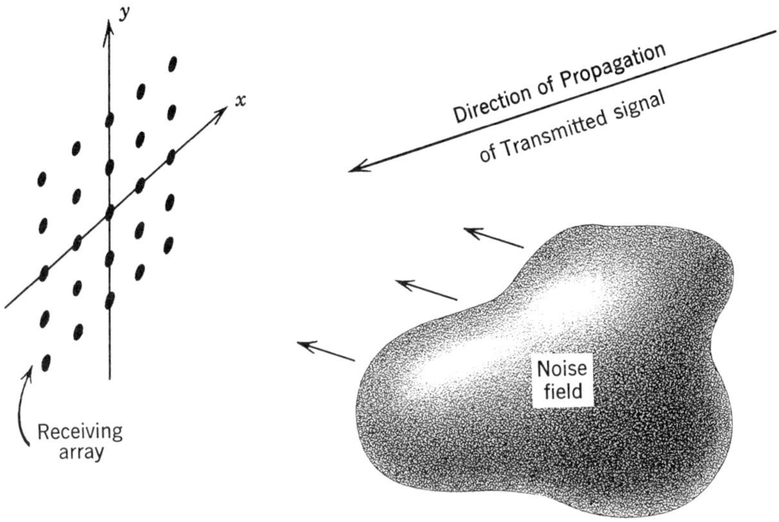

Fig. 5.13 A space-time system.

and

$$\mathbf{n}(t) = \begin{bmatrix} n_1(t) \\ n_2(t) \\ \vdots \\ n_M(t) \end{bmatrix}. \tag{136}$$

Then

$$\mathbf{r}(t) = \mathbf{s}(t, a(\tau)) + \mathbf{n}(t). \tag{137}$$

Here there is one message $a(t)$ to estimate and M waveforms are available to perform the estimation. We refer to this as a $1 \times M$-dimensional problem. The system we have shown is a frequency diversity system. Other obvious forms of diversity are space and polarization diversity.

A physical problem that is essentially a diversity system is discussed in the next case.

Case 3. A Space-Time System. In many sonar and radar problems the receiving system consists of an array of elements (Fig. 5.13). The received signal at the ith element consists of a signal component, $s_i(t, a(\tau))$, an external noise term $n_{Ei}(t)$, and a term due to the noise in the receiver element, $n_{Ri}(t)$. Thus the total received signal at the ith element is

$$r_i(t) = s_i(t, a(\tau)) + n_{Ri}(t) + n_{Ei}(t). \tag{138}$$

We define

$$n_i(t) = n_{Ri}(t) + n_{Ei}(t). \tag{139}$$

We see that this is simply a different physical situation in which we have M waveforms available to estimate a single message. Thus once again we have a $1 \times M$-dimension problem.

Case 4. $N \times M$-Dimensional Problems. If we take any of the multilevel modulation schemes of Case 1 and transmit them over a diversity channel, it is clear that we will have an $N \times M$-dimensional estimation problem. In this case the ith received signal, $r_i(t)$, has a component that depends on N messages, $a_j(t)$, $(j = 1, 2, \ldots, N)$. Thus

$$r_i(t) = s_i(t, \mathbf{a}(\tau)) + n_i(t), \qquad i = 1, 2, \ldots, M. \tag{140}$$

In matrix notation

$$\mathbf{r}(t) = \mathbf{s}(t, \mathbf{a}(\tau)) + \mathbf{n}(t). \tag{141}$$

These cases serve to illustrate the types of physical situations in which multidimensional estimation problems appear. We now formulate the model in general terms.

5.4.2 Problem Formulation†

Our first assumption is that the messages $a_i(t)$, $(i = 1, 2, \ldots, N)$, are sample functions from continuous, jointly Gaussian random processes. It is convenient to denote this set of processes by a single vector process $\mathbf{a}(t)$. (As before, we use the term vector and column matrix interchangeably.) We assume that the vector process has a zero mean. Thus it is completely characterized by an $N \times N$ covariance matrix,

$$\mathbf{K}_{\mathbf{a}}(t, u) \triangleq E[(\mathbf{a}(t)\, \mathbf{a}^T(u))]$$

$$= \begin{bmatrix} K_{a_1 a_1}(t, u) & K_{a_1 a_2}(t, u) & \cdots & K_{a_1 a_N}(t, u) \\ \vdots & & & \vdots \\ K_{a_N a_1}(t, u) & & \cdots & K_{a_N a_N}(t, u) \end{bmatrix}. \tag{142}$$

Thus the ijth element represents the cross-covariance function between the ith and jth messages.

The transmitted signal can be represented as a vector $\mathbf{s}(t, \mathbf{a}(\tau))$. This vector signal is deterministic in the sense that if a particular vector sample

† The multidimensional problem for no-memory signaling schemes and additive channels was first done in [9]. (See also [10].)

function $\mathbf{a}(\tau)$ is given, $\mathbf{s}(t, \mathbf{a}(\tau))$ will be uniquely determined. The transmitted signal is corrupted by an additive Gaussian noise $\mathbf{n}(t)$. The signal available to the receiver is an M-dimensional vector signal $\mathbf{r}(t)$,

$$\mathbf{r}(t) = \mathbf{s}(t, \mathbf{a}(\tau)) + \mathbf{n}(t), \qquad T_i \le t \le T_f, \tag{143}$$

or

$$\begin{bmatrix} r_1(t) \\ r_2(t) \\ \vdots \\ r_M(t) \end{bmatrix} = \begin{bmatrix} s_1(t, \mathbf{a}(\tau)) \\ s_2(t, \mathbf{a}(\tau)) \\ \vdots \\ s_M(t, \mathbf{a}(\tau)) \end{bmatrix} + \begin{bmatrix} n_1(t) \\ n_2(t) \\ \vdots \\ n_M(t) \end{bmatrix}, \qquad T_i \le t \le T_f. \tag{144}$$

The general model is shown in Fig. 5.14.

We assume that the M noise waveforms are sample functions from zero-mean jointly Gaussian random processes and that the messages and noises are statistically independent. (Dependent messages and noises can easily be included, cf. Problem 5.4.1.) We denote the M noises by a vector noise process $\mathbf{n}(t)$ which is completely characterized by an $M \times M$ covariance matrix $\mathbf{K_n}(t, u)$.

5.4.3 Derivation of Estimator Equations.

We now derive the equations for estimating a vector process. For simplicity we shall do only the no-memory modulation case here. Other cases are outlined in the problems.

The received signal is

$$\mathbf{r}(t) = \mathbf{s}(t, \mathbf{a}(t)) + \mathbf{n}(t), \qquad T_i \le t \le T_f, \tag{145}$$

where $\mathbf{s}(t, \mathbf{a}(t))$ is obtained by a no-memory transformation on the vector $\mathbf{a}(t)$. We also assume that $\mathbf{s}(t, \mathbf{a}(t))$ is differentiable with respect to each $a_i(t)$.

The first step is to expand $\mathbf{a}(t)$ in a vector orthogonal expansion.

$$\mathbf{a}(t) = \underset{K \to \infty}{\text{l.i.m.}} \sum_{r=1}^{K} a_r \boldsymbol{\psi}_r(t), \qquad T_i \le t \le T_f, \tag{146}$$

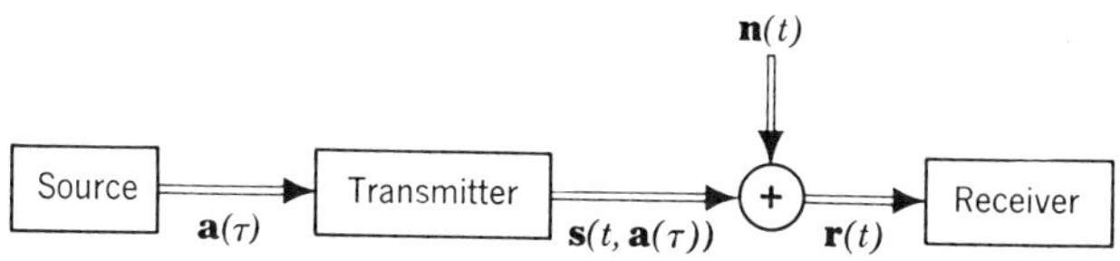

Fig. 5.14 The vector estimation model.

or

$$a_i(t) = \underset{K\to\infty}{\text{l.i.m.}} \sum_{r=1}^{K} a_r\, \psi_r^{(i)}(t), \qquad T_i \le t \le T_f, \tag{147}$$

where the $\boldsymbol{\psi}_r(t)$ are the vector eigenfunctions corresponding to the integral equation

$$\mu_k \boldsymbol{\psi}_k(t) = \int_{T_i}^{T_f} \mathbf{K_a}(t, u)\boldsymbol{\psi}_k(u)\, du, \qquad T_i \le t \le T_f. \tag{148}$$

This expansion was developed in detail in Section 3.7. Then we find $\hat{a}_r$ and define

$$\hat{\mathbf{a}}(t) = \underset{K\to\infty}{\text{l.i.m.}} \sum_{r=1}^{K} \hat{a}_r \boldsymbol{\psi}_r(t). \tag{149}$$

We have, however, already solved the problem of estimating a parameter a_r. By analogy to (72*d*) we have

$$\frac{\partial\,[\ln \Lambda(\mathbf{A}) + \ln p_{\mathbf{a}}(\mathbf{A})]}{\partial A_r} = \int_{T_i}^{T_f} \frac{\partial \mathbf{s}^T(z, \mathbf{a}(z))}{\partial A_r}\, [\mathbf{r_g}(z) - \mathbf{g}(z)]\, dz - \frac{A_r}{\mu_r}, \qquad (r = 1, 2, \ldots), \tag{150}$$

where

$$\mathbf{r_g}(z) \triangleq \int_{T_i}^{T_f} \mathbf{Q_n}(z, u)\, \mathbf{r}(u)\, du \tag{151}$$

and

$$\mathbf{g}(z) \triangleq \int_{T_i}^{T_f} \mathbf{Q_n}(z, u)\, \mathbf{s}(u, \mathbf{a}(u))\, du. \tag{152}$$

The left matrix in the integral in (150) is

$$\frac{\partial \mathbf{s}^T(t, \mathbf{a}(t))}{\partial A_r} = \left[\frac{\partial s_1(t, \mathbf{a}(t))}{\partial A_r} \,\vdots\, \cdots \,\vdots\, \frac{\partial s_M(t, \mathbf{a}(t))}{\partial A_r}\right]. \tag{153}$$

The first element in the matrix can be written as

$$\frac{\partial s_1(t, \mathbf{a}(t))}{\partial A_r} = \frac{\partial s_1(t, \mathbf{a}(t))}{\partial a_1(t)}\, \psi_r^{(1)}(t) + \frac{\partial s_1(t, \mathbf{a}(t))}{\partial a_2(t)}\, \psi_r^{(2)}(t) + \cdots + \frac{\partial s_1(t, \mathbf{a}(t))}{\partial a_N(t)}\, \psi_r^{(N)}(t). \tag{154}$$

Looking at the other elements, we see that if we define a derivative matrix

$$\mathbf{D}(t, \mathbf{a}(t)) \triangleq \nabla_{\mathbf{a}(t)}\{\mathbf{s}^T(t, \mathbf{a}(t))\} = \begin{bmatrix} \dfrac{\partial s_1(t, \mathbf{a}(t))}{\partial a_1(t)} & \cdots & \dfrac{\partial s_M(t, \mathbf{a}(t))}{\partial a_1(t)} \\ \vdots & \ddots & \vdots \\ \dfrac{\partial s_1(t, \mathbf{a}(t))}{\partial a_N(t)} & \cdots & \dfrac{\partial s_M(t, \mathbf{a}(t))}{\partial a_N(t)} \end{bmatrix} \tag{155}$$

we may write

$$\frac{\partial \mathbf{s}^T(t, \mathbf{a}(t))}{\partial A_r} = \boldsymbol{\psi}_r^T(t)\, \mathbf{D}(t, \mathbf{a}(t)), \tag{156}$$

so that

$$\frac{\partial\,[\ln \Lambda(\mathbf{A}) + \ln p_{\mathbf{a}}(\mathbf{A})]}{\partial A_r} = \int_{T_i}^{T_f} \boldsymbol{\psi}_r^T(z)\, \mathbf{D}(z, \mathbf{a}(z))[\mathbf{r}_{\mathbf{g}}(z) - \mathbf{g}(z)]\, dz - \frac{A_r}{\mu_r}. \tag{157}$$

Equating the right-hand side to zero, we obtain a necessary condition on the MAP estimate of A_r. Thus

$$\hat{a}_r = \mu_r \int_{T_i}^{T_f} \boldsymbol{\psi}_r^T(z)\, \mathbf{D}(z, \hat{\mathbf{a}}(z))[\mathbf{r}_{\mathbf{g}}(z) - \mathbf{g}(z)]\, dz. \tag{158}$$

Substituting (158) into (149), we obtain

$$\hat{\mathbf{a}}(t) = \int_{T_i}^{T_f} \left[\sum_{r=1}^{\infty} \mu_r \boldsymbol{\psi}_r(t) \boldsymbol{\psi}_r^T(z)\right] \mathbf{D}(z, \hat{\mathbf{a}}(z))[\mathbf{r}_{\mathbf{g}}(z) - \mathbf{g}(z)]\, dz. \tag{159}$$

We recognize the term in the bracket as the covariance matrix. Thus

$$\boxed{\hat{\mathbf{a}}(t) = \int_{T_i}^{T_f} \mathbf{K}_{\mathbf{a}}(t, z)\, \mathbf{D}(z, \hat{\mathbf{a}}(z))[\mathbf{r}_{\mathbf{g}}(z) - \mathbf{g}(z)]\, dz, \qquad T_i \le t \le T_f.} \tag{160}$$

As we would expect, the form of these equations is directly analogous to the one-dimensional case.

The next step is to find a lower bound on the mean-square error in estimating a vector random process.

5.4.4 Lower Bound on the Error Matrix

In the multidimensional case we are concerned with estimating the vector $\mathbf{a}(t)$. We can define an error vector

$$\hat{\mathbf{a}}(t) - \mathbf{a}(t) = \mathbf{a}_\epsilon(t), \tag{161}$$

which consists of N elements: $a_{\epsilon_1}(t), a_{\epsilon_2}(t), \ldots, a_{\epsilon_N}(t)$. We want to find the error correlation matrix.

Now,

$$\hat{\mathbf{a}}(t) - \mathbf{a}(t) = \sum_{i=1}^{\infty} (\hat{a}_i - a_i)\boldsymbol{\psi}_i(t) \triangleq \sum_{i=1}^{\infty} a_{\epsilon_i} \boldsymbol{\psi}_i(t). \tag{162}$$

Then, using the same approach as in Section 5.3, (68),

$$\begin{aligned}\mathbf{R}_{e_I} &\triangleq \frac{1}{T} E\left[\int_{T_i}^{T_f} \mathbf{a}_\epsilon(t)\, \mathbf{a}_\epsilon^T(t)\, dt\right] \\ &= \frac{1}{T} \lim_{K\to\infty} \int_{T_i}^{T_f} dt \sum_{i=1}^{K} \sum_{j=1}^{K} \boldsymbol{\psi}_i(t)\, E(a_{\epsilon_i} a_{\epsilon_j}) \boldsymbol{\psi}_j^T(t). \end{aligned} \tag{163}$$

We can lower bound the error matrix by a bound matrix $\mathbf{R}_B$ in the sense that the matrix $\mathbf{R}_{e_I} - \mathbf{R}_B$ is nonnegative definite. The diagonal terms in $\mathbf{R}_B$ represent lower bounds on the mean-square error in estimating the $a_i(t)$. Proceeding in a manner analogous to the one-dimensional case, we obtain

$$\mathbf{R}_B = \frac{1}{T}\int_{T_i}^{T_f} \mathbf{J}^{-1}(t, t)\, dt. \tag{164}$$

The kernel $\mathbf{J}^{-1}(t, x)$ is the inverse of the *information matrix kernel* $\mathbf{J}(t, x)$ and is defined by the matrix integral equation

$$\mathbf{J}^{-1}(t, x) + \int_{T_i}^{T_f} du \int_{T_i}^{T_f} dz\, \mathbf{J}^{-1}(t, u)\{E[\mathbf{D}(u, \mathbf{a}(u))\, \mathbf{Q_n}(u, z)\, \mathbf{D}^T(z, \mathbf{a}(z))]\}\mathbf{K_a}(z, x) = \mathbf{K_a}(t, x), \qquad T_i \le t, x \le T_f. \tag{165}$$

The derivation of (164) and (165) is quite tedious and does not add any insight to the problem. Therefore we omit it. The details are contained in [11].

As a final topic in our current discussion of multiple waveform estimation we develop an interesting interpretation of nonlinear estimation in the presence of noise that contains both a colored component and a white component.

5.4.5 Colored Noise Estimation

Consider the following problem:

$$r(t) = s(t, a(t)) + w(t) + n_c(t), \qquad T_i \le t \le T_f. \tag{166}$$

Here $w(t)$ is a white noise component ($N_0/2$, spectral height) and $n_c(t)$ is an independent colored noise component with covariance function $K_c(t, u)$. Then

$$K_n(t, u) = \frac{N_0}{2}\,\delta(t - u) + K_c(t, u). \tag{167}$$

The MAP estimate of $a(t)$ can be found from (31), (32), and (33)

$$\hat{a}(t) = \int_{T_i}^{T_f} K_a(t, u)\, \frac{\partial s(u, \hat{a}(u))}{\partial \hat{a}(u)}\, [r_g(u) - g(u)]\, du, \qquad T_i \le t \le T_f, \tag{168}$$

where

$$r(t) - s(t, \hat{a}(t)) = \int_{T_i}^{T_f} K_n(t, u)[r_g(u) - g(u)]\, du, \qquad T_i \le t \le T_f. \tag{169}$$

Substituting (167) into (169), we obtain

$$r(t) - s(t, \hat{a}(t)) = \int_{T_i}^{T_f} \left[\frac{N_0}{2}\,\delta(t - u) + K_c(t, u)\right][r_g(u) - g(u)]\, du. \tag{170}$$

We now want to demonstrate that the same estimate, $\hat{a}(t)$, is obtained if we estimate $a(t)$ and $n_c(t)$ *jointly*. In this case

$$\mathbf{a}(t) \triangleq \begin{bmatrix} a(t) \\ n_c(t) \end{bmatrix} \tag{171}$$

and

$$\mathbf{K_a}(t, u) = \begin{bmatrix} K_a(t, u) & 0 \\ 0 & K_c(t, u) \end{bmatrix}. \tag{172}$$

Because we are including the colored noise in the message vector, the only additive noise is the white component

$$K_n(t, u) = \frac{N_0}{2}\, \delta(t - u). \tag{173}$$

To use (160) we need the derivative matrix

$$\mathbf{D}(t, \mathbf{a}(t)) = \begin{bmatrix} \dfrac{\partial s(t, a(t))}{\partial a(t)} \\ 1 \end{bmatrix}. \tag{174}$$

Substituting into (160), we obtain two scalar equations,

$$\hat{a}(t) = \int_{T_i}^{T_f} \frac{2}{N_0} K_a(t, u) \frac{\partial s(u, \hat{a}(u))}{\partial \hat{a}(u)} [r(u) - s(u, \hat{a}(u)) - \hat{n}_c(u)]\, du, \qquad T_i \le t \le T_f, \tag{175}$$

$$\hat{n}_c(t) = \int_{T_i}^{T_f} \frac{2}{N_0} K_c(t, u)[r(u) - s(u, \hat{a}(u)) - \hat{n}_c(u)]\, du, \quad T_i \le t \le T_f. \tag{176}$$

Looking at (168), (170), and (175), we see that $\hat{a}(t)$ will be the same in both cases if

$$\int_{T_i}^{T_f} \frac{2}{N_0} [r(u) - s(u, \hat{a}(u)) - \hat{n}_c(u)]\left[\frac{N_0}{2}\, \delta(t - u) + K_c(t, u)\right] du = r(t) - s(t, \hat{a}(t)); \tag{177}$$

but (177) is identical to (176).

This leads to the following conclusion. Whenever there are independent white and colored noise components, we may always consider the colored noise as a message and jointly estimate it. The reason for this result is that the message and colored noise are independent and the noise enters into $r(t)$ in a linear manner. Thus the $N \times 1$ vector white noise problem includes all scalar colored noise problems in which there is a white noise component.

Before summarizing our results in this chapter we discuss the problem of estimating nonrandom waveforms briefly.

5.5 NONRANDOM WAVEFORM ESTIMATION

It is sometimes unrealistic to consider the signal that we are trying to estimate as a random waveform. For example, we may know that each time a particular event occurs the transmitted message will have certain distinctive features. If the message is modeled as a sample function of a random process then, in the processing of designing the optimum receiver, we may average out the features that are important. Situations of this type arise in sonar and seismic classification problems. Here it is more useful to model the message as an unknown, but nonrandom, waveform. To design the optimum processor we extend the maximum-likelihood estimation procedure for nonrandom variables to the waveform case. An appropriate model for the received signal is

$$r(t) = s(t, a(t)) + n(t), \qquad T_i \leq t \leq T_f, \tag{178}$$

where $n(t)$ is a zero-mean Gaussian process.

To find the maximum-likelihood estimate, we write the ln likelihood function and then choose the waveform $a(t)$ which maximizes it. The ln likelihood function is the limit of (72*b*) as $K \to \infty$.

$$\ln \Lambda(a(t)) = \int_{T_i}^{T_f} dt \int_{T_i}^{T_f} du\, s(t, a(t))\, Q_n(t, u)[r(u) - \tfrac{1}{2} s(u, a(u))], \tag{179}$$

where $Q_n(t, u)$ is the inverse kernel of the noise. For arbitrary $s(t, a(t))$ the minimization of $\ln \Lambda(a(t))$ is difficult. Fortunately, in the case of most interest to us the procedure is straightforward. This is the case in which the range of the function $s(t, a(t))$ includes all possible values of $r(t)$. An important example in which this is true is

$$r(t) = a(t) + n(t). \tag{180}$$

An example in which it is not true is

$$r(t) = \sin(\omega_c t + a(t)) + n(t). \tag{181}$$

Here all functions in the range of $s(t, a(t))$ have amplitudes less than one while the possible amplitudes of $r(t)$ are not bounded.

We confine our discussion to the case in which the range includes all possible values of $r(t)$. A necessary condition to minimize $\ln \Lambda(a(t))$ follows easily with variational techniques:

$$\int_{T_i}^{T_f} a_\epsilon(t)\, dt \left\{ \frac{\partial s(t, \hat{a}_{ml}(t))}{\partial \hat{a}_{ml}(t)} \int_{T_i}^{T_f} Q_n(t, u)[r(u) - s(u, \hat{a}_{ml}(u))]\, du \right\} = 0 \tag{182}$$

for every $a_\epsilon(t)$. A solution is

$$r(u) = s(u, \hat{a}_{ml}(u)), \qquad T_i \leq u \leq T_f, \tag{183}$$

because for every $r(u)$ there exists at least one $a(u)$ that could have mapped into it. There is no guarantee, however, that a unique inverse exists. Once again we can obtain a useful answer by narrowing our discussion. Specifically, we shall consider the problem given by (180). Then

$$\hat{a}_{ml}(u) = r(u). \tag{184}$$

Thus the maximum-likelihood estimate is simply the received waveform. It is an unbiased estimate of $a(t)$. It is easy to demonstrate that the maximum-likelihood estimate is efficient. Its variance can be obtained from a generalization of the Cramér-Rao bound or by direct computation. Using the latter procedure, we obtain

$$\sigma_I^2 \triangleq E\left\{\int_{T_i}^{T_f} [\hat{a}_{ml}(t) - a(t)]^2\, dt\right\}. \tag{185}$$

It is frequently convenient to normalize the variance by the length of the interval. We denote this normalized variance (which is just the average mean-square estimation error) as ξ_{ml}:

$$\xi_{ml} \triangleq E\left\{\frac{1}{T}\int_{T_i}^{T_f} [\hat{a}_{ml}(t) - a(t)]^2\, dt\right\}. \tag{186}$$

Using (180) and (184), we have

$$\xi_{ml} = \frac{1}{T}\int_{T_i}^{T_f} K_n(t, t)\, dt. \tag{187}$$

Several observations follow easily.

If the noise is white, the error is infinite. This is intuitively logical if we think of a series expansion of $a(t)$. We are trying to estimate an infinite number of components and because we have assumed no a priori information about their contribution in making up the signal we weight them equally. Because the mean-square errors are equal on each component, the equal weighting leads to an infinite mean-square error. Therefore to make the problem meaningful we must assume that the noise has finite energy over any finite interval. This can be justified physically in at least two ways:

1. The receiving elements (antenna, hydrophone, or seismometer) will have a finite bandwidth;

2. If we assume that we know an approximate frequency band that contains the signal, we can insert a filter that passes these frequencies without distortion and rejects other frequencies.†

† Up to this point we have argued that a white noise model had a good physical basis. The essential point of the argument was that if the noise was wideband compared with the bandwidth of the processors then we could consider it to be white. We tacitly assumed that the receiving elements mentioned in (1) had a much larger bandwidth than the signal processor. Now the mathematical model of the signal does not have enough structure and we must impose the bandwidth limitation to obtain meaningful results.

If the noise process is stationary, then

$$\xi_{ml} = \frac{1}{T}\int_{T_i}^{T_f} K_n(0)\,dt = K_n(0) = \int_{-\infty}^{\infty} S_n(\omega)\,\frac{d\omega}{2\pi}. \tag{188}$$

Our first reaction might be that such a crude procedure cannot be efficient. From the parameter estimation problem, however, we recall that a priori knowledge was not important when the measurement noise was small. The same result holds in the waveform case. We can illustrate this result with a simple example.

Example. Let $n'(t)$ be a white process with spectral height $N_0/2$ and assume that $T_i = -\infty$, $T_f = \infty$. We know that $a(t)$ does not have frequency components above W cps. We pass $r(t)$ through a filter with unity gain from $-W$ to $+W$ and zero gain outside this band. The output is the message $a(t)$ plus a noise $n(t)$ which is a sample function from a flat bandlimited process. The ML estimate of $a(t)$ is the output of this filter and

$$\xi_{ml} = N_0 W. \tag{189}$$

Now suppose that $a(t)$ is actually a sample function from a bandlimited random process $(-W, W)$ and spectral height P. If we used a MAP or an MMSE estimate, it would be efficient and the error would be given by (120),

$$\xi_{ms} = \xi_{map} = \frac{PN_0 W}{P + N_0/2}. \tag{190}$$

The normalized errors in the two cases are

$$\xi_{ml:n} = \frac{N_0 W}{P} \tag{191}$$

and

$$\xi_{ms:n} = \xi_{map:n} = \frac{N_0 W}{P}\left(1 + \frac{N_0}{2P}\right)^{-1}. \tag{192}$$

Thus the difference in the errors is negligible for $N_0/2P < 0.1$.

We see that in this example both estimation procedures assumed a knowledge of the signal bandwidth to design the processor. The MAP and MMSE estimates, however, also required a knowledge of the spectral heights. Another basic difference in the two procedures is not brought out by the example because of the simple spectra that were chosen. The MAP and MMSE estimates are formed by attenuating the various frequencies,

$$H_o(j\omega) = \frac{S_a(\omega)}{N_0/2 + S_a(\omega)}. \tag{193}$$

Therefore, unless the message spectrum is uniform over a fixed bandwidth, the message will be *distorted*. This distortion is introduced to reduce the total mean-square error, which is the sum of message and noise distortion.

On the other hand, the ML estimator never introduces any message distortion; the error is due solely to the noise. (For this reason ML estimators are also referred to as distortionless filters.)

In the sequel we concentrate on MAP estimation (an important exception is Section II.5.3); it is important to remember, however, that in many cases ML estimation serves a useful purpose (see Problem 5.5.2 for a further example.)

5.6 SUMMARY

In this chapter we formulated the problem of estimating a continuous waveform. The primary goal was to develop the equations that specify the estimates.

In the case of a single random waveform the MAP estimate was specified by two equations,

$$\hat{a}(t) = \int_{T_i}^{T_f} K_a(t, u) \frac{\partial s(u, \hat{a}(u))}{\partial \hat{a}(u)} [r_g(u) - g(u)]\, du, \qquad T_i \leq t \leq T_f, \quad (194)$$

where $[r_g(u) - g(u)]$ was specified by the equation

$$r(t) - s(t, \hat{a}(t)) = \int_{T_i}^{T_f} K_n(t, u)[r_g(u) - g(u)]\, du, \qquad T_i \leq t \leq T_f. \quad (195)$$

For the special case of white noise this reduced to

$$\hat{a}(t) = \frac{2}{N_0} \int_{T_i}^{T_f} K_a(t, u)[r(u) - s(u, \hat{a}(u))]\, du, \qquad T_i \leq t \leq T_f. \quad (196)$$

We then derived a bound on the mean-square error in terms of the trace of the information kernel,

$$\xi_I \geq \frac{1}{T} \int_{T_i}^{T_f} J^{-1}(t, t)\, dt. \quad (197)$$

For white noise this had a particularly simple interpretation.

$$\xi_I \geq \frac{N_0}{2R_{d_s}^*(0)} \frac{1}{T} \int_{T_i}^{T_f} h_o(t, t)\, dt, \quad (198)$$

where $h_o(t, u)$ satisfied the integral equation

$$K_a(t, u) = \frac{N_0}{2R_{d_s}^*(0)} h_o(t, u) + \int_{T_i}^{T_f} h_o(t, z)\, K_a(z, u)\, dz, \qquad T_i \leq t, u \leq T_f. \quad (199)$$

The function $h_o(t, u)$ was precisely the optimum processor for a *related* linear estimation problem. We showed that for linear modulation $\hat{a}_{\text{map}}(t)$ was efficient. For nonlinear modulation we shall find that $\hat{a}_{\text{map}}(t)$ is asymptotically efficient.

We then extended these results to the multidimensional case. The basic integral equations were logical extensions of those obtained in the scalar case. We also observed that a colored noise component could always be treated as an additional message and simultaneously estimated.

Finally, we looked at the problem of estimating a nonrandom waveform. The result for the problem of interest was straightforward. A simple example demonstrated a case in which it was essentially as good as a MAP estimate.

In subsequent chapters we shall study the estimator equations and the receiver structures that they suggest in detail. In Chapter 6 we study linear modulation and in Chapter II.2, nonlinear modulation.

5.7 PROBLEMS

Section P5.2 Derivation of Equations

Problem 5.2.1. If we approximate $a(t)$ by a K-term approximation $a_K(t)$, the inverse kernel $Q_{a_K}(t, u)$ is well-behaved. The logarithm of the likelihood function is

$$\ln \Lambda(a_K(t)) + \ln p_{\mathbf{a}}(\mathbf{A}) = \iint_{T_i}^{T_f} [s(t, a_K(t))]\, Q_n(t, u)[r(u) - \tfrac{1}{2}s(u, a_K(u))]\, du$$

$$- \tfrac{1}{2}\iint_{T_i}^{T_f} a_K(t)\, Q_{a_K}(t, u)\, a_K(u)\, dt\, du + \text{constant terms.}$$

1. Use an approach analogous to that in Section 3.4.5 to find $\hat{a}_K(t)$ [i.e., let $a_K(t) = \hat{a}_K(t) + \epsilon a_\epsilon(t)$].

2. Eliminate $Q_{a_K}(t, u)$ from the result and let $K \to \infty$ to get a final answer.

Problem 5.2.2. Let

$$r(t) = s(t, a(t)) + n(t), \qquad T_i \le t \le T_f,$$

where the processes are the same as in the text. Assume that

$$E[a(t)] = m_a(t).$$

1. Find the integral equation specifying $\hat{a}(t)$, the MAP estimate of $a(t)$.
2. Consider the special case in which

$$s(t, a(t)) = a(t).$$

Write the equations for this case.

Problem 5.2.3. Consider the case of the model in Section 5.2.1 in which

$$K_a(t, u) = \sigma_a^2, \qquad 0 \le t, u \le T.$$

1. What does this imply about $a(t)$.
2. Verify that (33) reduces to a previous result under this condition.

Problem 5.2.4. Consider the amplitude modulation system shown in Fig. P5.1. The Gaussian processes $a(t)$ and $n(t)$ are stationary with spectra $S_a(\omega)$ and $S_n(\omega)$, respectively. Let $T_i = -\infty$ and $T_f = \infty$.

1. Draw a block diagram of the optimum receiver.
2. Find $E[a_\epsilon^2(t)]$ as a function of $H(j\omega)$, $S_a(\omega)$, and $S_n(\omega)$.

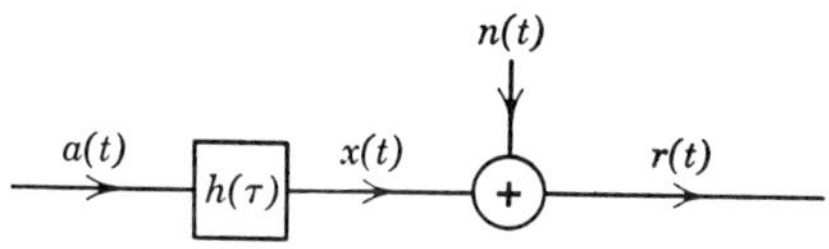

Fig. P5.1

Problem 5.2.5. Consider the communication system shown in Fig. P5.2. Draw a block diagram of the optimum nonrealizable receiver to estimate $a(t)$. Assume that a MAP interval estimate is required.

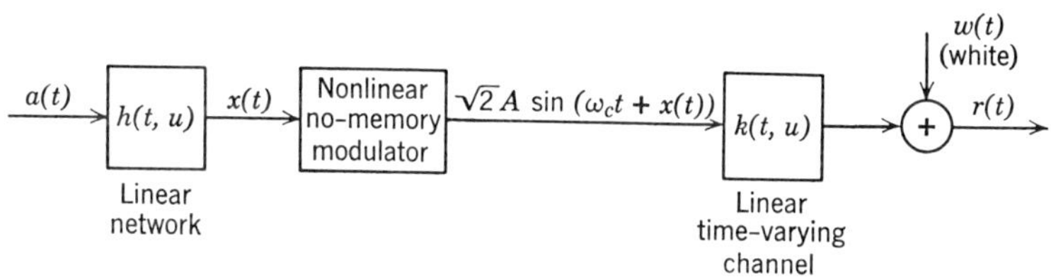

Fig. P5.2

Problem 5.2.6. Let

$$r(t) = s(t, a(t)) + n(t), \qquad -\infty < t < \infty,$$

where $a(t)$ and $n(t)$ are sample functions from zero-mean independent stationary Gaussian processes. Use the integrated Fourier transform method to derive the infinite time analog of (31) to (33) in the text.

Problem 5.2.7. In Chapter 4 (pp. 299–301) we derived the integral equation for the colored noise detection problem by using the idea of a sufficient statistic. Derive (31) to (33) by a suitable extension of this technique.

Section P5.3 Lower Bounds

Problem 5.3.1. In Chapter 2 we considered the case in which we wanted to estimate a linear function of the vector **A**,

$$d \triangleq g_d(\mathbf{A}) = \mathbf{G}_d\mathbf{A},$$

and proved that if $\hat{d}$ was unbiased then

$$E[(\hat{d} - d)^2] \geq \mathbf{G}_d \mathbf{J}^{-1} \mathbf{G}_d^T$$

[see (2.287) and (2.288)]. A similar result can be derived for random variables. Use the random variable results to derive (87).

Problem 5.3.2. Let

$$r(t) = s(t, a(t)) + n(t), \qquad T_i \leq t \leq T_f.$$

Assume that we want to estimate $a(T_f)$. Can you modify the results of Problem 5.3.1 to derive a bound on the mean-square *point* estimation error ?

$$\xi_P \triangleq E\{[\hat{a}(T_f) - a(T_f)]^2\}.$$

What difficulties arise in nonlinear *point* estimation ?

Problem 5.3.3. Let

$$r(t) = s(t, a(t)) + n(t), \qquad -\infty < t < \infty.$$

The processes $a(t)$ and $n(t)$ are statistically independent, stationary, zero-mean, Gaussian random processes with spectra $S_a(\omega)$ and $S_n(\omega)$ respectively. Derive a bound on the mean-square estimation error by using the integrated Fourier transform approach.

Problem 5.3.4. Let

$$r(t) = s(t, x(t)) + n(t), \qquad T_i \leq t \leq T_f,$$

where

$$x(t) = \int_{T_i}^{T_f} h(t, u)\, a(u)\, du, \qquad T_i \leq t \leq T_f,$$

and $a(t)$ and $n(t)$ are statistically independent, zero-mean Gaussian random processes.

1. Derive a bound on the mean-square interval error in estimating $a(t)$.
2. Consider the special case in which

$$s(t, x(t)) = x(t), \quad h(t, u) = h(t - u), \quad T_i = -\infty, \quad T_f = \infty,$$

and the processes are stationary. Verify that the estimate is efficient and express the error in terms of the various spectra.

Problem 5.3.5. Explain why a necessary and sufficient condition for an efficient estimate to exist in the waveform estimation case is that the modulation be linear [see (73)].

Problem 5.3.6. Prove the result given in (81) by starting with the definition of $\mathbf{J}^{-1}$ and using (80) and (82).

Section P5.4 Multidimensional Waveforms

Problem 5.4.1. The received waveform is

$$r(t) = s(t, a(t)) + n(t), \qquad 0 \leq t \leq T.$$

The message $a(t)$ and the noise $n(t)$ are sample functions from zero-mean, jointly Gaussian random processes.

$$E[a(t)\,a(u)] \triangleq K_{aa}(t, u),$$

$$E[a(t)\,n(u)] \triangleq K_{an}(t, u),$$

$$E[n(t)\,n(u)] \triangleq \frac{N_0}{2}\,\delta(t-u) + K_c(t, u).$$

Derive the integral equations that specify the MAP estimate $\hat{a}(t)$. *Hint.* Write a matrix covariance function $\mathbf{K_x}(t, u)$ for the vector $\mathbf{x}(t)$, where

$$\mathbf{x}(t) \triangleq \begin{bmatrix} a_K(t) \\ n(t) \end{bmatrix}.$$

Define an inverse matrix kernel,

$$\int_0^T \mathbf{Q_x}(t, u)\,\mathbf{K_x}(u, z)\,du = \mathbf{I}\,\delta(t-z).$$

Write

$$\ln \Lambda(\mathbf{x}(t)) = -\tfrac{1}{2}\iint_0^T [a_K(t) \,\vdots\, r(t) - s(t, a_K(t))]\mathbf{Q_x}(t, u)\begin{bmatrix} a_K(u) \\ r(u) - s(u, a_K(u)) \end{bmatrix} dt\,du,$$

$$-\tfrac{1}{2}\iint_0^T r(t)Q_{\mathbf{x},22}(t, u)\,r(u)\,dt\,du$$

$$-\tfrac{1}{2}\iint_0^T a_K(t)\,Q_a(t, u)a_K(u)\,dt\,du.$$

Use the variational approach of Problem 5.2.1.

Problem 5.4.2. Let

$$r(t) = s(t, a(t), \mathbf{B}) + n(t), \qquad T_i \le t \le T_f,$$

where $a(t)$ and $n(t)$ are statistically independent, zero-mean, Gaussian random processes. The vector $\mathbf{B}$ is Gaussian, $N(0, \mathbf{\Lambda_B})$, and is independent of $a(t)$ and $n(t)$. Find an equation that specifies the joint MAP estimates of $a(t)$ and $\mathbf{B}$.

Problem 5.4.3. In a PM/PM scheme the messages are phase-modulated onto sub-carriers and added:

$$z(t) = \sum_{j=1}^{N} \sqrt{2}\,g_j \sin\,[\omega_j t + \beta_j\,a_j(t)].$$

The sum $z(t)$ is then phase-modulated onto a main carrier.

$$s(t, \mathbf{a}(t)) = \sqrt{2P}\sin\,[\omega_c t + \beta_c z(t)].$$

The received signal is

$$r(t) = s(t, \mathbf{a}(t)) + w(t), \qquad -\infty < t < \infty.$$

The messages $a_i(t)$ are statistically independent with spectrum $S_a(\omega)$. The noise is independent of $\mathbf{a}(t)$ and is white $(N_0/2)$. Find the integral equation that specifies $\hat{\mathbf{a}}(t)$ and draw the block diagram of an unrealizable receiver. Simplify the diagram by exploiting the frequency difference between the messages and the carriers.

Problem 5.4.4. Let

$$\mathbf{r}(t) = \mathbf{a}(t) + \mathbf{n}(t), \qquad -\infty < t < \infty,$$

where $\mathbf{a}(t)$ and $\mathbf{n}(t)$ are independent Gaussian processes with spectral matrices $\mathbf{S_a}(\omega)$ and $\mathbf{S_n}(\omega)$, respectively.

1. Write (151), (152), and (160) in the frequency domain, using integrated transforms.
2. Verify that $\mathscr{F}[\mathbf{Q_n}(\tau)] = \mathbf{S_n}^{-1}(\omega)$.
3. Draw a block diagram of the optimum receiver. Reduce it to a single matrix filter.
4. Derive the frequency domain analogs to (164) and (165) and use them to write an error expression for this case.
5. Verify that exactly the same results (parts 1, 3, and 4) can be obtained *heuristically* by using ordinary Fourier transforms.

Problem 5.4.5. Let

$$\mathbf{r}(t) = \int_{-\infty}^{\infty} \mathbf{h}(t - \tau)\, a(\tau)\, d\tau + \mathbf{n}(t), \qquad -\infty < t < \infty,$$

where $\mathbf{h}(\tau)$ is a matrix filter with one input and N outputs. Repeat Problem 5.4.4.

Problem 5.4.6. Consider a simple five-element linear array with uniform spacing Δ. (Fig. P5.3). The message is a plane wave whose angle of arrival is θ and velocity of propagation is c. The output at the first element is

$$r_1(t) = a(t) + n_1(t), \qquad -\infty < t < \infty.$$

The output at the second element is

$$r_2(t) = a(t - \tau_\Delta) + n_2(t), \qquad -\infty < t < \infty,$$

where $\tau_\Delta = \Delta \sin \theta / c$. The other outputs follow in an obvious manner. The noises are statistically independent and white ($N_0/2$). The message spectrum is $S_a(\omega)$.

1. Show that this is a special case of Problem 5.4.5.
2. Give an intuitive interpretation of the optimum processor.
3. Write an expression for the minimum mean-square interval estimation error.

$$\xi_I = E[a_\epsilon^2(t)].$$

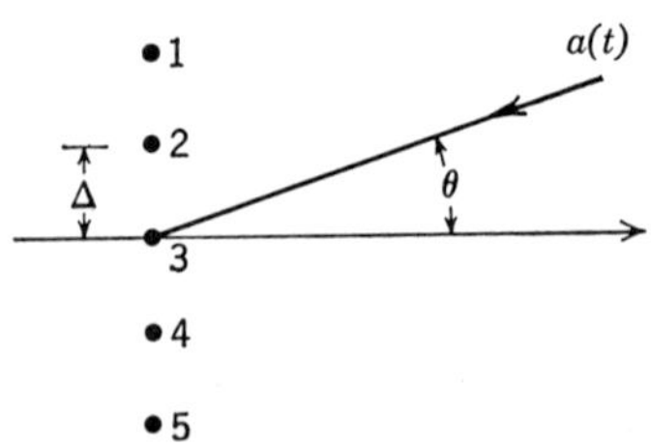

Fig. P5.3

Problem 5.4.7. Consider a zero-mean stationary Gaussian random process $a(t)$ with covariance function $K_a(\tau)$. We observe $a(t)$ in the interval (T_i, T_f) and want to estimate $a(t)$ in the interval (T_α, T_β), where $T_\alpha > T_f$.

1. Find the integral equation specifying $\hat{a}_{\text{map}}(t)$, $T_\alpha \le t \le T_\beta$.
2. Consider the special case in which

$$K_a(\tau) = \sigma_a{}^2 e^{-k|\tau|}, \qquad -\infty < \tau < \infty.$$

Verify that

$$\hat{a}_{\text{map}}(t_1) = a(T_f)e^{-k(t_1 - T_f)} \quad \text{for} \quad T_\alpha \le t_1 \le T_\beta.$$

Hint. Modify the procedure in Problem 5.4.1.

Section P5.5 Nonrandom Waveforms

Problem 5.5.1. Let

$$\mathbf{r}(t) = \mathbf{x}(t) + \mathbf{n}(t), \qquad -\infty < t < \infty,$$

where $\mathbf{n}(t)$ is a stationary, zero-mean, Gaussian process with spectral matrix $\mathbf{S_n}(\omega)$ and $\mathbf{x}(t)$ is a vector signal with finite energy. Denote the vector integrated Fourier transforms of the function as $\mathbf{Z_r}(\omega)$, $\mathbf{Z_x}(\omega)$, and $\mathbf{Z_n}(\omega)$, respectively [see (2.222) and (2.223)]. Denote the Fourier transform of $\mathbf{x}(t)$ as $\mathbf{X}(j\omega)$.

1. Write $\ln \Lambda(\mathbf{x}(t))$ in terms of these quantities.
2. Find $\hat{\mathbf{X}}_{\text{m}l}(j\omega)$.
3. Derive $\hat{\mathbf{X}}_{\text{m}l}(j\omega)$ heuristically using ordinary Fourier transforms for the processes.

Problem 5.5.2. Let

$$\mathbf{r}(t) = \int_{-\infty}^{\infty} \mathbf{h}(t - \tau)\, a(\tau)\, d\tau + \mathbf{n}(t), \qquad -\infty < t < \infty,$$

where $\mathbf{h}(\tau)$ is the impulse response of a matrix filter with one input and N outputs and transfer function $\mathbf{H}(j\omega)$.

1. Modify the results of Problem 5.5.1 to include this case.
2. Find $\hat{a}_{\text{m}l}(t)$.
3. Verify that $\hat{a}_{\text{m}l}(t)$ is unbiased.
4. Evaluate $\text{Var}\,[\hat{a}_{\text{m}l}(t) - a(t)]$.

REFERENCES

[1] D. C. Youla, "The Use of Maximum Likelihood in Estimating Continuously Modulated Intelligence Which Has Been Corrupted by Noise," *IRE Trans. Inform. Theory*, **IT-3**, 90–105 (March 1954).

[2] J. G. Lawton, private communication, dated August 20, 1962.

[3] J. G. Lawton, T. T. Chang, and C. J. Henrich, "Maximum Likelihood Reception of FM Signals," Detection Memo No. 25, Contract No. AF 30(602)-2210, Cornell Aeronautical Laboratory Inc., January 4, 1963.

[4] H. L. Van Trees, "Analog Communication Over Randomly Time-Varying Channels," presented at WESCON, August, 1964.

[5] L. L. Rauch, "Some Considerations on Demodulation of FM Signals," unpublished memorandum, 1961.
[6] L. L. Rauch, "Improved Demodulation and Predetection Recording," Research Report, Instrumentation Engineering Program, University of Michigan, February 28, 1961.
[7] H. L. Van Trees, "Bounds on the Accuracy Attainable in the Estimation of Random Processes," *IEEE Trans. Inform. Theory*, **IT-12**, No. 3, 298–304 (April 1966).
[8] N. H. Nichols and L. Rauch, *Radio Telemetry*, Wiley, New York, 1956.
[9] J. B. Thomas and E. Wong, "On the Statistical Theory of Optimum Demodulation," *IRE Trans. Inform. Theory*, **IT-6**, 420–425 (September 1960).
[10] J. K. Wolf, "On the Detection and Estimation Problem for Multiple Nonstationary Random Processes," Ph.D. Thesis, Princeton University, Princeton, New Jersey, 1959.
[11] H. L. Van Trees, "Accuracy Bounds in Multiple Process Estimation," Internal Memorandum, DET Group, M.I.T., April 1965.

6

Linear Estimation

In this chapter we shall study the linear estimation problem in detail. We recall from our work in Chapter 5 that in the linear modulation problem the received signal was given by

$$r(t) = c(t)\,a(t) + c_0(t) + n(t), \qquad T_i \le t \le T_f, \tag{1}$$

where $a(t)$ was the message, $c(t)$ was a deterministic carrier, $c_0(t)$ was a residual carrier component, and $n(t)$ was the additive noise. As we pointed out in Chapter 5, the effect of the residual carrier is not important in our logical development. Therefore we can assume that $c_0(t)$ equals zero for algebraic simplicity.

A more general form of linear modulation was obtained by passing $a(t)$ through a linear filter to obtain $x(t)$ and then modulating $c(t)$ with $x(t)$. In this case

$$r(t) = c(t)\,x(t) + n(t), \qquad T_i \le t \le T_f. \tag{2}$$

In Chapter 5 we defined linear modulation in terms of the derivative of the signal with respect to the message. An equivalent definition is the following:

Definition. The modulated signal is $s(t, a(t))$. Denote the component of $s(t, a(t))$ that does not depend on $a(t)$ as $c_0(t)$. If the signal $[s(t, a(t)) - c_0(t)]$ obeys superposition, then $s(t, a(t))$ is a linear modulation system.

We considered the problem of finding the maximum a posteriori probability (MAP) estimate of $a(t)$ over the interval $T_i \le t \le T_f$. In the case described by (1) the estimate $\hat{a}(t)$ was specified by two integral equations

$$\hat{a}(t) = \int_{T_i}^{T_f} K_a(t, u)\, c(u)[r_g(u) - g(u)]\, du, \qquad T_i \le t \le T_f, \tag{3}$$

and

$$r(t) - c(t)\,\hat{a}(t) = \int_{T_i}^{T_f} K_n(t, u)[r_g(u) - g(u)]\,du, \qquad T_i \le t \le T_f. \quad (4)$$

We shall study the solution of these equations and the properties of the resulting processors. Up to this point we have considered only *interval* estimation. In this chapter we also consider *point* estimators and show the relationship between the two types of estimators.

In Section 6.1 we develop some of the properties that result when we impose the linear modulation restriction. We shall explore the relation between the Gaussian assumption, the linear modulation assumption, the error criterion, and the structure of the resulting processor. In Section 6.2 we consider the special case in which the infinite past is available (i.e., $T_i = -\infty$), the processes of concern are stationary, and we want to make a minimum mean-square error point estimate. A constructive solution technique is obtained and its properties are discussed. In Section 6.3 we explore a different approach to point estimation. The result is a solution for the processor in terms of a feedback system. In Sections 6.2 and 6.3 we emphasize the case in which $c(t)$ is a constant. In Section 6.4 we look at conventional linear modulation systems such as amplitude modulation and single sideband. In the last two sections we summarize our results and comment on some related problems.

6.1 PROPERTIES OF OPTIMUM PROCESSORS

As suggested in the introduction, when we restrict our attention to linear modulation, certain simplifications are possible that were not possible in the general nonlinear modulation case.

The most important of these simplifications is contained in Property 1.

Property 1. The MAP interval estimate $\hat{a}(t)$ over the interval $T_i \le t \le T_f$, where

$$r(t) = c(t)\,a(t) + n(t), \qquad T_i \le t \le T_f, \quad (5)$$

is the received signal, can be obtained by using a *linear* processor.

Proof. A simple way to demonstrate that a linear processor can generate $\hat{a}(t)$ is to find an impulse response $h_o(t, u)$ such that

$$\hat{a}(t) = \int_{T_i}^{T_f} h_o(t, u)\,r(u)\,du, \qquad T_i \le t \le T_f. \quad (6)$$

First, we multiply (3) by $c(t)$ and add the result to (4), which gives

$$r(t) = \int_{T_i}^{T_f} [c(t)\,K_a(t, u)\,c(u) + K_n(t, u)][r_g(u) - g(u)]\,du, \qquad T_i \le t \le T_f. \quad (7)$$

We observe that the term in the bracket is $K_r(t, u)$. We rewrite (6) to indicate $K_r(t, u)$ explicitly. We also change t to x to avoid confusion of variables in the next step:

$$r(x) = \int_{T_i}^{T_f} K_r(x, u)[r_g(u) - g(u)]\, du, \qquad T_i \le x \le T_f. \tag{8}$$

Now multiply both sides of (8) by $h(t, x)$ and integrate with respect to x,

$$\int_{T_i}^{T_f} h(t, x)\, r(x)\, dx = \int_{T_i}^{T_f} [r_g(u) - g(u)]\, du \int_{T_i}^{T_f} h(t, x)\, K_r(x, u)\, dx. \tag{9}$$

We see that the left-hand side of (9) corresponds to passing the input $r(x)$, $T_i \le x \le T_f$ through a linear time-varying unrealizable filter. Comparing (3) and (9), we see that the output of the filter will equal $\hat{a}(t)$ if we require that the inner integral on the right-hand side of (9) equal $K_a(t,u)\,c(u)$ over the interval $T_i < u < T_f$, $T_i \le t \le T_f$. This gives the equation for the optimum impulse response.

$$K_a(t, u)\, c(u) = \int_{T_i}^{T_f} h_o(t, x)\, K_r(x, u)\, dx, \qquad T_i < u < T_f,\ T_i \le t \le T_f. \tag{10}$$

The subscript o denotes that $h_o(t, x)$ is the optimum processor. In (10) we have used a strict inequality on u. If $[r_g(u) - g(u)]$ does not contain impulses, either a strict or nonstrict equality is adequate. By choosing the inequality to be strict, however, we can find a continuous solution for $h_o(t, x)$. (See discussion in Chapter 3.) Whenever $r(t)$ contains a white noise component, this assumption is valid. As before, we define $h_o(t, x)$ at the end points by a continuity condition:

$$\begin{aligned} h_o(t, T_f) &\triangleq \lim_{x \to T_f^-} h_o(t, x), \\ h_o(t, T_i) &\triangleq \lim_{x \to T_i^+} h_o(t, x). \end{aligned} \tag{11}$$

It is frequently convenient to include the white noise component explicitly. Then we may write

$$K_r(x, u) = c(x)\, K_a(x, u)\, c(u) + K_c(x, u) + \frac{N_0}{2}\, \delta(x - u). \tag{12}$$

and (10) reduces to

$$K_a(t, u)\, c(u) = \frac{N_0}{2}\, h_o(t, u) + \int_{T_i}^{T_f} [c(x)\, K_a(x, u)\, c(u) + K_c(x, u)] h_o(t, x)\, dx,$$
$$T_i < u < T_f,\ T_i \le t \le T_f. \tag{13}$$

If $K_a(t, u)$, $K_c(t, u)$, and $c(t)$ are continuous square-integrable functions, our results in Chapter 4 guarantee that the integral equation specifying $h_o(t, x)$ will have a continuous square-integrable solution. Under these conditions (13) is also true for $u = T_f$ and $u = T_i$ because of our continuity assumption.

The importance of Property 1 is that it guarantees that the structure of the processor is linear and thus reduces the problem to one of finding the correct impulse response. A similar result follows easily for the case described by (2).

Property 1A. The MAP estimate $\hat{a}(t)$ of $a(t)$ over the interval $T_i \leq t \leq T_f$ using the received signal $r(t)$, where

$$r(t) = c(t)\,x(t) + n(t), \qquad T_i \leq t \leq T_f, \tag{14}$$

is obtained by using a linear processor.

The second property is one we have already proved in Chapter 5 (see p. 439). We include it here for completeness.

Property 2. The MAP estimate $\hat{a}(t)$ is also the minimum mean-square error interval estimate in the linear modulation case. (This results from the fact that the MAP estimate is efficient.)

Before solving (10) we shall discuss a related problem. Specifically, we shall look at the problem of estimating a waveform at a single point in time.

Point Estimation Model

Consider the typical estimation problem shown in Fig. 6.1. The signal available at the receiver for processing is $r(u)$. It is obtained by performing

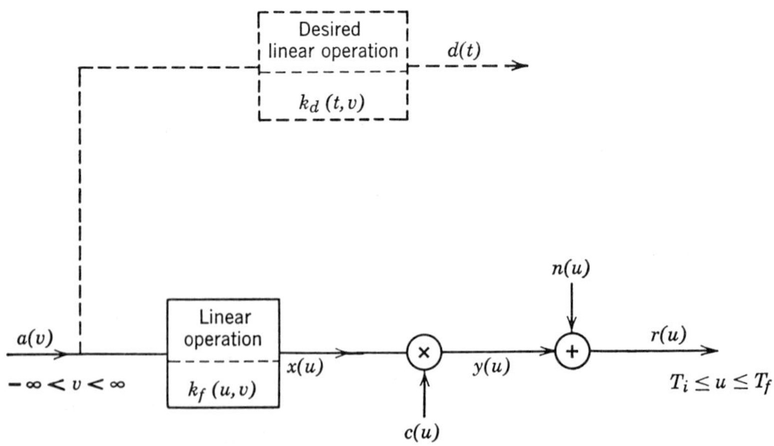

Fig. 6.1 Typical estimation problem.

a linear operation on $a(v)$ to obtain $x(u)$, which is then multiplied by a *known* modulation function. A noise $n(u)$ is added to the output $y(u)$ before it is observed. The dotted lines represent some linear operation (not necessarily time-invariant nor realizable) that we should like to perform on $a(v)$ *if* it were available (possibly for all time). The output is the desired signal $d(t)$ at some *particular* time t. The time t *may or may not* be included in the observation interval.

Common examples of desired signals are:

(i)
$$d(t) = a(t).$$

Here the output is simply the message. Clearly, *if* t were included in the observation interval, $x(u) = a(u)$, $n(u)$ were zero, and $c(u)$ were a constant, we could obtain the signal exactly. In general, this will not be the case.

(ii)
$$d(t) = a(t + \alpha).$$

Here, if α is a positive quantity, we wish to predict the value of $a(t)$ at some time in the future. Now, even in the absence of noise the estimation problem is nontrivial if $t + \alpha > T_f$. If α is a negative quantity, we shall want the value at some previous time.

(iii)
$$d(t) = \frac{d}{dt} a(t).$$

Here the desired signal is the derivative of the message. Other types of operations follow easily.

We shall assume that the linear operation is such that $d(t)$ is defined in the mean-square sense [i.e., if $d(t) = \dot{a}(t)$, as in (iii), we assume that $a(t)$ is a mean-square differentiable process]. Our discussion has been in the context of the linear modulation system in Fig. 6.1. We have not yet specified the statistics of the random processes. We describe the processes by the following assumption:

Gaussian Assumption. The message $a(t)$, the desired signal $d(t)$, and the received signal $r(t)$ are jointly Gaussian processes.

This assumption includes the linear modulation problem that we have discussed but avoids the necessity of describing the modulation system in detail. For algebraic simplicity we assume that the processes are zero-mean.

We now return to the optimum processing problem. We want to operate on $r(u)$, $T_i \le u \le T_f$ to obtain an estimate of $d(t)$. We denote this estimate as $\hat{d}(t)$ and choose our processor so that the quantity

$$\xi_P(t) \triangleq E\{[d(t) - \hat{d}(t)]^2\} = E[e^2(t)] \tag{15}$$

is minimized. First, observe that this is a *point* estimate (therefore the subscript P). Second, we observe that we are minimizing the mean-square error between the desired signal $d(t)$ and the estimate $\hat{d}(t)$.

We shall now find the optimum processor. The approach is as follows:

1. First, we shall find the optimum *linear* processor. Properties 3, 4, 5, and 6 deal with this problem. We shall see that the Gaussian assumption is not used in the derivation of the optimum linear processor.
2. Next, by including the Gaussian assumption, Property 7 shows that a linear processor is the best of *all possible* processors for the mean-square error criterion.
3. Property 8 demonstrates that under the Gaussian assumption the linear processor is optimum for a large class of error criteria.
4. Finally, Properties 9 and 10 show the relation between point estimators and interval estimators.

Property 3. The minimum mean-square *linear* estimate is the output of a linear processor whose impulse response is a solution to the integral equation

$$\boxed{K_{dr}(t, u) = \int_{T_i}^{T_f} h_o(t, \tau)\, K_r(\tau, u)\, d\tau, \qquad T_i < u < T_f.} \tag{16}$$

The proof of this property is analogous to the derivation in Section 3.4.5. The output of a linear processor can be written as

$$\hat{d}(t) = \int_{T_i}^{T_f} h(t, \tau)\, r(\tau)\, d\tau. \tag{17}$$

We assume that $h(t, \tau) = 0$, $\tau < T_i$, $\tau > T_f$. The mean-square error at time t is

$$\begin{aligned} \xi_P(t) &= \{E[d(t) - \hat{d}(t)]^2\} \\ &= E\left\{\left[d(t) - \int_{T_i}^{T_f} h(t, \tau)\, r(\tau)\, d\tau\right]^2\right\}. \end{aligned} \tag{18}$$

To minimize $\xi_P(t)$ we would go through the steps in Section 3.4.5 (pp. 198–204).

1. Let $h(t, \tau) = h_o(t, \tau) + \epsilon h_\epsilon(t, \tau)$.
2. Write $\xi_P(t)$ as the sum of the optimum error $\xi_{P_o}(t)$ and an incremental error $\Delta\xi(t, \epsilon)$.
3. Show that a necessary and sufficient condition for $\Delta\xi(t, \epsilon)$ to be greater than zero for $\epsilon \neq 0$ is the equation

$$E\left\{\left[d(t) - \int_{T_i}^{T_f} h_o(t, \tau)\, r(\tau)\, d\tau\right] r(u)\right\} = 0, \qquad T_i < u < T_f. \tag{19}$$

Bringing the expectation inside the integral, we obtain

$$K_{dr}(t, u) = \int_{T_i}^{T_f} h_o(t, \tau)\, K_r(\tau, u)\, d\tau, \qquad T_i < u < T_f, \tag{20}$$

which is the desired result. In Property 7A we shall show that the solution to (20) is unique iff $K_r(t, u)$ is positive-definite.

We observe that the only quantities needed to design the optimum linear processor for minimizing the mean-square error are the covariance function of the received signal $K_r(t, u)$ and the cross-covariance between the desired signal and the received signal, $K_{dr}(t, u)$. We emphasize that we have *not* used the Gaussian assumption.

Several special cases are important enough to be mentioned explicitly.

Property 3A. When $d(t) = a(t)$ and $T_f = t$, we have a realizable filtering problem, and (20) becomes

$$K_{ar}(t, u) = \int_{T_i}^{t} h_o(t, \tau)\, K_r(\tau, u)\, d\tau, \qquad T_i < u < t. \tag{21}$$

We use the term realizable because the filter indicated by (21) operates only on the past [i.e., $h_o(t, \tau) = 0$ for $\tau > t$].

Property 3B. Let $r(t) = c(t)\, x(t) + n(t) \triangleq y(t) + n(t)$. If the noise is white with spectral height $N_0/2$ and uncorrelated with $a(t)$, (20) becomes

$$K_{dy}(t, u) = \frac{N_0}{2}\, h_o(t, u) + \int_{T_i}^{T_f} h_o(t, \tau)\, K_y(\tau, u)\, d\tau, \qquad T_i \le u \le T_f. \tag{22}$$

Property 3C. When the assumptions of both 3A and 3B hold, and $x(t) = a(t)$, (20) becomes

$$K_a(t, u)\, c(u) = \frac{N_0}{2}\, h_o(t, u) + \int_{T_i}^{t} h_o(t, \tau)\, c(\tau)\, K_a(\tau, u)\, c(u)\, d\tau, \qquad T_i \le u \le t. \tag{23}$$

[The end point equalities were discussed after (13).]

Returning to the general case, we want to find an expression for the minimum mean-square error.

Property 4. The minimum mean-square error with the optimum linear processor is

$$\xi_{P_o}(t) \triangleq E[e_o^{\,2}(t)] = K_d(t, t) - \int_{T_i}^{T_f} h_o(t, \tau)\, K_{dr}(t, \tau)\, d\tau. \tag{24}$$

This follows by using (16) in (18). Hereafter we suppress the subscript *o* in the optimum error.

The error expressions for several special cases are also of interest. They all follow by straightforward substitution.

Property 4A. When $d(t) = a(t)$ and $T_f = t$, the minimum mean-square error is

$$\xi_P(t) = K_a(t, t) - \int_{T_i}^{t} h_o(t, \tau) K_{ar}(t, \tau) \, d\tau. \tag{25}$$

Property 4B. If the noise is white and uncorrelated with $a(t)$, the error is

$$\xi_P(t) = K_d(t, t) - \int_{T_i}^{T_f} h_o(t, \tau) K_{dy}(t, \tau) \, d\tau. \tag{26}$$

Property 4C. If the conditions of 4A and 4B hold and $x(t) = a(t)$, then

$$h_o(t, t) = \frac{2}{N_0} c(t) \, \xi_P(t). \tag{27}$$

If $c^{-1}(t)$ exists, (27) can be rewritten as

$$\xi_P(t) = \frac{N_0}{2} c^{-1}(t) \, h_o(t, t). \tag{28}$$

We may summarize the knowledge necessary to find the optimum linear processor in the following property:

Property 5. $K_r(t, u)$ and $K_{dr}(t, u)$ are the only quantities needed to find the MMSE point estimate when the processing is restricted to being linear. Any further statistical information about the processes cannot be used. All processes, Gaussian or non-Gaussian, with the same $K_r(t, u)$ and $K_{dr}(t, u)$ lead to the same processor and the same mean-square error if the processing is *required* to be linear.

Property 6. The error at time t using the optimum linear processor is uncorrelated with the input $r(u)$ at every point in the observation interval. This property follows directly from (19) by observing that the first term is the error using the optimum filter. Thus

$$E[e_o(t) \, r(u)] = 0, \qquad T_i < u < T_f. \tag{29}$$

We should observe that (29) can also be obtained by a simple heuristic geometric argument. In Fig. 6.2 we plot the desired signal $d(t)$ as a point in a vector space. The shaded plane area χ represents those points that can be achieved by a linear operation on the given input $r(u)$. We want to

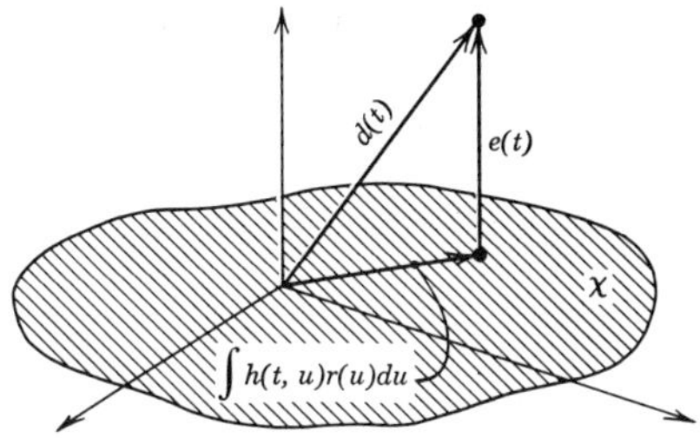

Fig. 6.2 Geometric interpretation of the optimum linear filter.

choose $\hat{d}(t)$ as the point in χ closest to $d(t)$. Intuitively, it is clear that we must choose the point directly under $d(t)$. Therefore the error vector is perpendicular to χ (or, equivalently, every vector in χ); that is, $e(t) \perp \int h(t, u)\, r(u)\, du$ for every $h(t, u)$.

The only difficulty is that the various functions are random. A suitable measure of the squared-magnitude of a vector is its mean-square value. The squared magnitude of the vector representing the error is $E[e^2(t)]$. Thus the condition of perpendicularity is expressed as an expectation:

$$E\left[e_o(t) \int_{T_i}^{T_f} h(t, u)\, r(u)\, du\right] = 0 \tag{30}$$

for every continuous $h(t, u)$; this implies

$$E[e_o(t)\, r(u)] = 0, \qquad T_i < u < T_f, \tag{31}$$

which is (29) [and, equivalently, (19)].†

Property 6A. If the processes of concern $d(t)$, $r(t)$, and $a(t)$ are jointly Gaussian, the error using the optimum linear processor is *statistically independent* of the input $r(u)$ at every point in the observation interval.

This property follows directly from the fact that uncorrelated Gaussian variables are statistically independent.

Property 7. When the Gaussian assumption holds, the optimum *linear* processor for minimizing the mean-square error is the best of *any* type. In other words, a nonlinear processor can not give an estimate with a smaller mean-square error.

Proof. Let $d_*(t)$ be an estimate generated by an arbitrary continuous processor operating on $r(u)$, $T_i \le u \le T_f$. We can denote it by

$$d_*(t) = f(t : r(u), T_i \le u \le T_f). \tag{32}$$

† Our discussion is obviously heuristic. It is easy to make it rigorous by introducing a few properties of linear vector spaces, but this is not necessary for our purposes.

Denote the mean-square error using this estimate as $\xi_*(t)$. We want to show that

$$\xi_*(t) \geq \xi_P(t), \tag{33}$$

with equality holding when the arbitrary processor is the optimum linear filter:

$$\begin{aligned}\xi_*(t) &= E\{[d_*(t) - d(t)]^2\} \\ &= E\{[d_*(t) - \hat{d}(t) + \hat{d}(t) - d(t)]^2\} \\ &= E\{[d_*(t) - \hat{d}(t)]^2\} + 2E\{[d_*(t) - \hat{d}(t)]e_o(t)\} + \xi_P(t). \end{aligned} \tag{34}$$

The first term is nonnegative. It remains to show that the second term is zero. Using (32) and (17), we can write the second term as

$$E\left\{\left[f(t:r(u), T_i \leq u \leq T_f) - \int_{T_i}^{T_f} h_o(t, u)\, r(u)\, du\right] e_o(t)\right\}. \tag{35}$$

This term is zero because $r(u)$ is *statistically independent* of $e_o(t)$ over the appropriate range, except for $u = T_f$ and $u = T_i$. (Because both processors are continuous, the expectation is also zero at the end point.) Therefore the optimum linear processor is as good as any other processor. The final question of interest is the uniqueness. To prove uniqueness we must show that the first term is strictly positive unless the two processors are equal. We discuss this issue in two parts.

Property 7A. First assume that $f(t:r(u), T_i \leq u \leq T_f)$ is a *linear* processor that is *not* equal to $h_o(t, u)$. Thus

$$f(t:r(u), T_i \leq u \leq T_f) = \int_{T_i}^{T_f} (h_o(t, u) + h_*(t, u))\, r(u)\, du, \tag{36}$$

where $h_*(t, u)$ represents the difference in the impulse responses.

Using (36) to evaluate the first term in (34), we have

$$E\{[d_*(t) - \hat{d}(t)]^2\} = \iint_{T_i}^{T_f} du\, dz\, h_*(t, u)\, K_r(u, z)\, h_*(t, z). \tag{37}$$

From (3.35) we know that if $K_r(u, z)$ is positive-definite the right-hand side will be positive for every $h_*(t, u)$ that is not identically zero. On the other hand, if $K_r(t, u)$ is only nonnegative definite, then from our discussion on p. 181 of Chapter 3 we know there exists an $h_*(t, u)$ such that

$$\int_{T_i}^{T_f} h_*(t, u)\, K_r(u, z)\, du = 0, \qquad T_i \leq z \leq T_f. \tag{38}$$

Because the eigenfunctions of $K_r(u, z)$ do not form a complete orthonormal set we can construct $h_*(t, u)$ out of functions that are orthogonal to $K_r(u, z)$.

Note that our discussion in 7A has not used the Gaussian assumption and that we have derived a necessary and sufficient condition for the uniqueness of the solution of (20). If $K_r(u, z)$ is not positive-definite, we can add an $h_*(t, u)$ satisfying (38) to any solution of (20) and still have a solution. Observe that the estimate $\hat{a}(t)$ is unique even if $K_r(u, z)$ is not positive-definite. This is because any $h_*(t, u)$ that we add to $h_o(t, u)$ must satisfy (38) and therefore cannot cause an output when the input is $r(t)$.

Property 7B. Now assume that $f(t{:}r(u), T_i \le u \le T_f)$ is a continuous nonlinear processor unequal to $h_o(t, u)$. Thus

$$f(t{:}r(u), T_i \le u \le T_f) = h_o(t, u) + f_*(t{:}r(u), T_i \le u \le T_f). \quad (39)$$

Then

$$E\{[d_*(t) - \hat{d}(t)]^2\} = E\int\int_{T_i}^{T_f} du\, dz\, f_*(t{:}r(u), T_i \le u \le T_f)\, f_*(t{:}r(z), T_i \le z \le T_f). \quad (40)$$

Because $r(u)$ is Gaussian and the higher moments factor, we can express the expectation on the right in terms of combinations of $K_r(u, z)$. Carrying out the tedious details gives the result that if $K_r(u, z)$ is positive-definite the expectation will be positive unless $f_*(t{:}r(z), T_i \le z \le T_f))$ is identically zero.

Property 7 is obviously quite important. It enables us to achieve two sets of results simultaneously by studying the linear processing problem.

1. If the Gaussian assumption holds, we are studying the best possible processor.
2. Even if the Gaussian assumption does not hold (or we cannot justify it), we shall have found the best possible linear processor.

In our discussion of waveform estimation we have considered only minimum mean-square error and MAP estimates. The next property generalizes the criterion.

Property 8A. Let $d_\epsilon(t)$ denote the error in estimating $d(t)$, using some estimate $\hat{d}(t)$.

$$e(t) = d(t) - \hat{d}(t). \quad (41)$$

The error is weighted with some cost function $C(e(t))$. The risk is the expected value of $C(e(t))$,

$$\mathcal{R}(\hat{d}(t), t) = E[C(e(t))] = E[C(d(t) - \hat{d}(t))]. \quad (42)$$

The Bayes point estimator is the estimate $\hat{d}_B(t)$ which minimizes the risk. If we assume that $C(e(t))$ is a symmetric convex upward function *and* the Gaussian assumption holds, the Bayes estimator is equal to the MMSE estimator.

$$\hat{d}_B(t) = \hat{d}_o(t). \tag{43}$$

Proof. The proof consists of three observations.

1. Under the Gaussian assumption the MMSE point estimator at any time (say t_1) is the conditional mean of the a posteriori density $p_{d_{t_1}|r(u)}[D_{t_1}|r(u): T_i \le u \le T_f]$. Observe that we are talking about a single random variable d_{t_1} so that this is a legitimate density. (See Problem 6.1.1.)
2. The a posteriori density is unimodal and symmetric about its conditional mean.
3. Property 1 on p. 60 of Chapter 2 is therefore applicable and gives the above conclusion.

Property 8B. If, in addition to the assumptions in Property 8A, we require the cost function to be *strictly* convex, then

$$\hat{d}_B(t) = \hat{d}_o(t) \tag{44}$$

is the unique Bayes point estimator.

This result follows from (2.158) in the derivation in Chapter 2.

Property 8C. If we replace the convexity requirement on the cost function with a requirement that it be a symmetric nondecreasing function such that

$$\lim_{X \to \infty} C(X) p_{d_{t_1}|r(u)}[X|r(u): T_i \le u \le T_f] = 0 \tag{45}$$

for all t_1 and $r(t)$ of interest, then (44) is still valid.

These properties are important because they guarantee that the processors we are studying in this chapter are optimum for a large class of criteria when the Gaussian assumption holds.

Finally, we can relate our results with respect to point estimators and MMSE and MAP interval estimators.

Property 9. A minimum mean-square error interval estimator is just a collection of point estimators. Specifically, suppose we observe a waveform $r(u)$ over the interval $T_i \le u \le T_f$ and want a signal $d(t)$ over the interval $T_\alpha \le t \le T_\beta$ such that the mean-square error averaged over the interval is minimized.

$$\xi_I \triangleq E\left\{\int_{T_\alpha}^{T_\beta} [d(t) - \hat{d}(t)]^2 \, dt\right\}. \tag{46}$$

Clearly, if we can minimize the expectation of the bracket for each t then ξ_I will be minimized. This is precisely what a MMSE point estimator does. Observe that the point estimator uses $r(u)$ over the entire observation interval to generate $\hat{d}(t)$. (Note that Property 9 is true for nonlinear modulation also.)

Property 10. Under the Gaussian assumption the minimum mean-square error point estimate and MAP point estimate are identical. This is just a special case of Property 8C. Because the MAP interval estimate is a collection of MAP point estimates, the interval estimates also coincide.

These ten properties serve as background for our study of the linear modulation case. Property 7 enables us to concentrate our efforts in this chapter on the *optimum linear processing* problem. When the Gaussian assumption holds, our results will correspond to the best possible processor (for the class of criterion described above). For arbitrary processes the results will correspond to the best linear processor.

We observe that all results carry over to the vector case with obvious modifications. Some properties, however, are used in the sequel and therefore we state them explicitly. *A typical* vector problem is shown in Fig. 6.3.

The message $\mathbf{a}(t)$ is a p-dimensional vector. We operate on it with a matrix linear filter which has p inputs and n outputs.

$$\mathbf{x}(u) = \int_{-\infty}^{\infty} \mathbf{k}_f(u, v)\, \mathbf{a}(v)\, dv, \qquad T_i \leq u \leq T_f. \tag{47}$$

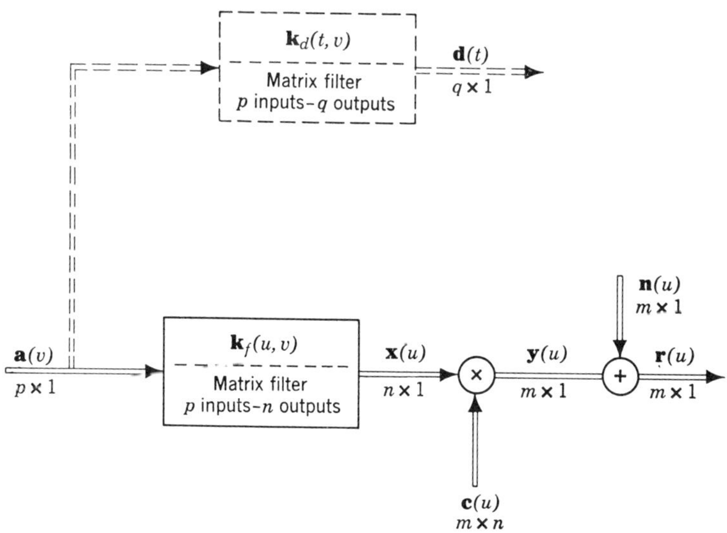

Fig. 6.3 Vector estimation problem.

The vector $\mathbf{x}(u)$ is multiplied by an $m \times n$ modulation matrix to give the m-dimensional vector $\mathbf{y}(t)$ which is transmitted over the channel. Observe that we have generated $\mathbf{y}(t)$ by a cascade of a linear operation with a memory and no-memory operation. The reason for this two-step procedure will become obvious later. The desired signal $\mathbf{d}(t)$ is a $q \times 1$-dimensional vector which is related to $\mathbf{a}(v)$ by a matrix filter with p inputs and q outputs Thus

$$\mathbf{d}(t) = \int_{-\infty}^{\infty} \mathbf{k}_d(t, v)\, \mathbf{a}(v)\, dv. \tag{48}$$

We shall encounter some typical vector problems later. Observe that p, q, m, and n, the dimensions of the various vectors, may all be different.

The desired signal $\mathbf{d}(t)$ has q components. Denote the estimate of the ith component as $\hat{d}_i(t)$. We want to minimize simultaneously

$$\xi_{P_i}(t) \triangleq E\{[d_i(t) - \hat{d}_i(t)]^2\}, \qquad i = 1, 2, \ldots, q. \tag{49}$$

The message $\mathbf{a}(v)$ is a zero-mean vector Gaussian process and the noise is an m-dimensional Gaussian random process. In general, we assume that it contains a white component $\mathbf{w}(t)$:

$$E[\mathbf{w}(t)\, \mathbf{w}^T(u)] \triangleq \mathbf{R}(t)\, \delta(t - u), \tag{50}$$

where $\mathbf{R}(t)$ is positive-definite. We assume also that the necessary covariance functions are known. We shall use the same property numbers as in the scalar case and add a V. We shall not restate the assumptions.

Property 3V.

$$\mathbf{K_{dr}}(t, u) = \int_{T_i}^{T_f} \mathbf{h}_o(t, \tau)\mathbf{K_r}(\tau, u)\, d\tau, \qquad T_i < u < T_f. \tag{51}$$

Proof. See Problem 6.1.2.

Property 3A-V.

$$\mathbf{K_{ar}}(t, u) = \int_{T_i}^{t} \mathbf{h}_o(t, \tau)\, \mathbf{K_r}(\tau, u)\, d\tau; \qquad T_i < u < t. \tag{52}$$

Property 4C-V.

$$\mathbf{h}_o(t, t) = \boldsymbol{\xi}_P(t)\, \mathbf{C}^T(t)\, \mathbf{R}^{-1}(t), \tag{53}$$

where $\xi_P(t)$ is the error covariance matrix whose elements are

$$\xi_{P_{ij}}(t) \triangleq E\{[a_i(t) - \hat{a}_i(t)][a_j(t) - \hat{a}_j(t)]\}. \tag{54}$$

(Because the errors are zero-mean, the correlation and covariance are identical.)

Proof. See Problem 6.1.3.

Other properties of the vector case follow by direct modification.

Summary

In this section we have explored properties that result when a linear modulation restriction is imposed. Although we have discussed the problem in the modulation context, it clearly has widespread applicability. We observe that if we let $c(t) = 1$ at certain instants of time and zero elsewhere, we will have the sampled observation model. This case and others of interest are illustrated in the problems (see Problems 6.1.4–6.1.9).

Up to this point we have restricted neither the processes nor the observation interval. In other words, the processes were stationary or nonstationary, the initial observation time T_i was arbitrary, and $T_f\,(\geq T_i)$ was arbitrary. Now we shall consider specific solution techniques. The easiest approach is by means of various special cases.

Throughout the rest of the chapter we shall be dealing with linear processors. In general, we do not specify explicitly that the Gaussian assumption holds. It is important to re-emphasize that in the absence of this assumption we are finding only the best *linear* processor (a nonlinear processor might be better). Corresponding to each problem we discuss there is another problem in which the processes are Gaussian, and for which the processor is the optimum of all processors for the given criterion.

It is also worthwhile to observe that the remainder of the chapter could have been studied directly after Chapter 1 if we had approached it as a "structured" problem and not used the Gaussian assumption. We feel that this places the emphasis incorrectly and that the linear processor should be viewed as a device that is generating the conditional mean. This viewpoint puts it into its proper place in the over-all statistical problem.

6.2 REALIZABLE LINEAR FILTERS: STATIONARY PROCESSES, INFINITE PAST: WIENER FILTERS

In this section we discuss an important case relating to (20). First, we assume that the final observation time corresponds to the time at which the estimate is desired. Thus $t = T_f$ and (20) becomes

$$K_{dr}(t, \sigma) = \int_{T_i}^{t} h_o(t, u)\, K_r(u, \sigma)\, du; \qquad T_i < \sigma < t. \tag{55}$$

Second, we assume that $T_i = -\infty$. This assumption means that we have the infinite past available to operate on to make our estimate. From a practical standpoint it simply means that the past is available beyond the significant memory time of our filter. In a later section, when we discuss finite T_i, we shall make some quantitative statements about how large $t - T_i$ must be in order to be considered infinite.

Third, we assume that the received signal is a sample function from a stationary process and that the desired signal and the received signal are jointly stationary. (In Fig. 6.1 we see that this implies that $c(t)$ is constant. Thus we say the process is unmodulated.†) Then we may write

$$K_{dr}(t - \sigma) = \int_{-\infty}^{t} h_o(t, u)\, K_r(u - \sigma)\, du, \qquad -\infty < \sigma < t. \tag{56}$$

Because the processes are stationary and the interval is infinite, let us try to find a solution to (56) which is time-invariant.

$$K_{dr}(t - \sigma) = \int_{-\infty}^{t} h_o(t - u)\, K_r(u - \sigma)\, du, \qquad -\infty < \sigma < t. \tag{57}$$

If we can find a solution to (57), it will also be a solution to (56). If $K_r(u - \sigma)$ is positive-definite, (56) has a unique solution. Thus, if (57) has a solution, it will be unique and will also be the only solution to (56). Letting $\tau = t - \sigma$ and $v = t - u$, we have

$$\boxed{K_{dr}(\tau) = \int_{0}^{\infty} h_o(v)\, K_r(\tau - v)\, dv, \qquad 0 < \tau < \infty,} \tag{58}$$

which is commonly referred to as the Wiener-Hopf equation. It was derived and solved by Wiener [1]. (The linear processing problem was studied independently by Kolmogoroff [2].)

6.2.1 Solution of Wiener-Hopf Equation

Our solution to the Wiener-Hopf equation is analogous to the approach by Bode and Shannon [3]. Although the amount of manipulation required is identical to that in Wiener's solution, the present procedure is more intuitive. We restrict our attention to the case in which the Fourier transform of $K_r(\tau)$, the input correlation function, is a rational function. This is not really a practical restriction because most spectra of interest can be approximated by a rational function. The general case is discussed by Wiener [1] but does not lead to a practical solution technique.

The first step in our solution is to observe that *if* $r(t)$ were white the solution to (58) would be trivial. If

$$K_r(\tau) = \delta(\tau), \tag{59}$$

then (58) becomes

$$K_{dr}(\tau) = \int_{0}^{\infty} h_o(v)\, \delta(\tau - v)\, dv, \qquad 0 < \tau < \infty, \tag{60}$$

† Our use of the term *modulated* is the opposite of the normal usage in which the message process *modulates* a carrier. The adjective *unmodulated* seems to be the best available.

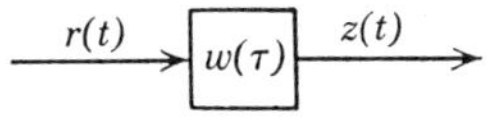

Fig. 6.4 Whitening filter.

and

$$
\begin{aligned}
h_o(\tau) &= K_{dr}(\tau), \qquad && \tau \geq 0, \\
&= 0, && \tau < 0,
\end{aligned}
\tag{61}
$$

where the value at $\tau = 0$ comes from our continuity restriction.

It is unlikely that (59) will be satisfied in many problems of interest. If, however, we could perform some preliminary operation on $r(t)$ to transform it into a white process, as shown in Fig. 6.4, the subsequent filtering problem in terms of the whitened process would be trivial. The idea of a whitening operation is familiar from Section 4.3 of Chapter 4. In that case the signal was deterministic and we whitened only the noise. In this case we whiten the entire input. In Section 4.3 we proved that *any* reversible operation could not degrade the over-all system performance. Now we also want the over-all processor to be a *realizable* linear filter. Therefore we show the following property:

Whitening Property. For all rational spectra there exists a realizable, time-invariant linear filter whose output $z(t)$ is a white process when the input is $r(t)$ *and* whose inverse is a realizable linear filter.

If we denote the impulse response of the whitening filter as $w(\tau)$ and the transfer function as $W(j\omega)$, then the property says:

(i) $$\iint_{-\infty}^{\infty} w(u)\, w(v)\, K_r(\tau - u - v)\, du\, dv = \delta(\tau), \qquad -\infty < \tau < \infty.$$

or

(ii) $$|W(j\omega)|^2 S_r(\omega) = 1.$$

If we denote the impulse response of the inverse filter as $w^{-1}(\tau)$, then

(iii) $$\int_{-\infty}^{\infty} w^{-1}(u - v)\, w(v)\, dv = \delta(u)$$

or

(iv) $$\mathcal{F}[w^{-1}(\tau)] = \frac{1}{W(j\omega)} = W^{-1}(j\omega)$$

and $w^{-1}(\tau)$ must be the impulse response of a realizable filter.

We derive this property by demonstrating a constructive technique for a simple example and then extending it to arbitrary rational spectra.

Example 1. Let

$$S_r(\omega) = \frac{2k}{\omega^2 + k^2}. \tag{62}$$

We want to choose the transfer function of the whitening filter so that it is realizable and the spectrum of its output $z(t)$ satisfies the equation

$$S_z(\omega) = S_r(\omega)|W(j\omega)|^2 = 1. \tag{63}$$

To accomplish this we divide $S_r(\omega)$ into two parts,

$$S_r(\omega) = \left(\frac{\sqrt{2k}}{j\omega + k}\right)\left(\frac{\sqrt{2k}}{-j\omega + k}\right) \triangleq [G^+(j\omega)][G^+(j\omega)]^*. \tag{64}$$

We denote the first term by $G^+(j\omega)$ because it is zero for negative time. The second term is its complex conjugate. Clearly, if we let

$$W(j\omega) = \frac{1}{G^+(j\omega)} = \frac{j\omega + k}{\sqrt{2k}}, \tag{65}$$

then (63) will be satisfied.

We observe that the whitening filter consists of a differentiator and a gain term in parallel. Because

$$W^{-1}(j\omega) = G^+(j\omega) = \frac{\sqrt{2k}}{j\omega + k}, \tag{66}$$

it is clear that the inverse is a realizable linear filter and therefore $W(j\omega)$ is a legitimate reversible operation. Thus we could operate on $z(t)$ in either of the two ways shown in Fig. 6.5 and, as we proved in Section 4.3, if we choose $h_o'(\tau)$ in an optimum manner the output of both systems will be $\hat{d}(t)$.

In this particular example the selection of $W(j\omega)$ was obvious. We now consider a more complicated example.

Example 2. Let

$$S_r(\omega) = \frac{c^2(j\omega + \alpha_1)(-j\omega + \alpha_1)}{(j\omega + \beta_1)(-j\omega + \beta_1)}. \tag{67}$$

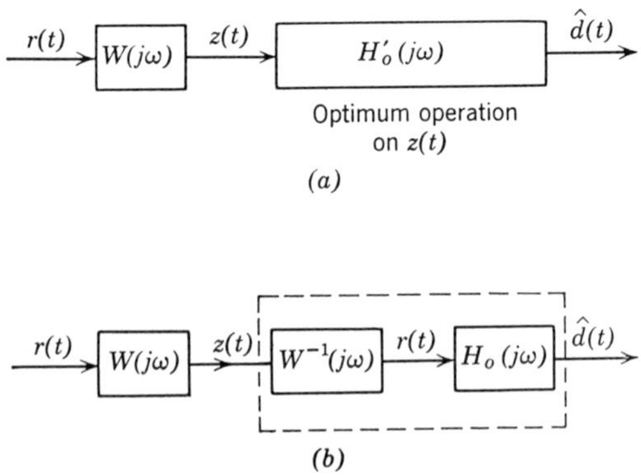

Fig. 6.5 Optimum filter: (*a*) approach No. 1; (*b*) approach No. 2.

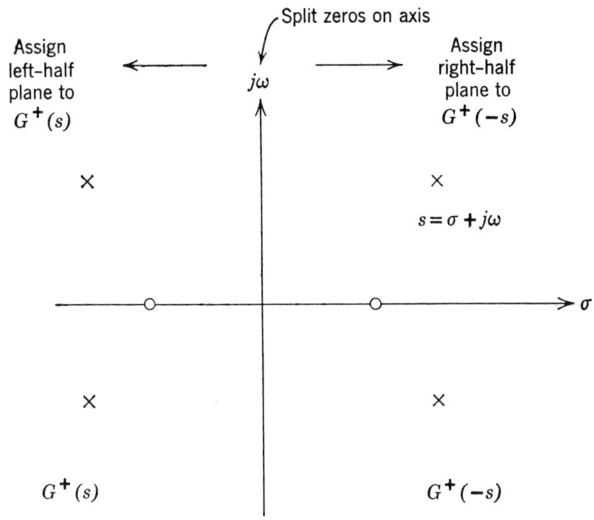

Fig. 6.6 A typical pole-zero plot.

We must choose $W(j\omega)$ so that

$$S_r(\omega) = |W^{-1}(j\omega)|^2 = |G^+(j\omega)|^2 \tag{68}$$

and both $W(j\omega)$ and $W^{-1}(j\omega)$ [or equivalently $G^+(j\omega)$ and $W(j\omega)$] are realizable. When discussing realizability, it is convenient to use the complex s-plane. We extend our functions to the entire complex plane by replacing $j\omega$ by s, where $s = \sigma + j\omega$. In order for $W(s)$ to be realizable, it cannot have any poles in the right half of the s-plane. Therefore we must assign the $(j\omega + \alpha_1)$ term to it. Similarly, for $W^{-1}(s)$ [or $G^+(s)$] to be realizable we assign to it the $(j\omega + \beta_1)$ term. The assignment of the constant is arbitrary because it adjusts only the white noise level. For simplicity we assume a unity level spectrum for $z(t)$ and divide the constant evenly. Therefore

$$G^+(j\omega) = c\,\frac{(j\omega + \alpha_1)}{(j\omega + \beta_1)}. \tag{69}$$

To study the general case we consider the pole-zero plot of the typical spectrum shown in Fig. 6.6. Assuming that this spectrum is typical, we then find that the procedure is clear. We factor $S_r(\omega)$ and assign all poles and zeros in the left half plane (and half of each pair of zeros on the axis) to $G^+(j\omega)$. The remaining poles and zeros will correspond exactly to the conjugate $[G^+(j\omega)]^*$. The fact that every rational spectrum can be divided in this manner follows directly from the fact that $S_r(\omega)$ is a real, even, nonnegative function of ω whose inverse transform is a correlation function. This implies the modes of behavior for the pole-zero plot shown in Fig. 6.7*a*–*c*:

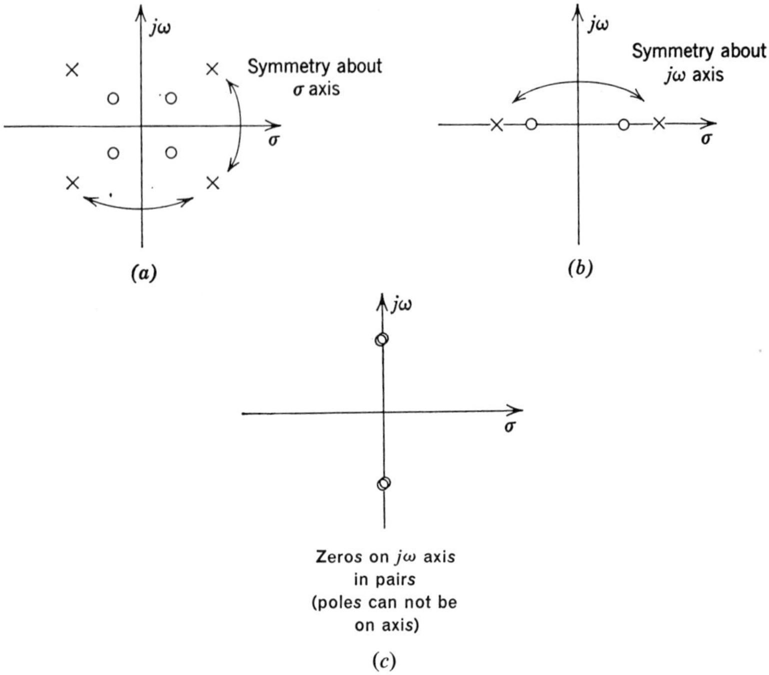

Fig. 6.7 Possible pole-zero plots in the s-plane.

1. Symmetry about the σ-axis. Otherwise $S_r(\omega)$ would not be real.
2. Symmetry about the $j\omega$-axis. Otherwise $S_r(\omega)$ would not also be even.
3. Any zeros on the $j\omega$-axis occur in pairs. Otherwise $S_r(\omega)$ would be negative for some value of ω.
4. No poles on the $j\omega$-axis. This would correspond to a $1/\omega^2$ term whose inverse is not the correlation function of a stationary process.

The verification of these properties is a straightforward exercise (see Problem 6.2.1).

We have now proved that we can always find a realizable, reversible whitening filter. The processing problem is now reduced to that shown in Fig. 6.8. We must design $H_o'(j\omega)$ so that it operates on $z(t)$ in such a way

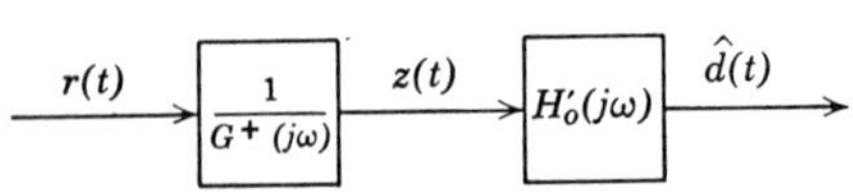

Fig. 6.8 Optimum filter.

that it produces the minimum mean-square error estimate of $d(t)$. Clearly, then, $h_o'(\tau)$ must satisfy (58) with r replaced by z,

$$K_{dz}(\tau) = \int_0^\infty h_o'(v)\, K_z(\tau - v)\, dv, \qquad 0 < \tau < \infty. \tag{70}$$

However, we have forced $z(t)$ to be white with unity spectral height. Therefore

$$h_o'(\tau) = K_{dz}(\tau), \qquad \tau \geq 0. \tag{71}$$

Thus, if we knew $K_{dz}(\tau)$, our solution would be complete. Because $z(t)$ is obtained from $r(t)$ by a linear operation, $K_{dz}(\tau)$ is easy to find,

$$K_{dz}(\tau) \triangleq E\left[d(t) \int_{-\infty}^{\infty} w(v)\, r(t - \tau - v)\, dv\right]$$
$$= \int_{-\infty}^{\infty} w(v)\, K_{dr}(\tau + v)\, dv = \int_{-\infty}^{\infty} w(-\beta)\, K_{dr}(\tau - \beta)\, d\beta. \tag{72}$$

Transforming,

$$S_{dz}(j\omega) = W^*(j\omega)\, S_{dr}(j\omega) = \frac{S_{dr}(j\omega)}{[G^+(j\omega)]^*}. \tag{73}$$

We simply find the inverse transform of $S_{dz}(j\omega)$, $K_{dz}(\tau)$, and retain the part corresponding to $\tau \geq 0$. A typical $K_{dz}(\tau)$ is shown in Fig. 6.9*a*. The associated $h_o'(\tau)$ is shown in Fig. 6.9*b*.

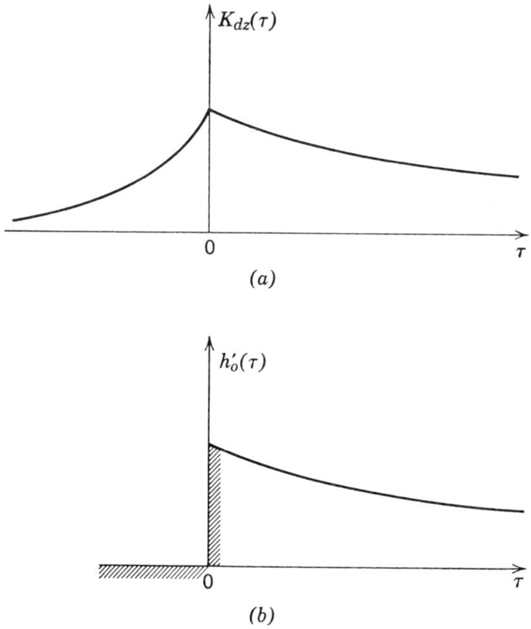

Fig. 6.9 Typical Functions: (*a*) a typical covariance function; (*b*) corresponding $h_o'(\tau)$.

We can denote the transform of $K_{dz}(\tau)$ for $\tau \geq 0$ by the symbol

$$[S_{dz}(j\omega)]_+ \triangleq \int_0^\infty K_{dz}(\tau)e^{-j\omega\tau}\,d\tau = \int_0^\infty h_o'(\tau)e^{-j\omega\tau}\,d\tau.\dagger \tag{74}$$

Similarly,

$$[S_{dz}(j\omega)]_- \triangleq \int_{-\infty}^{0^-} K_{dz}(\tau)e^{-j\omega\tau}\,d\tau. \tag{75}$$

Clearly,

$$S_{dz}(j\omega) = [S_{dz}(j\omega)]_+ + [S_{dz}(j\omega)]_-, \tag{76}$$

and we may write

$$H_o'(j\omega) = [S_{dz}(j\omega)]_+ = [W^*(j\omega)\,S_{dr}(j\omega)]_+ = \left[\frac{S_{dr}(j\omega)}{[G^+(j\omega)]^*}\right]_+. \tag{77}$$

Then the entire optimum filter is just a cascade of the whitening filter and $H_o'(j\omega)$,

$$\boxed{H_o(j\omega) = \left[\frac{1}{G^+(j\omega)}\right]\left[\frac{S_{dr}(j\omega)}{[G^+(j\omega)]^*}\right]_+.} \tag{78}$$

We see that by a series of routine, conceptually simple operations we have derived the desired filter. We summarize the steps briefly.

1. We factor the *input* spectrum into two parts. One term, $G^+(s)$, contains all the poles and zeros in the left half of the s-plane. The other factor is its mirror image about the $j\omega$-axis.
2. The cross-spectrum between $d(t)$ and $z(t)$ can be expressed in terms of the original cross-spectrum divided by $G^+(j\omega)$. This corresponds to a function that is nonzero for both positive and negative time. The realizable part of this function ($\tau \geq 0$) is $h_o'(\tau)$ and its transform is $H_o'(j\omega)$.
3. The transfer function of the optimum filter is a simple product of these two transfer functions. We shall see that the composite transfer function corresponds to a realizable system. Observe that we actually build the optimum linear filter as single system. The division into two parts is for conceptual purposes only.

Before we discuss the properties and implications of the solution, it will be worthwhile to consider a simple example to guarantee that we all agree on what (78) means.

Example 3. Assume that

$$r(t) = \sqrt{P}\,a(t) + n(t), \tag{79}$$

† In general, the symbol $[\sim]_+$ denotes the transform of the realizable part of the expression inside the bracket.

where $a(t)$ and $n(t)$ are uncorrelated zero-mean stationary processes and

$$S_a(\omega) = \frac{2k}{\omega^2 + k^2}. \tag{80}$$

[We see that $a(t)$ has unity power so that P is the transmitted power.]

$$S_n(\omega) = \frac{N_0}{2}. \tag{81}$$

The desired signal is

$$d(t) = a(t + \alpha), \tag{82}$$

where α is a constant.

By choosing α to be positive we have the prediction problem, choosing α to be zero gives the conventional filtering problem, and choosing α to be negative gives the filtering-with-delay problem.

The solution is a simple application of the procedure outlined in the preceding section:

$$S_r(\omega) = \frac{2kP}{\omega^2 + k^2} + \frac{N_0}{2} = \frac{N_0}{2}\frac{\omega^2 + k^2(1 + 4P/kN_0)}{\omega^2 + k^2}. \tag{83}$$

It is convenient to define

$$\Lambda = \frac{4P}{kN_0}. \tag{84}$$

(This quantity has a physical significance we shall discuss later. For the moment, it can be regarded as a useful parameter.) First we factor the spectrum

$$S_r(\omega) = \frac{N_0}{2}\frac{\omega^2 + k^2(1 + \Lambda)}{\omega^2 + k^2} = G^+(j\omega)\,[G^+(j\omega)]^*. \tag{85}$$

so

$$G^+(j\omega) = \left(\frac{N_0}{2}\right)^{1/2}\left(\frac{j\omega + k\sqrt{1 + \Lambda}}{j\omega + k}\right). \tag{86}$$

Now

$$\begin{aligned} K_{dr}(\tau) &= E[d(t)\, r(t - \tau)] = E\{a(t + \alpha)[\sqrt{P}\, a(t - \tau) + n(t - \tau)]\} \\ &= \sqrt{P}\, E[a(t + \alpha)\, a(t - \tau)] = \sqrt{P}\, K_a(\tau + \alpha). \end{aligned} \tag{87}$$

Transforming,

$$S_{dr}(j\omega) = \sqrt{P}\, S_a(\omega)e^{+j\omega\alpha} = \frac{2k\sqrt{P}\, e^{+j\omega\alpha}}{\omega^2 + k^2} \tag{88}$$

and

$$S_{dz}(j\omega) = \frac{S_{dr}(j\omega)}{[G^+(j\omega)]^*} = \frac{2k\sqrt{P}\, e^{+j\omega\alpha}}{\omega^2 + k^2}\cdot\frac{(-j\omega + k)}{\sqrt{N_0/2}(-j\omega + k\sqrt{1 + \Lambda})}. \tag{89}$$

To find the realizable part, we take the inverse transform:

$$K_{dz}(\tau) = \mathcal{F}^{-1}\left\{\frac{S_{dr}(j\omega)}{[G^+(j\omega)]^*}\right\} = \mathcal{F}^{-1}\left[\frac{2k\sqrt{P}\, e^{+j\omega\alpha}}{(j\omega + k)\sqrt{N_0/2}(-j\omega + k\sqrt{1 + \Lambda})}\right]. \tag{90}$$

The inverse transform can be evaluated easily (either by residues or a partial fraction expansion and the shifting theorem). The result is

$$K_{dz}(\tau) = \begin{cases} \dfrac{2\sqrt{P}}{\sqrt{N_0/2}}\, \dfrac{1}{1 + \sqrt{1 + \Lambda}}\, e^{-k(\tau + \alpha)}, & \tau + \alpha \geq 0, \\ \dfrac{2\sqrt{P}}{\sqrt{N_0/2}}\, \dfrac{1}{1 + \sqrt{1 + \Lambda}}\, e^{+k\sqrt{1 + \Lambda}\,(\tau + \alpha)}, & \tau + \alpha < 0. \end{cases} \tag{91}$$

The function is shown in Fig. 6.10.

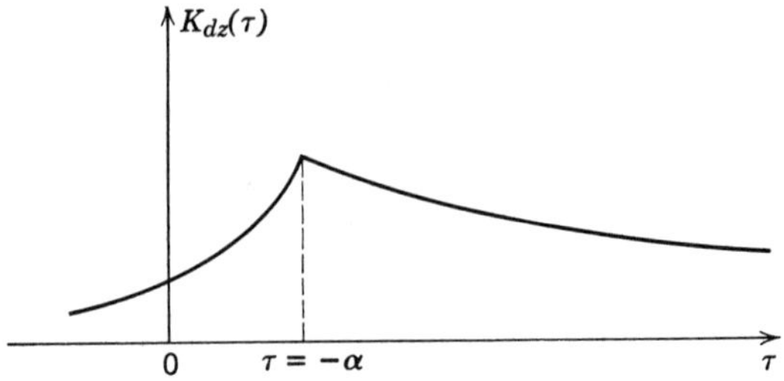

Fig. 6.10 Cross-covariance function.

Now $h_o'(\tau)$ depends on the value α. In other words, the amount of $K_{dz}(\tau)$ in the range $\tau \geq 0$ is a function of α. We consider three types of operations:

Case 1. $\alpha = 0$: filtering with zero delay. Letting $\alpha = 0$ in (91), we have

$$h_o'(\tau) = \frac{2\sqrt{P}}{\sqrt{N_0/2}} \frac{1}{1 + \sqrt{1 + \Lambda}} e^{-k\tau} u_{-1}(\tau), \tag{92}$$

or

$$H_o'(j\omega) = \frac{1}{1 + \sqrt{1 + \Lambda}} \frac{2\sqrt{P}}{\sqrt{N_0/2}} \frac{1}{j\omega + k}. \tag{93}$$

Then

$$H_0(j\omega) = \frac{H_o'(j\omega)}{G^+(j\omega)} = \frac{2\sqrt{P}}{(N_0/2)(1 + \sqrt{1 + \Lambda})} \frac{1}{j\omega + k\sqrt{1 + \Lambda}}. \tag{94}$$

We see that our result is intuitively logical. The amplitude of the filter response is shown in Fig. 6.11. The filter is a simple low-pass filter whose bandwidth varies as a function of k and Λ.

We now want to attach some physical significance to the parameter Λ. The bandwidth of the message process is directly proportional to k, as shown in Fig. 6.12*a*

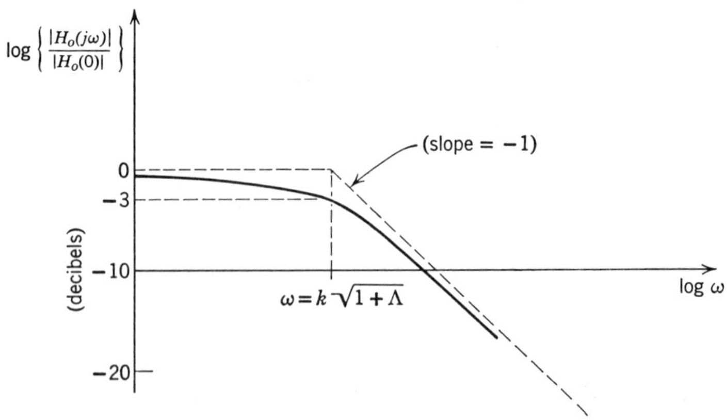

Fig. 6.11 Magnitude plot for optimum filter.

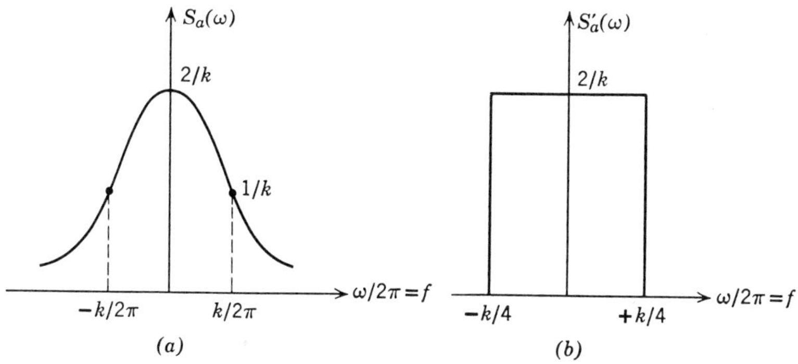

Fig. 6.12 Equivalent rectangular spectrum.

The 3-db bandwidth is k/π cps. Another common bandwidth measure is the equivalent rectangular bandwidth (ERB), which is the bandwidth of a rectangular spectrum with height $S_a(0)$ and the same total power of the actual message as shown in Fig. 6.12*b*. Physically, Λ is the signal-to-noise ratio in the message ERB. This ratio is most natural for most of our work. The relationship between Λ and the signal-to-noise ratio in the 3-db bandwidth depends on the particular spectrum. For *this particular case* $\Lambda_{3db} = (\pi/2)\Lambda$.

We see that for a fixed k the optimum filter bandwidth increases as Λ, the signal-to-noise ratio, increases. Thus, as $\Lambda \to \infty$, the filter magnitude approaches unity for all frequencies and it passes the message component without distortion. Because the noise is unimportant in this case, this is intuitively logical. On the other hand, as $\Lambda \to 0$, the filter 3-db point approaches k. The gain, however, approaches zero. Once again, this is intuitively logical. There is so much noise that, based on the mean square error criterion, the best filter output is zero (the mean value of the message).

Case 2. α is negative: filtering with delay. Here $h_o'(\tau)$ has the impulse response shown in Fig. 6.13. Transforming, we have

$$H_o'(j\omega) = \frac{2k\sqrt{P}}{\sqrt{N_0/2}} \left[\frac{e^{\alpha j\omega}}{(j\omega + k)(-j\omega + k\sqrt{1+\Lambda})} - \frac{e^{\alpha k\sqrt{1+\Lambda}}}{k(1+\sqrt{1+\Lambda})(-j\omega + k\sqrt{1+\Lambda})} \right] \tag{95}$$

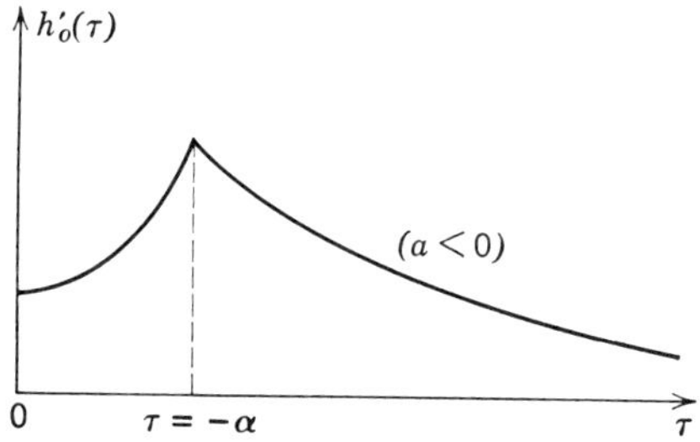

Fig. 6.13 Filtering with delay.

and

$$H_o(j\omega) = \frac{H_o'(j\omega)}{G^+(j\omega)} = \frac{2k\sqrt{P}}{N_0/2}\left\{\frac{e^{\alpha j\omega}}{[\omega^2 + k^2(1+\Lambda)]} - \frac{e^{\alpha k\sqrt{1+\Lambda}}(j\omega + k)}{k(1+\sqrt{1+\Lambda})[\omega^2 + k^2(1+\Lambda)]}\right\}. \tag{96a}$$

This can be rewritten as

$$H_o(j\omega) = \frac{2k\sqrt{P}\, e^{\alpha j\omega}}{(N_0/2)[\omega^2 + k^2(1+\Lambda)]}\left[1 - \frac{(j\omega + k)e^{\alpha(k\sqrt{1+\Lambda} - j\omega)}}{k(1+\sqrt{1+\Lambda})}\right]. \tag{96b}$$

We observe that the expression outside the bracket is just

$$\frac{S_{dr}(j\omega)}{S_r(\omega)}\, e^{\alpha j\omega}. \tag{97}$$

We see that when α is a large negative number the second term in the bracket is approximately zero. Thus $H_o(j\omega)$ approaches the expression in (97). This is just the ratio of the cross spectrum to the total input spectrum, with a delay to make the filter realizable.

We also observe that the impulse response in Fig. 6.13 is difficult to realize with conventional network synthesis techniques.

Case 3. α is positive: filtering with prediction. Here

$$h_o'(\tau) = \left(\frac{2\sqrt{P}}{\sqrt{N_0/2}}\frac{1}{1+\sqrt{1+\Lambda}}\, e^{-k\tau}\right)e^{-k\alpha}. \tag{98}$$

Comparing (98) with (92), we see that the optimum filter for prediction is just the optimum filter for estimating $a(t)$ multiplied by a gain $e^{-k\alpha}$, as shown in Fig. 6.14. The reason for this is that $a(t)$ is a first order wide-sense Markov process and the noise is white. We obtain a similar result for more general processes in Section 6.3.

Before concluding our discussion we amplify a point that was encountered in Case 1 of the example. One step of the solution is to find the realizable part of a function. Frequently it is unnecessary to find the time function and then retransform. Specifically, whenever $S_{dr}(j\omega)$ is a ratio of two polynomials in $j\omega$, we may write

$$\frac{S_{dr}(j\omega)}{[G^+(j\omega)]^*} = F(j\omega) + \sum_{i=1}^{N}\frac{c_i}{j\omega + p_i} + \sum_{j=1}^{N}\frac{d_j}{-j\omega + q_j}, \tag{99a}$$

where $F(j\omega)$ is a polynomial, the first sum contains all terms corresponding to poles in the left half of the s-plane (including the $j\omega$-axis), and the second sum contains all terms corresponding to poles in the right half of

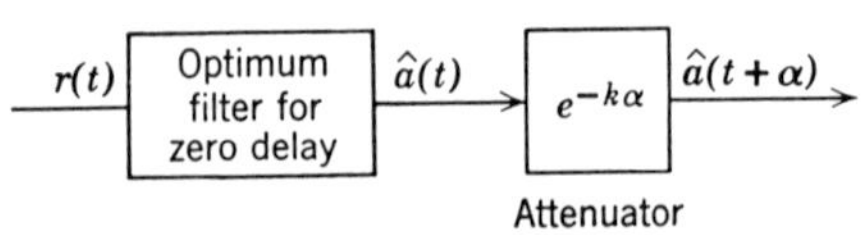

Fig. 6.14 Filtering with prediction.

the s-plane. In this expanded form the realizable part consists of the first two terms. Thus

$$\left[\frac{S_{dr}(j\omega)}{[G^+(j\omega)]^*}\right]_+ = F(j\omega) + \sum_{i=1}^{N} \frac{c_i}{j\omega + p_i}. \tag{99b}$$

The use of (99*b*) reduces the required manipulation.

In this section we have developed an algorithm for solving the Wiener-Hopf equation and presented a simple example to demonstrate the technique. Next we investigate the resulting mean-square error.

6.2.2 Errors in Optimum Systems

In order to evaluate the performance of the optimum linear filter we calculate the minimum mean-square error. The minimum mean-square error for the general case was given in (24) of Property 4. Because the processes are stationary and the filter is time-invariant, the mean-square error will not be a function of time. Thus (24) reduces to

$$\xi_P = K_d(0) - \int_0^\infty h_o(\tau)\, K_{dr}(\tau)\, d\tau. \tag{100}$$

Because $h_o(\tau) = 0$ for $\tau < 0$, we can equally well write (100) as

$$\xi_P = K_d(0) - \int_{-\infty}^\infty h_o(\tau)\, K_{dr}(\tau)\, d\tau. \tag{101}$$

Now

$$H_o(j\omega) = \frac{1}{G^+(j\omega)} \int_0^\infty K_{dz}(t) e^{-j\omega t}\, dt, \tag{102}$$

where

$$K_{dz}(t) = \mathcal{F}^{-1}\left\{\frac{S_{dr}(j\omega)}{[G^+(j\omega)]^*}\right\} = \frac{1}{2\pi}\int_{-\infty}^\infty \frac{S_{dr}(j\omega)}{[G^+(j\omega)]^*}\, e^{j\omega t}\, d\omega. \tag{103}$$

Substituting the inverse transform of (102) into (101), we obtain,

$$\xi_P = K_d(0) - \int_{-\infty}^\infty K_{dr}(\tau)\, d\tau \left[\frac{1}{2\pi}\int_{-\infty}^\infty e^{j\omega\tau}\, d\omega \cdot \frac{1}{G^+(j\omega)} \int_0^\infty K_{dz}(t) e^{-j\omega t}\, dt\right]. \tag{104}$$

Changing orders of integration, we have

$$\xi_P = K_d(0) - \int_0^\infty K_{dz}(t)\, dt \left[\frac{1}{2\pi}\int_{-\infty}^\infty e^{-j\omega t}\, d\omega\, \frac{1}{G^+(j\omega)} \int_{-\infty}^\infty K_{dr}(\tau) e^{j\omega\tau}\, d\tau\right]. \tag{105}$$

The part of the integral inside the brackets is just $K_{dz}^*(t)$. Thus, since $K_{dz}(t)$ is real,

$$\xi_P = K_d(0) - \int_0^\infty K_{dz}^{\,2}(t)\, dt. \tag{106}$$

The result in (106) is a convenient expression for the mean-square error. Observe that we must factor the input spectrum and perform an inverse transform in order to evaluate it. (The same shortcuts discussed above are applicable.)

We can use (106) to study the effect of α on the mean-square error. Denote the desired signal when $\alpha = 0$ as $d_0(t)$ and the desired signal for arbitrary α as $d_\alpha(t) \triangleq d_0(t + \alpha)$. Then

$$E[d_0(t)\, z(t - \tau)] = K_{d_0 z}(\tau) \triangleq \phi(\tau), \tag{107a}$$

and

$$E[d_\alpha(t)\, z(t - \tau)] = E[d_0(t + \alpha)\, z(t - \tau)] = \phi(\tau + \alpha). \tag{107b}$$

We can now rewrite (106) in terms of $\phi(\tau)$. Letting

$$K_{dz}(t) = \phi(t + \alpha) \tag{108a}$$

in (106), we have

$$\xi_P{}^\alpha = K_d(0) - \int_0^\infty \phi^2(t + \alpha)\, dt = K_d(0) - \int_\alpha^\infty \phi^2(u)\, du. \tag{108b}$$

Note that $\phi(u)$ is not a function of α. We observe that because the integrand is a positive quantity the error is monotone increasing with increasing α. Thus the smallest error is achieved when $\alpha = -\infty$ (infinite delay) and increases monotonely to unity as $\alpha \to +\infty$. This result says that for *any* desired signal the minimum mean-square error will decrease if we allow delay in the processing. The mean-square error for infinite delay provides a lower bound on the mean-square error for any finite delay and is frequently called the *irreducible error*. A more interesting quantity in some cases is the normalized error. We define the normalized error as

$$\xi_{Pn}{}^\alpha \triangleq \frac{\xi_P{}^\alpha}{K_d(0)}, \tag{109a}$$

or

$$\xi_{Pn}{}^\alpha = 1 - \frac{1}{K_d(0)} \int_\alpha^\infty \phi^2(u)\, du. \tag{109b}$$

We may now apply our results to the preceding example.

***Example 3* (*continued*).** For our example

$$\xi_{Pn}{}^\alpha = \begin{cases} 1 - \dfrac{8P}{N_0} \dfrac{1}{(1 + \sqrt{1 + \Lambda})^2} \left(\displaystyle\int_\alpha^0 dt\, e^{+2k\sqrt{1+\Lambda}\,t} + \int_0^\infty e^{-2kt}\, dt \right), & \alpha \le 0, \\ 1 - \dfrac{8P}{N_0} \dfrac{1}{(1 + \sqrt{1 + \Lambda})^2} \displaystyle\int_\alpha^\infty e^{-2kt}\, dt, & \alpha \ge 0. \end{cases} \tag{110}$$

Evaluating the integrals, we have

$$\xi_{Pn}{}^{\alpha} = \frac{1}{\sqrt{1+\Lambda}} + \frac{\Lambda e^{+2k\sqrt{1+\Lambda}\,\alpha}}{(1+\sqrt{1+\Lambda})^2\sqrt{1+\Lambda}}, \qquad \alpha \leq 0, \tag{111}$$

$$\xi_{Pn}{}^{0} = \frac{2}{1+\sqrt{1+\Lambda}}, \tag{112}$$

and

$$\xi_{Pn}{}^{\alpha} = \frac{2}{(1+\sqrt{1+\Lambda})} + \frac{\Lambda[1-e^{-2k\alpha}]}{(1+\sqrt{1+\Lambda})^2}, \qquad \alpha \geq 0. \tag{113}$$

The two limiting cases for (111) and (113) are $\alpha = -\infty$ and $\alpha = \infty$, respectively.

$$\xi_{Pn}{}^{-\infty} = \frac{1}{\sqrt{1+\Lambda}}. \tag{115}$$

$$\xi_{Pn}{}^{\infty} = 1, \tag{114}$$

A plot of $\xi_{Pn}{}^{\alpha}$ versus $(k\alpha)$ is shown in Fig. 6.15. Physically, the quantity $k\alpha$ is related to the reciprocal of the message bandwidth. If we define

$$\tau_c = \frac{1}{k}, \tag{116}$$

the units on the horizontal axis are α/τ_c, which corresponds to the delay measured in correlation times. We see that the error for a delay of one time constant is approximately the infinite delay error. Note that the error is not a symmetric function of α.

Before summarizing our discussion of realizable filters, we discuss the related problem of unrealizable filters.

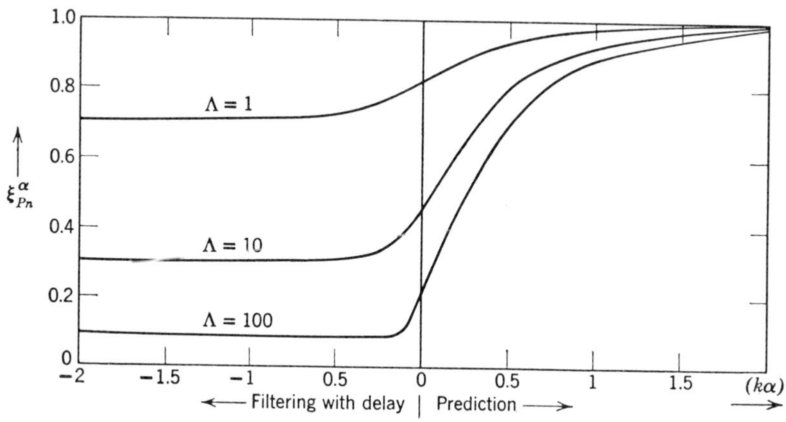

Fig. 6.15 Effect of time-shift on filtering error.

6.2.3 Unrealizable Filters

Instead of requiring the processor to be realizable, let us consider an optimum *unrealizable* system. This corresponds to letting $T_f > t$. In other words, we use the input $r(u)$ at times *later* than t to determine the estimate at t.

For the case in which $T_f > t$ we can modify (55) to obtain

$$K_{dr}(\tau) = \int_{t-T_f}^{\infty} h_o(t, t\text{-}v)\, K_r(\tau - v)\, dv, \qquad t - T_f < \tau < \infty. \tag{117}$$

In this case $h_o(t, t\text{-}v)$ is nonzero for all $v \geq t - T_f$. Because this includes values of v less than zero, the filter will be unrealizable. The case of most interest to us is the one in which $T_f = \infty$. Then (117) becomes

$$K_{dr}(\tau) = \int_{-\infty}^{\infty} h_{ou}(v)\, K_r(\tau - v)\, dv, \qquad -\infty < \tau < \infty. \tag{118}$$

We add the subscript u to emphasize that the filter is unrealizable. Because the equation is valid for all τ, we may solve by transforming

$$\boxed{H_{ou}(j\omega) = \frac{S_{dr}(j\omega)}{S_r(\omega)}.} \tag{119}$$

From Property 4, the mean-square error is

$$\xi_u = K_d(0) - \int_{-\infty}^{\infty} h_{ou}(\tau)\, K_{dr}(\tau)\, d\tau. \tag{120}$$

Note that ξ_u is a mean-square *point* estimation error. By Parseval's Theorem,

$$\xi_u = \frac{1}{2\pi}\int_{-\infty}^{\infty} [S_d(\omega) - H_{ou}(j\omega)\, S_{dr}^*(j\omega)]\, d\omega. \tag{121}$$

Substituting (119) into (121), we obtain

$$\boxed{\xi_u = \int_{-\infty}^{\infty} \frac{S_d(\omega)\, S_r(\omega) - |S_{dr}(j\omega)|^2}{S_r(\omega)} \frac{d\omega}{2\pi}.} \tag{122}$$

For the special case in which

$$\begin{aligned} d(t) &= a(t), \\ r(t) &= a(t) + n(t), \end{aligned} \tag{123}$$

and the message and noise are uncorrelated, (122) reduces to

$$\boxed{\xi_u = \int_{-\infty}^{\infty} \frac{S_n(\omega)\, S_a(\omega)}{S_a(\omega) + S_n(\omega)} \frac{d\omega}{2\pi}.} \tag{124}$$

In the example considered on p. 488, the noise is white. Therefore,

$$H_{ou}(j\omega) = \frac{\sqrt{P}\, S_a(\omega)}{S_r(\omega)}, \tag{125}$$

and

$$\xi_u = \frac{N_0}{2} \int_{-\infty}^{\infty} H_{ou}(j\omega) \frac{d\omega}{2\pi} = \frac{N_0}{2} \int_{-\infty}^{\infty} \frac{\sqrt{P}\, S_a(\omega)}{S_r(\omega)} \frac{d\omega}{2\pi}. \tag{126}$$

We now return to the general case. It is easy to demonstrate that the expression in (121) is also equal to

$$\xi_u = K_d(0) - \int_{-\infty}^{\infty} \phi^2(t)\, dt. \tag{127}$$

Comparing (127) with (107) we see that the effect of using an unrealizable filter is the same as allowing an infinite delay in the desired signal. This result is intuitively logical. In an unrealizable filter we allow ourselves (fictitiously, of course) to use the entire past and future of the input and produce the desired signal at the present time. A practical way to approximate this processing is to wait until more of the future input comes in and produce the desired output at a later time. In many, if not most, communications problems it is the unrealizable error that is a fundamental system limitation.

The essential points to remember when discussing unrealizable filters are the following:

1. The mean-square error using an unrealizable linear filter ($T_f = \infty$) provides a lower bound on the mean-square error for any realizable linear filter. It corresponds to the *irreducible* (or infinite delay) error that we encountered on p. 494. The computation of ξ_u (124) is usually easier than the computation of ξ_P (100) or (106). Therefore it is a logical preliminary calculation even if we are interested only in the realizable filtering problem.
2. We can build a realizable filter whose performance approaches the performance of the unrealizable filter by allowing delay in the output. We can obtain a mean-square error that is arbitrarily close to the irreducible error by increasing this delay. From the practical standpoint a delay of several times the reciprocal of the effective bandwidth of $[S_a(\omega) + S_n(\omega)]$ will usually result in a mean-square error close to the irreducible error.

We now return to the realizable filtering problem. In Sections 6.2.1 and 6.2.2 we devised an algorithm that gave us a constructive method for finding the optimum realizable filter and the resulting mean-square error. In other words, given the necessary information, we can always (conceptually, at least) proceed through a specified procedure and obtain the optimum filter and resulting performance. In practice, however, the algebraic complexity has caused most engineers studying optimum filters to use the one-pole spectrum as the canonic message spectrum. The lack of a closed-form mean-square error expression which did not require a spectrum factorization made it essentially impossible to study the effects of different message spectra.

In the next section we discuss a special class of linear estimation problems and develop a *closed-form* expression for the minimum mean-square error.

6.2.4 Closed-Form Error Expressions

In this section we shall derive some useful closed-form results for a special class of optimum linear filtering problems. The case of interest is when

$$r(u) = a(u) + n(u), \qquad -\infty < u \leq t. \tag{128}$$

In other words, the received signal consists of the message plus additive noise. The desired signal $d(t)$ is the message $a(t)$. We assume that the noise and message are uncorrelated. The message spectrum is rational with a finite variance. Our goal is to find an expression for the error that does not require spectrum factorization. The major results in this section were obtained originally by Yovits and Jackson [4]. It is convenient to consider white and nonwhite noise separately.

Errors in the Presence of White Noise. We assume that $n(t)$ is white with spectral height $N_0/2$. Although the result was first obtained by Yovits and Jackson, appreciably simpler proofs have been given by Viterbi and Cahn [5], Snyder [6], and Helstrom [7]. We follow a combination of these proofs. From (128)

$$S_r(\omega) = S_a(\omega) + \frac{N_0}{2}, \tag{129}$$

and

$$G^+(j\omega) = \left[S_a(j\omega) + \frac{N_0}{2}\right]^+. \tag{130}$$

From (78)

$$H_o(j\omega) = \frac{1}{[S_a(\omega) + N_0/2]^+}\left\{\frac{S_a(\omega)}{[S_a(\omega) + N_0/2]^-}\right\}_+ \tag{131†}$$

†To avoid a double superscript we introduce the notation

$$G^-(j\omega) = [G^+(j\omega)]^*.$$

Recall that conjugation in the frequency domain corresponds to reversal in the time domain. The time function corresponding to $G^+(j\omega)$ is zero for negative time. Therefore the time function corresponding to $G^-(j\omega)$ is zero for positive time.

or

$$H_o(j\omega) = \frac{1}{[S_a(\omega) + N_0/2]^+} \left\{ \frac{S_a(\omega) + N_0/2}{[S_a(\omega) + N_0/2]^-} - \frac{N_0/2}{[S_a(\omega) + N_0/2]^-} \right\}_+ . \tag{132}$$

Now, the first term in the bracket is just $[S_a(\omega) + N_0/2]^+$, which is realizable. Because the realizable part operator is linear, the first term comes out of the bracket without modification. Therefore

$$H_o(j\omega) = 1 - \frac{1}{[S_a(\omega) + N_0/2]^+} \left\{ \frac{N_0/2}{[S_a(\omega) + N_0/2]^-} \right\}_+ , \tag{133}$$

We take $\sqrt{N_0/2}$ out of the brace and put the remaining $\sqrt{N_0/2}$ inside the $[\cdot]^-$. The operation $[\cdot]^-$ is a factoring operation so we obtain $N_0/2$ inside.

$$H_o(j\omega) = 1 - \frac{\sqrt{N_0/2}}{[S_a(\omega) + N_0/2]^+} \left\{ \frac{1}{\left[\dfrac{S_a(\omega) + N_0/2}{N_0/2}\right]^-} \right\}_+ . \tag{134}$$

The next step is to prove that the realizable part of the term in the brace equals one.

Proof. Let $S_a(\omega)$ be a rational spectrum. Thus

$$S_a(\omega) = \frac{N(\omega^2)}{D(\omega^2)}, \tag{135}$$

where the denominator is a polynomial in ω^2 whose order is at least one higher than the numerator polynomial. Then

$$\frac{S_a(\omega) + N_0/2}{N_0/2} = \frac{N(\omega^2) + (N_0/2)\, D(\omega^2)}{(N_0/2)\, D(\omega^2)} \tag{136}$$

$$= \frac{D(\omega^2) + (2/N_0)\, N(\omega^2)}{D(\omega^2)} \tag{137}$$

$$= \prod_{i=1}^{n} \frac{\omega^2 + \alpha_i^2}{\omega^2 + \beta_i^2}. \tag{138}$$

Observe that there is no additional multiplier because the highest order term in the numerator and denominator are identical.

The α_i and β_i may always be chosen so that their real parts are positive. If any of the α_i or β_i are complex, the conjugate is also present. Inverting both sides of (138) and factoring the result, we have

$$\left\{ \left[\frac{S_a(\omega) + N_0/2}{N_0/2}\right]^- \right\}^{-1} = \prod_{i=1}^{n} \frac{(-j\omega + \beta_i)}{(-j\omega + \alpha_i)} \tag{139}$$

$$= \prod_{i=1}^{n} \left[1 + \frac{\beta_i - \alpha_i}{(-j\omega + \alpha_i)}\right]. \tag{140}$$

The transform of all terms in the product except the unity term will be zero for positive time (their poles are in the right-half s-plane). Multiplying the terms together corresponds to convolving their transforms. Convolving functions which are zero for positive time always gives functions that are zero for positive time. Therefore only the unity term remains when we take the realizable part of (140). This is the desired result. Therefore

$$\boxed{H_o(j\omega) = 1 - \frac{\sqrt{N_0/2}}{[S_a(j\omega) + N_0/2]^+}.} \tag{141}$$

The next step is to derive an expression for the error. From Property 4C (27–28) we know that

$$\xi_P = \frac{N_0}{2} \lim_{\tau \to t^-} h_o(t, \tau) = \frac{N_0}{2} \lim_{\epsilon \to 0^+} h_o(\epsilon) \triangleq \frac{N_0}{2} h_o(0^+) \tag{142}$$

for the time-invariant case. We also know that

$$\int_{-\infty}^{\infty} H_o(j\omega) \frac{d\omega}{2\pi} = \frac{h_o(0^+) + h_o(0^-)}{2} = \frac{h_o(0^+)}{2}, \tag{143}$$

because $h_0(\tau)$ is realizable. Combining (142) and (143), we obtain

$$\xi_P = N_0 \int_{-\infty}^{\infty} H_o(j\omega) \frac{d\omega}{2\pi}. \tag{144}$$

Using (141) in (144), we have

$$\xi_P = N_0 \int_{-\infty}^{\infty} \left(1 - \left\{\left[\frac{S_a(\omega) + N_0/2}{N_0/2}\right]^+\right\}^{-1}\right) \frac{d\omega}{2\pi}. \tag{145}$$

Using the conjugate of (139) in (145), we obtain

$$\xi_P = N_0 \int_{-\infty}^{\infty} \left[1 - \prod_{i=1}^{n} \frac{(j\omega + \beta_i)}{(j\omega + \alpha_i)}\right] \frac{d\omega}{2\pi}, \tag{146}$$

$$\xi_P = N_0 \int_{-\infty}^{\infty} \left[1 - \prod_{i=1}^{n} \left(1 + \frac{\beta_i - \alpha_i}{j\omega + \alpha_i}\right)\right] \frac{d\omega}{2\pi}. \tag{147}$$

Expanding the product, we have

$$\xi_P = N_0 \int_{-\infty}^{\infty} \left\{1 - \left[1 + \sum_{i=1}^{n} \frac{\beta_i - \alpha_i}{j\omega + \alpha_i} + \sum_{i=1}^{n} \sum_{j=1}^{n} \frac{\gamma_{ij}}{(j\omega + \alpha_i)(j\omega + \alpha_j)} + \sum_{i=1}^{n} \sum_{j=1}^{n} \sum_{k=1}^{n} \cdots\right]\right\} \frac{d\omega}{2\pi}, \tag{148}$$

$$\xi_P = N_0 \int_{-\infty}^{\infty} \sum_{i=1}^{n} \frac{(\alpha_i - \beta_i)}{(j\omega + \alpha_i)} \frac{d\omega}{2\pi} - N_0 \int_{-\infty}^{\infty} \left[\sum_{i=1}^{n} \sum_{j=1}^{n} \frac{\gamma_{ij}}{(j\omega + \alpha_i)(j\omega + \alpha_j)} + \sum_{i=1}^{n} \sum_{j=1}^{n} \sum_{k=1}^{n} \cdots\right] \frac{d\omega}{2\pi}. \tag{149}$$

The integral in the first term is just one half the sum of the residues (this result can be verified easily). We now show that the second term is zero. Because the integrand in the second term is analytic in the right half of the s-plane, the integral $[-\infty, \infty]$ equals the integral around a semicircle with infinite radius. All terms in the brackets, however, are at least of order $|s|^{-2}$ for large $|s|$. Therefore the integral on the semicircle is zero, which implies that the second term is zero. Therefore

$$\boxed{\xi_P = \frac{N_0}{2} \sum_{i=1}^{n} (\alpha_i - \beta_i).} \tag{150}$$

The last step is to find a closed-form expression for the sum of the residues. This follows by observing that

$$\int_{-\infty}^{\infty} \ln\left(\frac{\omega^2 + \alpha_i^2}{\omega^2 + \beta_i^2}\right) \frac{d\omega}{2\pi} = (\alpha_i - \beta_i). \tag{151}$$

(To verify this equation integrate the left-hand side by parts with $u = \ln[(\omega^2 + \alpha_i^2)/(\omega^2 + \beta_i^2)]$ and $dv = d\omega/2\pi$.)

Comparing (150), (151), and (138), we have

$$\boxed{\xi_P = \frac{N_0}{2} \int_{-\infty}^{\infty} \ln\left[1 + \frac{S_a(\omega)}{N_0/2}\right] \frac{d\omega}{2\pi},} \tag{152}$$

which is the desired result. Both forms of the error expressions (150) and (152) are useful. The first form is often the most convenient way to actually evaluate the error. The second form is useful when we want to find the $S_a(\omega)$ that minimizes ξ_P subject to certain constraints.

It is worthwhile to emphasize the importance of (152). In conventional Wiener theory to investigate the effect of various message spectra we had to actually factor the input spectrum. The result in (152) enables us to explore the error behavior directly. In later chapters we shall find it essential to the solution for the optimum pre-emphasis problem in angle modulation and other similar problems.

We observe in passing that the integral on the right-hand side is equal to twice the average mutual information (as defined by Shannon) between $r(t)$ and $a(t)$.

Errors for Typical Message Spectra. In this section we consider two families of message spectra. For each family we use (152) to compute the error when the optimum realizable filter is used. To evaluate the improvement obtained by allowing delay we also use (124) to calculate the unrealizable error.

Case 1. Butterworth Family. The message processes in the first class have spectral densities that are inverse Butterworth polynomials of order $2n$. From (3.98),

$$S_a(\omega:n) = \frac{2nP}{k} \frac{\sin(\pi/2n)}{1 + (\omega/k)^{2n}} \triangleq \frac{c_n}{1 + (\omega/k)^{2n}}. \tag{153}$$

The numerator is just a gain adjusted so that the power in the message spectrum is P.

Some members of the family are shown in Fig. 6.16. For $n = 1$ we have the one-pole spectrum of Section 6.2.1. The break-point is at $\omega = k$ rad/sec and the magnitude decreases at 6 db/octave above this point. For higher n the break-point remains the same, but the magnitude decreases at $6n$ db/octave. For $n = \infty$ we have a rectangular bandlimited spectrum

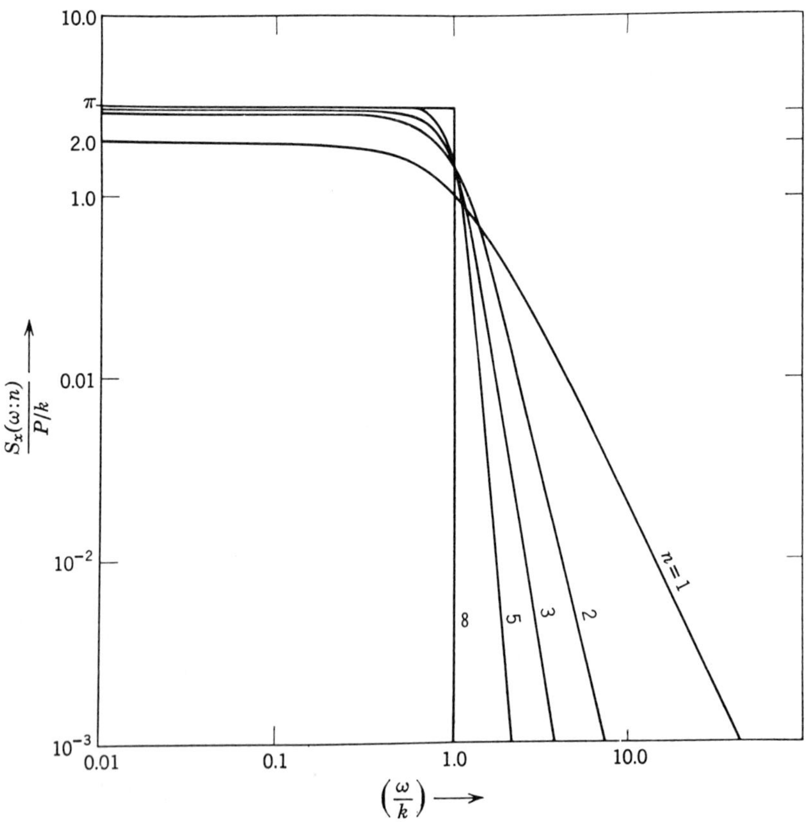

Fig. 6.16 Butterworth spectra.

[height $P\pi/k$; width k rad/sec]. We observe that if we use k/π cps as the reference bandwidth (double-sided) the signal-to-noise ratio will not be a function of n and will provide a useful reference:

$$\Lambda_B \triangleq \frac{P}{k/\pi \cdot N_0/2} = \frac{2\pi P}{kN_0}. \tag{154}$$

To find ξ_P we use (152):

$$\xi_P = \frac{N_0}{2} \int_{-\infty}^{\infty} \frac{d\omega}{2\pi} \ln \left[1 + \frac{2c_n/N_0}{1 + (\omega/k)^{2n}}\right]. \tag{155}$$

This can be integrated (see Problem 6.2.18) to give the following expression for the normalized error:

$$\xi_{Pn} = \frac{\pi}{\Lambda_B}\left(\sin\frac{\pi}{2n}\right)^{-1}\left[\left(1 + 2n\frac{\Lambda_B}{\pi}\sin\frac{\pi}{2n}\right)^{1/2n} - 1\right]. \tag{156}$$

Similarly, to find the unrealizable error we use (124):

$$\xi_u = \int_{-\infty}^{\infty} \frac{d\omega}{2\pi} \frac{c_n}{1 + (\omega/k)^{2n} + (2/N_0)c_n}. \tag{157}$$

This can be integrated (see Problem 6.2.19) to give

$$\xi_{un} = \left[1 + \frac{2n}{\pi}\Lambda_B \sin\frac{\pi}{2n}\right]^{(1/2n)-1}. \tag{158}$$

The reciprocal of the normalized error is plotted versus Λ_B in Fig. 6.17. We observe the vast difference in the error behavior as a function of n. The most difficult spectrum to filter is the one-pole spectrum. We see that asymptotically it behaves linearly whereas the bandlimited spectrum behaves exponentially. We also see that for $n = 3$ or 4 the performance is reasonably close to the bandlimited ($n = \infty$) case. Thus the one-pole message spectrum, which is commonly used as an example, has the worst error performance.

A second observation is the difference in improvement obtained by use of an unrealizable filter. For the one-pole spectrum

$$\xi_{un} > \tfrac{1}{2}\xi_{Pn}, \qquad (n = 1). \tag{159}$$

In other words, the maximum possible ratio is 2 (or 3 db). At the other extreme for $n = \infty$

$$\xi_{Pn} = \frac{1}{\Lambda_B}\ln(1 + \Lambda_B), \tag{160}$$

whereas

$$\xi_{un} = \frac{1}{1 + \Lambda_B}. \tag{161}$$

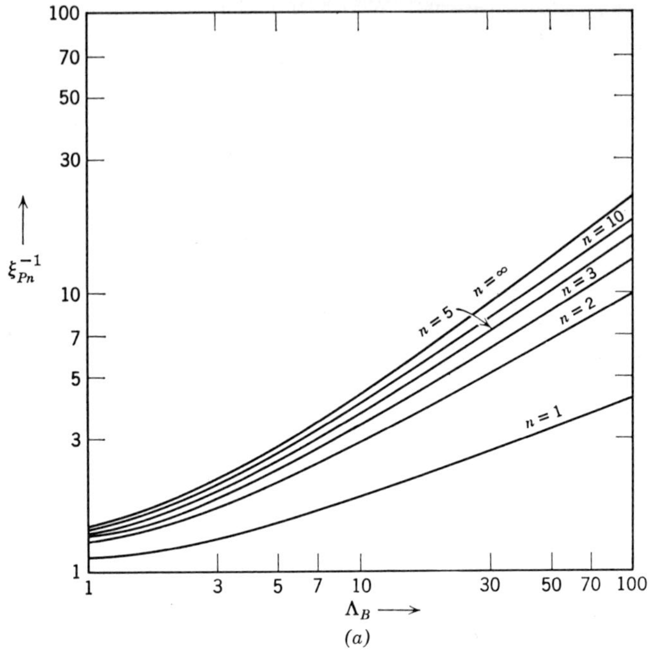

(a)

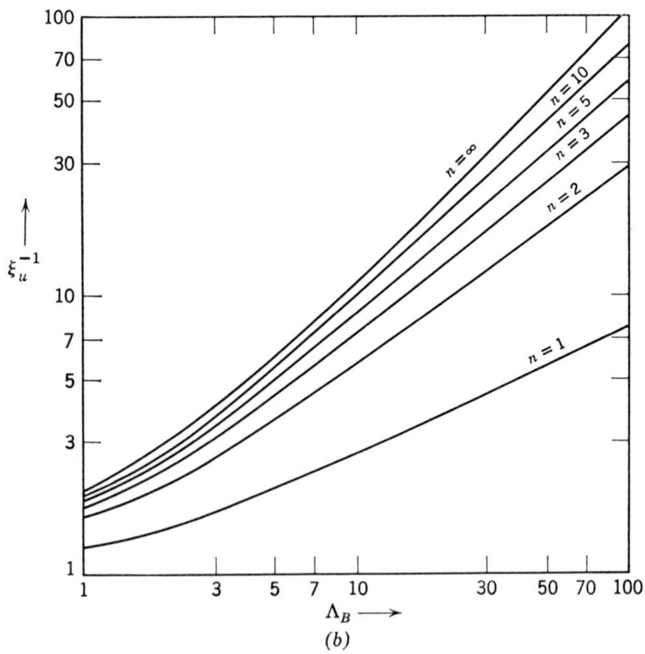

(b)

Fig. 6.17 Reciprocal of mean-square error, Butterworth spectra: (a) realizable; (b) unrealizable

Thus the ratio is

$$\frac{\xi_{un}}{\xi_{Pn}} = \frac{\Lambda_B}{1 + \Lambda_B} \frac{1}{\ln(1 + \Lambda_B)}. \tag{162}$$

For $n = \infty$ we achieve appreciable improvement for large Λ_B by allowing delay.

Case 2. Gaussian Family. A second family of spectra is given by

$$S_a(\omega:n) = \frac{2P\sqrt{\pi}\,\Gamma(n)}{k\sqrt{n}\,\Gamma(n-\frac{1}{2})} \frac{1}{(1 + \omega^2/nk^2)^n} \triangleq \frac{d_n}{(1 + \omega^2/nk^2)^n}, \tag{163}$$

obtained by passing white noise through n isolated one-pole filters. In the limit as $n \to \infty$, we have a Gaussian spectrum

$$\lim_{n \to \infty} S_a(\omega:n) = \frac{2\sqrt{\pi}}{k} Pe^{-\omega^2/k^2}. \tag{164}$$

The family of Gaussian spectra is shown in Fig. 6.18. Observe that for $n = 1$ the two cases are the same.

The expressions for the two errors of interest are

$$\xi_{Pn} = \frac{N_0}{2P} \int_{-\infty}^{\infty} \frac{d\omega}{2\pi} \ln\left[1 + \frac{2d_n/N_0}{(1 + \omega^2/nk^2)^n}\right] \tag{165}$$

and

$$\xi_{un} = \frac{1}{P} \int_{-\infty}^{\infty} \frac{d\omega}{2\pi} \frac{d_n}{(1 + \omega^2/nk^2)^n + (2/N_0)d_n}. \tag{166}$$

To evaluate ξ_{Pn} we rewrite (165) in the form of (150). For this case the evaluation of α_i and β_i is straightforward [53]. The results for $n = 1,2,3,$ and 5 are shown in Fig. 6.19*a*. For $n = \infty$ the most practical approach is to perform the integration numerically. We evaluate (166) by using a partial fraction expansion. Because we have already found the α_i and β_i, the residues follow easily. The results for $n = 1, 2, 3,$ and 5 are shown in Fig. 6.19*b*. For $n = \infty$ the result is obtained numerically. By comparing Figs. 6.17 and Fig. 6.19 we see that the Gaussian spectrum is more difficult to filter than the bandlimited spectrum. Notice that the limiting spectra in both families were nonrational (they were also not factorable).

In this section we have applied some of the closed-form results for the special case of filtering in the presence of additive white noise. We now briefly consider some other related problems.

Colored Noise and Linear Operations. The advantage of the error expression in (152) was its simplicity. As we proceed to more complex noise spectra, the results become more complicated. In almost all cases the error expressions are easier to evaluate than the expression obtained from the conventional Wiener approach.

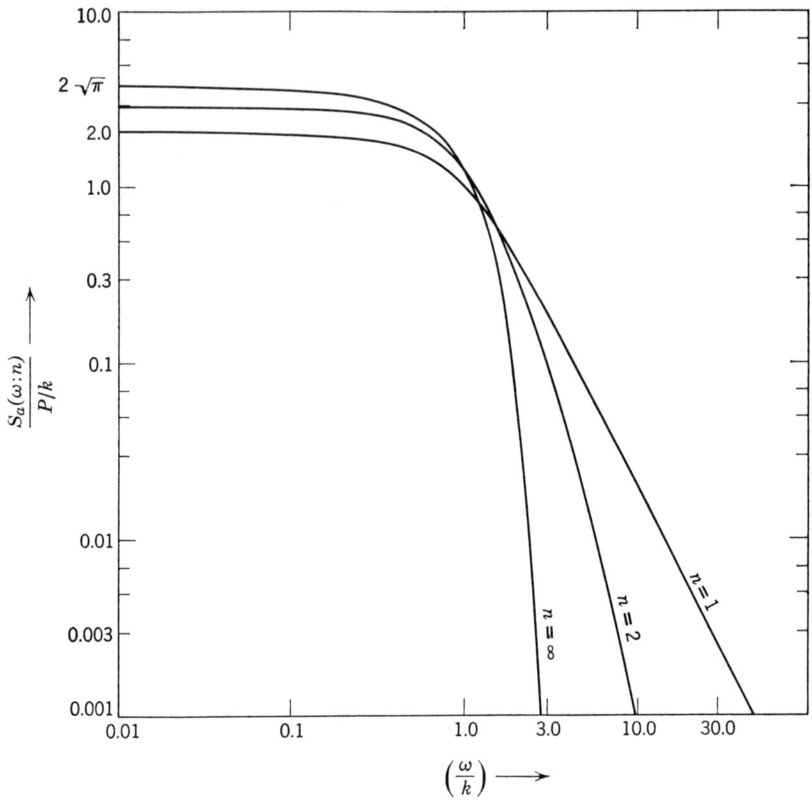

Fig. 6.18 Gaussian family.

For our purposes it is adequate to list a number of cases for which answers have been obtained. The derivations for some of the results are contained in the problems (see Problems 6.2.20 to 6.2.26).

1. The message $a(t)$ is transmitted. The additive noise has a spectrum that contains poles but not zeros.
2. The message $a(t)$ is passed through a linear operation whose transfer function contains only zeros before transmission. The additive noise is white.
3. The message $a(t)$ is transmitted. The noise has a polynomial spectrum

$$S_n(\omega) = N_0 + N_2\omega^2 + N_4\omega^4 + \cdots + N_{2n}\omega^{2n}.$$

4. The message $a(t)$ is passed through a linear operation whose transfer function contains *poles* only. The noise is white.

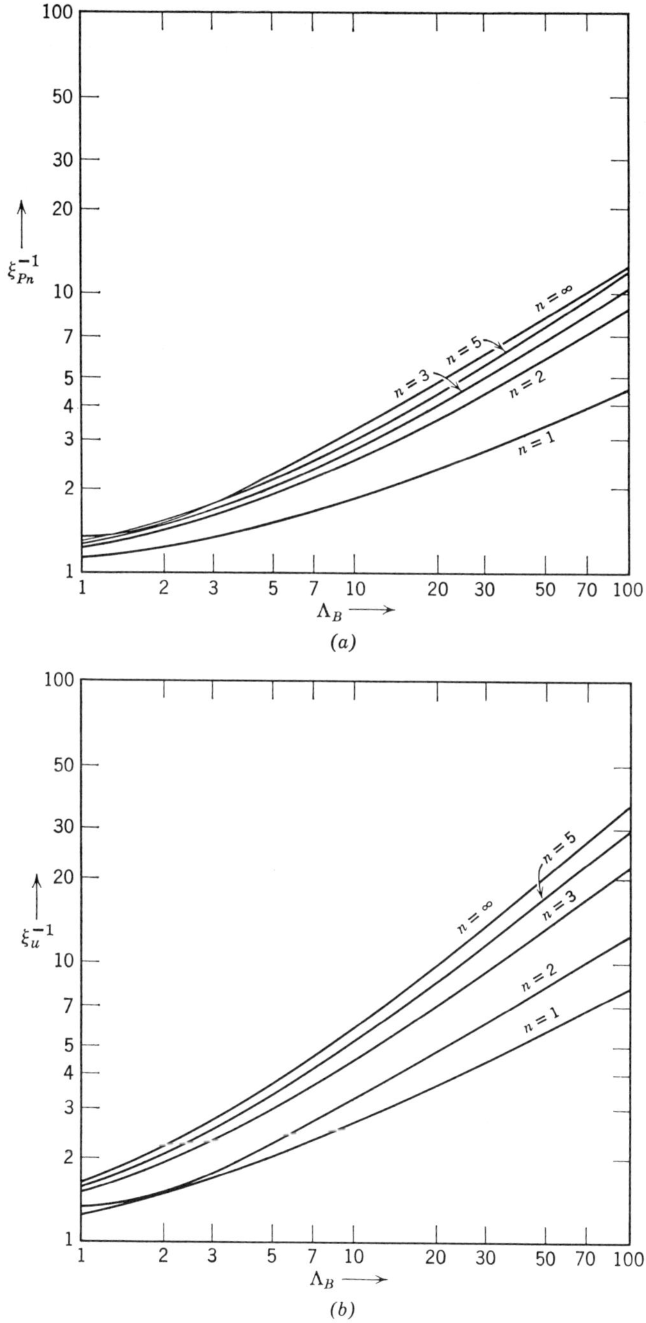

Fig. 6.19 Reciprocal of mean-square error: Gaussian family, (a) realizable; (b) unrealizable.

We observe that Cases 2 and 4 will lead to the same error expression as Cases 1 and 3, respectively.

To give an indication of the form of the answer we quote the result for typical problems from Cases 1 and 4.

Example (from Case 1). Let the additive noise $n(t)$ be uncorrelated with the message and have a spectrum,

$$S_n(\omega) = \frac{2c\sigma_n^2}{\omega^2 + c^2}. \tag{167}$$

Then

$$\xi_P = \sigma_n^2 \exp\left(-\frac{1}{\sigma_n^2}\int_{-\infty}^{\infty} S_n(\omega) \ln\left[1 + \frac{S_a(\omega)}{S_n(\omega)}\right]\frac{d\omega}{2\pi}\right). \tag{168}$$

This result is derived in Problem 6.2.20.

Example (from Case 4). The message $a(t)$ is integrated before being transmitted.

$$r(t) = \int_{-\infty}^{t} a(u)\, du + w(t). \tag{169}$$

Then

$$\xi_P = \frac{N_0}{6} I_1^3 + I_2, \tag{170}$$

where

$$I_1 = \int_{-\infty}^{\infty} \ln\left[1 + \frac{2S_a(\omega)}{\omega^2 N_0}\right]\frac{d\omega}{2\pi}, \tag{171}$$

$$I_2 = \frac{N_0}{2}\int_{-\infty}^{\infty} \omega^2 \ln\left[1 + \frac{2S_a(\omega)}{\omega^2 N_0}\right]\frac{d\omega}{2\pi}. \tag{172}$$

This result is derived in Problem 6.2.25.

It is worthwhile to point out that the *form* of the error expression depends only on the *form* of the noise spectrum or the linear operation. This allows us to vary the message spectrum and study the effects in an easy fashion.

As a final topic in our discussion of Wiener filtering, we consider optimum feedback systems.

6.2.5 Optimum Feedback Systems

One of the forms in which optimum linear filters are encountered in the sequel is as components in a feedback system. The modification of our results to include this case is straightforward.

We presume that the assumptions outlined at the beginning of Section 6.2 (pp. 481–482) are valid. In addition, we *require* the linear processor to have the form shown in Fig. 6.20. Here $g_l(\tau)$ is a linear filter. We are allowed to choose $g_l(\tau)$ to obtain the best $\hat{d}(t)$.

System constraints of this kind develop naturally in the control system context (see [9]). In Chapter II.2 we shall see how they arise as linearized

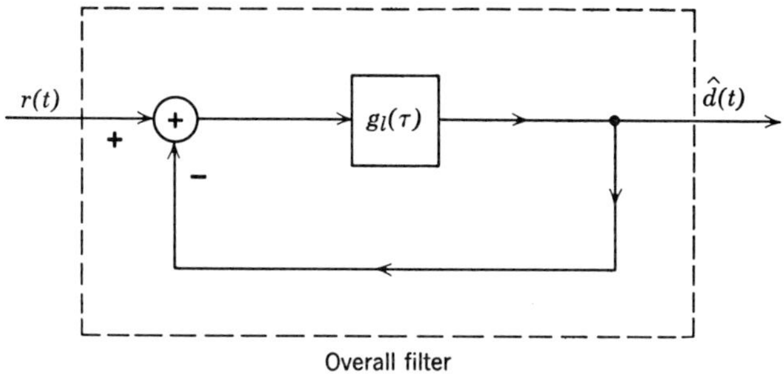

Fig. 6.20 Feedback system.

versions of demodulators. Clearly, we want the *closed* loop transfer function to equal $H_o(j\omega)$. We denote the loop filter that accomplishes this as $G_{lo}(j\omega)$. Now,

$$H_o(j\omega) = \frac{G_{lo}(j\omega)}{1 + G_{lo}(j\omega)} \tag{173}$$

Solving for $G_{lo}(j\omega)$,

$$G_{lo}(j\omega) = \frac{H_o(j\omega)}{1 - H_o(j\omega)}. \tag{174}$$

For the general case we evaluate $H_o(j\omega)$ by using (78) and substitute the result into (174).

For the special case in Section 6.2.4, we may write the answer directly. Substituting (141) into (174), we have

$$\boxed{G_{lo}(j\omega) = \left(\frac{2}{N_0}\right)^{1/2}\left\{\left[S_a(j\omega) + \frac{N_0}{2}\right]^+ - 1\right\}.} \tag{175}$$

We observe that $G_{lo}(j\omega)$ has the same *poles* as $G^+(j\omega)$ and is therefore a stable, realizable filter. We also observe that the poles of $G^+(j\omega)$ (and therefore the loop filter) are just the left-half-s-plane poles of the *message* spectrum.

We observe that the message can be visualized as the output of a linear filter when the input is white noise. The general rational case is shown in Fig. 6.21*a*. We control the power by adjusting the spectral height of $u(t)$:

$$E[u(t)\,u(\tau)] \triangleq q\,\delta(t - \tau). \tag{176}$$

The message spectrum is

$$S_a(\omega) = q\left|\frac{b_{n-1}(j\omega)^{n-1} + \cdots + b_0}{(j\omega)^n + p_{n-1}(j\omega)^{n-1} + \cdots + p_0}\right|^2. \tag{177}$$

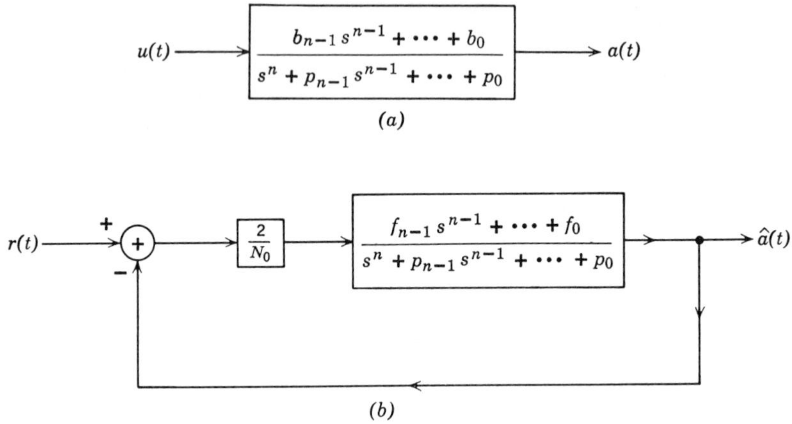

Fig. 6.21 Filters: (*a*) message generation; (*b*) canonic feedback filter.

The numerator must be at least one degree less than the denominator to satisfy the finite power assumption in Section 6.2.4. From (175) we see that the optimum loop filter has the same poles as the filter that could be used to generate the message. Therefore the loop filter has the form shown in Fig. 6.21*b*. It is straightforward to verify that the numerator of the loop filter is *exactly* one degree less than the denominator (see Problem 6.2.27). We refer to the structure in Fig. 6.21*b* as the canonic feedback realization of the optimum filter for rational message spectra [21]. In Section 6.3 we shall find that a general canonic feedback realization can be derived by for nonstationary processes and finite observation intervals.

Observe that to find the numerator we must still perform a factoring operation (see Problem 6.2.27). We can also show that the first coefficient in the numerator is $2\xi_P/N_0$ (see Problem 6.2.28).

A final question about feedback realizations of optimum linear filters concerns unrealizable filters. Because we have seen in Sections 6.2.2 and 6.2.3 [(108*b*) and (127)] that using an unrealizable filter (or allowing delay) always improves the performance, we should like to make provision for it in the feedback configuration. Previously we approximated unrealizable filters by allowing delay. Looking at Fig. 6.20, we see that this would not work in the case of $g_l(\tau)$ because its output is fed back in real time to become part of its input.

If we are willing to allow a postloop filter, as shown in Fig. 6.22, we can consider unrealizable operations. There is no difficulty with delay in the postloop filter because its output is *not* used in any other part of the system.

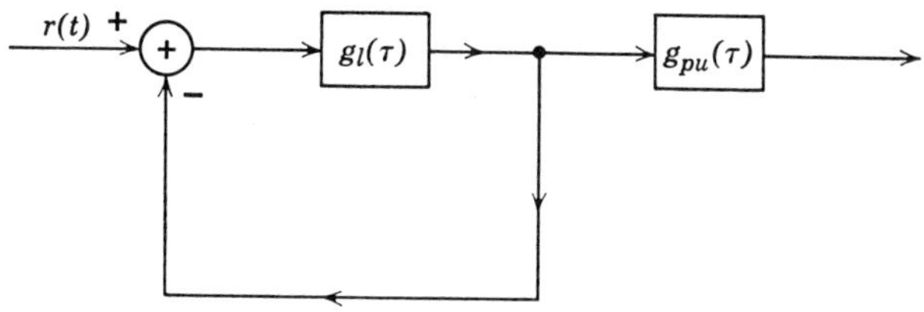

Fig. 6.22 Unrealizable postloop filter.

The expression for the optimum unrealizable postloop filter $G_{puo}(j\omega)$ follows easily. Because the cascade of the two systems must correspond to $H_{ou}(j\omega)$ and the closed loop to $H_o(j\omega)$, it follows that

$$G_{puo}(j\omega) = \frac{H_{ou}(j\omega)}{H_o(j\omega)}. \tag{178}$$

The resulting transfer function can be approximated arbitrarily closely by allowing delay.

6.2.6 Comments

In this section we discuss briefly some aspects of linear processing that are of interest and summarize our results.

Related Topics. *Multidimensional Problem.* Although we formulated the vector problem in Section 6.1, we have considered only the solution technique for the scalar problem. For the *unrealizable* case the extension to the vector problem is trivial. For the realizable case in which the message or the desired signal is a vector and the received waveform is a scalar the solution technique is an obvious modification of the scalar technique. For the realizable case in which the received signal is a vector, the technique becomes quite complex. Wiener outlined a solution in [1] which is quite tedious. In an alternate approach we factor the input spectral matrix. Techniques are discussed in [10] through [19].

Nonrational Spectra. We have confined our discussion to rational spectra. For nonrational spectra we indicated that we could use a rational approximation. A direct factorization is not always possible.

We can show that a necessary and sufficient condition for factorability is that the integral

$$\int_{-\infty}^{\infty} \left| \frac{\log S_r(\omega)}{1 + (\omega/2\pi)^2} \right| d\omega$$

must converge. Here $S_r(\omega)$ is the spectral density of the entire received waveform. This condition is derived and the implications are discussed in [1]. It is referred to as the *Paley-Wiener criterion.*

If this condition is not satisfied, $r(t)$ is termed a deterministic waveform. The adjective deterministic is used because we can *predict* the future of $r(t)$ exactly by using a *linear* operation on only the past data. A simple example of a deterministic waveform is given in Problem 6.2.39.

Both limiting message spectra in the examples in Section 6.2.4 were deterministic. This means, if the noise *were* zero, we would be able to predict the future of the message exactly. We can study this behavior easily by choosing some arbitrary prediction time α and looking at the prediction error as the index $n \to \infty$. For an arbitrary α we can make the mean-square prediction error less than any positive number by letting n become sufficiently large (see Problem 6.2.41).

In almost all cases the spectra of interest to us will correspond to non-deterministic waveforms. In particular, inclusion of white noise in $r(t)$ guarantees factorability.

Sensitivity. In the detection and parameter estimation areas we discussed the importance of investigating how sensitive the performance of the optimum system was with respect to the detailed assumptions of the model. Obviously, sensitivity is also important in linear modulation. In any particular case the technique for investigating the sensitivity is straightforward. Several interesting cases are discussed in the problems (6.2.31–6.2.33).

In the scalar case most problems are insensitive to the detailed assumptions. In the vector case we must exercise more care.

As before, a general statement is not too useful. The important point to re-emphasize is that we must always check the sensitivity.

Colored and White Noise. When trying to estimate the message $a(t)$ in the presence of noise containing both white and colored components, there is an interesting interpretation of the optimum filter.

Let

$$r(t) = a(t) + n_c(t) + w(t), \tag{179}$$

and

$$d(t) = a(t). \tag{180}$$

Now, it is clear that *if* we *knew* $n_c(t)$ the optimum processor would be that shown in Fig. 6.23*a*. Here $h_o(\tau)$ is just the optimum filter for $r'(t) = a(t) + w(t)$, which we have found before.

We do not know $n_c(t)$ because it is a sample function from a random process. A logical approach would be to estimate $n_c(t)$, subtract the estimate from $r(t)$, and pass the result through $h_o(\tau)$, as shown in Fig. 6.23*b*.

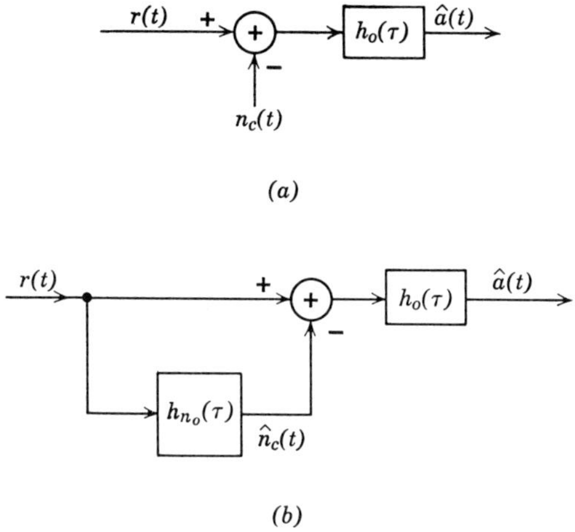

Fig. 6.23 Noise estimation.

We can show that the optimum system does exactly this (see Problem 6.2.34). (Note that $\hat{a}(t)$ and $\hat{n}_c(t)$ are coupled.) This is the same kind of intuitively pleasing result that we encountered in the detection theory area. The optimum processor does exactly what we would do if the disturbances were known exactly, only it uses estimates.

Linear Operations and Filters. In Fig. 6.1 we showed a *typical* estimation problem. With the assumptions in this section, it reduces to the problem shown in Fig. 6.24. The general results in (78), (119), and (122) are applicable. Because most interesting problems fall into the model in Fig. 6.24, it is worthwhile to state our results in a form that exhibits the effects of $k_d(\tau)$ and $k_f(\tau)$ explicitly. The desired relations for uncorrelated message and noise are

$$H_o(j\omega) = \frac{1}{[S_a(\omega)|K_f(j\omega)|^2 + S_n(\omega)]^+}\left[\frac{K_d(j\omega)\, S_a(\omega)\, K_f^*(j\omega)}{[S_a(\omega)|K_f(j\omega)|^2 + S_n(\omega)]^-}\right]_+ \qquad (181)$$

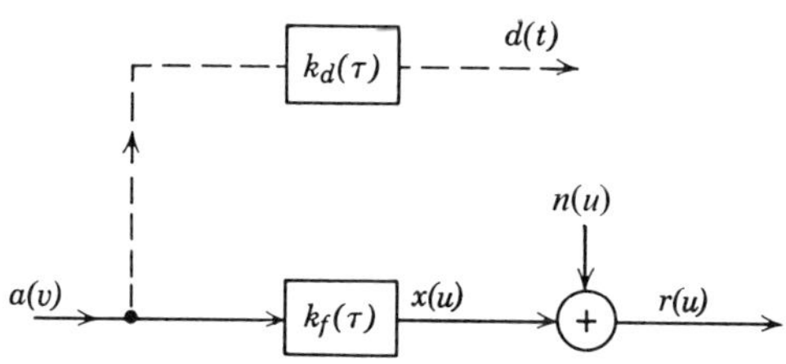

Fig. 6.24 Typical unmodulated estimation problem.

and

$$H_{ou}(j\omega) = \frac{K_d(j\omega)\, S_a(\omega)\, K_f^*(j\omega)}{S_a(\omega)|K_f(j\omega)|^2 + S_n(\omega)}. \tag{182}$$

In the realizable filtering case we must use (181) whenever $d(t) \neq a(t)$. By a simple counterexample (Problem 6.2.38) we can show that linear filtering and optimum realizable estimators *do not* commute in general. In other words, $\hat{d}(t)$ does not necessarily equal

$$\int_{-\infty}^{\infty} k_d(t - \tau)\, \hat{a}(\tau)\, d\tau$$

This result is in contrast to that obtained for MAP interval estimators. On the other hand, comparing (119) with (182), we see that linear filtering and optimum unrealizable ($T_f = \infty$) estimators *do* commute. The resulting error in the unrealizable case when $K_d(j\omega) = 1$ is

$$\xi_{uo} = \int_{-\infty}^{\infty} \frac{S_a(\omega)\, S_n(\omega)}{S_a(\omega)|K_f(j\omega)|^2 + S_n(\omega)} \frac{d\omega}{2\pi}. \tag{183}$$

This expression is obvious. For other $K_d(j\omega)$ see Problem 6.2.35. Some implications of (183) with respect to the prefiltering problem are discussed in Problem 6.2.36.

Remember that we assumed that $k_d(\tau)$ represents an "allowable" operation in the mean-square sense; for example, if the desired operation is a differentiation, we assume that $a(t)$ is a mean-square differentiable process (see Problem 6.2.37 for an example of possible difficulties when this assumption is not true).

Summary. We have discussed in some detail the problem of linear processing for stationary processes when the infinite past was available. The principal results are the following:

1. A constructive solution to the problem is given in (78):

$$H_o(j\omega) = \frac{1}{G^+(j\omega)} \left[\frac{S_{dr}(j\omega)}{[G^+(j\omega)]^*}\right]_+. \tag{78}$$

2. The effect of delay or prediction on the resulting error in an optimum linear processor is given in (108*b*). In all cases there is a monotone improvement as more delay is allowed. In many cases the improvement is sufficient to justify the resulting complexity.

3. The importance of the idea that an unrealizable filter can be approximated arbitrarily closely by allowing a processing delay. The advantage of the unrealizable concept is that the answer can almost always be easily obtained and represents a lower bound on the MMSE in any system.

4. A closed-form expression for the error in the presence of white noise is given in (152),

$$\xi_P = \frac{N_0}{2}\int_{-\infty}^{\infty} \ln\left[1 + \frac{S_a(\omega)}{N_0/2}\right]\frac{d\omega}{2\pi}. \tag{152}$$

5. The canonic filter structure for white noise shown in Fig. 6.21 enables us to relate the complexity of the optimum filter to the complexity of the message spectra by inspection.

We now consider another approach to the point estimation problem.

6.3 KALMAN-BUCY FILTERS

Once again the basic problem of interest is to operate on a received waveform $r(u)$, $T_i \le u \le t$, to obtain a minimum mean-square error *point* estimate of some desired waveform $d(t)$. In a simple scalar case the received waveform is

$$r(u) = c(u)\, a(u) + n(u), \qquad T_i \le u \le t, \tag{184}$$

where $a(t)$ and $n(t)$ are zero-mean random processes with covariance functions $K_a(t, u)$ and $(N_0/2)\,\delta(t - u)$, respectively, and $d(t) = a(t)$. The problem is much more general than this example, but the above case is adequate for motivation purposes.

The optimum processor consists of a linear filter that satisfies the equation

$$K_{ar}(t, \sigma) = \int_{T_i}^{t} h_o(t, \tau)\, K_r(\tau, \sigma)\, d\tau, \qquad T_i < \sigma < t. \tag{185}$$

In Section 6.2 we discussed a special case in which $T_i = -\infty$ and the processes were stationary. As part of the solution procedure we found a function $G^+(j\omega)$. We observed that if we passed white noise through a linear system whose transfer function was $G^+(j\omega)$ the output process had a spectrum $S_r(\omega)$. We also observed that in the white noise case, the filter could be realized in what we termed the canonic feedback filter form. The optimum loop filter had the same poles as the linear system whose output spectrum would equal $S_a(\omega)$ if the input were white noise. The only problem was to find the zeros of the optimum loop filter.

For the finite interval it is necessary to solve (185). In Chapter 4 we dealt with similar equations and observed that the conversion of the integral equation to a differential equation with a set of boundary conditions is a useful procedure.

We also observed in several examples that when the message is a scalar Markov process [recall that for a stationary Gaussian process this implies that the covariance had the form $A \exp(-B|t - u|)$] the results were simpler. These observations (plus a great deal of hindsight) lead us to make the following conjectures about an alternate approach to the problem that might be fruitful:

1. Instead of describing the processes of interest in terms of their covariance functions, characterize them in terms of the linear (possibly time-varying) systems that would generate them when driven with white noise.†
2. Instead of describing the linear system that generates the message in terms of a time-varying impulse response, describe it in terms of a differential equation whose solution is the message. The most convenient description will turn out to be a first-order vector differential equation.
3. Instead of specifying the optimum estimate as the output of a linear system which is specified by an integral equation, specify the optimum estimate as the solution to a differential equation whose coefficients are determined by the statistics of the processes. An obvious advantage of this method of specification is that even if we cannot solve the differential equation analytically, we can always solve it easily with an analog or digital computer.

In this section we make these observations more precise and investigate the results.

First, we discuss briefly the state-variable representation of linear, time-varying systems and the generation of random processes. Second, we derive a differential equation which is satisfied by the optimum estimate. Finally, we discuss some applications of the technique.

The original work in this area is due to Kalman and Bucy [23].

6.3.1 Differential Equation Representation of Linear Systems and Random Process Generation‡

In our previous discussions we have characterized linear systems by an impulse response $h(t, u)$ [or simply $h(\tau)$ in the time-invariant case].

† The advantages to be accrued by this characterization were first recognized and exploited by Dolph and Woodbury in 1949 [22].

‡ In this section we develop the background needed to solve the problems of immediate interest. A number of books cover the subject in detail (e.g., Zadeh and DeSoer [24], Gupta [25], Athans and Falb [26], DeRusso, Roy, and Close [27] and Schwartz and Friedland [28]. Our discussion is self-contained, but some results are stated without proof.

Implicit in this description was the assumption that the input was known over the interval $-\infty < t < \infty$. Frequently this method of description is the most convenient. Alternately, we can represent many systems in terms of a differential equation relating its input and output. Indeed, this is the method by which one is usually introduced to linear system theory. The impulse response $h(t, u)$ is just the solution to the differential equation when the input is an impulse at time u.

Three ideas of importance in the differential equation representation are presented in the context of a simple example.

The first idea of importance to us is the idea of initial conditions and state variables in dynamic systems. If we want to find the output over some interval $t_0 \leq t < t_1$, we must know not only the input over this interval but also a certain number of initial conditions that must be adequate to describe how any past inputs ($t < t_0$) affect the output of the system in the interval $t \geq t_0$.

We define the *state* of the system as the minimal amount of information about the effects of past inputs necessary to describe completely the output for $t \geq t_0$. The variables that contain this information are the *state variables*† There must be enough states that every input-output pair can be accounted for. When stated with more mathematical precision, these assumptions imply that, given the state of the system at t_0 and the input from t_0 to t_1, we can find both the *output* and the *state* at t_1. Note that our definition implies that the dynamic systems of interest are deterministic and realizable (future inputs cannot affect the output). If the state can be described by a finite-dimensional vector, we refer to the system as a finite-dimensional dynamic system. In this section we restrict our attention to finite-dimensional systems.

We can illustrate this with a simple example:

Example 1. Consider the *RC* circuit shown in Fig. 6.25. The output voltage $y(t)$ is related to the input voltage $u(t)$ by the differential equation

$$(RC)\,\dot{y}(t) + y(t) = u(t). \tag{186}$$

To find the output $y(t)$ in the interval $t \geq t_0$ we need to know $u(t)$, $t \geq t_0$, and the voltage across the capacitor at t_0. Thus a suitable state variable is $y(t)$.

The second idea is realizing (or simulating) a differential equation by using an analog computer. For our purposes we can visualize an analog computer as a system consisting of integrators, time-varying gains, adders, and nonlinear no-memory devices joined together to produce the desired input-output relation.

For the simple RC circuit example an analog computer realization is shown in Fig. 6.26. The initial condition $y(t_0)$ appears as a bias at the

† Zadeh and DeSoer [24]

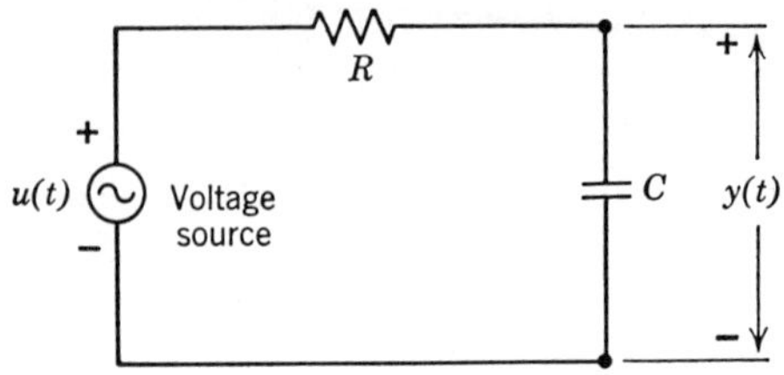

Fig. 6.25 An *RC* circuit.

output of the integrator. This biased integrator output is the state variable of the system.

The third idea is that of random process generation. If $u(t)$ is a random process or $y(t_0)$ is a random variable (or both), then $y(t)$ is a random process. Using the system described by (186), we can generate both nonstationary and stationary processes. As an example of a nonstationary process, let $y(t_0)$ be $N(0, \sigma_0)$, $u(t)$ be zero, and $k = 1/RC$. Then $y(t)$ is a zero-mean Gaussian random process with covariance function

$$K_y(t, u) = \sigma_0^{\,2} e^{-k(t+u-2t_0)}, \qquad t, u \geq t_0. \tag{187}$$

As an example of a stationary process, consider the case in which $u(t)$ is a sample function from a white noise process of spectral height q. If the input starts at $-\infty$ (i.e., $t_0 = -\infty$) and $y(t_0) = 0$, the output is a stationary process with a spectrum

$$S_y(\omega) = \frac{2k\sigma_y^{\,2}}{\omega^2 + k^2}, \tag{188}$$

where

$$q = 2\sigma_y^{\,2}/k. \tag{189}$$

We now explore these ideas in a more general context. Consider the system described by a differential equation of the form

$$y^{(n)}(t) + p_{n-1}\, y^{(n-1)}(t) + \cdots + p_0\, y(t) = b_0\, u(t), \tag{190}$$

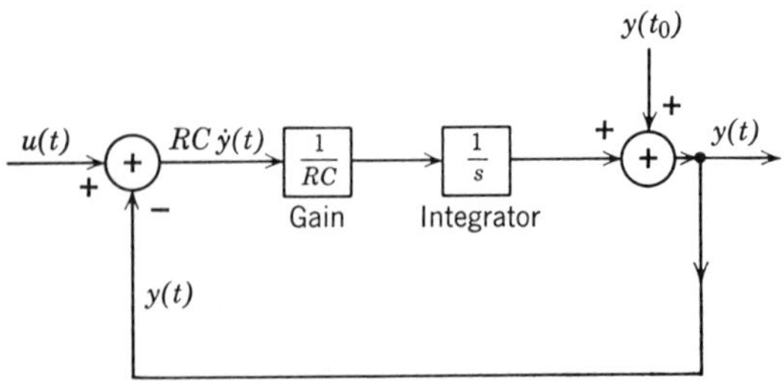

Fig. 6.26 An analog computer realization.

where $y^{(n)}(t)$ denotes the nth derivative of $y(t)$. Recall that to specify the solution to an nth-order equation we need the values of $y(t), \ldots, y^{(n-1)}(t)$ at t_0. This observation will be the key to finding the state representation for this system. The first step in finding an analog computer realization is to generate the terms on the left-hand side of the equation. This is shown in Fig. 6.27*a*. The next step is to interconnect these various quantities so that the differential equation is satisfied. The differential equation specifies the inputs to the summing point and gives the block diagram shown in Fig. 6.27*b*. Finally, we include the initial conditions by allowing

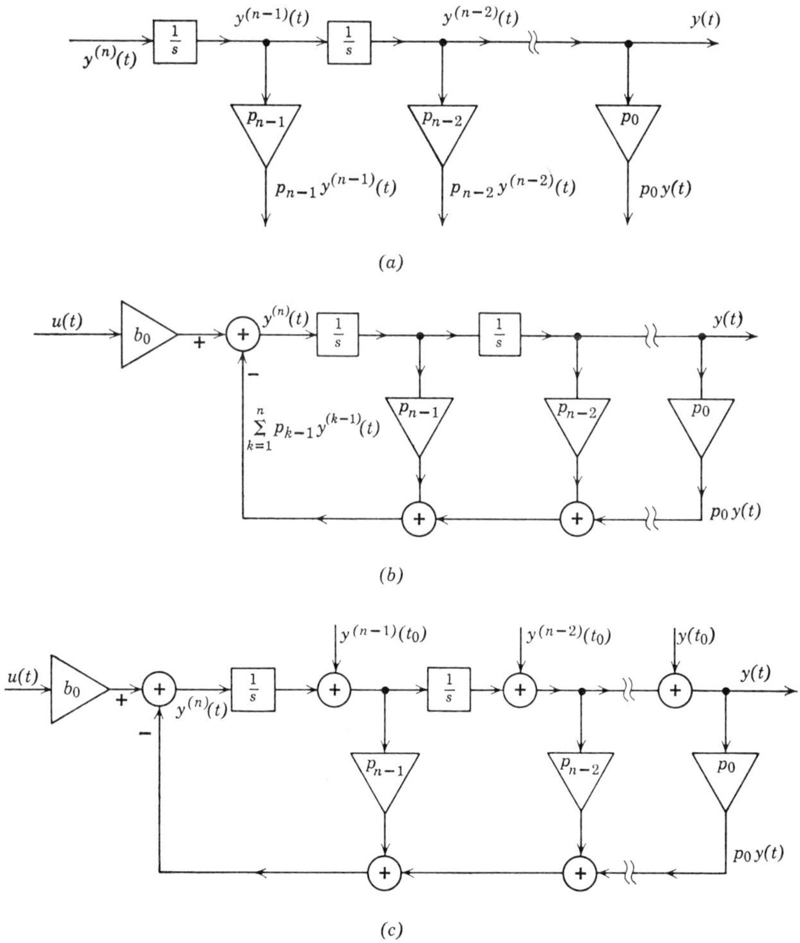

Fig. 6.27 Analog computer realization.

for a bias on the integrator outputs and obtain the realization shown in Fig. 6.27*c*. The state variables are the biased integrator outputs.

It is frequently easier to work with a first-order vector differential equation than an nth-order scalar equation. For (190) the transformation is straightforward. Let

$$\begin{aligned} x_1(t) &= y(t), \\ x_2(t) &= \dot{y}(t) = \dot{x}_1(t), \\ &\vdots \\ x_n(t) &= y^{(n-1)}(t) = \dot{x}_{n-1}(t). \end{aligned} \tag{191}$$

$$\begin{aligned} \dot{x}_n(t) = y^{(n)}(t) &= -\sum_{k=1}^{n} p_{k-1} y^{(k-1)}(t) + b_0 u(t) \\ &= -\sum_{k=1}^{n} p_{k-1} x_k(t) + b_0 u(t). \end{aligned} \tag{191}$$

Denoting the set of $x_i(t)$ by a column matrix, we see that the following first-order n-dimensional vector equation is equivalent to the nth-order scalar equation.

$$\frac{d\mathbf{x}(t)}{dt} \triangleq \dot{\mathbf{x}}(t) = \mathbf{F}\mathbf{x}(t) + \mathbf{G}u(t), \tag{192}$$

where

$$\mathbf{F} = \left[\begin{array}{c|ccccc} 0 & 1 & & & & \\ 0 & & 1 & & 0 & \\ 0 & & & 1 & & \\ \vdots & & 0 & & \ddots & \\ 0 & & & & & 1 \\ \hline -p_0 & -p_1 & -p_2 & -p_3 & \cdots & -p_{n-1} \end{array}\right] \tag{193}$$

and

$$\mathbf{G} = \begin{bmatrix} 0 \\ 0 \\ \vdots \\ 0 \\ b_0 \end{bmatrix}. \tag{194}$$

The vector $\mathbf{x}(t)$ is called the *state vector* for this linear system and (192) is called the *state equation* of the system. Note that the state vector $\mathbf{x}(t)$ we selected is not the only choice. Any nonsingular linear transformation of $\mathbf{x}(t)$ gives another state vector. The output $y(t)$ is related to the state vector by the equation

$$y(t) = \mathbf{C}\,\mathbf{x}(t), \tag{195}$$

where **C** is a $1 \times n$ matrix

$$\mathbf{C} = [1 \,\vdots\, 0 \,\vdots\, 0 \,\vdots\, 0 \cdots 0]. \tag{196}$$

Equation (195) is called the *output equation* of the system. The two equations (192) and (195) completely characterize the system.

Just as in the first example we can generate both nonstationary and stationary random processes using the system described by (192) and (195). For stationary processes it is clear (190) that we can generate any process with a rational spectrum in the form of

$$S_y(\omega) = \frac{k}{d_{2n}\omega^{2n} + d_{2n-2}\omega^{2n-2} + \cdots + d_0} \tag{197}$$

by letting $u(t)$ be a white noise process and $t_0 = -\infty$. In this case the state vector $\mathbf{x}(t)$ is a sample function from a vector random process and $y(t)$ is one component of this process.

The next more general differential equation is

$$\begin{aligned} y^{(n)}(t) + p_{n-1}\, y^{(n-1)}(t) + \cdots + p_0\, y(t) \\ = b_{n-1}\, u^{(n-1)}(t) + \cdots + b_0 u(t). \end{aligned} \tag{198}$$

The first step is to find an analog computer-type realization that corresponds to this differential equation. We illustrate one possible technique by looking at a simple example.

Example 2A. Consider the case in which $n = 2$ and the initial conditions are zero. Then (198) is

$$\ddot{y}(t) + p_1\, \dot{y}(t) + p_0\, y(t) = b_1\, \dot{u}(t) + b_0\, u(t). \tag{199}$$

Our first observation is that we want to avoid actually differentiating $u(t)$ because in many cases of interest it is a white noise process. Comparing the order of the highest derivatives on the two sides of (199), we see that this is possible. An easy approach is to assume that $\dot{u}(t)$ exists as part of the input to the first integrator in Fig. 6.28 and examine the consequences. To do this we rearrange terms as shown in (200):

$$[\ddot{y}(t) - b_1\, \dot{u}(t)] + p_1\, \dot{y}(t) + p_0\, y(t) = b_0\, u(t). \tag{200}$$

The result is shown in Fig. 6.28. Defining the state variables as the integrator outputs, we obtain

$$x_1(t) = y(t) \tag{201a}$$

and

$$x_2(t) = \dot{y}(t) - b_1\, u(t). \tag{201b}$$

Using (200) and (201), we have

$$\dot{x}_1(t) = x_2(t) + b_1\, u(t) \tag{202a}$$

$$\begin{aligned} \dot{x}_2(t) &= -p_0\, x_1(t) - p_1\,(x_2(t) + b_1\, u(t)) + b_0\, u(t) \\ &= -p_0\, x_1(t) - p_1\, x_2(t) + (b_0 - b_1 p_1)\, u(t). \end{aligned} \tag{202b}$$

We can write (202) as a vector state equation by defining

$$\mathbf{F} = \begin{bmatrix} 0 & 1 \\ -p_0 & -p_1 \end{bmatrix} \tag{203a}$$

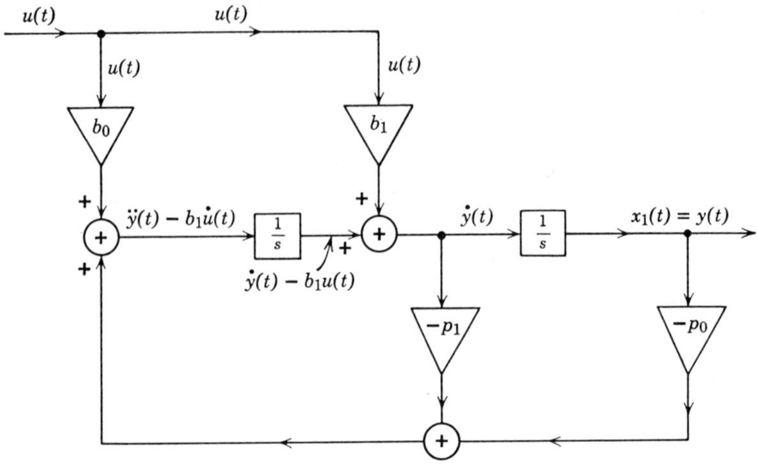

Fig. 6.28 Analog realization.

and

$$\mathbf{G} = \begin{bmatrix} b_1 \\ b_0 - p_1 b_1 \end{bmatrix}. \tag{203b}$$

Then

$$\mathbf{x}(t) = \mathbf{F}\,\mathbf{x}(t) + \mathbf{G}\,u(t). \tag{204a}$$

The output equation is

$$y(t) = [1 \,\vdots\, 0]\mathbf{x}(t) \triangleq \mathbf{C}\,\mathbf{x}(t). \tag{204b}$$

Equations 204*a* and 204*b* plus the initial condition $\mathbf{x}(t_0) = \mathbf{0}$ characterize the system.

It is straightforward to extend this particular technique to the nth order (see Problem 6.3.1). We refer to it as canonical realization No. 1. Our choice of state variables was somewhat arbitrary. To demonstrate this, we reconsider Example 2*A* and develop a different state representation.

Example 2B. Once again

$$\ddot{y}(t) + p_1\,\dot{y}(t) + p_0\,y(t) = b_1\,\dot{u}(t) + b_0\,u(t). \tag{205}$$

As a first step we draw the two integrators and the two paths caused by b_1 and b_0. This partial system is shown in Fig. 6.29*a*. We now want to introduce feedback paths and identify state variables in such a way that the elements in $\mathbf{F}$ and $\mathbf{G}$ will be one of the coefficients in the original differential equation, unity, or zero. Looking at Fig. 6.29*a*, we see that an easy way to do this is to feed back a weighted version of $x_1(t)$ $(= y(t))$ into each summing point as shown in Fig. 6.29*b*. The equations for the state variables are

$$x_1(t) = y(t), \tag{206}$$

$$\dot{x}_1(t) = x_2(t) - p_1\,y(t) + b_1\,u(t), \tag{207}$$

$$\dot{x}_2(t) = -p_0\,y(t) + b_0\,u(t). \tag{208}$$

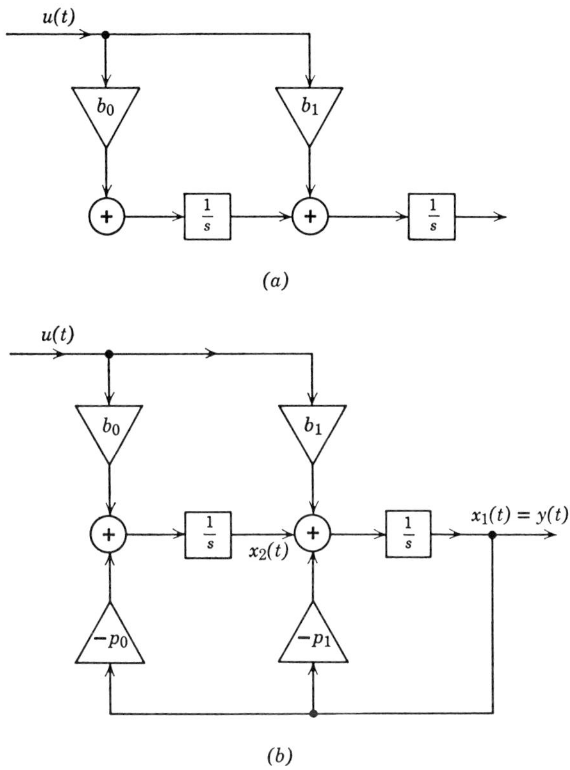

Fig. 6.29 Analog realization of (205).

The **F** matrix is

$$\mathbf{F} = \begin{bmatrix} -p_1 & +1 \\ -p_0 & 0 \end{bmatrix} \tag{209}$$

and the **G** matrix is

$$\mathbf{G} = \begin{bmatrix} b_1 \\ b_0 \end{bmatrix}. \tag{210}$$

We see that the system has the desired property.

The extension to the original nth-order differential equation is straightforward. The resulting realization is shown in Fig. 6.30. The equations for the state variables are

$$\begin{aligned} x_1(t) &= y(t), \\ x_2(t) &= \dot{x}_1(t) + p_{n-1}y(t) - b_{n-1}u(t), \\ &\vdots \\ x_n(t) &= \dot{x}_{n-1}(t) + p_1 y(t) - b_1 u(t), \\ \dot{x}_n(t) &= -p_0 y(t) + b_0 u(t). \end{aligned} \tag{211}$$

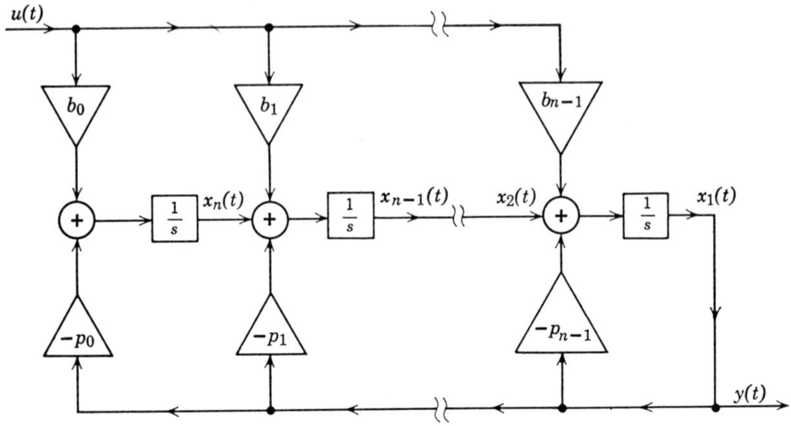

Fig. 6.30 Canonic realization No. 2: state variables.

The matrix for the vector differential equation is

$$\mathbf{F} = \left[\begin{array}{c|cccc} -p_{n-1} & 1 & 0 & & \\ -p_{n-2} & 0 & 1 & & 0 \\ & & & 1 & \\ \vdots & & & & \ddots \\ & & 0 & & \\ -p_1 & & & & 1 \\ \hline -p_0 & 0 & & \cdots & 0 \end{array}\right] \tag{212}$$

and

$$\mathbf{G} = \begin{bmatrix} b_{n-1} \\ b_{n-2} \\ \vdots \\ b_0 \end{bmatrix}. \tag{213}$$

We refer to this realization as canonical realization No. 2.

There is still a third useful realization to consider. The transfer function corresponding to (198) is

$$\frac{Y(s)}{X(s)} = \frac{b_{n-1}s^{n-1} + \cdots + b_0}{s^n + p_{n-1}s^{n-1} + \cdots + p_0} \triangleq H(s). \tag{214}$$

We can expand this equation in a partial fraction expansion

$$H(s) = \sum_{i=1}^{n} \frac{\alpha_i}{s - \lambda_i}, \tag{215}$$

where the λ_i are the roots of the denominator that are assumed to be distinct and the α_i are the corresponding residues. The system is shown in transform notation in Fig. 6.31*a*. Clearly, we can identify each subsystem output as a state variable and realize the over-all system as shown in Fig. 6.31*b*. The **F** matrix is diagonal.

$$\mathbf{F} = \begin{bmatrix} \lambda_1 & & & & \\ & \lambda_2 & & 0 & \\ & & \lambda_3 & & \\ & 0 & & \ddots & \\ & & & & \lambda_n \end{bmatrix} \tag{216}$$

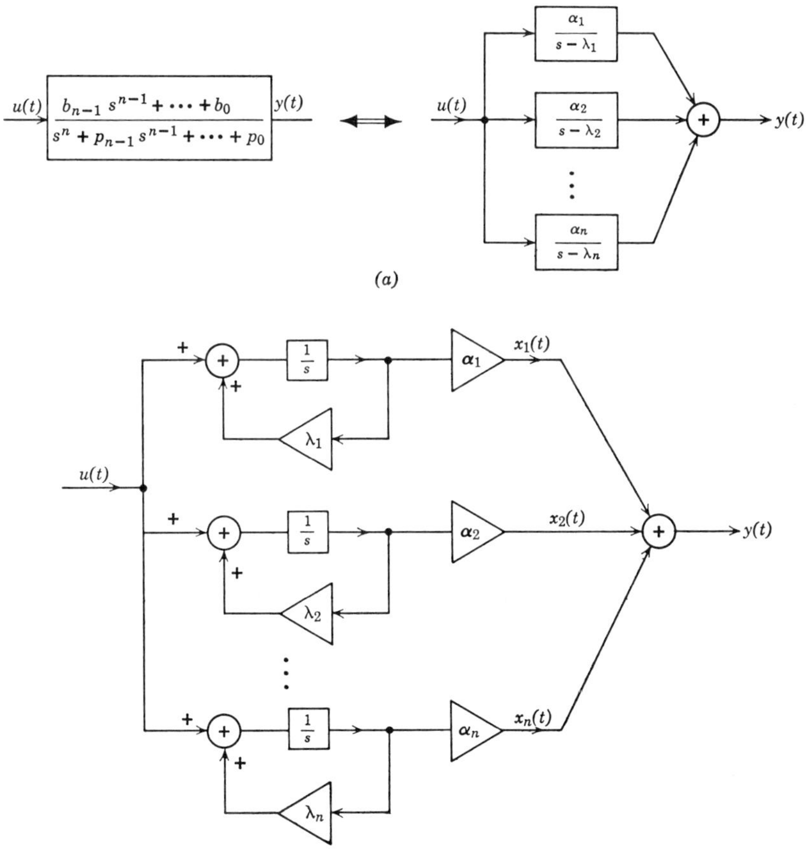

Fig. 6.31 Canonic realization No. 3: (*a*) transfer function; (*b*) analog computer realization.

and the elements in the **G** matrix are the residues

$$\mathbf{G} = \begin{bmatrix} \alpha_1 \\ \alpha_2 \\ \vdots \\ \alpha_n \end{bmatrix}. \tag{217}$$

Now the original output $y(t)$ is the sum of the state variables

$$y(t) = \sum_{i=1}^{n} x_i(t) = \mathbf{1}^T\mathbf{x}(t), \tag{218a}$$

where

$$\mathbf{1}^T \triangleq [1 \mid 1 \cdots 1]. \tag{218b}$$

We refer to this realization as canonical realization No. 3. (The realization for repeated roots is derived in Problem 6.3.2.)

Canonical realization No. 3 requires a partial fraction expansion to find **F** and **G**. Observe that the state equation consists of n *uncoupled* first-order scalar equations

$$\dot{x}_i = \lambda_i x_i(t) + \alpha_i u(t), \qquad i = 1, 2, \ldots, n. \tag{219}$$

The solution of this set is appreciably simpler than the solution of the vector equation. On the other hand, finding the partial fraction expansion may require some calculation whereas canonical realizations No. 1 and No. 2 can be obtained by inspection.

We have now developed three different methods for realizing a system described by an nth-order constant coefficient differential equation. In each case the state vector was different. The **F** matrices were different, but it is easy to verify that they all have the same eigenvalues. It is worthwhile to emphasize that even though we have labeled these realizations as canonical some other realization may be more desirable in a particular problem. Any nonsingular linear transformation of a state vector leads to a new state representation.

We now have the capability of generating *any* stationary random process with a rational spectrum and finite variance by exciting any of the three realizations with white noise. In addition we can generate a wide class of nonstationary processes.

Up to this point we have seen how to represent linear time-invariant systems in terms of a state-variable representation and the associated vector-differential equation. We saw that this could correspond to a physical realization in the form of an analog computer, and we learned how we could generate a large class of random processes.

The next step is to extend our discussion to include time-varying systems and multiple input-multiple output systems.

For time-varying systems we consider the vector equations

$$\frac{d\mathbf{x}(t)}{dt} = \mathbf{F}(t)\,\mathbf{x}(t) + \mathbf{G}(t)\,u(t), \tag{220a}$$

$$y(t) = \mathbf{C}(t)\,\mathbf{x}(t), \tag{220b}$$

as the basic representation.† The matrices $\mathbf{F}(t)$ and $\mathbf{G}(t)$ may be functions of time. By using a white noise input

$$E[u(t)\,u(\tau)] = q\,\delta(t - \tau), \tag{221}$$

we have the ability to generate some nonstationary random processes. It is worthwhile to observe that a nonstationary process can result even when $\mathbf{F}$ and $\mathbf{G}$ are constant and $\mathbf{x}(t_0)$ is deterministic. The Wiener process, defined on p. 195 of Chapter 3, is a good example.

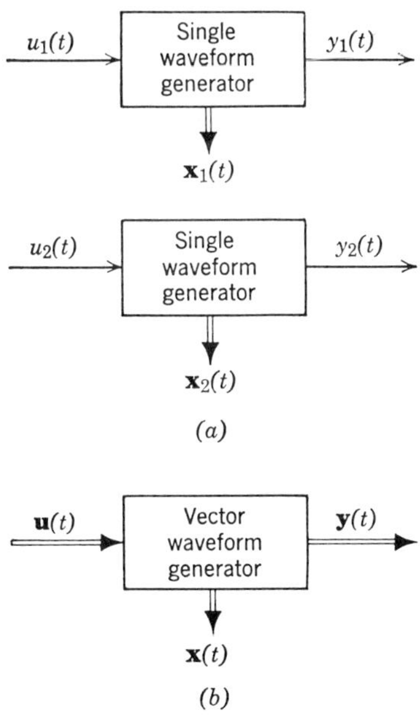

Fig. 6.32 Generation of two messages.

† The canonic realizations in Figs. 6.28 and 6.30 may still be used. It is important to observe that they do not correspond to the same nth-order differential equation as in the time-invariant case. See Problem 6.3.14.

Example 3. Here $\mathbf{F}(t) = 0$, $\mathbf{G}(t) = \sigma$, $\mathbf{C}(t) = 1$, and (220) becomes

$$\frac{dx(t)}{dt} = \sigma\, u(t). \tag{222}$$

Assuming that $x(0) = 0$, this gives the Wiener process.

Other specific examples of time-varying systems are discussed in later sections and in the problems.

The motivation for studying multiple input–multiple output systems follows directly from our discussions in Chapters 3, 4, and 5. Consider the simple system in Fig. 6.32 in which we generate two outputs $y_1(t)$ and $y_2(t)$. We assume that the state representation of system 1 is

$$\dot{\mathbf{x}}_1(t) = \mathbf{F}_1(t)\,\mathbf{x}_1(t) + \mathbf{G}_1(t)\,u_1(t), \tag{223}$$

$$y_1(t) = \mathbf{C}_1(t)\,\mathbf{x}_1(t), \tag{224}$$

where $\mathbf{x}_1(t)$ is an n-dimensional state vector. Similarly, the state representation of system 2 is

$$\dot{\mathbf{x}}_2(t) = \mathbf{F}_2(t)\,\mathbf{x}_2(t) + \mathbf{G}_2(t)\,u_2(t), \tag{225}$$

$$y_2(t) = \mathbf{C}_2(t)\,\mathbf{x}_2(t), \tag{226}$$

where $\mathbf{x}_2(t)$ is an m-dimensional state vector. A more convenient way to describe these two systems is as a single vector system with an $(n + m)$-dimensional state vector (Fig. 6.32*b*).

$$\mathbf{x}(t) = \begin{bmatrix} \mathbf{x}_1(t) \\ \hline \mathbf{x}_2(t) \end{bmatrix}, \tag{227}$$

$$\mathbf{F}(t) = \left[\begin{array}{c|c} \mathbf{F}_1(t) & \mathbf{0} \\ \hline \mathbf{0} & \mathbf{F}_2(t) \end{array}\right], \tag{228}$$

$$\mathbf{G}(t) = \left[\begin{array}{c|c} \mathbf{G}_1(t) & \mathbf{0} \\ \hline \mathbf{0} & \mathbf{G}_2(t) \end{array}\right], \tag{229}$$

$$\mathbf{u}(t) = \begin{bmatrix} u_1(t) \\ \hline u_2(t) \end{bmatrix}, \tag{230}$$

$$\mathbf{C}(t) = \left[\begin{array}{c|c} \mathbf{C}_1(t) & \mathbf{0} \\ \hline \mathbf{0} & \mathbf{C}_2(t) \end{array}\right], \tag{231}$$

and

$$\mathbf{y}(t) = \begin{bmatrix} y_1(t) \\ \hline y_2(t) \end{bmatrix}. \tag{232}$$

The resulting differential equations are

$$\dot{\mathbf{x}}(t) = \mathbf{F}(t)\,\mathbf{x}(t) + \mathbf{G}(t)\,\mathbf{u}(t), \tag{233}$$

$$\mathbf{y}(t) = \mathbf{C}(t)\,\mathbf{x}(t). \tag{234}$$

The driving function is a vector. For the message generation problem we assume that the driving function is a white process with a matrix covariance function

$$E[\mathbf{u}(t)\,\mathbf{u}^T(\tau)] \triangleq \mathbf{Q}\,\delta(t-\tau), \tag{235}$$

where $\mathbf{Q}$ is a nonnegative definite matrix. The block diagram of the generation process is shown in Fig. 6.33.

Observe that in general the initial conditions may be random variables. Then, to specify the second-moment characteristics we must know the covariance at the initial time

$$\mathbf{K}_{\mathbf{x}}(t_0, t_0) \triangleq E[\mathbf{x}(t_0)\,\mathbf{x}^T(t_0)] \tag{236}$$

and the mean value $E[\mathbf{x}(t_0)]$. We can also generate coupled processes by replacing the $\mathbf{0}$ matrices in (228), (229), or (231) with nonzero matrices.

The next step in our discussion is to consider the solution to (233). We begin our discussion with the *homogeneous time-invariant* case. Then (233) reduces to

$$\dot{\mathbf{x}}(t) = \mathbf{F}\mathbf{x}(t), \tag{237}$$

with initial condition $\mathbf{x}(t_0)$. If $\mathbf{x}(t)$ and $\mathbf{F}$ are scalars, the solution is familiar,

$$x(t) = e^{F(t-t_0)}x(t_0). \tag{238}$$

For the vector case we can show that (e.g. [27], [28], [29], or [30])

$$\mathbf{x}(t) = e^{\mathbf{F}(t-t_0)}\,\mathbf{x}(t_0), \tag{239}$$

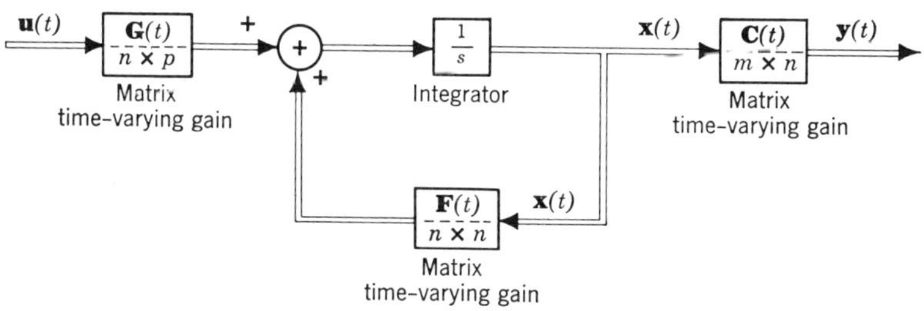

Fig. 6.33 Message generation process.

where $e^{\mathbf{F}t}$ is defined by the infinite series

$$e^{\mathbf{F}t} \triangleq \mathbf{I} + \mathbf{F}t + \frac{\mathbf{F}^2 t^2}{2!} + \cdots. \tag{240}$$

The function $e^{\mathbf{F}(t-t_0)}$ is denoted by $\boldsymbol{\phi}(t - t_0) \triangleq \boldsymbol{\phi}(\tau)$. The function $\boldsymbol{\phi}(t - t_0)$ is called the *state transition matrix* of the system. Two properties can easily be verified for the time-invariant case.

Property 11.† The state transition matrix satisfies the equation

$$\frac{d\boldsymbol{\phi}(t - t_0)}{dt} = \mathbf{F}\boldsymbol{\phi}(t - t_0) \tag{241}$$

or

$$\frac{d\boldsymbol{\phi}(\tau)}{d\tau} = \mathbf{F}\boldsymbol{\phi}(\tau). \tag{242}$$

[Use (240) and its derivative on both sides of (239).]

Property 12. The initial condition

$$\boldsymbol{\phi}(t_0 - t_0) = \boldsymbol{\phi}(0) = \mathbf{I} \tag{243}$$

follows directly from (239). The homogeneous solution can be rewritten in terms of $\boldsymbol{\phi}(t - t_0)$:

$$\mathbf{x}(t) = \boldsymbol{\phi}(t - t_0)\, \mathbf{x}(t_0). \tag{244}$$

The solution to (242) is easily obtained by using conventional Laplace transform techniques. Transforming (242), we have

$$s\,\boldsymbol{\Phi}(s) - \mathbf{I} = \mathbf{F}\,\boldsymbol{\Phi}(s), \tag{245}$$

where the identity matrix arises from the initial condition in (243). Rearranging terms, we have

$$[s\mathbf{I} - \mathbf{F}]\,\boldsymbol{\Phi}(s) = \mathbf{I} \tag{246}$$

or

$$\boldsymbol{\Phi}(s) = (s\mathbf{I} - \mathbf{F})^{-1}. \tag{247}$$

The state transition matrix is

$$\boldsymbol{\phi}(\tau) = \mathcal{L}^{-1}[\boldsymbol{\Phi}(s)] = \mathcal{L}^{-1}[(s\mathbf{I} - \mathbf{F})^{-1}]. \tag{248}$$

A simple example illustrates the technique.

Example 4. Consider the system described by (206–210). The transform of the transition matrix is,

$$\boldsymbol{\Phi}(s) = \left[s\begin{bmatrix}1 & 0\\ 0 & 1\end{bmatrix} - \begin{bmatrix}-p_1 & 1\\ -p_0 & 0\end{bmatrix}\right]^{-1}, \tag{249}$$

† Because we have to refer back to the properties at the beginning of the chapter we use a consecutive numbering system to avoid confusion.

$$\Phi(s) = \begin{bmatrix} s + p_1 & -1 \\ p_0 & s \end{bmatrix}^{-1}, \tag{250}$$

$$\Phi(s) = \frac{1}{s^2 + p_1 s + p_0} \begin{bmatrix} s & 1 \\ -p_0 & s + p_1 \end{bmatrix}. \tag{251}$$

To find $\phi(\tau)$ we take the inverse transform. For simplicity we let $p_1 = 3$ and $p_0 = 2$. Then

$$\phi(\tau) = \left[\begin{array}{c|c} 2e^{-2\tau} - e^{-\tau} & e^{-\tau} - e^{-2\tau} \\ \hline 2[e^{-2\tau} - e^{-\tau}] & 2e^{-\tau} - e^{-2\tau} \end{array}\right]. \tag{252}$$

It is important to observe that the complex natural frequencies involved in the solution are determined by the denominator of $\Phi(s)$. This is just the determinant of the matrix $s\mathbf{I} - \mathbf{F}$. Therefore these frequencies are just the roots of the equation

$$\det [s\mathbf{I} - \mathbf{F}] = 0. \tag{253}$$

For the time-varying case the basic concept of a state-transition matrix is still valid, but some of the above properties no longer hold. From the scalar case we know that $\phi(t, t_0)$ will be a function of two variables instead of just the difference between t and t_0.

Definition. The state transition matrix is defined to be a function of two variables $\phi(t, t_0)$ which satisfies the differential equation

$$\dot{\phi}(t, t_0) = \mathbf{F}(t)\phi(t, t_0) \tag{254a}$$

with initial condition $\phi(t_0, t_0) = \mathbf{I}$. The solution at any time is

$$\mathbf{x}(t) = \phi(t, t_0)\,\mathbf{x}(t_0). \tag{254b}$$

An analytic solution is normally difficult to obtain. Fortunately, in most of the cases in which we use the transition matrix an analytic solution is not necessary. Usually, we need only to know that it exists and that it has certain properties. In the cases in which it actually needs evaluation, we shall do it numerically.

Two properties follow easily:

$$\phi(t_2, t_0) = \phi(t_2, t_1)\phi(t_1, t_0), \qquad \text{for all } t_0, t_1, t_2 \tag{255a}$$

and

$$\phi^{-1}(t_1, t_0) = \phi(t_0, t_1). \tag{255b}$$

For the nonhomogeneous case the equation is

$$\dot{\mathbf{x}}(t) = \mathbf{F}(t)\,\mathbf{x}(t) + \mathbf{G}(t)\,\mathbf{u}(t). \tag{256}$$

The solution contains a homogeneous part and a particular part:

$$\mathbf{x}(t) = \phi(t, t_0)\,\mathbf{x}(t_0) + \int_{t_0}^{t} \phi(t, \tau)\,\mathbf{G}(\tau)\,\mathbf{u}(\tau)\,d\tau. \tag{257}$$

(Substitute (257) into (256) to verify that it is the solution.) The output $\mathbf{y}(t)$ is

$$\mathbf{y}(t) = \mathbf{C}(t)\,\mathbf{x}(t). \tag{258}$$

In our work in Chapters 4 and 5 and Section 6.1 we characterized time-varying linear systems by their impulse response $\mathbf{h}(t, \tau)$. This characterization assumes that the input is known from $-\infty$ to t. Thus

$$\mathbf{y}(t) = \int_{-\infty}^{t} \mathbf{h}(t, \tau)\,\mathbf{u}(\tau)\,d\tau. \tag{259}$$

For most cases of interest the effect of the initial condition $\mathbf{x}(-\infty)$ will disappear in (257). Therefore, we may set them equal to zero and obtain,

$$\mathbf{y}(t) = \mathbf{C}(t) \int_{-\infty}^{t} \boldsymbol{\phi}(t, \tau)\,\mathbf{G}(\tau)\,\mathbf{u}(\tau)\,d\tau. \tag{260}$$

Comparing (259) and (260), we have

$$\mathbf{h}(t, \tau) = \begin{cases} \mathbf{C}(t)\,\boldsymbol{\phi}(t, \tau)\,\mathbf{G}(\tau), & t \geq \tau, \\ \mathbf{0}, & \text{elsewhere.} \end{cases} \tag{261}$$

It is worthwhile to observe that the three matrices on the right will depend on the state representation that we choose for the system, but the matrix impulse response is unique. As pointed out earlier, the system is realizable. This is reflected by the $\mathbf{0}$ in (261).

For the time-invariant case

$$\mathbf{Y}(s) = \mathbf{H}(s)\,\mathbf{U}(s), \tag{262}$$

and

$$\mathbf{H}(s) = \mathbf{C}\,\boldsymbol{\Phi}(s)\mathbf{G}. \tag{263}$$

Equation 262 assumes that the input has a Laplace transform. For a stationary random process we would use the integrated transform (Section 3.6).

Most of our discussion up to this point has been valid for an arbitrary driving function $\mathbf{u}(t)$. We now derive some statistical properties of vector processes $\mathbf{x}(t)$ and $\mathbf{y}(t)$ for the specific case in which $\mathbf{u}(t)$ is a sample function of a vector white noise process.

$$E[\mathbf{u}(t)\,\mathbf{u}^T(\tau)] = \mathbf{Q}\,\delta(t - \tau). \tag{264}$$

Property 13. The cross correlation between the state vector $\mathbf{x}(t)$ of a system driven by a zero-mean white noise $\mathbf{u}(t)$ and the input $\mathbf{u}(\tau)$ is

$$\mathbf{K}_{\mathbf{xu}}(t, \tau) \triangleq E[\mathbf{x}(t)\,\mathbf{u}^T(\tau)]. \tag{265}$$

It is a discontinuous function that equals

$$\mathbf{K}_{\mathbf{xu}}(t, \tau) = \begin{cases} \mathbf{0}, & \tau > t, \\ \tfrac{1}{2}\mathbf{G}(t)\mathbf{Q}, & \tau = t, \\ \boldsymbol{\phi}(t, \tau)\,\mathbf{G}(\tau)\mathbf{Q}, & t_0 < \tau < t. \end{cases} \tag{266}$$

Proof. Substituting (257) into the definition in (265), we have

$$\mathbf{K}_{\mathbf{xu}}(t, \tau) = E\left\{\left[\boldsymbol{\phi}(t, t_0)\, \mathbf{x}(t_0) + \int_{t_0}^{t} \boldsymbol{\phi}(t, \alpha)\, \mathbf{G}(\alpha)\, \mathbf{u}(\alpha)\, d\alpha\right]\mathbf{u}^T(\tau)\right\}. \tag{267}$$

Bringing the expectation inside the integral and assuming that the initial state $\mathbf{x}(t_0)$ is independent of $\mathbf{u}(\tau)$ for $\tau > t_0$, we have

$$\begin{aligned}\mathbf{K}_{\mathbf{xu}}(t, \tau) &= \int_{t_0}^{t} \boldsymbol{\phi}(t, \alpha)\, \mathbf{G}(\alpha)\, E[\mathbf{u}(\alpha)\, \mathbf{u}^T(\tau)]\, d\alpha \\ &= \int_{t_0}^{t} \boldsymbol{\phi}(t, \alpha)\, \mathbf{G}(\alpha)\, \mathbf{Q}\, \delta(\alpha - \tau)\, d\alpha. \end{aligned} \tag{268}$$

If $\tau > t$, this expression is zero. If $\tau = t$ and we assume that the delta function is symmetric because it is the limit of a covariance function, we pick up only one half the area at the right end point. Thus

$$\mathbf{K}_{\mathbf{xu}}(t, t) = \tfrac{1}{2}\boldsymbol{\phi}(t, t)\, \mathbf{G}(t)\mathbf{Q}. \tag{269}$$

Using the result following (254*a*), we obtain the second line in (266).
If $\tau < t$, we have

$$\mathbf{K}_{\mathbf{xu}}(t, \tau) = \boldsymbol{\phi}(t, \tau)\, \mathbf{G}(\tau)\mathbf{Q}, \qquad \tau < t \tag{270a}$$

which is the third line in (266). A special case of (270*a*) that we shall use later is obtained by letting τ approach t from below.

$$\lim_{\tau \to t^-} \mathbf{K}_{\mathbf{xu}}(t, \tau) = \mathbf{G}(t)\mathbf{Q}. \tag{270b}$$

The cross correlation between the output vector $\mathbf{y}(t)$ and $\mathbf{u}(\tau)$ follows easily.

$$\mathbf{K}_{\mathbf{yu}}(t, \tau) = \mathbf{C}(t)\, \mathbf{K}_{\mathbf{xu}}(t, \tau). \tag{271}$$

Property 14. The variance matrix of the state vector $\mathbf{x}(t)$ of a system

$$\dot{\mathbf{x}}(t) = \mathbf{F}(t)\, \mathbf{x}(t) + \mathbf{G}(t)\, \mathbf{u}(t) \tag{272}$$

satisfies the differential equation

$$\dot{\boldsymbol{\Lambda}}_{\mathbf{x}}(t) = \mathbf{F}(t)\, \boldsymbol{\Lambda}_{\mathbf{x}}(t) + \boldsymbol{\Lambda}_{\mathbf{x}}(t)\, \mathbf{F}^T(t) + \mathbf{G}(t)\, \mathbf{Q}\, \mathbf{G}^T(t), \tag{273}$$

with the initial condition

$$\boldsymbol{\Lambda}_{\mathbf{x}}(t_0) = E[\mathbf{x}(t_0)\, \mathbf{x}^T(t_0)]. \tag{274}$$

[Observe that $\boldsymbol{\Lambda}_{\mathbf{x}}(t) = \mathbf{K}_{\mathbf{x}}(t, t)$.]

Proof.

$$\boldsymbol{\Lambda}_{\mathbf{x}}(t) \triangleq E[\mathbf{x}(t)\, \mathbf{x}^T(t)]. \tag{275}$$

Differentiating, we have

$$\frac{d\mathbf{\Lambda}_{\mathbf{x}}(t)}{dt} = E\left[\frac{d\mathbf{x}(t)}{dt}\,\mathbf{x}^T(t)\right] + E\left[\mathbf{x}(t)\,\frac{d\mathbf{x}^T(t)}{dt}\right]. \tag{276}$$

The second term is just the transpose of the first. [Observe that $\mathbf{x}(t)$ is not mean-square differentiable: therefore, we will have to be careful when dealing with (276).]

Substituting (272) into the first term in (276) gives

$$E\left[\frac{d\mathbf{x}(t)}{dt}\,\mathbf{x}^T(t)\right] = E\{[\mathbf{F}(t)\,\mathbf{x}(t) + \mathbf{G}(t)\,\mathbf{u}(t)]\,\mathbf{x}^T(t)\}. \tag{277}$$

Using Property 13 on the second term in (277), we have

$$E\left[\frac{d\mathbf{x}(t)}{dt}\,\mathbf{x}^T(t)\right] = \mathbf{F}(t)\,\mathbf{\Lambda}_{\mathbf{x}}(t) + \tfrac{1}{2}\mathbf{G}(t)\,\mathbf{Q}\,\mathbf{G}^T(t). \tag{278}$$

Using (278) and its transpose in (276) gives

$$\dot{\mathbf{\Lambda}}_{\mathbf{x}}(t) = \mathbf{F}(t)\,\mathbf{\Lambda}_{\mathbf{x}}(t) + \mathbf{\Lambda}_{\mathbf{x}}(t)\,\mathbf{F}^T(t) + \mathbf{G}(t)\,\mathbf{Q}\,\mathbf{G}^T(t), \tag{279}$$

which is the desired result.

We now have developed the following ideas:

1. State variables of a linear dynamic system;
2. Analog computer realizations;
3. First-order vector differential equations and state-transition matrices;
4. Random process generation.

The next step is to apply these ideas to the linear estimation problem.

Observation Model. In this section we recast the linear modulation problem described in the beginning of the chapter into state-variable terminology. The basic linear modulation problem was illustrated in Fig. 6.1. A state-variable formulation for a simpler special case is given in Fig. 6.34. The message $a(t)$ is generated by passing $y(t)$ through a linear system, as discussed in the preceding section. Thus

$$\dot{\mathbf{x}}(t) = \mathbf{F}(t)\,\mathbf{x}(t) + \mathbf{G}(t)\,u(t) \tag{280}$$

and

$$a(t) \triangleq x_1(t). \tag{281}$$

For simplicity in the explanation we have assumed that the message of interest is the first component of the state vector. The message is then modulated by multiplying by the carrier $c(t)$. In DSB-AM, $c(t)$ would be a sine wave. We include the carrier in the linear system by defining

$$\mathbf{C}(t) = [c(t) \,\vdots\, 0 \,\vdots\, 0 \quad \cdots \quad 0]. \tag{282}$$

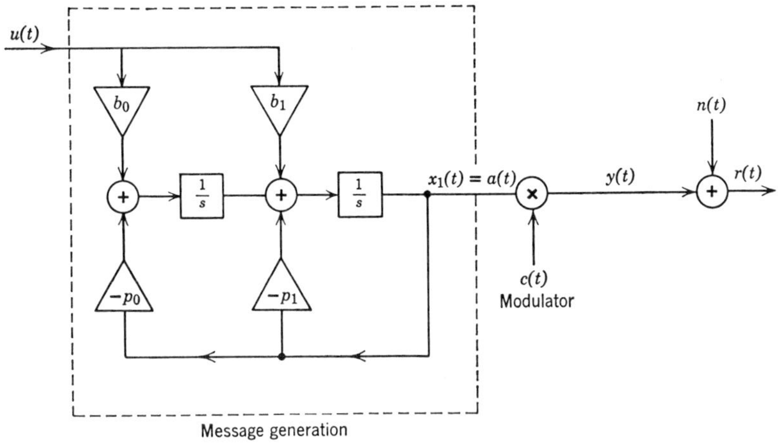

Fig. 6.34 A simple case of linear modulation in the state-variable formulation.

Then

$$y(t) = \mathbf{C}(t)\,\mathbf{x}(t). \tag{283}$$

We frequently refer to $\mathbf{C}(t)$ as the *modulation matrix.* (In the control literature it is called the *observation matrix.*) The waveform $y(t)$ is transmitted over an additive white noise channel. Thus

$$\begin{aligned} r(t) &= y(t) + w(t), \qquad && T_i \le t \le T_f \\ &= \mathbf{C}(t)\,\mathbf{x}(t) + w(t), \qquad && T_i \le t \le T_f, \end{aligned} \tag{284}$$

where

$$E[w(t)\,w(\tau)] = \frac{N_0}{2}\,\delta(t - \tau). \tag{285}$$

This particular model is too restrictive; therefore we generalize it in several different ways. Two of the modifications are fundamental and we explain them now. The others are deferred until Section 6.3.4 to avoid excess notation at this point.

Modification No. 1: Colored Noise. In this case there is a colored noise component $n_c(t)$ in addition to the white noise. We assume that the colored noise can be generated by driving a finite-dimensional dynamic system with white noise.

$$r(t) = \mathbf{C}_M(t)\,\mathbf{x}_M(t) + n_c(t) + w(t). \tag{286}$$

The subscript M denotes message. We can write (286) in the form

$$r(t) = \mathbf{C}(t)\,\mathbf{x}(t) + w(t) \tag{287}$$

by augmenting the message state vector to include the colored noise process. The new vector process $\mathbf{x}(t)$ consists of two parts. One is the vector process $\mathbf{x}_M(t)$ corresponding to the state variables of the system used to generate the message process. The second is the vector process $\mathbf{x}_N(t)$ corresponding to the state variables of the system used to generate the colored noise process. Thus

$$\mathbf{x}(t) \triangleq \begin{bmatrix} \mathbf{x}_M(t) \\ \hdashline \mathbf{x}_N(t) \end{bmatrix}. \tag{288}$$

If $\mathbf{x}_M(t)$ is n_1-dimensional and $\mathbf{x}_N(t)$ is n_2-dimensional, then $\mathbf{x}(t)$ has $(n_1 + n_2)$ dimensions. The modulation matrix is

$$\mathbf{C}(t) = [\mathbf{C}_M(t) \mid \mathbf{C}_N(t)]; \tag{289}$$

$\mathbf{C}_M(t)$ is defined in (286) and $\mathbf{C}_N(t)$ is chosen so that

$$n_c(t) = \mathbf{C}_N(t)\, \mathbf{x}_N(t). \tag{290}$$

With these definitions, we obtain (287). A simple example is appropriate at this point.

Example. Let

$$r(t) = \sqrt{2P}\, a(t) \sin \omega_c t + n_c(t) + w(t), \qquad -\infty < t, \tag{291}$$

where

$$S_a(\omega) = \frac{2k_a P_a}{\omega^2 + k_a^2}, \tag{292}$$

$$S_{n_c}(\omega) = \frac{2k_n P_n}{\omega^2 + k_n^2}, \tag{293}$$

and $n_c(t)$, $a(t)$, and $w(t)$ are uncorrelated. To obtain a state representation we let $t_0 = -\infty$ and assume that $a(-\infty) = n_c(-\infty) = 0$. We define the state vector $\mathbf{x}(t)$ as

$$\mathbf{x}(t) = \begin{bmatrix} a(t) \\ n_c(t) \end{bmatrix}. \tag{294}$$

Then,

$$\mathbf{F}(t) = \begin{bmatrix} -k_a & 0 \\ 0 & -k_n \end{bmatrix}, \tag{295}$$

$$\mathbf{G}(t) = \begin{bmatrix} 1 & 0 \\ 0 & 1 \end{bmatrix}, \tag{296}$$

$$\mathbf{Q} = \begin{bmatrix} 2k_a P_a & 0 \\ 0 & 2k_n P_n \end{bmatrix}, \tag{297}$$

and

$$\mathbf{C}(t) = [\sqrt{2P} \sin \omega_c t \mid 1]. \tag{298}$$

We see that **F**, **G**, and **Q** are diagonal because of the independence of the message and the colored noise and the fact that each has only one pole. In the general case of independent message and noise we can partition **F**, **G**, and **Q** and the off-diagonal partitions will be zero.

Modification No. 2: Vector Channels. The next case we need to include to get a general model is one in which we have multiple received waveforms. As we would expect, this extension is straightforward. Assuming m channels, we have a vector observation equation,

$$\mathbf{r}(t) = \mathbf{C}(t)\,\mathbf{x}(t) + \mathbf{w}(t). \tag{299}$$

where $\mathbf{r}(t)$ is m-dimensional. An example illustrates this model.

Example. A simple diversity system is shown in Fig. 6.35. Suppose $a(t)$ is a one-dimensional process. Then $\mathbf{x}(t) = a(t)$. The modulation matrix is $m \times 1$:

$$\mathbf{C}(t) = \begin{bmatrix} c_1(t) \\ c_2(t) \\ \vdots \\ c_m(t) \end{bmatrix}. \tag{300}$$

The channel noises are white with zero-means but may be correlated with one another. This correlation may be time-varying. The resulting covariance matrix is

$$E[\mathbf{w}(t)\,\mathbf{w}^T(u)] \triangleq \mathbf{R}(t)\,\delta(t - u), \tag{301}$$

where $\mathbf{R}(t)$ is positive-definite.

In general, $\mathbf{x}(t)$ is an n-dimensional vector and the channel is m-dimensional so that the modulation matrix is an $m \times n$ matrix. We assume that the channel noise $\mathbf{w}(t)$ and the white process $\mathbf{u}(t)$ which generates the message are uncorrelated.

With these two modifications our model is sufficiently general to include most cases of interest. The next step is to derive a differential equation for the optimum estimate. Before doing that we summarize the important relations.

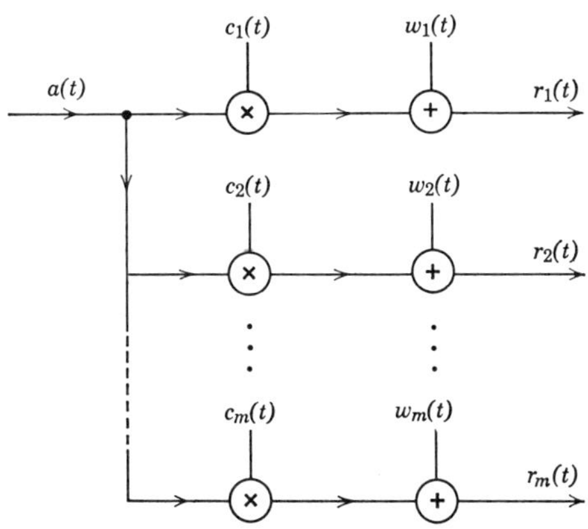

Fig. 6.35 A simple diversity system.

Summary of Model

All processes are assumed to be generated by passing white noise through a linear time-varying system. The processes are described by the vector-differential equation

$$\frac{d\mathbf{x}(t)}{dt} = \mathbf{F}(t)\,\mathbf{x}(t) + \mathbf{G}(t)\,\mathbf{u}(t), \tag{302}$$

where

$$E[\mathbf{u}(t)\,\mathbf{u}^T(\tau)] = \mathbf{Q}\,\delta(t-\tau). \tag{303}$$

and $\mathbf{x}(t_0)$ is specified either as a deterministic vector or as a random vector with known second-moment statistics.

The solution to (302) may be written in terms of a transition matrix:

$$\mathbf{x}(t) = \boldsymbol{\phi}(t, t_0)\,\mathbf{x}(t_0) + \int_{t_0}^{t} \boldsymbol{\phi}(t, \tau)\,\mathbf{G}(\tau)\,\mathbf{u}(\tau)\,d\tau. \tag{304}$$

The output process $\mathbf{y}(t)$ is obtained by a linear transformation of the state vector. It is observed after being corrupted by additive white noise. The received signal $\mathbf{r}(t)$ is described by the equation

$$\mathbf{r}(t) = \mathbf{C}(t)\,\mathbf{x}(t) + \mathbf{w}(t). \tag{305}$$

The measurement noise is white and is described by a covariance matrix:

$$E[\mathbf{w}(t)\,\mathbf{w}^T(u)] = \mathbf{R}(t)\,\delta(t-\tau). \tag{306}$$

Up to this point we have discussed only the second-moment properties of the random processes generated by driving linear dynamic systems with white noise. Clearly, if $\mathbf{u}(t)$ and $\mathbf{w}(t)$ are jointly Gaussian vector processes and if $\mathbf{x}(t_0)$ is a statistically independent Gaussian random vector, then the Gaussian assumption on p. 471 will hold. (The independence of $\mathbf{x}(t_0)$ is only convenient, not necessary.)

The next step is to show how we can modify the optimum linear filtering results we previously obtained to take advantage of this method of representation.

6.3.2 Derivation of Estimator Equations

In this section we want to derive a differential equation whose solution is the minimum mean-square estimate of the message (or messages). We recall that the MMSE estimate of a vector $\mathbf{x}(t)$ is a vector $\hat{\mathbf{x}}(t)$ whose components $\hat{x}_i(t)$ are chosen so that the mean-square error in estimating each component is minimized. In other words, $E[(\hat{x}_i(t) - x_i(t))^2]$, $i = 1, 2, \ldots, n$ is minimized. This implies that the sum of the mean-square errors, $E\{[\hat{\mathbf{x}}^T(t) - \mathbf{x}^T(t)][\mathbf{x}(t) - \hat{\mathbf{x}}(t)]\}$ is also minimized. The derivation is straightforward but somewhat lengthy. It consists of four parts.

1. Starting with the vector Wiener-Hopf equation (Property 3A–V) for realizable estimation, we derive a differential equation in t, with τ as a parameter, that the optimum filter $\mathbf{h}_o(t, \tau)$ must satisfy. This is (317).
2. Because the optimum estimate $\hat{\mathbf{x}}(t)$ is obtained by passing the received signal into the optimum filter, (317) leads to a differential equation that the optimum estimate must satisfy. This is (320). It turns out that all the coefficients in this equation are known except $\mathbf{h}_o(t, t)$.
3. The next step is to find an expression for $\mathbf{h}_o(t, t)$. Property 4B–V expresses $\mathbf{h}_o(t, t)$ in terms of the error matrix $\boldsymbol{\xi}_P(t)$. Thus we can equally well find an expression for $\boldsymbol{\xi}_P(t)$. To do this we first find a differential equation for the error $\mathbf{x}_\epsilon(t)$. This is (325).
4. Finally, because

$$\boldsymbol{\xi}_P(t) \triangleq E[\mathbf{x}_\epsilon(t)\, \mathbf{x}_\epsilon^T(t)], \tag{307}$$

we can use (325) to find a matrix differential equation that $\boldsymbol{\xi}_P(t)$ must satisfy. This is (330). We now carry out these four steps in detail.

Step 1. We start with the integral equation obtained for the optimum finite time point estimator [Property 3A–V, (52)]. We are estimating the entire vector $\mathbf{x}(t)$

$$\mathbf{K}_\mathbf{x}(t, \sigma)\, \mathbf{C}^T(\sigma) = \int_{T_i}^{t} \mathbf{h}_o(t, \tau)\, \mathbf{K}_\mathbf{r}(\tau, \sigma)\, d\tau, \qquad T_i < \sigma < t, \tag{308}$$

where

$$\mathbf{K}_\mathbf{r}(\tau, \sigma) = \mathbf{C}(\tau)\, \mathbf{K}_\mathbf{x}(\tau, \sigma)\mathbf{C}^T(\sigma) + \mathbf{R}(\tau)\, \delta(\tau - \sigma). \tag{309}$$

Differentiating both sides with respect to t, we have

$$\frac{\partial \mathbf{K}_\mathbf{x}(t, \sigma)}{\partial t}\, \mathbf{C}^T(\sigma) = \mathbf{h}_o(t, t)\, \mathbf{K}_\mathbf{r}(t, \sigma) + \int_{T_i}^{t} \frac{\partial \mathbf{h}_o(t, \tau)}{\partial t}\, \mathbf{K}_\mathbf{r}(\tau, \sigma)\, d\tau, \qquad T_i < \sigma < t. \tag{310}$$

First we consider the first term on the right-hand side of (310). For $\sigma < t$ we see from (309) that

$$\mathbf{K}_\mathbf{r}(t, \sigma) = \mathbf{C}(t)[\mathbf{K}_\mathbf{x}(t, \sigma)\mathbf{C}^T(\sigma)], \qquad \sigma < t. \tag{311}$$

The term inside the bracket is just the left-hand side of (308). Therefore,

$$\mathbf{h}_o(t, t)\, \mathbf{K}_\mathbf{r}(t, \sigma) = \int_{T_i}^{t} \mathbf{h}_o(t, t)\, \mathbf{C}(t)\, \mathbf{h}_o(t, \tau)\, \mathbf{K}_\mathbf{r}(\tau, \sigma)\, d\tau, \qquad \sigma < t. \tag{312}$$

Now consider the first term on the left-hand side of (310),

$$\frac{\partial \mathbf{K_x}(t, \sigma)}{\partial t} = E\left[\frac{d\mathbf{x}(t)}{dt}\, \mathbf{x}^T(\sigma)\right]. \tag{313}$$

Using (302), we have

$$\frac{\partial \mathbf{K_x}(t, \sigma)}{\partial t} = \mathbf{F}(t)\, \mathbf{K_x}(t, \sigma) + \mathbf{G}(t)\, \mathbf{K_{ux}}(t, \sigma), \tag{314}$$

but the second term is zero for $\sigma < t$ [see (266)]. Using (308), we see that

$$\mathbf{F}(t)\, \mathbf{K_x}(t, \sigma)\, \mathbf{C}^T(\sigma) = \int_{T_i}^{t} \mathbf{F}(t)\, \mathbf{h}_o(t, \tau)\, \mathbf{K_r}(\tau, \sigma)\, d\tau. \tag{315}$$

Substituting (315) and (312) into (310), we have

$$\mathbf{0} = \int_{T_i}^{t} \left[-\mathbf{F}(t)\, \mathbf{h}_o(t, \tau) + \mathbf{h}_o(t, t)\, \mathbf{C}(t)\, \mathbf{h}_o(t, \tau) + \frac{\partial \mathbf{h}_o(t, \tau)}{\partial t}\right] \mathbf{K_r}(\tau, \sigma)\, d\tau,$$

$$T_i < \sigma < t. \tag{316}$$

Clearly, if the term in the bracket is zero for all τ, $T_i \leq \tau \leq t$, (316) will be satisfied. Because $\mathbf{R}(t)$ is positive-definite the condition is also necessary; see Problem 6.3.19. Thus the differential equation satisfied by the optimum impulse response is

$$\frac{\partial \mathbf{h}_o(t, \tau)}{\partial t} = \mathbf{F}(t)\, \mathbf{h}_o(t, \tau) - \mathbf{h}_o(t, t)\mathbf{C}(t)\, \mathbf{h}_o(t, \tau). \tag{317}$$

Step 2. The optimum estimate is obtained by passing the input through the optimum filter. Thus

$$\hat{\mathbf{x}}(t) = \int_{T_i}^{t} \mathbf{h}_o(t, \tau)\, \mathbf{r}(\tau)\, d\tau. \tag{318}$$

We assumed in (318) that the MMSE realizable estimate of $\mathbf{x}(T_i) = \mathbf{0}$. Because there is no received data, our estimate at T_i is based on our a priori knowledge. If $\mathbf{x}(T_i)$ is a random variable with a mean-value vector $E[\mathbf{x}(T_i)]$ and a covariance matrix $\mathbf{K_x}(T_i, T_i)$, then the MMSE estimate is

$$\hat{\mathbf{x}}(T_i) = E[\mathbf{x}(T_i)].$$

If $\mathbf{x}(T_i)$ is a deterministic quantity, say $\mathbf{x}_D(T_i)$, then we may treat it as a random variable whose mean equals $\mathbf{x}_D(t)$

$$E[\mathbf{x}(T_i)] \triangleq \mathbf{x}_D(t)$$

and whose covariance matrix $\mathbf{K_x}(T_i, T_i)$ is identically zero.

$$\mathbf{K_x}(T_i, T_i) \triangleq \mathbf{0}.$$

In both cases (318) assumes $E[\mathbf{x}(T_i)] = \mathbf{0}$. The modification for other initial conditions is straightforward (see Problem 6.3.20). Differentiating (318), we have

$$\frac{d\hat{\mathbf{x}}(t)}{dt} = \mathbf{h}_o(t, t)\,\mathbf{r}(t) + \int_{T_i}^{t} \frac{\partial \mathbf{h}_o(t, \tau)}{\partial t}\,\mathbf{r}(\tau)\,d\tau. \tag{319}$$

Substituting (317) into the second term on the right-hand side of (319) and using (318), we obtain

$$\boxed{\frac{d\hat{\mathbf{x}}(t)}{dt} = \mathbf{F}(t)\,\hat{\mathbf{x}}(t) + \mathbf{h}_o(t, t)[\mathbf{r}(t) - \mathbf{C}(t)\,\hat{\mathbf{x}}(t)].} \tag{320}$$

It is convenient to introduce a new symbol for $\mathbf{h}_o(t, t)$ to indicate that it is only a function of one variable

$$\mathbf{z}(t) \triangleq \mathbf{h}_o(t, t). \tag{321}$$

The operations in (322) can be represented by the matrix block diagram of Fig. 6.36. We see that all the coefficients are known except $\mathbf{z}(t)$, but Property 4C–V (53) expresses $\mathbf{h}_o(t, t)$ in terms of the error matrix,

$$\boxed{\mathbf{z}(t) = \mathbf{h}_o(t, t) = \boldsymbol{\xi}_P(t)\mathbf{C}^T(t)\,\mathbf{R}^{-1}(t).} \tag{322}$$

Thus (320) will be completely determined if we can find an expression for $\boldsymbol{\xi}_P(t)$, the error covariance matrix for the optimum realizable point estimator.

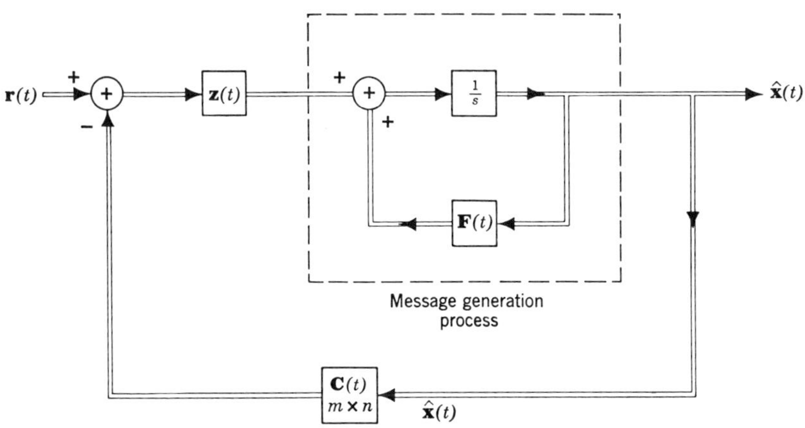

Fig. 6.36 Feedback estimator structure.

Step 3. We first find a differential equation for the error $\mathbf{x}_\epsilon(t)$, where

$$\mathbf{x}_\epsilon(t) \triangleq \mathbf{x}(t) - \hat{\mathbf{x}}(t). \tag{323}$$

Differentiating, we have

$$\frac{d\mathbf{x}_\epsilon(t)}{dt} = \frac{d\mathbf{x}(t)}{dt} - \frac{d\hat{\mathbf{x}}(t)}{dt}. \tag{324}$$

Substituting (302) for the first term on the right-hand side of (324), substituting (320) for the second term, and using (305), we obtain the desired equation

$$\frac{d\mathbf{x}_\epsilon(t)}{dt} = [\mathbf{F}(t) - \mathbf{z}(t)\,\mathbf{C}(t)]\mathbf{x}_\epsilon(t) - \mathbf{z}(t)\,\mathbf{w}(t) + \mathbf{G}(t)\,\mathbf{u}(t). \tag{325}$$

The last step is to derive a differential equation for $\boldsymbol{\xi}_P(t)$.

Step 4. Differentiating

$$\boldsymbol{\xi}_P(t) \triangleq E[\mathbf{x}_\epsilon(t)\,\mathbf{x}_\epsilon^T(t)], \tag{326}$$

we have

$$\frac{d\boldsymbol{\xi}_P(t)}{dt} = E\left[\frac{d\mathbf{x}_\epsilon(t)}{dt}\,\mathbf{x}_\epsilon^T(t)\right] + E\left[\mathbf{x}_\epsilon(t)\,\frac{d\mathbf{x}_\epsilon^T(t)}{dt}\right]. \tag{327}$$

Substituting (325) into the first term of (327), we have

$$E\left[\frac{d\mathbf{x}_\epsilon(t)}{dt}\,\mathbf{x}_\epsilon^T(t)\right] = E\{[\mathbf{F}(t) - \mathbf{z}(t)\,\mathbf{C}(t)]\,\mathbf{x}_\epsilon(t)\,\mathbf{x}_\epsilon^T(t) - \mathbf{z}(t)\,\mathbf{w}(t)\,\mathbf{x}_\epsilon^T(t) + \mathbf{G}(t)\,\mathbf{u}(t)\,\mathbf{x}_\epsilon^T(t)\}. \tag{328}$$

Looking at (325), we see $\mathbf{x}_\epsilon(t)$ is the state vector for a system driven by the weighted sum of two independent white noises $\mathbf{w}(t)$ and $\mathbf{u}(t)$. Therefore the expectations in the second and third terms are precisely the same type as we evaluated in Property 13 (second line of 266).

$$E\left[\frac{d\mathbf{x}_\epsilon(t)}{dt}\,\mathbf{x}_\epsilon^T(t)\right] = \mathbf{F}(t)\,\boldsymbol{\xi}_P(t) - \mathbf{z}(t)\,\mathbf{C}(t)\,\boldsymbol{\xi}_P(t) + \tfrac{1}{2}\mathbf{z}(t)\,\mathbf{R}(t)\,\mathbf{z}^T(t) + \tfrac{1}{2}\,\mathbf{G}(t)\mathbf{Q}\,\mathbf{G}^T(t). \tag{329}$$

Adding the transpose and replacing $\mathbf{z}(t)$ with the right-hand side of (322), we have

$$\boxed{\frac{d\boldsymbol{\xi}_P(t)}{dt} = \mathbf{F}(t)\,\boldsymbol{\xi}_P(t) + \boldsymbol{\xi}_P(t)\,\mathbf{F}^T(t) - \boldsymbol{\xi}_P(t)\,\mathbf{C}^T(t)\,\mathbf{R}^{-1}(t)\,\mathbf{C}(t)\,\boldsymbol{\xi}_P(t) + \mathbf{G}(t)\mathbf{Q}\,\mathbf{G}^T(t),} \tag{330}$$

which is called the *variance equation*. This equation and the initial condition

$$\boldsymbol{\xi}_P(T_i) = E[\mathbf{x}_\epsilon(T_i)\,\mathbf{x}_\epsilon^T(T_i)] \tag{331}$$

determine $\boldsymbol{\xi}_P(t)$ uniquely. Using (322) we obtain $\mathbf{z}(t)$, the gain in the optimal filter.

Observe that the variance equation does not contain the received signal. Therefore it may be solved before any data is received and used to determine the gains in the optimum filter. The variance equation is a matrix equation equivalent to n^2 scalar differential equations. However, because $\xi_P(t)$ is a symmetric matrix, we have $\frac{1}{2}n(n + 1)$ scalar nonlinear coupled differential equations to solve. In the general case it is impossible to obtain an explicit analytic solution, but this is unimportant because the equation is in a form which may be integrated using either an analog or digital computer.

The variance equation is a matrix Riccati equation whose properties have been studied extensively in other contexts (e.g., McLachlan [31]; Levin [32]; Reid [33], [34]; or Coles [35]). To study its behavior adequately requires more background than we have developed. Two properties are of interest. The first deals with the infinite memory, stationary process case (the Wiener problem) and the second deals with analytic solutions.

Property 15. Assume that T_i is fixed and that the matrices $\mathbf{F}$, $\mathbf{G}$, $\mathbf{C}$, $\mathbf{R}$, and $\mathbf{Q}$ are constant. Under certain conditions, as t increases there will be an initial transient period after which the filter gains will approach constant values. Looking at (322) and (330), we see that as $\dot{\xi}_P(t)$ approaches zero the error covariance matrix and gain matrix will approach constants. We refer to the problem when the condition $\dot{\xi}_P(t) = \mathbf{0}$ is true as the *steady-state estimation problem*.

The left-hand side of (330) is then zero and the variance equation becomes a set of $\frac{1}{2}n(n + 1)$ quadratic algebraic equations. The nonnegative definite solution is ξ_P.

Some comments regarding this statement are necessary.

1. How do we tell if the steady-state problem is meaningful? To give the *best* general answer requires notions that we have not developed [23]. A *sufficient* condition is that the message correspond to a stationary random process.

2. For small n it is feasible to calculate the various solutions and select the correct one. For even moderate n (e.g. $n = 2$) it is more practical to solve (330) numerically. We may start with some arbitrary nonnegative definite $\xi_P(T_i)$ and let the solution converge to the steady-state result (once again see [23], Theorem 4, p. 8, for a precise statement).

3. Once again we observe that we can generate $\xi_P(t)$ before the data is received or in real time. As a simple example of generating the variance using an analog computer, consider the equation:

$$\frac{d\xi_P(t)}{dt} = -2k\,\xi_P(t) - \frac{2}{N_0}\,\xi_P^2(t) + 2kP. \tag{332}$$

(This will appear in Example 1.) A simple analog method of generation is shown in Fig. 6.37. The initial condition is $\xi_P(T_i) = P$ (see discussion in the next paragraph).

4. To solve (or mechanize) the variance equation we must specify $\boldsymbol{\xi}_P(T_i)$. There are several possibilities.

   (a) The process may begin at T_i with a known value (i.e., zero variance) or with a random value having a known variance.

   (b) The process may have started at some time t_0 which is much earlier than T_i and reached a statistical steady state. In Property 14 on p. 533 we derived a differential equation that $\mathbf{\Lambda_x}(t)$ satisfied. If it has reached a statistical steady state, $\dot{\mathbf{\Lambda}}_\mathbf{x}(t) = \mathbf{0}$ and (273) reduces to

$$\mathbf{0} = \mathbf{F\Lambda_x} + \mathbf{\Lambda_x F}^T + \mathbf{GQG}^T. \tag{333a}$$

   This is an algebraic equation whose solution is $\mathbf{\Lambda_x}$. Then

$$\boldsymbol{\xi}_P(T_i) = \mathbf{\Lambda_x} \tag{333b}$$

   if the process has reached steady state before T_i. In order for the unobserved process to reach statistical steady state ($\dot{\mathbf{\Lambda}}_\mathbf{x}(t) = \mathbf{0}$), it is necessary and sufficient that the eigenvalues of $\mathbf{F}$ have negative real parts. This condition guarantees that the solution to (333) is nonnegative definite.

   In many cases the basic characterization of an unobserved stationary process is in terms of its spectrum $S_y(\omega)$. The elements

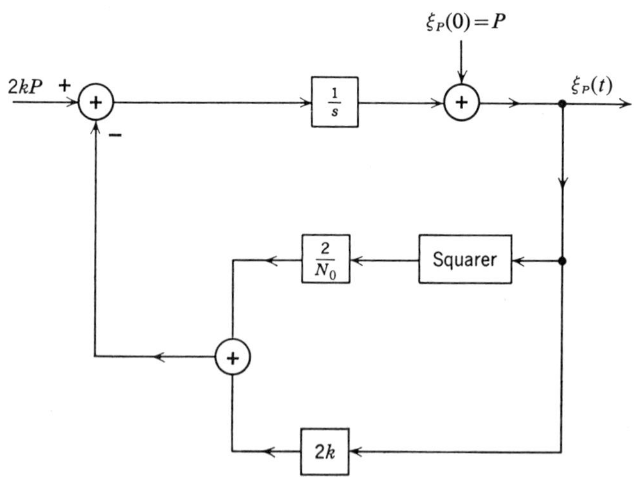

Fig. 6.37 Analog solution to variance equation.

in $\mathbf{\Lambda_x}$ follow easily from $S_y(\omega)$. As an example consider the state vector in (191),

$$x_i = \frac{d^{(i-1)}y(t)}{dt^{(i-1)}}, \qquad i = 1, 2, \ldots, n. \tag{334a}$$

If $y(t)$ is stationary, then

$$\Lambda_{\mathbf{x},11} = \int_{-\infty}^{\infty} S_y(\omega)\,\frac{d\omega}{2\pi} \tag{334b}$$

or, more generally,

$$\Lambda_{\mathbf{x},ik} = \int_{-\infty}^{\infty} (j\omega)^{i+k} S_y(\omega)\,\frac{d\omega}{2\pi}, \qquad \begin{aligned} i &= 1, 2, \ldots, n, \\ k &= 1, 2, \ldots, n. \end{aligned} \tag{334c}$$

Note that, for *this particular state vector*,

$$\Lambda_{\mathbf{x},ik} = 0 \quad \text{when} \quad i + k \text{ is odd}, \tag{334d}$$

because $S_y(\omega)$ is an even function.

A second property of the variance equation enables us to obtain analytic solutions in some cases (principally, the constant matrix, finite-time problem). We do not use the details in the text but some of the problems exploit them.

Property 16. The variance equation can be related to two simultaneous linear equations,

$$\begin{aligned} \frac{d\mathbf{v}_1(t)}{dt} &= \mathbf{F}(t)\,\mathbf{v}_1(t) + \mathbf{G}(t)\,\mathbf{QG}^T(t)\,\mathbf{v}_2(t), \\ \frac{d\mathbf{v}_2(t)}{dt} &= \mathbf{C}^T(t)\,\mathbf{R}^{-1}(t)\,\mathbf{C}(t)\,\mathbf{v}_1(t) - \mathbf{F}^T(t)\,\mathbf{v}_2(t), \end{aligned} \tag{335}$$

or, equivalently,

$$\frac{d}{dt}\begin{bmatrix} \mathbf{v}_1(t) \\ \mathbf{v}_2(t) \end{bmatrix} = \left[\begin{array}{c|c} \mathbf{F}(t) & \mathbf{G}(t)\,\mathbf{QG}^T(t) \\ \hline \mathbf{C}^T(t)\,\mathbf{R}^{-1}(t)\,\mathbf{C}(t) & -\mathbf{F}^T(t) \end{array}\right] \begin{bmatrix} \mathbf{v}_1(t) \\ \mathbf{v}_2(t) \end{bmatrix}. \tag{336}$$

Denote the transition matrix of (336) by

$$\mathbf{T}(t, T_i) = \left[\begin{array}{c|c} \mathbf{T}_{11}(t, T_i) & \mathbf{T}_{12}(t, T_i) \\ \hline \mathbf{T}_{21}(t, T_i) & \mathbf{T}_{22}(t, T_i) \end{array}\right]. \tag{337}$$

Then we can show [32],

$$\xi_P(t) = [\mathbf{T}_{11}(t, T_i)\,\xi_P(T_i) + \mathbf{T}_{12}(t, T_i)][\mathbf{T}_{21}(t, T_i)\,\xi_P(T_i) + \mathbf{T}_{22}(t, T_i)]^{-1}. \tag{338}$$

When the matrices of concern are constant, we can always find the transition matrix $\mathbf{T}$ (see Problem 6.3.21 for an example in which we find $\mathbf{T}$ by using Laplace transform techniques. As discussed in that problem, we must take the contour to the right of all the poles in order to include all the eigenvalues) of the coefficient matrix in (336).

In this section we have transformed the optimum linear filtering problem into a state variable formulation. All the quantities of interest are expressed as outputs of dynamic systems. The three equations that describe these dynamic systems are our principal results.

The Estimator Equation.

$$\frac{d\hat{\mathbf{x}}(t)}{dt} = \mathbf{F}(t)\,\hat{\mathbf{x}}(t) + \mathbf{z}(t)[\mathbf{r}(t) - \mathbf{C}(t)\,\hat{\mathbf{x}}(t)]. \tag{339}$$

The Gain Equation.

$$\mathbf{z}(t) = \boldsymbol{\xi}_P(t)\,\mathbf{C}^T(t)\,\mathbf{R}^{-1}(t). \tag{340}$$

The Variance Equation.

$$\frac{d\boldsymbol{\xi}_P(t)}{dt} = \mathbf{F}(t)\,\boldsymbol{\xi}_P(t) + \boldsymbol{\xi}_P(t)\,\mathbf{F}^T(t) - \boldsymbol{\xi}_P(t)\,\mathbf{C}^T(t)\,\mathbf{R}^{-1}(t)\,\mathbf{C}(t)\,\boldsymbol{\xi}_P(t) + \mathbf{G}(t)\,\mathbf{Q}\mathbf{G}^T(t). \tag{341}$$

To illustrate their application we consider some simple examples, chosen for one of three purposes:

1. To show an alternate approach to a problem that could be solved by conventional Wiener theory.
2. To illustrate a problem that could not be solved by conventional Wiener theory.
3. To develop a specific result that will be useful in the sequel.

6.3.3 Applications

In this section we consider some examples to illustrate the application of the results derived in Section 6.3.2.

Example 1. Consider the first-order message spectrum

$$S_a(\omega) = \frac{2kP}{\omega^2 + k^2}. \tag{342}$$

In this case $\mathbf{x}(t)$ is a scalar; $x(t) = a(t)$. If we assume that the message is not modulated and the measurement noise is white, then

$$r(t) = x(t) + w(t). \tag{343}$$

The necessary quantities follow by inspection:

$$\begin{aligned} \mathbf{F}(t) &= -k, \\ \mathbf{G}(t) &= 1, \\ \mathbf{C}(t) &= 1, \\ \mathbf{Q} &= 2kP, \\ \mathbf{R}(t) &= \frac{N_0}{2}. \end{aligned} \tag{344}$$

Substituting these quantities into (339), we obtain the differential equation for the optimum estimate:

$$\frac{d\hat{x}(t)}{dt} = -k\hat{x}(t) + z(t)[r(t) - \hat{x}(t)]. \tag{345}$$

The resulting filter is shown in Fig. 6.38. The value of the gain $z(t)$ is determined by solving the variance equation.

First, we assume that the estimator has reached steady state. Then the steady-state solution to the variance equation can be obtained easily. Setting the left-hand side of (341) equal to zero, we obtain

$$0 = -2k\xi_{P\infty} - \xi_{P\infty}^2 \frac{2}{N_0} + 2kP. \tag{346}$$

where $\xi_{P\infty}$ denotes the steady-state variance.

$$\xi_{P\infty} \triangleq \lim_{t\to\infty} \xi_P(t).$$

There are two solutions to (346); one is positive and one is negative. Because $\xi_{P\infty}$ is a mean-square error it must be positive. Therefore

$$\xi_{P\infty} = k\frac{N_0}{2}(-1 + \sqrt{1 + \Lambda}) \tag{347}$$

(recall that $\Lambda = 4P/kN_0$). From (340)

$$z(\infty) \triangleq z_\infty = \xi_{P\infty}R^{-1} = k(-1 + \sqrt{1 + \Lambda}). \tag{348}$$

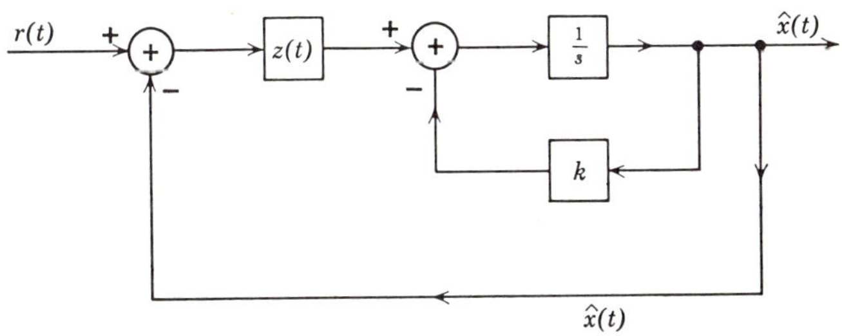

Fig. 6.38 Optimum filter: example 1.

Clearly, the filter must be equivalent to the one obtained in the example in Section 6.2. The closed loop transfer function is

$$H_o(j\omega) = \frac{z_\infty}{j\omega + k + z_\infty}. \tag{349}$$

Substituting (348) in (349), we have

$$H_o(j\omega) = \frac{k(\sqrt{1 + \Lambda} - 1)}{j\omega + k\sqrt{1 + \Lambda}}, \tag{350}$$

which is the same as (94).

The transient problem can be solved analytically or numerically. The details of the analytic solution are carried out in Problem 6.3.21 by using Property 16 on p. 545. The transition matrix is

$$\mathbf{T}(T_i + \tau, T_i) = \left[\begin{array}{c|c} \cosh(\gamma\tau) - \dfrac{k}{\gamma}\sinh(\gamma\tau) & \dfrac{2kP}{\gamma}\sinh(\gamma\tau) \\ \hline \dfrac{2}{N_0\gamma}\sinh(\gamma\tau) & \cosh(\gamma\tau) + \dfrac{k}{\gamma}\sinh(\gamma\tau) \end{array}\right], \tag{351}$$

where

$$\gamma \triangleq k\sqrt{1 + \Lambda}. \tag{352}$$

If we assume that the unobserved message is in a statistical steady state then $\hat{x}(T_i) = 0$ and $\xi_P(T_i) = P$. [(342) implies $a(t)$ is zero-mean.] Using these assumptions and (351) in (338), we obtain

$$\xi_P(t + T_i) = 2kP\left[\frac{(\gamma + k)e^{+\gamma t} + (\gamma + k)e^{-\gamma t}}{(\gamma + k)^2 e^{+\gamma t} - (\gamma - k)^2 e^{-\gamma t}}\right] = \frac{2kP}{\gamma + k}\left(\frac{1 + \left(\dfrac{\gamma - k}{\gamma + k}\right)e^{-2\gamma t}}{1 - \left(\dfrac{\gamma - k}{\gamma + k}\right)^2 e^{-2\gamma t}}\right) \tag{353}$$

As $t \to \infty$, we have

$$\lim_{t\to\infty} \xi_P(t + T_i) = \frac{2kP}{\gamma + k} = \frac{N_0}{2}[\gamma - k] = \xi_{P\infty}, \tag{354}$$

which agrees with (347). In Fig. 6.39 we show the behavior of the normalized error as a function of time for various values of Λ. The number on the right end of each curve is $\xi_P(1.2) - \xi_{P\infty}$. This is a measure of how close the error is to its steady-state value.

Example 2. A logical generalization of the one-pole case is the Butterworth family defined in (153):

$$S_a(\omega: n) = \frac{2nP}{k}\frac{\sin(\pi/2n)}{(1 + (\omega/k)^{2n})}. \tag{355}$$

To formulate this equation in state-variable terms we need the differential equation of the message generation process.

$$a^{(n)}(t) + p_{n-1}\, a^{(n-1)}(t) + \cdots + p_0\, a(t) = u(t). \tag{356}$$

The coefficients are tabulated for various n in circuit theory texts (e.g., Guillemin [37] or Weinberg [38]). The values for $k = 1$ are shown in Fig. 6.40*a*. The pole locations for various n are shown in Fig. 6.40*b*. If we are interested only in the

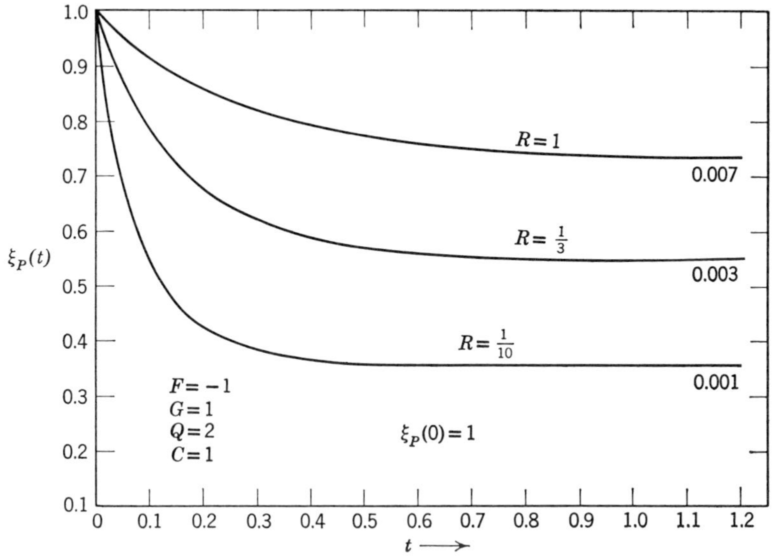

Fig. 6.39 Mean-square error, one-pole spectrum.

n	p_{n-1}	p_{n-2}	p_{n-3}	p_{n-4}	p_{n-5}	p_{n-6}	p_{n-7}
2	1.414	1.000					
3	2.000	2.000	1.000				
4	2.613	3.414	2.613	1.000			
5	3.236	5.236	5.236	3.236	1.000		
6	3.864	7.464	9.141	7.464	3.864	1.000	
7	4.494	10.103	14.606	14.606	10.103	4.494	1.000

$$a^{(n)}(t) + p_{n-1}a^{(n-1)}(t) + \cdots + p_0 a(t) = u(t)$$

Fig. 6.40 (*a*) Coefficients in the differential equation describing the Butterworth spectra [38].

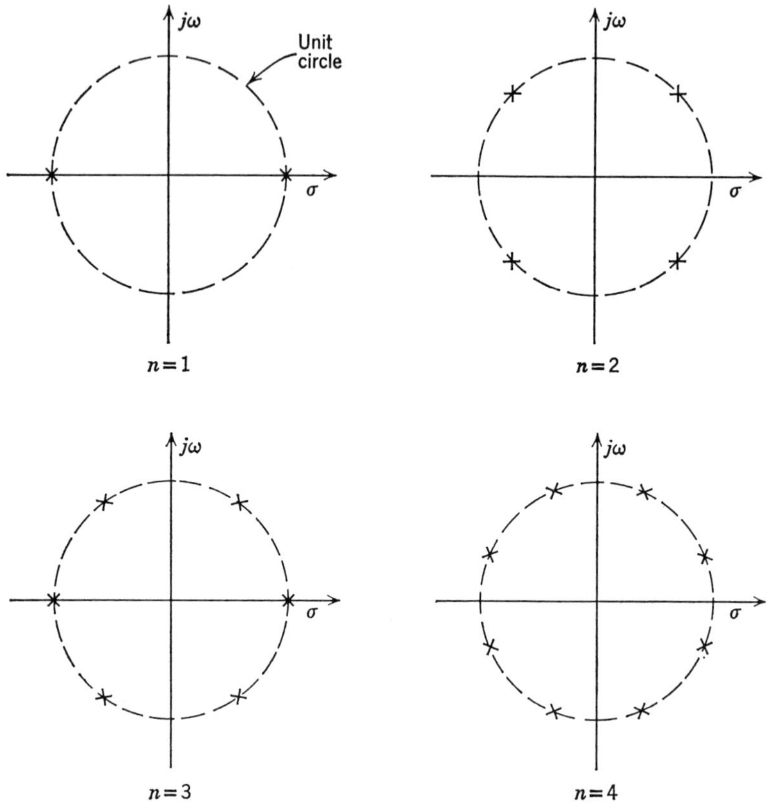

Poles are at $s = \exp\{j\pi(2m + n - 1)\} : m = 1, 2, \ldots, 2n$

Fig. 6.40 (*b*) pole plots, Butterworth spectra.

message process, we can choose any convenient state representation. An example is defined by (191),

$$
\begin{aligned}
x_1(t) &= a(t) \\
x_2(t) &= \dot{a}(t) = \dot{x}_1(t) \\
x_3(t) &= \ddot{a}(t) = \dot{x}_2(t) \\
&\vdots \\
x_n(t) &= a^{(n-1)}(t) = \dot{x}_{n-1}(t) \\
\dot{x}_n(t) &= -\sum_{k=1}^{n} p_{k-1}\, a^{(k-1)}(t) + u(t) \\
&= -\sum_{k=1}^{n} p_{k-1}\, x_k(t) + u(t)
\end{aligned}
\tag{357}
$$

The resulting **F** matrix for any n is given by using (356) and (193). The other quantities needed are

$$\mathbf{G}(t) = \begin{bmatrix} 0 \\ \hdashline 0 \\ \hdashline \vdots \\ \hdashline 1 \end{bmatrix} \tag{358}$$

$$\mathbf{C}(t) = [1 \;\vdots\; 0 \;\vdots\; \cdots \;\vdots\; 0] \tag{359}$$

$$\mathbf{Q} = 2nPk^{2n-1} \sin\left(\frac{\pi}{2n}\right) \tag{360}$$

$$\mathbf{R}(t) = \frac{N_0}{2}. \tag{361}$$

From (340) we observe that $\mathbf{z}(t)$ is an $n \times 1$ matrix,

$$\mathbf{z}(t) = \frac{2}{N_0} \begin{bmatrix} \xi_{11}(t) \\ \xi_{12}(t) \\ \vdots \\ \xi_{1n}(t) \end{bmatrix}, \tag{362}$$

$$\begin{aligned} \dot{\hat{x}}_1(t) &= \hat{x}_2(t) + \frac{2}{N_0}\, \xi_{11}(t)[r(t) - \hat{x}_1(t)] \\ \dot{\hat{x}}_2(t) &= \hat{x}_3(t) + \frac{2}{N_0}\, \xi_{12}(t)[r(t) - \hat{x}_1(t)] \\ &\vdots \\ \dot{\hat{x}}_n(t) &= -p_0\hat{x}_1(t) - p_1\hat{x}_2(t) - \cdots - p_{n-1}\,\hat{x}_n(t) + \frac{2}{N_0}\, \xi_{1n}(t)[r(t) - \hat{x}_1(t)]. \end{aligned} \tag{363}$$

The block diagram is shown in Fig. 6.41.

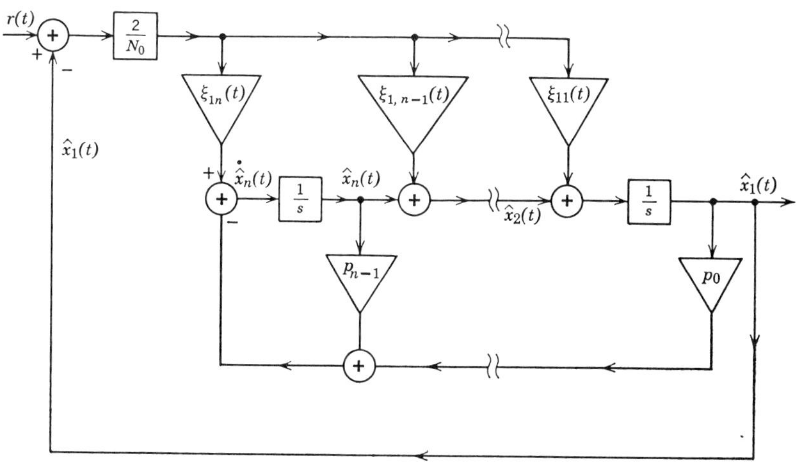

Fig. 6.41 Optimum filter: *n*th-order Butterworth message.

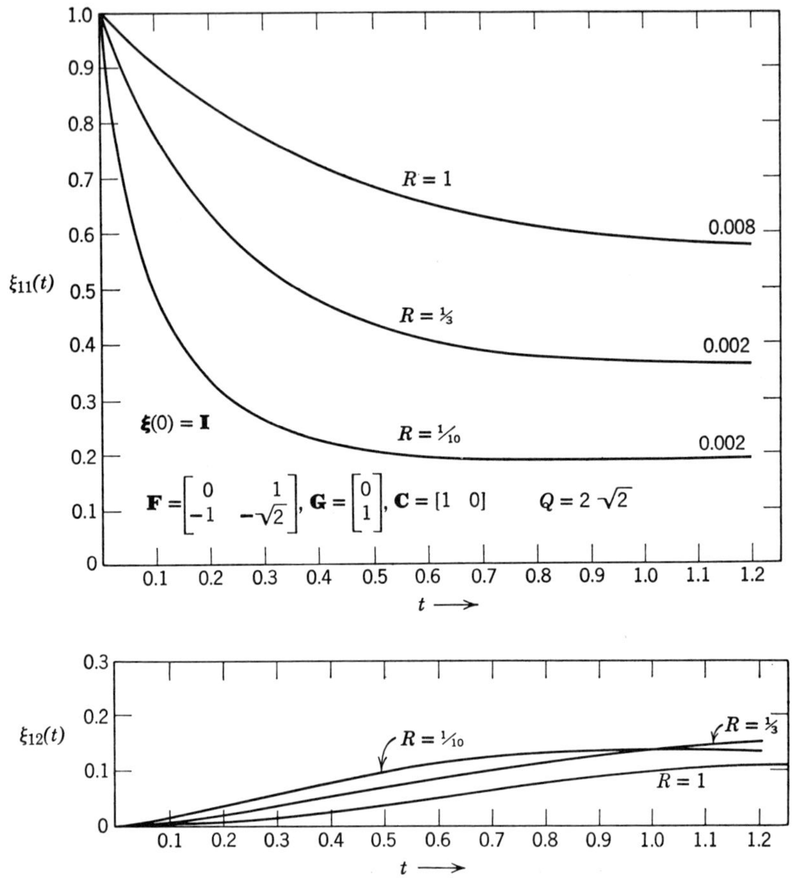

Fig. 6.42 (*a*) **Mean-square error, second-order Butterworth; (*b*) filter gains, second-order Butterworth.**

To find the values of $\xi_{11}(t), \ldots, \xi_{1n}(t)$, we solve the variance equation. This could be done analytically by using Property 16 on p. 545, but a numerical solution is much more practical. We assume that $T_i = 0$ and that the unobserved process is in a statistical steady state. We use (334) to find $\xi_P(0)$. Note that our choice of state variables causes (334*d*) to be satisfied. This is reflected by the zeros in $\boldsymbol{\xi}_P(0)$ as shown in the figures. In Fig. 6.42*a* we show the error as a function of time for the two-pole case. Once again the number on the right end of each curve is $\xi_P(1.2) - \xi_{P\infty}$. We see that for $t = 1$ the error has essentially reached its steady-state value. In Fig. 6.42*b* we show the term $\xi_{12}(t)$. Similar results are shown for the three-pole and four-pole cases in Figs. 6.43 and 6.44, respectively.† In all cases steady state is essentially reached by $t = 1$. (Observe that $k = 1$ so our time scale is normalized.) This means

† The numerical results in Figs. 6.39 and 6.41 through 6.44 are due to Baggeroer [36].

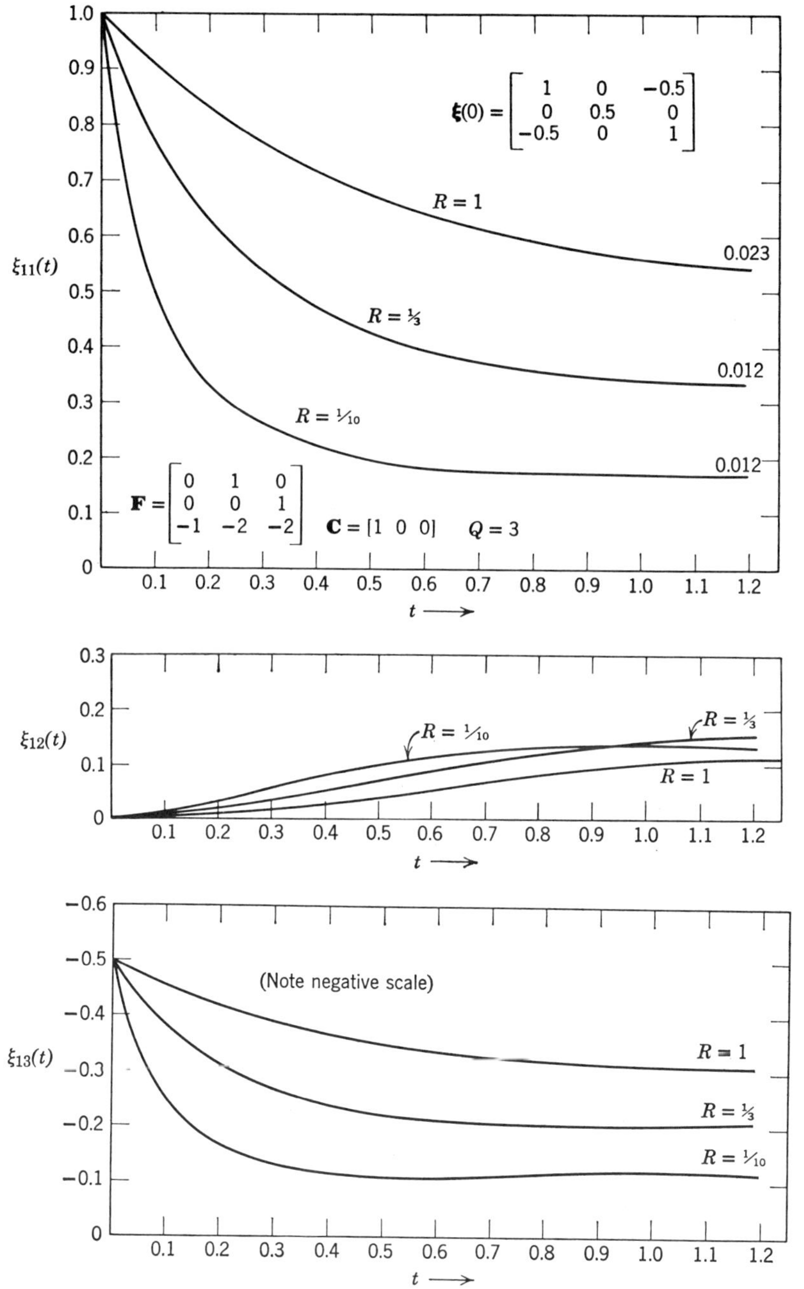

Fig. 6.43 (*a*) Mean-square error, third-order Butterworth; (*b*) filter gain, third-order Butterworth; (*c*) filter gain, third-order Butterworth.

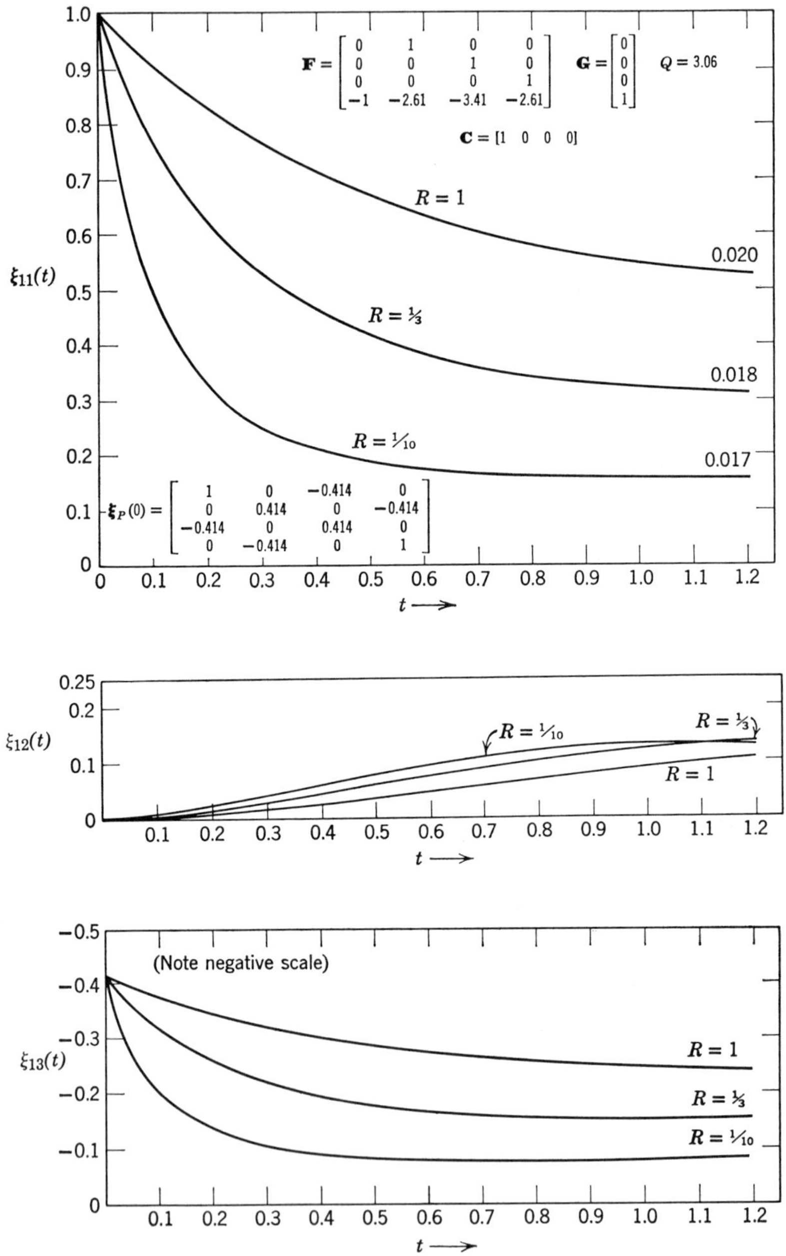

Fig. 6.44 (*a*) Mean-square error, fourth-order Butterworth; (*b*) filter gains, fourth-order Butterworth; (*c*) filter gains, fourth-order Butterworth.

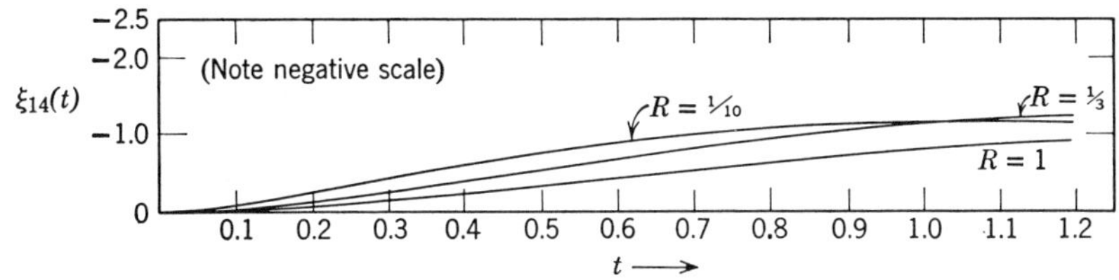

Fig. 6.44 (*d*) filter gains, fourth-order Butterworth.

that after $t = 1$ the filters are essentially time-invariant. (This does not imply that *all* terms in $\xi_P(t)$ have reached steady state.) A related question, which we leave as an exercise, is, "If we use a time-invariant filter designed for the steady-state gains, how will its performance during the initial transient compare with the optimum time-varying filter?" (See Problems 6.3.23 and 6.3.26.)

Example 3. The two preceding examples dealt with stationary processes. A simple nonstationary process is the Wiener process. It can be represented in differential equation form as

$$\dot{x}(t) = G\,u(t), \qquad x(0) = 0. \tag{364}$$

Observe that even though the coefficients in the differential equation are constant the process is nonstationary. If we assume that

$$r(t) = x(t) + w(t), \tag{365}$$

the estimator follows easily

$$\dot{\hat{x}}(t) = z(t)[r(t) - \hat{x}(t)], \tag{366}$$

where

$$z(t) = \frac{2}{N_0}\,\xi_P(t) \tag{367}$$

and

$$\dot{\xi}_P(t) = -\frac{2}{N_0}\,\xi_P{}^2(t) + G^2Q. \tag{368}$$

The transient problem can be solved easily by using Property 16 on p. 545 (see Problem 6.3.25). The result is

$$\xi_P(t) = \left(\frac{N_0}{2}\,G^2Q\right)^{1/2}\left(\frac{e^{\gamma t} - e^{-\gamma t}}{e^{\gamma t} + e^{-\gamma t}}\right) = \left(\frac{N_0G^2Q}{2}\right)^{1/2}\left(\frac{1 - e^{-2\gamma t}}{1 + e^{-2\gamma t}}\right), \tag{369}$$

where $\gamma = [2G^2Q|N_0]^{1/2}$. [Observe that (369) is not a limiting case of (353) because the initial condition is different.] As $t \to \infty$, the error approaches steady state.

$$\xi_{P\infty} = \left(\frac{N_0}{2}\,G^2Q\right)^{1/2} \tag{370}$$

[(370) can also be obtained directly from (368) by letting $\dot{\xi}_P(t) = 0$.]

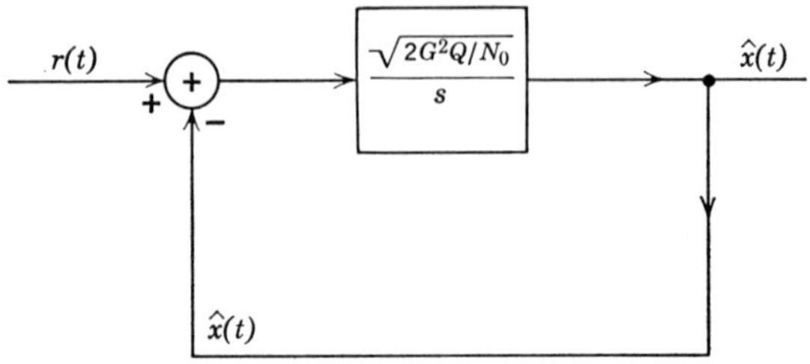

Fig. 6.45 Steady-state filter: Example 3.

The steady-state filter is shown in Fig. 6.45. It is interesting to observe that this problem is not included in the Wiener-Hopf model in Section 6.2. A heuristic way to include it is to write

$$S_x(\omega) = \frac{G^2Q}{\omega^2 + \epsilon^2}, \tag{371}$$

solve the problem by using spectral factorization techniques, and then let $\epsilon \to 0$. It is easy to verify that this approach gives the system shown in Fig. 6.45.

Example 4. In this example we derive a canonic receiver model for the following problem:

1. The message has a rational spectrum in which the order of the numerator as a function of ω^2 is at least one smaller than the order of the denominator. We use the state variable model described in Fig. 6.30. The **F** and **G** matrices are given by (212) and (213), respectively (Canonic Realization No. 2).
2. The received signal is scalar function.
3. The modulation matrix has unity in its first column and zero elsewhere. In other words, only the unmodulated message would be observed in the absence of measurement noise,

$$\mathbf{C}(t) = [1 \mid 0 \cdots 0]. \tag{372}$$

The equation describing the estimator is obtained from (339),

$$\dot{\hat{\mathbf{x}}}(t) = \mathbf{F}\hat{\mathbf{x}}(t) + \mathbf{z}(t)[r(t) - \hat{x}_1(t)] \tag{373}$$

and

$$\mathbf{z}(t) = \frac{2}{N_0}\, \boldsymbol{\xi}_P(t)\, \mathbf{C}^T(t). \tag{374}$$

As in Example 2, the gains are simply $2/N_0$ times the first row of the error matrix. The resulting filter structure is shown in Fig. 6.46. As $t \to \infty$, the gains become constant.

For the constant-gain case, by comparing the system inside the block to the two diagrams in Fig. 6.30 and 31a, we obtain the equivalent filter structure shown in Fig. 6.47.

Writing the loop filter in terms of its transfer function, we have

$$G_{lo}(s) = \frac{2}{N_0}\, \frac{\xi_{11}s^{n-1} + \cdots + \xi_{1n}}{s^n + p_{n-1}s^{n-1} + \cdots + p_0}. \tag{375}$$

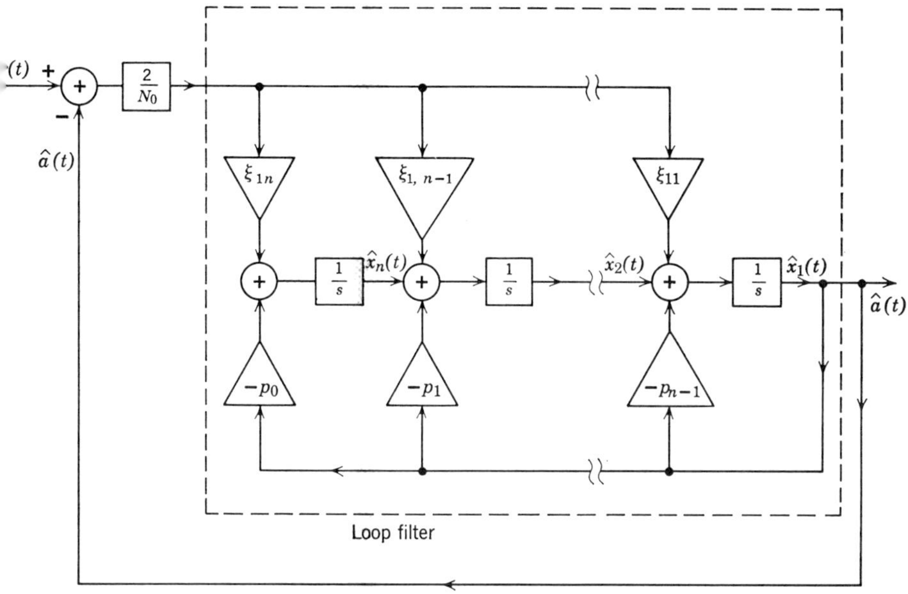

Fig. 6.46 Canonic estimator: stationary messages, statistical steady-state.

Thus the coefficients of the numerator of the loop-filter transfer function correspond to the first column in the error matrix. The poles (as we have seen before) are identical to those of the message spectrum.

Observe that we still have to solve the variance equation to obtain the numerator coefficients.

Example 5A [*23*]. As a simple example of the general case just discussed, consider the message process shown in Fig. 6.48*a*. If we want to use the canonic receiver structure we have just derived, we can redraw the message generator process as shown in Fig. 6.48*b*.

We see that

$$p_1 = k, \qquad p_0 = 0, \qquad b_1 = 0, \qquad b_0 = 1. \tag{376}$$

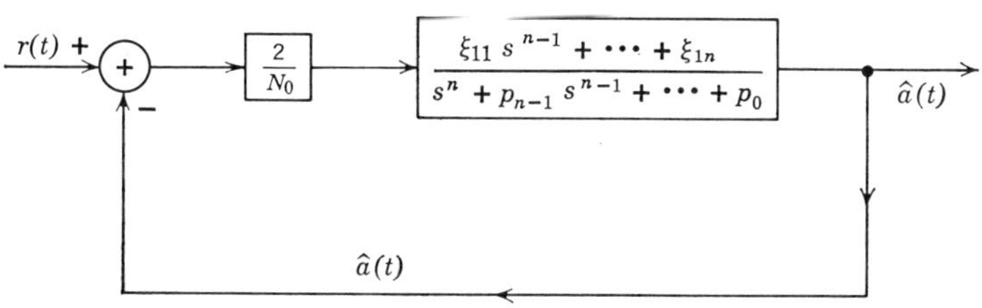

Fig. 6.47 Canonic estimator: stationary messages, statistical steady-state.

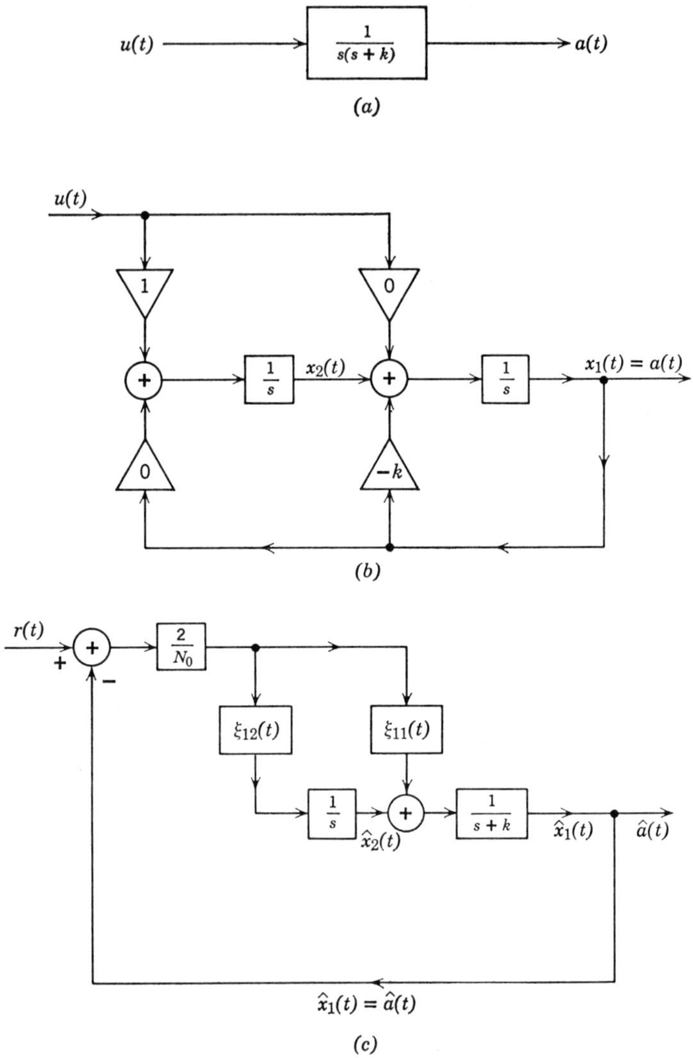

Fig. 6.48 Systems for example 5A: (*a*) message generator; (*b*) analog representation; (*c*) optimum estimator.

Then, using (212) and (213), we have

$$\mathbf{F} = \begin{bmatrix} -k & 1 \\ 0 & 0 \end{bmatrix}, \qquad \mathbf{G} = \begin{bmatrix} 0 \\ 1 \end{bmatrix}, \qquad \mathbf{C} = [1 \quad 0] \tag{377}$$

$\mathbf{Q} = q$, $\mathbf{R} = N_0/2$.
The resulting filter is just a special case of Fig. 6.47 as shown in Fig. 6.48.

The variance equation in the steady state is

$$2(-k\xi_{11} + \xi_{12}) - \frac{2}{N_0}\xi_{11}^2 = 0,$$

$$-k\xi_{12} + \xi_{22} - \frac{2}{N_0}\xi_{11}\xi_{12} = 0, \tag{378}$$

$$\frac{2}{N_0}\xi_{12}^2 = q.$$

Thus the steady-state errors are

$$\xi_{12} = \frac{N_0}{2}\left(\frac{2q}{N_0}\right)^{1/2}$$

$$\xi_{11} = \frac{N_0}{2}\left\{-k + \left[k^2 + 2\left(\frac{2q}{N_0}\right)^{1/2}\right]^{1/2}\right\}. \tag{379}$$

We have taken the positive definite solution. The loop filter for the steady-state case is

$$G_{lo}(s) = \frac{\left\{-k + \left[k^2 + 2\left(\frac{2q}{N_0}\right)^{1/2}\right]^{1/2}\right\}s + \left(\frac{2q}{N_0}\right)^{1/2}}{s(s+k)}. \tag{380}$$

Example 5B. An interesting example related to the preceding one is shown in Fig. 6.49. We now add A and B subscripts to denote quantities in Examples 5A and 5B, respectively. We see that, except for some constants, the output is the same as in Example 5A. The intermediate variable $x_{2_B}(t)$, however, did not appear in that realization.

We assume that the message of interest is $x_{2_B}(t)$. In Chapter II.2 we shall see the model in Fig. 6.49 and the resulting estimator play an important role in the FM problem. This is just a particular example of the general problem in which the message is subjected to a linear operation *before* transmission. There are two easy ways to solve this problem. One way is to observe that because we have already solved Example 5A we can use that result to obtain the answer. To use it we must express $x_{2_B}(t)$ as a linear transformation of $x_{1_A}(t)$ and $x_{2_A}(t)$, the state variables in Example 5A.

$$\beta\, x_{2_B}(t) = \dot{x}_{1_B}(t) \tag{381}$$

and

$$x_{1_B}(t) = x_{1_A}(t), \tag{382}$$

if we require

$$q_A = \beta^2 q_B. \tag{383}$$

Observing that

$$x_{2_A}(t) = k\, x_{1_A}(t) + \dot{x}_{1_A}(t), \tag{384}$$

we obtain

$$\beta\, x_{2_B}(t) = \dot{x}_{1_A}(t) = -kx_{1_A}(t) + x_{2_A}(t). \tag{385}$$

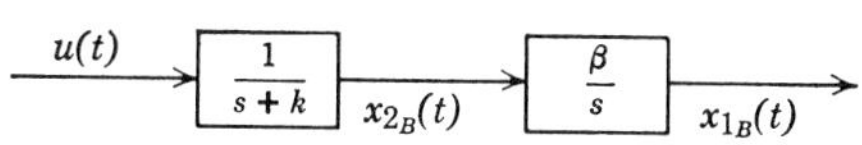

Fig. 6.49 Message generation, example 5B.

Now minimum mean-square error filtering commutes over *linear transformations.* (The proof is identical to the proof of 2.237.) Therefore

$$\hat{x}_{2_B}(t) = \frac{-1}{\beta}[k\,\hat{x}_{1_A}(t) + \hat{x}_{2_A}(t)]. \tag{386}$$

Observe that this is *not* equivalent to letting $\beta\hat{x}_{2_B}(t)$ equal the derivative of $\hat{x}_{1_A}(t)$. Thus

$$\hat{x}_{2_B}(t) \neq \frac{1}{\beta}\,\dot{\hat{x}}_{1_A}(t). \tag{387}$$

With these observations we may draw the optimum filter by modifying Fig. 6.48. The result is shown in Fig. 6.50. The error variance follows easily:

$$\beta^2\,\xi_{22_B}(t) = k^2\,\xi_{11_A}(t) - 2k\,\xi_{12_A}(t) + \xi_{22_A}(t). \tag{388}$$

Alternatively, if we had not solved Example 5A, we would approach the problem directly. We identify the message as one of the state variables. The appropriate matrices are

$$\mathbf{F} = \begin{bmatrix} 0 & \beta \\ 0 & -k \end{bmatrix}, \qquad \mathbf{G} = \begin{bmatrix} 0 \\ 1 \end{bmatrix}, \qquad \mathbf{C} = [1 \quad 0] \tag{389}$$

and

$$\mathbf{Q} = q_B. \tag{390}$$

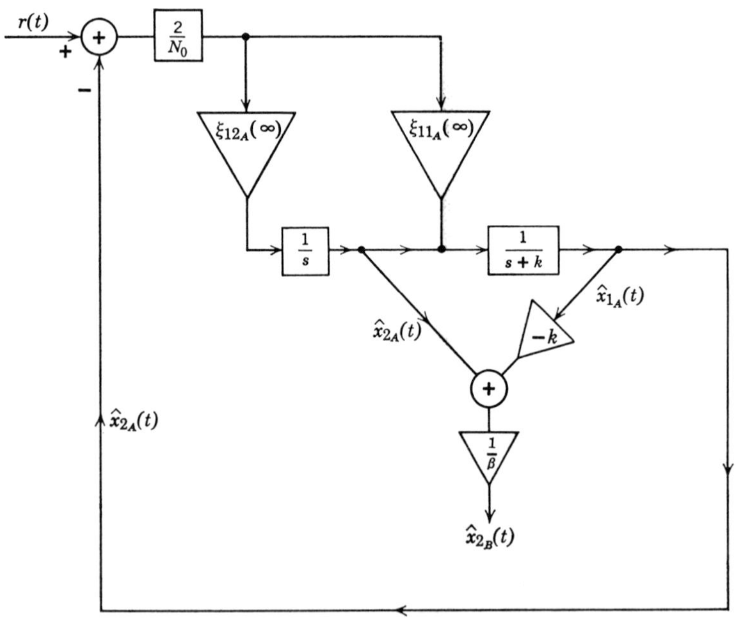

Fig. 6.50 Estimator for example 5B.

The structure of the estimator is shown in Fig. 6.51. The variance equation is

$$\dot{\xi}_{11}(t) = 2\beta\xi_{12}(t) - \frac{2}{N_0}\,\xi_{11}^2(t),$$

$$\dot{\xi}_{12}(t) = \beta\xi_{22}(t) - k\xi_{12}(t) - \frac{2}{N_0}\,\xi_{11}(t)\xi_{12}(t), \tag{391}$$

$$\dot{\xi}_{22}(t) = -2k\xi_{22}(t) - \frac{2}{N_0}\,\xi_{12}^2(t) + q_B.$$

Even in the steady state, these equations appear difficult to solve analytically. In this particular case we are helped by having just solved Example 5A. Clearly, $\xi_{11}(t)$ must be the same in both cases, if we let $q_A = \beta^2 q_B$. From (379)

$$\xi_{11,\infty} = \frac{kN_0}{2}\left\{-1 + \left[1 + \frac{2}{k^2}\left(\frac{2q_B\beta^2}{N_0}\right)^{1/2}\right]^{1/2}\right\}$$

$$\triangleq \frac{kN_0\kappa}{2}. \tag{392}$$

The other gain $\xi_{12,\infty}$ now follows easily

$$\xi_{12,\infty} = \frac{k^2N_0\kappa^2}{4\beta}. \tag{393}$$

Because we have assumed that the message of interest is $x_2(t)$, we can also easily calculate its error variance:

$$\xi_{22,\infty} = \frac{1}{2k}\left\{q_B - \frac{k^4\kappa^4N_0}{8\beta^2}\right\}. \tag{394}$$

It is straightforward to verify that (388) and (394) give the same result and that the block diagrams in Figs. 6.50 and 6.51 have identical responses between $r(t)$ and $\hat{x}_2(t)$. The internal difference in the two systems developed from the two different state representations we chose.

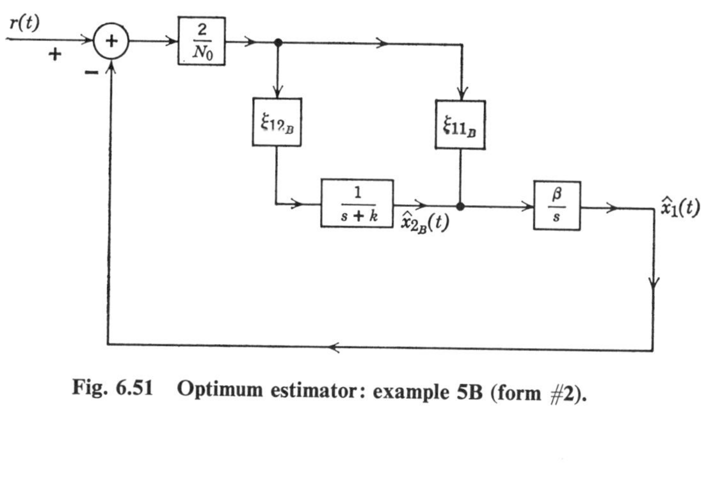

Fig. 6.51 Optimum estimator: example 5B (form #2).

Example 6. Now consider the same message process as in Example 1 but assume that the noise consists of the sum of a white noise and an uncorrelated colored noise,

$$n(t) = n_c(t) + w(t) \tag{395}$$

and

$$S_c(\omega) = \frac{2k_cP_c}{\omega^2 + k_c^2}. \tag{396}$$

As already discussed, we simply include $n_c(t)$ as a component in the state vector. Thus,

$$\mathbf{x}(t) \triangleq \begin{bmatrix} a(t) \\ n_c(t) \end{bmatrix}, \tag{397}$$

and

$$\mathbf{F}(t) = \begin{bmatrix} -k & 0 \\ 0 & -k_c \end{bmatrix}, \tag{398}$$

$$\mathbf{G}(t) = \begin{bmatrix} 1 & 0 \\ 0 & 1 \end{bmatrix}, \tag{399}$$

$$\mathbf{C}(t) = [1 \quad 1], \tag{400}$$

$$\mathbf{Q}(t) = \begin{bmatrix} 2kP & 0 \\ 0 & 2k_cP_c \end{bmatrix}, \tag{401}$$

and

$$\mathbf{R}(t) = \frac{N_0}{2}. \tag{402}$$

The gain matrix $\mathbf{z}(t)$ becomes

$$\mathbf{z}(t) = \frac{2}{N_0}\, \boldsymbol{\xi}_P(t)\, \mathbf{C}^T(t), \tag{403}$$

or

$$z_{11}(t) = \frac{2}{N_0}\, [\xi_{11}(t) + \xi_{12}(t)], \tag{404}$$

$$z_{21}(t) = \frac{2}{N_0}\, [\xi_{12}(t) + \xi_{22}(t)]. \tag{405}$$

The equation specifying the estimator is

$$\frac{d\hat{\mathbf{x}}(t)}{dt} = \mathbf{F}(t)\, \hat{\mathbf{x}}(t) + \mathbf{z}(t)[r(t) - \mathbf{C}(t)\, \hat{\mathbf{x}}(t)], \tag{406}$$

or in terms of the components

$$\frac{d\hat{x}_1(t)}{dt} = -k\hat{x}_1(t) + z_{11}(t)[r(t) - \hat{x}_1(t) - \hat{x}_2(t)], \tag{407}$$

$$\frac{d\hat{x}_2(t)}{dt} = -k_c\hat{x}_2(t) + z_{21}(t)[r(t) - \hat{x}_1(t) - \hat{x}_2(t)]. \tag{408}$$

The structure is shown in Fig. 6.52*a*. This form of the structure exhibits the symmetry of the estimation process. Observe that the estimates are coupled through the feedback path.

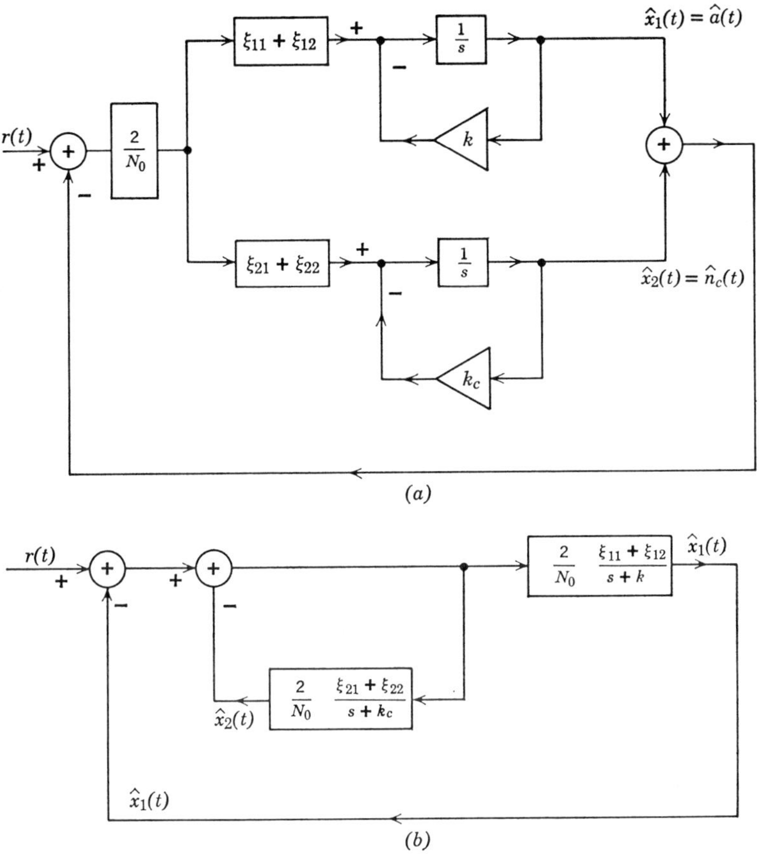

Fig. 6.52 Optimum estimator: colored and white noise (statistical steady-state).

An equivalent asymmetrical structure which shows the effect of the colored noise on the message is shown in Fig. 6.52*b*. To find the gains we must solve the variance equation. Substituting into (341), we find

$$\dot{\xi}_{11}(t) = -2k\xi_{11}(t) - \frac{2}{N_0}(\xi_{11}(t) + \xi_{12}(t))^2 + 2kP, \tag{409}$$

$$\dot{\xi}_{12}(t) = -(k + k_c)\xi_{12}(t) - \frac{2}{N_0}(\xi_{11}(t) + \xi_{12}(t))(\xi_{12}(t) + \xi_{22}(t)), \tag{410}$$

$$\dot{\xi}_{22}(t) = -2k_c\xi_{22}(t) - \frac{2}{N_0}(\xi_{12}(t) + \xi_{22}(t))^2 + 2k_cP_c. \tag{411}$$

Two comments are in order.

1. The system in Fig. 6.52*a* exhibits all of the essential features of the canonic structure for estimating a set of independent random processes whose sum is observed

in the presence of white noise (see Problem 6.3.31 for a derivation of the general canonical structure). In the general case, the coupling arises in exactly the same manner.

2. We are tempted to approach the case when there is no white noise ($N_0 = 0$) with a limiting operation. The difficulty is that the variance equation degenerates. A derivation of the receiver structure for the pure colored noise case is discussed in Section 6.3.4.

Example 7. The most common example of a multiple observation case in communications is a diversity system. A simplified version is given here. Assume that the message is transmitted over m channels with known gains as shown in Fig. 6.53. Each channel is corrupted by white noise. The modulation matrix is $m \times n$, but only the first column is nonzero:

$$\mathbf{C}(t) = \left[\begin{array}{c:c} c_2 & \\ c_2 & \mathbf{0} \\ \vdots & \\ c_m & \end{array}\right] \triangleq [\mathbf{c} \;\vdots\; \mathbf{0}]. \tag{412}$$

For simplicity we assume first that the channel noises are uncorrelated. Therefore $\mathbf{R}(t)$ is diagonal:

$$\mathbf{R}(t) = \begin{bmatrix} \frac{N_1}{2} & & & 0 \\ & \frac{N_2}{2} & & \\ & & \ddots & \\ 0 & & & \frac{N_m}{2} \end{bmatrix}. \tag{413}$$

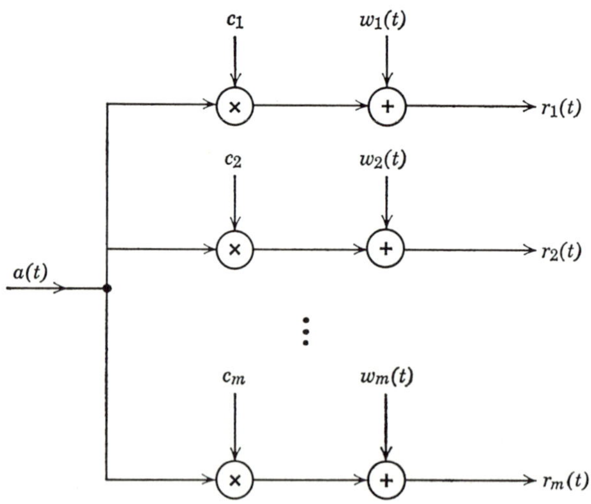

Fig. 6.53 Diversity system.

The gain matrix $\mathbf{z}(t)$ is an $n \times m$ matrix whose ijth element is

$$z_{ij}(t) = \frac{2c_j}{N_j}\,\xi_{i1}(t). \tag{414}$$

The general receiver structure is shown in Fig. 6.54a. We denote the input to the inner loop as $\mathbf{v}(t)$,

$$\mathbf{v}(t) = \mathbf{z}(t)[\mathbf{r}(t) - \mathbf{C}(t)\,\hat{\mathbf{x}}(t)]. \tag{415a}$$

Using (412)–(414), we have

$$v_i(t) = \xi_{i1}(t)\left[\sum_{j=1}^{m} \frac{2c_j}{N_j}\,r_j(t) - \left(\sum_{j=1}^{m} \frac{2c_j^2}{N_j}\right)\hat{x}_1(t)\right]. \tag{415b}$$

We see that the first term in the bracket represents a no-memory combiner of the received waveforms.

Denote the output of this combiner by $r_c(t)$,

$$r_c(t) = \sum_{j=1}^{m} \frac{2c_j}{N_j}\,r_j(t). \tag{416}$$

We see that it is precisely the maximal-ratio-combiner that we have already encountered in Chapter 4. The optimum filter may be redrawn as shown in Fig. 6.54b. We see that the problem is reduced to a single channel problem with the received waveform $r_c(t)$,

$$r_c(t) = \left(\sum_{j=1}^{m} \frac{2c_j^2}{N_j}\right) a(t) + \sum_{j=1}^{m} \frac{2c_j}{N_j}\,n_j(t). \tag{417a}$$

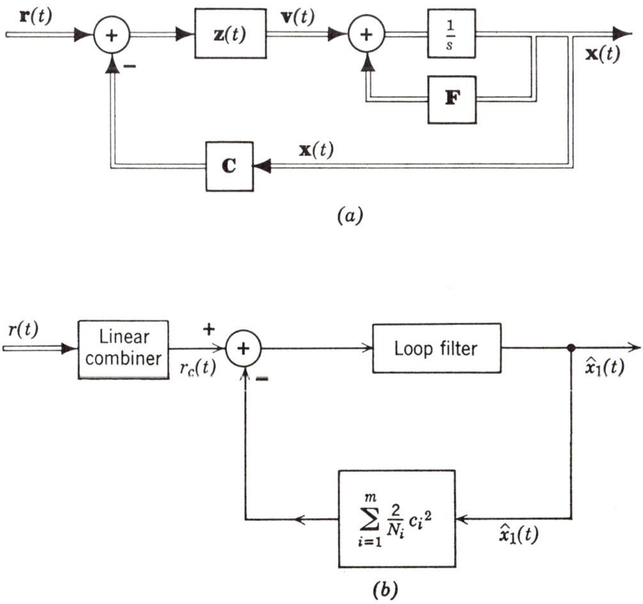

Fig. 6.54 Diversity receiver.

The modulation matrix is a scalar

$$c = \left(\sum_{j=1}^{m} \frac{2c_j^2}{N_j}\right). \tag{417b}$$

and the noise is

$$n_c(t) \triangleq \sum_{j=1}^{m} \frac{2c_j}{N_j} n_j(t). \tag{417c}$$

If the message $a(t)$ has unity variance then we have an effective power of

$$P_{ef} = \left(\sum_{j=1}^{m} \frac{2}{N_j} c_j^2\right)^2 \tag{418}$$

and an effective noise level of

$$\frac{N_{ef}}{2} = \sum_{j=1}^{m} \frac{2}{N_j} c_j^2. \tag{419}$$

Therefore all of the results for the single channel can be used with a simple scale change; for example, for the one-pole message spectrum in Example 1 we would have

$$\xi_{P\infty} = k\frac{N_{ef}}{2}(-1 + \sqrt{1 + \Lambda_{ef}}), \tag{420}$$

where

$$\Lambda_{ef} = \frac{2}{k}\sum_{j=1}^{m} \frac{2c_j^2}{N_j} = \frac{2}{k}\sum_{j=1}^{m} \frac{2P_j}{N_j}. \tag{421}$$

Similar results hold when $\mathbf{R}(t)$ is not diagonal and when the message is nonstationary.

These seven examples illustrate the problems encountered most frequently in the communications area. Other examples are included in the problems.

6.3.4 Generalizations

Several generalizations are necessary in order to include other problems of interest. We discuss them briefly in this section.

Prediction. In this case $\mathbf{d}(t) = \mathbf{x}(t + \alpha)$, where α is positive. We can show easily that

$$\hat{\mathbf{d}}(t) = \boldsymbol{\phi}(t + \alpha, t)\, \hat{\mathbf{x}}(t), \qquad \alpha > 0, \tag{422}$$

where $\boldsymbol{\phi}(t, \tau)$ is the transition matrix of the system,

$$\dot{\mathbf{x}}(t) = \mathbf{F}(t)\,\mathbf{x}(t) + \mathbf{G}(t)\,\mathbf{u}(t) \tag{423}$$

(see Problem 6.3.37).

When we deal with a constant parameter system,

$$\boldsymbol{\phi}(t + \alpha, t) = e^{\mathbf{F}\alpha}, \tag{424}$$

and (422) becomes

$$\hat{\mathbf{d}}(t) = e^{\mathbf{F}\alpha}\, \hat{\mathbf{x}}(t). \tag{425}$$

Filtering with Delay. In this case $\mathbf{d}(t) = \mathbf{x}(t + \alpha)$, but α is negative. From our discussions we know that considerable improvement is available and we would like to include it. The modification is not so straightforward as the prediction case. It turns out that the canonic receiver first finds the realizable estimate and then uses it to obtain the desired estimate. A good reference for this type of problem is Baggeroer [40]. The problem of estimating $\mathbf{x}(t_1)$, where t_1 is a point interior to a fixed observation interval, is also discussed in this reference. These problems are the state-variable counterparts to the unrealizable filters discussed in Section 6.2.3 (see Problem 6.6.4).

Linear Transformations on the State Vector. If $\mathbf{d}(t)$ is a linear transformation of the state variables $\mathbf{x}(t)$, that is,

$$\mathbf{d}(t) = \mathbf{k}_d(t)\,\mathbf{x}(t), \tag{426}$$

then

$$\hat{\mathbf{d}}(t) = \mathbf{k}_d(t)\,\hat{\mathbf{x}}(t). \tag{427}$$

Observe that $\mathbf{k}_d(t)$ is *not* a linear filter. It is a linear transformation of the state variables. This is simply a statement of the fact that minimum mean-square estimation and linear transformation commute. The error matrix follows easily,

$$\boldsymbol{\xi}_{\mathbf{d}}(t) \triangleq E[(\mathbf{d}(t) - \hat{\mathbf{d}}(t))(\mathbf{d}^T(t) - \hat{\mathbf{d}}^T(t))] = \mathbf{k}_d(t)\,\boldsymbol{\xi}_P(t)\,\mathbf{k}_d^{\,T}(t). \tag{428}$$

In Example 5B we used this technique.

Desired Linear Operations. In many cases the desired signal is obtained by passing $\mathbf{x}(t)$ or $\mathbf{y}(t)$ through a linear filter. This is shown in Fig. 6.55.

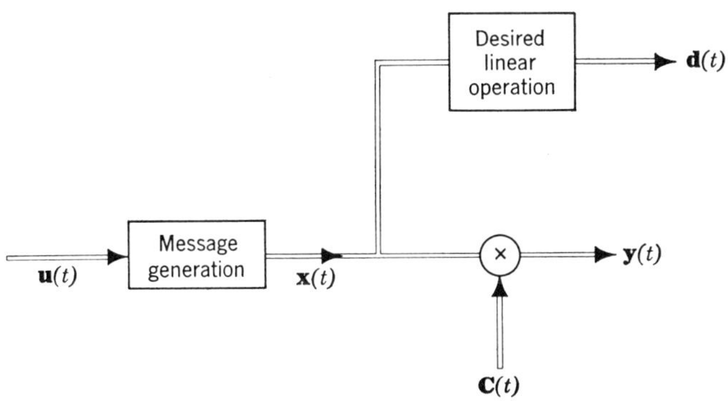

Fig. 6.55 Desired linear operations.

Three types of linear operations are of interest.

1. *Operations such as differentiation*; for example,

$$d(t) = \frac{d}{dt}\, x_1(t). \tag{429}$$

This expression is meaningful only if $x_1(t)$ is a sample function from a mean-square differentiable process.† If this is true, we can always choose $d(t)$ as one of the components of the message state vector and the results in Section 6.3.2 are immediately applicable. Observe that

$$\hat{d}(t) \neq \frac{d}{dt}\, \hat{x}_1(t). \tag{430}$$

In other words, linear filtering and realizable estimation do *not* commute. The result in (430) is obvious if we look at the estimator structure in Fig. 6.51.

2. *Improper operations*; for example, assume $y(t)$ is a scalar function and

$$d(t) = \int_0^\infty k_d(\tau)\, y(t - \tau)\, d\tau, \tag{431a}$$

where

$$K_d(j\omega) = \frac{j\omega + \alpha}{j\omega + \beta} = 1 + \frac{\alpha - \beta}{j\omega + \beta}. \tag{431b}$$

In this case the desired signal is the sum of two terms. The first term is $y(t)$. The second term is the output of a convolution operation on the past of $y(t)$. In general, an improper operation consists of a weighted sum of $y(t)$ and its derivatives plus an operation with memory. To get a state representation we must modify our results slightly. We denote the state vector of the dynamic system whose impulse response is $k_d(\tau)$ as $\mathbf{x}_d(t)$. (Here it is a scalar.) Then we have

$$\dot{x}_d(t) = -\beta x_d(t) + y(t) \tag{432}$$

and

$$d(t) = (\alpha - \beta)\, x_d(t) + y(t). \tag{433}$$

Thus the output equation contains an extra term. In general,

$$\dot{\mathbf{x}}_d(t) = \mathbf{F}_d(t)\, \mathbf{x}_d(t) + \mathbf{G}_d(t)\, \mathbf{x}(t), \tag{434}$$

$$\mathbf{d}(t) = \mathbf{C}_d(t)\, \mathbf{x}_d(t) + \mathbf{B}_d(t)\, \mathbf{x}(t). \tag{435}$$

Looking at (435) we see that if we *augment* the state-vector so that it contains both $\mathbf{x}_d(t)$ and $\mathbf{x}(t)$ then (427) will be valid. We define an augmented state vector

$$\mathbf{x}_a(t) \triangleq \begin{bmatrix} \mathbf{x}(t) \\ \hline \mathbf{x}_d(t) \end{bmatrix}. \tag{436}$$

† Note that we have worked with $\dot{\mathbf{x}}(t)$ when one of its components was not differentiable. However the output of the system always existed in the mean-square sense.

The equation for the augmented process is

$$\dot{\mathbf{x}}_a(t) = \begin{bmatrix} \mathbf{F}(t) & \vdots & \mathbf{0} \\ \hdashline \mathbf{G}_d(t) & \vdots & \mathbf{F}_d(t) \end{bmatrix} \mathbf{x}_a(t) + \begin{bmatrix} \mathbf{G}(t) \\ \hdashline \mathbf{0} \end{bmatrix} \mathbf{u}(t). \tag{437}$$

The observation equation is unchanged:

$$\mathbf{r}(t) = \mathbf{C}(t)\,\mathbf{x}(t) + \mathbf{w}(t); \tag{438}$$

However we must rewrite this in terms of the augmented state vector

$$\mathbf{r}(t) = \mathbf{C}_a(t)\,\mathbf{x}_a(t) + \mathbf{w}(t), \tag{439}$$

where

$$\mathbf{C}_a(t) \triangleq [\mathbf{C}(t) \mid \mathbf{0}]. \tag{440}$$

Now $\mathbf{d}(t)$ is obtained by a *linear* transformation on the augmented state vector. Thus

$$\hat{\mathbf{d}}(t) = \mathbf{C}_d(t)\,\hat{\mathbf{x}}_d(t) + \mathbf{B}_d(t)\,\hat{\mathbf{x}}(t) = [\mathbf{B}_d(t) \mid \mathbf{C}_d(t)][\hat{\mathbf{x}}_a(t)]. \tag{441}$$

(See Problem 6.3.41 for a simple example.)

3. *Proper operation*: In this case the impulse response of $k_d(\tau)$ does not contain any impulses or derivatives of impulses. The comments for the improper case apply directly by letting $\mathbf{B}_d(t) = \mathbf{0}$.

Linear Filtering Before Transmission. Here the message is passed through a linear filter before transmission, as shown in Fig. 6.56. All comments for the preceding case apply with obvious modification; for example, if the linear filter is an improper operation, we can write

$$\dot{\mathbf{x}}_f(t) = \mathbf{F}_f(t)\,\mathbf{x}_f(t) + \mathbf{G}_f(t)\,\mathbf{x}(t), \tag{442}$$

$$\mathbf{y}_f(t) = \mathbf{C}_f(t)\,\mathbf{x}_f(t) + \mathbf{B}_f(t)\,\mathbf{x}(t). \tag{443}$$

Then the augmented state vector is

$$\mathbf{x}_a(t) = \begin{bmatrix} \mathbf{x}(t) \\ \hdashline \mathbf{x}_f(t) \end{bmatrix}. \tag{444}$$

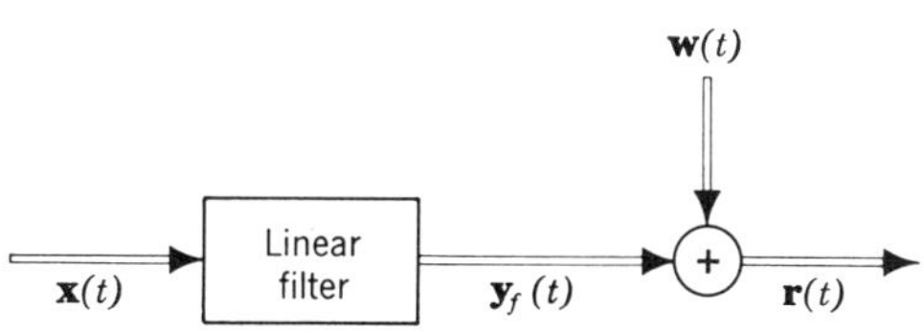

Fig. 6.56 Linear filtering before transmission.

The equation for the augmented process is

$$\dot{\mathbf{x}}_a(t) = \left[\begin{array}{c|c} \mathbf{F}(t) & \mathbf{0} \\ \hline \mathbf{G}_f(t) & \mathbf{F}_f(t) \end{array}\right] \mathbf{x}_a(t) + \left[\begin{array}{c} \mathbf{G}(t) \\ \hline \mathbf{0} \end{array}\right] \mathbf{u}(t). \tag{445}$$

The observation equation is modified to give

$$\mathbf{r}(t) = \mathbf{y}_f(t) + \mathbf{w}(t) = [\mathbf{B}_f(t) \mid \mathbf{C}_f(t)] \left[\begin{array}{c} \mathbf{x}(t) \\ \hline \mathbf{x}_f(t) \end{array}\right] + \mathbf{w}(t). \tag{446}$$

For the last two cases the key to the solution is the augmented state vector.

Correlation Between u(t) and w(t). We encounter cases in practice in which the vector white noise process $\mathbf{u}(t)$ that generates the message is correlated with the vector observation noise $\mathbf{w}(t)$. The modification in the derivation of the optimum estimator is straightforward.† Looking at the original derivation, we find that the results for the first two steps are unchanged. Thus

$$\dot{\hat{\mathbf{x}}}(t) = \mathbf{F}(t)\,\hat{\mathbf{x}}(t) + \mathbf{h}_o(t, t)[\mathbf{r}(t) - \mathbf{C}(t)\,\hat{\mathbf{x}}(t)]. \tag{447}$$

In the case of correlated $\mathbf{u}(t)$ and $\mathbf{w}(t)$ the expression for

$$\mathbf{z}(t) \triangleq \mathbf{h}_o(t, t) \triangleq \lim_{u \to t^-} \mathbf{h}_o(t, u) \tag{448}$$

must be modified. From Property 3A–V and the definition of $\boldsymbol{\xi}_P(t)$ we have

$$\boldsymbol{\xi}_P(t) = \lim_{u \to t^-} \left[\mathbf{K}_{\mathbf{x}}(t, u) - \int_{T_i}^{t} \mathbf{h}_o(t, \tau)\,\mathbf{K}_{\mathbf{rx}}(\tau, u)\, d\tau\right]. \tag{449}$$

Multiplying by $\mathbf{C}^T(t)$,

$$\begin{aligned} \boldsymbol{\xi}_P(t)\,\mathbf{C}^T(t) &= \lim_{u \to t^-} \left[\mathbf{K}_{\mathbf{x}}(t, u)\,\mathbf{C}^T(t) - \int_{T_i}^{t} \mathbf{h}_o(t, \tau)\,\mathbf{K}_{\mathbf{rx}}(\tau, u)\,\mathbf{C}^T(t) d\tau\right] \\ &= \lim_{u \to t^-} \left[\mathbf{K}_{\mathbf{xr}}(t, u) - \mathbf{K}_{\mathbf{xw}}(t, u) - \int_{T_i}^{t} \mathbf{h}_o(t, \tau)\,\mathbf{K}_{\mathbf{rx}}(\tau, u)\,\mathbf{C}^T(u)\, d\tau.\right] \end{aligned} \tag{450}$$

Now, the vector Wiener-Hopf equation implies that

$$\begin{aligned} \lim_{u \to t^-} \mathbf{K}_{\mathbf{xr}}(t, u) &= \lim_{u \to t^-} \left[\int_{T_i}^{t} \mathbf{h}_o(t, \tau)\,\mathbf{K}_{\mathbf{r}}(\tau, u)\, d\tau\right] \\ &= \lim_{u \to t^-} \left[\int_{T_i}^{t} \mathbf{h}_o(t, \tau)[\mathbf{K}_{\mathbf{rx}}(\tau, u)\,\mathbf{C}^T(u) + \mathbf{R}(\tau)\,\delta(\tau - u) \right. \\ &\qquad \left. + \mathbf{C}(\tau)\mathbf{K}_{\mathbf{xw}}(\tau, u)]\, d\tau\right]. \end{aligned} \tag{451}$$

† This particular case was first considered in [41]. Our derivation follows Collins [42].

Using (451) in (450), we obtain

$$\boldsymbol{\xi}_P(t)\,\mathbf{C}^T(t) = \lim_{u \to t^-} \left[\mathbf{h}_o(t, u)\mathbf{R}(u) + \int_{T_i}^{t} \mathbf{h}_o(t, \tau)\,\mathbf{C}(\tau)\,\mathbf{K}_{\mathbf{xw}}(\tau, u)\,d\tau - \mathbf{K}_{\mathbf{xw}}(t, u)\right]. \tag{452}$$

The first term is continuous. The second term is zero in the limit because $\mathbf{K}_{\mathbf{xw}}(\tau, u)$ is zero except when $u = t$. Due to the continuity of $\mathbf{h}_o(t, \tau)$ the integral is zero. The third term represents the effect of the correlation. Thus

$$\boldsymbol{\xi}_P(t)\mathbf{C}(t) = \mathbf{h}_o(t, t)\,\mathbf{R}(t) - \lim_{u \to t^-} \mathbf{K}_{\mathbf{xw}}(t, u). \tag{453}$$

Using Property 13 on p. 532, we have

$$\lim_{u \to t^-} \mathbf{K}_{\mathbf{xw}}(t, u) = \mathbf{G}(t)\,\mathbf{P}(t), \tag{454}$$

where

$$E[\mathbf{u}(t)\,\mathbf{w}^T(\tau)] \triangleq \delta(t - \tau)\,\mathbf{P}(t). \tag{455}$$

Then

$$\boxed{\mathbf{z}(t) \triangleq \mathbf{h}_o(t, t) = [\boldsymbol{\xi}_P(t)\,\mathbf{C}^T(t) + \mathbf{G}(t)\,\mathbf{P}(t)]\mathbf{R}^{-1}(t).} \tag{456}$$

The final step is to modify the variance equation. Looking at (328), we see that we have to evaluate the expectation

$$E\{[-\mathbf{z}(t)\,\mathbf{w}(t) + \mathbf{G}(t)\,\mathbf{u}(t)]\mathbf{x}_\epsilon^T(t)\}. \tag{457}$$

To do this we define a new white-noise driving function

$$\mathbf{v}(t) \triangleq -\mathbf{z}(t)\,\mathbf{w}(t) + \mathbf{G}(t)\,\mathbf{u}(t). \tag{458}$$

Then

$$\begin{aligned} E[\mathbf{v}(t)\mathbf{v}^T(\tau)] = {} & [\mathbf{z}(t)\mathbf{R}(t)\mathbf{z}^T(t) + \mathbf{G}(t)\mathbf{Q}\mathbf{G}^T(t)]\delta(t - \tau) \\ & - [\mathbf{z}(t)E[\mathbf{w}(t)\mathbf{u}^T(t)]\mathbf{G}^T(t) + \mathbf{G}(t)E[\mathbf{u}(t)\mathbf{w}^T(t)]\mathbf{z}^T(t)] \end{aligned} \tag{459}$$

or

$$\begin{aligned} E[\mathbf{v}(t)\,\mathbf{v}^T(\tau)] = {} & [\mathbf{z}(t)\,\mathbf{R}(t)\,\mathbf{z}^T(t) - \mathbf{z}(t)\,\mathbf{P}^T(t)\,\mathbf{G}^T(t) - \mathbf{G}(t)\,\mathbf{P}(t)\mathbf{z}^T(t) \\ & + \mathbf{G}(t)\,\mathbf{Q}\,\mathbf{G}^T(t)]\,\delta(t - \tau) \triangleq \mathbf{M}(t)\,\delta(t - \tau), \end{aligned} \tag{460}$$

and, using Property 13 on p. 532, we have

$$E[(-\mathbf{z}(t)\,\mathbf{w}(t) + \mathbf{G}(t)\,\mathbf{u}(t))\mathbf{x}_\epsilon^T(t)] = \tfrac{1}{2}\mathbf{M}(t). \tag{461}$$

Substituting into the variance equation (327), we have

$$\begin{aligned} \dot{\boldsymbol{\xi}}_P(t) = {} & \{\mathbf{F}(t)\boldsymbol{\xi}_P(t) - \mathbf{z}(t)\,\mathbf{C}(t)\boldsymbol{\xi}_P(t)\} + \{\sim\}^T \\ & + \mathbf{z}(t)\,\mathbf{R}(t)\,\mathbf{z}^T(t) - \mathbf{z}(t)\,\mathbf{P}^T(t)\,\mathbf{G}^T(t) \\ & - \mathbf{G}(t)\,\mathbf{P}(t)\,\mathbf{z}^T(t) + \mathbf{G}^T(t)\,\mathbf{Q}\,\mathbf{G}^T(t). \end{aligned} \tag{462}$$

Using the expression in (456) for $\mathbf{z}(t)$, (462) reduces to

$$\boxed{\begin{aligned}\dot{\boldsymbol{\xi}}_P(t) = {} & [\mathbf{F}(t) - \mathbf{G}(t)\,\mathbf{P}(t)\,\mathbf{R}^{-1}(t)\,\mathbf{C}(t)]\,\boldsymbol{\xi}_P(t) \\ & + \boldsymbol{\xi}_P(t)[\mathbf{F}^T(t) - \mathbf{C}^T(t)\,\mathbf{R}^{-1}(t)\mathbf{P}^T(t)\mathbf{G}(t)] \\ & - \boldsymbol{\xi}_P(t)\,\mathbf{C}^T(t)\,\mathbf{R}^{-1}(t)\,\mathbf{C}(t)\boldsymbol{\xi}_P(t) \\ & + \mathbf{G}(t)[\mathbf{Q} - \mathbf{P}(t)\,\mathbf{R}^{-1}(t)\mathbf{P}^T(t)]\,\mathbf{G}^T(t),\end{aligned}} \tag{463}$$

which is the desired result. Comparing (463) with the conventional variance equation (341), we see that we have exactly the same structure. The correlated noise has the same effect as changing $\mathbf{F}(t)$ and $\mathbf{Q}$ in (330). If we define

$$\mathbf{F}_{ef}(t) \triangleq \mathbf{F}(t) - \mathbf{G}(t)\,\mathbf{P}(t)\,\mathbf{R}^{-1}(t)\,\mathbf{C}(t) \tag{464}$$

and

$$\mathbf{Q}_{ef}(t) \triangleq \mathbf{Q} - \mathbf{P}(t)\,\mathbf{R}^{-1}(t)\,\mathbf{P}^T(t), \tag{465}$$

we can use (341) directly. Observe that the filter structure is identical to the case without correlation; only the time-varying gain $\mathbf{z}(t)$ is changed. This is the first time we have encountered a time-varying $\mathbf{Q}(t)$. The results in (339)–(341) are all valid for this case. Some interesting cases in which this correlation occurs are included in the problems.

Colored Noise Only. Throughout our discussion we have assumed that a nonzero white noise component is present. In the detection problem we encountered cases in which the removal of this assumption led to singular tests. Thus, even though the assumption is justified on physical grounds, it is worthwhile to investigate the case in which there is no white noise component. We begin our discussion with a simple example.

Example. The message process generation is described by the differential equation

$$\dot{x}_1(t) = F_1(t)\,x_1(t) + G_1(t)\,u_1(t). \tag{466}$$

The colored-noise generation is described by the differential equation

$$\dot{x}_2(t) = F_2(t)\,x_2(t) + G_2(t)\,u_2(t). \tag{467}$$

The observation process is the sum of these two processes:

$$y(t) = x_1(t) + x_2(t). \tag{468}$$

Observe that there is no white noise present. Our previous work with whitening filters suggests that in one procedure we could pass $y(t)$ through a filter designed so that the output due to $x_2(t)$ would be white. (Note that we whiten only the colored noise, not the entire input.) Looking at (468), we see that the desired output is $\dot{y}(t) - F_2(t)y(t)$. Denoting this new output as $r'(t)$, we have

$$\begin{aligned} r'(t) &\triangleq \dot{y}(t) - F_2(t)\,y(t) \\ &= \dot{x}_1(t) - F_2(t)\,x_1(t) + G_2(t)\,u_2(t), \end{aligned} \tag{469}$$

$$r'(t) = [F_1(t) - F_2(t)]x_1(t) + w'(t), \tag{470}$$

where

$$w'(t) \triangleq G_1(t)\, u_1(t) + G_2(t)\, u_2(t). \tag{471}$$

We now have the problem in a familiar format:

$$r'(t) = \mathbf{C}(t)\,\mathbf{x}(t) + w'(t), \tag{472}$$

where

$$\mathbf{C}(t) = [F_1(t) - F_2(t) \,\vdots\, 0]. \tag{473}$$

The state equation is

$$\begin{bmatrix} \dot{x}_1(t) \\ \dot{x}_2(t) \end{bmatrix} = \begin{bmatrix} F_1(t) & 0 \\ 0 & F_2(t) \end{bmatrix} \begin{bmatrix} x_1(t) \\ x_2(t) \end{bmatrix} + \begin{bmatrix} G_1(t) & 0 \\ 0 & G_2(t) \end{bmatrix} \begin{bmatrix} u_1(t) \\ u_2(t) \end{bmatrix}. \tag{474}$$

We observe that the observation noise $w'(t)$ is correlated with $\mathbf{u}(t)$.

$$E[\mathbf{u}(t)\, w'(\tau)] = \delta(t - \tau)\,\mathbf{P}(t), \tag{475}$$

so that

$$\mathbf{P}(t) = \begin{bmatrix} G_1(t)Q_1 \\ G_2(t)Q_2 \end{bmatrix}. \tag{476}$$

The optimal filter follows easily by using the gain equation (456) and the variance equation (463) derived in the last section. The general case† is somewhat more complicated but the basic ideas carry through (e.g., [43], [41], or Problem 6.3.45).

Sensitivity. In all of our discussion we have assumed that the matrices $\mathbf{F}(t)$, $\mathbf{G}(t)$, $\mathbf{C}(t)$, $\mathbf{Q}$, and $\mathbf{R}(t)$ were known exactly. In practice, the actual matrices may be different from those assumed. The sensitivity problem is to find the increase in error when the actual matrices are different. We assume the following model:

$$\dot{\mathbf{x}}_{\text{mo}}(t) = \mathbf{F}_{\text{mo}}(t)\,\mathbf{x}_{\text{mo}}(t) + \mathbf{G}_{\text{mo}}(t)\,\mathbf{u}_{\text{mo}}(t), \tag{477}$$

$$\mathbf{r}_{\text{mo}}(t) = \mathbf{C}_{\text{mo}}(t)\,\mathbf{x}_{\text{mo}}(t) + \mathbf{w}_{\text{mo}}(t). \tag{478}$$

The correlation matrices are $\mathbf{Q}_{\text{mo}}$ and $\mathbf{R}_{\text{mo}}(t)$. We denote the error matrix under the model assumptions as $\boldsymbol{\xi}_{\text{mo}}(t)$. (The subscript "mo" denotes model.) Now assume that the actual situation is described by the equations,

$$\dot{\mathbf{x}}_{\text{ac}}(t) = \mathbf{F}_{\text{ac}}(t)\,\mathbf{x}_{\text{ac}}(t) + \mathbf{G}_{\text{ac}}(t)\,\mathbf{u}_{\text{ac}}(t), \tag{479}$$

$$\mathbf{r}_{\text{ac}}(t) = \mathbf{C}_{\text{ac}}(t)\,\mathbf{x}_{\text{ac}}(t) + \mathbf{w}_{\text{ac}}(t), \tag{480}$$

† We have glossed over two important issues because the colored-noise-only problem does not occur frequently enough to warrant a lengthy discussion. The first issue is the minimal dimensionality of the problem. In this example, by a suitable choice of state variables, we can end up with a scalar variance equation instead of a 2×2 equation. We then retrieve $\hat{\mathbf{x}}(t)$ by a linear transformation on our estimate of the state variable for the minimal dimensional system and the received signal. The second issue is that of initial conditions in the variance equation. $\xi_P(0^-)$ may not equal $\xi_P(0^+)$. The interested reader should consult the two references listed for a more detailed discussion.

with correlation matrices $\mathbf{Q}_{ac}$ and $\mathbf{R}_{ac}(t)$. We want to find the actual error covariance matrix $\boldsymbol{\xi}_{ac}(t)$ for a system which is optimum under the model assumptions when the input is $\mathbf{r}_{ac}(t)$. (The subscript "ac" denotes actual.) The derivation is carried out in detail in [44]. The results are given in (484)–(487). We define the quantities:

$$\boldsymbol{\xi}_{ac}(t) \triangleq E[\mathbf{x}_{\epsilon_{ac}}(t)\, \mathbf{x}_{\epsilon_{ac}}^T(t)], \tag{481}$$

$$\boldsymbol{\xi}_{a\epsilon}(t) \triangleq E[\mathbf{x}_{ac}(t)\, \mathbf{x}_{\epsilon_{ac}}^T(t)], \tag{482}$$

$$\mathbf{F}_{\epsilon}(t) \triangleq \mathbf{F}_{ac}(t) - \mathbf{F}_{mo}(t). \tag{483a}$$

$$\mathbf{C}_{\epsilon}(t) \triangleq \mathbf{C}_{ac}(t) - \mathbf{C}_{mo}(t). \tag{483b}$$

The actual error covariance matrix is specified by three matrix equations:

$$\begin{aligned}\dot{\boldsymbol{\xi}}_{ac}(t) = {} & \{[\mathbf{F}_{mo}(t) - \boldsymbol{\xi}_{mo}(t)\, \mathbf{C}_{mo}{}^T(t)\, \mathbf{R}_{mo}{}^{-1}(t)\, \mathbf{C}_{mo}(t)]\, \boldsymbol{\xi}_{ac}(t) \\ & - [\mathbf{F}_{\epsilon}(t) - \boldsymbol{\xi}_{mo}(t)\, \mathbf{C}_{mo}{}^T(t)\, \mathbf{R}_{mo}{}^{-1}(t)\, \mathbf{C}_{\epsilon}(t)]\, \boldsymbol{\xi}_{a\epsilon}(t)\} \\ & + \{\sim\}^T + \mathbf{G}_{ac}(t)\, \mathbf{Q}_{ac}\, \mathbf{G}_{ac}{}^T(t) \\ & + \boldsymbol{\xi}_{mo}(t)\, \mathbf{C}_{mo}{}^T(t)\, \mathbf{R}_{mo}{}^{-1}(t)\, \mathbf{R}_{ac}(t)\, \mathbf{R}_{mo}{}^{-1}(t)\, \mathbf{C}_{mo}(t)\, \boldsymbol{\xi}_{mo}(t),\end{aligned} \tag{484}$$

$$\begin{aligned}\dot{\boldsymbol{\xi}}_{a\epsilon}(t) = {} & \mathbf{F}_{ac}(t)\, \boldsymbol{\xi}_{a\epsilon}(t) + \boldsymbol{\xi}_{a\epsilon}(t)\, \mathbf{F}_{mo}{}^T(t) \\ & - \boldsymbol{\xi}_{a\epsilon}(t)\, \mathbf{C}_{mo}{}^T(t)\, \mathbf{R}_{mo}{}^{-1}(t)\, \mathbf{C}_{mo}(t)\, \boldsymbol{\xi}_{mo}(t) \\ & - \boldsymbol{\Lambda}_{ac}(t)\, \mathbf{F}_{\epsilon}{}^T(t) + \boldsymbol{\Lambda}_{ac}(t)\, \mathbf{C}_{\epsilon}{}^T(t)\, \mathbf{R}_{mo}{}^{-1}(t)\, \mathbf{C}_{mo}(t)\, \boldsymbol{\xi}_{mo}(t) \\ & + \mathbf{G}_{ac}(t)\, \mathbf{Q}_{ac}\, \mathbf{G}_{ac}{}^T(t),\end{aligned} \tag{485}$$

where

$$\boldsymbol{\Lambda}_{ac}(t) \triangleq E[\mathbf{x}_{ac}(t)\, \mathbf{x}_{ac}{}^T(t)] \tag{486}$$

satisfies the familiar linear equation

$$\dot{\boldsymbol{\Lambda}}_{ac}(t) = \mathbf{F}_{ac}(t)\, \boldsymbol{\Lambda}_{ac}(t) + \boldsymbol{\Lambda}_{ac}(t)\, \mathbf{F}_{ac}{}^T(t) + \mathbf{G}_{ac}(t)\, \mathbf{Q}_{ac}\, \mathbf{G}_{ac}{}^T(t). \tag{487}$$

We observe that we can solve (487) then (485) and (484). In other words the equations are coupled in only one direction. Solving in this manner and assuming that the variance equation for $\boldsymbol{\xi}_{mo}(t)$ has already been solved, we see that the equations are linear and time-varying. Some typical examples are discussed in [44] and the problems.

Summary

With the inclusion of these generalizations, the feedback filter formulation can accommodate all the problems that we can solve by using conventional Wiener theory. (A possible exception is a stationary process with a nonrational spectrum. In theory, the spectral factorization techniques will work for nonrational spectra if they satisfy the Paley-Wiener criterion, but the actual solution is not practical to carry out in most cases of interest).

We summarize some of the advantages of the state-variable formulation.

1. Because it is a time-domain formulation, nonstationary processes and finite time intervals are easily included.
2. The form of the solution is such that it can be implemented on a computer. This advantage should not be underestimated. Frequently, when a problem is simple enough to solve analytically, our intuition is good enough so that the optimum processor will turn out to be only slightly better than a logically designed, but *ad hoc*, processor. However, as the complexity of the model increases, our intuition will start to break down, and the optimum scheme is frequently essential as a guide to design. If we cannot get quantitative answers for the optimum processor in an easy manner, the advantage is lost.
3. A third advantage is not evident from our discussion. The original work [23] recognizes and exploits heavily the duality between the estimation and control problem. This enables us to prove many of the desired results rigorously by using techniques from the control area.
4. Another advantage of the state-variable approach which we shall not exploit fully is its use in nonlinear system problems. In Chapter II.2 we indicate some of the results that can be derived with this approach.

Clearly, there are disadvantages also. Some of the more important ones are the following:

1. It appears difficult to obtain closed-form expressions for the error such as (152).
2. Several cases, such as unrealizable filters, are more difficult to solve with this formulation.

Our discussion in this section has served as an introduction to the role of the state-variable formulation. Since the original work of Kalman and Bucy a great deal of research has been done in the area. Various facets of the problem and related areas are discussed in many papers and books. In Chapters II.2, II.3, and II.4, we shall once again encounter interesting problems in which the state-variable approach is useful.

We now turn to the problem of amplitude modulation to see how the results of Sections 6.1 through 6.3 may be applied.

6.4 LINEAR MODULATION: COMMUNICATIONS CONTEXT

The general discussion in Section 6.1 was applicable to arbitrary linear modulation problems. In Section 6.2 we discussed solution techniques which were applicable to unmodulated messages. In Section 6.3 the general results were for linear modulations but the examples dealt primarily with unmodulated signals.

We now want to discuss a particular category of linear modulation that occurs frequently in communication problems. These are characterized by the property that the carrier is a waveform whose frequency is high compared with the message bandwidth. Common examples are

$$s(t, a(t)) = \sqrt{2P}\,[1 + ma(t)] \cos \omega_c t. \tag{488}$$

This is conventional AM with a residual carrier.

$$s(t, a(t)) = \sqrt{2P}\, a(t) \cos \omega_c t. \tag{489}$$

This is double-sideband suppressed-carrier amplitude modulation.

$$s(t, a(t)) = \sqrt{P}\,[a(t) \cos \omega_c t - \tilde{a}(t) \sin \omega_c t], \tag{490}$$

where $\tilde{a}(t)$ is related to the message $a(t)$ by a particular linear transformation (we discuss this in more detail on p. 581). By choosing the transformation properly we can obtain a single-sideband signal.

All of these systems are characterized by the property that $c(t)$ is essentially disjoint in frequency from the message process. We shall see that this leads to simplification of the estimator structure (or demodulator).

We consider several interesting cases and discuss both the realizable and unrealizable problems. Because the approach is fresh in our minds, let us look at a realizable point-estimation problem first.

6.4.1 DSB-AM: Realizable Demodulation

As a first example consider a double-sideband suppressed-carrier amplitude modulation system

$$s(t, a(t)) = [\sqrt{2P} \cos \omega_c t]\, a(t) = y(t). \tag{491}$$

We write this equation in the general linear modulation form by letting

$$c(t) \triangleq \sqrt{2P} \cos \omega_c t. \tag{492}$$

First, consider the problem of *realizable* demodulation in the presence of white noise. We denote the message state vector as $\mathbf{x}(t)$ and assume that the message $a(t)$ is its first component. The optimum estimate is specified by the equations,

$$\frac{d\hat{\mathbf{x}}(t)}{dt} = \mathbf{F}(t)\, \hat{\mathbf{x}}(t) + \mathbf{z}(t)[r(t) - \mathbf{C}(t)\, \hat{\mathbf{x}}(t)], \tag{493}$$

where

$$\mathbf{C}(t) = [c(t) \,\vdots\, 0 \,\vdots\, 0 \cdots 0] \tag{494}$$

and

$$\mathbf{z}(t) = \boldsymbol{\xi}_P(t)\, \mathbf{C}^T(t)\, \frac{2}{N_0}. \tag{495}$$

The block diagram of the receiver is shown in Fig. 6.57.

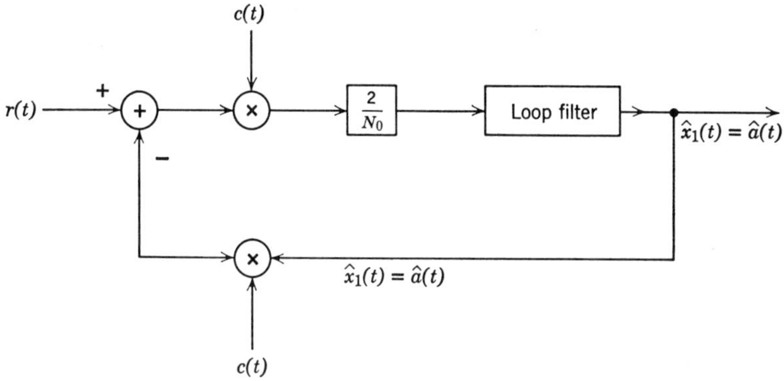

Fig. 6.57 DSB-AM receiver: form 1.

To simplify this structure we must examine the character of the loop filter and $c(t)$. First, let us *conjecture* that the loop filter will be low-pass with respect to ω_c, the carrier frequency. This is a logical conjecture because in an AM system the message is low-pass with respect to ω_c.

Now, let us look at what happens to $\hat{a}(t)$ in the feedback path when it is multiplied twice by $c(t)$. The result is

$$c^2(t)\,\hat{a}(t) = P(1 + \cos 2\omega_c t)\,\hat{a}(t). \tag{496}$$

From our original assumption, $\hat{a}(t)$ has no frequency components near $2\omega_c$. Thus, if the loop filter is low-pass, the term near $2\omega_c$ will not pass through and we could redraw the loop as shown in Fig. 6.58 to obtain the same output. We see that the loop is now operating at low-pass frequencies.

It remains to be shown that the resulting filter is just the conventional low-pass optimum filter. This follows easily by determining how $c(t)$ enters into the variance equation (see Problem 6.4.1).

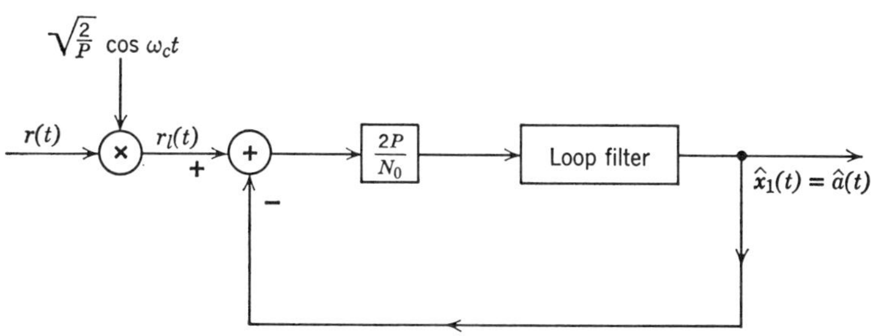

Fig. 6.58 DSB-AM receiver: final form.

We find that the resulting low-pass filter is identical to the unmodulated case. Because the modulator simply shifted the message to a higher frequency, we would expect the error variance to be the same. The error variance follows easily.

Looking at the input to the system and recalling that

$$r(t) = [\sqrt{2P} \cos \omega_c t]\, a(t) + n(t), \tag{497}$$

we see that the input to the loop is

$$r_l(t) = [\sqrt{P}\, a(t) + n_s(t)] + \text{double frequency terms}, \tag{498}$$

where $n_s(t)$ is the original noise $n(t)$ multiplied by $\sqrt{2/P} \cos \omega_c t$.

$$S_{n_s}(\omega) = \frac{N_0}{2P} \tag{499}$$

We see that this input is identical to that in Section 6.2. Thus the error expression in (152) carries over directly.

$$\xi_{Pn} = \frac{N_0}{2P} \int_{-\infty}^{\infty} \ln \left[1 + \frac{S_a(\omega)}{N_0/2P}\right] \frac{d\omega}{2\pi} \tag{500}$$

for DSB-SC amplitude modulation. The curves in Figs. 6.17 and 6.19 for the Butterworth and Gaussian spectra are directly applicable.

We should observe that the noise actually has the spectrum shown in Fig. 6.59 because there are elements operating at bandpass that the received waveform passes through before being available for processing. Because the filter in Fig. 6.58 is low-pass, the white noise approximation will be valid as long as the spectrum is flat over the effective filter bandwidth.

6.4.2 DSB-AM: Demodulation with Delay

Now consider the same problem for the case in which unrealizable filtering (or filtering with delay) is allowed. As a further simplification we

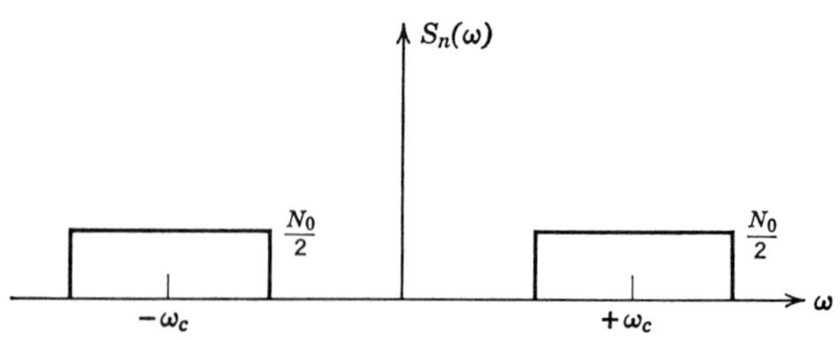

Fig. 6.59 Bandlimited noise with flat spectrum.

assume that the message process is stationary. Thus $T_f = \infty$ and $K_a(t, u) = K_a(t - u)$. In this case, the easiest approach is to assume the Gaussian model is valid and use the MAP estimation procedure developed in Chapter 5. The MAP estimator equation is obtained from (6.3) and (6.4),

$$\hat{a}_u(t) = \int_{-\infty}^{\infty} \frac{2}{N_0} K_a(t - u)\, c(u)[r(u) - c(u)\, \hat{a}(u)]\, du, \qquad -\infty < t < \infty. \tag{501}$$

The subscript u emphasizes the estimator is unrealizable. We see that the operation inside the integral is a convolution, which suggests the block

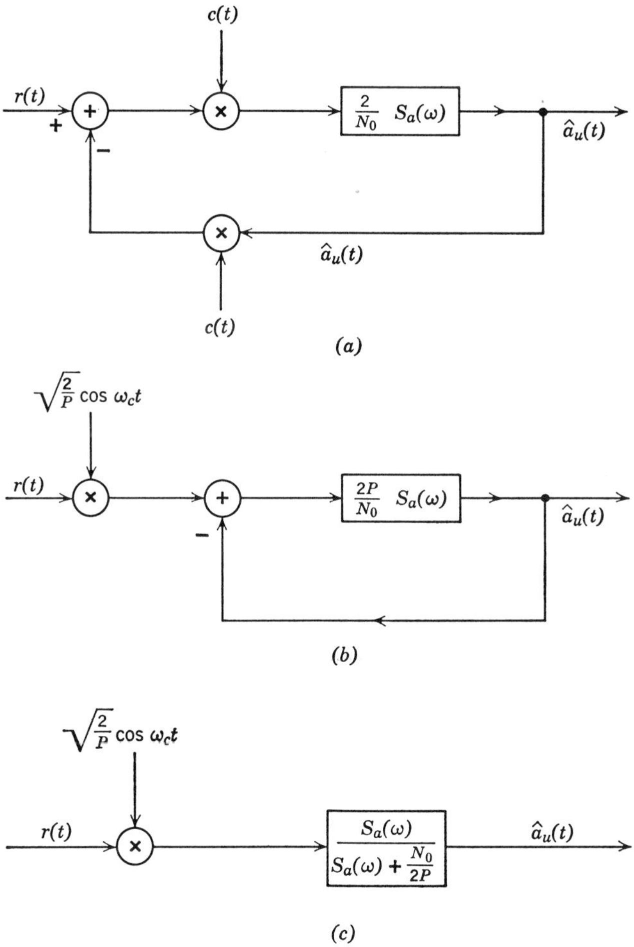

Fig. 6.60 Demodulation with delay.

diagram shown in Fig. 6.60*a*. Using an argument identical to that in Section 6.4.1, we obtain the block diagram in Fig. 6.60*b* and finally in Fig. 6.60*c*.

The optimum demodulator is simply a multiplication followed by an optimum unrealizable filter.† The error expression follows easily:

$$\xi_u = \int_{-\infty}^{\infty} \frac{S_a(\omega)(N_0/2P)}{S_a(\omega) + N_0/2P} \frac{d\omega}{2\pi}. \tag{502}$$

This result, of course, is identical to that of the unmodulated process case. As we have pointed out, this is because linear modulation is just a simple spectrum shifting operation.

6.4.3 Amplitude Modulation: Generalized Carriers

In most conventional communication systems the carrier is a sine wave. Many situations, however, develop when it is desirable to use a more general waveform as a carrier. Some simple examples are the following:

1. Depending on the spectrum of the noise, we shall see that different carriers will result in different demodulation errors. Thus in many cases a sine wave carrier is inefficient. A particular case is that in which additive noise is intentional jamming.
2. In communication nets with a large number of infrequent users we may want to assign many systems in the same frequency band. An example is a random access satellite communication system. By using different wideband orthogonal carriers, this assignment can be accomplished.
3. We shall see in Part II that a wideband carrier will enable us to combat randomly time-varying channels.

Many other cases arise in which it is useful to depart from sinusoidal carriers. The modification of the work in Section 6.4.1 is obvious. Let

$$s(t, a(t)) = c(t)\, a(t). \tag{503}$$

Looking at Fig. 6.60, we see that if

$$c^2(t)\, a(t) = k[a(t) + \text{a high frequency term}] \tag{504}$$

the succeeding steps are identical. If this is not true, the problem must be re-examined.

As a second example of linear modulation we consider single-sideband amplitude modulation.

† This particular result was first obtained in [46] (see also [45]).

6.4.4 Amplitude Modulation: Single-Sideband Suppressed-Carrier†

In a single-sideband communication system we modulate the carrier $\cos \omega_c t$ with the message $a(t)$. In addition, we modulate a carrier $\sin \omega_c t$ with a time function $\tilde{a}(t)$, which is linearly related to $a(t)$. Thus the transmitted signal is

$$s(t, a(t)) = \sqrt{P}\,[a(t) \cos \omega_c t - \tilde{a}(t) \sin \omega_c t]. \tag{505}$$

We have removed the $\sqrt{2}$ so that the transmitted power is the same as the DSB-AM case. Once again, the carrier is suppressed.

The function $\tilde{a}(t)$ is the Hilbert transform of $a(t)$. It corresponds to the output of a linear filter $h(\tau)$ when the input is $a(t)$. The transfer function of the filter is

$$H(j\omega) = \begin{cases} -j, & \omega > 0, \\ 0, & \omega = 0, \\ +j, & \omega < 0. \end{cases} \tag{506}$$

We can show (Problem 6.4.2) that the resulting signal has a spectrum that is entirely above the carrier (Fig. 6.61).

To find the structure of the optimum demodulator and the resulting performance we consider the case $T_f = \infty$ because it is somewhat easier. From (505) we observe that SSB is a mixture of a no-memory term and a memory term.

There are several easy ways to derive the estimator equation. We can return to the derivation in Chapter 5 (5.25), modify the expression for $\partial s(t, a(t))/\partial A_r$, and proceed from that point (see Problem 6.4.3). Alternatively, we can view it as a vector problem and jointly estimate $a(t)$ and $\tilde{a}(t)$ (see Problem 6.4.4). This equivalence is present because the transmitted signal contains $a(t)$ and $\tilde{a}(t)$ in a linear manner. Note that a state-

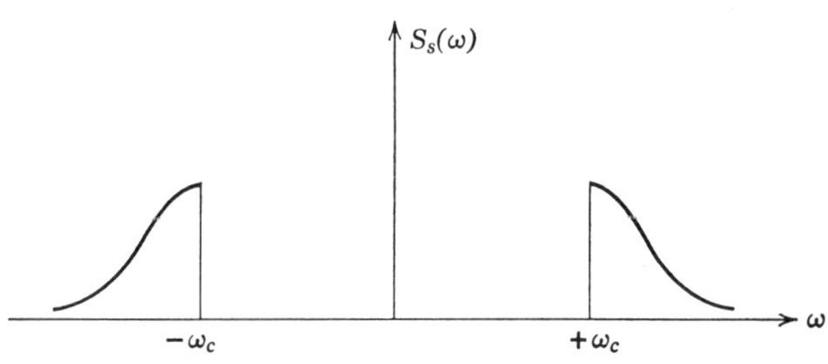

Fig. 6.61 SSB spectrum.

† We assume that the reader has heard of SSB. Suitable references are [47] to [49].

variable approach is not useful because the filter that generates the Hilbert transform (506) is not a finite-dimensional dynamic system.

We leave the derivation as an exercise and simply state the results. Assuming that the noise is white with spectral height $N_0/2$ and using (5.160), we obtain the estimator equations

$$\hat{a}(t) = \frac{2}{N_0}\int_{-\infty}^{\infty} \sqrt{P}\left\{\cos \omega_c z\, R_a(t, z) - \sin \omega_c z \left[\int_{-\infty}^{\infty} h(z - y) R_a(t,y)\, dy\right]\right\}$$
$$\{r(z) - \sqrt{P}\,[\hat{a}(z) \cos \omega_c z - \hat{\check{a}}(z) \sin \omega_c z]\}\, dz \tag{507}$$

and

$$\hat{\check{a}}(t) = \frac{2}{N_0}\int_{-\infty}^{\infty} \sqrt{P}\left\{\cos \omega_c z \left[\int_{-\infty}^{\infty} h(t - y) R_a(y, z)\, dy\right] - \sin \omega_c z\, R_a(t, y)\right\}$$
$$\{r(z) - \sqrt{P}\,[\hat{a}(z) \cos \omega_c z - \hat{\check{a}}(z) \sin \omega_c z]\}\, dz. \tag{508}$$

These equations look complicated. However, drawing the block diagram and using the definition of $H(j\omega)$ (506), we are led to the simple receiver in Fig. 6.62 (see Problem 6.4.5 for details).

We can easily verify that $n_s(t)$ has a spectral height of $N_0/2$. A comparison of Figs. 6.60 and 6.62 makes it clear that the mean-square performance of SSB-SC and DSB-SC are identical. Thus we may use other considerations such as bandwidth occupancy when selecting a system for a particular application.

These two examples demonstrate the basic ideas involved in the estimation of messages in the linear modulation systems used in conventional communication systems.

Two other systems of interest are double-sideband and single-sideband in which the carrier is *not* suppressed. The transmitted signal for the first case was given in (488). The resulting receivers follow in a similar manner.

From the standpoint of estimation accuracy we would expect that because part of the available power is devoted to transmitting a residual carrier the estimation error would increase. This qualitative increase is easy to demonstrate (see Problems 6.4.6, 6.4.7). We might ask why we

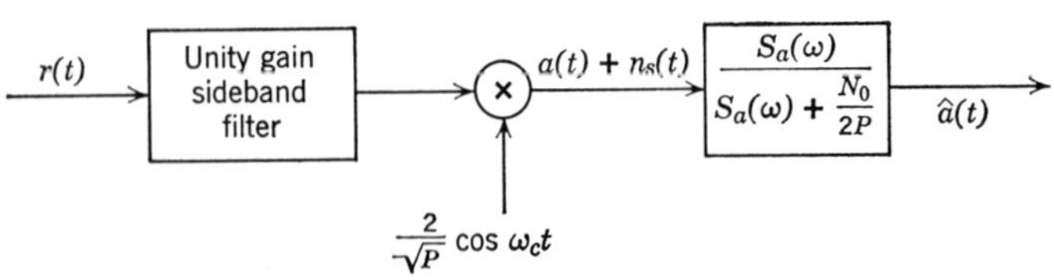

Fig. 6.62 SSB receiver.

would ever use a residual carrier. The answer, of course, lies in our model of the communication link.

We have assumed that $c(t)$, the modulation function (or carrier), is *exactly* known at the receiver. In other words, we assume that the oscillator at the receiver is *synchronized* in phase with the transmitter oscillator. For this reason, the optimum receivers in Figs. 6.58 and 6.60 are frequently referred to as *synchronous demodulators.* To implement such a demodulator in actual practice the receiver must be supplied with the carrier in some manner. In a simple method a pilot tune uniquely related to the carrier is transmitted. The receiver uses the pilot tone to construct a replica of $c(t)$. As soon as we consider systems in which the carrier is reconstructed from a signal sent from the transmitter, we encounter the question of an imperfect copy of $c(t)$. The imperfection occurs because there is noise in the channel in which we transmit the pilot tone. Although the details of reconstructing the carrier will develop more logically in Chapter II.2, we can illustrate the effect of a phase error in an AM system with a simple example.

Example. Let

$$r(t) = \sqrt{2P}\, a(t) \cos \omega_c t + n(t). \tag{509}$$

Assume that we are using the detector of Fig. 6.58 or 6.60. Instead of multiplying by exactly $c(t)$, however, we multiply by $\sqrt{2P} \cos (\omega_c t + \phi)$, where ϕ is phase angle that is a random variable governed by some probability density $p_\phi(\phi)$. We assume that ϕ is independent of $n(t)$.

It follows directly that for a given value of ϕ the effect of an imperfect phase reference is equivalent to a signal power reduction:

$$P_{\text{ef}} = P \cos^2 \phi. \tag{510}$$

We can then find an expression for the mean-square error (either realizable or nonrealizable) for the reduced power signal and average the result over $p_\phi(\phi)$. The calculations are conceptually straightforward but tedious (see Problems 6.4.8 and 6.4.9).

We can see intuitively that if ϕ is almost always small (say $|\phi| < 15°$) the effect will be negligible. In this case our model which assumes $c(t)$ is known exactly is a good approximation to the actual physical situation, and the results obtained from this model will accurately predict the performance of the actual system. A number of related questions arise:

1. Can we reconstruct the carrier without devoting any power to a pilot tone? This question is discussed by Costas [50]. We discuss it in the problem section of Chapter II.2.
2. If there is a random error in estimating $c(t)$, is the receiver structure of Fig. 6.58 or 6.60 optimum? The answer in general is "no". Fortunately, it is not too far from optimum in many practical cases.
3. Can we construct an optimum estimation theory that leads to practical receivers for the general case of a *random* modulation matrix, that is,

$$\mathbf{r}(t) = \mathbf{C}(t)\, \mathbf{x}(t) + \mathbf{n}(t), \tag{511}$$

where $\mathbf{C}(t)$ is random? We find that the practicality depends on the statistics of $C(t)$. (It turns out that the easiest place to answer this question is in the problem section of Chapter II.3.)

4. If synchronous detection is optimum, why is it not used more often? Here, the answer is complexity. In Problem 6.4.10 we compute the performance of a DSB residual-carrier system when a simple detector is used. For high-input SNR the degradation is minor. Thus, whenever we have a single transmitter and many receivers (e.g., commercial broadcasting), it is far easier to increase transmitter power than receiver complexity. In military and space applications, however, it is frequently easier to increase receiver complexity than transmitter power.

This completes our discussion of linear modulation systems. We now comment briefly on some of the results obtained in this chapter.

6.5 THE FUNDAMENTAL ROLE OF THE OPTIMUM LINEAR FILTER

Because we have already summarized the results of the various sections in detail it is not necessary to repeat the comments. Instead, we shall discuss briefly three distinct areas in which the techniques developed in this chapter are important.

Linear Systems. We introduced the topic by demonstrating that for linear modulation systems the MAP interval estimate of the message was obtained by processing $r(t)$ with a linear system. To consider point estimates we resorted to an approach that we had not used before. We required that the processor be a linear system and found the best possible linear system. We saw that if we constrained the structure to be linear then only the second moments of the processes were relevant. This is an example of the type mentioned in Chapter 1 in which a partial characterization is adequate because we employed a structure-oriented approach. We then completed our development by showing that a linear system was the best possible processor whenever the Gaussian assumption was valid. Thus all of our results in this chapter play a double role. They are the best processors under the Gaussian assumptions for the classes of criteria assumed and they are the best *linear* processors for any random process.

The techniques of this chapter play a fundamental role in two other areas.

Nonlinear Systems. In Chapter II.2 we shall develop optimum receivers for nonlinear modulation systems. As we would expect, these receivers are nonlinear systems. We shall find that the optimum linear filters we have derived in this chapter appear as components of the over-all nonlinear system. We shall also see that the model of the system with respect to its effect on the message is linear in many cases. In these cases the results in this chapter will be directly applicable. Finally, as we showed in Chapter 5 the demodulation error in a nonlinear system can be bounded by the error in some related linear system.

Detection of Random Processes. In Chapter II.3 were turn to the detection and estimation problem in the context of a more general model. We shall find that the linear filters we have discussed are components of the optimum detector (or estimator).

We shall demonstrate why the presence of an optimum linear filter should be expected in these two areas. When our study is completed the fundamental importance of optimum linear filters in many diverse contexts will be clear.

6.6 COMMENTS

It is worthwhile to comment on some related issues.

1. In Section 6.2.4 we saw that for stationary processes in white noise the realizable mean-square error was related to Shannon's mutual information. For the nonstationary, finite-interval case a similar relation may also be derived

$$\xi_P = 2\,\frac{N_0}{2}\,\frac{\partial I(T{:}r(t), a(t))}{\partial T}. \tag{512}$$

2. The discussion with respect to state-variable filters considered only the continuous-time case. We can easily modify the approach to include discrete-time systems. (The discrete system results were derived in Problem 2.6.15 of Chapter 2 by using a sequential estimation approach.)
3. Occasionally a problem is presented in which the input has a transient nonrandom component and a stationary random component. We may want to minimize the mean-square error caused by the random input while constraining the squared error due to transient component. This is a straightforward modification of the techniques discussed (e.g., [51]).
4. In Chapter 3 we discussed the eigenfunctions and eigenvalues of the integral equation,

$$\lambda\,\phi(t) = \int_{T_i}^{T_f} K_y(t, u)\,\phi(u)\,du, \qquad T_i \le t \le T_f. \tag{513}$$

For rational spectra we obtained solutions by finding the associated differential equation, solving it, and using the integral equation to evaluate the boundary conditions. From our discussion in Section 6.3 we anticipate that a computationally more efficient method could be found by using state-variable techniques. These techniques are developed in [52] and [54] (see also Problems 6.6.1–6.6.4) The specific results developed are:

(*a*) A solution technique for homogeneous Fredholm equations using state-variable methods. This technique enables us to find the eigenvalues and eigenfunctions of scalar and vector random processes in an efficient manner.

(*b*) A solution technique for nonhomogeneous Fredholm equations using state-variable methods. This technique enables us to find the function $g(t)$ that appears in optimum detector for the colored noise problem. It is also the key to the optimum signal design problem.

(*c*) A solution of the optimum unrealizable filter problem using state-variable techniques. This enables us to achieve the best possible performance using a given amount of input data.

The importance of these results should not be underestimated because they lead to solutions that can be evaluated easily with numerical technniques. We develop these techniques in greater detail in Part II and use them to solve various problems.

5. In Chapter 4 we discussed whitening filters for the problem of detecting signals in colored noise. In the initial discussion we did not require realizability. When we examined the infinite interval stationary process case (p. 312), we determined that a realizable filter could be found and one component interpreted as an optimum realizable estimate of the colored noise. A similar result can be derived for the finite interval nonstationary case (see Problem 6.6.5). This enables us to use state-variable techniques to find the whitening filter. This result will also be valuable in Chapter II.3.

6.7 PROBLEMS

P6.1 Properties of Linear Processors

Problem 6.1.1. Let

$$r(t) = a(t) + n(t), \qquad T_i \le t \le T_f,$$

where $a(t)$ and $n(t)$ are uncorrelated Gaussian zero-mean processes with covariance functions $K_a(t, u)$ and $K_n(t, u)$, respectively. Find $p_{a(t_1)|r(t):T_i \le t \le T_f}(A|r(t):T_i \le t \le T_f)$.

Problem 6.1.2. Consider the model in Fig. 6.3.

1. Derive Property 3V (51).
2. Specialize (51) to the case in which $\mathbf{d}(t) = \mathbf{x}(t)$.

Problem 6.1.3.

Consider the vector model in Fig. 6.3.

Prove that

$$\mathbf{h}_o(t, t)\,\mathbf{R}(t) = \boldsymbol{\xi}_P(t)\,\mathbf{C}^T(t).$$

Comment. Problems 6.1.4 to 6.1.9 illustrate cases in which the observation is a finite set of random variables. In addition, the observation noise is zero. They illustrate the simplicity that (29) leads to in linear estimation problems.

Problem 6.1.4. Consider a simple prediction problem. We observe $a(t)$ at a *single time*. The desired signal is

$$d(t) = a(t + \alpha),$$

where α is a positive constant. Assume that

$$E[a(t)] = 0,$$

$$E[a(t)\, a(u)] = K_a(t - u) \triangleq K_a(\tau).$$

1. Find the best linear MMSE estimate of $d(t)$.
2. What is the mean-square error?
3. Specialize to the case $K_a(\tau) = e^{-k|\tau|}$.
4. Show that, for the correlation function in part 3, the MMSE estimate would not change if the entire past were available.
5. Is this true for any other correlation function? Justify your answer.

Problem 6.1.5. Consider the following interpolation problem. You are given the values $a(0)$ and $a(T)$:

$$E[a(t)] = 0, \qquad -\infty < t < \infty,$$

$$E[a(t)a(u)] = K_a(t - u), \qquad -\infty < t, u < \infty.$$

1. Find the MMSE estimate of $a(t)$.
2. What is the resulting mean-square error?
3. Evaluate for $t = T/2$.
4. Consider the special case, $K_a(\tau) = e^{-k|\tau|}$, and evaluate the processor constants

Problem 6.1.6 [55]. We observe $a(t)$ and $\dot{a}(t)$. Let $d(t) = a(t + \alpha)$, where α is a positive constant.

1. Find the MMSE linear estimate of $d(t)$.
2. State the conditions on $K_a(\tau)$ for your answer to be meaningful.
3. Check for small α.

Problem 6.1.7 [55]. We observe $a(0)$ and $a(t)$. Let

$$d(t) = \int_0^t a(u)\, du.$$

1. Find the MMSE linear estimate of $d(t)$.
2. Check your result for $t \ll 1$.

Problem 6.1.8. Generalize the preceding model to $n + 1$ observations; $a(0)$, $a(t)$. $a(2t) \cdots a(nt)$.

$$d(t) = \int_0^{nt} a(u)\, du.$$

1. Find the equations which specify the optimum linear processor.
2. Find an explicit solution for $nt \ll 1$.

Problem 6.1.9. [55]. We want to reconstruct $a(t)$ from an infinite number of samples; $a(nT)$, $n = \cdots -1, 0, +1, \ldots$, using a MMSE linear estimate:

$$\hat{a}(t) = \sum_{n=-\infty}^{\infty} c_n(t)\, a(nT).$$

1. Find an expression that the coefficients $c_n(t)$ must satisfy.
2. Consider the special case in which

$$S_a(\omega) = 0 \qquad |\omega| > \frac{\pi}{T}.$$

Evaluate the coefficients.

3. Prove that the resulting mean-square error is zero. (Observe that this proves the sampling theorem for random processes.)

Problem 6.1.10. In (29) we saw that

$$E[e_o(t)\, r(u)] = 0, \qquad T_i < u < T_f.$$

1. In our derivation we assumed $h_o(t, u)$ was continuous and defined $h_o(t, T_i)$ and $h_o(t, T_f)$ by the continuity requirement. Assume $r(u)$ contains a white noise component. Prove

$$E[e_o(t) r(T_i)] \neq 0,$$
$$E[e_o(t) r(T_f)] \neq 0.$$

2. Now remove the continuity assumption on $h_o(t, u)$ and assume $r(u)$ contains a white noise component. Find an equation specifying an $h_o(t, u)$, such that

$$E[e_o(t) r(u)] = 0, \qquad T_i \leq u \leq T_f.$$

Are the mean-square errors for the filters in parts 1 and 2 the same? Why?

3. Discuss the implications of removing the white noise component from $r(u)$. Will $h_o(t, u)$ be continuous? Do we use strict or nonstrict inequalities in the integral equation?

P6.2 Stationary Processes, Infinite Past, (Wiener Filters)

Realizable and Unrealizable Filtering

Problem 6.2.1. We have restricted our attention to rational spectra. We write the spectrum as

$$S_r(\omega) = c\,\frac{(\omega - n_1)(\omega - n_2)\cdots(\omega - n_N)}{(\omega - d_1)(\omega - d_2)\cdots(\omega - d_M)}, \qquad n_i \neq d_j,$$

where N and M are even. We assume that $S_r(\omega)$ is integrable on the real line. Prove the following statements:

1. $S_r(\omega) = S_r^*(\omega)$.
2. c is real.
3. All n_i's and d_i's with nonzero imaginary parts occur in conjugate pairs.
4. $S_r(\omega) \geq 0$.
5. Any real roots of numerator occur with even multiplicity.
6. No root of the denominator can be real.
7. $N < M$.

Verify that these results imply all the properties indicated in Fig. 6.7.

Problem 6.2.2. Let

$$r(u) = a(u) + n(u), \qquad -\infty < u \leq t.$$

The waveforms $a(u)$ and $n(u)$ are sample functions from uncorrelated zero-mean processes with spectra

$$S_a(\omega) = \frac{2k\sigma_a^2}{\omega^2 + k^2}$$

and

$$S_n(\omega) = N_2\omega^2,$$

respectively.

1. The desired signal is $a(t)$. Find the realizable linear filter which minimizes the mean-square error.
2. What is the resulting mean-square error?
3. Repeat parts 1 and 2 for the case in which the filter may be unrealizable and compare the resulting mean-square errors.

Problem 6.2.3. Consider the model in Problem 6.2.2. Assume that

$$S_n(\omega) = N_0 + N_2\omega^2.$$

1. Repeat Problem 6.2.2.
2. Verify that your answers reduce to those in Problem 6.2.2 when $N_0 = 0$ and to those in the text when $N_2 = 0$.

Problem 6.2.4. Let

$$r(u) = a(u) + n(u), \qquad -\infty < u \leq t.$$

The functions $a(u)$ and $n(u)$ are sample functions from independent zero-mean Gaussian random processes.

$$S_a(\omega) = \frac{2k\sigma_a^2}{\omega^2 + k^2},$$

$$S_n(\omega) = \frac{2c\sigma_n^2}{\omega^2 + c^2}.$$

We want to find the MMSE point estimate of $a(t)$.

1. Set up an expression for the optimum processor.
2. Find an explicit expression for the special case

$$\sigma_n^2 = \sigma_a^2,$$
$$c = 2k.$$

3. Look at your answer in (2) and check to see if it is intuitively correct.

Problem 6.2.5. Consider the model in Problem 6.2.4. Now let

$$S_n(\omega) = \frac{N_0}{2} + \frac{2c\sigma_n^2}{\omega^2 + c^2}.$$

1. Find the optimum realizable linear filter (MMSE).
2. Find an expression for ξ_{Pn}.
3. Verify that the result in (1) reduces to the result in Problem 6.2.4 when $N_0 = 0$ and to the result in the text when $\sigma_n^2 = 0$.

Problem 6.2.6. Let

$$r(u) = a(u) + w(u), \qquad -\infty < u \leq t.$$

The processes are uncorrelated with spectra

$$S_a(\omega) = \frac{2\sqrt{2}\, P/k}{1 + (\omega^2/k^2)^2}$$

and

$$S_w(\omega) = \frac{N_0}{2}.$$

The desired signal is $a(t)$. Find the optimum realizable linear filter (MMSE).

Problem 6.2.7. The message $a(t)$ is passed through a linear network before transmission as shown in Fig. P6.1. The output $y(t)$ is corrupted by uncorrelated white noise ($N_0/2$). The message spectrum is $S_a(\omega)$.

$$S_a(\omega) = \frac{2k\sigma_a^2}{\omega^2 + k^2}.$$

1. A minimum mean-square error realizable estimate of $a(t)$ is desired. Find the optimum linear filter.
2. Find ξ_{Pn} as a function of α and $\Lambda \triangleq 4\sigma_a^2/kN_0$.
3. Find the value of α that minimizes ξ_{Pn}.
4. How do the results change if the zero in the prefilter is at $+k$ instead of $-k$.

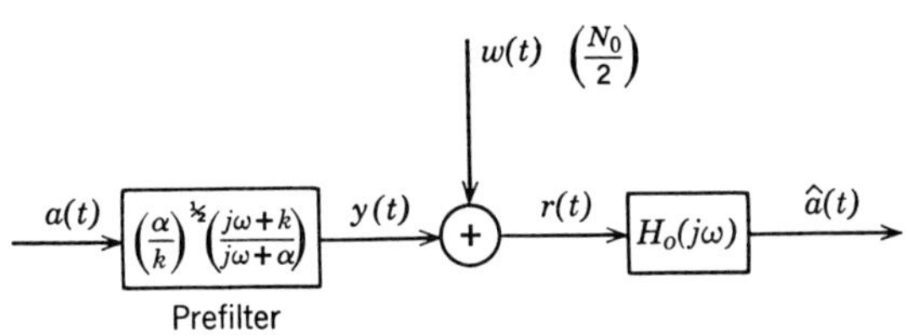

Fig. P6.1

Pure Prediction. The next four problems deal with pure prediction. The model is

$$r(u) = a(u), \qquad -\infty < u \leq t,$$

and

$$d(t) = a(t + \alpha),$$

where $\alpha \geq 0$. We see that there is no noise in the received waveform. The object is to *predict* $a(t)$.

Problem 6.2.8. Let

$$S_a(\omega) = \frac{2k}{\omega^2 + k^2}.$$

1. Find the optimum (MMSE) realizable filter.
2. Find the normalized prediction error ξ_{Pn}^{α}.

Problem 6.2.9. Let

$$S_a(\omega) = \frac{1}{(1 + \omega^2)^2}.$$

Repeat Problem 6.2.8.

Problem 6.2.10. Let

$$S_a(\omega) = \frac{1 + \omega^2}{1 + \omega^4}.$$

Repeat Problem 6.2.8.

Problem 6.2.11.

1. The received signal is $a(u)$, $-\infty < u \leq t$. The desired signal is

$$d(t) = a(t + \alpha), \qquad \alpha > 0.$$

Find $H_o(j\omega)$ to minimize the mean-square error

$$E\{[\hat{d}(t) - d(t)]^2\},$$

where

$$d(t) = \int_{-\infty}^{t} h_o(t - u)\, a(u)\, du.$$

The spectrum of $a(t)$ is

$$S_a(\omega) = \prod_{i=1}^{n} \frac{A^2}{(\omega^2 + k_i^2)},$$

where $k_i \neq k_j$; $i \neq j$ for $i = 1, \ldots, n$, $j = 1, \ldots, n$.

2. Now assume that the received signal is $a(u)$, $T_i \leq u \leq t$, where T_i is a *finite* number. Find $h_o(t, \tau)$ to minimize the mean-square error.

$$\hat{d}(t) = \int_{T_i}^{t} h_o(t, u)\, a(u)\, du.$$

3. Do your answers to parts 1 and 2 enable you to make any general statements about pure prediction problems in which the message spectrum has no zeros?

Problem 6.2.12. The message is generated as shown in Fig. P6.2, where $u(t)$ is a white noise process (unity spectral height) and α_i, $i = 1, 2$, and λ_i, $i = 1, 2$, are known positive constants. The additive white noise $w(t)(N_0/2)$ is uncorrelated with $u(t)$.

1. Find an expression for the linear filters whose outputs are the MMSE realizable estimates of $x_i(t)$, $i = 1, 2$.
2. Prove that

$$\hat{a}(t) = \sum_{i=1}^{2} \hat{x}_i(t).$$

3. Assume that

$$d(t) = \sum_{i=1}^{2} d_i\, x_i(t).$$

Prove that

$$\hat{d}(t) = \sum_{i=1}^{2} d_i\, \hat{x}_i(t).$$

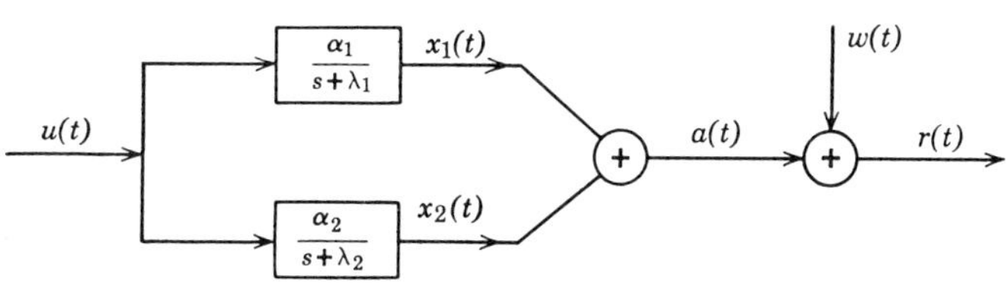

Fig. P6.2

Problem 6.2.13. Let

$$r(u) = a(u) + n(u), \qquad -\infty < u \leq t,$$

where $a(u)$ and $n(u)$ are uncorrelated random processes with spectra

$$S_a(\omega) = \frac{\omega^2}{\omega^4 + 1},$$

$$S_n(\omega) = \frac{1}{\omega^2 + \epsilon^2}.$$

The desired signal is $a(t)$. Find the optimum (MMSE) linear filter and the resulting error for the limiting case in which $\epsilon \to 0$. Sketch the magnitude and phase of $H_o(j\omega)$.

Problem 6.2.14. The received waveform $r(u)$ is

$$r(u) = a(u) + w(u), \qquad -\infty < u \leq t,$$

where $a(u)$ and $w(u)$ are uncorrelated random processes with spectra

$$S_a(\omega) = \frac{2k\sigma_a^2}{\omega^2 + k^2},$$

$$S_n(\omega) = \frac{N_0}{2}.$$

Let

$$d(t) \triangleq \int_t^{t+\alpha} a(u)\, du, \qquad \alpha > 0.$$

1. Find the optimum (MMSE) linear filter for estimating $d(t)$.
2. Find ξ_P^α.

Problem 6.2.15 (continuation). Consider the same model as Problem 6.2.14. Repeat that problem for the following desired signals:

1. $d(t) = \dfrac{1}{\alpha} \displaystyle\int_{t-\alpha}^{t} a(u)\, du, \qquad \alpha > 0.$
2. $d(t) = \dfrac{1}{\beta - \alpha} \displaystyle\int_{t+\alpha}^{t+\beta} a(u)\, du, \qquad \alpha > 0, \beta > 0, \beta \geq \alpha.$

What happens as $(\beta - \alpha) \to 0$?

3. $d(t) = \displaystyle\sum_{n=-1}^{+1} k_n\, a(t - n\alpha), \qquad \alpha > 0.$

Problem 6.2.16. Consider the model in Fig. P6.3. The function $u(t)$ is a sample function from a white process (unity spectral height). Find the MMSE realizable linear estimates, $\hat{x}_1(t)$ and $\hat{x}_2(t)$. Compute the mean-square errors and the cross correlation between the errors ($T_i = -\infty$).

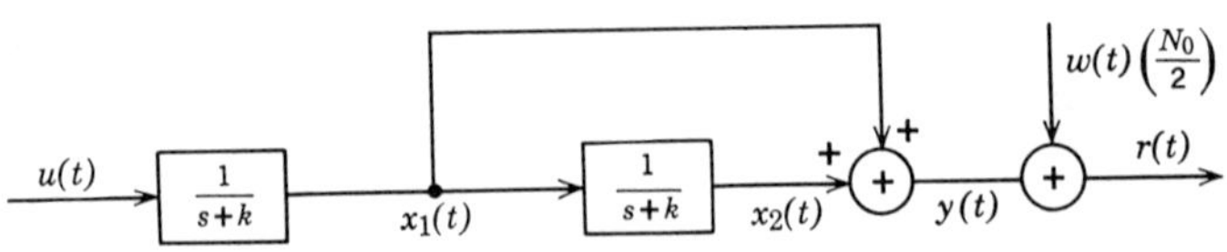

Fig. P6.3

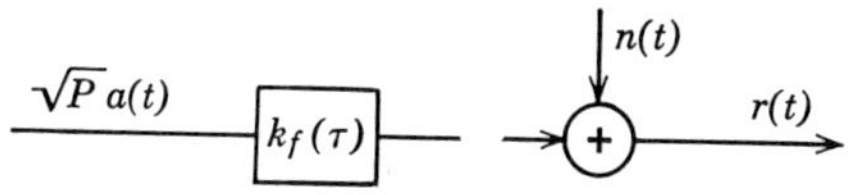

Fig. P6.4

Problem 6.2.17. Consider the communication problem in Fig. P6.4. The message $a(t)$ is a sample function from a stationary, zero-mean Gaussian process with unity variance. The channel $k_f(\tau)$ is a linear, time-invariant, not necessarily realizable system. The additive noise $n(t)$ is a sample function from a zero-mean white Gaussian process $(N_0/2)$.

1. We process $r(t)$ with the optimum unrealizable linear filter to find $\hat{a}(t)$. Assuming $\int_{-\infty}^{\infty} |K_f(j\omega)|^2(d\omega/2\pi) = 1$, find the $k_f(\tau)$ that minimizes the minimum mean-square error.
2. Sketch for

$$S_a(\omega) = \frac{2k}{\omega^2 + k^2}.$$

Closed Form Error Expressions

Problem 6.2.18. We want to integrate

$$\xi_P = \frac{N_0}{2}\int_{-\infty}^{\infty} \frac{d\omega}{2\pi} \ln\left[1 + \frac{2c_n/N_0}{1 + (\omega/k)^{2n}}\right].$$

1. Do this by letting $y = 2c_n/N_0$. Differentiate with respect to y and then integrate with respect to ω. Integrate the result from 0 to y.
2. Discuss the conditions under which this technique is valid.

Problem 6.2.19. Evaluate

$$\xi_u = \int_{-\infty}^{\infty} \frac{d\omega}{2\pi} \frac{c_n}{1 + (\omega/k)^{2n} + (2/N_0)c_n}.$$

Comment. In the next seven problems we develop closed-form error expressions for some interesting cases. In most of these problems the solutions are difficult. In all problems

$$r(u) = a(u) + n(u), \qquad -\infty < u \le t,$$

where $a(u)$ and $n(u)$ are uncorrelated. The desired signal is $a(t)$ and optimum (MMSE) linear filtering is used. The optimum realizable linear filter is $H_o(j\omega)$ and

$$G_o(j\omega) \triangleq 1 - H_o(j\omega).$$

Most of the results were obtained in [4].

Problem 6.2.20. Let

$$S_n(\omega) = \frac{N_0 a^2}{\omega^2 + a^2}.$$

Show that

$$H_o(j\omega) = 1 - k\,\frac{[S_n(\omega)]^+}{[S_a(\omega) + S_n(\omega)]^+},$$

where

$$k = \exp\left[\frac{2}{N_0 a}\int_0^\infty S_n(\omega)\ln\frac{S_n(\omega)}{S_a(\omega) + S_n(\omega)}\frac{d\omega}{2\pi}\right].$$

Problem 6.2.21. Show that if $\lim_{\omega\to\infty} S_n(\omega) \to 0$ then

$$\xi_P = 2\int_0^\infty \{S_n(\omega) - |G_o(j\omega)|^2[S_a(\omega) + S_n(\omega)]\}\frac{d\omega}{2\pi}.$$

Use this result and that of the preceding problem to show that for one-pole noise

$$\xi_P = \frac{N_0 a}{2}(1 - k^2).$$

Problem 6.2.22. Consider the case

$$S_n(\omega) = N_0 + N_2\omega^2 + N_4\omega^4.$$

Show that

$$|G_o(j\omega)|^2 = \frac{S_n(\omega) + K}{S_n(\omega) + S_a(\omega)},$$

where

$$\int_0^\infty \ln\left[\frac{S_n(\omega) + K}{S_n(\omega) + S_a(\omega)}\right] d\omega = 0$$

determines K.

Problem 6.2.23. Show that when $S_n(\omega)$ is a polynomial

$$\xi_P = -\frac{1}{\pi}\int_0^\infty d\omega\{S_n(\omega) - |G_o(j\omega)|^2[S_a(\omega) + S_n(\omega)] + S_n(\omega)\ln|G_o(j\omega)|^2\}.$$

Problem 6.2.24. As pointed out in the text, we can double the size of the class of problems for which these results apply by a simple observation. Figure P6.5*a* represents a typical system in which the message is filtered before transmission.

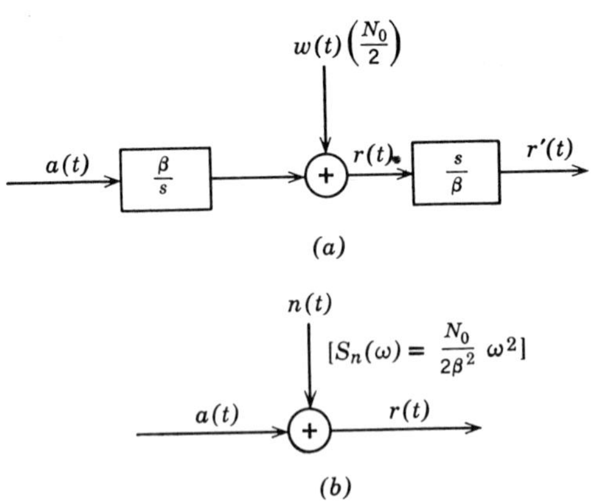

Fig. P6.5

Clearly the mean-square error in this system is identical to the error in the system in Fig. P6.5*b*. Using Problem 6.2.23, verify that

$$\xi_P = -\frac{1}{\pi}\int_0^\infty \left[\frac{N_0}{2\beta^2}\,\omega^2 - \left[\frac{N_0}{2\beta^2}\,\omega^2 + K\right] + \frac{N_0\omega^2}{2\beta^2}\ln|G_o(j\omega)|^2\right]d\omega$$

$$= \frac{1}{\pi}\int_0^\infty \left[K - \frac{N_0\omega^2}{2\beta^2}\ln\left(\frac{S_n(\omega) + K}{S_a(\omega) + S_n(\omega)}\right)\right]d\omega.$$

Problem 6.2.25 (continuation) [39]. Using the model of Problem 6.2.24, show that

$$\xi_P = \frac{N_0}{6}\,[f(0)]^3 + F(0),$$

where

$$f(0) = \int_{-\infty}^{\infty} \ln\left[1 + \frac{2S_a(\omega)}{\beta^2\omega^2 N_0}\right]\frac{d\omega}{2\pi}$$

and

$$F(0) = \int_{-\infty}^{\infty} \omega^2\,\frac{N_0}{2\beta^2}\ln\left[1 + \frac{2S_a(\omega)}{\beta^2\omega^2 N_0}\right]\frac{d\omega}{2\pi}.$$

Problem 6.2.26 [20]. Extend the results in Problem 6.2.20 to the case

$$S_n(\omega) = \frac{N_0}{2} + \frac{N_1 a^2}{\omega^2 + a^2}$$

to find $|G_o(j\omega)|^2$ and ξ_P.

FEEDBACK REALIZATIONS

Problem 6.2.27. Verify that the optimum loop filter is of the form indicated in Fig. 6.21*b*. Denote the numerator by $F(s)$.

1. Show that

$$F(s) = \left[\frac{2q}{N_0}\,B(s)\,B(-s) + P(s)\,P(-s)\right]^+ - P(s),$$

where $B(s)$ and $P(s)$ are defined in Fig. 6.21*a*.

2. Show that $F(s)$ is exactly one degree less than $P(s)$.

Problem 6.2.28. Prove

$$\xi_P = \frac{N_0}{2}\lim_{s\to\infty} s\,G_{lo}(s) = \frac{N_0}{2}\,f_{n-1}.$$

where $G_{lo}(s)$ is the optimum loop filter and f_{n-1} is defined in Fig. 6.21*b*.

Problem 6.2.29. In this problem we construct a realizable whitening filter. In Chapter 4 we saw that a conceptual unrealizable whitening filter may readily be obtained in terms of a Karhunen-Loève expansion. Let

$$r(u) = n_c(u) + w(u), \qquad -\infty < u \le t,$$

where $n_c(u)$ has a rational spectrum and $w(u)$ is an uncorrelated white noise process. Denote the optimum (MMSE) realizable linear filter for estimating $n_c(t)$ as $H_o(j\omega)$.

1. Prove that $1 - H_o(j\omega)$ is a realizable whitening filter. Draw a feedback realization of the whitening filter.

Hint. Recall the feedback structure of the optimum filter (173).

2. Find the inverse filter for $1 - H_o(j\omega)$. Draw a feedback realization of the inverse filter.

Problem 6.2.30 (continuation). What are the necessary and sufficient conditions for the inverse of the whitening filter found in Problem 6.2.29 to be stable?

GENERALIZATIONS

Problem 6.2.31. Consider the simple unrealizable filter problem in which

$$r(u) = a(u) + n(u), \qquad -\infty < u < \infty$$

and

$$d(t) = a(t).$$

Assume that we design the optimum unrealizable filter $H_{ou}(j\omega)$ using the spectrum $S_a(\omega)$ and $S_n(\omega)$. In practice the noise spectrum is

$$S_{np}(\omega) = S_{nd}(\omega) + S_{ne}(\omega).$$

1. Show that the mean-square error using $H_{ou}(j\omega)$ is

$$\xi_{up} = \xi_{uo} + \int_{-\infty}^{\infty} |H_{ou}(j\omega)|^2 \, S_{ne}(\omega) \frac{d\omega}{2\pi},$$

where *up* denotes unrealizable mean-square error in practice and *uo* denotes unrealizable mean-square error in the optimum filter when the design assumptions are exact.

2. Show that the change in error is

$$\Delta\xi_u = \int_{-\infty}^{\infty} \left[\frac{S_a(\omega)}{S_a(\omega) + S_{nd}(\omega)}\right]^2 S_{ne}(\omega) \frac{d\omega}{2\pi}.$$

3. Consider the case

$$S_{nd}(\omega) = \frac{N_0}{2},$$

$$S_{ne}(\omega) = \epsilon \frac{N_0}{2}.$$

The message spectrum is flat and bandlimited. Show that

$$\Delta\xi_u = \frac{\epsilon\Lambda}{(1 + \Lambda)^2},$$

where Λ is the signal-to-noise ratio in the message bandwidth.

Problem 6.2.32. Derive an expression for the change in the mean-square error in an optimum unrealizable filter when the actual message spectrum is different from the design message spectrum.

Problem 6.2.33. Repeat Problem 6.2.32 for an optimum realizable filter and white noise.

Problem 6.2.34. Prove that the system in Figs. 6.23*b* is the optimum realizable filter for estimating $a(t)$.

Problem 6.2.35. Derive (181) and (183) for arbitrary $K_d(j\omega)$.

Problem 6.2.36. The mean-square error using an optimum unrealizable filter is given by (183):

$$\xi_{uo} = \int_{-\infty}^{\infty} \frac{S_a(\omega)\, S_n(\omega)}{S_a(\omega)\, S_f(\omega) + S_n(\omega)} \frac{d\omega}{2\pi},$$

where $S_f(\omega) \triangleq |K_f(j\omega)|^2$.

1. Consider the following problem. Constrain the transmitted power

$$P = \int_{-\infty}^{\infty} S_a(\omega)\, S_f(\omega) \frac{d\omega}{2\pi}.$$

Find an expression for $S_f(\omega)$ that minimizes the mean-square error.
2. Evaluate the resulting mean-square error.

Problem 6.2.37. Let

$$r(u) = a(u) + n(u), \qquad -\infty < u \le t,$$

where $a(u)$ and $n(u)$ are uncorrelated. Let

$$S_a(\omega) = \frac{1}{1 + \omega^2}, \qquad S_n(\omega) = \epsilon^2.$$

The desired signal is $d(t) = (d/dt)\, a(t)$.

1. Find $H_o(j\omega)$.
2. Discuss the behavior of $H_o(j\omega)$ and ξ_P as $\epsilon \to 0$. Why is the answer misleading?

Problem 6.2.38. Repeat Problem 6.2.37 for the case

$$S_a(\omega) = \frac{1}{1 + \omega^4}, \qquad S_n(\omega) = \epsilon^4.$$

What is the important difference between the message random processes in the two problems? Verify that differentiation and optimum realizable filtering do not commute.

Problem 6.2.39. Let

$$a(u) = \cos(2\pi f u + \phi), \qquad -\infty < u \le t,$$

where ϕ and f are independent variables:

$$p_\phi(\phi) = \frac{1}{2\pi}, \qquad 0 \le \phi \le 2\pi$$

and

$$p_f(X) = 0, \qquad X \le 0.$$

1. Describe the resulting ensemble.
2. Prove that $S_a(f) = p_f(|f|)/4$.
3. Choose a $p_f(X)$ to make $a(t)$ a deterministic process (see p. 512). Demonstrate a linear predictor whose mean-square error is zero.
4. Choose a $p_f(X)$ to make $a(t)$ a nondeterministic process. Show that you can predict $a(t)$ with zero mean-square error by using a nonlinear predictor.

Problem 6.2.40. Let

$$g^+(\tau) = \mathcal{F}^{-1}[G^+(j\omega)].$$

Prove that the MMSE error for *pure* prediction is

$$\xi_P^\alpha = \int_0^\alpha [g^+(\tau)]^2\, d\tau.$$

Problem 6.2.41 [1]. Consider the message spectrum

$$S_a(\omega) = \left[\left(1 + \frac{\omega^2}{n}\right)^n\right]^{-1}.$$

1. Show that

$$g^+(\tau) = \frac{\tau^{n-1}\exp(-\tau\sqrt{n})}{n^{-n/2}(n-1)!}.$$

2. Show that (for large n)

$$\xi_P^\alpha \simeq \int_0^\alpha \frac{1}{2\pi} \exp\left[-2\left(t - \frac{n-1}{\sqrt{n}}\right)^2\right] dt.$$

3. Use part 2 to show that for any ϵ^2 and α we can make

$$\xi_P^\alpha < \epsilon^2$$

by increasing n sufficiently. Explain why this result is true.

Problem 6.2.42. The message $a(t)$ is a zero-mean process observed in the absence of noise. The desired signal $d(t) = a(t + \alpha)$, $\alpha > 0$.

1. Assume

$$K_a(\tau) = \frac{1}{\tau^2 + k^2}.$$

Find $\hat{d}(t)$ by using $a(t)$ and its derivatives. What is the mean-square error for $\alpha < k$?

2. Assume

$$K_a(\tau) = e^{-k\tau^2}.$$

Show that

$$\hat{a}(t + \alpha) = \sum_{n=0}^{\infty} \left[\frac{d^n}{dt^n}\, a(t)\right] \frac{\alpha^n}{n!},$$

and that the mean-square error is zero for all α.

Problem 6.2.43. Consider a simple diversity system,

$$r_1(t) = a(t) + n_1(t),$$

$$r_2(t) = a(t) + n_2(t),$$

where $a(t)$, $n_1(t)$, and $n_2(t)$ are independent zero-mean, stationary Gaussian processes with finite variances. We wish to process $r_1(t)$ and $r_2(t)$, as shown in Fig. P6.6. The spectra $S_{n_1}(\omega)$ and $S_{n_2}(\omega)$ are known; $S_a(\omega)$, however, is *unknown*. We require that the message $a(t)$ be undistorted. In other words, if $n_1(t)$ and $n_2(t)$ are zero, the output will be exactly $a(t)$.

1. What condition does this impose on $H_1(j\omega)$ and $H_2(j\omega)$?
2. We want to choose $H_1(j\omega)$ to minimize $E[n_c^2(t)]$, subject to the constraint that

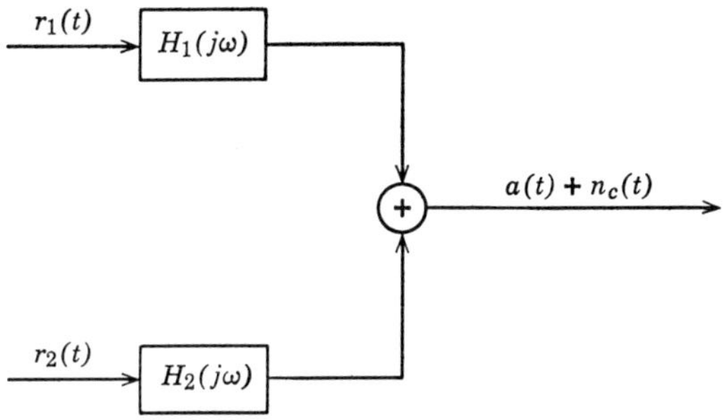

Fig. P6.6

$a(t)$ be reproduced exactly in the absence of input noise. The filters must be realizable and may operate on the infinite past. Find an expression for $H_{1_0}(j\omega)$ and $H_{2_0}(j\omega)$ in terms of the given quantities.

3. Prove that the $\hat{a}(t)$ obtained in part 2 is an unbiased, efficient estimate of the sample function $a(t)$. [Therefore $\hat{a}(t) = \hat{a}_{ml}(t)$.]

Problem 6.2.44. Generalize the result in Problem 6.2.43 to the n-input problem. Prove that any n-dimensional distortionless filter problem may be recast as an $(n-1)$-dimensional Wiener filter problem.

P6.3 Finite-time, Nonstationary Processes (Kalman-Bucy filters)

State-variable Representations

Problem 6.3.1. Consider the differential equation

$$y^{(n)}(t) + p_{n-1}y^{(n-1)}(t) + \cdots + p_0 y(t) = b_{n-1}u^{(n-1)}(t) + \cdots + b_0 u(t).$$

Extend Canonical Realization 1 on p. 522 to include this case. The desired $\mathbf{F}$ is

$$\mathbf{F} = \left[\begin{array}{c|cccc} 0 & 1 & & & 0 \\ \vdots & & & \ddots & \\ 0 & 0 & & & 1 \\ \hline -p_0 & -p_1 & & \cdots & -p_{n-1} \end{array}\right]$$

Draw an analog computer realization and find the $\mathbf{G}$ matrix.

Problem 6.3.2. Consider the differential equation in Problem 6.3.1. Derive Canonical Realization 3 (see p. 526) for the case of repeated roots.

Problem 6.3.3 [27]. Consider the differential equation

$$y^{(n)}(t) + p_{n-1}y^{(n-1)}(t) + \cdots + p_0 y(t) = b_{n-1}u^{(n-1)}(t) + \cdots + b_0 u(t).$$

1. Show that the system in Fig. P6.7 is a correct analog computer realization.
2. Write the vector differential equation that describes the system.

Problem 6.3.4. Draw an analog computer realization for the following systems:

$$\begin{aligned} 1. \quad \ddot{y}(t) + 3\dot{y}(t) + 4y(t) &= \dot{u}(t) + u(t), \\ 2. \quad \ddot{y}_1(t) + 3\dot{y}_1(t) + 2y_2(t) &= u_1(t) + 2\dot{u}_2(t) + 2u_2(t), \\ \ddot{y}_2(t) + 4\dot{y}_1(t) + 3y_2(t) &= 3u_2(t) + u_1(t). \end{aligned}$$

Write the associated vector differential equation.

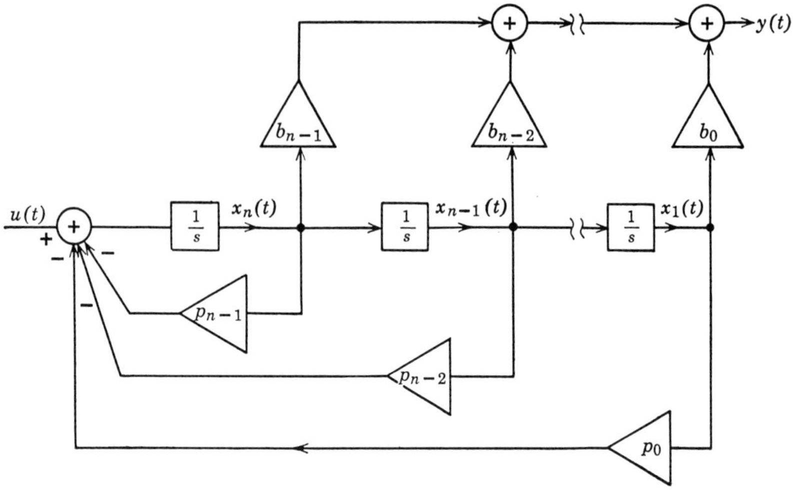

Fig. P6.7

Problem 6.3.5 [*27*]. Find the transfer function matrix and draw the transfer function diagram for the systems described below. Comment on the number of integrators required.

1. $\ddot{y}_1(t) + 3\dot{y}_1(t) + 2y_1(t) = \dot{u}_1(t) + 2u_1(t) + \dot{u}_2(t) + u_2(t),$
   $\dot{y}_2(t) + 2y_2(t) = -\dot{u}_1(t) - 2u_1(t) + u_2(t).$
2. $\dot{y}_1(t) + y_1(t) = u_1(t) + 2u_2(t),$
   $\ddot{y}_2(t) + 3\dot{y}_2(t) + 2y_2(t) = \dot{u}_2(t) + u_2(t) - u_1(t).$
3. $\ddot{y}_1(t) + 2\dot{y}_2(t) + y_1(t) = \dot{u}_1(t) + u_1(t) + u_2(t),$
   $\ddot{y}_2(t) + \dot{y}_1(t) + y_2(t) = u_2(t) + u_1(t).$
4. $\ddot{y}_1(t) + 3\dot{y}_1(t) + 2y_1(t) = 3\dot{u}_1(t) + 4\dot{u}_2(t) + 8u_2(t),$
   $\dot{y}_2(t) + 3y_2(t) - 4y_1(t) - \dot{y}_1(t) = \dot{u}_1(t) + 2\dot{u}_2(t) + 2u_2(t).$

Problem 6.3.6 [*27*]. Find the vector differential equations for the following systems, using the partial fraction technique.

1. $\ddot{y}(t) + 3\dot{y}(t) + 2y(t) = u(t),$
2. $\dddot{y}(t) + 4\ddot{y}(t) + 5\dot{y}(t) + 2y(t) = u(t),$
3. $\dddot{y}(t) + 4\ddot{y}(t) + 6\dot{y}(t) + 4y(t) = u(t),$
4. $\ddot{y}_1(t) - 10\dot{y}_2(t) + y_1(t) = u_1(t),$
   $\dot{y}_2(t) + 6y_2(t) = u_2(t).$

Problem 6.3.7. Compute $e^{\mathbf{F}t}$ for the following matrices:

1. $$\mathbf{F} = \begin{bmatrix} 1 & 1 \\ 1 & 1 \end{bmatrix}.$$

2. $$\mathbf{F} = \begin{bmatrix} 3 & 2 \\ -1 & 6 \end{bmatrix}.$$

3. $$\mathbf{F} = \begin{bmatrix} -2 & 5 \\ -4 & -3 \end{bmatrix}.$$

Problem 6.3.8. Compute $e^{\mathbf{F}t}$ for the following matrices:

1. $$\mathbf{F} = \begin{bmatrix} 0 & 1 & 0 \\ 0 & 0 & 1 \\ 0 & 0 & 0 \end{bmatrix}.$$

2. $$\mathbf{F} = \begin{bmatrix} 0 & 1 & 0 \\ 0 & 0 & 1 \\ 0 & 0 & -1 \end{bmatrix}.$$

3. $$\mathbf{F} = \begin{bmatrix} 0 & 1 & 0 \\ 0 & 0 & 1 \\ -6 & -11 & -6 \end{bmatrix}.$$

Problem 6.3.9. Given the system with state representation as follows,

$$\begin{aligned} \dot{\mathbf{x}}(t) &= \mathbf{F}\,\mathbf{x}(t) + \mathbf{G}\,u(t), \\ y(t) &= \mathbf{C}\,\mathbf{x}(t), \\ \mathbf{x}(0) &= \mathbf{0}. \end{aligned}$$

Let $U(s)$ and $Y(s)$ denote the Laplace transform of $u(t)$ and $y(t)$, respectively. We found that the transfer function was

$$\begin{aligned} H(s) = \frac{Y(s)}{U(s)} &= \mathbf{C}\,\mathbf{\Phi}(s)\mathbf{G} \\ &= \mathbf{C}(s\mathbf{I} - \mathbf{F})^{-1}\mathbf{G}. \end{aligned}$$

Show that the poles of $H(s)$ are the eigenvalues of the matrix $\mathbf{F}$.

Problem 6.3.10. Consider the circuit shown in Fig. P6.8. The source is turned on at $t = 0$. The current $i(0^-)$ and the voltage across the capacitor $v_c(0^-)$ are both zero. The observed quantity is the voltage across R.

1. Write the vector differential equations that describe the system and an equation that describes the observation process.
2. Draw an analog computer realization of the circuit.

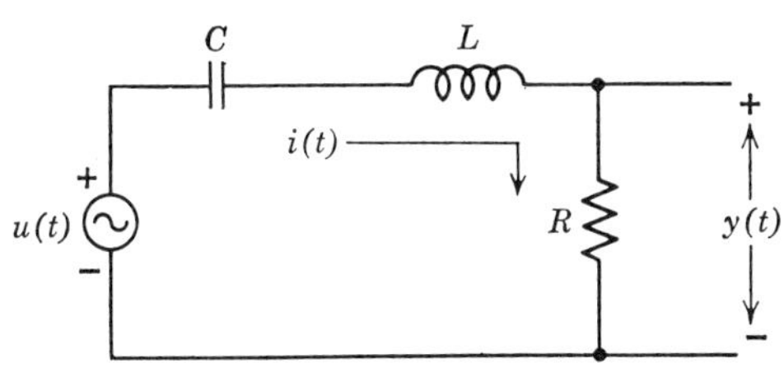

Fig. P6.8

Problem 6.3.11. Consider the control system shown in Fig. P6.9. The output of the system is $a(t)$. The two inputs, $b(t)$ and $n(t)$, are sample functions from zero-mean, uncorrelated, stationary random processes. Their spectra are

$$S_b(\omega) = \frac{2\sigma_b^2 k}{\omega^2 + k^2}$$

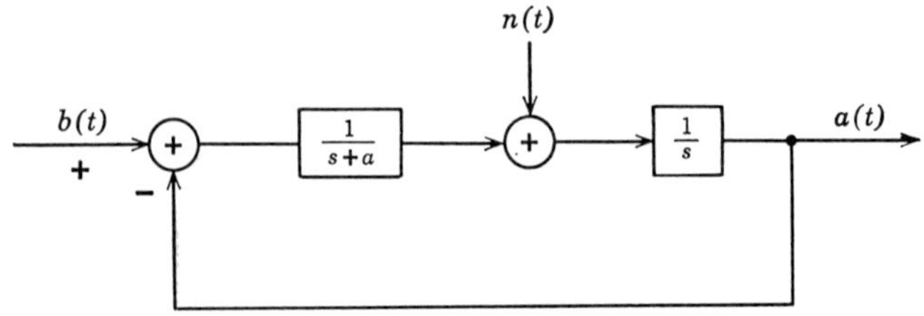

Fig. P6.9

and

$$S_n(\omega) = \frac{N_0}{2}.$$

Write the vector differential equation that describes a mathematically equivalent system whose input is a vector white noise $\mathbf{u}(t)$ and whose output is $a(t)$.

Problem 6.3.12. Consider the discrete multipath model shown in Fig. P6.10. The time delays are assumed known. The channel multipliers are independent, zero-mean processes with spectra

$$S_{b_j}(\omega) = \frac{2k_j\sigma_j^2}{\omega^2 + k_j^2}, \qquad \text{for } j = 1,2,3.$$

The additive white noise is uncorrelated and has spectral height $N_0/2$. The input signal $s(t)$ is a known waveform.

1. Write the state and observation equations for the process.
2. Indicate how this would be modified if the channel gains were correlated.

Problem 6.3.13. In the text we considered in detail state representations for time invariant systems.

Consider the time varying system

$$\ddot{y}(t) + p_1(t)\,\dot{y}(t) + p_0(t)\,y(t) = b_1(t)\,\dot{u}(t) + b_0(t)\,u(t).$$

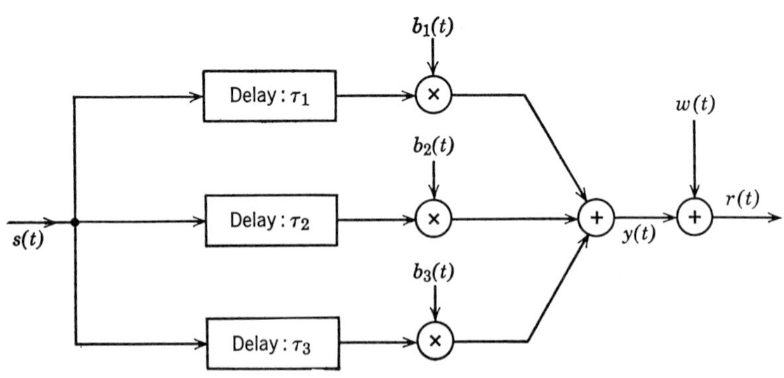

Fig. P6.10

Show that this system has a state representation of the same form as that in Example 2.

$$\frac{d}{dt}\mathbf{x}(t) = \begin{bmatrix} 0 & 1 \\ -p_0(t) & -p_1(t) \end{bmatrix} \mathbf{x}(t) + \begin{bmatrix} h_1(t) \\ h_2(t) \end{bmatrix} u(t),$$

$$y(t) = [1 \quad 0]\,\mathbf{x}(t) = x_1(t),$$

where $h_1(t)$ and $h_2(t)$ are functions that you must find.

Problem 6.3.14 [27]. Given the system defined by the time-varying differential equation

$$y^{(n)}(t) + \sum_{k=0}^{n-1} p_{n-k}(t)\, y^{(k)}(t) = \sum_{k=0}^{n} b_{n-k}(t)\, u^{(k)}(t),$$

Show that this system has the state equations

$$\begin{bmatrix} \dot{x}_1(t) \\ \dot{x}_2(t) \\ \cdot \\ \cdot \\ \cdot \\ \dot{x}_n(t) \end{bmatrix} = \begin{bmatrix} 0 & 1 & 0 & \cdots & 0 \\ 0 & 0 & 1 & \cdots & 0 \\ \cdot & \cdot & \cdot & \cdots & \cdot \\ \cdot & \cdot & \cdot & \cdots & \cdot \\ \cdot & \cdot & \cdot & \cdots & 1 \\ -p_n & -p_{n-1} & \cdot & \cdots & -p_1 \end{bmatrix} \begin{bmatrix} x_1(t) \\ x_2(t) \\ \cdot \\ \cdot \\ \cdot \\ x_n(t) \end{bmatrix} + \begin{bmatrix} g_1(t) \\ g_2(t) \\ \cdot \\ \cdot \\ \cdot \\ g_n(t) \end{bmatrix} u(t)$$

$$y(t) = x_1(t) + g_0(t)\, u(t),$$

where

$$g_0(t) = b_0(t),$$

$$g_i(t) = b_i(t) - \sum_{r=0}^{i-1} \sum_{m=0}^{i-r} \binom{n+m-i}{n-i} p_{i-r-m}(t)\, g_r^{(m)}(t).$$

Problem 6.3.15. Demonstrate that the following is a solution to (273).

$$\mathbf{\Lambda_x}(t) = \mathbf{\Phi}(t, t_0)\left[\mathbf{\Lambda_x}(t_0) + \int_{t_0}^{t} \mathbf{\Phi}(t_0, \tau)\mathbf{G}(\tau)\,\mathbf{Q}(\tau)\,\mathbf{G}^T(\tau)\mathbf{\Phi}^T(t_0, \tau)\, d\tau\right]\mathbf{\Phi}^T(t, t_0),$$

where $\mathbf{\Phi}(t, t_0)$ is the fundamental transition matrix; that is,

$$\frac{d}{dt}\mathbf{\Phi}(t, t_0) = \mathbf{F}(t)\,\mathbf{\Phi}(t, t_0),$$

$$\mathbf{\Phi}(t_0, t_0) = \mathbf{I}.$$

Demonstrate that this solution is unique.

Problem 6.3.16. Evaluate $\mathbf{K_y}(t, \tau)$ in terms of $\mathbf{K_x}(t, t)$ and $\mathbf{\Phi}(t, \tau)$, where

$$\begin{aligned} \dot{\mathbf{x}}(t) &= \mathbf{F}(t)\,\mathbf{x}(t) + \mathbf{G}(t)\,\mathbf{u}(t), \\ \mathbf{y}(t) &= \mathbf{C}(t)\,\mathbf{x}(t), \\ E[\mathbf{u}(t)\,\mathbf{u}^T(\tau)] &= \mathbf{Q}\,\delta(t-\tau). \end{aligned}$$

Problem 6.3.17. Consider the first-order system defined by

$$\frac{dx(t)}{dt} = -k(t)\,x(t) + g(t)\,u(t).$$

$$y(t) = x(t).$$

1. Determine a general expression for the transition matrix for this system.
2. What is $h(t, \tau)$ for this system?
3. Evaluate $h(t, \tau)$ for

$$k(t) = k(1 + m \sin(\omega_0 t)),$$
$$g(t) = 1.$$

4. Does this technique generalize to vector equations?

Problem 6.3.18. Show that for constant parameter systems the steady-state variance of the unobserved process is given by

$$\lim_{t\to\infty} \mathbf{K_x}(t, t) = \int_0^\infty e^{+\mathbf{F}\tau}\mathbf{G}\mathbf{Q}\mathbf{G}^T e^{\mathbf{F}^T\tau}\, d\tau,$$

where

$$\dot{\mathbf{x}}(t) = \mathbf{F}\mathbf{x}(t) + \mathbf{G}\mathbf{u}(t),$$
$$E[\mathbf{u}(t)\, \mathbf{u}^T(\tau)] = \mathbf{Q}\, \delta(t - \tau),$$

or, equivalently,

$$\lim_{t\to\infty} \mathbf{K_x}(t, t) = \frac{1}{2\pi j}\int_{-j\infty}^{j\infty} [s\mathbf{I} - \mathbf{F}]^{-1}\mathbf{G}\mathbf{Q}\mathbf{G}^T[-s\mathbf{I} - \mathbf{F}^T]^{-1}\, ds.$$

Problem 6.3.19. Prove that the condition in (317) is necessary when $\mathbf{R}(t)$ is positive-definite.

Problem 6.3.20. In this problem we incorporate the effect of nonzero means into our estimation procedure. The equations describing the model are (302)–(306).

1. Assume that $\mathbf{x}(T_i)$ is a Gaussian random vector

$$E[\mathbf{x}(T_i)] \triangleq \mathbf{m}(T_i) \neq \mathbf{0},$$

and

$$E\{[\mathbf{x}(T_i) - \mathbf{m}(T_i)][\mathbf{x}^T(T_i) - \mathbf{m}^T(T_i)]\} = \mathbf{K_x}(T_i, T_i).$$

It is statistically independent of $\mathbf{u}(t)$ and $\mathbf{w}(t)$. Find the vector differential equations that specify the MMSE estimate $\hat{\mathbf{x}}(t)$, $t \geq T_i$.

2. Assume that $\mathbf{m}(T_i) = \mathbf{0}$. Remove the zero-mean assumption on $\mathbf{u}(t)$,

$$E[\mathbf{u}(t)] = \mathbf{m_u}(t),$$

and

$$E\{[\mathbf{u}(t) - \mathbf{m_u}(t)][\mathbf{u}^T(\tau) - \mathbf{m_u}^T(\tau)]\} = \mathbf{Q}(t)\delta(t - \tau).$$

Find the vector differential equations that specify $\hat{\mathbf{x}}(t)$.

Problem 6.3.21. Consider Example 1 on p. 546. Use Property 16 to derive (351). Remember that when using the Laplace transform technique the contour must be taken to the right of all the poles.

Problem 6.3.22. Consider the second-order system illustrated in Fig. P6.11, where

$$E[u(t)\, u(\tau)] = 2Pab(a + b)\, \delta(t - \tau),$$

$$E[w(t)\, w(\tau)] = \frac{N_0}{2}\, \delta(t - \tau).$$

(a, b are possibly complex conjugates.) The state variables are

$$x_1(t) = y(t),$$

$$x_2(t) = \dot{y}(t).$$

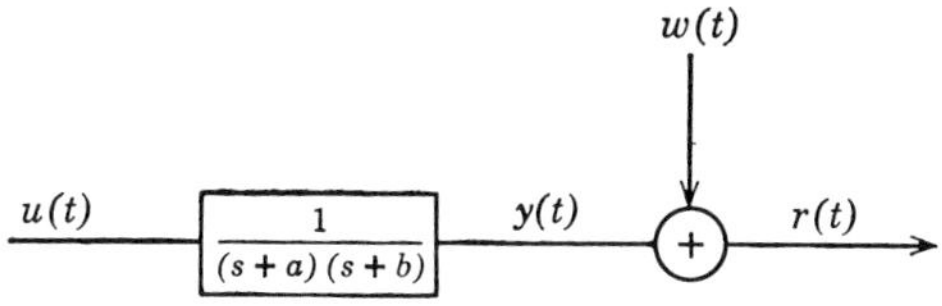

Fig. P6.11

1. Write the state equation and the output equation for the system.
2. For this state representation determine the steady state variance matrix $\mathbf{\Lambda_x}$ of the unobserved process. In other words, find

$$\mathbf{\Lambda_x} = \lim_{t \to \infty} E[\mathbf{x}(t)\, \mathbf{x}^T(t)],$$

where $\mathbf{x}(t)$ is the state vector of the system.

3. Find the transition matrix $\mathbf{T}(t, T_i)$ for the equation,

$$\frac{d\mathbf{T}(t, T_i)}{dt} = \begin{bmatrix} \mathbf{F} & \mathbf{GQG}^T \\ \mathbf{C}^T\mathbf{R}^{-1}\mathbf{C} & -\mathbf{F}^T \end{bmatrix} \mathbf{T}(t, T_i), \qquad \text{[text (336)]}$$

by using Laplace transform techniques. (Depending on the values of a, b, q, and $N_0/2$, the exponentials involved will be real, complex, or both.)

4. Find $\xi_P(t)$ when the initial condition is

$$\xi_P(T_i) = \mathbf{\Lambda_x}.$$

Comment. Although we have an analytic means of determining $\xi_P(t)$ for a system of any order, this problem illustrates that numerical means are more appropriate.

Problem 6.3.23. Because of its time-invariant nature, the optimal linear filter, as determined by Wiener spectral factorization techniques, will lead to a nonoptimal estimate when a finite observation interval is involved. The purpose of this problem is to determine how much we degrade our estimate by using a Wiener filter when the observation interval is finite. Consider the first order system.

$$\dot{x}(t) = -kx(t) + u(t),$$

where

$$\begin{aligned} r(t) &= x(t) + w(t), \\ E[u(t)\, u(\tau)] &= 2kP\, \delta(t - \tau), \\ E[w(t)\, w(\tau)] &= \frac{N_0}{2}\, \delta(t - \tau), \\ E[x(0)] &= 0, \\ E[x^2(0)] &= P_0. \\ T_i &= 0. \end{aligned}$$

1. What is the variance of error obtained by using Kalman-Bucy filtering?
2. Show that the steady-state filter (i.e., the Wiener filter) is given by

$$H_o(j\omega) = \frac{4kP/N_0}{(k + \gamma)(j\omega + \gamma)},$$

where $\gamma = k(1 + 4P/kN_0)^{1/2}$. Denote the output of the Wiener filter as $\hat{x}_{w_o}(t)$.

3. Show that a state representation for the Wiener filter is

$$\dot{\hat{x}}_{w_o}(t) = -\gamma\,\hat{x}_{w_o}(t) + \frac{4Pk}{N_0(k+\gamma)}\,r(t),$$

where

$$\hat{x}_{w_o}(0) = 0.$$

4. Show that the error for this system is

$$\dot{\epsilon}_{w_o}(t) = -\gamma\,\epsilon_{w_o}(t) - u(t) + \frac{4Pk}{N_0(k+\gamma)}\,w(t),$$

$$\epsilon_{w_o}(0) = -x(0).$$

5. Define

$$\xi_{w_o}(t) = E[\epsilon_{w_o}^{\,2}(t)].$$

Show that

$$\dot{\xi}_{w_o}(t) = -2\gamma\xi_{w_o}(t) + \frac{4kP\gamma}{\gamma + k}$$

$$\xi_{w_o}(0) = P_0$$

and verify that

$$\xi_{w_o}(t) = \xi_{P\infty}(1 - e^{-2\gamma t}) + P_o e^{-2\gamma t}.$$

6. Plot the ratio of the mean-square error using the Kalman-Bucy filter to the mean-square error using the Wiener filter. (Note that both errors are a function of time.)

$$\beta(t) = \frac{\xi_P(t)}{\xi_{w_o}(t)} \quad \text{for} \quad \gamma = 1.5k,\ 2k, \text{ and } 3k.$$

$$P_0 = 0,\ 0.5P, \text{ and } P.$$

Note that the expression for $\xi_P(t)$ in (353) is only valid for $P_0 = P$. Is your result intuitively correct?

Problem 6.3.24.

Consider the following system:

$$\dot{\mathbf{x}}(t) = \mathbf{F}\mathbf{x}(t) + \mathbf{G}u(t)$$

$$y(t) = \mathbf{C}\,\mathbf{x}(t)$$

where

$$\mathbf{F} = \begin{bmatrix} 0 & 1 & 0 & 0 \\ 0 & 0 & 1 & 0 \\ 0 & 0 & 0 & 1 \\ -p_0 & -p_1 & -p_2 & -p_3 \end{bmatrix}, \qquad \mathbf{G} = \begin{bmatrix} 0 \\ 0 \\ 0 \\ 1 \end{bmatrix}, \qquad \mathbf{C} = [1\,0\,0\,0].$$

$$E[u(t)\,u(\tau)] = Q\,\delta(t - \tau).$$

Find the steady-state covariance matrix, that is,

$$\lim_{t\to\infty} \mathbf{K_x}(t, t),$$

for a fourth-order Butterworth process using the above representation.

$$S_a(\omega) = \frac{8\sin(\pi/16)}{1 + \omega^8}$$

Hint. Use the results on p. 545.

Problem 6.3.25. Consider Example 3 on p. 555. Use Property 16 to solve (368).

Problem 6.3.26 (continuation). Assume that the steady-state filter shown in Fig. 6.45 is used. Compute the transient behavior of the error variance for this filter. Compare it with the optimum error variance given in (369).

Problem 6.3.27. Consider the system shown in Fig. P6.12*a* where

$$E[u(t)\,u(\tau)] = \sigma^2\,\delta(t-\tau),$$

$$E[w(t)\,w(\tau)] = \frac{N_0}{2}\,\delta(t-\tau).$$

$$a(T_i) = \dot{a}(T_i) = 0.$$

1. Find the optimum linear filter.
2. Solve the steady-state variance equation.
3. Verify that the "pole-splitting" technique of conventional Wiener theory gives the correct answer.

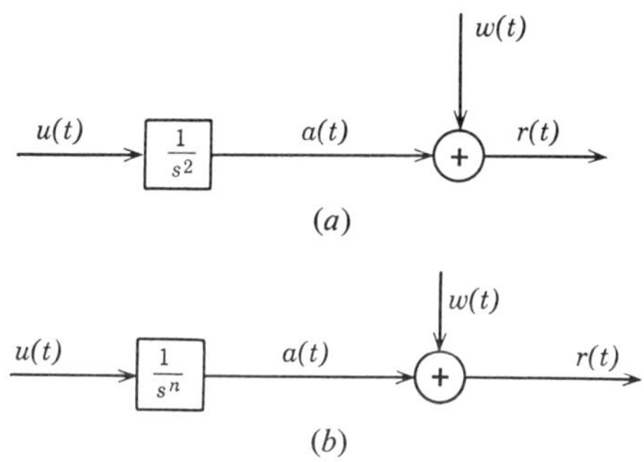

Fig. P6.12

Problem 6.3.28 (continuation). A generalization of Problem 6.3.27 is shown in Fig. P6.12*b*. Repeat Problem 6.3.27.

Hint. Use the tabulated characteristics of Butterworth polynomials given in Fig. 6.40.

Problem 6.3.29. Consider the model in Problem 6.3.27. Define the state-vector as

$$\mathbf{x}(t) = \begin{bmatrix} x_1(t) \\ x_2(t) \end{bmatrix} \triangleq \begin{bmatrix} a(t) \\ \dot{a}(t) \end{bmatrix}, \qquad \mathbf{x}(T_i) = \mathbf{0}.$$

1. Determine $\mathbf{K_x}(t, u) = E[\mathbf{x}(t)\,\mathbf{x}^T(u)]$.
2. Determine the optimum realizable filter for estimating $\mathbf{x}(t)$ (calculate the gains analytically).
3. Verify that your answer reduces to the answer in Problem 6.3.27 as $t \rightarrow \infty$.

Problem 6.3.30. Consider Example 5*A* on p. 557. Write the variance equation for arbitrary time ($t \geq 0$) and solve it.

Problem 6.3.31. Let

$$r(t) + \sum_{i=1}^{n} k_i a_i(t) + w(t),$$

where the $a_i(t)$ are statistically independent messages with state representations,

$$\dot{\mathbf{x}}_i(t) = \mathbf{F}_i(t)\,\mathbf{x}_i(t) + \mathbf{G}_i(t)\,\mathbf{u}_i(t),$$
$$a_i(t) = \mathbf{C}_i(t)\,\mathbf{x}_i(t),$$

and $w(t)$ is white noise $(N_0/2)$. Generalize the optimum filter in Fig. 6.52*a* to include this case.

Problem 6.3.32. Assume that a particle leaves the origin at $t = 0$ and travels at a constant but unknown velocity. The observation is corrupted by additive white Gaussian noise of spectral height $N_0/2$. Thus

$$r(t) = vt + w(t), \qquad t \geq 0.$$

Assume that

$$E(v) = 0,$$
$$E(v^2) = \sigma^2,$$

and that v is a Gaussian random variable.

1. Find the equation specifying the MAP estimate of vt.
2. Find the equation specifying the MMSE estimate of vt.

Use the techniques of Chapter 4 to solve this problem.

Problem 6.3.33. Consider the model in Problem 6.3.32. Use the techniques of Section 6.3 to solve this problem.

1. Find the minimum mean-square error linear estimate of the message

$$a(t) \triangleq vt.$$

2. Find the resulting mean-square error.
3. Show that for large t

$$\xi_P(t) \simeq \left(\frac{3N_0}{2t}\right)^{1/2}.$$

Problem 6.3.34 (continuation).

1. Verify that the answers to Problems 6.3.32 and 6.3.33 are the same.
2. Modify your estimation procedure in Problem 6.3.32 to obtain a maximum likelihood estimate (assume that v is an unknown nonrandom variable).
3. Discuss qualitatively when the a priori knowledge is useful.

Problem 6.3.35 (continuation).

1. Generalize the model of Problem 6.3.32 to include an arbitrary polynomial message:

$$a(t) = \sum_{i=1}^{K} v_i t^i,$$

where

$$E(v_i) = 0,$$
$$E(v_i v_j) = \sigma_i^2\,\delta_{ij}.$$

2. Solve for $k = 0$, 1, and 2.

Problem 6.3.36. Consider the second-order system shown in Fig. P6.13.

$$E[u(t)\,u(\tau)] = Q\,\delta(t - \tau),$$

$$E[w(t)\,w(\tau)] = \frac{N_0}{2}\,\delta(t - \tau),$$

$$x_1(t) = y(t),$$
$$x_2(t) = \dot{y}(t).$$

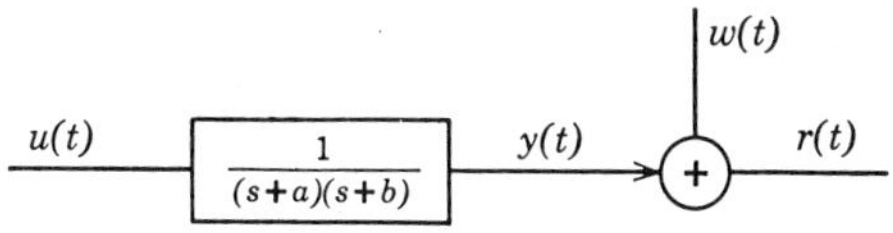

Fig. P6.13

1. Write the state equation and determine the steady state solution to the covariance equation by setting $\dot{\boldsymbol{\xi}}_P(t) = \mathbf{0}$.
2. Do the values of a, b, Q, N_0 influence the roots we select in order that the covariance matrix will be positive definite?
3. In general there are eight possible roots. In the a,b-plane determine which root is selected for any particular point for fixed Q and N_0.

Problem 6.3.37. Consider the prediction problem discussed on p. 566.

1. Derive the result stated in (422). Recall $\mathbf{d}(t) = \mathbf{x}(t+\alpha)$; $\alpha > 0$.
2. Define the prediction covariance matrix as

$$\boldsymbol{\xi}_P{}^{\alpha} \triangleq E\{[\hat{\mathbf{d}}(t) - \mathbf{d}(t)][\hat{\mathbf{d}}^T(t) - \mathbf{d}^T(t)]\}$$

Find an expression for $\boldsymbol{\xi}_P{}^{\alpha}$. Verify that your answer has the correct behavior for $\alpha \to \infty$.

Problem 6.3.38 (continuation). Apply the result in part 2 to the message and noise model in Example 3 on p. 494. Verify that the result is identical to (113).

Problem 6.3.39 (continuation). Let

$$r(u) = a(u) + w(u), \qquad -\infty < u \le t$$

and

$$d(t) = a(t + \alpha).$$

The processes $a(u)$ and $w(u)$ are uncorrelated with spectra

$$S_a(\omega) = \frac{2\sqrt{2}\, P/k}{1 + (\omega^2/k^2)^2},$$

$$S_n(\omega) = \frac{N_0}{2}.$$

Use the result of Problem 6.3.37 to find $E[(\hat{d}(t) - d(t))^2]$ as a function of α.
Compare your result with the result in Problem 6.3.38. Would you expect that the prediction error is a monotone function of n, the order of the Butterworth spectra?

Problem 6.3.40. Consider the following optimum realizable filtering problem:

$$r(u) = a(u) + w(u), \qquad 0 \le u \le t$$

$$S_a(\omega) = \frac{1}{(\omega^2 + k^2)^2},$$

$$S_w(\omega) = \frac{N_0}{2},$$

$$S_{aw}(\omega) = 0.$$

The desired signal $d(t)$ is

$$d(t) = \frac{da(t)}{dt}.$$

We want to find the optimum linear filter by using state-variable techniques.

1. Set the problem up. Define explicitly the state variables you are using and *all* matrices.
2. Draw an explicit block diagram of the optimum receiver. (Do not use matrix notation here.)
3. Write the variance equation as a set of scalar equations. *Comment* on how you would solve it.
4. Find the steady-state solution by letting $\dot{\xi}_P(t) = 0$.

Problem 6.3.41. Let

$$r(u) = a(u) + w(u), \qquad 0 \le u \le t,$$

where $a(u)$ and $n(u)$ are uncorrelated processes with spectra

$$S_a(\omega) = \frac{2k\sigma_a^2}{\omega^2 + k^2}$$

and

$$S_n(\omega) = \frac{N_0}{2}.$$

The desired signal is obtained by passing $a(t)$ through a linear system whose transfer function is

$$K_d(j\omega) = \frac{-j\omega + k}{j\omega + \beta}$$

1. Find the optimum linear filter to estimate $d(t)$ and the variance equation.
2. Solve the variance equation for the steady-state case.

Problem 6.3.42. Consider the model in Problem 6.3.41. Let

$$K_d(j\omega) = \left(\frac{1}{j\omega + \beta}\right).$$

Repeat Problem 6.3.41.

Problem 6.3.43. Consider the model in Problem 6.3.41. Let

$$d(t) = \frac{1}{\beta - \alpha}\int_{t+\alpha}^{t+\beta} a(u)\, du, \qquad \alpha > 0,\ \beta > 0,\ \beta > \alpha.$$

1. Does this problem fit into any of the cases discussed in Section 6.3.4 of the text?
2. Demonstrate that you can solve it by using state variable techniques.
3. What is the basic reason that the solution in part 2 is possible?

Problem 6.3.44. Consider the following pre-emphasis problem shown in Fig. P6.14.

$$S_a(\omega) = \frac{2p_1\sigma_a^2}{\omega^2 + p_1^2},$$

$$S_w(\omega) = \frac{N_0}{2},$$

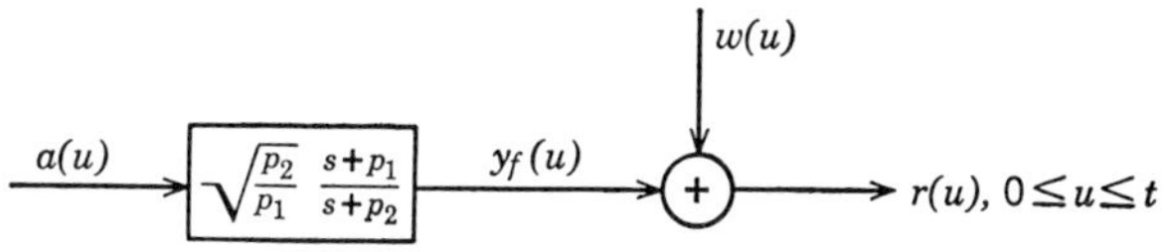

Fig. P6.14

and the processes are uncorrelated.

$$d(t) = a(t).$$

1. Find the optimum realizable linear filter by using a state-variable formulation.
2. Solve the variance equation for $t \rightarrow \infty$. (You may assume that a statistical steady state exists.) Observe that $a(u)$, not $y_f(t)$, is the message of interest.

Problem 6.3.45. Estimation in Colored Noise. In this problem we consider a simple example of state-variable estimation in colored noise. Our approach is simpler than that in the text because we are required to estimate only one state variable. Consider the following system.

$$\begin{aligned}\dot{x}_1(t) &= -k_1\, x_1(t) + u_1(t),\\ \dot{x}_2(t) &= -k_2\, x_2(t) + u_2(t).\\ E[u_1(t)\, u_2(\tau)] &= 0,\\ E[u_1(t)\, u_1(\tau)] &= 2k_1 P_1\, \delta(t-\tau),\\ E[u_2(t)\, u_2(\tau)] &= 2k_1 P_2\, \delta(t-\tau),\\ E[x_1^2(0)] &= P_1,\\ E[x_2^2(0)] &= P_2,\\ E[x_1(0)\, x_2(0)] &= 0.\end{aligned}$$

We observe

$$r(t) = x_1(t) + x_2(t);$$

that is, no white noise is present in the observation. We want to apply whitening concepts to estimating $x_1(t)$ and $x_2(t)$. First we generate a signal $r'(t)$ which has a white component.

1. Define a linear transformation of the state variables by

$$\begin{bmatrix} y_1(t) \\ y_2(t) \end{bmatrix} = \frac{1}{2}\begin{bmatrix} 1 & -1 \\ 1 & 1 \end{bmatrix}\begin{bmatrix} x_1(t) \\ x_2(t) \end{bmatrix} = \frac{1}{2}\begin{bmatrix} x_1(t) - x_2(t) \\ r(t) \end{bmatrix}.$$

Note that one of our new state variables, $y_2(t)$, is $\frac{1}{2}r(t)$; therefore, it is known at the receiver.

Find the state equations for $y(t)$.

2. Show that the new state equations may be written as

$$\begin{aligned}\dot{y}_1(t) &= -k' y_1(t) + [u'(t) + m_r(t)]\\ r'(t) &= \dot{y}_2(t) + k' y_2(t) = C' y_1(t) + w'(t),\end{aligned}$$

where

$$k' = \frac{(k_1 + k_2)}{2},$$

$$C' = \frac{-(k_1 - k_2)}{2},$$

$$u'(t) = \tfrac{1}{2}[u_1(t) - u_2(t)],$$

$$w'(t) = \tfrac{1}{2}[u_1(t) + u_2(t)],$$

$$m_r(t) = C' y_2(t).$$

Notice that $m_r(t)$ is a known function at the receiver; therefore, its effect upon $y_1(t)$ is known. Also notice that our new observation signal $r'(t)$ consists of a linear modulation of $y_1(t)$ plus a white noise component.

3. Apply the estimation with correlated noise results to derive the Kalman-Bucy realizable filter equations.

4. Find equations which relate the error variance of $\hat{x}_1(t)$ to

$$E[(\hat{y}_1(t) - y_1(t))^2] \triangleq P_y(t).$$

5. Specialize your results to the case in which $k_1 = k_2$. What does the variance equation become in the limit as $t \to \infty$? Is this intuitively satisfying?

6. Comment on how this technique generalizes for higher dimension systems when there is no white noise component present in the observation. In the case of multiple observations can we ever have a singular problem: i.e., perfect estimation?

Comment. The solution to the unrealizable filter problem was done in a different manner in [56].

Problem 6.3.46. Let

$$\begin{aligned} \dot{a}(\tau) &= -k_{mo}a(\tau) + u(\tau), \qquad 0 \le \tau \le t, \\ r(\tau) &= a(\tau) + w(\tau), \qquad 0 \le \tau \le t, \end{aligned}$$

where

$$E[a(0)] = 0, \; E[a^2(0)] = \sigma^2/2k_{mo},$$

$$K_u(t, \tau) = \sigma^2\, \delta(t - \tau),$$

$$K_w(t, \tau) = \frac{N_0}{2}\, \delta(t - \tau).$$

Assume that we are processing $r(\tau)$ by using a realizable filter which is optimum for the above message model. The actual message process is

$$\dot{a}(\tau) = -k_{ac}\, a(\tau) + u(\tau), \qquad 0 \le \tau \le t.$$

Find the equations which specify $\xi_{ac}(t)$, the actual error variance.

P6.4 Linear Modulation, Communications Context

Problem 6.4.1. Write the variance equation for the DSB-AM example discussed on p. 576. Draw a block diagram of the system to generate it and verify that the high-frequency terms can be ignored.

Problem 6.4.2. Let

$$s(t, a(t)) = \sqrt{P}\,[a(t)\cos(\omega_c t + \theta) - \tilde{a}(t)\sin(\omega_c t + \theta)],$$

where

$$\tilde{A}(j\omega) = H(j\omega)\, A(j\omega).$$

$H(j\omega)$ is specified by (506) and θ is independent of $a(t)$ and uniformly distributed $(0, 2\pi)$. Find the power-density spectrum of $s(t, a(t))$ in terms of $S_a(\omega)$.

Problem 6.4.3. In this problem we derive the integral equation that specifies the optimum estimate of a SSB signal [see (505, 506)]. Start the derivation with (5.25) and obtain (507).

Problem 6.4.4. Consider the model in Problem 6.4.3. Define

$$\mathbf{a}(t) = \begin{bmatrix} a(t) \\ \tilde{a}(t) \end{bmatrix}.$$

Use the vector process estimation results of Section 5.4 to derive (507).

Problem 6.4.5.

1. Draw the block diagram corresponding to (508).
2. Use block diagram manipulation and the properties of $H(j\omega)$ given in (506) to obtain Fig. 6.62.

Problem 6.4.6. Let

$$s(t, a(t)) = \left(\frac{P}{1 + m^2}\right)^{1/2} [1 + ma(t)] \cos \omega_c t,$$

where

$$S_a(\omega) = \frac{2k}{\omega^2 + k^2}.$$

The received waveform is

$$r(t) = s(t, a(t)) + w(t), \qquad -\infty < t < \infty$$

where $w(t)$ is white ($N_0/2$). Find the optimum unrealizable demodulator and plot the mean-square error as a function of m.

Problem 6.4.7 (continuation). Consider the model in Problem 6.4.6. Let

$$S_a(\omega) = \begin{cases} \dfrac{1}{2W}, & |\omega| \le 2\pi W, \\ 0, & \text{elsewhere.} \end{cases}$$

Find the optimum unrealizable demodulator and plot the mean-square error as a function of m.

Problem 6.4.8. Consider the example on p. 583. Assume that

$$p_\phi(\theta) = \frac{e^{\Lambda_m \cos\theta}}{2\pi I_0(\Lambda_m)}, \qquad -\pi \le \theta \le \pi,$$

and

$$S_a(\omega) = \frac{2k}{\omega^2 + k^2}.$$

1. Find an expression for the mean-square error using an unrealizable demodulator designed to be optimum for the known-phase case.
2. Approximate the integral in part 1 for the case in which $\Lambda_m \gg 1$.

Problem 6.4.9 (continuation). Consider the model in Problem 6.4.8. Let

$$S_a(\omega) = \begin{cases} \dfrac{1}{2W}, & |\omega| \le 2\pi W, \\ 0, & \text{elsewhere.} \end{cases}$$

Repeat Problem 6.4.8.

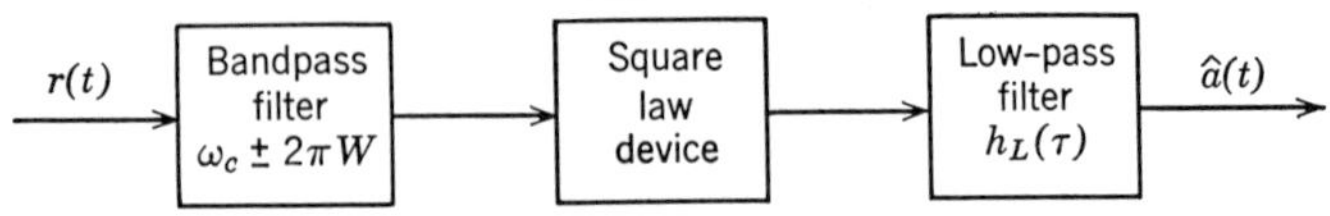

Fig. P6.15

Problem 6.4.10. Consider the model in Problem 6.4.7. The demodulator is shown in Fig. P6.15. Assume $m \ll 1$ and

$$S_a(\omega) = \begin{cases} \dfrac{1}{2W}, & 2\pi W_1 \le |\omega| \le 2\pi(W_1 + W), \\ 0, & \text{elsewhere,} \end{cases}$$

where $2\pi(W_1 + W) \ll \omega_c$.

Choose $h_L(\tau)$ to minimize the mean-square error. Calculate the resulting error ξ_P.

P6.6 Related Issues

In Problems 6.6.1 through 6.6.4, we show how the state-variable techniques we have developed can be used in several important applications. The first problem develops a necessary preliminary result. The second and third problems develop a solution technique for homogeneous and nonhomogeneous Fredholm equations (either vector or scalar). The fourth problem develops the optimum unrealizable filter. A complete development is given in [54]. The model for the four problems is

$$\begin{aligned} \dot{\mathbf{x}}(t) &= \mathbf{F}(t)\mathbf{x}(t) + \mathbf{G}(t)\mathbf{u}(t) \\ \mathbf{y}(t) &= \mathbf{C}(t)\mathbf{x}(t) \\ E[\mathbf{u}(t)\mathbf{u}^T(\tau)] &= \mathbf{Q}\delta(t - \tau) \end{aligned}$$

We use a function $\boldsymbol{\xi}(t)$ to agree with the notation of [54]. It is not related to the variance matrix $\boldsymbol{\xi}_P(t)$.

Problem 6.6.1. Define the linear functional

$$\boldsymbol{\xi}(t) = \int_{T_i}^{T_f} \mathbf{K_x}(t, \tau)\, \mathbf{s}(\tau)\, d\tau,$$

where $\mathbf{s}(t)$ is a bounded vector function.

We want to show that when $\mathbf{K_x}(t, \tau)$ is the covariance matrix for a state-variable random process $\mathbf{x}(t)$ we can represent this functional as the solution to the differential equations

$$\frac{d\boldsymbol{\xi}(t)}{dt} = \mathbf{F}(t)\,\boldsymbol{\xi}(t) + \mathbf{G}(t)\,\mathbf{Q}\mathbf{G}^T(t)\,\boldsymbol{\eta}(t)$$

and

$$\frac{d\boldsymbol{\eta}(t)}{dt} = -\mathbf{F}^T(t)\,\boldsymbol{\eta}(t) - \mathbf{s}(t),$$

with the boundary conditions

$$\boldsymbol{\eta}(T_f) = \mathbf{0},$$

and

$$\boldsymbol{\xi}(T_i) = \mathbf{P}_0\boldsymbol{\eta}(T_i),$$

where

$$\mathbf{P}_0 = \mathbf{K_x}(T_i, T_i).$$

1. Show that we can write the above integral as

$$\boldsymbol{\xi}(t) = \int_{T_i}^{t} \boldsymbol{\Phi}(t, \tau)\, \mathbf{K_x}(\tau, \tau)\, \mathbf{s}(\tau)\, d\tau + \int_{t}^{T_f} \mathbf{K_x}(t, t)\, \boldsymbol{\Phi}^T(\tau, t)\, \mathbf{s}(\tau)\, d\tau.$$

Hint. See Problem 6.3.16.

2. By using Leibnitz's rule, show that

$$\frac{d\boldsymbol{\xi}(t)}{dt} = \mathbf{F}(t)\, \boldsymbol{\xi}(t) + \mathbf{G}(t)\mathbf{Q}\mathbf{G}^T(t) \int_{t}^{T_f} \boldsymbol{\Phi}^T(\tau, t)\, \mathbf{s}(\tau)\, d\tau.$$

Hint. Note that $\mathbf{K_x}(t, t)$ satisfies the differential equation

$$\frac{d\mathbf{K_x}(t, t)}{dt} = \mathbf{F}(t)\, \mathbf{K_x}(t, t) + \mathbf{K_x}(t, t)\mathbf{F}^T(t) + \mathbf{G}(t)\mathbf{Q}\mathbf{G}^T(t), \qquad \text{(text 273)}$$

with $\mathbf{K_x}(T_i, T_i) = \mathbf{P}_0$, and $\boldsymbol{\Phi}^T(\tau, t)$ satisfies the adjoint equation; that is,

$$\frac{d\boldsymbol{\Phi}^T(\tau, t)}{dt} = -\mathbf{F}^T(t)\, \boldsymbol{\Phi}^T(\tau, t),$$

with $\boldsymbol{\Phi}(T_i, T_i) = \mathbf{I}$.

3. Define a second functional $\boldsymbol{\eta}(t)$ by

$$\boldsymbol{\eta}(t) = \int_{t}^{T_f} \boldsymbol{\Phi}^T(\tau, t)\, \mathbf{s}(\tau)\, d\tau.$$

Show that it satisfies the differential equation

$$\frac{d\boldsymbol{\eta}(t)}{dt} = -\mathbf{F}^T(t)\, \boldsymbol{\eta}(t) - \mathbf{s}(t).$$

4. Show that the differential equations must satisfy the two independent boundary conditions

$$\boldsymbol{\eta}(T_f) = \mathbf{0},$$
$$\boldsymbol{\xi}(T_i) = \mathbf{P}_i \boldsymbol{\eta}(T_i).$$

5. By combining the results in parts 2, 3, and 4, show the desired result.

Problem 6.6.2. Homogeneous Fredholm Equation. In this problem we derive a set of differential equations to determine the eigenfunctions for the homogeneous Fredholm equation. The equation is given by

$$\int_{T_i}^{T_f} \mathbf{K_y}(t, \tau)\boldsymbol{\phi}(\tau)\, d\tau = \lambda \boldsymbol{\phi}(t), \qquad T_i \le t \le T_f,$$

or

$$\boldsymbol{\phi}(t) = \frac{1}{\lambda} \mathbf{C}(t) \int_{T_i}^{T_f} \mathbf{K_x}(t, \tau)\, \mathbf{C}^T(\tau)\boldsymbol{\phi}(\tau)\, d\tau, \qquad \text{for } \lambda > 0.$$

Define

$$\boldsymbol{\xi}(t) = \int_{T_i}^{T_f} \mathbf{K_x}(t, \tau)\mathbf{C}^T(\tau)\boldsymbol{\phi}(\tau)\, d\tau,$$

so that

$$\boldsymbol{\phi}(t) = \frac{1}{\lambda} \mathbf{C}(t)\, \boldsymbol{\xi}(t).$$

1. Show that $\boldsymbol{\xi}(t)$ satisfies the differential equations

$$\frac{d}{dt}\begin{bmatrix}\boldsymbol{\xi}(t)\\ \boldsymbol{\eta}(t)\end{bmatrix} = \left[\begin{array}{c|c}\mathbf{F}(t) & \mathbf{G}(t)\mathbf{Q}\mathbf{G}^T(t)\\ \hline \dfrac{-\mathbf{C}^T(t)\,\mathbf{C}(t)}{\lambda} & -\mathbf{F}^T(t)\end{array}\right]\begin{bmatrix}\boldsymbol{\xi}(t)\\ \boldsymbol{\eta}(t)\end{bmatrix},$$

with

$$\boldsymbol{\xi}(T_i) = \mathbf{P}_0\boldsymbol{\eta}(T_i),$$
$$\boldsymbol{\eta}(T_f) = \mathbf{0}.$$

(Use the results of Problem 6.6.1.)

2. Show that to have a nontrivial solution which satisfies the boundary conditions we need

$$\det\left[\boldsymbol{\Psi}_{\eta\xi}(T_f, T_i:\lambda)\mathbf{P}_0 + \boldsymbol{\Psi}_{\eta\eta}(T_f, T_i:\lambda)\right] = 0,$$

where $\boldsymbol{\Psi}(t, T_i:\lambda)$ is given by

$$\frac{d}{dt}\left[\begin{array}{c|c}\boldsymbol{\Psi}_{\xi\xi}(t, T_i:\lambda) & \boldsymbol{\Psi}_{\xi\eta}(t, T_i:\lambda)\\ \hline \boldsymbol{\Psi}_{\eta\xi}(t, T_i:\lambda) & \boldsymbol{\Psi}_{\eta\eta}(t, T_i:\lambda)\end{array}\right]$$

$$= \left[\begin{array}{c|c}\mathbf{F}(t) & \mathbf{G}(t)\,\mathbf{Q}\mathbf{G}^T(t)\\ \hline \dfrac{-\mathbf{C}^T(t)\,\mathbf{C}(t)}{\lambda} & -\mathbf{F}^T(t)\end{array}\right]\left[\begin{array}{c|c}\boldsymbol{\Psi}_{\xi\xi}(t, T_i:\lambda) & \boldsymbol{\Psi}_{\xi\xi}(t, T_i:\lambda)\\ \hline \boldsymbol{\Psi}_{\eta\xi}(t, T_i:\lambda) & \boldsymbol{\Psi}_{\eta\eta}(t, T_i:\lambda)\end{array}\right],$$

and $\boldsymbol{\Psi}(T_i, T_i:\lambda) = \mathbf{I}$. The values of λ which satisfy this equation are the eigenvalues.

3. Show that the eigenfunctions are given by

$$\boldsymbol{\phi}(t, T_i:\lambda) = \frac{\mathbf{C}(t)}{\lambda}\left[\boldsymbol{\Psi}_{\xi\eta}(t, T_i:\lambda)\,\mathbf{P}_0 + \boldsymbol{\Psi}_{\xi\eta}(t, T_i:\lambda)\right]\boldsymbol{\eta}(T_i)$$

where $\boldsymbol{\eta}(T_i)$ satisfies the orthogonality relationship

$$\left[\boldsymbol{\Psi}_{\eta\xi}(T_f, T_i:\lambda)\,\mathbf{P}_0 + \boldsymbol{\Psi}_{\eta\eta}(T_f, T_i:\lambda)\right]\boldsymbol{\eta}(T_i) = \mathbf{0}.$$

Problem 6.6.3. Nonhomogeneous Fredholm Equation. In this problem we derive a set of differential equations to determine the solution to the nonhomogeneous Fredholm equation. This equation is given by

$$\int_{T_i}^{T_f} \mathbf{K_y}(t, \tau)\,\mathbf{g}(\tau)\,d\tau + \sigma\,\mathbf{g}(t) = \mathbf{s}(t), \qquad T_i \le t \le T_f,\ \sigma > 0.$$

1. If we define $\boldsymbol{\xi}(t)$ as

$$\boldsymbol{\xi}(t) = \int_{T_i}^{T_f} \mathbf{K_x}(t, \tau)\,\mathbf{C}^T(\tau)\,\mathbf{g}(\tau)\,d\tau,$$

show that we may write the nonhomogeneous equation as

$$\mathbf{g}(t) = \frac{1}{\sigma}\left[\mathbf{s}(t) - \mathbf{C}(t)\,\boldsymbol{\xi}(t)\right].$$

2. Using Problem 6.6.1, show that $\boldsymbol{\xi}(t)$ satisfies the differential equations

$$\frac{d}{dt}\begin{bmatrix}\boldsymbol{\xi}(t)\\ \boldsymbol{\eta}(t)\end{bmatrix} = \left[\begin{array}{c|c}\mathbf{F}(t) & \mathbf{G}(t)\,\mathbf{Q}\mathbf{G}^T(t)\\ \hline \dfrac{\mathbf{C}^T(t)\,\mathbf{C}(t)}{\sigma} & -\mathbf{F}^T(t)\end{array}\right]\begin{bmatrix}\boldsymbol{\xi}(t)\\ \boldsymbol{\eta}(t)\end{bmatrix} - \left[\begin{array}{c}\mathbf{0}\\ \hline \dfrac{\mathbf{C}^T(t)\,\mathbf{s}(t)}{\sigma}\end{array}\right],$$

with

$$\boldsymbol{\xi}(T_i) = \mathbf{P}_0\,\boldsymbol{\eta}(T_i),$$
$$\boldsymbol{\eta}(T_f) = \mathbf{0}.$$

Comment. In general we can replace σ by an arbitrary positive-definite time-varying matrix ($\mathbf{R}(t)$) and the derivation is valid with obvious modifications.

Problem No. 6.6.4. ***Unrealizable Filters.*** In this problem we show how the nonhomogeneous Fredholm equation may be used to determine the optimal unrealizable filter structure. For algebraic simplicity we assume $\mathbf{r}(t)$ is a scalar.

$$\mathbf{R}(t) = N_0/2.$$

1. Show that the integral equation specifying the optimum unrealizable filter for estimating $\mathbf{x}(t)$ at any point t in the interval $[T_i, T_f]$ is

$$\mathbf{K_x}(t, \tau)\,\mathbf{C}^T(\tau) = \int_{T_i}^{T_f} \mathbf{h}_o(t, \sigma)\,K_r(\sigma, \tau)\,d\sigma, \qquad T_i \le t \le T_f, \qquad T_i < \tau < T_f. \quad (1)$$

2. Using the inverse kernel of $K_r(t, \tau)$ [Chapter 4, (4.161)] show that

$$\hat{\mathbf{x}}(t) = \int_{T_i}^{T_f} \mathbf{K_x}(t, \tau)\,\mathbf{C}^T(\tau)\left(\int_{T_i}^{T_f} Q_r(\tau, \sigma)\,r(\sigma)\,d\sigma\right), \qquad T_i \le t \le T_f. \quad (2)$$

3. As in Chapter 5, define the term in parentheses as $r_g(\tau)$. We note that $r_g(\tau)$ solves the nonhomogeneous Fredholm equation when the input is $r(t)$. Using Problem No. 6.6.3, show that $r_g(t)$ is given by

$$r_g(t) = \frac{2}{N_0}\,(r(t) - \mathbf{C}(t)\boldsymbol{\xi}(t)), \quad (3)$$

where

$$\frac{d\boldsymbol{\xi}(t)}{dt} = \mathbf{F}(t)\boldsymbol{\xi}(t) + \mathbf{G}(t)\mathbf{Q}\mathbf{G}^T(t)\boldsymbol{\eta}_1(t), \quad (4a)$$

$$\frac{d\boldsymbol{\eta}_1(t)}{dt} = \frac{2}{N_0}\,\mathbf{C}^T(t)\mathbf{C}(t)\boldsymbol{\xi}(t) - \mathbf{F}^T(t)\boldsymbol{\eta}_1(t) - \frac{2}{N_0}\,\mathbf{C}^T(t)r(t), \quad (4b)$$

and

$$\boldsymbol{\xi}(T_i) = \mathbf{K_x}(T_i, T_i)\boldsymbol{\eta}_1(T_i), \quad (5a)$$

$$\boldsymbol{\eta}_1(T_f) = \mathbf{0}. \quad (5b)$$

4. Using the results of Problem No. 6.6.1, show that $\hat{\mathbf{x}}(t)$ satisfies the differential equation,

$$\frac{d\hat{\mathbf{x}}(t)}{dt} = \mathbf{F}(t)\hat{\mathbf{x}}(t) + \mathbf{G}(t)\mathbf{Q}\mathbf{G}^T(t)\boldsymbol{\eta}_2(t), \quad (6a)$$

$$\frac{d\boldsymbol{\eta}_2(t)}{dt} = -\mathbf{C}^T(t)r_g(t) - \mathbf{F}^T(t)\boldsymbol{\eta}_2(t), \quad (6b)$$

where

$$\hat{\mathbf{x}}(T_i) = \mathbf{K_x}(T_i, T_i)\boldsymbol{\eta}_2(T_i), \quad (7a)$$

$$\boldsymbol{\eta}_2(T_f) = \mathbf{0}. \quad (7b)$$

5. Substitute (3) into (6*b*). Show that

$$\boldsymbol{\eta}_1(t) = \boldsymbol{\eta}_2(t), \quad (8a)$$

and

$$\hat{\mathbf{x}}(t) = \boldsymbol{\xi}(t). \quad (8b)$$

6. Show that the differential equation structure for the optimum unrealizable estimate of $\hat{\mathbf{x}}(t)$ is

$$\frac{d\hat{\mathbf{x}}(t)}{dt} = \mathbf{F}(t)\hat{\mathbf{x}}(t) + \mathbf{G}(t)\mathbf{Q}\mathbf{G}^T(t)\boldsymbol{\eta}(t), \tag{9a}$$

$$\frac{d\boldsymbol{\eta}(t)}{dt} = \mathbf{C}^T(t)\frac{2}{N_0}\mathbf{C}(t)\hat{\mathbf{x}}(t) - \mathbf{F}^T(t)\boldsymbol{\eta}(t) - \mathbf{C}^T(t)\frac{2}{N_0}r(t), \tag{9b}$$

where

$$\hat{\mathbf{x}}(T_i) = \mathbf{K}_{\mathbf{x}}(T_i, T_i)\boldsymbol{\eta}(T_i), \tag{10a}$$

$$\boldsymbol{\eta}(T_f) = \mathbf{0}. \tag{10b}$$

Comments

1. We have two n-dimensional linear vector differential equations to solve.
2. The performance given by the unrealizable error covariance matrix is not part of the filter structure.
3. By letting T_f be a variable we can determine a differential equation structure for $\hat{\mathbf{x}}(T_f)$ as a function of T_f. These equations are just the Kalman-Bucy equations for the optimum realizable filter.

Problem 6.6.5. In Problems 6.2.29 and 6.2.30 we discussed a realizable whitening filter for infinite intervals and stationary processes. In this problem we verify that these results generalize to finite intervals and nonstationary processes.

Let

$$r(\tau) = n_c(\tau) + w(\tau), \qquad T_i \le \tau \le t,$$

where $n_c(\tau)$ can be generated as the output of a dynamic system,

$$\dot{\mathbf{x}}(t) = \mathbf{F}(t)\mathbf{x}(t) + \mathbf{G}(t)u(t)$$
$$n_c(t) = \mathbf{C}(t)\mathbf{x}(t),$$

driven by white noise $u(t)$.

Show that the process

$$r'(t) = r(t) - \hat{n}_c(t),$$
$$= r(t) - \mathbf{C}(t)\hat{\mathbf{x}}(t),$$

is white.

Problem 6.6.6. Sequential Estimation. Consider the convolutional encoder in Fig. P6.16.

1. What is the "state" of this system?
2. Show that an appropriate set of state equations is

$$\mathbf{x}_{n+1} = \begin{bmatrix} x_{1,n+1} \\ x_{2,n+1} \\ x_{3,n+1} \end{bmatrix} = \begin{bmatrix} 0 & 1 & 0 \\ 0 & 0 & 1 \\ 0 & 0 & 0 \end{bmatrix} \begin{bmatrix} x_{1,n} \\ x_{2,n} \\ x_{3,n} \end{bmatrix} \oplus \begin{bmatrix} 0 \\ 0 \\ 1 \end{bmatrix} u_n$$

$$\mathbf{y}_{n+1} = \begin{bmatrix} 1 & 0 & 0 \\ 1 & 1 & 1 \end{bmatrix} \begin{bmatrix} x_{1,n+1} \\ x_{2,n+1} \\ x_{3,n+1} \end{bmatrix}$$

(all additions are modulo 2).

Assume that the process u is composed of a sequence of independent binary random variables with

$$E(u_n = 0) = P_u,$$
$$E(u_n = 1) = 1 - P_u.$$

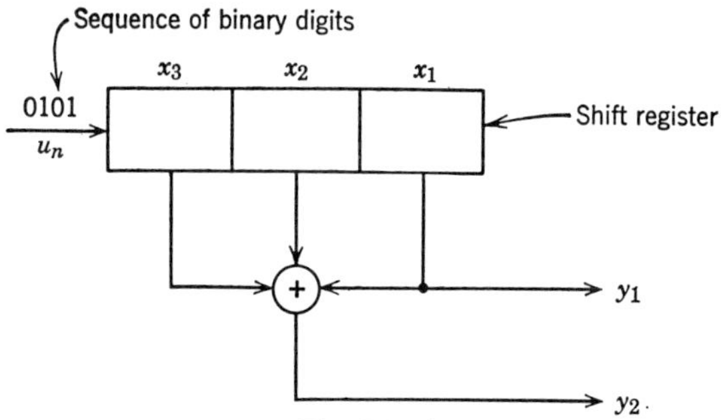

Fig. P6.16

In addition, assume that the components of **y** are sent over two independent identical binary symmetric channels such that

$$\mathbf{r}_n = \mathbf{y} \oplus \mathbf{w},$$

where

$$E(w_{n,1} = 0) = P_w,$$
$$E(w_{n,1} = 1) = 1 - P_w.$$

Finally, let the sequence of measurements

$$\mathbf{r}_1, \mathbf{r}_2, \ldots, \mathbf{r}_n, \text{ be denoted by } \mathbf{z}_n.$$

3. Show that the a posteriori probability density $p_{\mathbf{x}_{n+1}|\mathbf{z}_{n+1}}(\mathbf{X}_{n+1}|\mathbf{Z}_{n+1})$ satisfies the following recursive relationship:

$$p_{\mathbf{x}_{n+1}|\mathbf{z}_{n+1}}(\mathbf{X}_{n+1}|\mathbf{Z}_{n+1}) = \frac{p_{\mathbf{r}_{n+1}|\mathbf{z}_{n+1}}(\mathbf{R}_{n+1}|\mathbf{X}_{n+1}) \sum_{\mathbf{X}_n} p_{\mathbf{x}_{n+1}|\mathbf{x}_n}(\mathbf{X}_{n+1}|\mathbf{X}_n) p_{\mathbf{x}_n|\mathbf{z}_n}(\mathbf{X}_n|\mathbf{Z}_n)}{\sum_{\mathbf{X}_{n+1}} p_{\mathbf{r}_{n+1}|\mathbf{x}_{n+1}}(\mathbf{R}_{n+1}|\mathbf{X}_{n+1}) \sum_{\mathbf{X}_n} p_{\mathbf{x}_{n+1}|\mathbf{x}_n}(\mathbf{X}_{n+1}|\mathbf{X}_n) p_{\mathbf{x}_n|\mathbf{z}_n}(\mathbf{X}_n|\mathbf{Z}_n)}$$

where $\sum_{\mathbf{X}_n}$ denotes the sum over all possible states.

4. How would you design the MAP receiver that estimates $\mathbf{x}_{n+1}$? What must be computed for each receiver estimate?

5. How does this estimation procedure compare with the discrete Kalman filter?

6. How does the complexity of this procedure increase as the length of the convolutional encoder increases?

REFERENCES

[1] N. Wiener, *The Extrapolation, Interpolation, and Smoothing of Stationary Time Series*, Wiley, New York, 1949.

[2] A. Kolmogoroff, "Interpolation and Extrapolation," *Bull. Acad. Sci., U.S.S.R.*, Ser. Math., **5**, 3–14 (1941).

[3] H. W. Bode and C. E. Shannon, "A Simplified Derivation of Linear Least-squares Smoothing and Prediction Theory," *Proc. IRE*, **38**, 417–426 (1950).

[4] M. C. Yovits and J. L. Jackson, "Linear Filter Optimization with Game Theory Considerations," *IRE National Convention Record*, Pt. 4, 193–199 (1955).

[5] A. J. Viterbi and C. R. Cahn, "Optimum Coherent Phase and Frequency

Demodulation of a Class of Modulating Spectra," *IEEE Trans. Space Electronics and Telemetry*, **SET-10**, No. 3, 95–102 (September 1964).

[6] D. L. Snyder, "Some Useful Expressions for Optimum Linear Filtering in White Noise," *Proc. IEEE*, **53**, No. 6, 629–630 (1965).

[7] C. W. Helstrom, *Statistical Theory of Signal Detection*, Pergamon, New York, 1960.

[8] W. C. Lindsay, "Optimum Coherent Amplitude Demodulation," *Proc. NEC*, **20**, 497–503 (1964).

[9] G. C. Newton, L. A. Gould, and J. F. Kaiser, *Analytical Design of Linear Feedback Controls*, Wiley, New York, 1957.

[10] E. Wong and J. B. Thomas, "On the Multidimensional Filtering and Prediction Problem and the Factorization of Spectral Matrices," *J. Franklin Inst.*, **8**, 87–99, (1961).

[11] D. C. Youla, "On the Factorization of Rational Matrices," *IRE Trans. Inform. Theory*, **IT-7**, 172–189 (July 1961).

[12] R. C. Amara, "The Linear Least Squares Synthesis of Continuous and Sampled Data Multivariable Systems," Report No. 40, Stanford Research Laboratories, 1958.

[13] R. Brockett, "Spectral Factorization of Rational Matrices," *IEEE Trans. Inform. Theory* (to be published).

[14] M. C. Davis, "On Factoring the Spectral Matrix," Joint Automatic Control Conference, Preprints, pp. 549–466, June 1963; also, *IEEE Transactions on Automatic Control*, 296–305, October 1963.

[15] R. J. Kavanaugh, "A Note on Optimum Linear Multivariable Filters," *Proc. Inst. Elec. Engs.*, **108**, Pt. C, 412–217, Paper No. 439M, 1961.

[16] N. Wiener and P. Massani, "The Prediction Theory of Multivariable Stochastic Processes," *Acta Math.*, **98** (June 1958).

[17] H. C. Hsieh and C. T. Leondes, "On the Optimum Synthesis of Multipole Control Systems in the Wiener Sense," *IRE Trans.* on Automatic Control, **AC-4**, 16–29 (1959).

[18] L. G. McCracken, "An Extension of Wiener Theory of Multi-Variable Controls," *IRE International Convention Record*, Pt. 4, p. 56, 1961.

[19] H. C. Lee, "Canonical Factorization of Nonnegative Hermetian Matrices," *J. London Math. Soc.*, **23**, 100–110 (1948).

[20] M. Mohajeri, "Closed-form Error Expressions," M.Sc. Thesis, Dept. of Electrical Engineering, M.I.T., 1968.

[21] D. L. Snyder, "The State-Variable Approach to Continuous Estimation," M.I.T., Sc.D. Thesis, February 1966.

[22] C. L. Dolph and M. A. Woodbury, "On the Relation Between Green's Functions and Covariances of Covariances of Certain Stochastic Processes and its Application to Unbiased Linear Prediction," *Trans. Am. Math. Soc.*, 1948.

[23] R. E. Kalman and R. S. Bucy, "New Results in Linear Filtering and Prediction Theory," *ASME J. Basic Engr.*, March 1961.

[24] L. A. Zadeh and C. A. DeSoer, *Linear System Theory*, McGraw-Hill, New York, 1963.

[25] S. C. Gupta, *Transform and State Variable Methods in Linear Systems*, Wiley, New York, 1966.

[26] M. Athans and P. L. Falb, *Optimal Control*, McGraw-Hill, New York, 1966.

[27] P. M. DeRusso, R. J. Roy, and C. M. Close, *State Variables for Engineers*, Wiley, New York, 1965.

[28] R. J. Schwartz and B. Friedland, *Linear Systems*, McGraw-Hill, New York, 1965.
[29] E. A. Coddington and N. Levinson, *Theory of Ordinary Differential Equations*, McGraw-Hill, New York, 1955.
[30] R. E. Bellman, *Stability Theory of Differential Equations*, McGraw-Hill, New York, 1953.
[31] N. W. McLachlan, *Ordinary Non-linear Differential Equations in Engineering and Physical Sciences*, Clarendon Press, Oxford, 1950.
[32] J. J. Levin, "On the Matrix Riccati Equation," *Proc. Am. Math. Soc.*, **10**, 519–524 (1959).
[33] W. T. Reid, "Solutions of a Riccati Matrix Differential Equation as Functions of Initial Values," *J. Math. Mech.*, **8**, No. 2 (1959).
[34] W. T. Reid, "A Matrix Differential Equation of Riccati Type," *Am. J. Math.*, **68**, 237–246 (1946); Addendum, ibid., **70**, 460 (1948).
[35] W. J. Coles, "Matrix Riccati Differential Equations," *J. Soc. Indust. Appl. Math.*, **13**, No. 3 (September 1965).
[36] A. B. Baggeroer, "Solution to the Ricatti Equation for Butterworth Processes," Internal Memo., Detection and Estimation Theory Group, M.I.T., 1966.
[37] E. A. Guillemin, *Synthesis of Passive Networks*, Wiley, New York, 1957.
[38] L. Weinberg, *Network Analysis and Synthesis*, McGraw-Hill, New York.
[39] D. L. Snyder, "Optimum Linear Filtering of an Integrated Signal in White Noise," *IEEE Trans. Aerospace Electron. Systems*, **AES-2**, No. 2, 231–232 (March 1966).
[40] A. B. Baggeroer, "Maximum A Posteriori Interval Estimation," WESCON, Paper No. 7/3, August 23–26, 1966.
[41] A. E. Bryson and D. E. Johansen, "Linear Filtering for Time-Varying Systems Using Measurements Containing Colored Noise," *IEEE Transactions on Automatic Control*, **AC-10**, No. 1 (January 1965).
[42] L. D. Collins, "Optimum Linear Filters for Correlated Messages and Noises," Internal Memo. 17, Detection and Estimation Theory Group, M.I.T., October 28, 1966.
[43] A. B. Baggeroer and L. D. Collins, "Maximum A Posteriori Estimation in Correlated Noise," Internal Memo., Detection and Estimation Theory Group, M.I.T., October 25, 1966.
[44] H. L. Van Trees, "Sensitivity Equations for Optimum Linear Filters," Internal Memorandum, Detection and Estimation Theory Group, M.I.T., February 1966.
[45] J. B. Thomas, "On the Statistical Design of Demodulation Systems for Signals in Additive Noise," TR-88, Stanford University, August 1955.
[46] R. C. Booton, Jr., and M. H. Goldstein, Jr., "The Design and Optimization of Synchronous Demodulators," IRE Wescon Convention Record, Part 2, 154–170, 1957.
[47] "Single Sideband Issue," *Proc. IRE*, **44**, No. 12, (December 1956).
[48] J. M. Wozencraft, *Lectures on Communication System Theory*, E. J. Baghdady, ed., McGraw-Hill, New York, Chapter 12, 1961.
[49] M. Schwartz, *Information Transmission, Modulation, and Noise*, McGraw-Hill Co., New York, 1959.
[50] J. P. Costas, "Synchronous Communication," *Proc. IRE*, **44**, 1713–1718 (December 1956).
[51] R. Jaffe and E. Rechtin, "Design and Performance of Phase-Lock Circuits

Capable of Near-optimum Performance over a Wide Range of Input Signal and Noise Levels," *IRE Trans. Inform. Theory*, **IT-1**, 66–76 (March 1955).

[52] A. B. Baggeroer, "A State Variable Technique for the Solution of Fredholm Integral Equations," *Proc. 1967 Intl. Conf. Communications*, Minneapolis, Minn., June 12–14, 1967.

[53] A. B. Baggeroer and L. D. Collins, "Evaluation of Minimum Mean Square Error for Estimating Messages Whose Spectra Belong to the Gaussian Family," Internal Memo, Detection and Estimation Theory Group, M.I.T., R.L.E., July 12, 1967.

[54] A. B. Baggeroer, "A State-Variable Approach to the Solution of Fredholm Integral Equations," M.I.T., R.L.E. Technical Report 459, July 15, 1967.

[55] A. Papoulis, *Probability, Random Variables, and Stochastic Processes*, McGraw-Hill Co., New York, 1965.

[56] A. E. Bryson and M. Frazier, "Smoothing for Linear and Nonlinear Dynamic Systems," USAF Tech. Rpt. ASD-TDR-63-119 (February 1963).

7

Discussion

The next step in our development is to apply the results in Chapter 5 to the nonlinear estimation problem. In Chapter 4 we studied the problem of estimating a single parameter which was contained in the observed signal in a nonlinear manner. The discrete frequency modulation example we discussed illustrated the type of difficulties met. We anticipate that even more problems will appear when we try to estimate a sample function from a random process. This conjecture turns out to be true and it forces us to make several approximations in order to arrive at a satisfactory answer. To make useful approximations it is necessary to consider specific nonlinear estimation problems rather than the general case. In view of this specialization it seems appropriate to pause and summarize what we have already accomplished. In Chapter 2 of Part II we shall discuss nonlinear estimation. The remainder of Part II discusses random channels, radar/sonar signal processing, and array processing. In this brief chapter we discuss three issues.

1. The problem areas that we have studied. These correspond to the first two levels in the hierarchy outlined in Chapter 1.
2. The problem areas that remain to be covered when our development is resumed in Part II.
3. Some areas that we have encountered in our discussion that we do not pursue in Part II.

7.1 SUMMARY

Our initial efforts in this book were divided into developing background in the area of classical detection and estimation theory and random process representation. Chapter 2 developed the ideas of binary and M-ary hypothesis testing for Bayes and Neyman–Pearson criteria. The fundamental role of the likelihood ratio was developed. Turning to estimation theory, we studied Bayes estimation for random parameters and maximum-likelihood estimation for nonrandom parameters. The idea of likelihood functions, efficient estimates, and the Cramér-Rao inequality were of central importance in this discussion. The close relation between detection and estimation theory was emphasized in the composite hypothesis testing problem. A particularly important section from the standpoint of the remainder of the text was the general Gaussian problem. The model was simply the finite-dimensional version of the signals in noise problems which were the subject of most of our subsequent work. A section on performance bounds which we shall need in Chapter II.3 concluded the discussion of the classical problem.

In Chapter 3 we reviewed some of the techniques for representing random processes which we needed to extend the classical results to the waveform problem. Emphasis was placed on the Karhunen-Loéve expansion as a method for obtaining a set of uncorrelated random variables as a representation of the process over a finite interval. Other representations, such as the Fourier series expansion for periodic processes, the sampling theorem for bandlimited processes, and the integrated transform for stationary processes on an infinite interval, were also developed. As preparation for some of the problems that were encountered later in the text, we discussed various techniques for solving the integral equation that specified the eigenfunctions and eigenvalues. Simple characterizations for vector processes were also obtained. Later, in Chapter 6, we returned to the representation problem and found that the state-variable approach provided another method of representing processes in terms of a set of random variables. The relation between these two approaches is further developed in [1] (see Problem 6.6.2 of Chapter 6).

The next step was to apply these techniques to the solution of some basic problems in the communications and radar areas. In Chapter 4 we studied a large group of basic problems. The simplest problem was that of detecting one of two known signals in the presence of additive white Gaussian noise. The matched filter receiver was the first important result. The ideas of linear estimation and nonlinear estimation of parameters were derived. As pointed out at that time, these problems corresponded to engineering systems commonly referred to as uncoded PCM, PAM, and PFM, respectively. The extension to nonwhite Gaussian noise followed easily. In this case it was necessary to solve an integral equation to find the functions used in the optimum receiver, so we developed techniques to obtain these solutions. The method for dealing with unwanted parameters such as a random amplitude or random phase angle was derived and some typical cases were examined in detail. Finally, the extensions to multiple-parameter estimation and multiple-channel systems were accomplished. At this point in our development we could design and analyze digital communication systems and time-sampled analog communication systems operating in a reasonably general environment.

The third major area was the estimation of continuous waveforms. In Chapter 5 we approached the problem from the viewpoint of *maximum a posteriori probability* estimation. The primary result was a pair of integral equations that the optimum estimate had to satisfy. To obtain solutions we divided the problem into linear estimation which we studied in Chapter 6 and nonlinear estimation which we shall study in Chapter II.2.

Although MAP interval estimation served as a motivation and introduction to Chapter 6, the actual emphasis in the chapter was on minimum mean-square point estimators. After a brief discussion of general properties we concentrated on two classes of problems. The first was the optimum realizable point estimator for the case in which the processes were stationary and the infinite past was available. Using the Wiener-Hopf spectrum factorization techniques, we derived an algorithm for obtaining the form expressions for the errors. The second class was the optimum realizable point estimation problem for finite observation times and possibly nonstationary processes. After characterizing the processes by using state-variable methods, a differential equation that implicitly specified the optimum estimator was derived. This result, due to Kalman and Bucy, provided a computationally feasible way of finding optimum processors for complex systems. Finally, the results were applied to the types of linear modulation scheme encountered in practice, such as DSB-AM and SSB.

It is important to re-emphasize two points. The linear processors resulted from our initial Gaussian assumption and were the best processors of any kind. The use of the minimum mean-square error criterion was not a restriction because we showed in Chapter 2 that when the a posteriori density is Gaussian the conditional mean is the optimum estimate for any convex (upward) error criterion. Thus the results of Chapter 6 are of far greater generality than might appear at first glance.

7.2 PREVIEW OF PART II

In Part II we deal with four major areas. The first is the solution of the nonlinear estimation problem that was formulated in Chapter 5. In Chapter II.2 we study angle modulation in detail. The first step is to show how the integral equation that specifies the optimum MAP estimate suggests a demodulator configuration, commonly referred to as a phase-lock loop (PLL). We then demonstrate that under certain signal and noise conditions the output of the phase-lock loop is an asymptotically efficient estimate of the message. For the simple case which arises when the PLL is used for carrier synchronization an exact analysis of the performance in the nonlinear region is derived. Turning to frequency modulation, we investigate the design of optimum demodulators in the presence of bandwidth and power constraints. We then compare the performance of these optimum demodulators with conventional limiter-discriminators. After designing the optimum receiver the next step is to return to the transmitter and modify it to improve the overall system performance. This modification, analogous to the pre-emphasis problem in conventional frequency modulation, leads to an optimum angle modulation system. Our results in this chapter and Chapter I.4. give us the background to answer the following question: If we have an analog message to transmit, should we (a) sample and quantize it and use a digital transmission system, (b) sample it and use a continuous amplitude-time discrete system, or (c) use a continuous analog system? In order to answer this question we first use the rate-distortion results of Shannon to derive bounds on how well *any* system could do. We then compare the various techniques discussed above with these bounds. As a final topic in this chapter we show how state-variable techniques can be used to derive optimum nonlinear demodulators.

In Chapter II.3 we return to a more general problem. We first consider the problem of observing a received waveform $r(t)$ and deciding to which of two random processes it belongs. This type of problem occurs naturally in radio astronomy, scatter communication, and passive sonar. The likelihood ratio test leads to a quadratic receiver which can be realized in several mathematically equivalent ways. One receiver implementation, the estimator-correlator realization, contains the optimum linear filter as a component and lends further importance to the results in Chapter I.6. In a parallel manner we consider the problem of estimating parameters contained in the covariance function of a random process. Specific problems encountered in practice, such as the estimation of the center frequency of a bandpass process or the spectral width of a process, are discussed in detail. In both the detection and estimation cases particularly simple solutions are obtained when the processes are stationary and the observation times are long. To complete the hierarchy of problems outlined in Chapter 1 we study the problem of transmitting an analog message over a randomly time-varying channel. As a specific example, we study an angle modulation system operating over a Rayleigh channel.

In Chapter II.4 we show how our earlier results can be applied to solve detection and parameter estimation problems in the radar-sonar area. Because we are interested in narrow-band signals, we develop a representation for them by using complex envelopes. Three classes of target models are considered. The simplest is a slowly fluctuating point target whose range and velocity are to be estimated. We find that the issues of accuracy, ambiguity, and resolution must all be considered when designing signals and processors. The radar ambiguity function originated by Woodward plays a central role in our discussion. The targets in the next class are referred to as singly spread. This class includes fast-fluctuating point targets which spread the transmitted signal in frequency and slowly fluctuating dispersive targets which spread the signal in time (range). The third class consists of doubly spread targets which spread the signal in both time and frequency. Many diverse physical situations, such as reverberation in active sonar systems, communication over scatter channels, and resolution in mapping radars, are included as examples. The overall development provides a unified picture of modern radar-sonar theory.

The fourth major area in Part II is the study of multiple-process and multivariable process problems. The primary topic in this area is a detailed study of array processing in passive sonar (or seismic) systems. Optimum processors for typical noise fields are analyzed from the standpoint of signal waveform estimation accuracy, detection performance, and beam patterns. Two other topics, multiplex transmission systems and multivariable systems (e.g., continuous receiving apertures or optical systems), are discussed briefly.

In spite of the length of the two volumes, a number of interesting topics have been omitted. Some of them are outlined in the next section.

7.3 UNEXPLORED ISSUES

Several times in our development we have encountered interesting ideas whose complete discussion would have taken us too far afield. In this section we provide suitable references for further study.

Coded Digital Communication Systems. The most important topic we have not discussed is the use of coding techniques to reduce errors in systems that transmit sequences of digits. Shannon's classical information theory results indicate how well we can do, and a great deal of research has been devoted to finding ways to approach these bounds. Suitable texts in this area are Wozencraft and Jacobs [2], Fano [3], Gallager [4], Peterson [5], and Golomb [6]. Bibliographies of current papers appear in the "Progress in Information Theory" series [7] and [8].

Sequential Detection and Estimation Schemes. All our work has dealt with a fixed observation interval. Improved performance can frequently be obtained if a variable length test is allowed. The fundamental work in this area is due to Wald [10]. It was applied to the waveform problem by

Peterson, Birdsall, and Fox [11] and Bussgang and Middleton [12]. Since that time a great many papers have considered various aspects of the subject (e.g. [13] to [22]).

Nonparametric Techniques. Throughout our discussion we have assumed that the random variables and processes have known density functions. The nonparametric statistics approach tries to develop tests that do not depend on the density function. As we might expect, the problem is more difficult, and even in the classical case there are a number of unsolved problems. Two books in the classical area are Fraser [23] and Kendall [24]. Other references are given on pp. 261–264 of [25] and in [26]. The progress in the waveform case is less satisfactory. (A number of models start by sampling the input and then use a known classical result.) Some recent interesting papers include [27] to [32].

Adaptive and Learning Systems. Ever-popular adjectives are the words "adaptive" and "learning." By a suitable definition of adaptivity many familiar systems can be described as adaptive or learning systems. The basic notion is straightforward. We want to build a system to operate efficiently in an unknown or changing environment. By allowing the system to change its parameters or structure as a function of its input, the performance can be improved over that of a fixed system. The complexity of the system depends on the assumed model of the environment and the degrees of freedom that are allowed in the system. There are a large number of references in this general area. A representative group is [33] to [50] and [9].

Pattern Recognition. The problem of interest in this area is to recognize (or classify) patterns based on some type of imperfect observation. Typical areas of interest include printed character recognition, target classification in sonar systems, speech recognition, and various medical areas such as cardiology. The problem can be formulated in either a statistical or non-statistical manner. Our work would be appropriate to a statistical formulation.

In the statistical formulation the observation is usually reduced to an N-dimensional vector. If there are M possible patterns, the problem reduces to the finite dimensional M-hypothesis testing problem of Chapter 2. If we further assume a Gaussian distribution for the measurements, the general Gaussian problem of Section 2.6 is directly applicable. Some typical pattern recognition applications were developed in the problems of Chapter 2. A more challenging problem arises when the Gaussian assumption is not valid; [8] discusses some of the results that have been obtained and also provides further references.

Discrete Time Processes. Most of our discussion has dealt with observations that were sample functions from continuous time processes. Many of the results can be reformulated easily in terms of discrete time processes. Some of these transitions were developed in the problems.

Non-Gaussian Noise. Starting with Chapter 4, we confined our work to Gaussian processes. As we pointed out at the end of Chapter 4, various types of result can be obtained for other forms of interference. Some typical results are contained [51] to [57].

Receiver-to-Transmitter Feedback Systems. In many physical situations it is reasonable to have a feedback link from the receiver to the transmitter. The addition of the link provides a new dimension in system design and frequently enables us to achieve efficient performance with far less complexity than would be possible in the absence of feedback. Green [58] describes the work in the area up to 1961. Recent work in the area includes [59] to [63] and [76].

Physical Realizations. The end product of most of our developments has been a block diagram of the optimum receiver. Various references discuss practical realizations of these block diagrams. Representative systems that use the techniques of modern communication theory are described in [64] to [68].

Markov Process-Differential Equation Approach. With the exception of Section 6.3, our approach to the detection and estimation problem could be described as a "covariance function-impulse response" type whose success was based on the fact that the processes of concern were Gaussian. An alternate approach could be based on the Markovian nature of the processes of concern. This technique might be labeled the "state-variable-differential equation" approach and appears to offer advantages in many problems: [69] to [75] discuss this approach for some specific problems.

Undoubtedly there are other related topics that we have not mentioned, but the foregoing items illustrate the major ones.

REFERENCES

[1] A. B. Baggeroer, "Some Applications of State Variable Techniques in Communications Theory," Sc.D. Thesis, Dept. of Electrical Engineering, M.I.T., February 1968.

[2] J. M. Wozencraft and I. M. Jacobs, *Principles of Communication Engineering*, Wiley, New York, 1965.

[3] R. M. Fano, *The Transmission of Information*, The M.I.T. Press and Wiley, New York, 1961.

[4] R. G. Gallager, *Notes on Information Theory*, Course 6.574, M.I.T.

[5] W. W. Peterson, *Error-Correcting Codes*, Wiley, New York, 1961.

[6] S. W. Golomb (ed.), *Digital Communications with Space Applications*, Prentice-Hall, Englewood Cliffs, New Jersey, 1964.

[7] P. Elias, A. Gill, R. Price, P. Swerling, L. Zadeh, and N. Abramson; "Progress in Information Theory in the U.S.A., 1957–1960," *IRE Trans. Inform. Theory*, **IT-7**, No. 3, 128–143 (July 1960).

[8] L. A. Zadeh, N. Abramson, A. V. Balakrishnan, D. Braverman, M. Eden, E. A. Feigenbaum, T. Kailath, R. M. Lerner, J. Massey, G. E. Mueller, W. W. Peterson, R. Price, G. Sebestyen, D. Slepian, A. J. Thomasian, G. L. Turin, "Report on Progress in Information Theory in the U.S.A., 1960–1963," *IEEE Trans. Inform. Theory*, **IT-9**, 221–264 (October 1963).

[9] M. E. Austin, "Decision-Feedback Equalization for Digital Communication Over Dispersive Channels," M.I.T. Sc.D. Thesis, May 1967.

[10] A. Wald, *Sequential Analysis*, Wiley, New York, 1947.

[11] W. W. Peterson, T. G. Birdsall, and W. C. Fox, "The Theory of Signal Detectability," *IRE Trans. Inform. Theory*, **PGIT-4**, 171 (1954).

[12] J. Bussgang and D. Middleton, "Optimum Sequential Detection of Signals in Noise," *IRE Trans. Inform. Theory*, **IT-1**, 5 (December 1955).

[13] H. Blasbalg, "The Relationship of Sequential Filter Theory to Information Theory and Its Application to the Detection of Signals in Noise by Bernoulli Trials," *IRE Trans. Inform. Theory*, **IT-3**, 122 (June 1957); Theory of Sequential Filtering and Its Application to Signal Detection and Classification, doctoral dissertation, Johns Hopkins University, 1955; "The Sequential Detection of a Sine-wave Carrier of Arbitrary Duty Ratio in Gaussian Noise," *IRE Trans. Inform. Theory*, **IT-3**, 248 (December 1957).

[14] M. B. Marcus and P. Swerling, "Sequential Detection in Radar With Multiple Resolution Elements," *IRE Trans. Inform. Theory*, **IT-8**, 237–245 (April 1962).

[15] G. L. Turin, "Signal Design for Sequential Detection Systems With Feedback," *IEEE Trans. Inform. Theory*, **IT-11**, 401–408 (July 1965); "Comparison of Sequential and Nonsequential Detection Systems with Uncertainty Feedback," *IEEE Trans. Inform. Theory*, **IT-12**, 5–8 (January 1966).

[16] A. J. Viterbi, "The Effect of Sequential Decision Feedback on Communication over the Gaussian Channel," *Information and Control*, **8**, 80–92 (February 1965).

[17] R. M. Gagliardi and I. S. Reed, "On the Sequential Detection of Emerging Targets," *IEEE Trans. Inform. Theory*, **IT-11**, 260–262 (April 1965).

[18] C. W. Helstrom, "A Range Sampled Sequential Detection System," *IRE Trans. Inform. Theory*, **IT-8**, 43–47 (January 1962).

[19] W. Kendall and I. S. Reed, "A Sequential Test for Radar Detection of Multiple Targets," *IRE Trans. Inform. Theory*, **IT-9**, 51 (January 1963).

[20] M. B. Marcus and J. Bussgang, "Truncated Sequential Hypothesis Tests," The RAND Corp., RM-4268-ARPA, Santa Monica, California, November 1964.

[21] I. S. Reed and I. Selin, "A Sequential Test for the Presence of a Signal in One of k Possible Positions," *IRE Trans. Inform. Theory*, **IT-9**, 286 (October 1963).

[22] I. Selin, "The Sequential Estimation and Detection of Signals in Normal Noise, I," *Information and Control*, **7**, 512–534 (December 1964); "The Sequential Estimation and Detection of Signals in Normal Noise, II," *Information and Control*, **8**, 1–35 (January 1965).

[23] D. A. S. Fraser, *Nonparametric Methods in Statistics*, Wiley, New York, 1957.

[24] M. G. Kendall, *Rank Correlation Methods*, 2nd ed., Griffin, London, 1955.

[25] E. L. Lehmann, *Testing Statistical Hypotheses*, Wiley, New York, 1959.

[26] I. R. Savage, "Bibliography of nonparametric statistics and related topics," *J. American Statistical Association*, **48**, 844–906 (1953).

[27] M. Kanefsky, "On sign tests and adaptive techniques for nonparametric detection," Ph.D. dissertation, Princeton University, Princeton, New Jersey, 1964.

[28] S. S. Wolff, J. B. Thomas, and T. Williams, "The Polarity Coincidence Correlator: a Nonparametric Detection Device," *IRE Trans. Inform. Theory*, **IT-8**, 1–19 (January 1962).

[29] H. Ekre, "Polarity Coincidence Correlation Detection of a Weak Noise Source," *IEEE Trans. Inform. Theory*, **IT-9**, 18–23 (January 1963).

[30] M. Kanefsky, "Detection of Weak Signals with Polarity Coincidence Arrays," *IEEE Trans. Inform. Theory*, **IT-12** (April 1966).

[31] R. F. Daly, "Nonparametric Detection Based on Rank Statistics," *IEEE Trans. Inform. Theory*, **IT-12** (April 1966).

[32] J. B. Millard and L. Kurz, "Nonparametric Signal Detection—An Application of the Kolmogorov-Smirnov, Cramer-Von Mises Tests," *IEEE Trans. Inform. Theory*, **IT-12** (April 1966).

[33] C. S. Weaver, "Adaptive Communication Filtering," *IEEE Trans. Inform. Theory*, **IT-8**, S169–S178 (September 1962).

[34] A. M. Breipohl and A. H. Koschmann, "A Communication System with Adaptive Decoding," Third Symposium on Discrete Adaptive Processes, 1964 National Electronics Conference, Chicago, 72–85, October 19–21.

[35] J. G. Lawton, "Investigations of Adaptive Detection Techniques," Cornell Aeronautical Laboratory Report, No. RM-1744-S-2, Buffalo, New York, November, 1964.

[36] R. W. Lucky, "Techniques for Adaptive Equalization of Digital Communication System," *Bell System Tech. J.*, 255–286 (February 1966).

[37] D. T. Magill, "Optimal Adaptive Estimation of Sampled Stochastic Processes," Stanford Electronics Laboratories, TR No. 6302–2, Stanford University, California, December, 1963.

[38] K. Steiglitz and J. B. Thomas, "A Class of Adaptive Matched Digital Filters," Third Symposium on Discrete Adaptive Processes, 1964 National Electronic Conference, pp. 102–115, Chicago, Illinois, October, 1964.

[39] H. L. Groginsky, L. R. Wilson, and D. Middleton, "Adaptive Detection of Statistical Signals in Noise," *IEEE Trans. Inform. Theory*, **IT-12** (July 1966).

[40] Y. C. Ho and R. C. Lee, "Identification of Linear Dynamic Systems," Third Symposium on Discrete Adaptive Processes, 1964 National Electronic Conference, pp. 86–101, Chicago, Illinois, October, 1964.

[41] L. W. Nolte, "An Adaptive Realization of thc Optimum Receiver for a Sporadic-Recurrent Waveform in Noise," First IEEE Annual Communication Convention, pp. 593–398, Boulder, Colorado, June, 1965.

[42] R. Esposito, D. Middleton, and J. A. Mullen, "Advantages of Amplitude and Phase Adaptivity in the Detection of Signals Subject to Slow Rayleigh Fading," *IEEE Trans. Inform. Theory*, **IT-11**, 473–482 (October 1965).

[43] J. C. Hancock and P. A. Wintz, "An Adaptive Receiver Approach to the Time Synchronization Problem," *IEEE Trans. Commun. Systems*, **COM-13**, 90–96 (March 1965).

[44] M. J. DiToro, "A New Method of High-Speed Adaptive Serial Communication

Through Any Time-Variable and Dispersive Transmission Medium," First IEEE Annual Communication Convention, pp. 763–767, Boulder, Colorado, June, 1965.

[45] L. D. Davisson, "A Theory of Adaptive Filtering," *IEEE Trans. Inform. Theory*, **IT-12**, 97–102 (April 1966).

[46] H. J. Scudder, "Probability of Error of Some Adaptive Pattern-Recognition Machines," *IEEE Trans. Inform. Theory*, **IT-11**, 363–371 (July 1965).

[47] C. V. Jakowitz, R. L. Shuey, and G. M. White, "Adaptive Waveform Recognition," TR60-RL-2353E, General Electric Research Laboratory, Schenectady, New York, September, 1960.

[48] E. M. Glaser, "Signal Detection by Adaptive Filters," *IRE Trans.*, **IT-7**, No. 2, 87–98 (April 1961).

[49] D. Boyd, "Several Adaptive Detection Systems," M.Sc. Thesis, Department of Electrical Engineering, M.I.T., July, 1965.

[50] H. J. Scudder, "Adaptive Communication Receivers," *IEEE Trans. Inform. Theory*, **IT-11**, No. 2, 167–174 (April 1965).

[51] V. R. Algazi and R. M. Lerner, "Optimum Binary Detection in Non-Gaussian Noise," *IEEE Trans. Inform. Theory*, **IT-12**, (April 1966).

[52] B. Reiffen and H. Sherman, "An Optimum Demodulator for Poisson Processes; Photon Source Detectors," *Proc. IEEE*, **51**, 1316–1320 (October 1963).

[53] L. Kurz, "A Method of Digital Signalling in the Presence of Additive Gaussian and Impulsive Noise," 1962 *IRE International Convention Record*, Pt. 4, pp. 161–173.

[54] R. M. Lerner, "Design of Signals," in *Lectures on Communication System Theory*, E. J. Baghdady, ed., McGraw-Hill, New York, Chapter 11, 1961.

[55] D. Middleton, "Acoustic Signal Detection by Simple Correlators in the Presence of Non-Gaussian Noise," *J. Acoust. Soc. Am.*, **34**, 1598–1609 (October 1962).

[56] J. R. Feldman and J. N. Faraone, "An analysis of frequency shift keying systems," *Proc. National Electronics Conference*, **17**, 530–542 (1961).

[57] K. Abend, "Noncoherent optical communication," Philco Scientific Lab., Blue Bell, Pa., Philco Rept. 312 (February 1963).

[58] P. E. Green, "Feedback Communication Systems," *Lectures on Communication System Theory*, E. J. Baghdady (ed.), McGraw-Hill, New York, 1961.

[59] J. K. Omura, *Signal Optimization for Channels with Feedback*, Stanford Electronics Laboratories, Stanford University, California, Technical Report No. 7001–1, August 1966.

[60] J. P. M. Schalkwijk and T. Kailath, "A Coding Scheme for Additive Noise Channels with Feedback—Part I: No Bandwidth Constraint," *IEEE Trans. Inform. Theory*, **IT-12**, April 1966.

[61] J. P. M. Schalkwijk, "Coding for a Class of Unknown Channels," *IEEE Trans. Inform. Theory*, **IT-12** (April 1966).

[62] M. Horstein, "Sequential Transmission using Noiseless Feedback," *IEEE Trans. Inform. Theory*, **IT-9**, No. 3, 136–142 (July 1963).

[63] T. J. Cruise, "Communication using Feedback," Sc.D. Thesis, Dept. of Electrical Engr., M.I.T. (to be published, 1968).

[64] G. L. Turin, "Signal Design for Sequential Detection Systems with Feedback," *IEEE Trans. Inform. Theory*, **IT-11**, 401–408 (July 1965).

[65] J. D. Belenski, "The application of correlation techniques to meteor burst communications," presented at the 1962 WESCON Convention, Paper No. 14.4.

[66] R. M. Lerner, "A Matched Filter Communication System for Complicated Doppler Shifted Signals," *IRE Trans. Inform. Theory*, **IT-6**, 373–385 (June 1960).

[67] R. S. Berg, et al., "The DICON System," MIT Lincoln Laboratory, Lexington, Massachusetts, Tech. Report No. 251, October 31, 1961.

[68] C. S. Krakauer, "Serial-Parallel Converter System 'SEPATH'," presented at the 1961 IRE WESCON Convention, Paper No. 36/2.

[69] H. L. Van Trees, "Detection and Continuous Estimation: The Fundamental Role of the Optimum Realizable Linear Filter," 1966 WESCON Convention, Paper No. 7/1, August 23–26, 1966.

[70] D. L. Snyder, "A Theory of Continuous Nonlinear Recursive-Filtering with Application of Optimum Analog Demodulation," 1966 WESCON Convention, Paper No. 7/2, August 23–26, 1966.

[71] A. B. Baggeroer, "Maximum A Posteriori Interval Estimation," 1966 WESCON Convention, Paper No. 7/3, August 23–26, 1966.

[72] J. K. Omura, "Signal Optimization for Additive Noise Channels with Feedback," 1966 WESCON Convention, Paper No. 7/4, August 23–26, 1966.

[73] F. C. Schweppe, "A Modern Systems Approach to Signal Design," 1966 WESCON Convention, Paper 7/5, August 23–26, 1966.

[74] M. Athans, F. C. Schweppe, "On Optimal Waveform Design Via Control Theoretic Concepts I: Formulations," MIT Lincoln Laboratory, Lexington, Massachusetts (submitted to Information and Control); "On Optimal Waveform Design Via Control Theoretic Concepts II: The On-Off Principle," MIT Lincoln Laboratory, Lexington, Massachusetts, JA-2789, July 5, 1966.

[75] F. C. Schweppe, D. Gray, "Radar Signal Design Subject to Simultaneous Peak and Average Power Constraints," *IEEE Trans. Inform. Theory*, **II-12**, No. 1, 13–26 (1966).

[76] T. J. Cruise, "Achievement of Rate-Distortion Bound Over Additive White Noise Channel Utilizing a Noiseless Feedback Channel," Proceedings of the IEEE, **55**, No. 4 (April 1967).

Appendix:
A Typical Course Outline

It would be presumptious of us to tell a professor how to teach a course at this level. On the other hand, we have spent a great deal of time experimenting with different presentations in search of an efficient and pedagogically sound approach. The concepts are simple, but the great amount of detail can cause confusion unless the important issues are emphasized. The following course outline is the result of these experiments; it should be useful to an instructor who is using the book or teaching this type of material for the first time.

The course outline is based on a 15-week term of three hours of lectures a week. The homework assignments, including reading and working the assigned problems, will take 6 to 15 hours a week. The prerequisite assumed is a course in random processes. Typically, it should include Chapters 1 to 6, 8, and 9 of Davenport and Root or Chapters 1 to 10 of Papoulis. Very little specific material in either of these references is used, but the student needs a certain level of sophistication in applied probability theory to appreciate the subject material.

Each lecture unit corresponds to a one-and-one-half-hour lecture and contains a topical outline, the corresponding text material, and additional comments when necessary. Each even-numbered lecture contains a problem assignment. A set of solutions for this collection of problems is available.

In a normal term we get through the first 28 lectures, but this requires a brisk pace and leaves no room for making up background deficiencies. An ideal format would be to teach the material in two 10-week quarters. The expansion from 32 to 40 lectures is easily accomplished (probably without additional planning). A third alternative is two 15-week terms, which would allow time to cover more material in class and reduce the homework load. We have not tried either of the last two alternatives, but student comments indicate they should work well.

One final word is worthwhile. There is a great deal of difference between reading the text and being able to apply the material to solve actual problems of interest. This book (and therefore presumably any course using it) is designed to train engineers and scientists to solve new problems. The only way for most of us to acquire this ability is by practice. Therefore any effective course must include a fair amount of problem solving and a critique (or grading) of the students' efforts.

	Lecture 1 **pp. 1–18**
Chapter 1	Discussion of the physical situations that give rise to detection, estimation, and modulation theory problems
1.1	Detection theory, 1–5 Digital communication systems (known signal in noise), 1–2 Radar/sonar systems (signal with unknown parameters), 3 Scatter communication, passive sonar (random signals), 4 Show hierarchy in Fig. 1.4, 5 Estimation theory, 6–8 PAM and PFM systems (known signal in noise), 6 Range and Doppler estimation in radar (signal with unknown parameters), 7 Power spectrum parameter estimation (random signal), 8 Show hierarchy in Fig. 1.7, 8 Modulation theory, 9–12 Continuous modulation systems (AM and FM), 9 Show hierarchy in Fig. 1.10, 11
1.2	Various approaches, 12–15 Structured versus nonstructured, 12–15 Classical versus waveform, 15
1.3	Outline of course, 15–18

	Lecture 2	**pp. 19–33**
Chapter 2	Formulation of the hypothesis testing problem, 19–23	

2.2.1	Decision criteria (Bayes), 23–30
	    Necessary inputs to implement test; a priori probabilities and costs, 23–24
	    Set up risk expression and find LRT, 25–27
	    Do three examples in text, introduce idea of sufficient statistic, 27–30
	Minimax test, 31–33
	Minimum Pr(ϵ) test, idea of maximum a posteriori rule, 30	

	Problem Assignment 1 **1.** 2.2.1 **2.** 2.2.2	

	Lecture 3 **pp. 33–52**
	Neyman–Pearson tests, 33–34 Fundamental role of LRT, relative unimportance of the criteria, 34 Sufficient statistics, 34 Definition, geometric interpretation

2.2.2	Performance, idea of ROC, 36–46
	  Example 1 on pp. 36–38; Bound on $\text{erfc}_*(X)$, (72)
	Properties of ROC, 44–48
	  Concave down, slope, minimax

2.3	M-Hypotheses, 46–52
	  Set up risk expression, do $M = 3$ case and demonstrate that the decision space has at most two dimensions; emphasize that regardless of the observation space dimension the decision space has at most $M - 1$ dimensions, 52
	  Develop idea of maximum a posteriori probability test (109)

Comments

The randomized tests discussed on p. 43 are not important in the sequel and may be omitted in the first reading.

	Lecture 4	**pp. 52–62**

2.4	Estimation theory, 52
	Model, 52–54
	  Parameter space, question of randomness, 53
	  Mapping into observation space, 53
	  Estimation rule, 53
	Bayes estimation, 54–63
	  Cost functions, 54
	  Typical single-argument expressions, 55
	   Mean-square, absolute magnitude, uniform, 55
	  Risk expression, 55
	  Solve for $\hat{a}_{ms}(\mathbf{R})$, $\hat{a}_{map}(\mathbf{R})$, and $\hat{a}_{mae}(\mathbf{R})$, 56–58
	  Linear example, 58–59
	  Nonlinear example, 62	

Problem Assignment 2

1.	2.2.10	**5.**	2.3.3
2.	2.2.15	**6.**	2.3.5
3.	2.2.17	**7.**	2.4.2
4.	2.3.2	**8.**	2.4.3

	Lecture 5	**pp. 60–69**

	Bayes estimation (*continued*)
	  Convex cost criteria, 60–61
	  Optimality of $\hat{a}_{ms}(\mathbf{R})$, 61	

2.4.2	Nonrandom parameter estimation, 63–73
	  Difficulty with direct approach, 64
	  Bias, variance, 64
	Maximum likelihood estimation, 65
	Bounds
	  Cramér-Rao inequality, 66–67
	  Efficiency, 66
	  Optimality of $\hat{a}_{ml}(\mathbf{R})$ when efficient estimate exists, 68
	Linear example, 68–69	

	Lecture 6 **pp. 69–98**
	Nonrandom parameter estimation (*continued*) Nonlinear example, 69 Asymptotic results, 70–71 Intuitive explanation of when C–R bound is accurate, 70–71 Bounds for random variables, 72–73
2.4.3, 2.4.4 **2.5**	Assign the sections on multiple parameter estimation and composite hypothesis testing for reading, 74–96
2.6	General Gaussian problem, 96–116 Definition of a Gaussian random vector, 96 Expressions for $M_{\mathbf{r}}(j\mathbf{v})$, $p_{\mathbf{r}}(\mathbf{R})$, 97 Derive LRT, define quadratic forms, 97–98
	Problem Assignment 3 **1.** 2.4.9 **5.** 2.6.1 **2.** 2.4.12 *Optional* **3.** 2.4.27 **6.** 2.5.1 **4.** 2.4.28

	Lecture 7	**pp. 98–133**
2.6	General Gaussian problem (*continued*)	
	Equal covariance matrices, unequal mean vectors, 98–107	
	Expression for d^2, interpretation as distance, 99–100	
	Diagonalization of **Q**, eigenvalues, eigenvectors, 101–107	
	Geometric interpretation, 102	
	Unequal covariance matrices, equal mean vectors, 107–116	
	Structure of LRT, interpretation as estimator, 107	
	Diagonal matrices with identical components, χ^2 density, 107–111	
	Computational problem, motivation of performance bounds, 116	
2.7	Assign section on performance bounds as reading, 116–133	

Comments

1. (p. 99) Note that $\text{Var}[l \mid H_1] = \text{Var}[l \mid H_0]$ and that d^2 completely characterizes the test because l is Gaussian on both hypotheses.
2. (p. 119) The idea of a "tilted" density seems to cause trouble. Emphasize the motivation for tilting.

	Lecture 8 **pp. 166–186**
Chapter 3	

3.1, 3.2	Extension of results to waveform observations
	Deterministic waveforms
	    Time-domain and frequency-domain characterizations, 166–169
	    Orthogonal function representations, 169–174
	    Complete orthonormal sets, 171
	    Geometric interpretation, 172–174

3.3, 3.3.1	Second-moment characterizations, 174–178 Positive definiteness, nonnegative definiteness, and symmetry of covariance functions, 176–177

3.3.3	Gaussian random processes, 182
	    Difficulty with usual definitions, 185
	    Definition in terms of a linear functional, 183
	    Jointly Gaussian processes, 185
	    Consequences of definition; joint Gaussian density at any set of times, 184–185

	Problem Assignment 4 **1.** 2.6.2 *Optional* **2.** 2.6.4. **5.** 2.7.1 **3.** 2.6.8 **6.** 2.7.2 **4.** 2.6.10

	Lecture 9 **pp. 178–194**
3.3.2	Orthogonal representation for random processes, 178–182 Choice of coefficients to minimize mean-square representation error, 178 Choice of coordinate system, 179 Karhunen-Loève expansion, 179 Properties of integral equations, 180–181 Analogy with finite-dimensional case in Section 2.6 The following comparison should be made:

Gaussian definition $z = \mathbf{g}^T\mathbf{x}$	$z = \int_0^T g(u)\, x(u)\, du$
Symmetry $K_{ij} = K_{ji}$	$K_x(t, u) = K_x(u, t)$
Nonnegative definiteness $\mathbf{x}^T\mathbf{K}\mathbf{x} \geq 0$	$\int_0^T dt \int_0^T du\, x(t)\, K_x(t, u)\, x(u) \geq 0$
Coordinate system $\mathbf{K}\boldsymbol{\phi} = \lambda\boldsymbol{\phi}$	$\int_0^T K_x(t, u)\, \phi(u)\, du = \lambda\phi(t) \quad 0 \leq t \leq T$
Orthogonality $\boldsymbol{\phi}_i^T\boldsymbol{\phi}_j = \delta_{ij}$	$\int_0^T \phi_i(t)\, \phi_j(t)\, dt = \delta_{ij}$

	Mercer's theorem, 181 Convergence in mean-square sense, 182
3.4, 3.4.1, 3.4.2	Assign Section 3.4 on integral equation solutions for reading, 186–194

	Lecture 10	**pp. 186–226**
3.4.1	Solution of integral equations, 186–196	
3.4.2	Basic technique, obtain differential equation, solve, and satisfy boundary conditions, 186–191	
3.4.3	Example: Wiener process, 194–196	
3.4.4	White noise and its properties, 196–198 Impulsive covariance function, flat spectrum, orthogonal representation, 197–198	
3.4.5	Optimum linear filter, 198–204 This derivation illustrates variational procedures; the specific result is needed in Chapter 4; the integral equation (144) should be emphasized because it appears in many later discussions; a series solution in terms of eigenvalues and eigenfunctions is adequate for the present	
3.4.6–3.7	Assign the remainder of Chapter 3 for reading; Sections 3.5 and 3.6 are not used in the text (some problems use the results); Section 3.7 is not needed until Section 4.5 (the discussion of vector processes can be avoided until Section 6.3, when it becomes essential), 204–226	
	Problem Assignment 5 **1.** 3.3.1 **5.** 3.4.4 **2.** 3.3.6 **6.** 3.4.6 **3.** 3.3.19 **7.** 3.4.8 **4.** 3.3.22	

Comment

At the beginning of the derivation on p. 200 we assumed $h(t, u)$ was continuous. Whenever there is a white noise component in $r(t)$ (143), this restriction does not affect the performance. When $K_r(t, u)$ does not contain an impulse, a discontinuous filter may perform better. The optimum filter satisfies (138) for $0 \leq u \leq T$ if discontinuities are allowed.

	Lecture 11 **pp. 239–257**
Chapter 4	
4.1	Physical situations in which detection problem arises Communication, radar/sonar, 239–246
4.2	Detection of known signals in additive white Gaussian noise, 246–271 Simple binary detection, 247 Sufficient statistic, reduction to scalar problem, 248 Applicability of ROC in Fig. 2.9 with $d^2 = 2E/N_0$, 250–251 Lack of dependence on signal shape, 253 General binary detection, 254–257 Coordinate system using Gram-Schmidt, 254 LRT, reduction to single sufficient statistic, 256 Minimum-distance rule; "largest-of" rule, 257 Expression for d^2, 256 Optimum signal choice, 257

	Lecture 12 **pp. 257–271**
4.2.1	*M*-ary detection in white Gaussian noise, 257 Set of sufficient statistics; Gram-Schmidt procedure leads to at most *M* (less if signals have some linear dependence); Illustrate with PSK and FSK set, 259 Emphasize equal a priori probabilities and minimum $\Pr(\epsilon)$ criterion. This leads to minimum-distance rules. Go through three examples in text and calculate $\Pr(\epsilon)$ Example 1 illustrates symmetry and rotation, 261 Example 3 illustrates complexity of exact calculation for a simple signal set. Derive bound and discuss accuracy, 261–264 Example 4 illustrates the idea of transmitting sequences of digits and possible performance improvements, 264–267 Sensitivity, 267–271 Functional variation and parameter variation; this is a mundane topic that is usually omitted; it is a crucial issue when we try to implement optimum systems and must measure the quantities needed in the mathematical model
	Problem Assignment 6 **1.** 4.2.4 **5.** 4.2.16 **2.** 4.2.6 **6.** 4.2.23 **3.** 4.2.8 **4.** 4.2.9

	Lecture 13 **pp. 271–278**
4.2.2	Estimation of signal parameters Derivation of likelihood function, 274 Necessary conditions on $\hat{a}_{map}(r(t))$ and $\hat{a}_{ml}(r(t))$, 274 Generalization of Cramér-Rao inequality, 275 Conditions for efficient estimates, 276 Linear estimation, 271–273 Simplicity of solution, 272 Relation to detection problem, 273
4.2.3	Nonlinear estimation, 273–286 Optimum receiver for estimating arrival time (equivalently, the PPM problem), 276 Intuitive discussion of system performance, 277

	Lecture 14	**pp. 278–289**

4.2.3	Nonlinear estimation (*continued*), 278–286
	Pulse frequency modulation (PFM), 278
	  An approximation to the optimum receiver (interval selector followed by selection of local maximum), 279
	  Performance in weak noise, effect of βT product, 280
	  Threshold analysis, using orthogonal signal approximation, 280–281
	  Bandwidth constraints, 282
	  Design of system under threshold and bandwidth constraints, 282
	  Total mean-square error, 283–285	

4.2.4	Summary: known signals in white noise, 286–287	
4.3	Introduction to colored noise problem, 287–289 Model, observation interval, motivation for including white noise component	
	Problem Assignment 7 **1.** 4.2.25 **3.** 4.2.28 **2.** 4.2.26	

Comments on Lecture 15 (*lecture is on p. 651*)

1. As in Section 3.4.5, we must be careful about the endpoints of the interval. If there is a white noise component, we may choose a continuous $g(t)$ without affecting the performance. This leads to an integral equation on an open interval. If there is no white component, we must use a closed interval and the solution will usually contain singularities.

2. On p. 296 it is useful to emphasize the analogy between an inverse kernel and an inverse matrix.

	Lecture 15 **pp. 289–333**
4.3	Colored noise problem
4.3.1, 4.3.2, 4.3.3	Possible approaches: Karhunen-Loéve expansion, prewhitening filter, generation of sufficient statistic, 289
	Reversibility proof (this idea is used several times in the text), 289–290
	Whitening derivation, 290–297
	Define whitening filter, $h_w(t, u)$, 291
	Define inverse kernel $Q_n(t, u)$ and function for correlator $g(t)$, 292
	Derive integral equations for above functions, 293–297
	Draw the three realizations for optimum receiver (Fig. 4.38), 293
	Construction of $Q_n(t, u)$
	Interpretation as impulse minus optimum linear filter, 294–295
	Series solutions, 296–297
4.3.4	Performance, 301–307
	Expression for d^2 as quadratic form and in terms of eigenvalues, 302
	Optimum signal design, 302–303
	Singularity, 303–305
	Importance of white noise assumption and effect of removing it
4.3.8	Duality with known channel problem, 331–333
4.3.5–4.3.8	Assign as reading,
	1. Estimation (4.3.5)
	2. Solution to integral equations (4.3.6)
	3. Sensitivity (4.3.7)
	4. Known linear channels (4.3.8)
	Problem Set 7 (*continued*)
	4. 4.3.4 **7.** 4.3.12
	5. 4.3.7 **8.** 4.3.21
	6. 4.3.8

	Lecture 16 **pp. 333–377**
4.4	Signals with unwanted parameters, 333–366 Example of random phase problem to motivate the model, 336 Construction of LRT by integrating out unwanted parameters, 334 Models for unwanted parameters, 334
4.4.1	Random phase, 335–348 Formulate bandpass model, 335–336 Go to $\Lambda(r(t))$ by inspection, define quadrature sufficient statistics L_c and L_s, 337 Introduce phase density (364), motivate by brief phaselock loop discussion, 337 Obtain LRT, discuss properties of ln $I_0(x)$, point out it can be eliminated here but will be needed in diversity problems, 338–341 Compute ROC for uniform phase case, 344 Introduce Marcum's Q function Discuss extension to binary and M-ary case, signal selection in partly coherent channels
4.4.2	Random amplitude and phase, 349–366 Motivate Rayleigh channel model, piecewise constant approximation, possibility of continuous measurement, 349–352 Formulate in terms of quadrature components, 352 Solve general Gaussian problem, 352–353 Interpret as filter-squarer receiver and estimator-correlator receiver, 354 Apply Gaussian result to Rayleigh channel, 355 Point out that performance was already computed in Chapter 2 Discuss Rician channel, modifications necessary to obtain receiver, relation to partially-coherent channel in Section 4.4.1, 360–364
4.5–4.7	Assign Section 4.6 to be read before next lecture; Sections 4.5 and 4.7 may be read later, 366–377

Lecture 16 *(continued)*

Problem Assignment 8

1.	4.4.3	**5.**	4.4.29
2.	4.4.5	**6.**	4.4.42
3.	4.4.13	**7.**	4.6.6
4.	4.4.27		

	Lecture 17 **pp. 370–460**
4.6	Multiple-parameter estimation, 370–374 Set up a model and derive MAP equations for the colored noise case, 374 The examples can be left as a reading assignment but the MAP equations are needed for the next topic
Chapter 5	Continuous waveform estimation, 423–460
5.1, 5.2	Model of problem, typical continuous systems such as AM, PM, and FM; other problems such as channel estimation; linear and nonlinear modulation, 423–426 Restriction to no-memory modulation, 427 Definition of $\hat{a}_{\text{map}}(r(t))$ in terms of an orthogonal expansion, complete equivalence to multiple parameter problem, 429–430 Derivation of MAP equations (31–33), 427–431 Block diagram interpretation, 432–433 Conditions for an efficient estimate to exist, 439
5.3–5.6	Assign remainder of Chapter 5 for reading, 433–460

	Lecture 18	**pp. 467–481**
Chapter 6	Linear modulation	
6.1	Model for linear problem, equations for MAP interval estimation, 467–468	
	Property 1:	MAP estimate can be obtained by using linear processor; derive integral equation, 468
	Property 2:	MAP and MMSE estimates coincide for linear modulation because efficient estimate exists, 470
	Formulation of linear *point* estimation problem, 470	
	Gaussian assumption, 471	
	Structured approach, linear processors	
	Property 3:	Derivation of optimum linear processor, 472
	Property 4:	Derivation of error expression [emphasize (27)], 473
	Property 5:	Summary of information needed, 474
	Property 6:	Optimum error and received waveform are uncorrelated, 474
	Property 6A:	In addition, $e_o(t)$ and $r(u)$ are statistically independent under Gaussian assumption, 475
	Optimality of linear filter under Gaussian assumption, 475–477	
	Property 7:	Prove no other processor could be better, 475
	Property 7A:	Conditions for uniqueness, 476
	Generalization of criteria (Gaussian assumption), 477	
	Property 8:	Extend to convex criteria; uniqueness for strictly convex criteria; extend to monotone increasing criteria, 477–478
	Relationship to interval and MAP estimators	
	Property 9:	Interval estimate is a collection of point estimates, 478
	Property 10:	MAP point estimates and MMSE point estimates coincide, 479

	Lecture 18 ***(continued)***
	Summary, 481 Emphasize interplay between structure, criteria, and Gaussian assumption Point out that the optimum linear filter plays a central role in nonlinear modulation (Chapter II.2) and detection of random signals in random noise (Chapter II.3) **Problem Assignment 9** **1.** 5.2.1 **3.** 6.1.1 **2.** 5.3.5 **4.** 6.1.4

Lecture 19 **pp. 481–515**

6.2 Realizable linear filters, stationary processes, infinite time (Wiener-Hopf problem)

Modification of general equation to get Wiener-Hopf equation, 482

Solution of Wiener-Hopf equation, rational spectra, 482

Whitening property; illustrate with one-pole example and then indicate general case, 483–486

Demonstrate a unique spectrum factorization procedure, 485

Express $H_o'(j\omega)$ in terms of original quantities; define "realizable part" operator, 487

Combine to get final solution, 488

Example of one-pole spectrum plus white noise, 488–493

Desired signal is message shifted in time

Find optimum linear filter and resulting error, 494

Prove that the mean-square error is a monotone increasing function of α, the prediction time; emphasize importance of filtering with delay to reduce the mean-square error, 493–495

Unrealizable filters, 496–497

Solve equation by using Fourier transforms

Emphasize that error performance is easy to compute and bounds the performance of a realizable system; unrealizable error represents ultimate performance; filters can be approximated arbitrarily closely by allowing delay

Closed-form error expressions in the presence of white noise, 498–508

Indicate form of result and its importance in system studies

Assign derivation as reading

Discuss error behavior for Butterworth family, 502

Assign the remainder of Section 6.2 as reading, 508–515

Problem Assignment 9 (*continued*)

5. 6.2.1 **7.** 6.2.7

6. 6.2.3 **8.** 6.2.43

	Lecture 20* **pp. 515–538**
6.3	State-variable approach to optimum filters (Kalman-Bucy problem), 515–575

6.3.1	Motivation for differential equation approach, 515
	State variable representation of system, 515–526
	    Differential equation description
	    Initial conditions and state variables
	    Analog computer realization
	    Process generation
	Example 1: First-order system, 517
	Example 2: System with poles only, introduction of vector differential equation, vector block diagrams, 518
	Example 3: General linear differential equation, 521
	Vector-inputs, time-varying coefficients, 527
	System observation model, 529
	State transition matrix, $\boldsymbol{\phi}(t, \tau)$, 529
	    Properties, solution for time-invariant case, 529–531
	    Relation to impulse response, 532
	Statistical properties of system driven by white noise
	    Properties 13 and 14, 532–534
	Linear modulation model in presence of white noise, 534
	Generalizations of model, 535–538

	Problem Assignment 10 **1.** 6.3.1 **4.** 6.3.9 **2.** 6.3.4 **5.** 6.3.12 **3.** 6.3.7(a, b) **6.** 6.3.16

*Lectures 20–22 may be omitted if time is a limitation and the material in Lectures 28–30 on the radar/sonar problem is of particular interest to the audience. For graduate students the material in 20–22 should be used because of its fundamental nature and importance in current research.

	Lecture 21	**pp. 538–546**
6.3.2	Derivation of Kalman-Bucy estimation equations	
	Step 1:	Derive the differential equation that $\mathbf{h}_o(t, \tau)$ satisfies, 539
	Step 2:	Derive the differential equation that $\hat{\mathbf{x}}(t)$ satisfies, 540
	Step 3:	Relate $\xi_P(t)$, the error covariance matrix, and $\mathbf{h}_o(t, t)$, 542
	Step 4:	Derive the variance equation, 542
	Properties of variance equation	
	Property 15:	Steady-state solution, relation to Wiener filter, 543
	Property 16:	Relation to two simultaneous linear vector, equations; analytic solution procedure for constant coefficient case, 545

	Lecture 22 **pp. 546–586**
6.3.3	Applications to typical estimation problems, 546–566 Example 1: One-pole spectrum, transient behavior, 546 Example 3: Wiener process, relation to pole-splitting, 555 Example 4: Canonic receiver for stationary messages in single channel, 556 Example 5: FM problem; emphasize that optimum realizable estimation commutes with linear transformations, *not* linear filtering (this point seems to cause confusion unless discussed explicitly), 557–561 Example 7: Diversity system, maximal ratio combining, 564–565
6.3.4	Generalizations, 566–575 List the eight topics in Section 6.3.4 and explain why they are of interest; assign derivations as reading Compare state-variable approach to conventional Wiener approach, 575
6.4	Amplitude modulation, 575–584 Derive synchronous demodulator; assign the remainder of Section 6.4 as reading
6.5–6.6	Assign remainder of Chapter 6 as reading. Emphasize the importance of optimum linear filters in other areas
	Problem Assignment 11 **1.** 6.3.23 **4.** 6.3.37 **2.** 6.3.27 **5.** 6.3.43 **3.** 6.3.32 **6.** 6.3.44

	Lecture 23*

Chapter II–2	Nonlinear modulation
	Model of angle modulation system
	  Applications; synchronization, analog communication
	  Intuitive discussion of what optimum demodulator should be

II–2.2	MAP estimation equations

II–2.3	Derived in Chapter I–5, specialize to phase modulation
	  Interpretation as unrealizable block diagram
	  Approximation by realizable loop followed by unrealizable postloop filter
	  Derivation of linear model, loop error variance constraint
	Synchronization example
	  Design of filters
	Nonlinear behavior
	  Cycle-skipping
	  Indicate method of performing exact nonlinear analysis

*The problem assignments for Lectures 23–32 will be included in Appendix 1 of Part II.

	Lecture 24
II–2.5	Frequency modulation Design of optimum demodulator Optimum loop filters and postloop filters Signal-to-noise constraints Bandwidth constraints Indicate comparison of optimum demodulator and conventional limiter-discriminator Discuss other design techniques
II–2.6	Optimum angle modulation Threshold and bandwidth constraints Derive optimum pre-emphasis filter Compare with optimum FM systems

	Lecture 25

II–2.7	Comparison of various systems for transmitting analog messages
	Sampled and quantized systems
	    Discuss simple schemes such as binary and M-ary signaling
	    Derive expressions for system transmitting at channel capacity
	Sampled, continuous amplitude systems
	    Develop PFM system, use results from Chapter I–4, and compare with continuous FM system
	Bounds on analog transmission
	    Rate-distortion functions
	    Expression for Gaussian sources
	    Channel capacity formulas
	    Comparison for infinite-bandwidth channel of continuous modulation schemes with the bound
	    Bandlimited message and bandlimited channel
	      Comparison of optimum FM with bound
	      Comparison of simple companding schemes with bound
	Summary of analog message transmission and continuous waveform estimation

	Lecture 26
Chapter II–3	Gaussian signals in Gaussian noise
3.2.1	Simple binary problem, white Gaussian noise on H_0 and H_1, additional colored Gaussian noise on H_1 Derivation of LRT using Karhunen-Loéve expansion Various receiver realizations Estimator-correlator Filter-squarer Structure with optimum realizable filter as component (this discussion is most effective when Lectures 20–22 are included; it should be mentioned, however, even if they were not studied) Computation of bias terms Performance bounds using $\mu(s)$ and tilted probability densities (at this point we must digress and develop the material in Section 2.7 of Chapter I–2). Interpretation of $\mu(s)$ in terms of realizable filtering errors Example: Structure and performance bounds for the case in which additive colored noise has a one-pole spectrum

	Lecture 27

3.2.2	General binary problem
	    Derive LRT using whitening approach
	    Eliminate explicit dependence on white noise
	    Singularity
	    Derive $\mu(s)$ expression
	Symmetric binary problems
	    $\Pr(\epsilon)$ expressions, relation to Bhattacharyya distance
	    Inadequacy of signal-to-noise criterion

	Lecture 28
Chapter II–3	Special cases of particular importance
3.2.3	Separable kernels Time diversity Frequency diversity Eigenfunction diversity Optimum diversity
3.2.4	Coherently undetectable case Receiver structure Show how $\mu(s)$ degenerates into an expression involving d^2
3.2.5	Stationary processes, long observation times Simplifications that occur in receiver structure Use the results in Lecture 26 to show when these approximations are valid Asymptotic formulas for $\mu(s)$ Example: Do same example as in Lecture 26 Plot P_D versus kT for various E/N_0 ratios and P_F's Find the optimum kT product (this is continuous version of the optimum diversity problem)
	Assign remainder of Chapter II–3 (Sections 3.3–3.6) as reading

	Lecture 29
Chapter II–4	Radar-sonar problem
4.1	Representation of narrow-band signals and processes Typical signals, quadrature representation, complex signal representation. Derive properties: energy, correlation, moments; narrow-band random processes; quadrature and complex waveform representation. Complex state variables Possible target models; develop target hierarchy in Fig. 4.6
4.2	Slowly-fluctuating point targets System model Optimum receiver for estimating range and Doppler Develop time-frequency autocorrelation function and radar ambiguity function Examples: Rectangular pulse Ideal ambiguity function Sequence of pulses Simple Gaussian pulse Effect of frequency modulation on the signal ambiguity function Accuracy relations

	Lecture 30
4.2.4	Properties of autocorrelation functions and ambiguity functions. Emphasize: Property 3: Volume invariance Property 4: Symmetry Property 6: Scaling Property 11: Multiplication Property 13: Selftransform Property 14: Partial volume invariances Assign the remaining properties as reading
4.2.5	Pseudo-random signals Properties of interest Generation using shift-registers
4.2.6	Resolution Model of problem, discrete and continuous resolution environments, possible solutions: optimum or "conventional" receiver Performance of conventional receiver, intuitive discussion of optimal signal design Assign the remainder of Section 4.2.6 and Section 4.2.7 as reading.

	Lecture 31

4.3	Singly spread targets (or channels)
	    Frequency spreading—delay spreading
	Model for Doppler-spread channel
	    Derivation of statistics (output covariance)
	    Intuitive discussion of time-selective fading
	Optimum receiver for Doppler-spread target (simple example)
	
	Assign the remainder of Section 4.3 as reading
	
	Doubly spread targets
	    Physical problems of interest: reverberation, scatter communication
	    Model for doubly spread return, idea of target or channel scattering function
	Reverberation (resolution in a dense environment)
	    Conventional or optimum receiver
	    Interaction between targets, scattering function, and signal ambiguity function
	    Typical target-reverberation configurations
	      Optimum signal design
	
	Assign the remainder of Chapter II–4 as reading

	Lecture 32
Chapter II–5	
5.1	Physical situations in which multiple waveform and multiple variable problems arise Review vector Karhunen-Loéve expansion briefly (Chapter I–3)
5.3	Formulate array processing problem for sonar
5.3.1	Active sonar Consider single-signal source, develop array steering Homogeneous noise case, array gain Comparison of optimum space-time system with conventional space-optimum time system Beam patterns Distributed noise fields Point noise sources
5.3.2	Passive sonar Formulate problem, indicate result. Assign derivation as reading Assign remainder of Chapter II–5 and Chapter II–6 as reading

Glossary

In this section we discuss the conventions, abbreviations, and symbols used in the book.

CONVENTIONS

The following conventions have been used:

1. Boldface roman denotes a vector or matrix.
2. The symbol $| \quad |$ means the magnitude of the vector or scalar contained within.
3. The determinant of a square matrix $\mathbf{A}$ is denoted by $|\mathbf{A}|$ or det $\mathbf{A}$.
4. The script letters $\mathcal{F}(\cdot)$ and $\mathcal{L}(\cdot)$ denote the Fourier transform and Laplace transform respectively.
5. Multiple integrals are frequently written as,

$$\int d\tau\, f(\tau) \int dt\, g(t, \tau) \triangleq \int f(\tau) \left\{\int dt\, g(t, \tau)\right\} d\tau,$$

that is, an integral is inside all integrals to its left unless a multiplication is specifically indicated by parentheses.

6. $E[\cdot]$ denotes the statistical expectation of the quantity in the bracket. The overbar $\bar{x}$ is also used infrequently to denote expectation.
7. The symbol $\otimes$ denotes convolution.

$$x(t) \otimes y(t) \triangleq \int_{-\infty}^{\infty} x(t - \tau) y(\tau)\, d\tau$$

8. Random variables are lower case (e.g., x and $\mathbf{x}$). Values of random variables and nonrandom parameters are capital (e.g., X and $\mathbf{X}$). In some estimation theory problems much of the discussion is valid for both random and nonrandom parameters. Here we depart from the above conventions to avoid repeating each equation.

9. The probability density of x is denoted by $p_x(\cdot)$ and the probability distribution by $P_x(\cdot)$. The probability of an event A is denoted by $\Pr[A]$. The probability density of x, given that the random variable a has a value A, is denoted by $P_{x|a}(X|A)$. When a probability density depends on non-random parameter A we also use the notation $p_{x|a}(X|A)$. (This is non-standard but convenient for the same reasons as 8.)

10. A vertical line in an expression means "such that" or "given that"; that is $\Pr[A|x \le X]$ is the probability that event A occurs given that the random variable x is less than or equal to the value of X.

11. Fourier transforms are denoted by both $F(j\omega)$ and $F(\omega)$. The latter is used when we want to emphasize that the transform is a real-valued function of ω. The form used should always be clear from the context.

12. Some common mathematical symbols used include,

(i) $\propto$	proportional to
(ii) $t \to T^-$	t approaches T from below
(iii) $A + B \triangleq A \cup B$	A or B or both
(iv) l.i.m.	limit in the mean
(v) $\int_{-\infty}^{\infty} d\mathbf{R}$	an integral over the same dimension as the vector
(vi) $\mathbf{A}^T$	transpose of $\mathbf{A}$
(vii) $\mathbf{A}^{-1}$	inverse of $\mathbf{A}$
(viii) $\mathbf{0}$	matrix with all zero elements
(ix) $\binom{N}{k}$	binomial coefficient $\left(= \frac{N!}{k!\,(N-k)!}\right)$
(x) $\triangleq$	defined as
(xi) $\int_{\Omega} d\mathbf{R}$	integral over the set Ω

ABBREVIATIONS

Some abbreviations used in the text are:

ML	maximum likelihood
MAP	maximum a posteriori probability
PFM	pulse frequency modulation
PAM	pulse amplitude modulation
FM	frequency modulation
DSB-SC-AM	double-sideband–suppressed carrier–amplitude modulation
DSB-AM	double sideband-amplitude modulation

PM	phase modulation
NLNM	nonlinear no-memory
FM/FM	two-level frequency modulation
MMSE	minimum mean-square error
ERB	equivalent rectangular bandwidth
UMP	uniformly most powerful
ROC	receiver operating characteristic
LRT	likelihood ratio test

SYMBOLS

The principal symbols used are defined below. In many cases the vector symbol is an obvious modification of the scalar symbol and is not included.

A_a	actual value of parameter
A_i	sample at t_i
$\tilde{a}(t)$	Hilbert transform of $a(t)$
$\hat{a}_{\text{abs}}$	minimum absolute error estimate of a
$\hat{a}_{\text{map}}$	maximum a posteriori probability estimate of a
$\hat{a}_{\text{m}l}$	maximum likelihood estimate of A
$\hat{a}_{\text{m}l}(t)$	maximum likelihood estimate of $a(t)$
$\hat{a}_{\text{ms}}$	minimum mean-square estimate of a
α	amplitude weighting of specular component in Rician channel
α	constraint on P_F (in Neyman-Pearson test)
α	delay or prediction time (in context of waveform estimation)
B	constant bias
$B(A)$	bias that is a function of A
$\mathbf{B}_d(t)$	matrix in state equation for desired signal
β	parameter in PFM and angle modulation
C	channel capacity
$C(a_\epsilon)$	cost of an estimation error, a_ϵ
$C(\hat{a}, a)$	cost of estimating a when a is the actual parameter
$C(d_\epsilon(t))$	cost function for point estimation
C_F	cost of a false alarm (say H_1 when H_0 is true)
C_{ij}	cost of saying H_i is true when H_j is true
C_M	cost of a miss (say H_0 when H_1 is true)
C_∞	channel capacity, infinite bandwidth
$\mathbf{C}(t)$	modulation (or observation) matrix
$\mathbf{C}_d(t)$	observation matrix, desired signal

$\mathbf{C}_M(t)$	message modulation matrix
$\mathbf{C}_N(t)$	noise modulation matrix
χ	parameter space
χ_a	parameter space for a
χ_θ	parameter space for θ
χ^2	chi-square (description of a probability density)
$D(\omega^2)$	denominator of spectrum
d	desired function of parameter
d	performance index parameter on ROC for Gaussian problems
$\hat{d}$	estimate of desired function
$d(t)$	desired signal
$\hat{d}(t)$	estimate of desired signal
d_a	actual performance index
$\hat{d}_B(t)$	Bayes point estimate
d_f	parameter in FM system (frequency deviation)
$\hat{d}_o(t)$	optimum MMSE estimate
$d_s(t, a(t))$	derivative of $s(t, a(t))$ with respect to $a(t)$
$d_\epsilon(t)$	error in desired point estimate
$d_*(t)$	output of arbitrary nonlinear operation
δ	phase of specular component (Rician channel)
Δ	interval in PFM detector
Δd	change in performance index
Δd_x	desired change in d
ΔN	change in white noise level
Δ_n	constraint on covariance function error
$\mathbf{\Delta m}$	mean difference vector (i.e., vector denoting the difference between two mean vectors)
$\mathbf{\Delta Q}$	matrix denoting difference between two inverse covariance matrices
E	energy (no subscript when there is only one energy in the problem)
E_a	expectation over the random variable a only
$E_e(N)$	energy in error waveform (as a function of the number of terms in approximating series)
E_I	energy in interfering signal
E_i	energy on ith hypothesis
$\bar{E}_r$	expected value of received energy
E_t	transmitted energy
E_y	energy in $y(t)$

E_1, E_0	energy of signals on H_1 and H_0 respectively
E_ϵ	energy in error signal (sensitivity context)
$e_N(t)$	error waveform
ϵ_I	interval error
ϵ_T	total error
$\text{erf}\,(\cdot)$	error function (conventional)
$\text{erf}_*\,(\cdot)$	error function (as defined in text)
$\text{erfc}\,(\cdot)$	complement of error function (conventional)
$\text{erfc}_*\,(\cdot)$	complement of error function (as defined in text)
η	(eta) threshold in likelihood ratio test
$E(\cdot)$	expectation operation (also denoted by $\overline{(\cdot)}$ infrequently)
F	function to minimize or maximize that includes Lagrange multiplier
$f(t)$	envelope of transmitted signal
$f(t)$	function used in various contexts
$f(t: r(u), T_i \leq u \leq T_f)$	nonlinear operation on $r(u)$ (includes linear operation as special case)
f_c	oscillator frequency ($\omega_c = 2\pi f_c$)
$f_\Delta(t)$	normalized difference signal
$\mathbf{F}$	matrix in differential equation
$\mathbf{F}(t)$	time-varying matrix in differential equation
$\mathbf{F}_d(t)$	matrix in equation describing desired signal
$G^+(j\omega)$	factor of $S_r(\omega)$ that has all of the poles and zeros in LHP (and $\frac{1}{2}$ of the zeros on $j\omega$-axis). Its transform is zero for negative time.
$g(t)$	function in colored noise correlator
$g(t, A), g(t, \mathbf{A})$	function in problem of estimating A (or $\mathbf{A}$) in colored noise
$g(\lambda_i)$	a function of an eigenvalue
$g_h(t)$	homogeneous solution
$g_l(\tau)$	filter in loop
$g_{lo}(\tau), G_{lo}(j\omega)$	impulse response and transfer function optimum loop filter
$g_{pu}(\tau)$	unrealizable post-loop filter
$g_{puo}(\tau), G_{puo}(j\omega)$	optimum unrealizable post-loop filter
$g_\delta(t)$	impulse solution
$g_\Delta(t)$	difference function in colored noise correlator
g_λ	a weighted sum of $g(\lambda_i)$
$g_\infty(t), G_\infty(j\omega)$	infinite interval solution

$\mathbf{G}$	matrix in differential equation
$\mathbf{G}(t)$	time-varying matrix in differential equation
$\mathbf{G}_d$	linear transformation describing desired vector $\mathbf{d}$
$\mathbf{G}_d(t)$	matrix in differential equation for desired signal
$\mathbf{g}(t)$	function for vector correlator
$\mathbf{g_d}(\mathbf{A})$	nonlinear transformation describing desired vector $\mathbf{d}$
$\Gamma(x)$	Gamma function
γ	parameter ($\gamma = k\sqrt{1 + \Lambda}$)
γ	threshold for arbitrary test (frequently various constants absorbed in γ)
γ_a	factor in nonlinear modulation problem which controls the error variance
$H_0, H_1, \ldots, H_i$	hypotheses in decision problem
$h(t, u)$	impulse response of time-varying filter (output at t due to impulse input at u)
$h_{\text{ch}}(t, u)$	channel impulse response
$h_L(t)$	low pass function (envelope of bandpass filter)
$h_o(t, u)$	optimum linear filter
$h'_o(\tau), H'_o(j\omega)$	optimum processor on whitened signal: impulse response and transfer function, respectively
$h_{ou}(\tau), H_{ou}(j\omega)$	optimum unrealizable filter (impulse response and transfer function)
$h_w(t, u)$	whitening filter
$h_\epsilon(t, u)$	arbitrary linear filter
$h_*(t, u)$	linear filter in uniqueness discussion
$\mathbf{H}$	linear matrix transformation
$h_o(t, u)$	optimum linear matrix filter
$I_o(\cdot)$	modified Bessel function of 1st kind and order zero
I_1, I_2	integrals
I_Γ	incomplete Gamma function
$\mathbf{I}$	identity matrix
$J(t, u)$	information kernel
J^{ij}	elements in $\mathbf{J}^{-1}$
$J^{-1}(t, u)$	inverse information kernel
J_{ij}	elements in information matrix
$J_K(t, u)$	kth term approximation to information kernel
$\mathbf{J}$	information matrix (Fisher's)
$\mathbf{J}_D$	data component of information matrix
$\mathbf{J}_P$	a priori component of information matrix

$\mathbf{J}_T$	total information matrix
$\mathfrak{J}(\omega)$	transform of $J(\tau)$
$\mathfrak{J}^{-1}(\omega)$	transform of $J^{-1}(\tau)$
$K_{na}(t, u)$	actual noise covariance (sensitivity discussion)
$K_{ne}(t, u)$	effective noise covariance
$K_{n\epsilon}(t, u)$	error in noise covariance (sensitivity discussion)
$K_x(t, u)$	covariance function of $x(t)$
k	Boltzmann's constant
$k(t, r(u))$	operation in reversibility proof
$\mathbf{K}$	covariance matrix
$\mathbf{k}_d(t)$	linear transformation of $\mathbf{x}(t)$
$\mathbf{k}_d(t, v)$	matrix filter with p inputs and q outputs relating $\mathbf{a}(v)$ and $\mathbf{d}(t)$
$\mathbf{k}_f(u, v)$	matrix filter with p inputs and n outputs relating $\mathbf{a}(v)$ and $\mathbf{x}(u)$
$l(\mathbf{R}), l$	sufficient statistic
$l(A)$	likelihood function
l_a	actual sufficient statistic (sensitivity problem)
l_c, l_s	sufficient statistics corresponds to cosine and sine components
$\boldsymbol{l}$	a set of sufficient statistics
Λ	a parameter which frequently corresponds to a signal-to-noise ratio in message ERB
$\Lambda(\mathbf{R})$	likelihood ratio
$\Lambda(r_K(t))$	likelihood ratio
$\Lambda(r_K(t), A)$	likelihood function
Λ_B	signal-to-noise ratio in reference bandwidth for Butterworth spectra
Λ_{ef}	effective signal-to-noise ratio
Λ_g	generalized likelihood ratio
Λ_m	parameter in phase probability density
Λ_{3db}	signal-to-noise ratio in 3-db bandwidth
$\mathbf{\Lambda_x}$	covariance matrix of vector $\mathbf{x}$
$\mathbf{\Lambda_x}(t)$	covariance matrix of state vector $(= \mathbf{K_x}(t, t))$
λ	Lagrange multiplier
λ_i	eigenvalue of matrix or integral equation
λ_i^{ch}	eigenvalues of channel quadratic form
λ_i^T	total eigenvalue
ln	natural logarithm
$\log_a$	logarithm to the base a

$M_x(jv)$, $M_{\mathbf{x}}(j\mathbf{v})$	characteristic function of random variable x (or $\mathbf{x}$)
$m_x(t)$	mean-value function of process
$\mathbf{M}$	matrix used in colored noise derivation
$\mathbf{m}$	mean vector
$\mu(s)$	exponent of moment-generating function
N	dimension of observation space
N	number of coefficients in series expansion
$N(m, \sigma)$	Gaussian (or Normal) density with mean m and standard deviation σ
$N(\omega^2)$	numerator of spectrum
N_{ef}	effective noise level
N_0	spectral height (joules)
$n(t)$	noise random process
$n_c(t)$	colored noise (does not contain white noise)
$n_{Ei}(t)$	external noise
n_i	ith noise component
$n_{Ri}(t)$	receiver noise
$n_*(t)$	noise component at output of whitening filter
$\hat{n}_{c_r}(t)$	MMSE realizable estimate of colored noise component
$\hat{n}_{c_u}(t)$	MMSE unrealizable estimate of colored noise component
$\mathbf{N}$	noise correlation matrix numbers)
n, $\mathbf{n}$	noise random variable (or vector variable)
$\xi_{a\epsilon}$	cross-correlation between error and actual state vector
ξ_I	expected value of interval estimation error
$\xi_{ij}(t)$	elements in error covariance matrix
$\xi_{\mathrm{m}l}$	variance of ML interval estimate
$\xi_P(t)$	expected value of *realizable* point estimation error
$\xi_{Pi}(t)$	variance of error of point estimate of ith signal
$\xi_{Pn}(t)$	normalized realizable point estimation error
ξ_{Pn}^{α}	normalized error as function of prediction (or lag) time
$\xi_{P\infty}$	expected value of point estimation error, statistical steady state
ξ_u	optimum unrealizable error
ξ_{un}	normalized optimum unrealizable error
$\xi_*(t)$	mean-square error using nonlinear operation
$\boldsymbol{\xi}_{\mathrm{ac}}$	actual covariance matrix
$\boldsymbol{\xi}_d(t)$	covariance matrix in estimating $d(t)$
$\boldsymbol{\xi}_{P\infty}$	steady-state error covariance matrix

ω_c	carrier frequency (radians/second)
ω_D	Doppler shift
P	power
$Pr(\epsilon)$	probability of error
P_D	probability of detection (a conditional probability)
P_{ef}	effective power
P_F	probability of false alarm (a conditional probability)
P_i	a priori probability of ith hypothesis
P_M	probability of a miss (a conditional probability)
$P_D(\theta)$	probability of detection for a particular value of θ
p	operator to denote d/dt (used infrequently)
p_0	fixed probability of interval error in PFM problems
$p_{\mathbf{r}\|H_i}(\mathbf{R}\|H_i)$	probability density of $\mathbf{r}$, given that H_i is true
$p_{x_t}(X_t)$ or $p_{x_t}(X: t)$	probability density of a random process at time t
$\phi(t)$	eigenfunction
$\phi_i(t)$	ith coordinate function, ith eigenfunction
$\phi_x(s)$	moment generating function of random variable x
$\phi(t)$	phase of signal
$\psi_L(t)$	low pass phase function
$\mathbf{P}(t)$	cross-correlation matrix between input to message generator and additive channel noise
$\boldsymbol{\phi}(t, \tau)$	state transition matrix, time-varying system
$\boldsymbol{\phi}(t - t_0) \triangleq \boldsymbol{\phi}(\tau)$	state transition matrix, time-invariant system
$\Pr[\cdot]$, $\Pr(\cdot)$	probability of event in brackets or parentheses
$Q(\alpha, \beta)$	Marcum's Q function
$Q_n(t, u)$	inverse kernel
q	height of scalar white noise drive
$\mathbf{Q}$	covariance matrix of vector white noise drive (Section 6.3)
$\mathbf{Q}$	inverse of covariance matrix $\mathbf{K}$
$\mathbf{Q}_1$, $\mathbf{Q}_0$	inverse of covariance matrix $\mathbf{K}_1$, $\mathbf{K}_0$
$\mathbf{Q}_n(u, z)$	inverse matrix kernel
R	rate (digits/second)
$R_x(t, u)$	correlation function
$\mathcal{R}$	risk
$\mathcal{R}(d(t), t)$	risk in point estimation
$\mathcal{R}_{abs}$	risk using absolute value cost function
$\mathcal{R}_B$	Bayes risk

$\mathcal{R}_F$	risk using fixed test
$\mathcal{R}_{\text{ms}}$	risk using mean-square cost function
$\mathcal{R}_{\text{unf}}$	risk using uniform cost function
$r(t)$	received waveform (denotes both the random process and a sample function of the process)
$r_c(t)$	combined received signal
$r_g(t)$	output when inverse kernel filter operates on $r(t)$
$r_K(t)$	K term approximation
$r_*(t)$	output of whitening filter
$r_{*a}(t)$	actual output of whitening filter (sensitivity context)
$r_{**}(t)$	output of $S_Q(\omega)$ filter (equivalent to cascading two whitening filters)
ρ_{ij}	normalized correlation $s_i(t)$ and $s_j(t)$ (normalized signals)
ρ_{12}	normalized covariance between two random variables
$\mathbf{R}(t)$	covariance matrix of vector white noise $\mathbf{w}(t)$
$\mathbf{R}$	correlation matrix of errors
$\mathbf{R}_{\epsilon_I}$	error correlation matrix, interval estimate
$\mathbf{r}$, $\mathbf{R}$	observation vector
$R_{on}{}^1[\cdot, \cdot]$	radial prolate spheroidal function
$S(j\omega)$	Fourier transform of $s(t)$
$S_c(\omega)$	spectrum of colored noise
$S_{on}[\cdot, \cdot]$	angular prolate spheroidal function
$S_Q(\omega)$	Fourier transform of $Q(\tau)$
$S_r(\omega)$	power density spectrum of received signal
$S_x(\omega)$	power density spectrum
$S_{\epsilon_o}(j\omega)$	transform of optimum error signal
$s(t)$	signal component in $r(t)$, no subscript when only one signal
$s(t, A)$	signal depending on A
$s(t, a(t))$	modulated signal
$s_a(t)$	actual $s(t)$ (sensitivity context)
$s_I(t)$	interfering signal
$s_{Ia}(t)$	actual interfering signal (sensitivity context)
s_i	coefficient in expansion of $s(t)$
s_i	ith signal component
$s_r(t, \theta)$	received signal component
$s_t(t)$	signal transmitted
$s_0(t)$	signal on H_0
$s_1(t)$	signal on H_1
$s_1(t, \boldsymbol{\theta})$, $s_0(t, \boldsymbol{\theta})$	signal with unwanted parameters
$s_\epsilon(t)$	error signal (sensitivity context)

$s_\Delta(t)$	difference signal $(\sqrt{E_1}\, s_1(t) - \sqrt{E_0}\, s_0(t))$
$s_\Omega(t)$	random signal
$s_{\epsilon *}(t)$	whitened difference signal
$s_*(t)$	signal component at output of whitening filter
$s_{*\epsilon}(t)$	output of whitening filter due to signal error
σ^2	variance
${\sigma_1}^2$, ${\sigma_0}^2$	variance on H_1, H_0
${\sigma_{\epsilon_i}}^2$	error variance
$\mathbf{s}(t)$	vector signal
T_e	effective noise temperature
θ, $\boldsymbol{\theta}$	unwanted parameter
$\hat{\theta}$	phase estimate
θ_a	actual phase in binary system
$\theta_{\text{ch}}(t)$	phase of channel response
$\hat{\boldsymbol{\theta}}_1$	estimate of $\boldsymbol{\theta}_1$
$\mathbf{T}(t, \tau)$	transition matrix
$[\ ]^T$	transpose of matrix
$u_{-1}(t)$	unit step function
$u(t)$, $\mathbf{u}(t)$	input to system
V	variable in piecewise approximation to $V_{\text{ch}}(t)$
$V_{\text{ch}}(t)$	envelope of channel response
$\mathbf{v}(t)$	combined drive for correlated noise case
$\mathbf{v}_1(t)$, $\mathbf{v}_2(t)$	vector functions in Property 16 of Chapter 6
W	bandwidth parameter (cps)
$W(j\omega)$	transfer function of whitening filter
W_{ch}	channel bandwidth (cps) single-sided
$W^{-1}(j\omega)$	transform of inverse of whitening filter
$w(t)$	white noise process
$w(\tau)$	impulse response of whitening filter
$\mathbf{W}$	a matrix operation whose output vector has a diagonal covariance matrix
$x(t)$	input to modulator
$x(t)$	random process
$\hat{x}(t)$	estimate of random process
$\mathbf{x}$	random vector
$\mathbf{x}(t)$	state vector
$\mathbf{x}_a(t)$	augmented state vector

$\mathbf{x}_{\mathrm{ac}}$	actual state vector
$\mathbf{x}_d(t)$	state vector for desired operation
$\mathbf{x}_f(t)$	prefiltered state vector
$\mathbf{x}_M(t)$	state vector, message
$\mathbf{x}_{mo}$	state vector in model
$\mathbf{x}_N(t)$	state vector, noise
$y(t)$	output of differential equation
$y(t)$	portion of $r(t)$ not needed for decision
$y(t)$	transmitted signal
$\mathbf{y}$	vector component of observation that is not needed for decision
$y = s(A)$	nonlinear function of parameter A
Z	observation space
$Z_c(\omega)$	integrated cosine transform
$Z_s(\omega)$	integrated sine transform
Z_1, Z_2	subspace of observation space
$z(t)$	output of whitening filter
$\mathbf{z}(t)$	gain matrix in state-variable filter ($\triangleq \mathbf{h}_o(t, t)$)

Author Index

Chapter and reference numbers are in brackets; page numbers follow.

Subject Index